UNIVERSITY OF SUNDERLAND

WITHDRAWN

Microcontroller Programming
The Microchip PIC®

Microcontroller Programming
The Microchip PIC®

Julio Sanchez
Minnesota State University, Mankato

Maria P. Canton
South Central College, North Mankato, Minnesota

CRC Press
Taylor & Francis Group
Boca Raton   London   New York

CRC Press is an imprint of the
Taylor & Francis Group, an informa business

CRC Press
Taylor & Francis Group
6000 Broken Sound Parkway NW, Suite 300
Boca Raton, FL 33487-2742

© 2007 by Taylor & Francis Group, LLC
CRC Press is an imprint of Taylor & Francis Group, an Informa business

No claim to original U.S. Government works
Printed in the United States of America on acid-free paper
10 9 8 7 6 5 4 3 2

International Standard Book Number-10: 0-8493-7189-9 (Hardcover)
International Standard Book Number-13: 978-0-8493-7189-9 (Hardcover)

This book contains information obtained from authentic and highly regarded sources. Reprinted material is quoted with permission, and sources are indicated. A wide variety of references are listed. Reasonable efforts have been made to publish reliable data and information, but the author and the publisher cannot assume responsibility for the validity of all materials or for the consequences of their use.

No part of this book may be reprinted, reproduced, transmitted, or utilized in any form by any electronic, mechanical, or other means, now known or hereafter invented, including photocopying, microfilming, and recording, or in any information storage or retrieval system, without written permission from the publishers.

For permission to photocopy or use material electronically from this work, please access www.copyright.com (http://www.copyright.com/) or contact the Copyright Clearance Center, Inc. (CCC) 222 Rosewood Drive, Danvers, MA 01923, 978-750-8400. CCC is a not-for-profit organization that provides licenses and registration for a variety of users. For organizations that have been granted a photocopy license by the CCC, a separate system of payment has been arranged.

Trademark Notice: Product or corporate names may be trademarks or registered trademarks, and are used only for identification and explanation without intent to infringe.

Visit the Taylor & Francis Web site at
http://www.taylorandfrancis.com

and the CRC Press Web site at
http://www.crcpress.com

Table of Contents

Preface xv

Chapter 1 - Basic Electronics — 1

1.0 The Atom — 1
1.1 Isotopes and Ions — 2
1.2 Static Electricity — 3
1.3 Electrical Charge — 4
    1.3.1 Voltage — 4
    1.3.2 Current — 4
    1.3.3 Power — 5
    1.3.4 Ohm's Law — 5
1.4 Electrical Circuits — 6
    1.4.1 Types of Circuits — 6
1.5 Circuit Elements — 8
    1.5.1 Resistors — 9
    1.5.2 Revisiting Ohm's Law — 9
    1.5.3 Resistors in Series and Parallel — 10
    1.5.4 Capacitors — 12
    1.5.5 Capacitors in Series and in Parallel — 13
    1.5.6 Inductors — 14
    1.5.7 Transformers — 15
1.6 Semiconductors — 15
    1.6.1 Integrated Circuits — 16
    1.6.2 Semiconductor Electronics — 16
    1.6.3 P-Type and N-Type Silicon — 17
    1.6.4 The Diode — 17

Chapter 2 - Number Systems — 19

2.0 Counting — 19
    2.0.1 The Tally System — 19
    2.0.2 Roman Numerals — 20
2.1 The Origins of the Decimal System — 20
    2.1.1 Number Systems for Digital-Electronics — 22
    2.1.2 Positional Characteristics — 22
    2.1.3 Radix or Base of a Number System — 23

2.2 Types of Numbers 23
    2.2.1 Whole Numbers 24
    2.2.2 Signed Numbers 24
    2.2.3 Rational, Irrational, and Imaginary Numbers 24
2.3 Radix Representations 25
    2.3.1 Decimal versus Binary Numbers 25
    2.3.2 Hexadecimal and Octal 26
2.4 Number System Conversions 27
    2.4.1 Binary-to-ASCII-Decimal 28
    2.4.2 Binary-to-Hexadecimal Conversion 29
    2.4.3 Decimal-to-Binary Conversion 29

Chapter 3 - Data Types and Data Storage   33

3.0 Electronic-Digital Machines 33
3.1 Character Representations 33
    3.1.1 ASCII 34
    3.1.2 EBCDIC and IBM 36
    3.1.3 Unicode 36
3.2 Storage and Encoding of Integers 37
    3.2.1 Signed and Unsigned Representations 37
    3.2.2 Word Size 38
    3.2.3 Byte Ordering 39
    3.2.4 Sign-Magnitude Representation 40
    3.2.5 Radix Complement Representation 41
3.3 Encoding of Fractional Numbers 44
    3.3.1 Fixed-Point Representations 45
    3.3.2 Floating-Point Representations 46
    3.3.3 Standardized Floating-Point Representations 47
    3.3.4 IEEE 754 Single Format 48
    3.3.5 Encoding and Decoding Floating-Point Numbers 50
3.4 Binary-Coded Decimals (BCD) 51
    3.4.1 Floating-Point BCD 52

Chapter 4 - Digital Logic, Arithmetic, and Conversions   55

4.0 Microcontroller Logic and Arithmetic 55
    4.0.1 CPU Flags 55
    4.0.2 Word Size 56
4.1 Logical Instructions 56
    4.1.1 Logical AND 57
    4.1.2 Logical OR 57
    4.1.3 Logical XOR 57
    4.1.4 Logical NOT 58
4.2 Microcontroller Arithmetic 58
    4.2.1 Unsigned and Two's Complement Arithmetic 58
    4.2.2 Operations on Decimal Numbers 60
4.3 Bit Manipulations and Auxiliary Operations 62
    4.3.1 Bit Shift and Rotate 62
    4.3.2 Comparison Operations 63
    4.3.3 Other Support Operations 63

4.4 Unsigned Binary Arithmetic 64
    4.4.1 Multi-byte Unsigned Addition 64
    4.4.2 Unsigned Multiplication 65
    4.4.3 Unsigned Division 67
4.5 Signed Binary Arithmetic 67
    4.5.1 Overflow Detection in Signed Arithmetic 69
    4.5.2 Sign Extension Operations 70
    4.5.3 Multi-byte Signed Operations 71
4.6 Data Format Conversions 72
    4.6.1 BCD Digits to ASCII Decimal 72
    4.6.2 Unsigned Binary to ASCII Decimal Digits 73
    4.6.3 ASCII Decimal String to Unsigned Binary 73
    4.6.4 Unsigned Binary to ASCII Hexadecimal Digits 75
    4.6.6 Signed Numerical Conversions 76

Chapter 5 - Circuits and Logic Gates   77

5.0 Digital Circuits 77
5.1 The Diode Revisited 78
    5.1.1 The Light-Emitting Diode (LED) 79
5.2 The Transistor 81
    5.2.1 Bipolar Transistor 81
    5.2.2 MOS Transistor 83
5.3 Logic Gates 84
5.4 Transistor-Transistor Logic 85
    5.4.1 Inverter Gates 86
    5.4.2 The AND Gate 87
    5.4.3 The NAND Gate 87
    5.4.4 The OR Gate 88
    5.4.5 The NOR Gate 88
    5.4.6 Positive and Negative Logic 89
    5.4.7 The XOR Gate 90
    5.4.8 Schmitt Trigger Inverter 91
5.5 Other TTL Logic Families 93
5.6 CMOS Logic Gates 93

Chapter 6 - Circuit Components   95

6.0 Power Supplies 95
6.1 Clocked Logic and Flip-flops 96
    6.1.1 The RS Flip-flop 96
    6.1.2 Clocked RS Flip-flop 98
    6.1.3 The D Flip-flop 99
    6.1.4 The Edge-triggered D Flip-flop 100
    6.1.5 Preset and Clear Signals 101
    6.1.6 D Flip-flop Waveform Action 102
    6.1.7 Flip-flop Applications 103
6.2 Clocks 103
    6.2.1 Clock Waveforms 104
    6.2.2 The TTL Clock 105
    6.2.3 The 555 Timer 106

6.2.4 Microcontroller Clocks	106
6.3 Frequency Dividers and Counters	107
6.3.1 Frequency Dividers	107
6.3.2 The JK Flip-flop Counter	107
6.3.3 Ripple Counters	108
6.3.4 Decoding Gates	110
6.3.5 Synchronous Counters	110
6.3.6 Counter ICs	112
6.3.7 Shift Registers	113
6.4 Multiplexers and Demultiplexers	115
6.4.1 Multiplexers	115
6.4.2 Demultiplexers	118
6.4.3 Multiplexer and Demultiplexer ICs	118
6.5 Input Devices	118
6.5.1 Switches	118
6.5.2 Switch Contact Bounce	120
6.5.3 Keypads	121
6.6 Output Devices	122
6.6.1 Seven-segment LED	122
6.6.2 Liquid Crystal Displays	124
6.6.3 LCD Technologies	125

Chapter 7 - The Microchip PIC   129

7.0 The PICMicro Microcontroller	129
7.0.1 Programming the PIC	130
PIC Programmers	131
Development Boards	131
7.0.2 Prototyping the PIC Circuit	132
7.1 PIC Architecture	134
7.1.1 Baseline PIC Family	134
PIC10 Devices	135
PIC12 Devices	135
PIC14 Devices	138
7.1.2 Mid-range PIC Family	138
PIC16 Devices	139
7.1.3 High-Performance PIC Family	139
PIC18 Devices	139

Chapter 8 - Mid-range PIC Architecture   141

8.0 Processor Architecture and Design	142
8.0.1 Harvard Architecture	142
8.0.2 RISC CPU Design	143
8.0.3 Single-word Instructions	143
8.0.4 Instruction Format	144
8.0.5 Mid-Range Device Versions	145
8.1 The Mid-range Core Features	145
8.1.1 Oscillator	145
8.1.2 System Reset	147
8.1.3 Interrupts	148
8.2 Mid-Range CPU and Instruction Set	149

Contents

      8.2.1 Mid-Range Instruction Set   149
      8.2.2 STATUS and OPTION Registers   151
   8.3 EEPROM Data Storage   153
      8.3.1 EEPROM in Mid-Range PICs   153
   8.4 Data Memory Organization   154
      8.4.1 The w Register   154
      8.4.2 The Data Registers   154
          Memory Banks   154
          The SFRs   155
          The GPRs   157
      8.4.3 Indirect Addressing   158
   8.5 Mid-range I/O and Peripheral Modules   158
      8.5.1 I/O Ports   159
      8.5.2 Timer Modules   160
      8.5.3 Capture-and-Compare Module   160
      8.5.4 Master Synchronous Serial Port (MSSP) Module   161
      8.5.5 USART Module   161
      8.5.6 A/D Module   161

Chapter 9 - PIC Programming: Tools and Techniques   163

   9.0 Microchip's MPLAB   163
      9.0.1 Embedded Systems   164
   9.1 Integrated Development Environment   165
      9.1.1 Installing MPLAB   165
      9.1.2 Creating the Project   167
      9.1.3 Project Build Options   169
      9.1.4 Building the Project   169
   9.2 Simulators and Debuggers   170
      9.2.1 MPLAB SIM   171
      9.2.2 MPLAB Hardware Debuggers   172
      9.2.3 A "Quick-and-Dirty" Debugger   174
   9.3 Programmers   174
   9.4 Engineering PIC Software   175
      9.4.1 Using Program Comments   176
          Program Header   176
          Commented Banners   177
          Commented Bitmaps   178
      9.4.2 Defining Data Elements   179
          The cblock Directive   179
      9.4.3 Banking Techniques   180
          The banksel Directive   180
          Bank Selection Macros   180
          Deprecated Banking Instructions   181
      9.4.4 Processor and Configuration Controls   182
          Configuration Bits   182
      9.4.5 Naming Conventions   184
      9.4.6 Errorlevel Directive   186
   9.5 Pseudo Instructions   186

Chapter 10 - Programming Essentials: Input and Output   189

   10.0 16F84A Programming Template   189
   10.1 Introducing the 16F84A   191
      10.1.1 Template Circuit for 16F84A   191
      10.1.2 Power Supplies   191
          Voltage Regulator   192
      10.1.3 Comparisons in PIC Programming   193
          The Infamous PIC Carry Flag   194
   10.2 Simple Circuits and Programs   194
      10.2.1 A Single LED Circuit   194
          LED Flasher Program   196
      10.2.2 LED/Pushbutton Circuit   199
      10.2.3 Multiple LED Circuit   202
   10.3 Programming the Seven-segment LED   204
   10.4 A Demonstration Board   206
      10.4.1 PCB Images for Demo Board   206
      10.4.2 TestDemo1 Program   208

Chapter 11 - Interrupts   211

   11.0 Interrupts on the 16F84   211
      11.0.1 The Interrupt Control Register   211
      11.0.2 The OPTION Register   212
   11.1 Interrupt Sources   213
      11.1.1 Port-B External Interrupt   214
      11.1.2 Timer0 Interrupt   214
      11.1.3 Port-B Line Change Interrupt   215
          Multiple External Interrupts   217
      11.1.4 EEPROM Data Write Interrupt   217
   11.2 Interrupt Handlers   217
      11.2.1 Context Saving Operations   218
          Saving w and STATUS Registers   218
   11.3 Interrupt Programming   218
      11.3.1 Programming the External Interrupt   219
          RB0 Interrupt Initialization   220
          RB0 Interrupt Service Routine   221
      11.3.2 Wakeup from SLEEP Using the RB0 Interrupt   222
          The SleepDemo Program   223
      11.3.3 Port-B Bits 4-7 Status Change Interrupt   224
          RB4-7 Interrupt Initialization   225
          RB4-7 Change Interrupt Service Routine   227
   11.4 Sample Programs   229
      11.4.1 The RB0Int Program   229
      11.4.2 The SleepDemo Program   232
      11.4.3 The RB4to7Int Program   235

Chapter 12 - Timers and Counters   241

   12.0 The 16F84 Timer0 Module   241
      12.0.1 Timer0 Operation   241

Timer0 Interrupt	242
Timer0 Prescaler	242
12.1 Delays Using Timer0	243
12.1.1 Long Delay Loops	244
How Accurate the Delay?	245
The Black-Ammerman Method	245
12.2 Timer0 as a Counter	246
12.3 Timer0 Programming	247
12.3.1 Programming a Counter	247
A Timer/Counter Test Circuit	248
The Tmr0Counter Program	248
12.3.2 Timer0 as a Simple Delay Timer	250
12.3.3 Measured Time Lapse	252
Interrupt-driven Timer	255
12.4 The Watchdog Timer	259
12.4.1 Watchdog Timer Programming	260
12.5 Sample Programs	260
12.5.1 The Tmr0Counter program	260
12.5.2 The Timer0 Program	263
12.5.3 The LapseTimer Program	265
12.5.4 The LapseTmrInt Program	269

Chapter 13 - LCD Interfacing and Programming   275

13.0 LCD Features and Architecture	275
13.0.1 LCD Functions and Components	276
Internal Registers	276
Busy Flag	276
Address Counter	276
Display Data RAM (DDRAM)	276
Character Generator ROM (CGROM)	276
Character Generator RAM (CGRAM)	277
Timing Generation Circuit	277
Liquid Crystal Display Driver Circuit	278
Cursor/Blink Control Circuit	278
13.0.2 Connectivity and Pin-Out	278
13.1 Interfacing with the HD44780	279
13.1.1 Busy Flag or Timed Delay Options	280
13.1.2 Contrast Control	281
13.1.3 Display Backlight	281
13.1.4 Display Memory Mapping	281
13.2 HD44780 Instruction Set	283
13.2.1 Instruction Set Overview	283
Clearing the Display	283
Return home	284
Entry mode set	284
Display and Cursor ON/OFF	284
Cursor/display shift	284
Function set	285
Set CGRAM address	285
Set DDRAM address	285
Read busy flag and Address register	285

Write data	285
Read data	286
13.2.2 A 16F84 8-bit Data Mode Circuit	286
13.3 LCD Programming	287
13.3.1 Defining Constants and Variables	287
Using MPLAB Data Directives	289
13.3.2 LCD Initialization	290
Function Set Command	290
Display Off	291
Display and Cursor On	291
Set Entry Mode	292
Cursor and Display Shift	292
Clear Display	293
13.3.3 Auxiliary Operations	293
Time Delay Routine	293
Pulsing the E Line	295
Reading the Busy Flag	295
Bit Merging Operations	296
13.3.4 Text Data Storage and Display	298
Generating and Storing a Text String	299
Displaying the Text String	301
13.3.5 Data Compression Techniques	302
4-bit Data Transfer Mode	302
Master/Slave Systems	304
13.4 Sample Programs	306
13.4.1 LCDTest1	306
13.4.2 LCDTest2 Program	316
13.4.3 LCDTest3 Program	327

Chapter 14 - Communications   339

14.0 PIC Communications Overview	339
14.1 Serial Data Transmission	340
14.1.1 Asynchronous Serial Transmission	340
14.1.2 Synchronous Serial Transmission	342
14.1.3 PIC Serial Communications	342
14.1.4 The RS-232-C Standard	343
Essential Concepts	344
The Serial Bit Stream	344
Parity Testing	345
Connectors and Wiring	345
The Null Modem	346
The Null Modem Cable	347
14.1.5 The EIA-485 Standard	349
EIA-485 in PIC-based Systems	350
14.2 Parallel Data Transmission	350
14.2.1 PIC Parallel Slave Port (PSP)	351
14.3 PIC "Free-style" Serial Programming	351
14.3.1 PIC-to-PIC Serial Communications	352
PIC-to-PIC Serial Communications Circuits	352
PIC-to-PIC Serial Communications Programs	354
14.3.2 Program Using Shift Register ICs	360

The 74HC165 Parallel-to-Serial Shift Register	361
74HC164 Serial-to-Parallel Shift Register	364
14.4 PIC Protocol-based Serial Programming	366
14.4.1 RS-232-C Communications on the 16F84	366
The RS-232-C Transceiver IC	367
PIC to PC Communications	368
An RS-232-C TTY Board	368
A 16F84A UART Emulation	369
An LCD Scrolling Routine	371
14.4.2 RS-232-C Communications on the 16F87x	375
The 16F87x USART Module	376
The USART Baud Rate Generator	376
16F87x USART Asynchronous Transmitter	379
16F87x USART Asynchronous Receiver	380
PIC-to-PC RS-232-C Communications Circuit	381
16F877 PIC Initialization Code	381
USART Receive and Transmit Routines	384
The USART Receive Interrupt	386
14.5 Sample Programs	389
14.5.1 SerialSnd Program	389
14.5.2 SerialRcv Program	394
14.5.3 Serial6465 Program	400
14.5.4 TTYUsart Program	404
14.5.5 SerComLCD Program	420
14.5.6 SerIntLCD Program	438

Chapter 15 - Data EEPROM Programming   459

15.0 PIC Internal EEPROM Memory	460
15.0.1 EEPROM Programming on the 16F84	460
Reading EEPROM Data Memory on the 16F84	460
16F84 EEPROM Data Memory Write	461
16F84 EEPROM Demonstration Program	462
15.0.2 EEPROM Programming on the 16F87x	465
Reading EEPROM Data Memory on the 16F87x	467
Writing to EEPROM Data Memory in the 16F87x	467
GFR Access Issue in the 16F87x	469
15.0.3 16F87x EEPROM Circuit and Program	469
15.1 EEPROM Devices and Interfaces	475
15.1.1 The I2C Serial Interface	476
15.1.2 I2C Communications	476
15.1.3 EEPROM Communications Conditions	477
15.1.4 EEPROM Write Operation	478
15.1.5 EEPROM Read Operation	478
15.1.6 I2C EEPROM Devices	479
15.1.7 PIC Master Synchronous Serial Port (MSSP)	480
MSSP in Master Mode	482
15.1.8 I2C Serial EEPROM Programming on the 16F877	486
IC2 Initialization Procedure	486
I2C Write Byte Procedure	488
I2C Read Byte Procedure	490
15.2 Sample Programs	492

15.2.1 EECounter Program	492
15.2.2 Ser2EEP Program	504
15.2.3 I2CEEP Program	521

Chapter 16 - Analog to Digital and Realtime Clocks   543

16.0 A/D Converters	544
16.0.1 Converter Resolution	544
16.0.2 ADC Implementation	545
16.1 A/D Integrated Circuits	546
16.1.1 ADC0331 Sample Circuit and Program	547
16.2 PIC On-Board A/D Hardware	549
16.2.1 A/D Module on the 16F87x	549
The ADCON0 Register	550
The ADCON1 Register	552
SLEEP Mode Operation	554
16.2.2 A/D Module Sample Circuit and Program	554
16.3 Realtime Clocks	558
16.3.1 The NJU6355 Realtime Clock	558
16.3.2 RTC Demonstration Circuit and Program	560
BCD Conversion Procedures	565
16.4 Sample Programs	568
16.4.1 ADF84 Program	568
16.4.2 A2DinLCD Program	580
16.4.3 RTC2LCD Program	595

Appendix A - Resistor Color Codes   613

Appendix B - Building Your Own Circuit Boards   615

Appendix C - Mid-range Instruction Set   621

Appendix D - Supplementary Programs   659

Index   795

Preface

There are two sides to the computer revolution: one is represented by the PC on your desktop and the second one by the device that remote-controls your TV, monitors and operates your car engine, and allows you to set up your answering machine and your microwave oven. At the core of the PC you find a microprocessor, while at the heart of a self-contained programmable device (also called an embedded system) is a microcontroller.

Microcontrollers are virtually everywhere in our modern society. They are found in automobiles, airplanes, toys, kitchen appliances, computers, TVs and VCRs, phones and answering machines, space telescopes, and practically every electronic digital device that furnishes an independent functionality to its user. In this sense a microcontroller is a self-contained computer system that includes a processor, memory, and some way of communicating with the outside world, all in a single chip that can be smaller than a postage stamp.

A microcontroller (sometimes called an MCU) is actually a computer on a chip. Essentially it is a control device and its design places emphasis on being self-sufficient and inexpensive. The typical microcontroller contains all the components and features necessary to perform its functions, such as a central processor, input/output facilities, timers, RAM memory for storing program data and executable code, and a clock or oscillator that provides a timing beat. In addition, some microcontrollers include a variety of additional modules and circuits. Some common ones are serial and parallel communications, analog-to-digital converters, realtime clocks, and flash memory.

Engineers, inventors, experimenters, students, and device designers in general deal with microcontrollers on an everyday basis. In fact, interest in microcontrollers is not limited to electrical, electronic, and computer engineers. Mechanical and automotive engineers, among many others, often design devices or components that contain microcontrollers. The system that controls the hatch of a ballistic missile silo and the one that operates the doglike toy that barks and rolls on its back, both contain microcontrollers.

The Microchip PIC

Microcontrollers include an enormous array of models and variations of general- and special-purpose devices. Discussing all of them in a single volume would have forced a superficial scope. Even the products of a single manufacturer can have a mind-boggling variety, which sometimes include hundreds of different MCU models in a half-dozen families, all with very different applications and features.

For this reason we have focused the book on a single type of microcontroller: the Microchip PIC. Not only are the PIC the most used and best known microcontrollers, they are also the best supported. In fact, PIC system design and programming has become a powerful specialization with a large number of professional and amateur specialists. There are hundreds of WEB sites devoted to PIC-related topics. An entire cottage industry of PIC software and hardware has flourished around this technology.

For practical reasons we have limited the book's scope to 8-bit PICs. In fact, the book concentrates on a particular type of 8-bit PIC known as the mid-range family. We have chosen this approach partly because of space limitations and partly due to the fact that 16- and 32-bit microcontrollers (sometimes called external memory microcontrollers) are more related to microprocessor technology than to the topic at hand.

The Book's Design

The book is intended as a resource kit for PIC microcontroller programming. But programming microcontrollers is a different paradigm from microprocessor programming. PIC programming requires a set of skills and a knowledge base quite different from the one needed by a computer programmer. The reason is that the designer/programmer is responsible for the entire system. A typical embedded system has no DOS, Windows, or UNIX software to handle the operational and housekeeping chores. Thus, the PIC programmer provides all the functionality needed by the application with very little assistance from other programs. This makes the microcontroller programmer an application developer, a system's programmer, and an input/output specialist, all at the same time.

For these reasons, the microcontroller programmer must be familiar with a host of computer science topics, including low-level data representations, binary arithmetic, computer organization, input/output programming, concurrency and scheduling, memory management, timing operations, and system functions. At the same time, he or she must be quite conversant with digital electronics and circuit design since the object of the program is a hardware device.

In the first six chapters of the book we have attempted to provide the necessary background both in digital electronics and in computer science. Chapters 7, 8, and 9 are an overview of PIC architecture and programming tools. The remainder of the book deals with programming the various functions, modules, and devices. The appendices contain supplementary materials and expand the coding contained in the text. Readers familiar with electronics and circuit design can skip over Chapters 1, 5, and 6. Those well versed in computer science can do the same with Chapters 2, 3, and 4.

Mapleton, Minnesota　　　　　　　　　　　　　　　　　　　　Julio Sanchez

June 28, 2006　　　　　　　　　　　　　　　　　　　　　　　Maria P. Canton

Additional Material

Additional material is available from the CRC Web site:

www.crcpress.com

Under the menu Electronic Products (located on the left side of the screen), click on Downloads & Updates. A list of books in alphabetical order with Web downloads will appear. Locate this book by a search, or scroll down to it. After clicking on the book title, a brief summary of the book will appear. Go to the bottom of this screen and click on the hyperlinked "Download" that is in a zip file.

Chapter 1

Basic Electronics

1.0 The Atom

Until the end of the nineteenth century it was assumed that matter was composed of small, indivisible particles called *atoms*. The work of J.J. Thompson, Daniel Rutheford, and Neils Bohr proved that atoms were complex structures that contained both positive and negative particles. The positive ones were called *protons* and the negative ones *electrons*.

Several models of the atom were proposed: the one by Thompson assumed that there were equal numbers of protons and electrons inside the atom and that these elements were scattered at random, as in the leftmost drawing in Figure 1-1. Later, in 1913, Daniel Rutheford's experiments led him to believe that atoms contained a heavy central positive nucleus with the electrons scattered randomly. So he modified Thompson's model as shown in the center drawing. Finally, Neils Bohr theorized that electrons had different energy levels, as if they moved around the nucleus in different orbits, like planets around a sun. The rightmost drawing represents this orbital model.

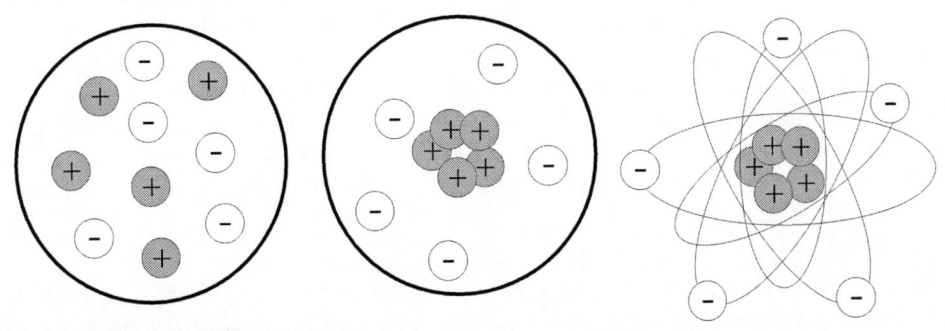

Figure 1-1 Models of the Atom

Investigations also showed that the normal atom is electrically neutral. Protons (positively charged particles) have a mass of 1.673×10^{-24} grams. Electrons (negatively charged particles) have a mass of 9.109×10^{-28} grams. Furthermore, the orbital model of the atom is not actually valid since orbits have little meaning at the atomic level. A more accurate representation is based on concentric spherical shells about the nucleus. An active area of research deals with atomic and sub-atomic structures.

The number of protons in an atom determines its atomic number; for example, the hydrogen atom has a single proton and an atomic number of 1, helium has 2 protons, carbon has 6, and uranium has 92. But when we compare the ratio of mass to electrical charge in different atoms we find that the nucleus must be made up of more than protons. For example, the helium nucleus has twice the charge of the hydrogen nucleus, but four times the mass. The additional mass is explained by assuming that there is another particle in the nucleus, called a neutron, which has the same mass as the proton but no electrical charge. Figure 1-2 shows a model of the helium atom with two protons, two electrons, and two neutrons.

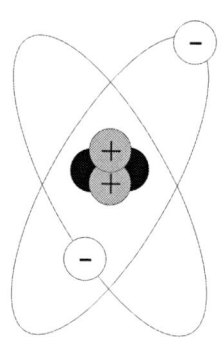

Figure 1-2 Model of the Helium Atom

1.1 Isotopes and Ions

But nature is not always consistent with such neat models. Whereas in a neutral atom, the number of protons in the atomic nucleus exactly matches the number of electrons, the number of protons need not match the number of neutrons. For example, most hydrogen atoms have a single proton, but no neutrons, while a small percentage have one neutron, and an even smaller one have two neutrons. In this sense, atoms of an element that contains different number of neutrons are isotopes of the element; for example water (H_2O) containing hydrogen atoms with two neutrons (deuterium) is called "heavy water."

An atom that is electrically charged due to an excess or deficiency of electrons is called an *ion*. When the dislodged elements are one or more electrons the atom takes a positive charge. In this case it is called a *positive ion*. When a stray electron combines with a normal atom the result is called a *negative ion*.

1.2 Static Electricity

Free electrons can travel through matter or remain at rest on a surface. When electrons are at rest, the surface is said to have a static electrical charge that can be positive or negative. When electrons are moving in a stream-like manner we call this movement an *electrical current*. Electrons can be removed from a surface by means of friction, heat, light, or a chemical reaction. In this case the surface becomes positively charged.

The ancient Greeks discovered that when amber was rubbed with wool the amber became electrically charged and would attract small pieces of material. In this case, the charge is a positive one. Friction can cause other materials, such as hard rubber or plastic, to become negatively charged. Observing objects that have positive and negative charges we note that like charges repel and unlike charges attract each other, as shown in Figure 1-3.

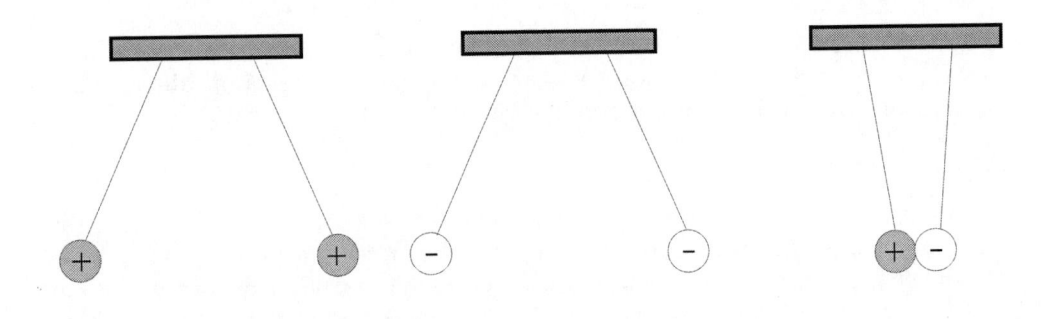

Figure 1-3 Like and Unlike Charges

Friction causes loosely-held electrons to be transferred from one surface to the other. This results in a net negative charge on the surface that has gained electrons, and a net positive charge on the surface that has lost electrons. If there is no path for the electrons to take to restore the balance of electrical charges, these charges remain until they gradually leak off. If the electrical charge continues building it eventually reaches the point where it can no longer be contained. In this case it discharges itself over any available path, as is the case with lightning.

Static electricity does not move from one place to another. While some interesting experiments can be performed with it, it does not serve the practical purpose of providing energy to do sustained work.

Static electricity certainly exists, and under certain circumstances we must allow for it and account for its possible presence, but it will not be the main theme of these pages.

1.3 Electrical Charge

Physicists often resort to models and theories to describe and represent some force that can be measured in the real world. But very often these models and representations are no more than concepts that fail to physically represent the object. In this sense, no one knows exactly what gravity is, or what is an electrical charge. Gravity, which can be felt and measured, is the force between masses.

By the same token, bodies in "certain electrical conditions" also exert measurable forces on one another. The term "electrical charge" was coined to explain these observations.

Three simple postulates or assumptions serve to explain all electrical phenomena:

1. Electrical charge exists and can be measured. Charge is measured in Coulombs, a unit named for the French scientist Charles Agustin Coulomb.
2. Charge can be positive or negative.
3. Charge can neither be created nor destroyed. If two objects with equal amounts of positive and negative charge are combined on some object, the resulting object will be electrically neutral and will have zero net charge.

1.3.1 Voltage

Objects with opposite charges attract, that is, they exert a force upon each other that pulls them together. In this case, the magnitude of the force is proportional to the product of the charge on each mass. Like gravity, electrical force depends inversely on the distance squared between the two bodies; the closer the bodies the greater the force. Consequently, it takes energy to pull apart objects that are positively and negatively charged, in the same manner that it takes energy to raise a big mass against the pull of gravity.

The potential that separate objects with opposite charges have for doing work is called *voltage*. Voltage is measured in units of *volts* (V). The unit is named for the Italian scientist *Alessandro Volta*.

The greater the charge and the greater the separation, the greater the stored energy, or voltage. By the same token, the greater the voltage, the greater the force that drives the charges together.

Voltage is always measured between two points that represent the positive and negative charges. In order to compare voltages of several charged bodies a common reference point is necessary. This point is usually called "ground."

1.3.2 Current

Electrical charge flows freely in certain materials, called *conductors*, but not in others, called *insulators*. Metals and a few other elements and compounds are good conductors, while air, glass, plastics, and rubber are insulators. In addition, there is a third category of materials called semiconductors; sometimes they seem to be good conductors but much less so other times. Silicon and Germanium are two such semiconductors. We discuss semiconductors in the context of integrated circuits later in the book.

Figure 1-4 shows two connected, oppositely charged bodies. The force between them has the potential for work; therefore, there is voltage. If the two bodies are connected by a conductor, as in the illustration, the positive charge moves along the wire to the other sphere. On the other end, the negative charge flows out on the wire towards the positive side. In this case, positive and negative charges combine to neutralize each other until there are no charge differences between any points in the system.

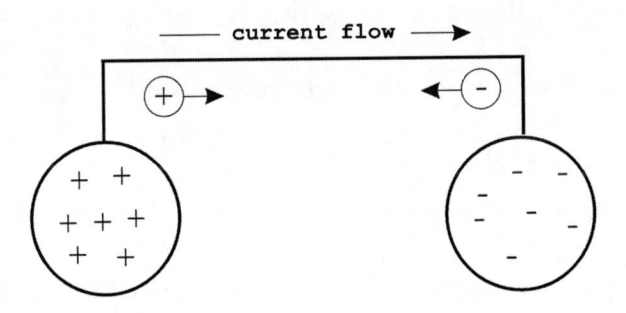

Figure 1-4 Connected Opposite Charges

The flow of an electrical charge is called a current. Current is measured in *amperes* (a), also called *amps*, after Andre Ampere, a French mathematician and physicist. An ampere is defined as a flow of one Coulomb of charge in one second.

Electrical current is directional; therefore, a positive current is the flow current from a positive point A to a negative point B. However, most current results from the flow of negative-to-positive charges.

1.3.3 Power

Current flowing through a conductor produces heat. The heat is the result of the energy that comes from the charge traveling across the voltage difference. The work involved in producing this heat is *electrical power*. Power is measured in units of *watts* (W), named after the Englishman *James Watt*, who invented the steam engine.

1.3.4 Ohm's Law

The relationship between voltage, current, and power is described by *Ohm's Law*, named after the German physicist *Georg Simon Ohm*. Using equipment of his own creation, Ohm determined that the current that flows through a wire is proportional to its cross-sectional area and inversely proportional to its length. This allowed defining the relationship between voltage, current, and power, as expressed by the equation:

$$P = V \times I$$

Where P represents the power in watts, V is the voltage in volts, and I is the current in amperes. Ohm's Law can also be formulated in terms of voltage, current, and resistance as shown later in this chapter.

1.4 Electrical Circuits

An *electrical network* is an interconnection of electrical elements. An *electrical circuit* is a network in a closed loop, giving a return path for the current. A network is a connection of two or more simple elements, and may not necessarily be a circuit.

Although there are several types of electrical circuits they all have some of the following elements:

1. A power source, which can be a battery, alternator, etc., produces an electrical potential.
2. Conductors, in the form of wires or circuit boards, provide a path for the current.
3. Loads, in the form of devices such as lamps, motors, etc., use the electrical energy to produce some form of work.
4. Control devices, such as potentiometers and switches, regulate the amount of current flow or turn it on and off.
5. Protection devices, such as fuses or circuit breakers, prevent damage to the system in case of overload.
6. A common ground.

Figure 1-5 shows a simple circuit that contains all of these elements.

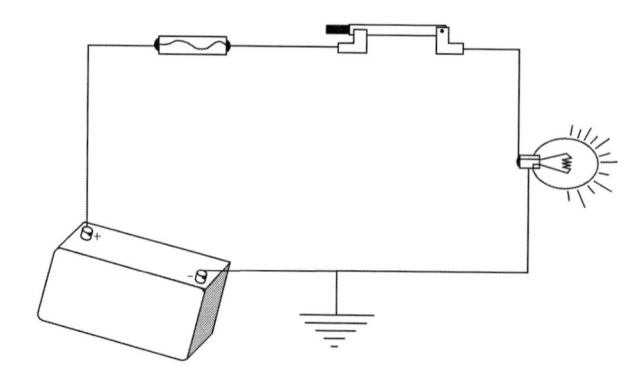

Figure 1-5 Simple Circuit

1.4.1 Types of Circuits

There are three common types of circuits: series, parallel, and series-parallel. The circuit type is determined by how the components are connected. In other words, by how the circuit elements, power source, load, and control and protection devices are interconnected. The simplest circuit is one in which the components offer a single current path. In this case, although the loads may be different, the amount of current flowing through each one is the same. Figure 1.6 shows a series circuit with two light bulbs.

Basic Electronics

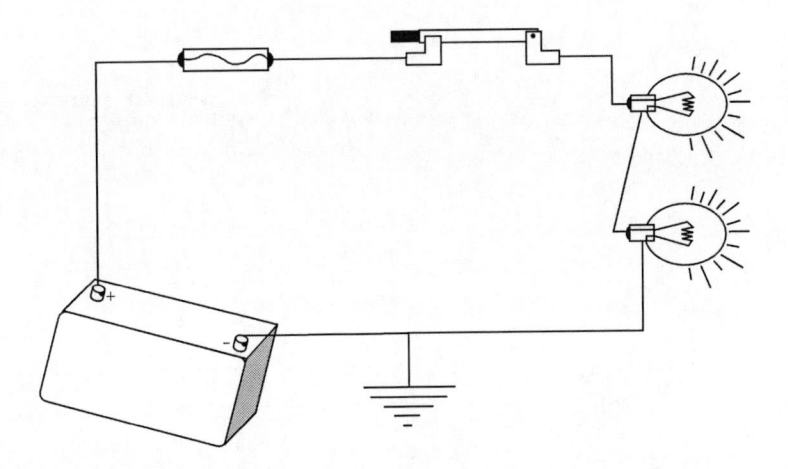

Figure 1-6 Series Circuit

In the *series circuit* in Figure 1-6 if one of the light bulbs burn out, the circuit flow is interrupted and the other one will not light. Some Christmas lights are wired in this manner, and if a single bulb fails the whole string will not light.

In a *parallel circuit* there is more than one path for current flow. Figure 1-7 shows a circuit wired in parallel.

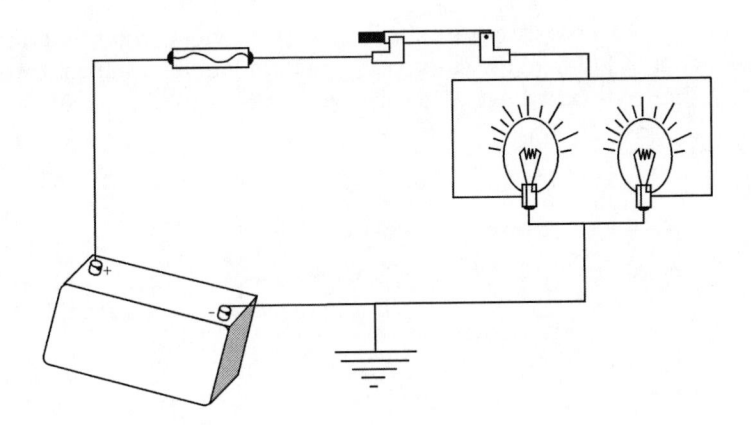

Figure 1-7 Parallel Circuit

In the circuit of Figure 1-7, if one of the light bulbs burns out, the other one will still light. Also, if the load is the same in each circuit branch, so will be the current flow in that branch. By the same token, if the load in each branch is different, so will be the current flow in each branch.

The series-parallel circuit has some components wired in series and others in parallel. Therefore, the circuit shares the characteristics of both series and parallel circuits. Figure 1-8 shows the same parallel circuit to which a series *rheostat* (dimmer) has been added in series.

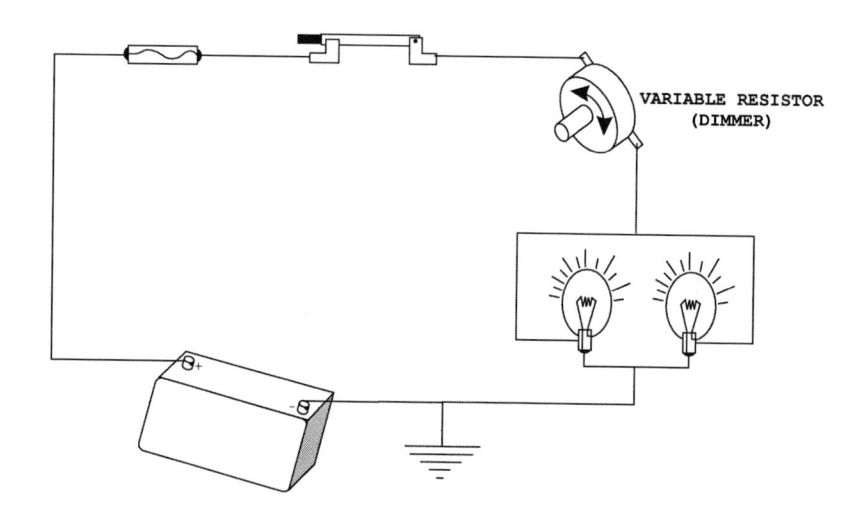

Figure 1-8 Series-Parallel Circuit

In the circuit of Figure 1-8 the two light bulbs are wired in parallel, so if one fails the other one will not. However, the rheostat (dimmer) is wired in series with the circuit, so its action affects both light bulbs.

1.5 Circuit Elements

So far we have represented circuits using a pictorial style. Circuit diagrams are more often used since they achieve the same purpose with much less artistic effort and are easier to read. Figure 1-9 is a diagrammatic representation of the circuit in Figure 1-8.

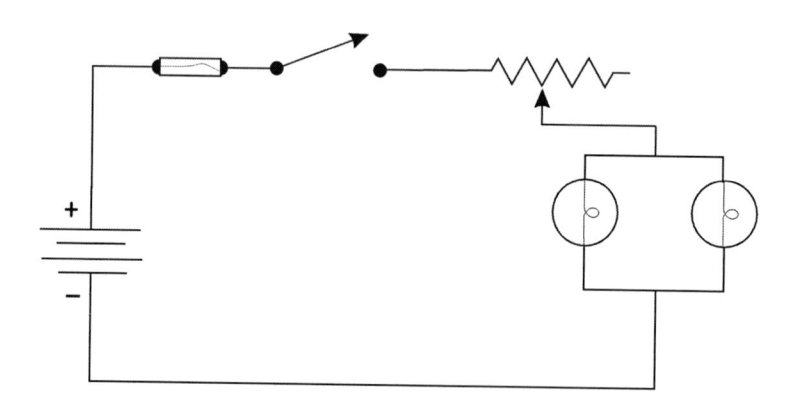

Figure 1-9 Diagram of a Series-Parallel Circuit

Certain components are commonly used in electrical circuits. These include power sources, resistors, capacitors, inductors, and several forms of semiconductor devices.

Basic Electronics

1.5.1 Resistors

If the current flow from, say, a battery is not controlled, a *short-circuit* takes place and the wires can melt or the battery may even explode. *Resistors* provide a way of controlling the flow of current from a source. A resistor is to current flow in an electrical circuit as a valve is to water flow: both elements "resist" flow. Resistors are typically made of materials that are poor conductors. The most common ones are made from powdered carbon and some sort of binder. Such carbon composition resistors usually have a dark-colored cylindrical body with a wire lead on each end. Color bands on the body of the resistor indicate its value, measured in ohms and represented by the Greek letter ω The color code for resistor bands can be found in Appendix A.

The potentiometer and the rheostat are variable resistors. When the knob of a potentiometer or rheostat is turned, a slider moves along the resistance element and reduces or increases the resistance. A potentiometer is used as a dimmer in the circuits of Figure 1-8 and Figure 1-9. The photoresistor or photocell is composed of a light sensitive material whose resistance decreases when exposed to light. Photoresistors can be used as light sensors.

1.5.2 Revisiting Ohm's Law

We have seen how Ohm's Law describes the relationship between voltage, current, and power. The law is reformulated in terms of resistance so as to express the relationship between voltage, current, and resistance, as follows:

In this case V represents voltage, I is the current, and R is the resistance in the circuit. Ohm's Law equation can be manipulated in order to find current or resistance in terms of the other variables, as follows:

$$V = I \times R$$

Note that the voltage value in Ohm's Law refers to the voltage across the resistor, in other words, the voltage between the two terminal wires. In this sense the voltage is actually produced by the resistor, since the resistor is restricting the flow of charge much as a valve or nozzle restricts the flow of water. It is the restriction created by the resistor that forms an excess of charge with respect to the other side of the circuit. The charge difference results in a voltage between the two points. Ohm's Law is used to calculate the voltage if we know the resistor value and the current flow.

$$I = \frac{V}{R}$$
$$R = \frac{V}{I}$$

V=IR

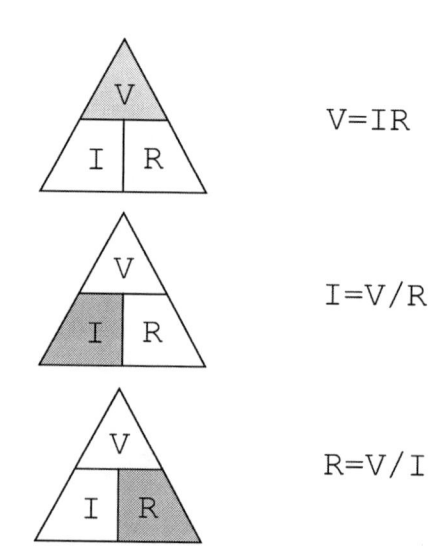

I=V/R

R=V/I

Figure 1-10 Ohm's Law Pyramid

A popular mnemonics for Ohm's Law consists of drawing a pyramid with the voltage symbol at the top and current and resistance in the lower level. Then, it is easy to solve for each of the values by observing the position of the other two symbols in the pyramid, as shown in Figure 1-10.

1.5.3 Resistors in Series and Parallel

When resistors are in series the total resistance equals the sum of the individual resistances. The diagram in Figure 1-11 shows two resistors (R1 and R2) wired in series in a simple circuit.

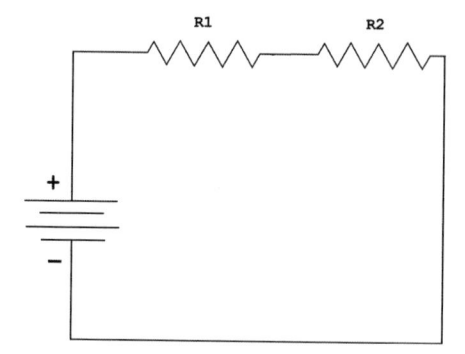

Figure 1-11 Resistors in Series

In Figure 1-11 the total resistance (RT) is calculated by adding the resistance values of R1 and R2, thus, RT = R1 + R2.

In terms of water flow, a series of partially closed valves in a pipe add up to slow the flow of water.

Resistors can also be connected in parallel, as shown in Figure 1-12.

Basic Electronics

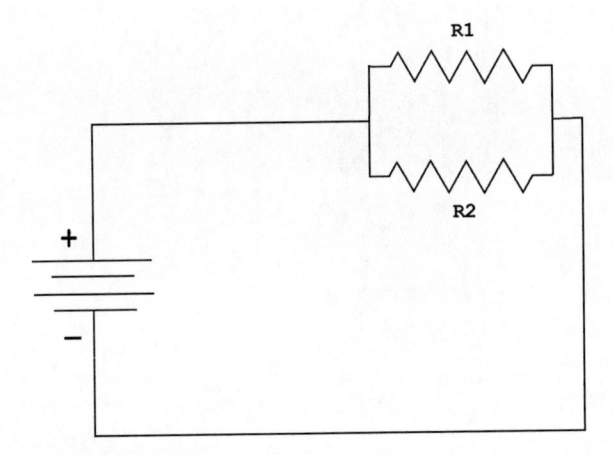

Figure 1-12 Resistors in Parallel

When resistors are placed in parallel, the combination has less resistance than any one of the resistors. If the resistors have different values, then more current flows through the path of least resistance. The total resistance in a parallel circuit is obtained by dividing the product of the individual resistors by their sum, as in the formula:

$$RT = \frac{R1 \times R2}{R1 + R2}$$

If more than two resistors are connected in parallel, then the formula can be expressed as follows:

$$RT = \frac{1}{\frac{1}{R1} + \frac{1}{R2} + \frac{1}{R3}\ldots}$$

Also note that the diagram representation of resistors in parallel can have different appearances. For example, the circuit in Figure 1-13 is electrically identical to the one in Figure 1-12.

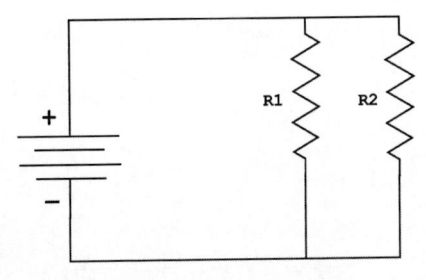

Figure 1-13 Alternative Circuit of Parallel Resistors

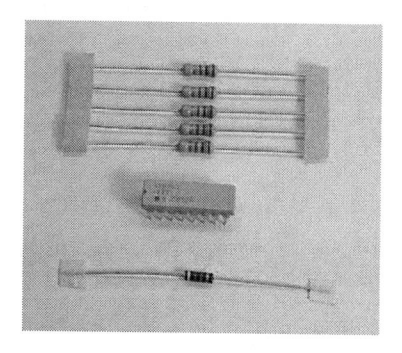

Figure 1-14 Resistors

Figure 1-14 shows several commercial resistors. The integrated circuit at the center of the image combines eight resistors of the same value. These devices are convenient when the circuit design calls for several identical resistors. The color-coded cylindrical resistors in the image are made of carbon

Appendix A contains the color codes used in identifying resistors whose surface area does not allow printing its value.

1.5.4 Capacitors

An element often used in the control of the flow of an electrical charge is a *capacitor*. The name originated in the notion of a "capacity" to store charge. In that sense a capacitor functions as a small battery. Capacitors are made of two conducting surfaces separated by an insulator. A wire lead is usually connected to each surface. Two large metal plates separated by air would perform as a capacitor. More frequently capacitors are made of thin metal foils separated by a plastic film or another form of solid insulator. Figure 1-15 shows a circuit which contains both a capacitor and a resistor.

In Figure 1-15 charge flows from the battery terminals, along the conductor wire, onto the capacitor plates. Positive charges collect on one plate and negative charges on the other plate. The initial current is limited only by the resistance of the wires and by the resistor in the circuit. As charge builds up on the plates, charge repulsion resists the flow and the current is reduced. At some point the repulsive force from charge on the plates is strong enough to balance the force from charge on the battery, and the current stops.

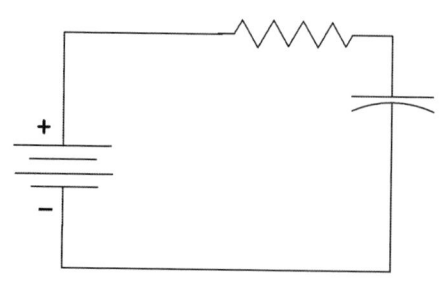

Figure 1-15 Capacitor Circuit

Basic Electronics

The existence of charges on the capacitor plates means there must be a voltage between the plates. When the current stops this voltage is equal to the voltage in the battery. Since the points in the circuit are connected by conductors, then they have the same voltage, even if there is a resistor in the circuit. If the current is zero, there is no voltage across the resistor, according to Ohm's law.

The amount of charge on the plates of the capacitor is a measure of the value of the capacitor. This "capacitance" is measured in farads (f), named in honor of the English scientist Michael Faraday.

The relationship is expressed by the equation:

$$C = \frac{Q}{V}$$

where C is the capacitance in farads, Q is the charge in Coulombs, and V is the voltage. Capacitors of one farad or more are rare. Generally capacitors are rated in microfarads (μf), one-millionth of a farad, or picofarads (pf), one-trillionth of a farad.

Consider the circuit of Figure 1-15 after the current has stabilized. If we now remove the capacitor from the circuit it still holds a charge on its plates. That is, there is a voltage between the capacitor terminals. In one sense, the charged capacitor appears somewhat like a battery. If we were to short-circuit the capacitor's terminals a current would flow as the positive and negative charges neutralize each other. But unlike a battery, the capacitor does not replace its charge. So the voltage drops, the current drops, and finally there is no net charge and no voltage difference anywhere in the circuit.

1.5.5 Capacitors in Series and in Parallel

Like resistors, capacitors can be joined together in series and in parallel. Connecting two capacitors in parallel results in a bigger capacitance value, since there is a larger plate area. Thus, the formula for total capacitance (CT) in a parallel circuit containing capacitors C1 and C2 is:

$$CT = C1 + C2$$

Note that the formula for calculating capacitance in parallel is similar to the one for calculating series resistance. By the same token, where several capacitors are connected in series the formula for calculating the total capacitance is:

$$CT = \frac{1}{\frac{1}{C1} + \frac{1}{C2} + \frac{1}{C3} \ldots}$$

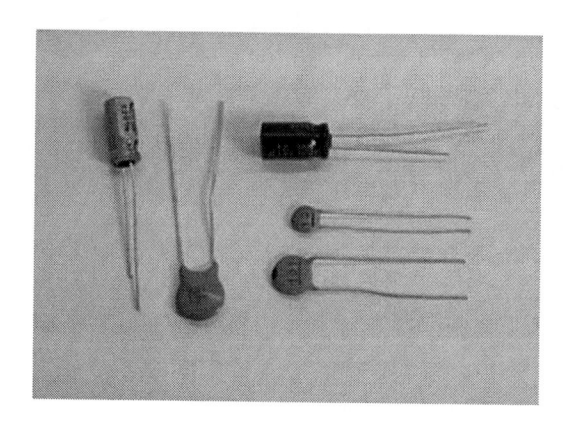

Figure 1-16 Assorted Commercial Capacitors

Note that the total capacitance of a connection in series is lower than for any capacitor in the series, considering that for a given voltage across the entire group there is less charge on each plate.

There are several types of commercial capacitors, including mylar, ceramic, disk, and electrolytic. Figure 1-16 shows several commercial capacitors.

1.5.6 Inductors

Inductors are the third type of basic circuit components. An inductor is a coil of wire with many windings. The wire windings are often made around a core of a magnetic material, such as iron. The properties of inductors are derived from magnetic rather than electric forces.

When current flows through a coil it produces a magnetic field in the space outside the wire. This makes the coil behave just like a natural, permanent magnet. Moving a wire through a magnetic field generates a current in the wire, and this current will flow through the associated circuit. Since it takes mechanical energy to move the wire through the field, then it is the mechanical energy that is transformed into electrical energy. A generator is a device that converts mechanical to electrical energy by means of induction. An electric motor is the opposite of a generator. In the motor electrical energy is converted to mechanical energy by means of induction.

The current in an inductor is similar to the voltage across a capacitor. In both cases it takes time to change the voltage from an initially high current flow. Such induced voltages can be very high and can damage other circuit components, so it is common to connect a resistor or a capacitor across the inductor to provide a current path to absorb the induced voltage. In combination inductors behave just like resistors: inductance adds in series. By the same token, parallel connection reduces induction. Induction is measured in henrys (h), but more commonly in mh, and μh.

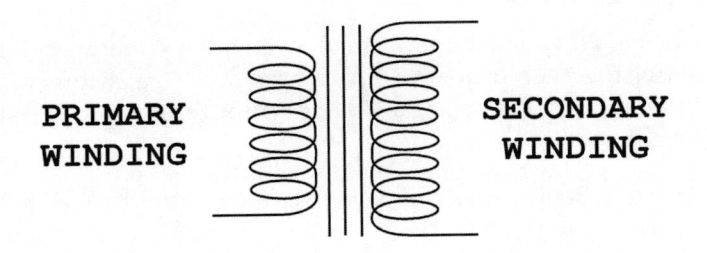

Figure 1-17 Transformer Schematics

1.5.7 Transformers

The *transformer* is an induction device that changes voltage or current levels. The typical transformer has two or more windings wrapped around a core made of laminated iron sheets. One of the windings, called the primary, receives a fluctuating current. The other winding, called the secondary, produces a current induced by the primary. Figure 1-17 shows the schematics of a transformer.

The device in Figure 1-17 is a step-up transformer. This is determined by the number of windings in the primary and secondary coils. The ratio of the number of turns in each winding determines the voltage increase. A transformer with an equal number of turns in the primary and secondary transfers the current unaltered. This type of device is sometimes called an isolation transformer. A transformer with less turns in the secondary than in the primary is a step-down transformer and its effect is to reduce the primary voltage at the secondary.

Transformers require an alternating or fluctuating current since it is the fluctuations in the current flow in the primary that induce a current in the secondary. The ignition coil in an automobile is a transformer that converts the low-level battery voltage to the high voltage level necessary to produce a spark.

1.6 Semiconductors

The name *semiconductor* stems from the property of some materials that act either as a conductor or as an insulator depending on certain conditions. Several elements are classified as semiconductors including Silicon, Zinc, and Germanium. Silicon is the most widely used semiconductor material because it is easily obtained.

In the ultra-pure form of silicon the addition of minute amounts of certain impurities (called *dopants*) alters the atomic structure of the silicon. This determines whether the Silicon can then be made to act as a conductor or as a nonconductor, depending upon the polarity of an electrical charge applied to it.

In the early days of radio, receivers required a device called a rectifier to detect signals. Ferdinand Braun used the rectifying properties of the galena crystal, a semiconductor material composed of lead sulfide, to create a "cat's whisker" diode that served this purpose. This was the first semiconductor device.

1.6.1 Integrated Circuits

Until 1959, electronic components performed a single function; therefore, many of them had to be wired together to create a functional circuit. Transistors were individually packaged in small cans. Packaging and hand wiring the components into circuits was extremely inefficient.

In 1959, at Fairchild Semiconductor, Jean Hoerni and Robert Noyce developed a process which made it possible to diffuse various layers onto the surface of a silicon wafer, while leaving a layer of protective oxide on the junctions. By allowing the metal interconnections to be evaporated onto the flat transistor surface the process replaced hand wiring. By 1961, nearly 90% of all the components manufactured were integrated circuits.

1.6.2 Semiconductor Electronics

To understand the workings of semiconductor devices we need to re-consider the nature of the electrical charge. Electrons are one of the components of atoms, and atoms are the building blocks of all matter. Atoms bond with each other to form molecules. Molecules of just one type of atom are called elements. In this sense gold, oxygen, and plutonium are elements since they all consist of only one type of atom. When a molecule contains more than one atom it is known as a compound. Water, which has both hydrogen and oxygen atoms, is a compound. Figure 1-18 represents an orbital model of an atom with five protons and three electrons.

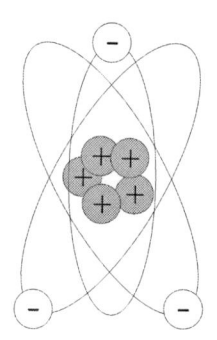

Figure 1-18 Orbital Model of the Boron Atom with its Valence Electrons

In Figure 1-18, protons carry positive charge and electrons carry negative charge. Neutrons, not represented in the illustration, are not electrically charged. Atoms that have the same number of protons and electrons have no net electrical charge.

Electrons that are far from the nucleus are relatively free to move around since the attraction from the positive charge in the nucleus is weak at large distances. In fact, it takes little force to completely remove an outer electron from an atom, leaving an ion with a net positive charge. A free electron can move at speeds approaching the speed of light (approximately 186,282 miles per second).

Electric current takes place in metal conductors due to the flow of free electrons. Because electrons have negative charge, the flow is in a direction opposite to the

Basic Electronics

positive current. Free electrons traveling through a conductor drift until they hit other electrons attached to atoms. These electrons are then dislodged from their orbits and replaced by the formerly free electrons. The newly freed electrons then start the process anew.

1.6.3 P-Type and N-Type Silicon

Semiconductor devices are made primarily of silicon. Pure silicon forms rigid crystals because of its four outermost electrons. Since it contains no free electrons it is not a conductor. But silicon can be made conductive by combining it with other elements (doping) such as boron and phosphorus. The boron atom has three outer valence electrons (Figure 1-18) and the phosphorus atom has five. When three silicon atoms and one phosphorus atom bind together, creating a structure of four atoms, there is an extra electron and a net negative charge.

The combination of silicon and phosphorous, with the extra phosphorus electron, is called an N-type silicon. In this case the N stands for the extra negative electron. The extra electron donated by the phosphorus atom can easily move through the crystal; therefore N-type silicon can carry an electrical current.

When a boron atom combines with a cluster of silicon atoms there is a deficiency of one electron in the resulting crystal. Silicon with a deficient electron is called P-type silicon (P stands for positive). The vacant electron position is sometimes called a "hole." An electron from another nearby atom can "fall" into this hole, thereby moving the hole to a new location. In this case, the hole can carry a current in the P-type silicon.

1.6.4 The Diode

Both P-type and N-type silicon conduct electricity. In either case, the conductivity is determined by the proportion of holes or the surplus of electrons. By forming some P-type silicon in a chip of N-type silicon it is possible to control electron flow so that it takes place in a single direction. This is the principle of the diode, and the p-n action is called a pn-junction.

A diode is said to have a forward bias if it has a positive voltage across it from the P- to N-type material. In this condition, the diode acts rather like a good conductor, and current can flow, as in Figure 1-19.

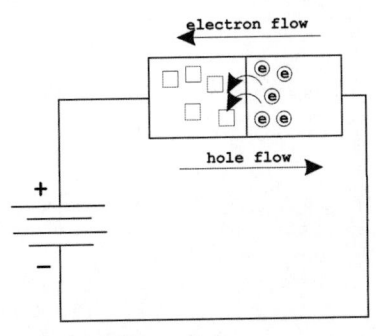

Figure 1-19 A Forward Biased Diode

If the polarity of the voltage applied to the silicon is reversed, then the diode is *reverse-biased* and appears nonconducting. This nonsymmetric behavior is due to the properties of the *pn-junction*. The fact that a diode acts like a one-way valve for current is a very useful characteristic. One application is to convert *alternating current* (AC) into *direct current* (DC). Diodes are so often used for this purpose that they are sometimes called rectifiers.

Chapter 2

Number Systems

In order to perform more efficient digital operations on numeric data, mathematicians have devised systems and structures that differ from those used traditionally. This chapter presents the background material necessary for understanding and using the number systems and numeric data storage structures employed in digital devices.

2.0 Counting

The fundamental application of a number system is counting. A stone-age hunter uses his or her fingers to show other members of the tribe how many mammoths were spotted at the bottom of the ravine. In this manner the hunter is able to transmit a unique type of information that does not relate to the species, size, or color of the animals, but to their numbers. Our minds have the ability to capture this notion of "oneness" independently from other properties of objects.

The most primitive method of counting consists of using objects to represent degrees of oneness. The stone-age hunter uses fingers to represent individual mammoth. Alternatively, the hunter could have resorted to pebbles, sticks, lines on the ground, or scratches on the cave wall to show how many units there were of the object.

2.0.1 The Tally System

The tally system probably originated from notches on a stick or scratches on a cave wall. In its simplest form each scratch, notch, or line represents an object. The method is so simple and intuitive that we still resort to it occasionally. Tallying requires no knowledge of quantity and no elaborate symbols. Had there been 12 mammoth in the ravine the cave wall would have appeared as follows:

$$| | | | | | | | | | | |$$

A logical evolution of the tally system consists of grouping the marks. Since we have five fingers in each hand, the 12 mammoth may have been grouped as follows:

$$| | | | | \ \ | | | | | \ \ | |$$

Perhaps a primitive mathematical genius added one final sophistication to the tally system. By drawing one tally line diagonally the visualization is further improved, as in this familiar style:

$$\cancel{||||} \quad \cancel{||||} \quad ||$$

2.0.2 Roman Numerals

Roman numerals show how a simple graphical tally system evolved into a symbolic numeric representation. The first five digits were encoded with the symbols:

I, II, III, IIII, and V

The Roman symbol V is conceivably a simplification of the tally encoding using a diagonal line to complete the grouping.

Table 2.1
Symbols in the Roman Numeration System

ROMAN	DECIMAL
I	1
V	5
X	10
L	50
C	100
D	500
M	1000

The Roman numeral system is based on an add-subtract rule whereby the elements of a number, read left-to-right, are either added or subtracted to the previous sum according to its value. Thereby the decimal number 1994 is represented in Roman numerals as follows:

$$\begin{aligned}
\text{MCMXCIV} &= M + (C - M) + (X - C) + (I - V) \\
&= 1000 + (1000 - 100) + (100 - 10) + (5 - 1) \\
&= 1000 + 900 + 90 + 4 \\
&= 1994
\end{aligned}$$

The uncertainty in the positional value of each digit, the absence of a symbol for zero, and the fact that some numbers require either one or two symbols (I, IV, V, IX, and X) complicate the rules of arithmetic using Roman numerals.

2.1 The Origins of the Decimal System

The one element of our civilization which has transcended all cultural and social differences is our decimal system of numbers. While mankind is yet to agree on the most desirable political order, on generally acceptable rules of moral behavior, or on a universal language, the Hindu-Arabic numerals have been adopted by practically all the nations and cultures of the world.

Number Systems

By the 9th century A.D. the Arabs were using a ten-symbol positional system of numbers which included the special symbol for 0. The Latin title of the first book on the subject of "Indian numbers" is *Liber Algorismi de Numero Indorum*. The author is the Arab mathematician *al-Khowarizmi*.

In spite of the evident advantages of this number system its adoption in Europe took place only after considerable debate and controversy. Many scholars of the time still considered Roman numerals to be easier to learn and more convenient for operations on the *abacus*. The supporters of the Roman numeral system, called abacists, engaged in intellectual combat with the algorists, who were in favor of the Hindu-Arabic numerals as described by al-Khowarizmi. For several centuries abacists and algorists debated about the advantages of their systems, with the Catholic church often siding with the abacists. This controversy explains why the Hindu-Arabic numerals were not accepted into general use in Europe until the beginning of the 16th century.

It is sometimes said that the reason for there being ten symbols in the Hindu-Arabic numerals is related to the fact that we have ten fingers. However, if we make a one-to-one correlation between the Hindu-Arabic numerals and our fingers, we find that the last finger must be represented by a combination of two symbols, 10. Also, one Hindu-Arabic symbol, 0, cannot be matched to an individual finger. In fact, the decimal system of numbers, as used in a positional notation that includes a zero digit, is a refined and abstract scheme which should be considered one of the greatest achievements of human intelligence. We will never know for certain if the Hindu-Arabic numerals are related to the fact that we have ten fingers, but its profoundness and usefulness clearly transcends this biological fact.

The most significant feature of the Hindu-Arabic numerals is the presence of a special symbol, 0, which by itself represents no quantity. Nevertheless, the special symbol 0 is combined with the other ones. In this manner the nine other symbols are reused to represent larger quantities. Another characteristic of decimal numbers is that the value of each digit depends on its position in a digit string. This positional characteristic, in conjunction with the use of the special symbol 0 as a placeholder, allows the following representations:

```
   1 = one
  10 = ten
 100 = hundred
1000 = thousand
```

The result is a counting scheme where the value of each symbol is determined by its column position. This positional feature requires the use of the special symbol, 0, which does not correspond to any unit-amount, but is used as a place-holder in multicolumn representations. We must marvel at the intelligence, capability for abstraction, and even the sense of humor of the mind that conceived a counting system that has a symbol that represents nothing. We must also wonder about the evolution of mathematics, science, and technology had this system not been invented. One intriguing question is whether a positional counting system that includes the zero symbol is a natural and predictable step in the evolution of our mathematical thought, or whether its invention was a stroke of genius that could have been missed for the next two thousand years.

2.1.1 Number Systems for Digital-Electronics

The computers built in the United States during the early 1940s operated on decimal numbers. However, in 1946, von Neumann, Burks, and Goldstine published a trend-setting paper titled Preliminary Discussion of the Logical Design of an Electronic Computing Instrument, in which they state:

> *"In a discussion of the arithmetic organs of a computing machine one is naturally led to a consideration of the number system to be adopted. In spite of the long-standing tradition of building digital machines in the decimal system, we must feel strongly in favor of the binary system for our device."*

In their paper, von Neumann, Burks, and Goldstine also consider the possibility of a computing device that uses binary-coded decimal numbers. However, the idea is discarded in favor of a pure binary encoding. The argument is that binary numbers are more compact than binary-coded decimals. Later in this book you will see that binary-coded decimal numbers (called BCD) are used today in some types of computer calculations.

In 1941, Konrad Zuse, a German who had done pioneering work in computing machines, released the first programmable computer designed to solve complex engineering equations. The machine, called the Z3, was controlled by perforated strips of discarded movie film and used the binary number system.

The use of the binary number system in digital calculators and computers was made possible by previous research on number systems and on numerical representations, starting with an article by G.W. Leibnitz published in Paris in 1703. Researchers concluded that it is possible to count and perform arithmetic operations using any set of symbols as long as the set contains at least two symbols, one of which must be zero.

In digital electronics the binary symbol 1 is equated with the electronic state ON, and the binary symbol 0 with the state OFF. The two symbols of the binary system can also represent conducting and nonconducting states, positive or negative, or any other bi-valued condition. It was the binary system that presented the Hindu-Arabic decimal number system with the first challenge in 800 years. In digital-electronics two steady states are easier to implement and more reliable than a ten-digit encoding.

2.1.2 Positional Characteristics

All modern number systems, including decimal, hexadecimal, and binary, are positional and include the digit zero. It is the positional feature that is used to determine the total value of a multi-digit representation. For example, the digits in the decimal number 4359 have the following positional weights:

Number Systems

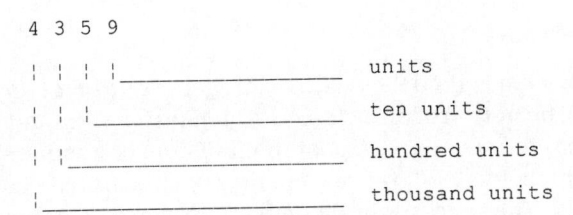

```
4 3 5 9
| | | |_____  units
| | |_____  ten units
| |_____  hundred units
|_____  thousand units
```

The total value is obtained by adding the column weights of each unit:

```
      4000  ---  4 thousand units
       300  ---  3 hundred units
 +      50  ---  5 ten units
         9  ---  9 unit
      ----
      4359
```

2.1.3 Radix or Base of a Number System

In any positional number system the weight of each column is determined by the total number of symbols in the set, including zero. This is called the base or radix of the system. The base of the decimal system is 10 and the base of the binary system is 2. The positional value or weight (P) of a digit in a multi-digit number is determined by the formula:

$$P = d \times B^c$$

where d is the digit, B is the base or radix, and c is the zero-based column number, starting from right to left. Note that the increase in column weight from right to left is purely conventional. You could construct a number system in which the column weights increase in the opposite direction. In fact, in the original Hindu notation the most significant digit was placed at the right.

In radix-positional terms a decimal number can be expressed as a sum of digits by the formula:

$$\sum_{i=-m}^{n} d_i \times 10^i$$

where i is the system's range and n is its limit.

2.2 Types of Numbers

By the adoption of special representations for different types of numbers the usefulness of a positional number system can be extended beyond the simple counting function.

2.2.1 Whole Numbers

The digits of a number system, called the positive integers or *natural numbers*, are an ordered set of symbols. The notion of an *ordered set* means that the numerical symbols are assigned a predetermined sequence. A positional system of numbers also requires the special digit zero which, by itself, represents the absence of oneness, or nothing, and thus is not included in the set of natural numbers. However, 0 assumes a cardinal function when it is combined with other digits, for instance, 10 or 30. The *whole numbers* are the set of natural numbers, including the number zero.

2.2.2 Signed Numbers

A number system can also encode direction. We generally use the + and - signs to represent opposite numerical directions. The typical illustration for a set of signed numbers is as follows:

```
-9 -8 -7 -6 -5 -4 -3 -2 -1   0 +1 +2 +3 +4 +5 +6 +7 +8 +9

negative numbers   <-        zero      -> positive numbers
```

The number zero, which separates the positive and the negative numbers, has no sign of its own, although in some binary encodings we can end up with a negative and a positive zero.

2.2.3 Rational, Irrational, and Imaginary Numbers

A number system also represents parts of a whole. For example, when a carpenter cuts one board into two boards of equal length we can represent the result with the fraction 1/2; the fraction 1/2 represents one of the two parts which make up the object. Rational numbers are those expressed as a ratio of two integers, for example, 1/2, 2/3, 5/248. Note that this use of the word rational is related to the mathematical concept of a ratio, and not to reason.

The denominator of a rational number expresses the number of potential parts. In this sense 2/5 indicates two of five possible parts. There is no reason why the number 1 cannot be used to indicate the number of potential parts, for example 2/1, 128/1. In this case the ratio x/1 indicates x elements of an undivided part. Therefore, it follows that x/1 = x. The implication is that the set of rational numbers includes the integers, since an integer can be expressed as a ratio by using a unit denominator.

But not all non-integer numbers can be written as an exact ratio of two integers. The discovery of the first irrational number is usually associated with the investigation of a right triangle by the Greek mathematician *Pythagoras* (approximately 600 BC). The *Pythagorean Theorem* states that in any right triangle the square of the longest side (hypotenuse) is equal to the sum of the squares of the other two sides.

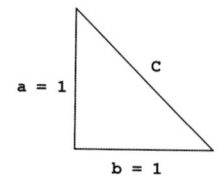

```
         c
a = 1
       b = 1
```

Number Systems

For this triangle, the Pythagorean theorem states that

$$a^2 + b^2 = c^2$$
$$2 = c^2$$
$$2 = c \times c$$
$$c = \sqrt{2}$$

Therefore, the length of the hypotenuse in a right triangle with unit sides is a number that, when multiplied by itself, gives 2. This number (approximately 1.414213562) cannot be expressed as the exact ratio of two integers. Other irrational numbers are the square roots of 3 and 5, as well as the mathematical constants π and e.

The set of numbers that includes the natural numbers, the whole numbers, and the rational and irrational numbers is called the real numbers. Most common mathematical problems are solved using real numbers. However, during the investigation of squares and roots we notice that there can be no real number whose square is negative. Mathematicians of the 18th century extended the number system to include operations with roots of negative numbers. They did this by defining an imaginary unit as follows:

$$i = \sqrt{-1}$$

The imaginary unit makes possible a new set of numbers, called complex numbers, that consist of a real part and an imaginary part. One of the uses of complex numbers is in finding the solution of a quadratic equation. Complex numbers are also useful in vector analysis, graphics, and in solving many engineering, scientific, and mathematical problems.

2.3 Radix Representations

The radix of a number system is the number of symbols in the set, including zero. Thus, the radix of the decimal system is 10, and the radix of the binary system is 2. Digital electronics is based on circuits that can be in one of two stable states. Therefore, a number system based on two symbols is better suited for work in digital electronics, since each state can be represented by a digit.

2.3.1 Decimal versus Binary Numbers

The binary system of numbers uses two symbols, 1 and 0. It is the simplest possible set of symbols with which we can count and perform arithmetic. Most of the difficulties in learning and using the binary system arise from this simplicity. Figure 2.1 shows sixteen groups of four electronic cells each in all possible combinations of two states.

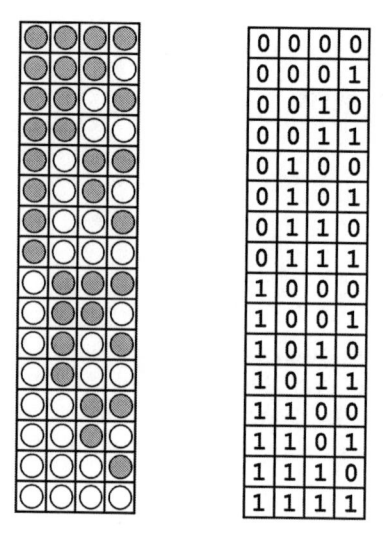

Figure 2-1 Electronic Cells and Binary Numbers

It is interesting to note that binary numbers match the physical state of each electronic cell. If we think of each cell as a miniature light bulb, then the binary number 1 can be used to represent the state of a charged cell (light ON) and the binary number 0 to represent the state of an uncharged cell (light OFF).

2.3.2 Hexadecimal and Octal

Binary numbers are convenient in digital electronics; however, one of their drawbacks is the number of symbols required to encode a large value. For example, the number 9134 is represented in four decimal digits. However, the binary equivalent 10001110101110 requires fourteen digits. In addition, large binary numbers are difficult to remember.

One possible way of compensating for these limitations of binary numbers is to use individual symbols to represent groups of binary digits. For example, a group of three binary numbers allows eight possible combinations. In this case, we can use the decimal digits 0 to 7 to represent each possible combination of three binary digits. This grouping of three binary digits gives rise to the following table:

```
         binary          octal

         0 0 0             0

         0 0 1             1

         0 1 0             2

         0 1 1             3

         1 0 0             4

         1 0 1             5

         1 1 0             6

         1 1 1             7
```

Number Systems

The *octal* encoding serves as a shorthand representation for groups of 3-digit binary numbers.

Hexadecimal numbers (base 16) are used for representing values encoded in four binary digits. Since there are only ten decimal digits, the hexadecimal system borrows the first six letters of the alphabet (A, B, C, D, E, and F). The result is a set of sixteen symbols, as follows:

$$0\ 1\ 2\ 3\ 4\ 5\ 6\ 7\ 8\ 9\ A\ B\ C\ D\ E\ F$$

Most modern computers are designed with memory cells, registers, and data paths in multiples of four binary digits. Table 2.2 lists some common units of memory storage.

Table 2.2
Units of Memory Storage

UNIT	BITS	HEX DIGITS	HEX RANGE
Nibble	4	1	0 to F
Byte	8	2	0 to FF
Word	16	4	0 to FFFF
Doubleword	32	8	0 to FFFFFFFF

In most digital-electronic devices memory addressing is organized in multiples of four binary digits. Here again, the hexadecimal number system provides a convenient way to represent addresses. Table 2.3 lists some common memory addressing units and their hexadecimal and decimal range.

Table 2.3
Units of Memory Addressing

UNIT	DATA PATH IN BITS	ADDRESS RANGE DECIMAL	HEX
1 paragraph	4	0 to 15	0-F
1 page	8	0 to 255	0-FF
1 kilobyte	16	0 to 65,535	0-FFFF
1 megabyte	20	0 to 1,048,575	0-FFFFF
4 gigabytes	32	0 to 4,294,967,295	0-FFFFFFFF

2.4 Number System Conversions

We use decimal numbers in our everyday life because they meaningfully represent common units used in the real world. To state that a certain historical event took place in the year 7C6 hexadecimal would convey little information to the average person. However, in computer systems based on two-state electronic cells binary representations are more convenient. Also note that hexadecimal and octal numbers are handy shorthand for representing groups of binary digits.

Numerical conversions between positional systems of different radices are based on the number of symbols in the respective sets and on the positional value (weight) of each column. But methods used for manual conversions are not always suitable for machine conversions, as we will see in the forthcoming sections.

2.4.1 Binary-to-ASCII-Decimal

To manually convert a binary number to its decimal equivalent we take into account the positional weight of each binary digit, as shown in Figure 2-2.

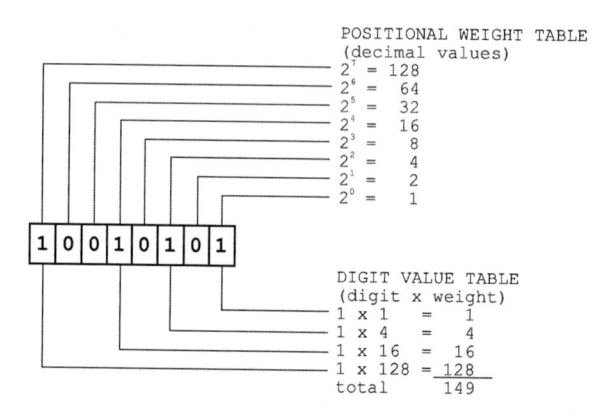

```
                              POSITIONAL WEIGHT TABLE
                              (decimal values)
                              2^7 = 128
                              2^6 =  64
                              2^5 =  32
                              2^4 =  16
                              2^3 =   8
                              2^2 =   4
                              2^1 =   2
                              2^0 =   1

        |1|0|0|1|0|1|0|1|
                              DIGIT VALUE TABLE
                              (digit x weight)
                              1 x   1 =    1
                              1 x   4 =    4
                              1 x  16 =   16
                              1 x 128 =  128
                              total      149
```

Figure 2-2 Binary to ASCII Decimal Conversion Example

The positional weight table in Figure 2-2 lists the decimal value of each binary column. These weights are powers of the system's base (2 in the binary system). In the digit value table, also in Figure 2-2, the decimal values of the binary columns holding a 1 digit are added. The sum of the weights of all the one-digits in the operand is the decimal equivalent of the binary number. In this case 10010101 binary = 149 decimal.

The method in Figure 2-2, although useful in manual conversions, is not an algorithm for computer conversions. Figure 2-3 is a flowchart of a low-level binary-to-decimal conversion routine.

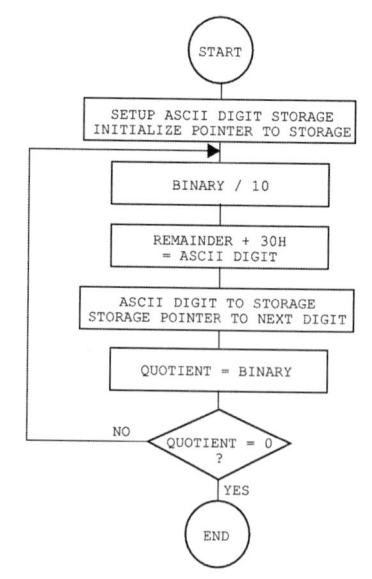

```
                    ( START )
                        |
          +-----------------------------+
          | SETUP ASCII DIGIT STORAGE   |
    +---->| INITIALIZE POINTER TO STORAGE|
    |     +-----------------------------+
    |                   |
    |     +-----------------------------+
    |     |        BINARY / 10          |
    |     +-----------------------------+
    |                   |
    |     +-----------------------------+
    |     |      REMAINDER + 30H        |
    |     |       = ASCII DIGIT         |
    |     +-----------------------------+
    |                   |
    |     +-----------------------------+
    |     |  ASCII DIGIT TO STORAGE     |
    |     | STORAGE POINTER TO NEXT DIGIT|
    |     +-----------------------------+
    |                   |
    |     +-----------------------------+
    |     |     QUOTIENT = BINARY       |
    |     +-----------------------------+
    |                   |
    |    NO       < QUOTIENT = 0 ? >
    +-------------      |
                       YES
                        |
                    (  END  )
```

Figure 2-3 Flowchart for a Binary to ASCII Decimal Conversion

Number Systems

The algorithm for the processing in Figure 2-3 can be written as follows:

1. Set up and initialize a string storage area (sometimes called a buffer) to hold the ASCII decimal digits of the result. Set up the buffer pointer to the right-most digit position of the result.
2. Obtain the remainder of the value divided by 10.
3. Add 30H to remainder digit to convert to ASCII representation.
4. Store remainder digit in buffer and index the buffer pointer to the preceding digit.
5. Quotient of division by 10 becomes the new binary value.
6. End conversion routine if quotient is equal to 0. Otherwise, continue at step 2.

Note that the numerical digits are located from 30H to 39H in the ASCII table. This makes is easy to convert a binary digit to ASCII simply by adding 30H. Likewise, an ASCII digit is converted to binary by subtracting 30H.

2.4.2 Binary-to-Hexadecimal Conversion

The method described in Section 2.4.1 for a binary to ASCII decimal conversion can be adapted to other radices by representing the positional weight of each binary digit in the number system to which the conversion is to be made. In the case of a binary to ASCII hexadecimal conversion the positional weight of each binary digit is a hexadecimal value. Figure 2-4 shows the conversion of the binary value 10010101 into hexadecimal by using the corresponding positional weights.

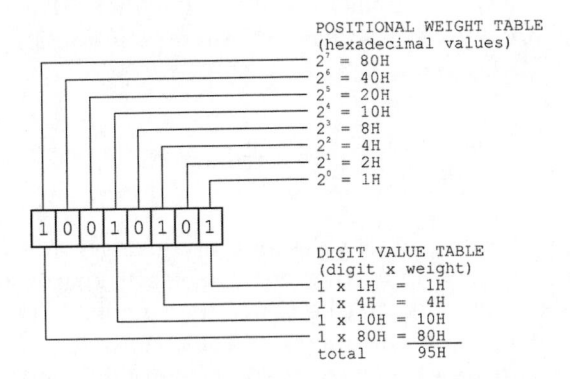

```
                        POSITIONAL WEIGHT TABLE
                        (hexadecimal values)
                        2^7 = 80H
                        2^6 = 40H
                        2^5 = 20H
                        2^4 = 10H
                        2^3 = 8H
                        2^2 = 4H
                        2^1 = 2H
                        2^0 = 1H
   ┌─┬─┬─┬─┬─┬─┬─┬─┐
   │1│0│0│1│0│1│0│1│
   └─┴─┴─┴─┴─┴─┴─┴─┘
                        DIGIT VALUE TABLE
                        (digit x weight)
                        1 x  1H =   1H
                        1 x  4H =   4H
                        1 x 10H =  10H
                        1 x 80H =  80H
                        total      95H
```

Figure 2-4 Binary to ASCII Hexadecimal Conversion Example

The machine conversion binary to ASCII hexadecimal is similar to the binary to ASCII decimal algorithm described previously. In the case of the conversion into ASCII hexadecimal digits the buffer need only hold four ASCII characters, since a 16-bit binary cannot exceed the value FFFFH. In the case of binary to ASCII hex the divisor for obtaining the digits is 16 instead of 10.

2.4.3 Decimal-to-Binary Conversion

Longhand conversion of decimal into binary can be performed by using the positional weights to find the binary 1-digits and then subtracting this positional weight from the decimal value. The process is shown in Figure 2-5.

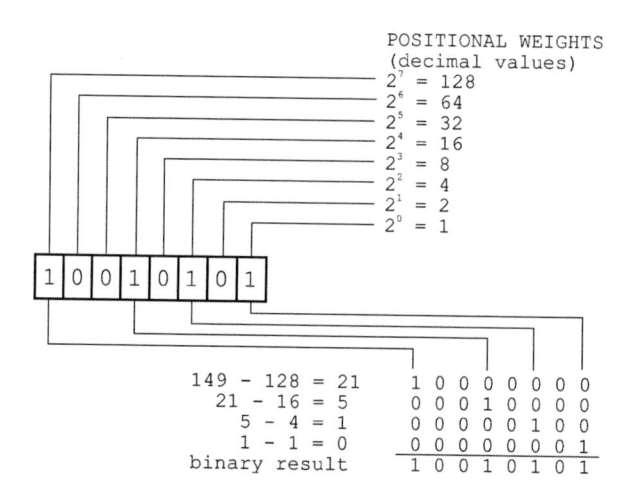

```
                                    POSITIONAL WEIGHTS
                                      (decimal values)
                                      2⁷ = 128
                                      2⁶ = 64
                                      2⁵ = 32
                                      2⁴ = 16
                                      2³ = 8
                                      2² = 4
                                      2¹ = 2
                                      2⁰ = 1

              | 1 | 0 | 0 | 1 | 0 | 1 | 0 | 1 |


              149 - 128 = 21      1 0 0 0 0 0 0 0
               21 -  16 =  5      0 0 0 1 0 0 0 0
                5 -   4 =  1      0 0 0 0 0 1 0 0
                1 -   1 =  0      0 0 0 0 0 0 0 1
              binary result       1 0 0 1 0 1 0 1
```

Figure 2-5 Example of Decimal to Binary Conversion

In the example of Figure 2-5 we start with the decimal value 149. Since the highest power of 2 smaller than 149 is 128, which corresponds to bit 7, we set bit 7 in the result and perform the subtraction:

$$149 - 128 = 21$$

At this point the highest positional weight smaller than 21 is 16, which corresponds to bit 4. Therefore we set bit 4, and perform the subtraction:

$$21 - 16 = 5$$

The remaining steps in the conversion can be seen in the illustration. The conversion is finished when the result of the subtraction is 0.

Suppose there is a numerical value in the form of a string of ASCII decimal, octal, or hexadecimal digits. In order for a processor to perform simple arithmetic operations on such data, the data must first be converted to binary. The binary value is then loaded into machine registers or memory cells. However, methods suited for manual conversion do not always make a good computer algorithm. Figure 2.6 shows two decimal-to-binary conversion algorithms that are suited for machine coding.

Using the first method of Figure 2-6, the individual decimal digits are multiplied by their corresponding positional values. The final result is obtained by adding all the partial products. Although this method is frequently used, it has the disadvantage that a different multiplier is used during each iteration (1, 10, 100, 1000). The second method in Figure 2-6 starts with the high-order ASCII-decimal digit. The calculations consist of multiplying an accumulated value by 10. Initially, this accumulated value is set to 0. After multiplication by 10, the value of the digit is added to the accumulated value. The following algorithm is based on the second method in Figure 2-6.

Number Systems

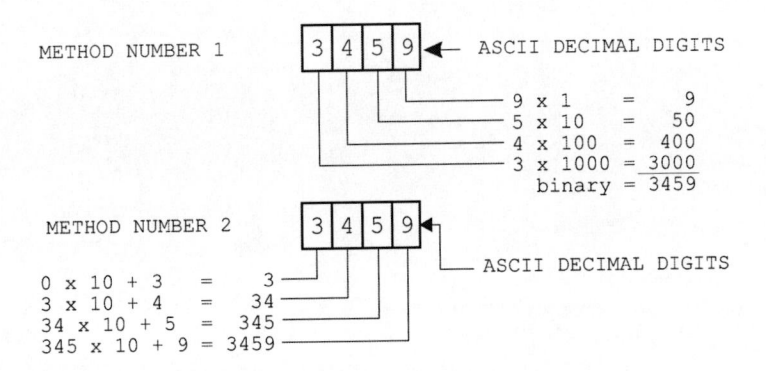

```
METHOD NUMBER 1       3 4 5 9  ← ASCII DECIMAL DIGITS
                              9 x 1    =     9
                              5 x 10   =    50
                              4 x 100  =   400
                              3 x 1000 =  3000
                                binary =  3459

METHOD NUMBER 2       3 4 5 9
                              ASCII DECIMAL DIGITS
   0 x 10 + 3   =    3
   3 x 10 + 4   =   34
  34 x 10 + 5   =  345
 345 x 10 + 9   = 3459
```

Figure 2-6 Machine Conversion of ASCII Decimal to Binary

1. Set up and initialize to binary zero a storage location for holding the value accumulated during conversion. Set up a pointer to the highest order ASCII digit in the source string.
2. Test the ASCII digit for a value in the range 0 to 9. End of routine if the ASCII digit is not in this range.
3. Subtract 30H from ASCII decimal digit.
4. Multiply accumulated value by 10.
5. Add digit to accumulated value.
6. Increment the pointer to the next digit and continue at step 2.

   Figure 2-7 is a flowchart of the conversion algorithm.

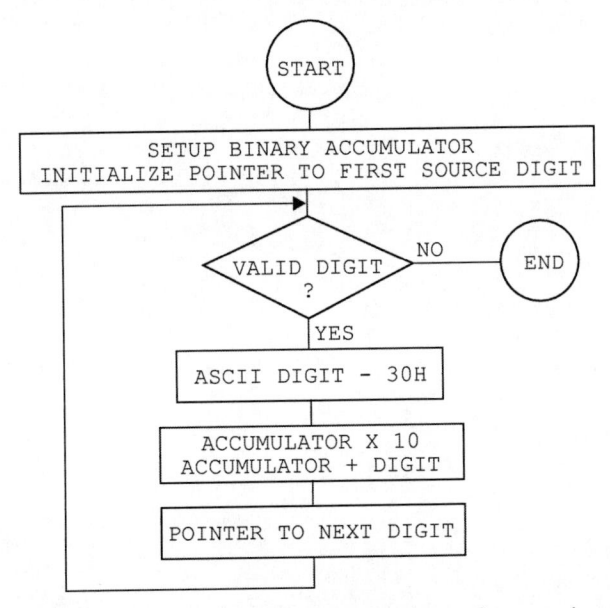

```
                    ( START )
                        │
        ┌───────────────────────────────────┐
        │      SETUP BINARY ACCUMULATOR     │
        │ INITIALIZE POINTER TO FIRST SOURCE DIGIT │
        └───────────────────────────────────┘
                        │
           ┌────────────┘
           │       ╱╲
           │      ╱  ╲       NO
           │     ╱VALID╲──────────( END )
           │     ╲DIGIT╱
           │      ╲ ? ╱
           │       ╲╱
           │        │ YES
           │   ┌─────────────────┐
           │   │ ASCII DIGIT - 30H│
           │   └─────────────────┘
           │        │
           │   ┌─────────────────┐
           │   │ ACCUMULATOR X 10 │
           │   │ ACCUMULATOR + DIGIT│
           │   └─────────────────┘
           │        │
           │   ┌─────────────────┐
           │   │POINTER TO NEXT DIGIT│
           │   └─────────────────┘
           └────────┘
```

Figure 2-7 Flowchart for ASCII to Machine Register Conversion

Chapter 3

Data Types and Data Storage

In this chapter we review the various encodings and formats used for representing character and numeric data in digital systems. Tha character formats are used for encoding the letters, symbols, and control codes of the various alphabets. The numeric formats allow representing binary numbers as signed and unsigned integers in several forms, binary floating-point numbers, and decimal floating-point numbers, usually called *binary-coded decimals* or *BCD*.

3.0 Electronic-Digital Machines

The mechanization of arithmetic is often traced back to the abacus, slide rule, mechanical calculators, and punch card machines. The work of *John von Neumann* at Princeton's *Institute for Advanced Study and Research* marks the first highlight in the design and construction of a digital-electronic calculating machine. In von Neumann's design, data and instructions are stored in a common memory area. An alternative approach, known as Harvard architecture, was discarded at first but has recently been re-validated and is in use in several microcontroller families.

The calculating power of the first computer was approximately 2000 operations per second, while previous electro-mechanical devices were capable of performing only 3 or 4 operations. Today's digital machines can execute more than 1 billion instructions per second. Technological advances and miniaturization techniques have reduced the cost and size of computing machinery.

3.1 Character Representations

Over the years, data representation issues have often been determined by the various conventions used by the different hardware manufacturer. Machines have had different word lengths and different character sets and have used various schemes for storing character and data. Fortunately, in microprocessor and microcontroller design, the encoding of character data has not been subject to major disagreements.

Historically, the methods used to represent characters have varied widely, but the basic approach has always been to choose a fixed number of bits and then map the various bit combinations to the various characters. Clearly, the number of bits of the storage format limits the total number of distinct characters that can be represented. In this manner, the 6-bit codes used on a number of earlier computing machines allow representing 64 characters. This range allows including the uppercase letters, the decimal digits, some special characters, but not the lowercase letters. Computer manufacturers that used the 6-bit format often argued that their customers had no need for lower-case letters. Nowadays 7- and 8-bit codes that allow representing the lower-case letters have been adopted almost universally.

Most of the world (except IBM) has standardized character representations by using the *ISO (International Standards Organization)* code. ISO exists in several national variants; the one used in the United States is called ASCII, which stands for *American Standard Code for Information Interchange.* All microcomputers and microcontrollers use ASCII as the code for character representation.

3.1.1 ASCII

ASCII is a character encoding based on the English alphabet. ASCII was first published as a standard in 1967 and was last updated in 1986. The first 33 codes, referred to as non-printing codes, are mostly obsolete control characters. The remaining 95 printable characters (starting with the space character) include the common characters found in a standard keyboard, the decimal digits, and the upper- and lower-case characters of the English alphabet. Table 3.1 lists the ASCII characters in decimal, hexadecimal, and binary.

Table 3.1
ASCII Character Representation

DECIMAL	HEX	BINARY	VALUE	
000	000	00000000	annual	(Null character)
001	001	00000001	SOH	(Start of Header)
002	002	00000010	STX	(Start of Text)
003	003	00000011	ETX	(End of Text)
004	004	00000100	EOT	(End of Transmission)
005	005	00000101	ENQ	(Enquiry)
006	006	00000110	ACK	(Acknowledgment)
007	007	00000111	BEL	(Bell)
008	008	00001000	BS	(Backspace)
009	009	00001001	HT	(Horizontal Tab)
010	00A	00001010	LF	(Line Feed)
011	00B	00001011	VT	(Vertical Tab)
012	00C	00001100	FF	(Form Feed)
013	00D	00001101	CR	(Carriage Return)
014	00E	00001110	SO	(Shift Out)
015	00F	00001111	SI	(Shift In)
016	010	00010000	DLE	(Data Link Escape)
017	011	00010001	DC1	(XON)(Device Control 1)
018	012	00010010	DC2	(Device Control 2)
019	013	00010011	DC3	(XOFF)(Device Control 3)
020	014	00010100	DC4	(Device Control 4)
021	015	00010101	NAK	(- Acknowledge)
022	016	00010110	SYN	(Synchronous Idle)

(continues)

Data Types and Data Storage

Table 3.1

ASCII Character Representation (conitnued)

DECIMAL	HEX	BINARY	VALUE	
000	000	00000000	annual	(Null character)
023	017	00010111	ETB	(End of Trans. Block)
024	018	00011000	CAN	(Cancel)
025	019	00011001	EM	(End of Medium)
026	01A	00011010	SUB	(Substitute)
027	01B	00011011	ESC	(Escape)
028	01C	00011100	FS	(File Separator)
029	01D	00011101	GS	(Group Separator)
030	01E	00011110	RS	(Request to Send)
031	01F	00011111	US	(Unit Separator)
032	020	00100000	SP	(Space)
033	021	00100001	!	(exclamation mark)
034	022	00100010	"	(double quote)
035	023	00100011	#	(number sign)
036	024	00100100	$	(dollar sign)
037	025	00100101	%	(percent)
038	026	00100110	&	(ampersand)
039	027	00100111	'	(single quote)
040	028	00101000	(	(left/opening parenthesis)
041	029	00101001	)	(right/closing parenthesis)
042	02A	00101010	*	(asterisk)
043	02B	00101011	+	(plus)
044	02C	00101100	,	(comma)
045	02D	00101101	-	(minus or dash)
046	02E	00101110	.	(dot)
047	02F	00101111	/	(forward slash)
048	030	00110000	0	(decimal digits ...)
049	031	00110001	1	
050	032	00110010	2	
051	033	00110011	3	
052	034	00110100	4	
053	035	00110101	5	
054	036	00110110	6	
055	037	00110111	7	
056	038	00111000	8	
057	039	00111001	9	
058	03A	00111010	:	(colon)
059	03B	00111011	;	(semi-colon)
060	03C	00111100	<	(less than)
061	03D	00111101	=	(equal sign)
062	03E	00111110	>	(greater than)
063	03F	00111111	?	(question mark)
064	040	01000000	@	(AT symbol)
065	041	01000001	A	
066	042	01000010	B	
067	043	01000011	C	
...				
090	05A	01011010	Z	
091	05B	01011011	[	(left/opening bracket)
092	05C	01011100	\	(back slash)
093	05D	01011101	]	(right/closing bracket)

(continues)

Table 3.1

ASCII Character Representation (conitnued)

DECIMAL	HEX	BINARY	VALUE	
094	05E	01011110	^	(circumflex)
095	05F	01011111	_	(underscore)
096	060	01100000	`	(accent)
097	061	01100001	a	
098	062	01100010	b	
099	063	01100011	c	
...				
122	07A	01111010	z	
123	07B	01111011	{	(left/opening brace)
124	07C	01111100	|	(vertical bar)
125	07D	01111101	}	(right/closing brace)
126	07E	01111110	~	(tilde)
127	07F	01111111	DEL	(delete)

3.1.2 EBCDIC and IBM

In spite of ASCII's general acceptance, IBM continues to use EBCDIC (*Extended Binary Coded Decimal Interchange Code*) for character encoding. IBM mainframes and midrange systems such as the AS/400 use a wholly incompatible character set primarily designed for punched cards.

EBCDIC uses the full eight bits available to it, so there is no place left to implement parity checking. On the other hand, EBCDIC has a wider range of control characters than ASCII.

EBCDIC character encoding is based on Binary Coded Decimal (BCD), which we discuss later in this chapter. There are four main blocks in the EBCDIC code page:

1. The range 0000 0000 to 0011 1111 is reserved for control characters.
2. The range 0100 0000 to 0111 1111 is for punctuation.
3. The range 1000 0000 to 1011 1111 is for lowercase characters.
4. The range 1100 0000 to 1111 1111 is for uppercase characters and numbers.

Actually, microprocessor and microcontroller design need not address how character data is encoded. Usually a set of instructions allows manipulating 8-bit quantities, but the processor need not be concerned with what the encodings represent. On the other hand, some mainframe processors do have instructions that manipulate character codes. For example, the EDIT instruction on the IBM 370 implements the kind of picture conversion that appears in COBOL programs.

3.1.3 Unicode

One of the limitations of the ASCII code is that eight bits are not enough for representing characters sets in languages such as Japanese or Chinese which use large character sets. This has led to the development of encodings which allow representing large character sets. *Unicode* has been proposed as a universal character encoding standard that can be used for representation of text for computer processing.

Unicode attempts to provide a consistent way of encoding multilingual text and thus make it possible to exchange text files internationally. The design of Unicode is based on the ASCII code, but goes beyond the Latin alphabet to which ASCII is limited. The Unicode Standard provides the capacity to encode all of the characters used for the written languages of the world. Like ASCII, Unicode assigns each character a unique numeric value and name. Unicode uses three encoding forms that use a common repertoire of characters. These forms allow encoding as many as a million characters.

The three encoding forms of the Unicode Standard allow the same data to be transmitted in a byte, word, or double word format, that is, in 8-, 16- or 32-bits per character.

- UTF-8 is a way of transforming all Unicode characters into a variable length encoding of bytes. In this format the Unicode characters corresponding to the familiar ASCII set have the same byte values as ASCII. By the same token, Unicode characters transformed into UTF-8 can be used with existing software.
- UTF-16 is designed to balance efficient access to characters with economical use of storage. It is reasonably compact and all the heavily used characters fit into a single 16-bit code unit, while all other characters are accessible via pairs of 16-bit code units.
- UTF-32 is used where memory space is no concern, but fixed width, single code unit access to characters is desired. In UTF-32 each Unicode character is represented by a single 32-bit code.

3.2 Storage and Encoding of Integers

The Indian mathematician Pingala first described binary numbers in the fifth century B.C. The modern system of binary numbers first appears in the work of Gottfried Leibniz during the seventeenth century. During the mid-nineteenth century the British logician George Boole described a logical system which used binary numbers to represent logical true and false. In 1937, Claude Shannon published his master's thesis that used Boolean algebra and binary arithmetic to implement electronic relays and switches. The thesis paper entitled *A Symbolic Analysis of Relay and Switching Circuits* is usually considered to be the origin of modern digital circuit design.

Also in 1937, George Stibitz completed a relay-based computer which could perform binary addition. The Bell Labs *Complex Number Computer*, also designed by Stibitz, was completed in January 1940. The system was demonstrated to the *American Mathematical Society* in September 1940. The attendants included John Von Neumann, John Mauchly, and Norbert Wiener. In 1945, von Neumann wrote a seminal paper in which he stated that binary numbers were the ideal computational format.

3.2.1 Signed and Unsigned Representations

For unsigned integers there is little doubt that the binary representation is ideal. Successive bits indicate powers of 2, with the most significant bit at the left and the least significant one on the right, as is customary in decimal representations. Figure 3-1 shows the digit weights and the conventional bit numbering in the binary encoding.

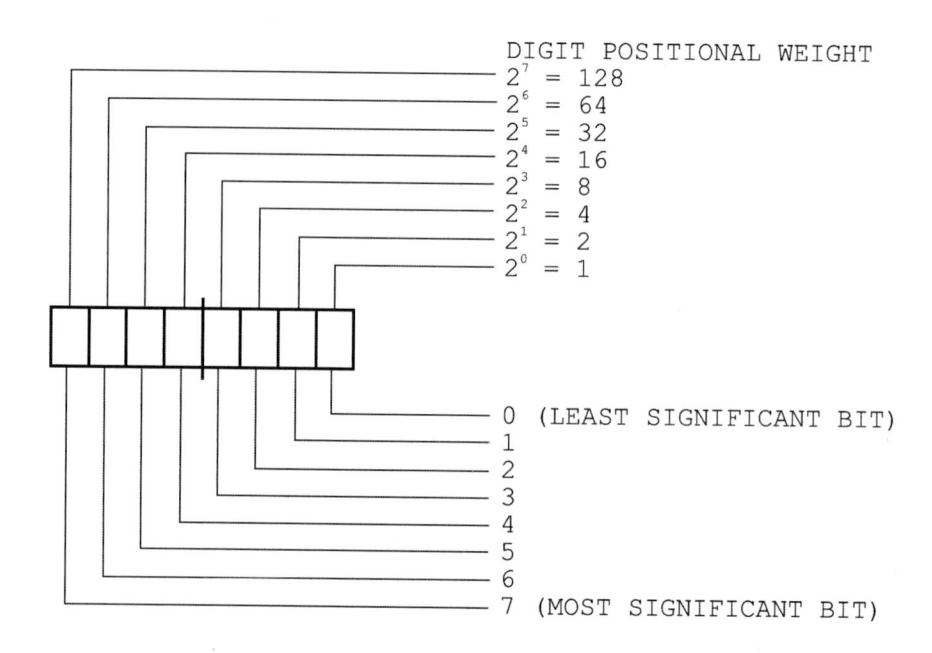

```
                                    DIGIT POSITIONAL WEIGHT
                                    2⁷ = 128
                                    2⁶ = 64
                                    2⁵ = 32
                                    2⁴ = 16
                                    2³ = 8
                                    2² = 4
                                    2¹ = 2
                                    2⁰ = 1

        [ | | | | | | | | ]
                                    0 (LEAST SIGNIFICANT BIT)
                                    1
                                    2
                                    3
                                    4
                                    5
                                    6
                                    7 (MOST SIGNIFICANT BIT)
```

Figure 3-1 Binary Digit Weights and Numbering

In order to perform arithmetic operations, the digital machine must be capable of storing and retrieving numerical data. Numerical data is stored in standard formats, designed to minimize space and optimize processing. Historically, numeric data was stored in data structures devised to fit the characteristics of a specific machine, or the preferences of its designers. It was in 1985 that the *Institute of Electrical and Electronics Engineers* (IEEE) and the *American National Standards Institute* (ANSI) formally approved mathematical standards for encoding and storing numerical data in digital devices.

The electronic and physical mechanisms used for storing data have evolved with technology. One common feature of many devices, from punched tape to integrated circuits, is that the encoding is represented in two possible states. In paper tape the two states are holes or no holes, while in electronic media they are usually the presence or absence of an electrical charge.

Data stored in processor registers, in magnetic media, in optical devices, or in punched tape is usually encoded in binary. Thus, the programmer and the operator can usually ignore the physical characteristics of the storage medium. In other words, the bit pattern 10010011 can be encoded as holes in a strip of paper tape, as magnetic charges on a mylar-coated disk, as positive voltages in an integrated circuit memory cell, or as minute craters on the surface of the CD. In all cases 10010011 represents the decimal number 147.

3.2.2 Word Size

In electronic digital devices the bistable states are represented by a *binary digit*, or *bit*. Circuit designers group several individual cells to form a unit of storage that holds

Data Types and Data Storage

several bits. In a particular machine the basic unit of data storage is called the word size. Word size in computers often ranges from 8 to 128 bits, in powers of 2. Microcontrollers and other digital devices sometimes use word-sizes that are determined by their specific architectures. For example, some PIC microcontrollers use a 14-bit word size.

In most digital machines the smallest unit of storage individually addressable is eight bits (one *byte*). Individual bits are not directly addressable and must be manipulated as part of larger units of data storage.

3.2.3 Byte Ordering

The storage of a single-byte integer can be done according to the scheme in Figure 3-1. However, the maximum value that can be represented in eight bits is the decimal number 255. To represent larger binary integers requires additional storage area. Since memory is usually organized in byte-size units, any decimal number larger than 255 requires more than one byte of storage. In this case the encoding is padded with the necessary leading zeros. Figure 3-2 is a representation of the decimal number 21,141 stored in two consecutive data bytes.

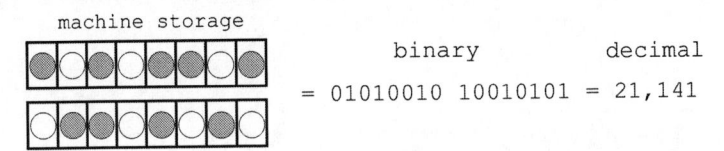

```
       machine storage
                               binary            decimal
                          = 01010010 10010101 = 21,141
```

Figure 3-2 Representation of an Unsigned Integer

One issue related to using multiple memory bytes to encode binary integers is the successive layout of the various byte-size units. In other words, does the representation store the most significant byte at the lowest numbered memory location, or viceversa. For example, when a 32-bit binary integer is stored in a 32-bit storage area we can follow the conventional pattern of placing the low-order bit on the right-hand side and the high-order bit on the left, as we did in Figure 3-1. However, if the 32-bit number is to be stored into four byte size memory cells, then two possible storage schemes are possible, as shown in Figure 3-3.

LOW-TO-LOW STORAGE SCHEME

HIGH-TO-LOW STORAGE SCHEME

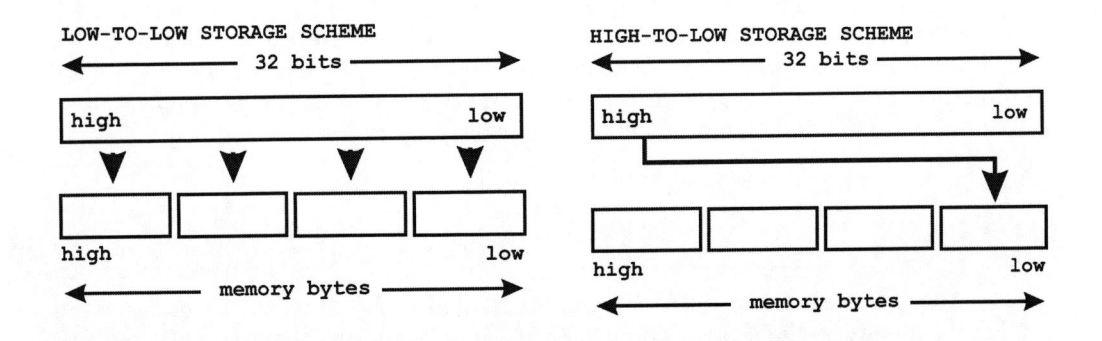

Figure 3-3 Byte Ordering Schemes

In the low-to-low storage scheme the low-order 8-bits of the operand are stored in the low-order memory byte, the next group of 8-bits are moved to the following memory byte in low-to-high order, and so on. Conceivably, this scheme can be described by saying that the "little end" of the operand is stored first, that is, in lowest memory. According to this notion, the storage scheme is described as the *little-endian* format. If the "big-end" of the operand, that is, the highest valued bits, is stored in the low memory addresses then the byte ordering is said to be in *big-endian* format. Some Intel processors (like those of 80x86 family) follow the little-endian format. Some Motorola processors (like those of the 68030 family) follow the big-endian format, while others (such as the MIPS 2000) can be configured to store data in either format.

In many situations the programmer needs to be aware of the byte-ordering scheme; for example, to retrieve memory data into processor registers so as to perform multi-byte arithmetic, or to convert data stored in one format to the other one. This last operation is a simple byte-swap. For example, if the hex value 01020304 is stored in four consecutive memory cells in low-to-high order (little-endian format) it appears in memory (low-to-high) as the values 04030201. Converting this data to the big-endian format consists of swapping the individual bytes so that they are stored in the order 01010304. Figure 3-4 is a diagram of a byte swap operation.

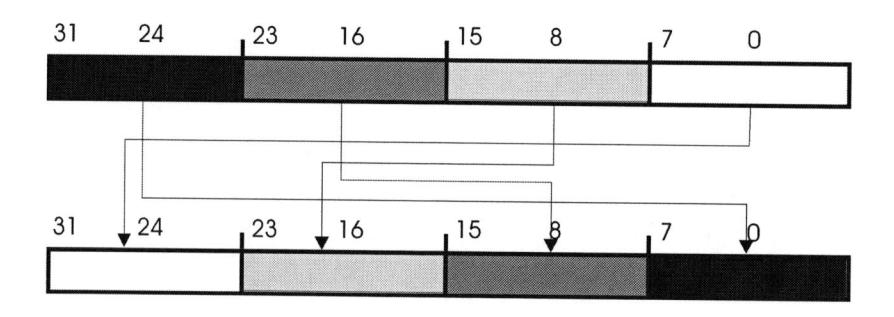

Figure 3-4 Data Format Conversion by Byte Swapping

3.2.4 Sign-Magnitude Representation

Representing signed numbers requires differentiating between positive and negative magnitudes. One possible scheme is to devote one bit to represent the sign. Typically the high-order bit is set (1) to denote negatives and reset (0) to denote positives. Using this convention the decimal numbers 93 and -93 are represented as follows:

```
01011101 binary = 93 decimal
11011101 binary = -93 decimal
|
|----------- sign bit
```

This way of designating negative numbers, called a *sign-magnitude* representation, corresponds to the conventional way in which we write negative and positive numbers longhand, that is, we precede the number by its sign. Sign-magnitude representation has the following characteristics:

Data Types and Data Storage 41

1. The absolute value of positive and negative numbers is the same.
2. Positive numbers can be distinguished from negative numbers by examining the high-order bit.
3. There are two possible representations for zero, one negative (10000000B) and one positive (00000000B).

But a major limitation of sign-magnitude representation is that the processing required to perform addition is different from that for subtraction. Complicated rules are required for the addition of signed numbers. For example, considering two operands labeled x and y, the following rules must be observed for performing signed addition:

1. If x and y have the same sign, they are added directly and the result is given the common sign.
2. If x is larger than y, then y is subtracted from x and the result is given the sign of x.
3. If y is larger than x then x is subtracted from y and the result is given the sign of y.
4. If either x or y is 0 or -0 the result is the non-zero element.
5. If both x and y are -0, then the sum is 0.

However, there are other numeric representations that avoid this situation. A consequence of sign-magnitude representation is that, in some cases, it is necessary to take into account the magnitude of the operands in order to determine the sign of the result. Also, the presence of an encoding for negative zero reduces the numerical range of the representation and is, for most practical uses, an unnecessary complication. An important limitation of using the high-order bit for representing the sign is the resulting halving of the numerical range.

3.2.5 Radix Complement Representation

The *radix complement* of a number is defined as the difference between the number and the next integer power of the base that is larger than the number. In decimal numbers the radix complement is called the *ten's complement*. In the binary system the radix complement is called the *two's complement*. For example, the radix complement of the decimal number 89 (ten's complement) is calculated as follows:

```
    100  = higher power of 10
-    89
    ----
     11  = ten's complement of 89
```

The use of radix complements to simplify machine subtraction operations can best be seen in an example. The operation $x = a - b$ with the following values:

```
a =  602
b =  353
           602
         - 353
           ---
x =        249
```

Note that in the process of performing longhand subtraction we had to perform two borrow operations. Now consider that the radix complement (ten's complement) of 353 is:

```
1000 - 353 = 647
```

Using complements we can reformulate subtraction as the addition of the ten's complement of the subtrahend, as follows:

```
      602
  +   647
  -------
     1249
     |_____ discarded digit
```

The result is adjusted by discarding the digit that overflows the number of digits in the operands.

In performing longhand decimal arithmetic there is little advantage in replacing subtraction with ten's complement addition. The work of calculating the ten's complement cancels out any other possible benefit. However, in binary arithmetic the use of radix complements entails significant computational advantages because binary machines can calculate complements efficiently.

The two's complement of a binary number is obtained in the same manner as the ten's complement of a decimal number, that is, by subtracting the number from an integer power of the base that is larger than the number. For example, the two's complement of the binary number 101 is:

```
    1000B  =  2^3 =  8 decimal (higher power of 2)
  -  101B  =         5 decimal
    -----             ---------
     011B  =         3 decimal
```

While the two's complement of 10110B is calculated as follows:

```
  100000B  =  2^5 = 32 decimal (higher power of 2)
  - 10110B =        22 decimal
   ------            ---------
   01010B           10 decimal
```

You can perform the binary subtraction of 11111B (31 decimal) minus 10110B (22 decimal) by finding the two's complement of the subtrahend, adding the two operands, and discarding any overflow digit, as follows:

```
           11111B  =  31 decimal
        +  01010B  =  10 decimal (two's complement of 22)
          -------
           101001B
 discard_____|
           01001B  =  9 decimal (31 minus 22 = 9)
```

In addition to the radix complement representation, there is a diminished radix representation that is often useful. This encoding, sometimes called the *radix-minus-one form*, is created by subtracting 1 from an integer power of the base that is larger than the number, then subtracting the operand from this value. In the decimal

Data Types and Data Storage

system the diminished radix representation is sometimes called the *nine's complement*. This is due to the fact that an integer power of ten, minus one, results in one or more 9-digits. In the binary system the diminished radix representation is called the one's complement. The nine's complement of the decimal number 76 is calculated as follows:

```
    100  = next highest integer power of 10

     99  = 100 minus 1
  -  76
    ----
     23  = nine's complement of 89
```

The one's complement of a binary number is obtained by subtracting the number from an integer power of the base that is larger than the number, minus one. For example, the one's complement of the binary number 101 (5 decimal) can be calculated as follows:

```
    1000B  =  2^3 = 8 decimal

     111B  =  1000B minus 1 =  7 decimal
  -  101B                      5 decimal
    ------                    ----------
     010B  =                   2 decimal
```

An interesting feature of one's complement is that it can be obtained changing every 1 binary digit to a 0 and every 0 binary digit to a 1. In this example 010B is the one's complement of 101B. In this context the 0 binary digit is often said to be the complement of the 1 binary digit, and vice versa. Most modern computers contain an instruction that inverts all the digits of a value by changing all 1 digits into 0, and all 0 digits into 1. The operation is also known as *logical negation*.

Furthermore, the two's complement can be obtained by adding 1 to the one's complement of a number. Therefore, instead of calculating

```
           100000B
        -   10110B
           -------
            01010B
```

we can find the two's complement of 10110B as follows:

```
      10110B  = number
      01001B  = change 0 to 1 and 1 to 0 (one's complement)
  +       1B    then add 1
     ---------
      01010B  = two's complement
```

This algorithm provides a convenient way of calculating the two's complement in a machine equipped with a complement instruction. Finally, the two's complement can be obtained by subtracting the operand from zero and discarding the overflow.

The radix complement of a number is the difference between the number and an integer power of the base that is larger than the number. Following this rule, we calculate the radix complement of the binary number 10110 as follows:

```
     100000B  =  2^5 = 32 decimal
  -   10110B  =        22 decimal
     -------           ----------
      01010B           10 decimal
```

However, the machine calculation of the two's complement of the same value often produces a different result, for example:

```
  100000000B  =  28 =  256 decimal
-  00010110B  =         22 decimal
  ----------             ----------
   11101010B             234 decimal
```

The difference is due to the fact that in the longhand method we have used the next higher integer power of the base compared to the value of the subtrahend (in this case 100000B) while the machine calculations use the next higher integer power of the base compared to the operand's word size, which is normally either 8 or 16 bits. In this example the operand's word size is eight bits and the next highest integer power of 2 is 100000000B. In either case, the results from two's complement subtraction are valid as long as the minuend is an integer power of the base that is larger than the subtrahend.

For example, to perform the binary subtraction of 00011111B (31 decimal) minus 00010110B (22 decimal) we can find the two's complement of the subtrahend and add, discarding any overflow digit, as follows:

```
           00011111B  =  31 decimal
        +  11101010B  =  234 decimal (two's complement of 22)
           ----------
          100001001B
discard____|
           00001001B  =  9 decimal (31 minus 22 = 9)
```

In addition to the simplification of subtraction, two's complement arithmetic has the advantage that there is no representation for negative 0. It can be argued that there are cases in which a negative zero notation could be useful, but in fact this is usually unnecessary. While both the two's complement and the one's complement schemes can be used to implement binary arithmetic, system designers usually prefer the two's complement.

3.3 Encoding of Fractional Numbers

In any positional number system the weight of each integer digit is determined by the formula:

$$P = d * B^C$$

where d is the digit, B is the base or radix, and C is the zero-based column number, starting from right to left. Therefore, the value of a multi-digit positive integer to n digits can be expressed as a sum of the digit values:

$$d_n*B^n + d_{n-1}*B^{n-1} + d_{n-2}*B^{n-2} + ... + d_0*B^0$$

where d is the value of the digit and B is the base or radix of the number system. This representation can be extended to represent fractional values. Recalling that we can

Data Types and Data Storage

extend the sequence to the right of the radix point, as follows:

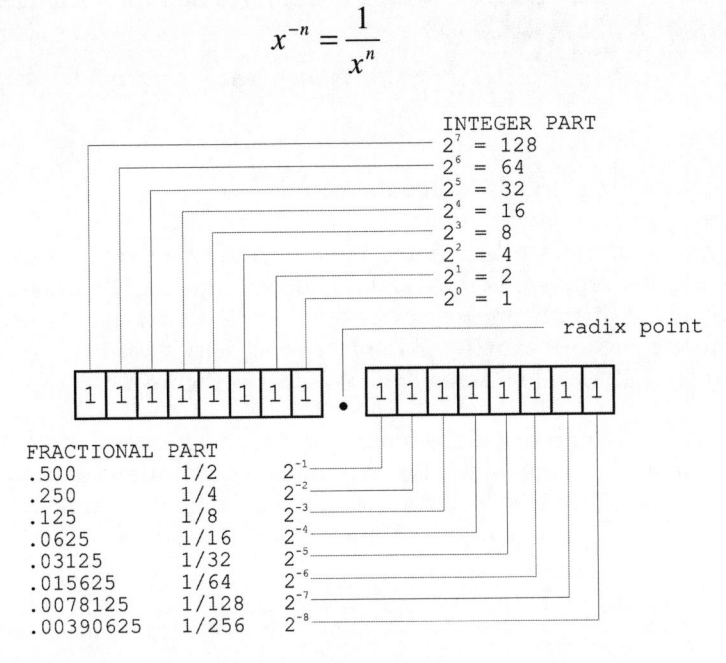

$$x^{-n} = \frac{1}{x^n}$$

```
                                       INTEGER PART
                                        2⁷ = 128
                                        2⁶ = 64
                                        2⁵ = 32
                                        2⁴ = 16
                                        2³ = 8
                                        2² = 4
                                        2¹ = 2
                                        2⁰ = 1
                                                       radix point

           | 1 | 1 | 1 | 1 | 1 | 1 | 1 | 1 | . | 1 | 1 | 1 | 1 | 1 | 1 | 1 | 1 |

FRACTIONAL PART
 .500          1/2       2⁻¹
 .250          1/4       2⁻²
 .125          1/8       2⁻³
 .0625         1/16      2⁻⁴
 .03125        1/32      2⁻⁵
 .015625       1/64      2⁻⁶
 .0078125      1/128     2⁻⁷
 .00390625     1/256     2⁻⁸
```

Figure 3-5 Positional Weights in a Binary Fraction

In the decimal system the value of each digit to the right of the decimal point is calculated as 1/10, 1/100, 1/1000, and so on. The value of each successive digit of a binary fraction is the reciprocal of a power of 2; therefore, the sequence is: 1/2, 1/4, 1/8, 1/16, ... Figure 3-5 shows the positional weight of the integer and fractional digits in a binary number.

In Chapter 2 we used the positional weights of the binary digits to convert a binary number to its decimal equivalent. A similar method can be used to convert the fractional part of a binary number. Using the decimal equivalents shown in Figure 3-5 we convert the binary fraction .10101 to a decimal fraction as follows

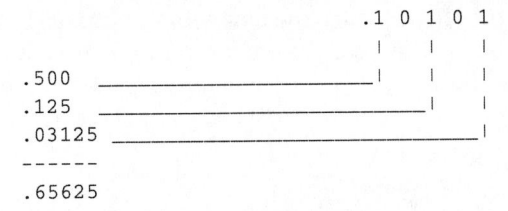

```
                           . 1 0 1 0 1
                             |   |   |
   .500    _____|   |   |
   .125    _____|   |
   .03125  _____|
   ------
   .65625
```

3.3.1 Fixed-Point Representations

The encoding and storage of fractional numbers (also called real numbers) in binary form presents several difficulties. The first one is related to the representation of the *radix point*. Since there are only two symbols in the binary set, and both are used to represent the numerical value of the number, there is no other symbol available for the decimal point.

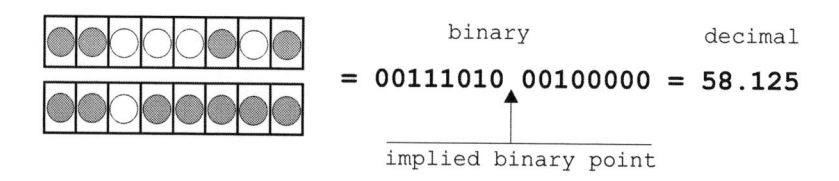

```
                                binary              decimal
                     = 00111010 00100000 = 58.125
                                ↑
                           ─────────────
                           implied binary point
```

Figure 3-6 Binary Fixed-Point Representation

One possible solution is to predefine the digit field that represents the integer part and the one that represents the fractional part. For example, if a real number is to be encoded in two data bytes we can assign the high-order byte to encode the integer part and the low-order byte for the fractional part. In this case, the positive decimal number 58.125 could be encoded as shown in Figure 3-6.

In Figure 3-6 we assumed that the binary point is positioned between the eighth and the ninth digit of the encoding. Fixed-point representations assume that whatever distribution of digits is selected for the integer and the fractional part of the representation is maintained in all cases. This is the greatest limitation of the fixed-point formats.

Suppose we want to store the value 312.250. This number is represented in binary as follows:

```
312  = 100111000
.250 = .01
```

In this case, the total number of binary digits required for the binary encoding is 11. The number can be physically stored in a 16-digit structure (as the one in Figure 3-6) leaving five cells to spare. However, since the fixed-point format we have adopted assigns eight cells to the integer part of the number, 312.250 cannot be encoded because the integer part requires nine binary digits. In spite of this limitation, the-fixed point format was the only one used in early computers.

3.3.2 Floating-Point Representations

An alternative to fixed-point is not to assume that the radix point has a fixed position in the encoding, but to allow it to float, hence the name *floating-point*. The idea of separately encoding the position of the radix point originated in scientific notation, where a number is written as a base greater than or equal to 1 and smaller than 10, multiplied by a power of 10. For example, the value 310.25 in scientific notation is written:

$$3.1025 \times 10^2$$

A number in scientific notation has a real part and an exponent part. Using the terminology of logarithms these two parts are sometimes called the *mantissa* and the *characteristic*. The following simplification of scientific notation is often used in computer work:

```
3.1025 E2
```

Data Types and Data Storage

In the computer version of scientific notation the multiplication symbol and the base are implied. The letter E, which is used to signal the start of the exponent part of the representations, accounts for the name "exponential form." Numbers smaller than 1 can be represented in scientific notation or in exponential form by using negative powers. For example, the number .0004256 can be written:

$$4.256 \times 10^{-4}$$

or as

```
4.256 E-4
```

Floating-point representations provide a more efficient use of the machine's storage space. For example, the numerical range of the fixed point encoding shown in Figure 3-6 is from 255.99609375 to 0.00390625. To improve this range we can re-assign the sixteen bits of storage so that four bits are used for encoding the exponent and twelve bits for the fractional part, called the significand. In this case the encoded number appears as follows:

```
0000 000000000000
+--+ +----------+
      |_____   12-bit fractional part
                   (significand)
 |_____  4-bit exponent part
```

If we were to use the first bit of the exponent to indicate the sign of the exponent, then the range of the remaining three digits would be 0 to 7. Note that the sign of the exponent indicates the direction in which the decimal point is to be moved; this is unrelated to the sign of the number. In this example, the fractional part (or significand) could hold values in the range 1,048,575 to 1. The combined range of exponent and significand allows representing decimal numbers in the range 4095 to 0.00000001 that considerably exceeds the range in the same storage space in fixed-point format.

3.3.3 Standardized Floating-Point Representations

Both the significand and the exponent of a floating-point number can be stored as an integer, in sign-magnitude, or in radix complement form. The number of bits assigned to each field varies according to the range and the precision required. For example, the computers of the CDC 6000, 7000, and CYBER series used a 96-digit significand with an 11-digit exponent, while the PDP 11 series used 55-digit significands and 8-digit exponents in their extended precision formats.

Variations, incompatibilities, and inconsistencies in floating-point formats led to the development of a standard format. In March and July 1985, the *Computer Society of the Institute of Electric and Electronic Engineers* (IEEE) and the *American National Standards Institute* (ANSI) approved a standard for binary floating-point arithmetic (*ANSI/IEEE Standard 754-1985*). This standard establishes four formats for encoding binary floating-point numbers. Table 3.1 summarizes the characteristics of these formats.

Table 3.1
ANSI/IEEE Floating Point Formats

PARAMETER	SINGLE	SINGLE EXTENDED	DOUBLE	DOUBLE EXTENDED
total bits	32	43	64	79
significand bits	24	32	53	64
maximum exponent	+127	1023	1023	16383
minimum exponent	-126	1022	-1022	16382
exponent width	8	11	11	15
exponent bias	+127	---	+1023	---

3.3.4 IEEE 754 Single Format

Figure 3-7 shows the IEEE floating-point single format.

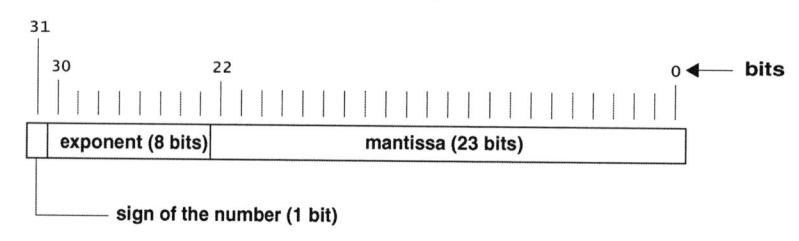

Figure 3-7 IEEE Floating-Point Single Format

If a floating-point encoding is to allow the representation of signed numbers it must devote one binary digit to encode the number's sign. In the IEEE 754 single format in Figure 3-7 the high-order bit represents the sign of the number. A value of 1 indicates a negative number.

The exponent of a binary floating-point number represents the integer power of the base with which the significand must be multiplied. The exponent can be stored in integer, sign magnitude, or radix complement representations. The IEEE 754 standard for floating-point arithmetic establishes that the exponent be stored in biased form, although the bias is not defined in all formats defined in the standard.

The word *bias*, in this context, means a constant that is added to the exponent in order to determine its final value. The term *excess-n notation* has also been used in this context. The constant is usually calculated to be approximately one-half the numerical range of the exponent field. For example, the IEEE single format devotes eight digits for the exponent field (see Figure 3-7). The numerical range of eight binary digits is 0 to 255 decimal and one-half of this range is approximately 127. Adding the constant 127 to all positive exponents places them in the range 127 to 255. The lower half of the range (1 to 126) is used for negative exponents. A 0-value in the exponent field is reserved to encode zero and *denormals*. Denormals are a special type of number discussed in the following paragraph. Table 3.2 shows the values of the exponent and the biased representation in the IEEE single format for floating-point numbers.

Table 3.2
Interpretation of Exponent in the IEEE Single Format

BIASED EXPONENT	SIGN OF NUMBER	TRUE EXPONENT	SIGNIFICAND	CLASS
0000 0000	+	-	00 ... 00	positive zero
	-	-	00 ... 00	negative zero
			11 ... 11 to 00 ... 01	denormals
0000 0001 to 0111 1111	-	-126 to 0	00 ... 00 to 11 ... 11	normals
1000 0000 to 1111 1110	-	1 to 127	00 ... 00 to 11 ... 11	normals
1111 1111	+	-	00 ... 00	+ infinity
	-		00 ... 00	- infinity
	-		10 ... 00	indefinite
	-		00 ... 01 to 11 ... 11	not-a-number

Note in Table 3.2 that the exponent value 00000000B is used to represent zero and denormal numbers. Denormals, or denormalized numbers, occur when the exponent of the number is too small to represent in the corresponding floating-point format. On the other hand, the exponent 11111111B is used to encode numbers that are too large for the single format, or to represent error conditions. The exponent range 00000001B to 11111110B (decimal values 1 to 254) is used to represent *normal* numbers, that is, numbers that are within the range of the format.

In IEEE 754 floating-point formats the high bit of the exponent field does not encode the sign, as is the case in the sign-magnitude form. Instead, the bias 127 scheme, mentioned previously, is used to represent negative and positive exponents. Negative exponents are in the range 1 to 127 (see Table 3.2) and positive exponents are in the range 128 to 254. In contrast with fixed point conventions, the high bit of the exponent is set to indicate a positive exponent, and is zero to indicate negative exponent. The main advantage of a biased exponent is that the numbers can be compared bitwise, from left to right, to determine the larger one. The number's true exponent is obtained by subtracting the bias.

The third field of the floating-point representation is known by several names: fractional part, *mantissa*, *characteristic*, and significand (see Figure 3-7). The word *significand* is the one most commonly used in the literature. Like the exponent, the significand can be stored as an integer, or in sign-magnitude or radix complement representations.

A floating-point binary number is said to be in *normalized form* when the first digit of its significand is 1. An un-normalized binary floating-point number can be normalized by successively shifting the digits of the significand to the left, while simultaneously subtracting one from the exponent. This process is continued until the high-order bit of the significand is a binary 1. The process does not change the value of the number, since shifting the significand bits to the left effectively multiplies the number by 2, while subtracting one from the exponent divides the number by 2. Clearly, the value of a number does not change if it is multiplied and divided by the same value. Also, note that normalization applies to the entire encoded number since it requires adjustments of both the exponent and the significand. Therefore, it is not correct to speak of a normalized significand or a normalized mantissa; we should refer to the significand of a normalized floating-point number.

One advantage of the normalized form is that the significand contains a maximum number of significant bits. However, addition and subtraction of floating-point numbers require that both operands have the same exponent. Therefore, before performing these operations it is often necessary to shift the significand digits to the right or to the left so that the exponents are equal.

The IEEE standard takes advantage of the fact that a normalized significand of a binary floating-point starts with a 1-digit. In the single- and double-precision formats this leading bit of the significand is assumed, in effect doubling the range of the representation. Not so in the extended formats, in which the digit must be explicitly coded. Note that this assumption is not valid if the exponent is all zeros. A zero exponent and a non-zero significand indicate a *denormal*, as shown in Table 3.2. Also, the use of an implicit bit makes necessary a special representation for zero (see Table 3.2). This special zero must be handled separately during arithmetic operations.

3.3.5 Encoding and Decoding Floating-Point Numbers

The formats in the IEEE 754 standard for binary floating-point arithmetic were designed to provide maximum storage capacity and processing efficiency. For example, the exponent in the IEEE single format, stored in biased form, takes up eight bits; however, these eight bits do not fall on a byte boundary. The exponent bits take up seven bit positions in the high-order byte, and one bit position in the next byte, as shown in Figure 3-7. In the same IEEE single encoding the significand takes up seven bits of the second byte as well as the third and fourth bytes. The sign of the number is the high-order bit of the high-order byte. Figure 3-8 shows the number 127.375 stored in the IEEE floating-point single format.

The encoding in Figure 3-8 is interpreted as follows:

```
sign of number = 0 (positive)
biased exponent = 10000101B = 133 decimal
real exponent = 133 - bias = 133 - 127 = 6
significand = 1.1111110 11000000 00000000 (adding explicit digit)
significand is adjusted by moving the radix point six places
to the right
new significand = 1111111.01100...000
```

The significand bits are intepreted as follows::

```
integer part = 1111111 = 127
fractional part = .01100..00 = .375
```

Data Types and Data Storage

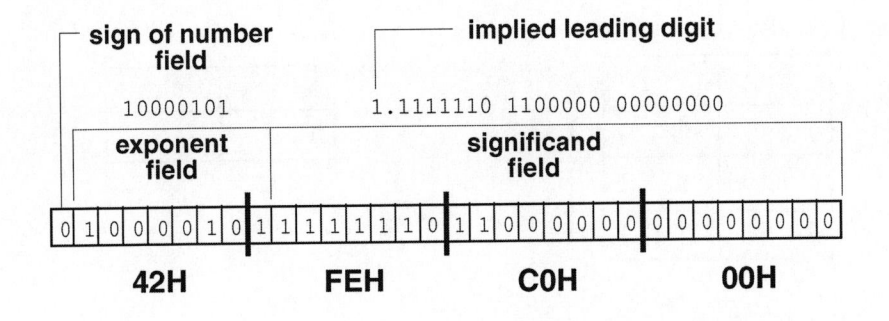

```
        sign of number                    implied leading digit
            field
         10000101              1.1111110 1100000 00000000
        exponent                     significand
         field                         field
   |0|1|0|0|0|0|1|0|1|1|1|1|1|1|1|0|1|1|0|0|0|0|0|0|0|0|0|0|0|0|0|0|
         42H                FEH              C0H              00H
```

MEMORY LAYOUT OF 127.375
IN LITTLE-ENDIAN FORMAT

LOW ADDRESS

0	0	0	0	0	0	0	0	00H
1	1	0	0	0	0	0	0	C0H
1	1	1	1	1	1	1	0	FEH
0	1	0	0	0	0	1	0	42H

HIGH ADDRESS

MEMORY LAYOUT MAP FOR
IEEE SINGLE FORMAT

LOW ADDRESS

legend:
s = sign bit
e = exponent bit
m = mantissa bit

m 16	m 17	m 18	m 19	m 20	m 21	m 22	m 23
m 8	m 9	m 10	m 11	m 12	m 13	m 14	m 15
e 8	m 1	m 2	m 3	m 4	m 5	m 6	m 7
s	e 1	e 2	e 3	e 4	e 5	e 6	e 7

HIGH ADDRESS

Figure 3-8 Encoding of the Number 127.375 in IEEE Single Format

```
bit value: 11111110-11000000-00000000 = 16,695,296
           |------| |---------------|
                |           |
                |           |_____ fractional part
                |_____ integer part

number: 127.375
```

3.4 Binary-Coded Decimals (BCD)

Floating-point encodings are the most efficient format for storing numerical data in a digital device and binary arithmetic is the fastest way to perform numerical calculations. But other representations are also useful. BCD (binary-coded decimal) is a way of representing decimal digits in binary form. There are two common ways of encoding decimal digits in binary format. One is known as the *packed BCD* format and the other one as *unpacked*. In the unpacked format each BCD digit is stored in one byte. In packed form two BCD digits are encoded per byte. The unpacked BCD format does not use the four high-order bits of each byte, which is wasted storage space. On the other hand, the unpacked format facilitates conversions and arithmetic operations on some machines. Figure 3.9 shows the memory storage of a packed and unpacked BCD number.

UNPACKED BCD

0 0 0 0	0 0 1 0	2
0 0 0 0	0 0 1 1	3
0 0 0 0	0 1 1 1	7
0 0 0 0	1 0 0 1	9

PACKED BCD

0 0 1 0	0 0 1 1	23
0 1 1 1	1 0 0 1	79

Figure 3-9 Packed and Unpacked BCD

3.4.1 Floating-Point BCD

Unlike the floating-point binary numbers, binary-coded decimal representations and BCD arithmetic have not been explicitly described in a formal standard. Each machine or software package stores and manipulates BCD numbers in a unique and often incompatible way. Some machines include packed decimal formats, which are sign-magnitude BCD representations. These integer formats are useful for conversions and input-output operations. For performing arithmetic calculations a floating-point BCD encoding is required. This approach provides all the advantages of floating-point as well as the accuracy of decimal encodings.

The BCD floating-point format which we call BCD12 is shown Figure 3-8.

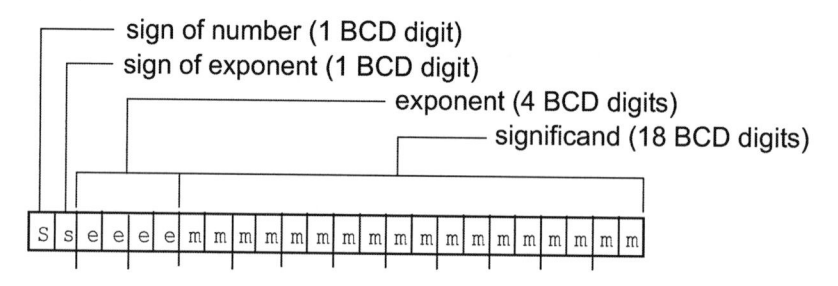

Figure 3-10 Map of the BCD12 Format

BCD12 requires 12 bytes of storage and is described as follows:

1. The sign of the number (S) is encoded in the left-most packed BCD digit. Therefore, the first four bits are either 0000B (positive number) or 0001B (negative number).

2. The sign of the exponent is represented in the four low-order bits of the first byte. The sign of the exponent is also encoded in one packed BCD digit. As is the case with the sign of the number field, the sign of the exponent is either 0000B (positive exponent) or 0001B (negative exponent)

3. The following two bytes encode the exponent in four packed BCD digits. The decimal range of the exponent is 0000 to 9999.

4. The remaining nine bytes are devoted to the significand field, consisting of 18 packed BCD digits. Positive and negative numbers are represented with a significand normal-

Data Types and Data Storage

ized to the range 1.00...00 to 9.00...99. The decimal point following the first significand digit is implied. The special value 0 has an all-zero significand.

5. The special value FF hexadecimal in the number's sign byte indicates an invalid number.

The structure of the BCD12 format is described in Table 3.4.

Table 3.4
Field Structure of the BCD12 Format

CODE	FIELD NAME	BITS WIDE	BCD DIGITS	RANGE
S	sign of number	4	1	0 - 1 (BCD)
S	sign of exponent	4	1	0 - 1 (BCD)
E	exponent	16	4	0 - 9999
M	significand	72	18	0 - 99..99 (18 digits)
	Format size	96 (12 bytes)		

Notes:
1. The significand is scaled (normalized) to a number in the range 1.00..00 to 9.99..99.
2. The encoding for the value zero (0.00..00) is a special case.
3. The special value FFH in the sign byte indicates an invalid number.

The BCD12 format, as is the case in all BCD encodings, does not make ideal use of the available storage space. In the first place, each packed BCD digit requires four bits, which in binary could serve to encode six additional combinations. At a byte level the wasted space is of 100 encodings (BCD 0 to 99) out of a possible 256 (0 to FFH). The sign field in the BCD12 format is wasteful since only one binary digit is actually required for storing the sign. Regarding efficient use of storage, BCD formats cannot compete with floating-point binary encodings. The advantages of BCD representations are a greater ease of conversion into decimal forms, and the possibility of using the processors' BCD arithmetic instructions.

Chapter 4

Digital Logic, Arithmetic, and Conversions

This chapter is about the fundamental arithmetic and logical operations of digital machines. It serves as a background for developing processing routines which involve decisions, data filtering and processing, and number crunching. Here we discuss logical and arithmetic operations in general, that is, without reference to any individual processor. There are so many different hardware versions of microcontrollers that it is not feasible to develop an actual routine for each device. On the other hand, once the logic is understood, the actual coding is a simple process of finding a way of implementing it in a specific instruction set. The chapter also includes material related to data type conversions since these operations are closely related to the other material in this chapter.

4.0 Microcontroller Logic and Arithmetic

All microcontrollers contain instructions to perform arithmetic and logic transformations on binary or decimal operands. These instructions can be classified into three groups:

1. Logical instructions. Sometimes these are called Boolean operators. The group includes instructions with mnemonics such as AND, NOT, OR, and XOR. They perform the logical functions that correspond to their names.

2. Arithmetic instructions. Typically this group of instructions performs integer addition and subtraction. Occasionally, the instruction set includes multiplication and division. The operands can be signed or unsigned binary and binary coded decimal numbers.

3. Auxiliary and bit manipulation instructions. This group includes instructions to shift and rotate bits, to compare operands, to test, set, and reset individual binary digits, and to perform various auxiliary operations.

4.0.1 CPU Flags

All microcontrollers are equipped with a special register that reflects the current processing status. This register, sometimes called the *status register* or the *flags register*, contains individual bits, usually called *flags*, that are meaningful during the execution of logic and arithmetic operations. The most common flags are:

1. The *zero flag*. This flag is set if a previous operation produces a value of zero.
2. The *carry/overflow flag*. This flag is set if there has been a carry or a *borrow-out* of the high-order bit of the operand.
3. The *half-carry* or *digit-carry flag*. This flag is set if there has been a carry or a *borrow-out* of the low-order nibble of the operand.

Not all instructions affect all the flags. For example, loading a zero constant into a register may be said to produce a zero value; however, such an instruction may or may not affect the zero flag, according to the implementation on each particular device. More powerful and sophisticated microcontrollers sometimes implement other flags, such as flags to indicate a negative operand, an arithmetic overflow, or an interrupt.

4.0.2 Word Size

The *word-size* of a computer or a digital device refers to the number of bits used in storing data and in moving data in and out of the various machine units. In other words, a machine's word-size is the native data unit for a particular architecture. In this manner we speak of the Pentium having a 32-bit word size or the PIC16x84 having an 8-bit word-size for data operations and 14-bit program words.

In the context of digital arithmetic and logic the data word-size determines the processing capabilities of each device. For example, a machine with an 8-bit word-size can perform unsigned addition of operands whose sum does not exceed the decimal value 255, since 255 is the largest unsigned integer that can be stored in eight bits. However, a machine with 16-bit words can perform unsigned additions up to a sum of 65,535 since it is the largest number that can be stored in 16 bits.

Therefore, the coding of numerical routines is determined by the word size of the machine or device. A device with 8-bit word-size requires multi-byte arithmetic to perform addition that exceeds a sum of 255, while a machine with a 16-bit word can do direct addition up to the sum 65,535. Considering that most popular microcontrollers have 8-bit word-sizes, we assume this limit in the arithmetic and logic algorithms and routines developed in this chapter.

4.1 Logical Instructions

The logical instructions include the *Boolean operators*, AND, OR, NOT, and XOR, as well as instructions to shift and rotate individual bits.

The logical instructions operate on a bit-by-bit basis; therefore, in the AND, OR, NOT, and XOR there is no interaction between bits. The action performed by the logical instructions is as follows:

1. AND, OR, and XOR logically combine each bit in the source operand with the corresponding bit in the destination operand. The result does not affect the neighboring bits.
2. The NOT operator inverts all bits in the destination operand.

Digital Logic, Arithmetic, and Conversions

These actions explain the term *bitwise operation* sometimes used to describe the instructions.

4.1.1 Logical AND

The AND instruction performs a bitwise logical AND of two operands. This determines that a bit in the result is set if and only if the corresponding bits are set in both operands. A frequent use of the AND operation is to clear one or more bits without affecting the remaining ones. This action is possible because ANDing with a 0 bit always clears the result bit and ANDing with a 1 bit preserves the original value of the first operand.

For example, if we have the binary coded decimal number 34 packed into a single byte, we can isolate the four low-order bits as follows:

```
          hexadecimal           binary
          34                    0011 0100
    AND   0F                    0000 1111    mask
          ------------          ---------
          04                    0000 0100
```

The second operand, in this case 0FH, is called a *mask*. The AND operation preserves the 1-bits in the mask and clears the bits that are 0. Consequently, the mask 00000001B clears the seven high-order bits and preserves the original value of the low-order bit.

4.1.2 Logical OR

The OR operation performs the bitwise logical inclusive OR of two operands. After a logical OR, a bit in the result is set if one or both of the corresponding bits in the operands were set. A frequent use for the OR is to selectively set one or more bits. The action takes place because ORing with a 1-bit always sets the result bit, while ORing with a 0-bit preserves the original value in the first operand.

For example, to set the high-order bit (bit number 7) we can OR with a 1 bit, as follows:

```
          hexadecimal           binary
          34                    0011 0100
    OR    80                    1000 0000    mask
          ----                  ---------
          B4                    1011 0100
```

The OR operation sets the bits that are 1 in the mask and preserves the bits that are masked 0.

4.1.3 Logical XOR

The XOR operator performs the bitwise logical exclusive OR of the two operands. Therefore, a bit in the result is set if the corresponding bits in the operands have opposite values. For this reason, XORing a value with itself always generates a zero result since all bits necessarily have the same value. On the other hand, XORing with a 1-bit inverts the value of the other operand, since 0 XOR 1 is 1 and 1 XOR 1 is 0. This toggling action of XORing with a 1 bit generates identical bitwise results as the NOT operation, but by selecting the XOR mask, the programmer can control which bits of the operand are inverted and which are preserved.

In this manner it is possible to invert the four high-order bits of an operand by XORing with a mask that has these bits set. If the four low-order bits of the mask are clear, then the original values of the bits in the other operand are preserved in the result. For example:

```
            hexadecimal          binary
               55                0101 0101
    XOR        F0                1111 0000  mask
               ----              ---------
               A5                1010 0101
```

In the previous example, the XOR operation inverts the bits that are 1 in the mask and preserves the bits that are masked 0. Consequently, the XOR mask 11110000B inverts the four high-order bits.

4.1.4 Logical NOT

In contrast with the other logical operators which require two operands, the NOT instruction acts on a single value. Its action is consistent with a Boolean NOT function, which converts all 1-bits to 0 and all 0-bits to 1. Arithmetically, the result is the one's complement of the original value. This instruction can be useful in obtaining the two's complement representation by performing the logical NOT and then adding one to the results.

4.2 Microcontroller Arithmetic

Microcontrollers are not designed for intensive numeric processing; therefore, they are not equipped with many arithmetic operators usually found in microprocessors. A typical mid-range microcontroller has instructions to add and subtract integers and perhaps to increment and decrement. Hardware multiplication is rarely available and even more so is division. Likewise, there is usually no hardware support for decimal and floating-point arithmetic. For this reason the microcontroller programmer is often challenged to provide most arithmetic and data processing operations in software.

In this discussion we assume a mid-range microcontroller, such as the PIC 16f8x. These devices contain primitives for adding and subtracting integers, shifting and rotating bits, incrementing and decrementing machine registers, some support for decimal operations and conversions, as well as the basic logic primitives AND, OR, XOR, and NOT. Multiplication and division operators, as well as floating-point operators, are not available in the mid-range devices.

4.2.1 Unsigned and Two's Complement Arithmetic

In Chapter 3 we discussed the various representations for signed and unsigned binary and decimal numbers. Arithmetic operations of unsigned operands are the simplest. In this case we assume that the encoding always represents a positive number and that all bits relate to the number's magnitude.

Digital Logic, Arithmetic, and Conversions

Unsigned arithmetic can be binary or decimal. In a machine with 8-bit words binary arithmetic on unsigned numbers use the entire range of the format. This is true even when the primitive operations are valid in two's complement form; in fact, it is one of the great advantages of two's complement representation. Table 4.1 shows a 4-bit binary in several numeric formats.

Table 4.1

Interpretations of 4-bit Binary Numbers

BINARY	1'S COMPLEMENT	DECIMAL VALUES 2'S COMPLEMENT	UNSIGNED
0111	7	7	7
0110	6	6	6
0101	5	5	5
0100	4	4	4
0011	3	3	3
0010	2	2	2
0001	1	1	1
0000	0	0	0
1111	-0	-1	15
1110	-1	-2	14
1101	-2	-3	13
1100	-3	-4	12
1011	-4	-5	11
1010	-5	-6	10
1001	-6	-7	9
1000	-7	-8	8

Assume a machine with a 4-bit word size and consider addition of two unsigned numbers:

```
    BINARY    DECIMAL
    0111      7
  + 0110      6
    ------    ----
    1101      13
```

Note, in the previous example, that if the encoding were in two's complement form, the addition of the positive values 6 plus 7 would produce a result that overflows the capacity of the representation. In 4-bit two's complement representation there is no way of encoding the value 13.

The question that arises is: in a device that performs two's complement addition, must we always assume that the operands are in two's complement form? The answer is: no. Signed addition of two's complement operands and the unsigned addition of integer operands can be performed with identical processing and by the same electronic circuitry. It is the software that must take into account the encoding of the operands in order to interpret the results. For example, in the 4-binary digit device previously considered, the two's complement addition of the values 6 and 7 produce an overflow, which can be detected by observing the change in the high-order bit (the sign bit) of the result. Therefore, in this case, the result of the addition operation is invalid. However, if the same decimal values represent unsigned operands, then the addition of 7 plus 6 produce the valid result 13. In either case the binary values of the operands, as well as the result, are the same.

Microcontrollers usually support the fundamental operations of addition and subtraction on signed and unsigned integer operands with a single primitive operation. The addition and subtraction operators in low- and mid-range devices allow two operands. The more powerful microcontrollers support addition and subtraction of three operands, which is useful in implementing multi-digit routines. In either case, the software determines if the result is signed or unsigned by interpreting the changes in the high-order bit of the operands and by evaluating the status flags if these are available.

4.2.2 Operations on Decimal Numbers

Although microcontrollers are binary devices, the instruction set often includes operations for performing arithmetic on binary coded decimal numbers. In Chapter 3 we saw that BCD numbers can be stored in packed or unpacked form. In packed format two BCD digits are contained in each byte. The low-order BCD digit takes up bits 0 to 3 and the high-order BCD digit takes up bits 4 to 7. Unpacked BCD digits are stored one digit per byte; in this case the high-order nibble is unused. The packed and unpacked binary coded decimal formats can be seen in Figure 3-9.

Microcontroller designers usually adopt the packed BCD format for representing decimal operands. One advantage of packed BCDs is that the two decimal digits encoded in a single byte can be represented as hexadecimal digits. For example, the values H24 and H99 represent the packed BCD digits 24 and 99 respectively. Note that each hex digit is preceded by the letter H to indicate radix 16. In actual microcontroller programming other ways are often used for representing numbers in hexadecimal notation.

The addition and subtraction of decimal numbers represented in packed BCD can be performed with binary primitive operations, complemented with some additional adjustments. In some cases the addition of two BCD numbers in packed format may produce a valid result, for example:

```
           H23          H31          H56
     +     H12          H38          H22
           ----         ----         ----
           H35          H69          H78
```

In the previous examples the results are valid because the sum of each digit does not exceed the range of the BCD format. However, the following additions do not produce valid BCD results:

```
           H33          H31          H56
     +     H27          H59          H27
           ----         ----         ----
           H5A          H8A          H7D
```

In the case of the first operation the valid BCD result would be: 33 + 27 = 60, in the second one 31 + 59 = 90, and in the third one 56 + 27 = 83. A simple adjustment corrects the error, as follows:

Digital Logic, Arithmetic, and Conversions

```
      H33           H31           H56
   +  H27        +  H59        +  H27
      ---           ---           ---
      H5A           H8A           H7D
   +  H 6        +  H 6        +  H 6
      ---           ---           ---
      H60           H90           H83
```

In all three cases adding 6 to the previous sum produces the expected result. The logic for deciding when the value 6 must be added is simple: if the sum of the low-order digit is greater than 9 or if the sum produced a carry out of the low-order nibble, then add 6 to the sum to perform the decimal adjustment. Some high-end microcontrollers contain a primitive instruction that executes the decimal adjustment automatically, that is, without having to test the sum. However, this instruction is not available in low- and mid-range devices.

Also note that the largest number that can be encoded in packed BCD format is the decimal 99. When adding two BCD digits the high-order digit of the sum cannot be greater than 9. If so, then the capacity of the format has been exceeded and the result cannot be adjusted by the simple addition of 6. Here again, a multi-byte processing routine can be developed in order to accommodate the result of BCD addition when the sum exceeds a single byte.

Many microcontrollers are equipped with a flag that indicates overflow from binary digit number 3. This flag, sometimes called the *digit carry* or the *half carry flag*, can be used to detect that a calculation has overflowed the storage capacity of four binary digits. The availability of this flag simplifies the logic necessary for adjusting binary addition of decimal operands since the value 6 must be added when the digit in the low-order nibble is larger than 9, or when there has been a carry to the next digit. The following flowchart shows this processing.

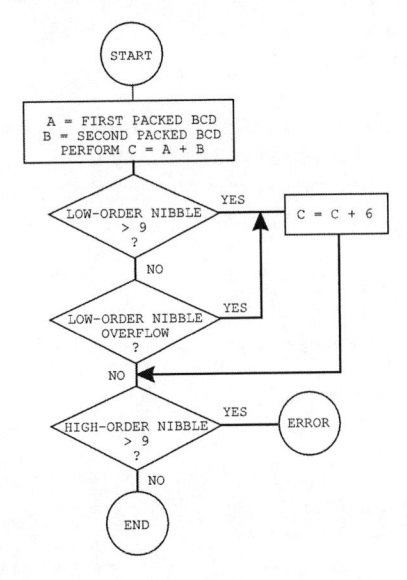

Figure 4-1 Flowchart for Two-byte BCD Addition

4.3 Bit Manipulations and Auxiliary Operations

In addition to basic logic and arithmetic, microcontrollers contain primitive operators to manipulate individual bits, to compare operands, to make decision based on the state of individual bits and flags, and to convert data to other formats. As always, presence or absence of some of these operations, as well as their degree of power and sophistication, varies with the individual microcontroller. In the following subsections we describe the most commonly available primitives.

4.3.1 Bit Shift and Rotate

The fundamental operators to shift and rotate are useful in developing BCD and binary arithmetic routines. One interesting use of bit shifting is in implementing binary multiplication and division routines.

Shift operations consist of transposing to the left or right all the bits in the operand. In microcontrollers the operand is usually a processor register. For example, after a right shift operation all the bits in the value 01110101B (75H) are moved one position to the right, resulting in the value 00111010B (3AH). Note that on a right shift the right-most bit disappears and a zero comes into the high-order bit. By the same token, in a left shift the high-order bit disappears and a zero comes into the low-order bit. Figure 4-2 shows the action of a left-shift operation.

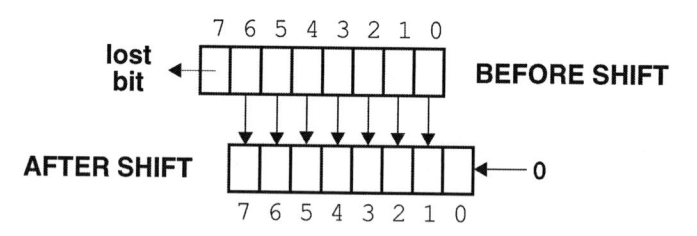

Figure 4-2 Left Shift Operation

The rotate operation differs from the shift in that in the rotate the low-order bit is either a copy of the high-order bit or of the carry flag. In the first case the operation is a pure rotate, in the second case the rotate is referred to as *rotate-through-carry*. Figure 4-3 shows the action of a *left-rotate-through-carry* flag.

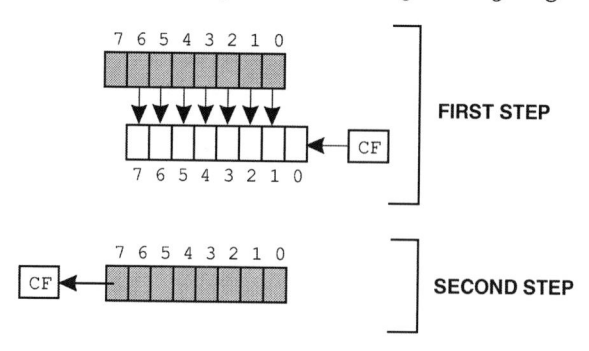

Figure 4-3 Rotate-through-carry Left Operation

Digital Logic, Arithmetic, and Conversions

Note in Figure 4-3 that the contents of the carry flag are first copied to the low-order bit of the destination operand, then the individual bits of the source (in gray in the illustration) are shifted left and moved to the destination. Finally the high-order bit of the source is copied to the carry flag.

There are several possible variations of the rotate operation. The Intel microprocessors distinguish between arithmetic and logic rotates. In the arithmetic rotation the high-order bit is preserved in order to maintain the sign of the operand. The rotate shown in Figure 4-3 is the one most common in microcontroller hardware. Clearing the carry flag before the rotate takes place makes the operation identical to a shift.

4.3.2 Comparison Operations

An interesting property of subtraction is its use in finding the relative size of two operands. This interesting action of subtraction is based on the following logic:

1. If the result of a subtraction is zero, then both operands were of the same size.

2. If the result of a subtraction is a positive number, then the subtrahend was smaller than the minuend.

3. If the result of a subtraction is a negative number, the subtrahend was larger than the minuend.

In a binary/digital device the result of a subtraction operation can be determined by observing the flags. If the zero flag is set, then the operands were the same (case 1, above). If the carry flag is set, then the subtrahend was larger than the minuend (case 3, above). If neither the carry nor the zero flag is set, then resulting subtrahend was smaller than the minuend (case 2, above). Since all microcontrollers offer some mechanism for re-directing execution according to the state of the flags, a program can use subtraction to make these decisions.

The one objection to the use of subtraction in comparing the size of two operands is that the process will change one of them. To use subtraction in comparison operations the programmer has to find some way of preserving the minuend. Alternatively, some devices contain a comparison operator that sets the flags as if a subtraction had taken place but without changing the operands. High-end microcontrollers are equipped with dedicated comparison operators, but the middle- and low-range devices usually are not.

4.3.3 Other Support Operations

Mid- and high-range microcontrollers contain other auxiliary bitwise, arithmetic, and logic operators that can be useful to the programmer. These include instructions to:

1. Increment and decrement operands

2. Clear registers or storage locations

3. Swap nibbles

4. Clear and set individual bits

5. Test individual bits

Usually instructions to increment and decrement and to test individual bits are also capable of redirecting execution according to the result. For example, a special decrement can be followed by a jump if decrementing sets the zero flag. Or a bit test instruction can include a jump that is taken if the tested bit is set or reset.

4.4 Unsigned Binary Arithmetic

Since microcontrollers are not used in data processing, microcontroller programming does not usually require the development of powerful or sophisticated numerical routines. At the same time, because microcontrollers often lack primitive support for even the most essential calculations, the programmer makes up for this deficiency. For example, mid-range PIC microcontrollers contain primitive instructions for signed and unsigned addition and subtraction of byte-size operands. Unsigned addition and subtraction operations that exceed one byte, as well as unsigned multiplication and division, must be provided in software.

In unsigned arithmetic all bits of the binary encoding are interpreted as magnitude bits and all numbers are positive. Addition of unsigned binary numbers is limited by the machine's word size. For example, a mid-range PIC microcontroller performs unsigned addition on 8-bit operands. An overflow of the sum is reported by the carry flag set. In this case the carry flag clear indicates that the sum is within the storage capacity of the format. In unsigned arithmetic processing, routines for extending operations to multiple bytes are straightforward and relatively simple.

4.4.1 Multi-byte Unsigned Addition

Many microcontrollers are one-byte machines, so operands and results for arithmetic operations must be contained within eight bits. The largest unsigned value that can be represented in a single byte is the decimal number 255. But often applications require adding operands that are larger than a single byte and storing results that exceed this limit. In these cases multi-byte routines become necessary.

The simplest case is the addition of two unsigned byte-size operands whose sum exceeds 255 decimal. This case requires storing the result in a two-byte area and detecting those cases in which there is a carry into the high-order byte. In this case the largest possible operands for byte addition are the hexadecimal numbers FF. Addition is as follows:

```
          Binary:
        1 1 1 1 1 1 1 1
      + 1 1 1 1 1 1 1 1
        ---------------
        1 1 1 1 1 1 1 0
   C <=
```

In this example the symbol C <= represents a carry out of the high-order bit, the case when the sum exceeds the capacity of a single byte. In hexadecimal, the sum of HFF + HFF = H1FE. You can add two byte-size operands into a two-byte storage area by using byte addition to determine the low-order byte of the result and testing for a carry out of the high-order bit. If there is a carry, then the high-order byte of the result is 1; otherwise the high-order byte is 0.

Digital Logic, Arithmetic, and Conversions

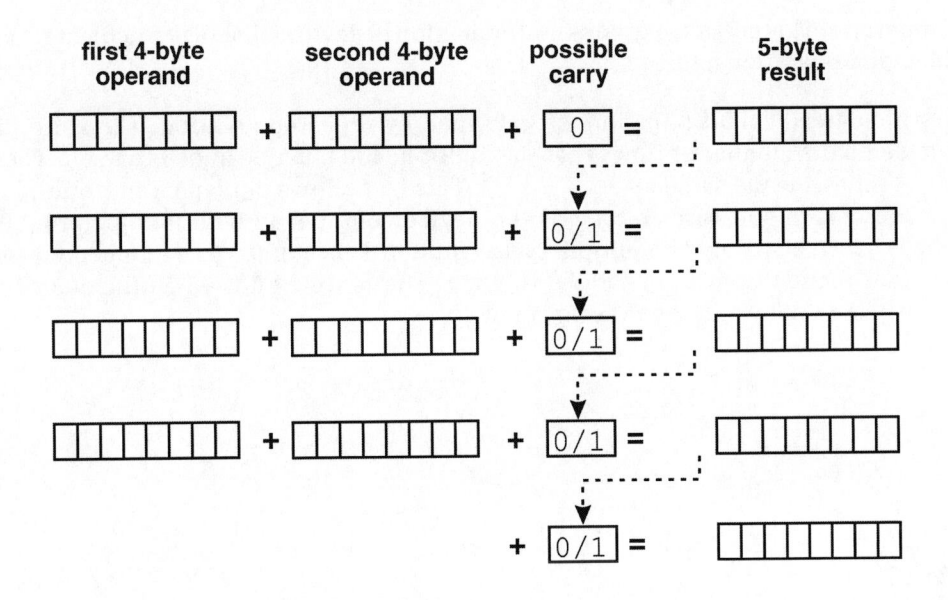

```
   first 4-byte        second 4-byte       possible         5-byte
    operand              operand            carry           result

  [ | | | | ]   +   [ | | | | ]    +   [  0  ]  =    [ | | | | | ]

  [ | | | | ]   +   [ | | | | ]    +   [ 0/1 ]  =    [ | | | | | ]

  [ | | | | ]   +   [ | | | | ]    +   [ 0/1 ]  =    [ | | | | | ]

  [ | | | | ]   +   [ | | | | ]    +   [ 0/1 ]  =    [ | | | | | ]

                                    +   [ 0/1 ]  =    [ | | | | | ]
```

Figure 4-4 Unsigned Multi-byte Addition

The same logic can be generalized to add more than two byte-size operands as long as the storage area for the result exceeds the size of the operands by one byte. For example, two word-size operands (16 bits each) can be added into a 3-byte (24-bit) storage area, or two double-word operands (32 bits) into a 5-byte storage area. The general algorithm for multi-byte addition is shown in Figure 4-4.

The case shown in Figure 4-4 consists of adding two, 4-byte operands into a 5-byte sum. The addition of the first two operands assumes that there is no carry. In the remaining stages there can be a possible carry from the previous stage if the sum of the two byte-size operands, plus the previous carry, exceed the storage capacity of eight bits. The last byte of the result is determined solely by the possible carry from the previous stage.

In Figure 4-4 we see that multi-byte addition requires the sum of three values in all stages except the first and the last one. Some high-end microcontrollers have addition operators that accept a three-byte operand. Others have special addition opcodes that automatically add-in the carry flag. The latter operators are referred to as *add-with-carry*. However, in most low- and mid-range devices the software must take care of incrementing the sum if there is a carry from the previous stage. The actual multi-byte addition routines are developed in the context of programming the various microcontrollers, discussed later in this book.

4.4.2 Unsigned Multiplication

The case for multiplication cannot be generalized since high-end microcontrollers usually contain one or more multiplication operators; this is not the case in low- and mid-range devices. In the first case implementation is simply by using the corresponding operator. This section explains multiplication in devices that lack a dedicated multiplication operation code.

Arithmetically, multiplication is performed by repeated addition. The multiplier represents the number of times that the multiplicand must be added to itself. Therefore, 3 times 4 is the same as 3 + 3 + 3 + 3. This fact allows implementing multiplication routines in software as long as the device contains an addition operator. The logic is based on using the multiplier as a counter. This counter is decremented each time that the multiplicand is added to itself. The routine ends when the counter is exhausted, as shown in the flowchart in Figure 4-5.

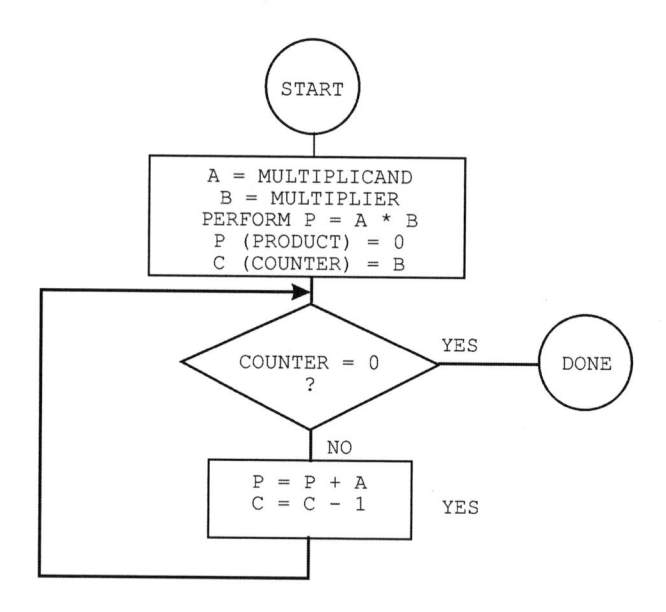

```
                    START

        A = MULTIPLICAND
        B = MULTIPLIER
        PERFORM P = A * B
        P (PRODUCT) = 0
        C (COUNTER) = B

                              YES
        COUNTER = 0  ───────────────  DONE
            ?
              NO
        P = P + A
        C = C - 1          YES
```

Figure 4-5 Unsigned Multiplication Flowchart

The beauty of the repeated addition algorithm is its simplicity and its main shortcoming is its slowness. An alternative way of performing multiplication is by shifting the bits of the operand. This method is based on the properties of a binary positional system, in which the value of each digit is a successive power of 2. Therefore, by shifting all digits to the left, the value 0001B (1 decimal) successively becomes 0010B (2 decimal), 0100B (4 decimal), 1000B (8 decimal), and so on.

Binary multiplication by means of bit shifting has the downside that the multiplier must be a power of 2. Otherwise, the software must shift by a power of 2 that is smaller than the multiplier and then add the multiplier as many times as necessary to complete the product. In this manner, to multiply by 5 we can shift left twice and add once the value of the multiplicand. To multiply by 7 we would shift left twice and then add three times the value of the multiplicand. As the multiplier gets larger and more distant from the smaller power of 2, the number of addition operations required is also larger, and the effectiveness of the algorithm diminishes.

A third approach is based on the manipulations performed during longhand multiplication. For example, the multiplication of 00101101B (45 decimal) by 01101101B (109 decimal) can be expressed as a series of products and shifts, in the following manner:

```
                        0 0 1 0 1 1 0 1 B = 45 decimal
        times           0 1 1 0 1 1 0 1 B = 109 decimal
                        -----------------
                        0 0 1 0 1 1 0 1
                      0 0 0 0 0 0 0 0
                    0 0 1 0 1 1 0 1
                  0 0 1 0 1 1 0 1
                0 0 0 0 0 0 0 0
              0 0 1 0 1 1 0 1
            0 0 1 0 1 1 0 1
          0 0 0 0 0 0 0 0
          -------------------------------
          0 0 1 0 0 1 1 0 0 1 0 1 0 0 1 B = 4905 decimal
```

The actual calculations in this method of binary multiplication are quite simple since the product by a 0 digit is zero and the product by a 1 digit is the multiplicand itself. Consequently, the multiplication routine simply tests each digit in the multiplier. If the digit is zero no action need be performed; if the digit is one, the multiplicand is shifted left and added into an accumulator.

The storage allocation to hold the product of a multiplication operation is not the same as that to hold the sum. In multi-byte addition one additional byte is required to hold the sum. In multiplication the storage allocation must be twice the size of the operands. For example, byte multiplication requires a two-byte storage, while multiplying two double-byte operands requires a four-byte storage allocation.

4.4.3 Unsigned Division

If multiplication can be reduced to repeated addition, then division can be conceptualized as repeated subtraction. In the case of division, the quotient (result) is the number of times the divisor must be subtracted from the dividend before zero or a negative value results from the subtraction. The flowchart in Figure 4-6 (in the following page) shows the logic steps in unsigned division.

In Figure 4-6 note that the logic tests for a zero divisor, since division by zero is mathematically undefined. Also, since the operation is unsigned, the result cannot be negative; therefore, the divisor must be larger than the dividend. Finally, the logic must consider the case in which subtracting the divisor from the reminder produces a negative value, in which case an adjustment is necessary to produce a valid quotient. This adjustment avoids the need for searching for a trial divisor, as in the case in the common longhand division algorithm. In machine code the negative result is detected as an overflow (carry flag set) from the subtraction.

4.5 Signed Binary Arithmetic

In two's complement and sign-magnitude representations the high-order bit represents the sign of the operand, while its magnitude is represented in the remaining bits. Therefore, in the case of signed numbers, a carry out of the high-order bit is meaning-

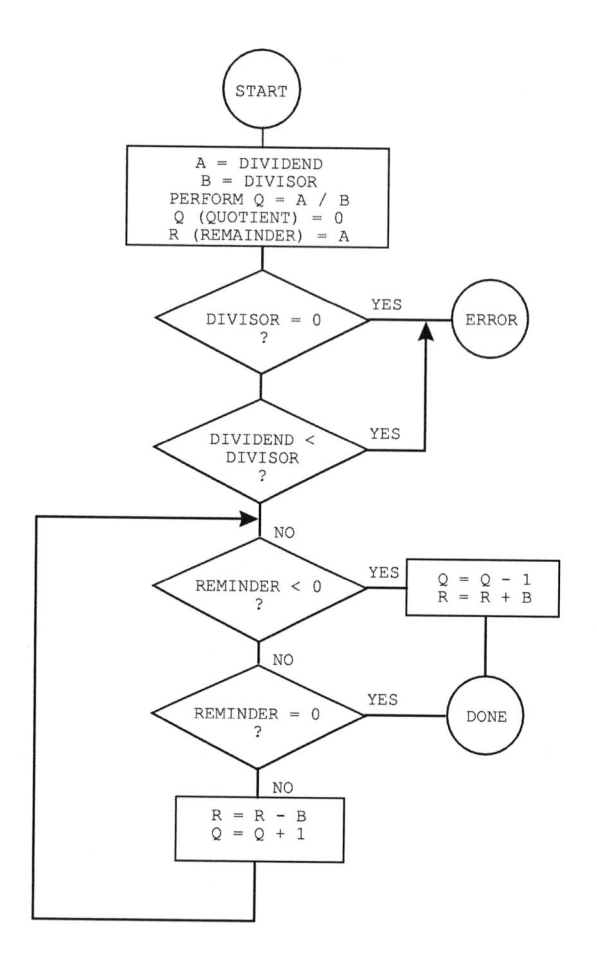

Figure 4-6 Unsigned Division Flowchart

less since the high-order bit is not a magnitude bit. For example, consider the following operation in an 8-bit device that performs unsigned and two's complement addition:

```
      80   =   0101 0000B
  +   90   =   0101 1010B
      ------------------
     170   =   1010 1010B
```

If the operands are assumed to be in unsigned binary format the result is valid. However, if the operands (the decimal values 80 and 90) are assumed to be positive numbers in two's complement form, then the result is invalid since the positive number 170 cannot be represented in an 8-bit two's complement encoding.

Clearly, multi-byte operations on signed representations cannot be performed identically as with unsigned operands. Table 4.2 shows the unsigned and two's complement representations of one-byte numbers.

Digital Logic, Arithmetic, and Conversions

Table 4.2
Signed and Unsigned Representations of One-Byte Numbers

BINARY	2'S COMPLEMENT	UNSIGNED
0000 0000	0	0
0000 0001	1	1
0000 0010	2	2
0000 0011	3	3
.	.	.
.	.	.
.	.	.
0111 1111	127	127
1000 0000	-128	128
1000 0001	-127	129
1000 0010	-126	130
1000 0011	-125	131
.	.	.
1111 1110	-2	254
1111 1111	-1	255

4.5.1 Overflow Detection in Signed Arithmetic

In unsigned addition the carry flag is magnitude-related. It is set when there is a carry out of the high-order bit of the destination operand, which takes place when its capacity has been exceeded. This is usually described as an overflow condition. However, a carry out of the high-order bit of the result is not always meaningful in signed arithmetic. For example, suppose the following two's complement addition:

```
       Decimal      binary
       127          0111 1111
 +     127          0111 1111
       ----         ---------
       ??           1111 1110
```

In this case the sum clearly exceeds the capacity of the format, since the largest positive value that can be represented in a two's complement 8-bit format is 127 (see Table 4-2). However, the operation did not generate a carry out of the high-order bit. Therefore, the carry flag could not have been used to detect the overflow error in this case.

Now consider the addition of two negative numbers in two's complement form:

```
       Decimal      binary
       -4           1111 1100
 +     -5           1111 1011
       ----         ---------
       -9           C <= 1111 0111
```

In this case the addition operation generated a carry out of the high-order bit; however, the sum is arithmetically correct. In fact, any addition of negative operands in two's complement notation generates a carry out of the most significant bit.

These two examples show that the carry flag, by itself, cannot be used to detect an error or no-error condition in two's complement arithmetic. Detecting an overflow condition in two's complement representations requires observing the carry into the high-order bit of the encoding as well as the carry out. In both previous examples we note that there was a carry into the high-order bit of the result. However, in the first case there was no carry out. The general rule is: *two's complement overflow takes place when the carry into and the carry out of the high-order bit have opposite values.* Figure 4-7 is a flowchart to detect overflow in signed arithmetic.

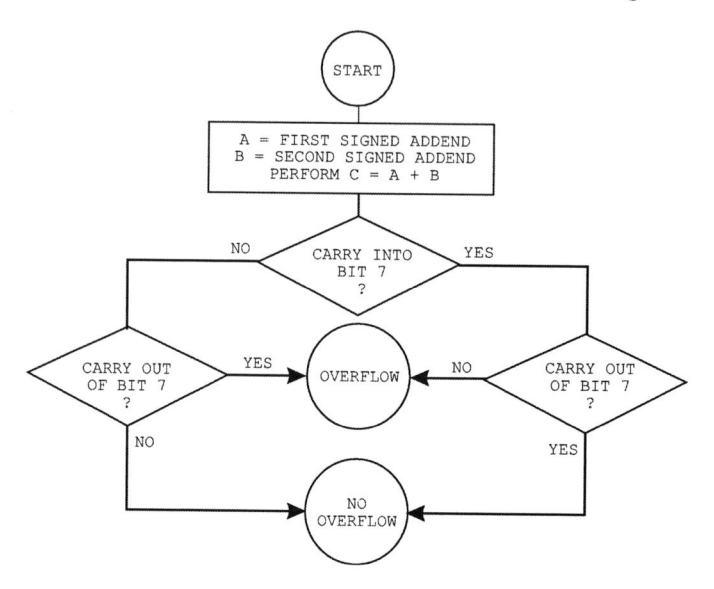

Figure 4-7 Detecting Overflow in Two's Complement Arithmetic

Most microprocessors and some high-end microcontrollers contain hardware facilities for detecting signed arithmetic overflow. In some cases the hardware support consists of a single overflow flag that is set whenever the result of an arithmetic operation exceeds the capacity of the format. In other cases, as in the PIC 18CXX2 family, the status register contains a negative bit flag that indicates a 1-bit in the sign bit position, as well as an overflow bit that is set whenever there is an overflow from the magnitude bits (0 to 6) into the sign bit (bit 7) of the destination operand. In this device, software can test one or both of these flags to detect two's complement overflow.

In low- and mid-range devices, with no hardware support for signed arithmetic, detecting a two's complement overflow is by no means simple. Without a hardware flag to report a carry condition into a particular bit position, software is confronted with several possible alternatives, but none is simple or straightforward.

4.5.2 Sign Extension Operations

Observing the carry into and the carry out of the most significant bit is a valid way of detecting overflow of a two's complement arithmetic operation. In theory, the logic described in the flowchart of Figure 4-7 can be implemented in devices without hard-

Digital Logic, Arithmetic, and Conversions

ware support for signed overflow; however, the processing is complicated and therefore costly in execution time. An alternative approach is to ensure that the format has sufficient capacity to store the arithmetic result. The rule developed previously lets us determine that, for addition and subtraction, the destination format must have at least one more byte than the operands. In multiplication, the destination operand must be at least twice the size of the source operands.

A simple mechanism for extending the capacity of two's complement encoding is called *sign extension*. The process consists of copying the sign bit into the high-order bit positions of the extended encoding. For example, to extend a two's complement 8-bit number into 16 bits, copy the sign bit of the original value (bit number 7) into all the bits of the extended byte. The process is shown in Figure 4-8 for both positive and negative operands.

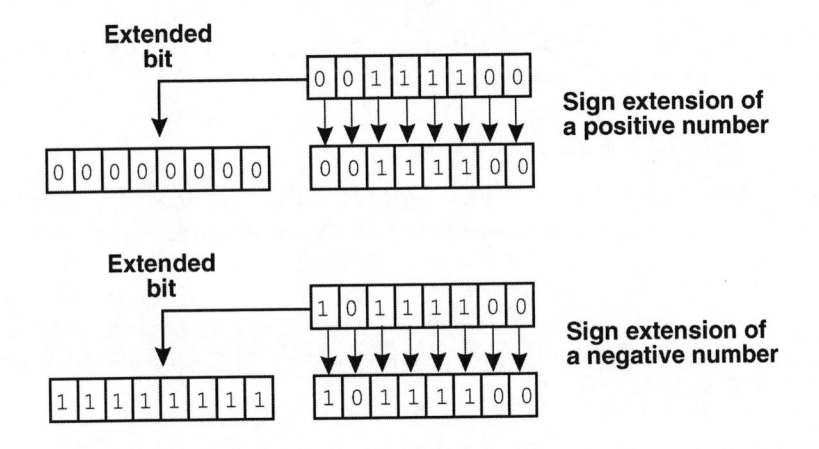

Figure 4-8 Sign Extension of Two's Complement Numbers

4.5.3 Multi-byte Signed Operations

Signed operations on two's complement numbers encoded in multiple bytes can be performed using the processor's arithmetic primitives. Consider the addition of the numbers -513 and -523, each one encoded in 16-bit two's complement form:

```
decimal           binary
                  HOB         LOB
-513              1111 1101   1111 1111
-523              1111 1101   1111 0101
-----             ---------------------
-1036             1111 1011   1111 0100
```

In the preceding example, adding the low-order bytes produces the sum shown, plus a carry. Adding the high-order bytes plus the carry, and discarding the overflow, produces the sum of high-order bytes shown above. The result is the correct value in two's complement form. The fact that the result did not overflow the capacity of the 16-bit format can be ascertained by observing that there was a carry into the fifteenth digit but also a carry out. Carry in and carry out of the sign bit is one of the conditions for no overflow in the flowchart of Figure 4-7.

4.6 Data Format Conversions

Quite often code needs to convert data into and from different numeric formats; for example, to display ASCII digits in an output device, or to convert numeric keyboard input in ASCII into binary or BCD encodings for processing. In this section we consider the logic for the following cases:

1. BCD digits to ASCII decimal
2. Binary to string of ASCII decimal digits
3. String of ASCII decimal digits to binary
4. Binary to string of ASCII hexadecimal digits

As in the previous cases, implementation of these conversions is device-dependent and varies in the different hardware.

4.6.1 BCD Digits to ASCII Decimal

Packed BCD digits are encoded in one digit-per nibble, as shown in Section 4.2.2. Thus, each digit is a binary value in the range 0 to 9. Converting each digit to ASCII consists of isolating each nibble and then changing the binary into an ASCII representation. Note in Table 3.1 that the numeric ASCII digits start at 30H for the digit zero and extend to 39H for the digit 9. For this reason converting a numeric digit from binary into ASCII consists of adding 30H. By the same token, subtracting 30H converts a single ASCII digit to binary.

Assume four packed BCD digits in two consecutive memory bytes, labeled A and B, where A holds the two low-order digits; also, a four-digit storage buffer to which the variable P is a pointer. The conversion algorithm can be described as follows:

1. Initialize buffer pointer P to the first storage location.
2. Copy digit A to temporary digit T.
3. Mask out four high-order bits of T.
4. Add 30H to value in T and store in buffer by pointer P.
5. Bump buffer pointer to next digit storage.
6. Copy digit A to temporary digit T.
7. Mask out four low-order bits in T.
8. Shift four high-order bits to the right by 4 bits.
9. Add 30H to value in T and store in buffer by pointer P.
10. Bump buffer pointer to next digit.
11. Proceed with digit B in the same manner.

4.6.2 Unsigned Binary to ASCII Decimal Digits

Often we hold an unsigned binary number in memory or a machine register and need to display its value to some ASCII-based output device. The process requires converting the binary value to a string of ASCII decimal digits. The number of decimal digits depends on the number of bits in the binary representation. A one-byte unsigned binary requires three ASCII decimal digits since the value ranges from 0 to 255. A two-byte unsigned binary requires a string of five ASCII decimal digits since the range of a two-byte representation is from 0 to 65,535, and so on. The storage area for the ASCII digits is sometimes referred to as a *buffer*.

The process of converting binary to ASCII decimal consists of dividing the binary by 10 to obtain each decimal digit, then adding 30H to the remainder in order to turn the digit into ASCII. The process continues until the original dividend is reduced to zero, as shown in the flowchart of Figure 4-9.

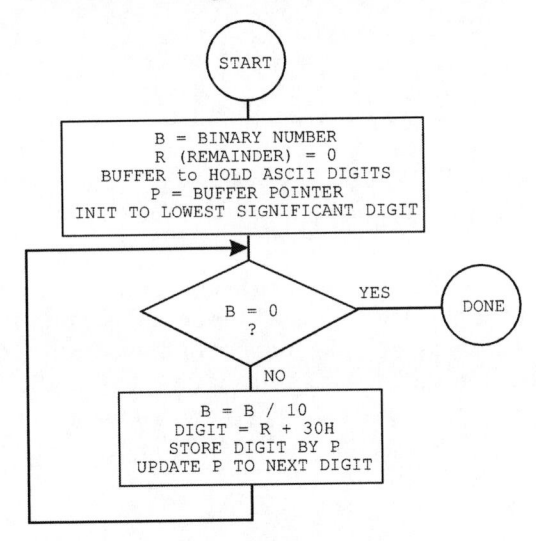

Figure 4-9 Unsigned Binary to ASCII Decimal String

4.6.3 ASCII Decimal String to Unsigned Binary

Another conversion operation frequently needed in software is the transformation of a string of ASCII decimal digits into binary. This type of conversion typically arises when the program needs to receive input which must later be processed by the device. For example, the user enters a numeric value from a keyboard and the application must process this data in binary form.

In designing the conversion routine we must first delimit the value range of the input data so as to allocate a sufficiently large binary format to store the result. For example, code can store in a single unsigned byte a binary in the range 0 to 255, but it requires two bytes to store one in the range 0 to 65,535. Once the binary storage size is determined, the conversion logic is based on converting each ASCII digit to binary, high-to-low, and adding its value to a previous sum multiplied by 10. The following flowchart describes the conversion logic.

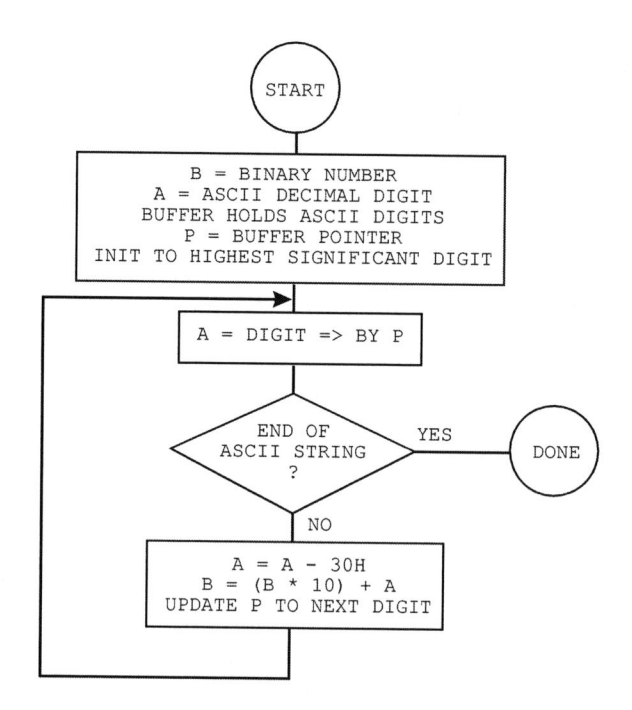

```
                        START
                          │
                          ▼
        ┌─────────────────────────────────┐
        │      B = BINARY NUMBER          │
        │   A = ASCII DECIMAL DIGIT       │
        │   BUFFER HOLDS ASCII DIGITS     │
        │      P = BUFFER POINTER         │
        │ INIT TO HIGHEST SIGNIFICANT DIGIT│
        └─────────────────────────────────┘
                          │
                          ▼
        ┌─────────────────────────────────┐
   ┌───▶│       A = DIGIT => BY P         │
   │    └─────────────────────────────────┘
   │                      │
   │                      ▼
   │                 ╱ END OF ╲     YES
   │                ╱  ASCII   ╲─────────▶ DONE
   │                ╲  STRING  ╱
   │                 ╲   ?    ╱
   │                  ╲──────╱
   │                     │ NO
   │                     ▼
   │    ┌─────────────────────────────────┐
   │    │        A = A - 30H              │
   │    │      B = (B * 10) + A           │
   │    │    UPDATE P TO NEXT DIGIT       │
   │    └─────────────────────────────────┘
   │                     │
   └─────────────────────┘
```

Figure 4-10 Decimal String to Unsigned Binary

The logic in the flowchart of Figure 4-10 assumes that there is some way of detecting the end of the string of ASCII digits. This could be a terminator character embedded in the string or a counter for the number of digits. Here again we use a buffer pointer that is initialized to the least significant digit in the ASCII string. The ASCII digit is converted to binary by subtracting 30H and then added to the previous sum, multiplied by 10. For example, assume the ASCII string of the decimal digits 564.

```
STRING = '564'

FIRST ITERATION:
STEP 1: B = 0
        P => HIGH DIGIT IN STRING '564'
STEP 2: A = '5'
STEP 3: END OF STRING?
        NO
STEP 4: A = 4 ('5' - 30H = 5)
        B = (0 * 10) + A = 5
        P TO NEXT LOWER DIGIT
SECOND ITERATION:
STEP 2: A = '6'
STEP 3: END OF STRING?
        NO
STEP 4: A = 6 ('6' - 30H = 6)
        B = (5 * 10) + A = 56
        P TO NEXT LOWER DIGIT
THIRD ITERATION:
STEP 2: A = '4'
STEP 3: END OF STRING?
        NO
STEP 4: A = 4 ('4' - 30H = 4)
        B = (56 * 10) + A = 564
```

```
          P TO NEXT LOWER DIGIT
FOURTH ITERATION:
STEP 2: A = ??
STEP 3: END OF STRING?
        YES
RESULT: B = 564
```

4.6.4 Unsigned Binary to ASCII Hexadecimal Digits

Converting a binary number to a string of ASCII hex digits is quite similar to converting from binary to an ASCII decimal string, as described in Section 4.6.2. Here again, the digit space to allocate for the ASCII string depends on the size of the binary operand. An 8-bit binary is represented in two ASCII hex digits, a 16-bit binary into four ASCII hex digits, and so on.

The process of converting binary to ASCII hexadecimal consists of dividing the binary by 16 to obtain each hex digit. If the remaining hexadecimal digit is in the range 0 to 9 we add 30H to turn it into the corresponding ASCII digit. If it is in the range A to F then we must add 40H to convert into ASCII. The process continues until the original dividend is reduced to zero, as shown in the flowchart of Figure 4-11.

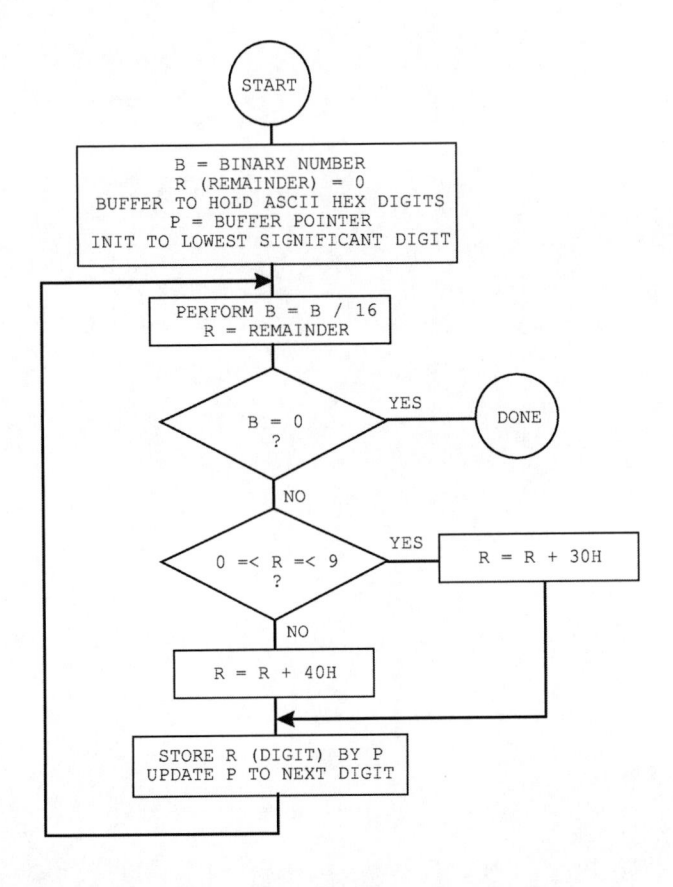

Figure 4-11 Unsigned Binary to ASCII Hexadecimal String

4.6.6 Signed Numerical Conversions

Conversion routines that use signed operands are usually a variation of the unsigned ones described in previous sections. Although logic can be developed that directly encodes to and from two's complement format, the more convenient approach is to determine the sign of the operand then use unsigned conversion for the digit values. For example, to convert a signed binary in two's complement form into a string of ASCII decimal digits the logic first determines if the binary operand is negative or positive. If a positive number, then the unsigned conversion routine can be used directly. If the binary operand is a negative number, the minus sign is placed in the storage buffer. Then the two's complement binary is converted to an unsigned number so that the ASCII digits can be obtained with the conversion routine described in Section 4.6.2.

Chapter 5

Circuits and Logic Gates

In Chapter 1 we covered basic electronics and elementary circuit components such as resistors, capacitors, inductors, transformers, and simple semiconductors. In this chapter we expand these topics and introduce new ones so as to provide a basic background in digital electronics and in the electronic circuits that are often used in microcontroller-based systems. The chapter also contains information on some of the simpler electronic devices often found in electronic circuit boards, such as diodes, LEDs, and logic gates. Chapter 6 covers other circuit components including switches, seven-segment displays, LCDs (liquid crystal displays), buzzers, motors, and flip-flops.

5.0 Digital Circuits

Digital circuits are the basic building blocks from which microprocessors, microcontrollers, computer systems, and virtually all digital electronic devices are constructed. These building blocks are essential and perform elementary functions. A single device can contain thousands of these primitive components. Knowing about these elementary building blocks is necessary if you are to design or program digital circuitry.

Understanding these components requires viewing them at the proper level of abstraction. To understand a simple digital device you must know how the simpler transistors that make up the device operate. To understand how a shift register works it is useful to visualize it in term of the logic gates from which it is built. Similarly, once you understand how counters and registers work it is easy to grasp how a complex large-scale integrated circuit, such as a serial port, operates.

Fortunately, at any given level of abstraction, it is not necessary to consider every single device of that class, because knowing about one or two representative devices is usually sufficient. For example, once you understand the operation of a few different logic gates you can assume that others work in a similar manner. So we start by explaining the basic facts about diodes and transistors, then we consider logic gates that are built from transistors, then the more complex circuits that are built from elementary logic gates, and so on.

5.1 The Diode Revisited

Chapter 1 concluded with a brief discussion of diodes and p-type and n-type silicon junctions. The diode acts as a very useful one-way valve for electrical current and is one of the most powerful developments in semiconductor physics.

But in order to use the diode it is not necessary to comprehend the physical and electrical principles that make it work. Rather, the diode can be treated as a device made from two pieces of silicon and it has the property of passing current in one direction.

When a voltage is applied to the diode that makes the n-type end more positive than its p-type end, electrons flow from the n to the p direction, but not from the p to n direction. In this manner the diode behaves as a one-way filter that allows electrons to flow in one direction but not in the other one. Figure 5-1 shows the p-n junction in a diode and its electrical symbol.

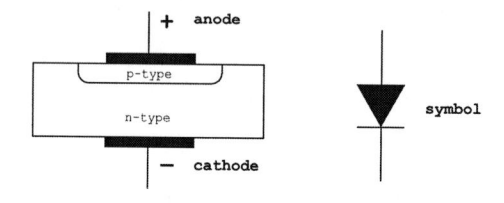

Figure 5-1 Diode Construction and Symbol

The general convention is that current flows from positive to negative, although in reality electrons flow from negative to positive. Benjamin Franklin is usually held responsible for this erroneous convention. Therefore, current in the diode in Figure 5-1 flows from the anode to the cathode, but not vice versa.

The electrical symbol for a diode, in Figure 5-1, resembles an arrow pointing in the direction of current flow. When the anode voltage of a diode exceeds the cathode voltage the diode is said to be *forward-biased*. A forward-biased diode acts like a short circuit. To prevent too much current from flowing a resistor is usually inserted in series with the diode, as in Figure 5-2.

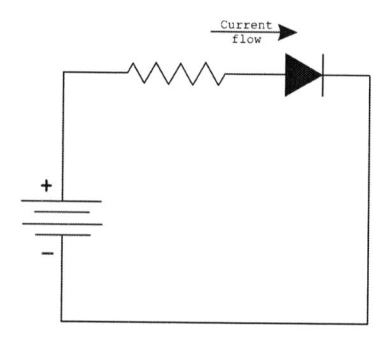

Figure 5-2 Diode and Resistor in a Circuit

Circuits and Logic Gates

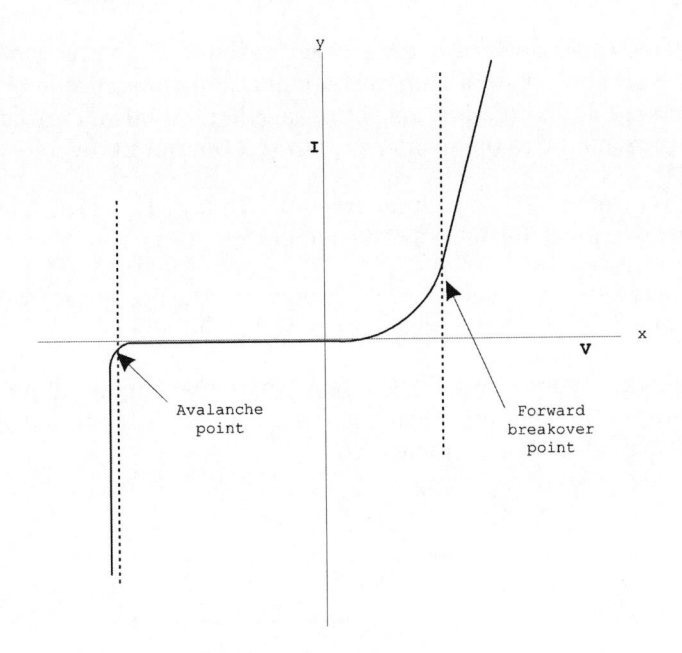

Figure 5-3 I/V Plot in a Diode

The diode's behavior can be also be represented by a curve that shows current-versus-voltage, sometimes called an *I/V curve*. If the voltage is represented on the abscissa of the Cartesian coordinate plane (x-axis) and current on the ordinate (y-axis), then the plot resembles the one in Figure 5-3.

In Figure 5-3 the current is non-linear; that is, it becomes very large if a positive voltage difference across the diode exceeds about 0.6 volts. This point is called the *forward breakover point*. If the diode is reverse-biased and the voltage is progressively increased, a point is reached in which the junction suddenly begins to conduct. This is called the *avalanche point*. The effect is similar to an internal short and the diode can be destroyed. Note that the I-V plot of a resistor is quite different from that of a diode. Since the resistor obeys Ohm's Law its I/V curve would be a straight line.

The typical diode, such as the ones used in logic and display circuits, can handle a current of 10 to 20 milliamps. For a 5-volt supply a 300-ohm series resistor limits the current through the diode to a reasonable value.

5.1.1 The Light-Emitting Diode (LED)

One of the most useful types of diodes is an LED (light emitting diode). The LED produces light when it is forward-biased. The most common LEDs have a distinctive red color, although they may be amber, green, blue, or white.

The LED is a semiconductor device that emits incoherent light when forward-biased. The color of the light depends on the chemical composition of the semiconducting material. The first practical LEDs were developed in 1962. LEDs are used in many electronic devices to signal the presence of an electric current. Like any diode, the LED consists of a chip of semiconducting material impregnated with impurities to create a p-n junction. As is the case in all diodes, current flows easily from the p-side, or anode, to the n-side, or cathode, but not in reverse.

The first LEDs were made of gallium arsenide. Today LEDs are made of a variety of materials so as to produce light of different colors.

Advances in materials science have made possible the production of devices with ever shorter wavelengths, producing light in a variety of colors.

Because LEDs are diodes they light only with positive electrical polarity, that is, when forward-biased. When the polarity is reversed very little or no light is emitted by the LED. Figure 5-4 shows a typical LED.

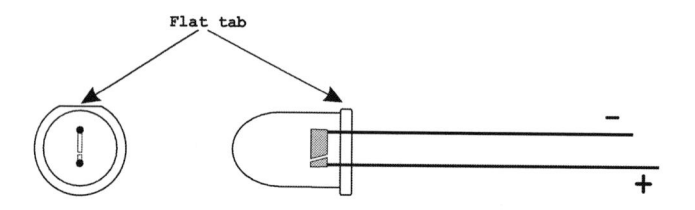

Figure 5-4 A Typical LED

The correct polarity of a new LED can usually be determined by observing that the longest terminal is the anode. If the terminals have been altered, then it is risky to try to determine polarity by observing the LED's internals. Although in most LEDs the larger internal tab is the cathode, there are others in which it is not. A more dependable clue to the LED's polarity is the flat tab on the LED's base, which indicates the cathode, as in Figure 5-4.

Ratings vary among the different sizes and types of LEDs. Most LEDs are rated to operate between 1.7 and 3.8 volts and at currents of 10 to 40 mA. The light-emitting capacity of an LED is measured in *megacandela* or *mcd*. Small commercial LEDs range from 10 to about 5000 mcd.

Once the LED's ratings and circuit's voltage are known it is necessary to calculate the value of the series resistor so that the current does not exceed the LED's capacity. For example, the series resistor for wiring a commercial red LED rated at 2.6 VDC and 28 mA on a 5 volt circuit is calculated as follows:

STEP 1: Calculate the voltage across the resistor by subtracting the LED's forward voltage from the supply voltage, in this case:

STEP 2: Apply Ohm's Law to calculate the required resistor:

The electronic symbol for an LED is somewhat similar to that for a diode, as shown in Figure 5-5.

Circuits and Logic Gates

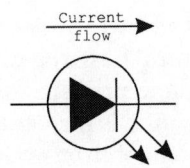

Figure 5-5 Electrical Symbol for LED

As a simple experiment, connect an LED in series with a 330-ohm resistor to a 5-volt power supply and see how light is emitted for one orientation of the diode and not emitted for the other. This little circuit makes a convenient probe for logic circuits. If the LED's cathode is touched to some point in a circuit, the LED lights up if the voltage at that point is less than about one or two volts. The LED remains dark if the voltage is greater than this value. This is a 1-bit binary digital voltmeter.

In addition to LEDs, there are logic diodes such as the 1N4148 or its equivalent, the IN914. These are used simply to ensure that the current in some circuit can flow in only one direction. There are also much heftier diodes that are used to manufacture the DC power supplies needed for computers and other electronic equipment. They take the 110-volt AC that comes out of wall outlets and converts it to a unidirectional DC voltage.

5.2 The Transistor

The transistor is a solid state semiconductor device that is used for signal amplification, voltage stabilization, switching, signal modulation, and many other functions. It can be considered as a variable valve which controls the current it draws from a voltage source. Transistors are manufactured as individual components or as part of an integrated circuit. Transistors come in two basic varieties: bipolar and MOS.

5.2.1 Bipolar Transistor

The bipolar transistor was the first type of transistor to be commercially mass-produced. The terminals of a bipolar transistor are named emitter, base, and collector. Physically, the bipolar transistor consists of two n-type regions separated by a thin p-type region or, alternatively, by two p-type regions separated by a thin n-type. When a transistor has two n-type regions, the device is called an NPN transistor. One of the n-type regions is called the collector, the other the emitter, and the central p-type region the base. The NPN bipolar transistor is shown in Figure 5-6.

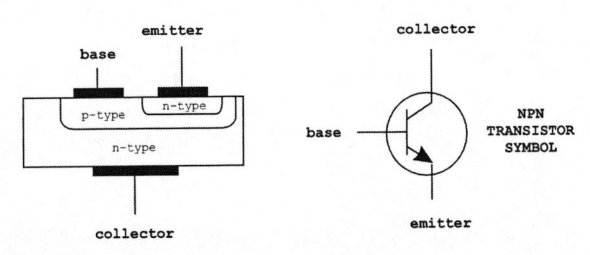

Figure 5-6 Bipolar NPN Transistor andSymbol

Simply, a bipolar transistor consists of two diodes connected back to back so that they share a common end. Since the central base region between the collector and emitter is very thin, the device has the unique property of serving as an amplifier. When the transistor's base-to-emitter p-n junction is forward-biased (this could be called the p-n diode) it creates a low resistance in the thin base region. This allows a much larger current to flow from the collector to the emitter. If the base-emitter current is turned off, then the collector-emitter current is also completely turned off. In this case the transistor is said to be cut off.

Over a given range, the collector-emitter current is directly proportional to the base-emitter current. In this manner the transistor amplifies small currents into larger ones, as in radios and other sound amplifying applications. For larger base currents the transistor acts as if there were nearly a short circuit between the collector and the emitter. In this case the transistor is said to be in saturation.

The effect is that a positive voltage on the base turns on the transistor and pulls the output low (to about 0.5 volts). When this voltage is removed, the transistor is turned off and the output is high (+5 volts). The action is that of a current controlled switch, as shown in Figure 5-7.

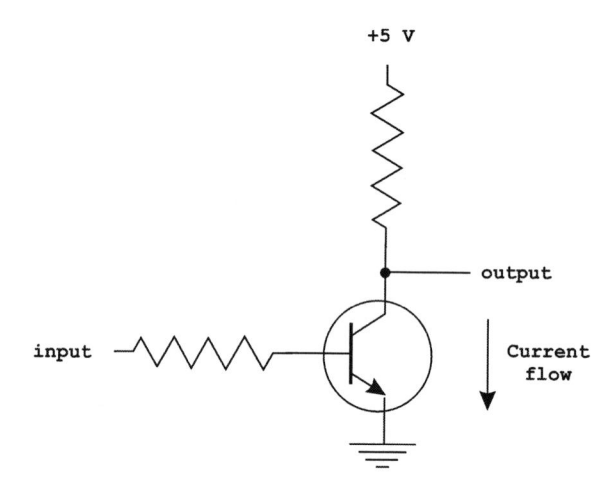

Figure 5-7 NPN Transistor Used as a Switch

The circuit in Figure 5-7 operates as follows: if the input voltage is held at zero volts, the p-n base-emitter junction has no current flowing through it and the output voltage is +5 volts. However, if the input voltage is raised to any value between +2 and +5 volts, a base-emitter current flows. This in turn allows a collector-emitter current to flow and the output voltage is pulled down to ground (typically between 0.5 and 1 V).

An alternative architecture for a bipolar transistor is called PNP. In this case the n-type silicon is sandwiched between two p-types, as shown in Figure 5-8.

Circuits and Logic Gates

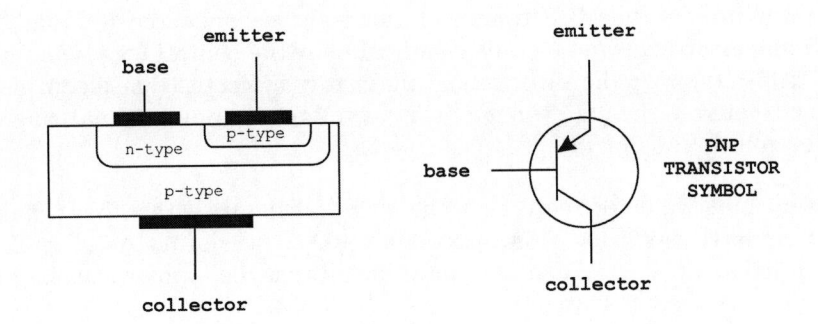

Figure 5-8 PNP Transistor and Symbol

The PNP transistor in Figure 5-8 works in the same way as the NPN transistor, except that in the PNP design the base has to have a negative voltage with respect to the emitter in order to turn on the transistor.

5.2.2 MOS Transistor

The second major type of transistor is the *metal oxide semiconductor transistor*, or MOS. It consists of two separate n-type regions embedded in p-type silicon. Alternatively, the MOS can consist of two p-type regions embedded in n-type silicon. In the first case the device is called an *n-channel MOS* (or NMOS) transistor; in the second case it is called a *PMOS*.

One of the two n-type regions is called the *source*, and the other is called the *drain*. An area between the source and the drain consists of a metal contact separated from the p-type body by a thin layer of non-conductive silicon dioxide. This area is called the *gate*. When a positive voltage is applied to the gate, the electric field attracts a thin layer of electrons into the p-type region underneath the gate. This provides a low resistance path between the two n-type regions. Figure 5-9 shows the construction of an NMOS transistor and the symbols for the NMOS and PMOS.

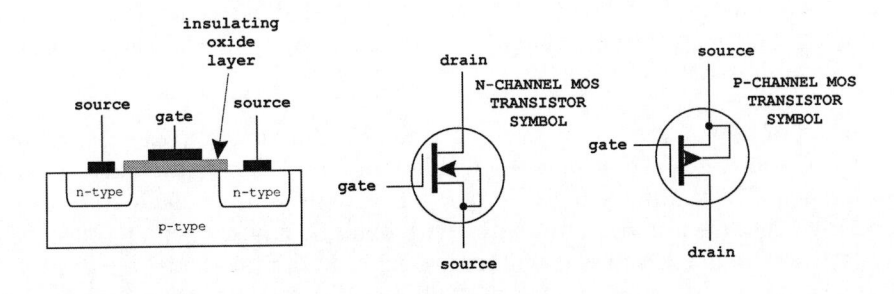

Figure 5-9 MOS Transistor and Symbols

In construction of the MOS transistor the body is connected internally to the source. In the electrical symbols this is indicated by the central wire with an arrow. In the NMOS transistor the direction of the arrow indicates that electrons in the body are attracted to the gate when a positive voltage is applied. This same voltage repels electrons in the PMOS.

One of the most valuable features of the MOS transistors is that they require very small currents to turn on. This makes the MOS transistors behave like voltage-controlled switches in a digital circuit. Recall that the bipolar transistors operate as current-controlled switches.

5.3 Logic Gates

A logic gate is an electronic device that takes one or more binary signals as inputs and produces a binary output that is a logical function of the input or inputs. The basic logical operations of AND, OR, XOR, and NOT were covered in Section 4.1. Although logic gates can be made from electromagnetic relays, mechanical switches, or optical components, nowadays they are normally implemented using diodes and transistors.

Charles Babbage's Analytical Engine, developed around 1837, used mechanical logic gates based on gears. Electromagnetic relays were later used for logic gates, and these were eventually replaced by vacuum tubes, since Lee De Forest's modification of the Fleming valve can be used as an AND logic gate. In 1937, Claude E. Shannon wrote a thesis paper that introduced the use of Boolean algebra in the analysis and design of switching circuits. The first modern electronic gate was invented by Walther Bothe in 1924, for which he received part of the 1954 Nobel Prize in physics.

The primitive types of gate are AND, OR, and NOT; in addition, the XOR gate offers an alternative version of the OR. The other Boolean operations can be implemented by combining the three primitive types. However, for convenience, other combined types have been developed. These are called NAND (NOT plus AND), NOR (NOT plus OR), and XNOR (XOR plus NOT). The advantage of these secondary logic gates is that they require fewer circuit elements for a given function. In fact, the NAND gate is the simplest of all gates, except for the NOT gate. Also, a NAND can implement both a NOT and an OR function; therefore, it can replace AND, OR, and NOT. The NAND gate is the only type actually needed in a real system. Programmable logic arrays very often contain nothing but NAND gates. The symbols for logic gates are shown in Figure 5-10.

The notion of a binary signal is accomplished by allowing it to be in only one of two states. These states are designated as high and low. Conventionally, we represent a high signal with binary digit "1" and a low with a binary digit "0." True and false and high and low are also associated with binary signals, binary 0 representing false or low and binary 1 representing true or high. In digital electronics, voltage is used to encode binary 0 and 1. A voltage of about 0.5 volts (actually 0 to 0.8 volts) is interpreted as logic 0 and a voltage of about 3.5 volts (actually 2.4 to 5.0 volts) is interpreted as logic 1. Voltages from 0.8 to 2.4 volts are not allowed. This voltage convention is referred to as TTL (*transistor-transistor logic*).

Circuits and Logic Gates

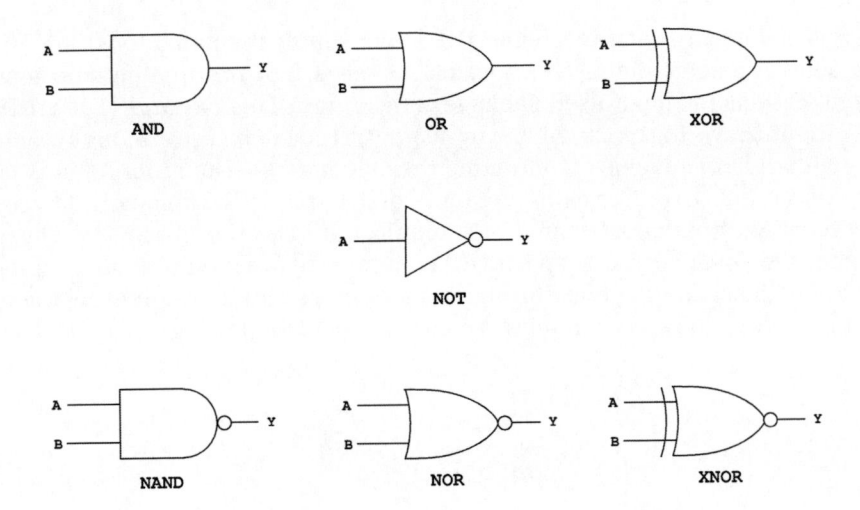

Figure 5-10 Logic Gate Symbols

5.4 Transistor-Transistor Logic

Transistor-Transistor Logic is a class of digital circuits built from bipolar transistors and resistors. TTL is used in a popular family of integrated circuits originally developed by Texas Instruments in 1962. These are known as the 7400 series of ICs. Components of the 7400 family are used in computers, industrial controllers, music synthesizers, and electronic test and measurement instruments. TTL provided a low-cost digital option to the expensive analog methods of the day.

TTL integrated circuit are available to perform the following functions:

1. logic gates such as AND, OR, NAND, NOR, and XOR
2. flip-flops
3. latch elements
4. counters and adders
5. shift registers
6. timing circuits
7. data bus drivers and buffers
8. display drivers
9. multifunction logic
10. memory
11. programmable logic arrays

The TTL logic gates require a 5-volt DC power supply regulated to within 5% (5V ± 0.25V). They are available in a variety of packages. For prototyping and manually built applications the most-used package is the *dual inline package* (DIP). DIP integrated circuits have 14, 16, 20, 24, 28, or 40 pins arranged in a row along two sides of a rectangular plastic package containing the silicon chip. The ground pin is usually the last in the first row. For example, pin 7 of a 14-pin DIP. The 5-volt power pin is usually the highest numbered pin. For example, pin 14 on the 14-pin DIP. The pins in a DIP package are spaced 0.1" apart. DIP components are mounted on a printed circuit board by inserting the pins through a set of holes and then soldering the pins in place. Figure 5-11 shows two integrated circuits in a DIP package.

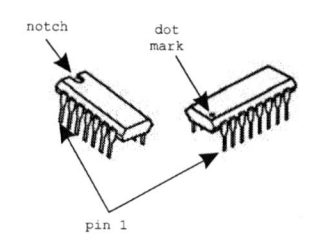

Figure 5-11 Dual Inline Packaged (DIP) Integrated Circuits

One objection to DIP packages is that they become too large if more than 40 pins are required. The *pin grid array* (PGA) is one solution to this problem. PGAs are square packages with an array of pins coming out of the bottom. Microprocessors are sometimes implemented in PGAs. Surface mount technologies are often used in commercial electronic boards, since they allow fitting more circuitry into a smaller space. The pins on surface mount packages are bent out horizontally and soldered to the top surface of the board. In surface mount ICs the pin spacing is .05 inch or less. But surface mount packages are difficult to handle outside of commercial production, since the smaller pins spacing require special soldering irons and inspection microscopes. For non-manufactured projects, such as the ones in this book, the DIP is the most suitable.

In the present context we discuss TTL logic gates furnished as integrated circuits constructed using semiconductor electronics. The part number of logic gate ICs is in the format 74XXX, where XXX refers to the specific gate implementation.

5.4.1 Inverter Gates

IC number 7404 is a hex inverter. Here the term hex refers to the six inverters included in the circuit. Figure 5-12 shows the schematics of the 7404 hex inverter.

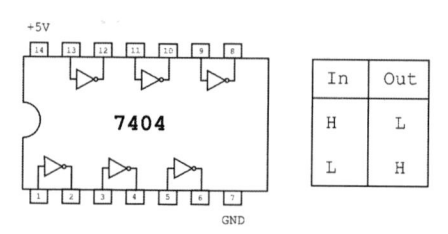

In	Out
H	L
L	H

Figure 5-12 7404 TTL Hex Inverter IC

Circuits and Logic Gates 87

The function of a given logic gate can be shown in a truth table, such as the one in Figure 5-12. The truth table lists the outputs (high or low) for the given inputs (also high or low).

5.4.2 The AND Gate

The 7408 is an AND gate that includes four individual gates per package. The 7408 AND gate and its truth table are shown in Figure 5-13.

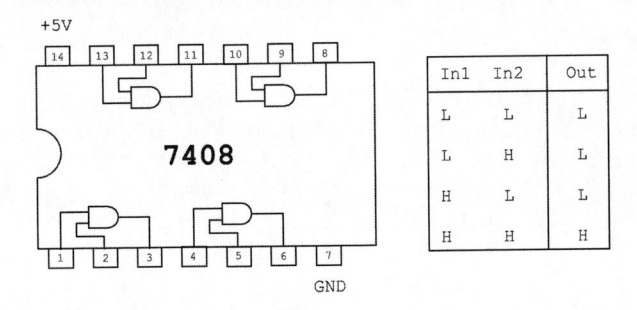

In1	In2	Out
L	L	L
L	H	L
H	L	L
H	H	H

Figure 5-13 7408 Quad 2-input AND Gate

The AND gate in Figure 5-13 is described as quad 2-input. Four individual AND gates are contained in the circuit and each one has two input lines. The gate logic corresponds to the Boolean AND: if both inputs are high, then the output is high, otherwise the output is low. If the input lines are designated as A and B, and the output as Y, then the AND operation can be expressed in the equation:

$$Y = \overline{A \bullet B}$$

In this case the dot operator represents the AND function, not arithmetic multiplication.

5.4.3 The NAND Gate

A variation of the AND gate is the 7400 NAND gate. In this case the AND operation is replaced with the inverted AND, or NAND. Thus, if both inputs are high, the output is low, otherwise the output is high. The 7400 NAND gate is shown in Figure 5-14.

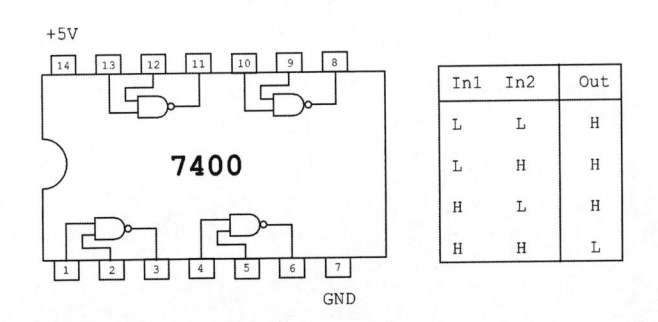

In1	In2	Out
L	L	H
L	H	H
H	L	H
H	H	L

Figure 5-14 7400 Quad 2-input NAND Gate

Note that in the NAND gate of Figure 5-14, the AND symbol has been replaced with the NAND symbol as in Figure 5-10. The logic equation for the NAND gate is the combination of the AND and NOT operations, as follows:

$$Y = \overline{A \bullet B}$$

Here the vertical bar over the equation's right-hand side indicates negation.

5.4.4 The OR Gate

The 7432 OR gate performs the Boolean OR of the two input lines. If either line *A* or line *B* is high, then line *Y* is high, otherwise line *Y* is low. Figure 5-15 is a diagram of the 7432 quad 2-input OR gate.

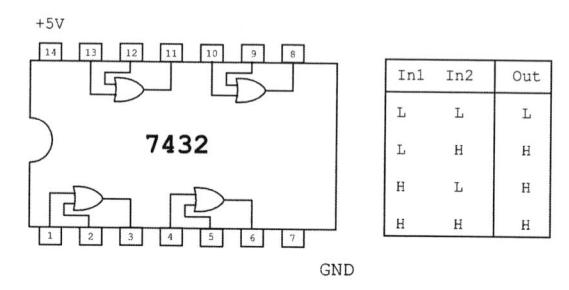

In1	In2	Out
L	L	L
L	H	H
H	L	H
H	H	H

Figure 5-15 7432 Quad 2-Input OR Gate

The equation of the OR operation with two inputs is:

$$Y = A + B$$

The plus sign in the previous equation indicates the Boolean OR operation, not arithmetic addition.

5.4.5 The NOR Gate

Another version of the OR gate is the 7402 NOR quad 2-input NOR gate. Here the Boolean OR is negated, as shown in Figure 5-16.

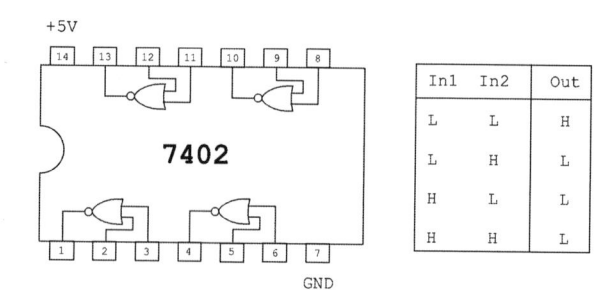

In1	In2	Out
L	L	H
L	H	L
H	L	L
H	H	L

Figure 5-16 7402 Quad 2-input NOR gate

Circuits and Logic Gates

The equation for the NOR gate consists of negating the inputs of the OR gate, as follows:

$$Y = \overline{A + B}$$

The gates shown in this section contain two input lines, labeled A and B. Logic gates are also available that contain more than two inputs, for example, the 7410 is a three 3-input NAND gate. Other logic gates with 3, 4, and 8 inputs are available. For example, the 7410 is a three 3-input NAND gate, the 7420 a two 4-input NAND gate, and the 7430 is a single 8-input NAND gate.

5.4.6 Positive and Negative Logic

The gates discussed so far assume that logic high is regarded as true and logic low as false. This is called *positive logic*. If we were to invert these assumptions so that logic high is regarded as true and logic low as false we would have a system based on *negative logic*. In this case the AND and the OR functions would be exchanged in regards to positive logic.

Digital circuit designers can often reduce the number of required integrated circuits by switching between positive and negative logic. For example, if an extra AND gate is available but the circuit requires an OR gate, the AND gate can be used by assuming negative logic. Circuit diagrams can be shown to use positive or negative logic by the position of the inverting circles. By convention, a circle on the input lines indicates negative logic and positive logic if it is placed on the output line. Figure 5-17 shows the equivalent circuits for gates using positive and negative logic.

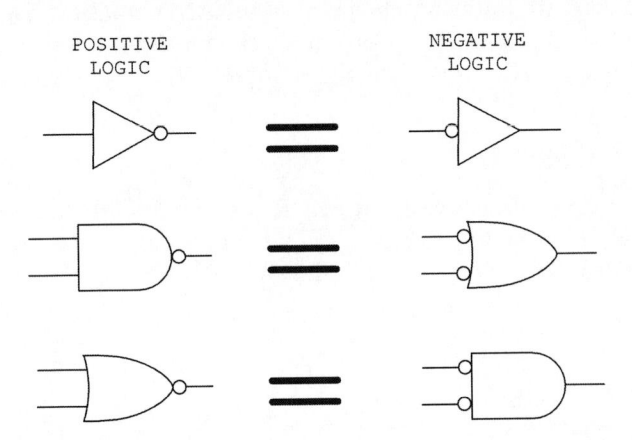

Figure 5-17 Circuit Symbols for Positive and Negative Logic Gates

The position of the inverting circles in the circuit diagrams is consistent with the notion that inverting the inputs changes the gate function. Thus, a negative logic AND gate functions as an OR gate and vice versa. In this manner a circle on the input line is read as the signal on that line being *active low*. An active low signal is asserted as true when it is electrically low. For example, the output of the 2-input NAND gates of Figure 5-14 is low when inputs A and B are both high, as shown in the circuit truth table. But the NAND gate can also be interpreted as a negative logic OR gate as in Figure 5-17. The one logic operation that is the same in negative or positive logic is NOT, as also shown in Figure 5-17.

An alternative explanation of positive and negative logic can be based on the truth table for the Boolean OR, as follows:

```
   A    B  |   Y
--------- |-----
   T    T  |   T
   T    F  |   T
   F    T  |   T
   F    F  |   F
```

In binary form and positive logic the OR truth table is as follows:

```
   A    B  |   Y
--------- |----
   1    1  |   1
   1    0  |   1
   0    1  |   1
   0    0  |   0
```

If we now invert the binary values in the second table, the results are as follows:

```
   A    B  |   Y
--------- |-----
   0    0  |   0
   0    1  |   0
   1    0  |   0
   1    1  |   1
```

Note that the last truth table matches that of the AND function. Thus, by inverting the truth table for the logical OR we produced the truth table for logical AND, validating the previous assertion that a negative logic AND gate is equivalent to an OR gate.

5.4.7 The XOR Gate

The last elementary logic gate is called the XOR or exclusive OR gate. In the XOR function the output is high if the inputs have opposite values, otherwise the output is low. Figure 5-18 shows the 7486 quad 2-input XOR gate and the corresponding truth table.

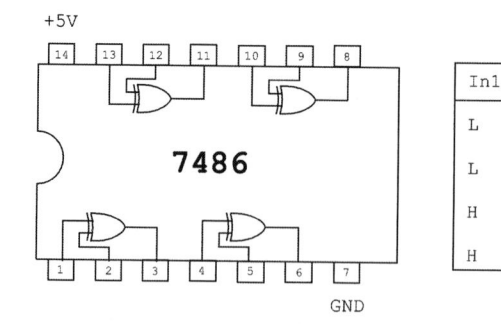

In1	In2	Out
L	L	L
L	H	H
H	L	H
H	H	L

Figure 5-18 7486 Quad 2-input XOR gate

Circuits and Logic Gates

Since an XOR gate's output is high if the inputs are different it can serve as a difference detector for logic levels. The following equation expresses the XOR relationship for two inputs (labeled A and B) and one output (labeled Y).

$$Y = A \oplus B$$

The logic symbol for XOR is the symbol for OR (+) enclosed in a circle. The XOR function can also be expressed in terms of Boolean OR and AND operations, as in the following equation:

$$Y = (A \bullet \overline{B}) + (\overline{A} \bullet B)$$

Figure 5-19 is the circuit diagram for the XOR gate constructed from OR, AND, and NOR gates.

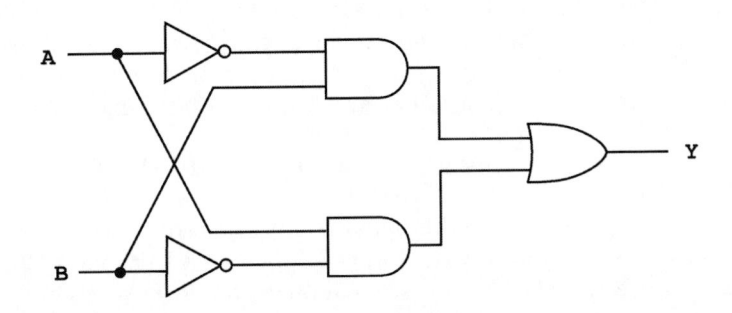

Figure 5-19 XOR Gate Circuit Diagram

Note in the XOR truth table in Figure 5-18 that if one of the inputs is forced high, then the gate functions as an inverter for the other input. Also, note in the truth table for the NOR gate (Figure 5-16) that if both inputs are low or high, then the circuit also functions as an inverter. If the inputs of a NOR gate are tied together the gate performs as an inverter. Often a circuit designer can take advantage of these identities in order to use an available gate for a function other than the one originally intended, thus saving having to use an additional IC.

5.4.8 Schmitt Trigger Inverter

Digital signals used in operating electronic devices consist of pulses. Conceptually, the pulses instantly fluctuate between a high and a low voltage level, ideally generating a square waveform. But signal noise in a circuit often contaminates the waveform into a non-rectangular shape. This noise can be the cause of circuit problems. For example, consider a plot of output voltage versus input voltage for a simple inverter, as shown in Figure 5-20.

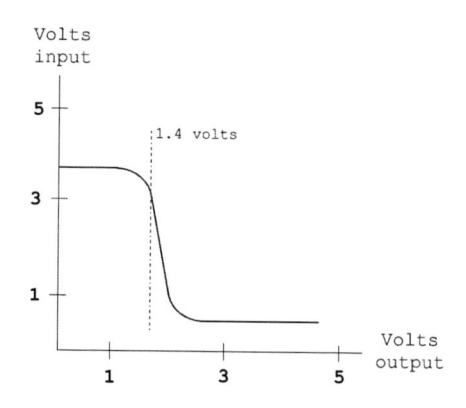

Figure 5-20 TTL Input and Output Voltage

In Figure 5-20 you notice that as the input voltage is raised from zero, the output voltage stays high. However, when the input voltage reaches about 1.4 volts (dashed line in Figure 5-20) the output switches sharply from high to low. Now suppose there is noise on the input line and that this noise causes the voltage to go above 1.4 volts and then below this value. In this case, the inverter's output may also switch its logic state several times during the transition period following the voltage fluctuations.

One common solution to this problem is based on a property of physical systems called *hysteresis*. Although the term derives from a Greek work meaning deficiency, it can also be related to "history." In this sense hysteresis refers to the property of an object that does not instantly follow the forces applied to it, but reacts "historically" to these forces. In other words, the new state depends on the object's immediate history. Adding hysteresis to a circuit's input makes it so that the point at which the output changes state depends on the current state of the output. For example, if output is high, it does not go low until the input voltage is raised above 1.7 volts. On the other hand, once the output goes low it will not change back to high until the input falls below 0.9 volt. This "lag" before changing to a new state makes the output much less susceptible to being inadvertently switched by noise. Figure 5-21 shows a plot of the input versus the output currents on a circuit with hysteresis.

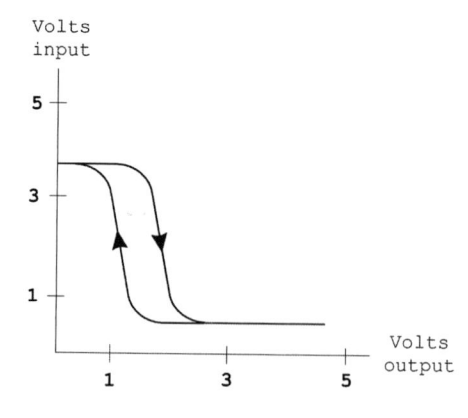

Figure 5-21 Effect of Hysteresis in an Inverter Circuit

Logic gates whose input has hysteresis are often known as a *Schmitt trigger*. Figure 5-22 shows a 7414 hex Schmitt trigger inverter.

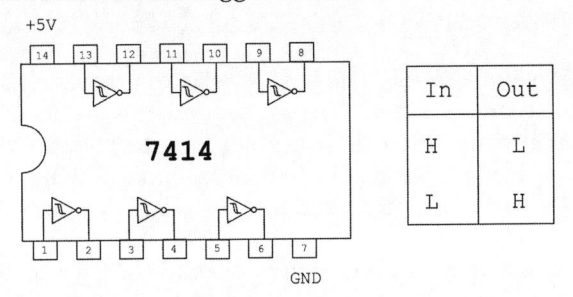

Figure 5-22 7414 Hex Schmitt Trigger Inverter

Note in Figure 5-22 that there is a small hysteresis curve inside each inverter symbol. This indicates that the inverters are the Schmitt triggers.

5.5 Other TTL Logic Families

In 1971 a major advance in TTL logic occurred with the introduction of TTL devices that incorporate *Schottky* diodes. They are based on the property of aluminum to act much like a p-type semiconductor when in contact with n-type silicon. The Schottky diode acts like an ordinary p-n diode except that it has a faster response time and the voltage drop is about 0.3 volts instead of 0.6 volts. When a Schottky diode is connected between the base and the collector of a bipolar transistor, the transistor is prevented from going into saturation. The Schottky diode/transistor combination, known as a Schottky transistor, has a significantly faster switching speed. Schottky TTL logic devices have part numbers 74SXXX and give three times the speed of standard TTL using only twice the power.

By increasing the resistor sizes, low-power Schottky TTL was developed giving the same speed as standard TTL, but using only 1/5 the power. These devices, whose part numbers are in the format 74LSXXX, were the standard TTL logic parts for many years. In 1980, more sophisticated Schottky-type logic circuits using smaller, higher performance transistors were developed by Texas Instruments. These are the advanced Schottky and advanced low-power Schottky logic families. Their part numbers are 74ASXXX and 74ALSXXX respectively.

5.6 CMOS Logic Gates

Around the same time that the original TTL circuits using bipolar transistors were introduced, a line of logic circuits using *CMOS* (complementary metal-oxide semiconductor) technology became available. A line of TTL-compatible CMOS ICs have part numbers 74XXX. TTL series pinouts are also available with part numbers 74CXXX.

CMOS logic circuits have two significant advantages over TTL. In the first place, CMOS circuits operate with very low power dissipation. A CMOS input requires virtually no current to remain at a given logic level. In fact, the entire circuit draws insignificant current when it is not switching between logic levels. In CMOS, power is consumed only during switching, while bipolar logic power dissipation is only weakly dependent on the switching rate. At low switching rates, CMOS provides huge savings in power dissipation.

A second advantage of CMOS is the smaller size of the circuits. Since no resistors and only two simple types of transistors are needed, the resulting logic gates require less area on a silicon wafer than their bipolar counterparts. The combined advantages of less power consumption and less area make CMOS the choice for *VLSI (very large scale integration)* integrated circuits such as microprocessors.

However, there are also significant drawbacks to CMOS which have prevented it from completely replacing bipolar logic. One of them is that CMOS circuits have slower switching speeds and propagation delays compared to bipolar circuits. The original CMOS logic gates had switching speeds that were about five to ten times slower than the 74XXX bipolar logic gates. High-speed CMOS, introduced in 1980, have improved processing technology and smaller transistor sizes, resulting in higher switching speeds and improved output drive current capability.

The CMOS 74HCT parts are completely TTL-compatible and can be freely intermixed with bipolar TTL parts. The 74HC series, on the other hand, have a logic transition threshold of 2.5 V when using a 5-volt power supply, compared to the 1.4 volts TTL standard. Since CMOS outputs have 5-volt and 0-volt logic levels, a 2.5-volt threshold provides better noise immunity than TTL; however, 74HC series parts cannot be mixed with standard TTL parts. For this reason, in mixed circuits, it is preferable to use the 74HCT parts.

An advanced CMOS technology family was introduced in 1985 having part numbers 74ACXXX. The TTL-compatible versions have part numbers 74ACTXXX. These new ICs have about double the speed of HC and HCT with yet another increase in drive power. The result is that the propagation delays for 74ACT parts approach those of bipolar TTL, although they are not quite equal to the fastest TTL families. To further increase CMOS speeds manufacturers turned to a process known as *BiCMOS*, which uses a mixture of bipolar and MOS transistors on the same chip. By strategically placing bipolar transistors at critical points in the circuit, the switching speed can be improved with only a small increase in power dissipation. The most popular BiCMOS logic family is the 74FCTXXX (fast CMOS) series of logic ICs.

Still another drawback to CMOS logic is that the circuits are susceptible to static electricity. The static discharge of the human body in a dry environment can destroy a CMOS transistor. Although protective diodes on CMOS circuit inputs provide some protection to static breakdown, all CMOS circuits are susceptible. For this reason ICs and circuits boards should be stored in conductive pouches and not handled until you have discharged yourself by touching a good electrical ground.

Chapter 6

Circuit Components

This chapter covers the most common general-purpose circuit components often found in microcontroller boards. Some simpler circuit devices such as diodes, LEDs, and logic gates, were discussed in chapter 5. Here we mention other common components including power supplies, switches, clocks and timers, flip-flops, decoders, seven-segment displays, and liquid crystal displays (LCDs). Other components sometimes found in microcontroller-based digital circuits are not discussed either because of their simplicity (buzzers and relays), their passive nature (connectors, adapters, batteries, and wiring), or their specialized features (motors, digital-to-analog and analog-to-digital converters, and memory).

6.0 Power Supplies

Standard logic circuits usually require a power source of +5 VDC. One possible source of +5 VDC is in one or more batteries. A D- cell battery generates 1.5 volts, so three of them can be connected in series to produce 4.5 VDC. An alternative power source can be from the standard wall outlet. Household electrical service in the United States is in the form of 110 volt AC (alternating current) power. Actually, 110 volts is the root mean square value of a sine wave that oscillates 60 times per second from about +155 volts to about -155 volts. The circuitry required to convert 110 VAC into 5 VDC is known as a *power supply*.

To obtain +5 VDC from 110 VAC requires scaling down the voltage and converting alternating current to direct current. In addition, most power supplies include a voltage regulator component that ensures that the circuit voltage is exactly +5 volts. The circuit in Figure 6-1 (in the following page) is a regulated 5-VDC power supply. The transformer reduces the household voltage from 110 to about 12 VAC. The diodes rectify the input to an oscillating signal of about +12 VDC. The 100mF electrolytic capacitor smoothes out the oscillation producing a largely DC voltage with little ripple. The 7805 is a voltage regulator that accepts an input voltage from about 8 volts to about 35 volts and produces a constant 5V output. Voltage regulator ICs are Zener diodes with a precise, reverse-biased breakdown voltage.

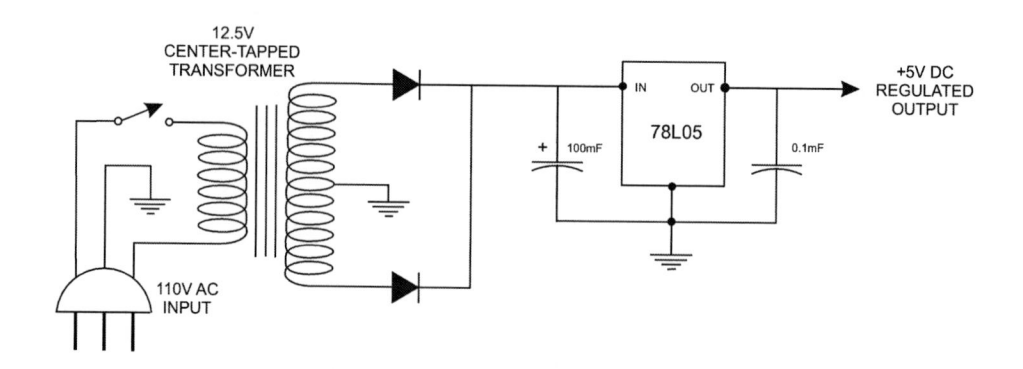

Figure 6-1 Regulated +5 VDC Power Supply

The 7805 is usually mounted on a metal base with a drilled hole so that a heat sink can be attached to it. With a heat sink the 7805 can produce up to 1 amp output. Figure 6-2 shows a 7805 voltage regulator IC.

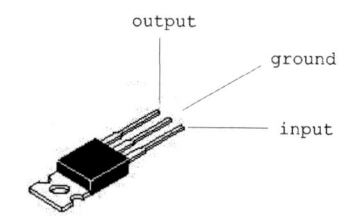

Figure 6-2 7805 Voltage Regulator IC

6.1 Clocked Logic and Flip-flops

In the digital circuits considered so far the outputs are entirely determined by the inputs to these circuits. In other words, if the inputs change so do the outputs. However, we often need a digital component whose output remains unchanged even if there is a change in input, for example, to store a binary number. A *flip-flop* is such a circuit since it performs as a 1-bit memory that stores either the value 0 or 1.

6.1.1 The RS Flip-flop

A circuit is said to be *bistable* if it has two, and only two, stable states. For example, a toggle switch which can be either OPEN or CLOSED is a bistable device. In a sense the toggle switch has memory since it remains in any one of the two positions until changed.

A flip-flop is an electronic circuit with two stable states, since its output is either 0 or +5 VDC. In this context we say that a flip-flop is set if it stores a binary 1 and reset otherwise. The RS designation refers to the Reset and Set stages. The flip-flop can also be said to have memory since its output remains set or reset until it is intentionally changed. When the flip-flop output is 0 VDC it can be regarded as storing a logic 0 and when its output is +5 VDC as storing a logic 1.

Circuit Components

Flip-flops can be constructed using primary logic gates. One possibility is using two NAND gates, as in Figure 6-3.

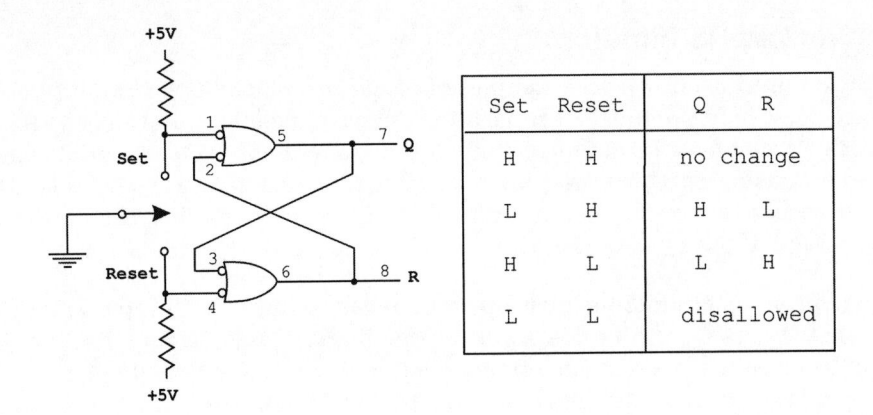

Set	Reset	Q	R
H	H	no change	
L	H	H	L
H	L	L	H
L	L	disallowed	

Figure 6-3 NOR Gate-based RS Flip-flop

Recall from Chapter 5 that a NAND gate is equivalent to a negative logic OR gate; this makes the flip-flop easier to understand. Looking at Figure 6-3, first consider that the Set input is pulled low by flipping the switch counterclockwise and sending the input to ground. In this case the output of the upper gate (1) is forced high since the gate's output goes high if either input 1 or input 2 is low. Since the Reset input to the lower gate is high (4), then neither input of the lower gate (3 or 4) is low and its output is low. Note that input 3 is low because the bubble on the lower OR gate inverts the value fed back from the upper OR gate. Now the feedback line from the lower gate (6) sends low to input 2 on the upper gate, which is inverted by the upper gate bubble. So both inputs to the upper gate are high, determining that the upper gate's output remains high even when the Set input returns to a logic high, as would be the case if the switch were turned back to the neutral position. Thus, the Q output of the flip-flop stays high (and the inverted Q output remains low). When the flip-flop is in this state, it is set. The flip-flop is placed in the cleared state by momentarily pulling the Reset input low. This forces the lower gate's output to be high and the upper gate's to be low.

The action of the flip-flop in Figure 6-3 is consistent with the description of a device with two steady states, labeled Set and Reset, and controlled by two corresponding input lines. Once a device is in either state, it remains in that state until the opposite state is enabled, thus "remembering" its set or reset status. The rotary switch mechanism ensures that the device will have either two high input lines or one high and one low. The condition of two low input lines is not allowed in this flip-flop, as shown in the truth table.

All mechanical switches used in electronic devices contain a spring of some sort. It is this spring that maintains the switch's contact in either position, but it also makes the switch electrically *"bounce"* whenever it is activated. Although the bounce only takes a few milliseconds, the logic level can change between high and low several times during this period. If an RS flip-flop is connected to the switch, the first contact switches the flip-flop and subsequent ones have no effect, thus effectively "debouncing" the switch.

6.1.2 Clocked RS Flip-flop

So far the circuits discussed are examples of *combinatorial* or *asynchronous logic*. If we ignore a few nanoseconds of propagation delay, in combinatorial circuits the outputs change as soon as the inputs change. Although in theory you can build complex logic circuits using combinatorial logic, it is more convenient to use clocked logic pulses to ensure high reliability and noise immunity. Circuits that use clocked impulses are said to use *synchronous logic*.

In synchronous circuits unconstrained changes in logic gate outputs are not allowed. Instead, the logic is designed so that logic level changes can progress through the circuitry one stage at a time under control of a clock. Between the clock pulses that cause changes to take place, the temporary state of the system is stored in memory elements or flip-flops.

In clocked or synchronous logic all the gates in the system change outputs at the same time. The output state of each gate depends only on the state of the gate inputs at the time of the clock pulse. In combinatorial circuits the gates may briefly "see" the wrong logic level and cause incorrect operation of the circuit. With clocked logic, the gate outputs "settle down" during the time between clock pulses so that only valid logic levels are present by the time the next clock pulse arrives.

The RS flip-flop in Figure 6-3 is not suitable for use in a clocked logic circuit because its output changes immediately whenever the Set or Reset inputs change. However, the circuit can be made into a clocked RS flip-flop by adding two NAND gates, as shown in Figure 6-4.

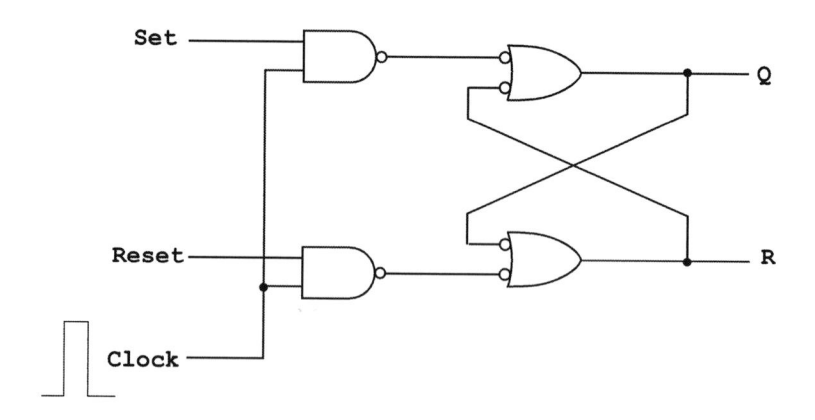

Figure 6-4 Clocked RS Flip-flop

Circuit Components

In the clocked flip-flop of Figure 6-4 the Set and Reset inputs can change at any time, but those changes are ignored by the flip-flop except during the interval when the logic high of a clock pulse is present. During the clock pulse the state of the Set or Reset line is stored by the flip-flop.

6.1.3 The D Flip-flop

One of the objections to the flip-flops in Figures 6-3 and 6-4 is that there are two data input lines, labeled Set and Reset in the illustrations. One possible solution is to only use one of the inputs by connecting an inverter between the Set line and the Reset input. The circuits for a D Flip-flop are shown in Figure 6-5.

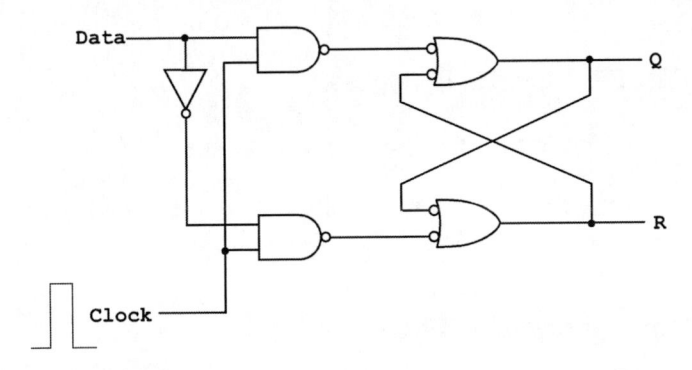

Figure 6-5 The D Flip-flop

The name D (or data) flip-flop originates in the fact that it contains a single data line. The *D flip-flop* is also called a transparent latch, or a *D latch*. In the D flip-flop the state of the input line, called the D input, is stored in the flip-flop when a clock pulse occurs. An advantage of this design is that the disallowed state (see Figure 6-3), in which both Set and Reset are simultaneously low, cannot be reached accidentally.

A flip-flop can be used for storing binary data. To visualize how this can be done, imagine four D flip-flops driven by the same clock signal. When the clock goes high, input data is loaded into the flip-flops and appears at the output. When the clock goes low, the output retains the data. For example, consider four data inputs, as follows:

$$D_0 D_1 D_2 D_3 = 0101$$

When the clock signal goes high, these four bits are loaded into the D latches, resulting in the output:

$$Q_0 Q_1 Q_2 Q_3 = 0101$$

This operation is represented in Figure 6-6.

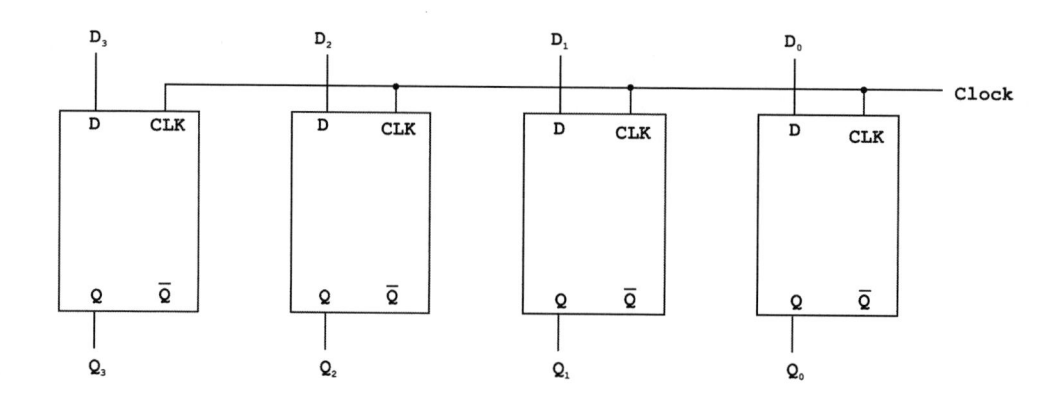

Figure 6-6 4 Data Bits Stored in D Latches

In the 4-bit D latch of Figure 6-6 the output data is stored as soon as the clock goes low. For as long as the clock is low, the D values can change without affecting the Q values. The 7475 IC contains four D flip-flops and is called a *quad bistable latch*. This circuit is well suited for handling 4-bit data bits simultaneously (one nibble).

6.1.4 The Edge-triggered D Flip-flop

The D flip-flop or transparent latch is available in several versions in addition to the 7475. Although the pure D flip-flop is a useful IC, for some applications it has the drawback that outputs follow the D input during the entire time that the clock line is high. In some circuits it would be ideal to have a flip-flop that stores data at a unique point in time. The edge-triggered D-type flip-flop approaches this behavior. In this device, the flip-flop stores the state of the data line at the instant the clock signal makes a transition from low to high and ignores it otherwise. Figure 6-7 shows an edge-triggered D flip-flop.

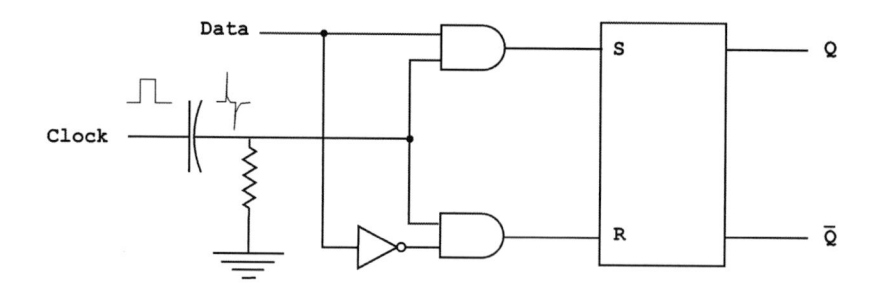

Figure 6-7 Edge-Triggered D Latch

The circuit in Figure 6-7 is sometimes called an *RC differentiated clock input latch*. In this case RC stands for the resistor/capacitor combination at the input of the D latch. By design, the RC time constant is made smaller than the clock's pulse width. This determines that the capacitor fully charges when the clock goes high,

Circuit Components

producing a narrow positive voltage spike across the resistor. Later, the trailing edge of the pulse results in a narrow negative spike, enabling the AND gates for a brief period. The effect is to activate the AND gates only during the positive spike; the negative spike does nothing in this circuit. The result is equivalent to sampling the value of D for an instant. At this point in time, D and its complement hit the flip-flop inputs, forcing Q to set or reset (unless Q is already equal to D).

6.1.5 Preset and Clear Signals

The use of flip-flops in digital circuits usually requires some way of placing the signals in a known state. In this sense a *Preset* signal is used to make sure that the Set line is high, and a *Clear* signal to make sure that the Reset line is high. Alternatively, these signals are referred to as *Preset R* and *Preset S*. Figure 6-8 shows how the Preset and Clear functions can be implemented in an RS flip-flop.

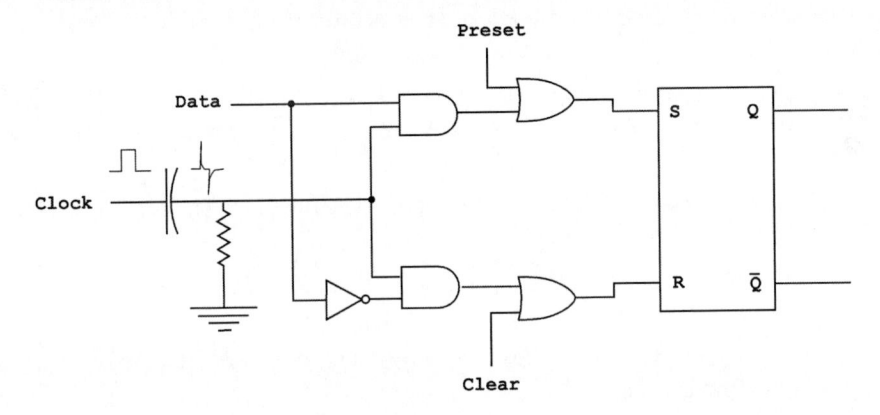

Figure 6-8 Implementing Preset and Clear

The OR gates in the circuit of Figure 6-8 allow selectively setting the S or the R lines of the edge-triggered D flip-flop. The Preset and Clear signals are called *asynchronous inputs* since they activate the R or S lines of the flip-flop independently of the clock. The D input, on the other hand, is *synchronous* since it has an effect only when the clock edge signal is high. Figure 6-9 shows the electrical symbol for a positive edge-triggered flip-flop with active high Preset and Clear lines.

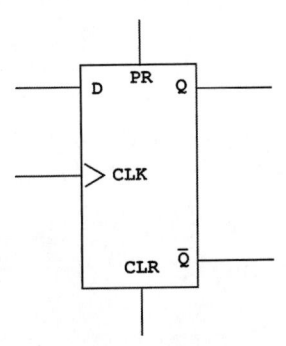

Figure 6-9 D-Type Edge-triggered Flip-flop Symbol

In the normal mode of operation, a D-type flip-flop has the Set and Clear inputs high (not active), so that a transition of the clock input from low to high (called a *positive edge*) clocks the value of D into Q and the inverse of D into not-Q. The clock transition is required; D can do anything it wants to, but nothing happens to Q and not-Q until a positive edge occurs on the clock line.

6.1.6 D Flip-flop Waveform Action

An easy way of understanding the interaction of the various signals in a clocked RS flip-flop is by means of a *waveform diagram*. The reference circuit is the one in Figure 6-8. This includes a clock signal, a data input line, Preset and Clear lines, Set and Reset input lines into the flip-flop, and Q and not-Q output lines. The signals are described as follows:

1. The clock signal (CLK) is a square wave that oscillates between a high and a low state. It provides a synchronized beat that coordinates the various digital devices present in the circuit.

2. The data signal is used as a single input line into the flip-flop. Setting the data signal high also sets high the flip-flop's Set line. A low data signal makes the flip-flop Reset line high.

3. The Set signal or line is one of the two inputs into the flip-flop. The other one is the Reset line.

4. The Preset line is used to make the flip-flop set line active. The Clear signal has the effect of setting high the Reset line into the flip-flop.

5. The Q and not-Q lines provide the flip-flop output. Q is high if the Set line is high; otherwise the not-Q line is high.

Figure 6-10 is a waveform diagram for a clocked RS flip-flop.

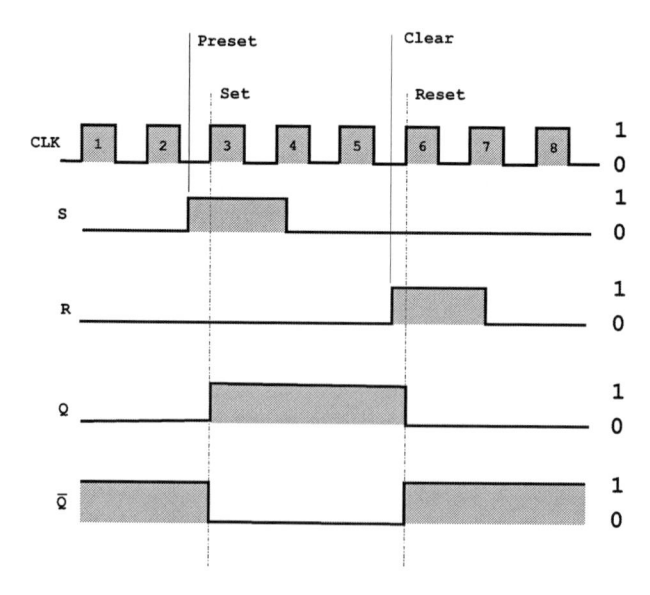

Figure 6-10 Waveform Diagram for Clocked RS Flip-flop

In reference to Figure 6-10 note that the clock input (at the top of the illustration) provides the synchronization beat for the flip-flop inputs (R and S) and the outputs (Q and not-Q). However, the Preset and Clear signals are asynchronous, that is, they operate independently of the clock pulse. Therefore, when the Preset line is set high, the S input line into the flip-flop immediately follows. However, the Q output line must wait until the next rising clock pulse, which corresponds to the dot-dash line labeled Set in the illustration. Similarly, the Clear signal immediately sets the R line; however, the not-Q output is not set until the next rising clock pulse. Note that during clock pulse number 4 both the R and S lines are held low. This corresponds to the hold state and the output on lines Q and not-Q remains unchanged.

6.1.7 Flip-flop Applications

The D-type flip-flop finds many uses in digital technology. Perhaps the most obvious one is as a memory. The flip-flop stores the value clocked into it from the D line; said value can be read on the output lines Q and not-Q. A type of memory known as *static RAM* is implemented as a large array of flip-flops with address decoding circuitry that allows selecting which flip-flop is being accessed by a read or write operation. Processors and microcontrollers contain many flip-flops, usually in the form of registers, which are just a group of 8, 16, 32, or 64 flip-flops. Flags are also flip-flops that are set or cleared by the results of the CPU's internal operations.

Digital devices interface with the outside world by means of input and output ports. These elements are implemented as flip-flops. For example, supporting the logic requires turning on a LED so as to signal that some event has occurred. To achieve this, a data line from the digital device can be connected to the D input of a flip-flop. Then a pulse is sent on another line to the clock input. When the clock pulse goes from low to high, the state of the data line at that instant is clocked into the flip-flop. This state remains on the Q output until a new value is clocked in. Another example is the 74374 IC, which contains 8 flip-flops in a single 20-pin DIP package. The chip is called an *octal latch* because data is latched into all eight flip-flops all at once by a single clock line.

D-type flip-flops are also used in implementing digital interfaces; for example, to have a digital device read in data from some external source, such as a switch. Each time new data is produced by the switch, a flip-flop is set and the output of this flip-flop is connected to an interrupt request line (IRQ) on the device. When the IRQ line goes high, the microcontroller saves its current state and branches off to an input routine that takes some action according to the state of the switch; for example, turns on a LED if the switch is high. To prevent the microcontroller from getting interrupted again by the same input, the same signal is also used to clear the flip-flop until the next data byte comes along.

6.2 Clocks

A clock signal consists of a sequence of regularly spaced pulses, typically in the form of a square wave. Digital devices use the rising or the falling edges of the square wave to run logic circuits. Clocks provide the heartbeat without which the system would cease to function.

6.2.1 Clock Waveforms

In a digital device, such as a microcontroller system, the clock provides a periodic waveform that is used as a synchronizing signal. Although the typical clock waveform is depicted as a square wave (as in Figure 6-10) it need not be perfectly symmetrical. In fact, a series of positive or negative waves could serve as a timing pulse in a digital circuit. The one requirement of a clock pulse is that it be perfectly periodic.

The basic timing interval for a digital circuit, which is equal to one full waveform period, is called the *clock cycle*. This determines that all logic elements in the circuit, including gates and flip-flops, complete their transitions in a complete clock cycle or less.

We can assume that the ideal clock produces a perfectly square waveform that is absolutely stable, as the one shown in Figure 6-11.

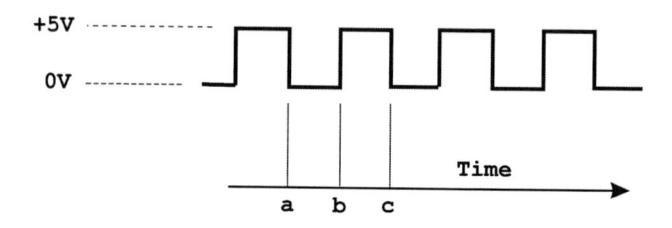

Figure 6-11 Ideal Waveform

A stable and uniform waveform reaches exactly the same voltage every time the clock is high; for example, +5 volts. By the same token, every time the clock signal goes low the voltage level must be the same, typically 0 volts. In addition, the clock signal must remain at the high and low levels for the same time and the time between each high and low cycle must be exactly the same. This last element is usually called the *frequency stability* of the clock. In Figure 6-11 the frequency stability refers to the time it takes for the signal to transition from point a to point c during each clock cycle. In practice, the stability and uniformity of the clock signal are more important than the absolute value. For example, it is usually acceptable that the high voltage level of the clock signal be 4.8 volts instead of 5 volts, as long as the 4.8 volts level is exactly reproduced at every clock cycle. Figure 6-11 shows an ideal waveform.

Another characteristic of the clock signal is the time required for clock levels to change from high to low and vice versa. Ideally this transition could be represented by a vertical line, as in Figure 6-11. This would mean that the transition is instantaneous, which is not achievable in actual circuits. In practice some time is required for the waveform to transition from low to high and vice versa. So the actual graph of the waveform, as can be seen in an oscilloscope, shows a slightly sloping side. Customarily, the actual measurement of the transition time is referred to as the 10 and 90 percent points. For example, in a 5 volt waveform, the rise time is the time it takes for the voltage to go from 0.5 to 4.5 volts, which are the 10 and 90 percent points for that waveform.

Circuit Components

6.2.2 The TTL Clock

A much used TTL-compatible clock can be built around a 7404 hex inverter IC such as the one in Figure 5-12. The idea is to use two inverters to build a two-stage amplifier with an overall shift of 360 degrees. The output signal at one of the inverters is fed back, through a crystal, to the first inverter; this determines that the circuit oscillates at a frequency determined by the crystal. Thus, the frequency of this clock signal is determined by the crystal: values between 1 and 20 MHz are common. The TTL clock circuit is shown in Figure 6-12.

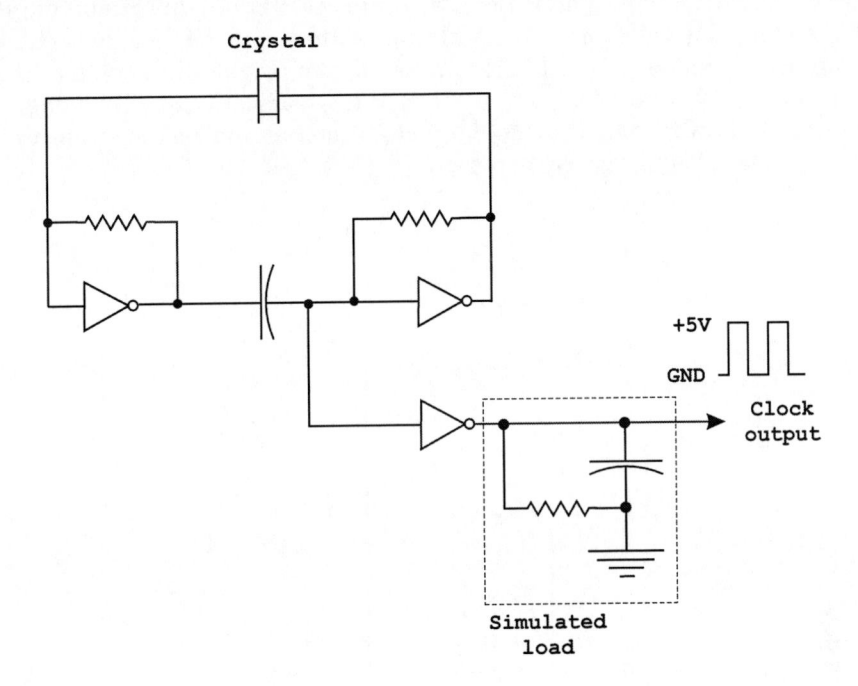

Figure 6-12 TTL Clock Circuit

The crystal in the circuit of Figure 6-12 makes the frequency of oscillation very stable. The third inverter is used as an output buffer and allows driving the load simulated by the RC circuit.

The clocks used in digital systems need to be stable and uniform so that the frequency is the same and each pulse is the same as every other one. To achieve this, a narrow band frequency-selective filter whose center frequency does not change is required. Quartz crystals are a good choice since they provide a stable, precision oscillation. A quartz crystal is actually a thin piece of polished crystalline quartz with contacts plated on each surface and a lead attached to each contact. Quartz is a piezoelectric material, which means that there is one particular electrical frequency that excites the crystal's resonance. It is this narrow resonant frequency that is used to build a frequency-selective filter whose center frequency changes very little as the components age or with changes in temperature. Crystal oscillators are available with frequencies that range from 10 KHz up to 600 MHz or more. They are typically housed in small metal cases with the frequency printed on the outside.

6.2.3 The 555 Timer

One of the most versatile timer ICs is the TTL-compatible *555 timer*. This chip can be used to make many different kinds of oscillators, pulse generators, and timers. As an oscillator, the 555 can be made to produce square, sawtooth, or triangle waves, and its frequency can be modulated by an external input. Although the 555 is not a TTL part, its output is TTL-compatible when it is used with a 5-volt power supply.

The 555 timer has two distinct output levels that continuously switch back and forth between two unstable states. Because of this oscillation, the circuit output is a periodic, rectangular waveform. The fact that neither output is stable accounts for the circuit being astable or bistable. The frequency of oscillation as well as the duty cycle are accurately controlled by two external resistors and a single timing capacitor. Figure 6-13 shows the logic symbol for a 555 timer as well as the wiring to implement an asymmetric square wave generator.

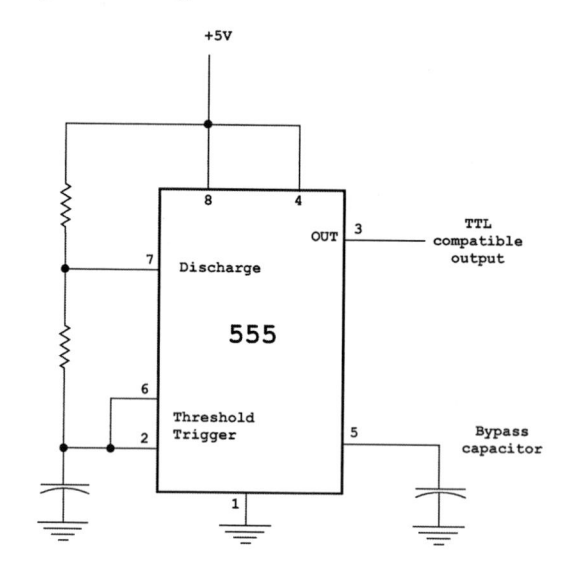

Figure 6-13 555 Timer as a Square Wave Generator.

6.2.4 Microcontroller Clocks

Microcontrollers, like most digital components, require a synchronizing timing pulse provided by some form of clocking device.

There are five common ways of implementing a timer in a microcontroller:

1. Internal clock
2. RC network
3. Crystal oscillator
4. Ceramic resonator
5. External oscillator

The selection depends on the specific microcontroller, the circuit requirements, and the cost of each available option. The least expensive option is the *resistor/capacitor oscillator circuit* (*RC network*). The disadvantages are its slow speed and inherent inaccuracies. Some of the newer generations of microcontrollers come equipped with an internal RC oscillator that operate as a programmable timer. Typical speeds are 4 MHz with a 1.5 percent error. The actual use and implementation of microcontroller clocks is discussed in relation to each specific device.

6.3 Frequency Dividers and Counters

Frequency dividers and *counters* are actually the same circuitry used in different ways. Counters are one of the most useful and versatile digital devices. Counters can be used to count the number of clock cycles and as an instrument for measuring time and therefore period or frequency. The two different types of counters are synchronous and asynchronous.

6.3.1 Frequency Dividers

Circuit designers often needed to reduce the frequency of a wave clock signal. One easy way of doing it is to divide the frequency by two, which is done by feeding back the not-Q output of a D-type flip-flop to its data line. Figure 6-14 shows a divide-by-2 circuit and its effect on the resulting wave.

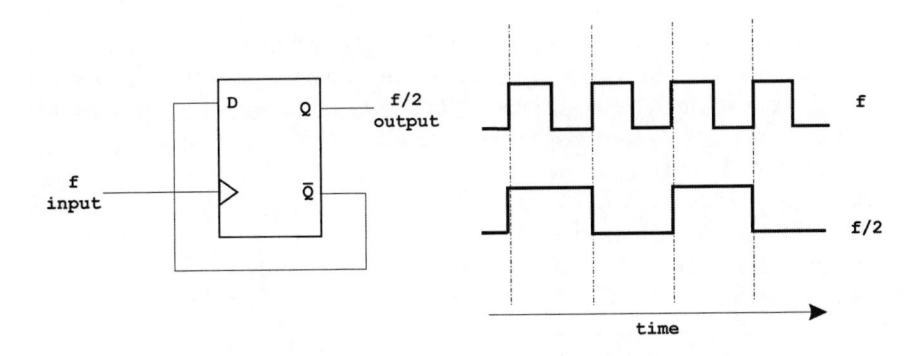

Figure 6-14 A Divide-by-two Circuit

In the circuit of Figure 6-14 the frequency division occurs because each input clock rising edge toggles the flip-flop's output. When the Q output goes low, the not-Q line goes high and the high feedback signal is fed back to the data line, thus canceling out the next high wave of the f signal.

6.3.2 The JK Flip-flop Counter

One type of specialized flip-flop that we did not cover in Section 6.2 is the JK flip-flop. The JK flip-flop is an ideal component to build a circuit that keeps track of the number of positive or negative clock edges on the input clock. The name of this flip-flop relates to the two variables, J and K, that are used as inputs to the circuit. Figure 6-15 shows one possible circuit implementation for the JK flip-flop.

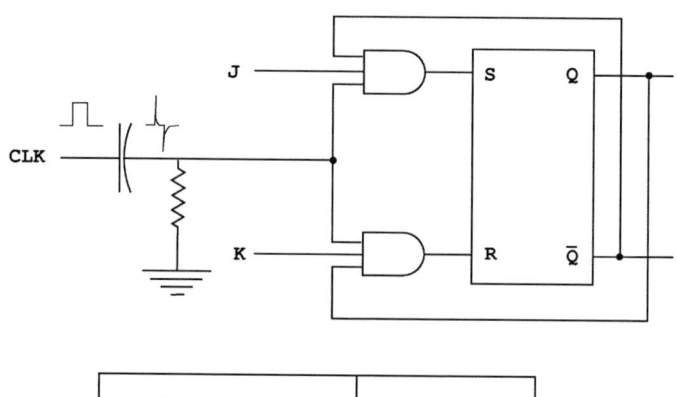

Figure 6-15 A JK Flip-Flop Circuit

CLK	J	K	Q
X	L	L	last state
H	L	H	L
H	H	L	H
H	H	H	toggle

In Figure 6-15 the RC circuit converts the rectangular wave clock pulse into a narrow spike. The three-input AND gates make the circuit positive-edge-triggered. When J and K are low, both AND gates are disabled; therefore, clock pulses have no effect. This corresponds to the first entry in the truth table. When J is low and K is high (second entry in the truth table) the upper gate is disabled, so the flip-flop cannot be set; it must be reset. When Q is high, the lower gate passes a Reset trigger as soon as the next positive clock edge arrives. This forces Q to become low (the same second entry in the truth table). Therefore, J low and K high means that the next positive clock edge resets the flip-flop.

When J is high and K is low (third entry in the truth table) the lower gate is disabled, so it is impossible to reset the flip-flop. However, the flip-flop can be reset when Q is low because not-Q is high; therefore, the upper gate passes a Set trigger on the next positive clock edge. This drives Q into the high state (the third entry in the truth table). As you can see, J = 1 and K = 0 means that the next positive clock edge sets the flip-flop (unless Q is already high). When J and K are both high it is possible to set or reset the flip-flop. If Q is high, the lower gate passes a RESET trigger on the next positive clock edge. On the other hand, when Q is low, the upper gate passes a SET trigger on the next positive clock edge. Either way, Q changes to the complement of the last state (see last entry in the truth table). Therefore, when J = 1 and K = 1 the flip-flop will toggle on the next positive clock edge.

6.3.3 Ripple Counters

The simplest of all counters is called a *ripple counter*. A two-bit ripple counter can be constructed by wiring together two divide-by-two circuits, as in Figure 6-16.

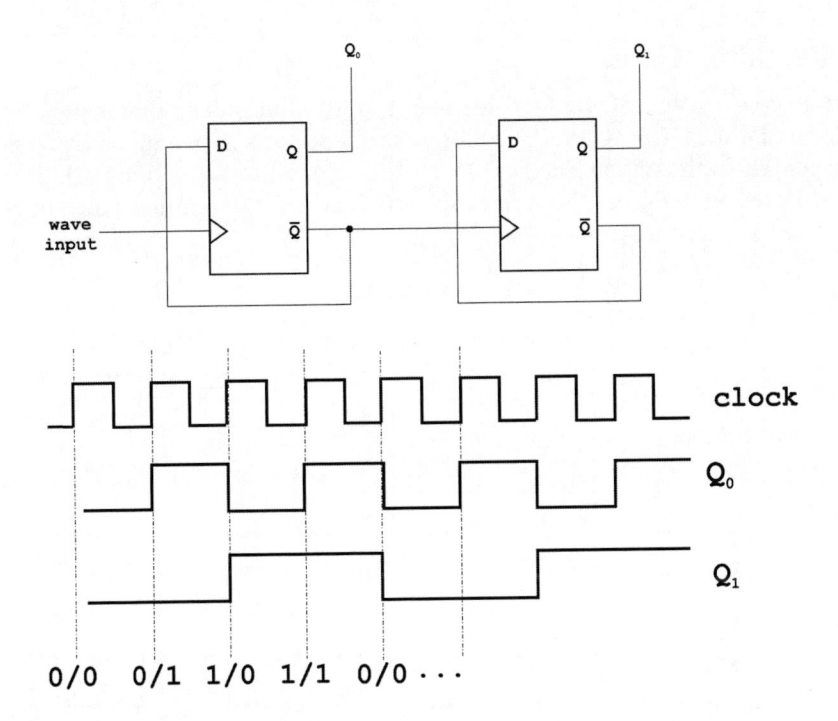

Figure 6-16 Two-Bit Ripple Counter

Stringing together two divide-by-two circuits, as in Figure 6-16, produces a divide-by-four circuit. Stringing together three flip-flops produces a divide-by-eight circuit, four flip-flops create a divide-by-sixteen circuit, and so on. The counting action of the connected flip-flops is based on the fact that each flip-flop changes state before triggering the next one in line. Thus, each stage performs as a bit in a binary counter, the first stage being the LSB and the last stage the MSB. Since the preceding flip-flop acts as a clock for the next one in line, the flip-flop to the right toggles each time its neighbor to the left goes low. In Figure 6-15 the signal labeled Q_0 is the LSB of a two-bit counter, while the signal labeled Q_1 is the most significant bit.

In this design each flip-flop is triggered by the previous one; thus the count is said to "ripple" down the device. One objection to the ripple counter is that the change in each output is determined by the previous output in the flip-flop chain; this produces a few nanoseconds of time lag from output line to output line. This cumulative settling time is why these counters are called serial or asynchronous.

Note that the ripple counter of Figure 6-16 uses the not-Q line to drive the following flip-flop. If a ripple counter is wired so that the Q line drives each next stage, then the transitions take place not when the previous waveform goes low, but when it goes high. The result is that the counter counts down instead of up. In other words, in the down counter, the count is reduced by one during each clock transition. Commercial counters, such as the 74193, can be made to operate as up-counters or down-counters by selecting the corresponding input line.

6.3.4 Decoding Gates

A decoding gate is a way of connecting the output of a counter so that it signals a given state. For example, if four D-type flip-flops are wired so as to produce a four-bit ripple counter similar to the one in Figure 6-15, the counter represents binary digits 0000 to 1111. If we wanted to detect the value 1101 (16 decimal) the resulting circuit could be designed as in Figure 6-17.

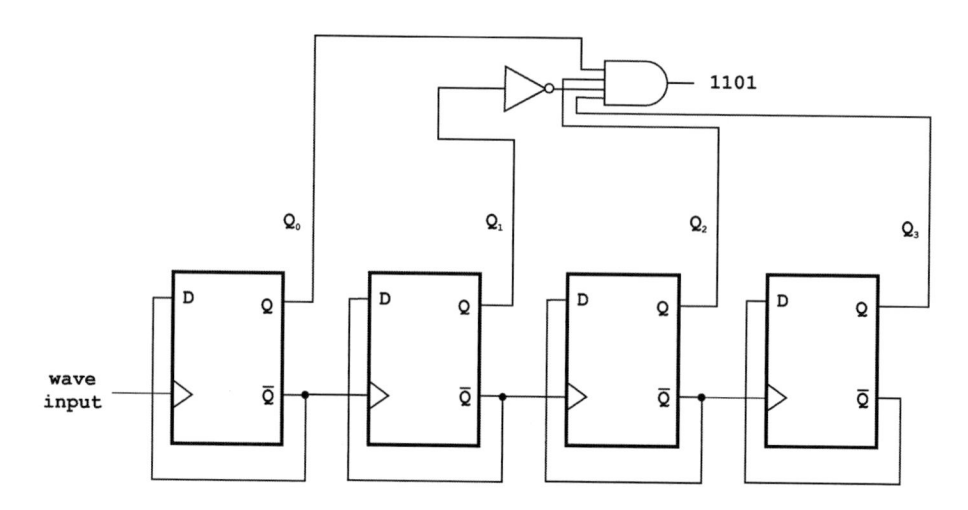

Figure 6-17 Decoding Gate

The circuit of Figure 6-17 uses a NOR gate to invert the value of bit number 1. Then the AND gate serves to trigger the output when bits 0, 2, and 3 are high and bit 1 is low. This corresponds to the binary value 1011.

6.3.5 Synchronous Counters

Although the ripple counter is the simplest one, it has the previously mentioned disadvantage that each flip-flop has to wait for its neighbor to switch states. This means that in a ripple counter the delay times are additive, and also that the total "settling" time for the counter is approximately the delay multiplied by the total number of flip-flops. In addition, with ripple counters the resulting delay creates the possibility of glitches occurring at the output of decoding gates. These problems can be overcome by the use of a synchronous or parallel counter.

By observing how counting takes place in binary numbers, a counter in which each flip-flop is triggered at every clock beat can be built. Binary counting has the property that when a bit changes from high to low (1 to 0) it sends a toggle command to its neighbor to the left. So assuming that the low-order bit changes consecutively from one state to its complement, and starting from all bits initialized to 0, binary counting can be visualized as in Figure 6-18.

Circuit Components

```
0000
0001
 ▼
0010
0011
▼▼
0100
0101
 ▼
0110
0111
▼▼▼
1000
1001
 ▼
1010
1011
▼▼
1100
1101
 ▼
1110
1111
```

Figure 6-18 The Binary Counting Mechanism

Note in Figure 6-18 that the arrows indicate the transition from high to low, which is the command for the column to the left to change to its complement (toggle). Using this property of binary counting, it is possible to wire four JK flip-flops so that every high-to-low transition of a flip-flop triggers its higher-order neighbor to toggle its state. Figure 6-19 shows such a system.

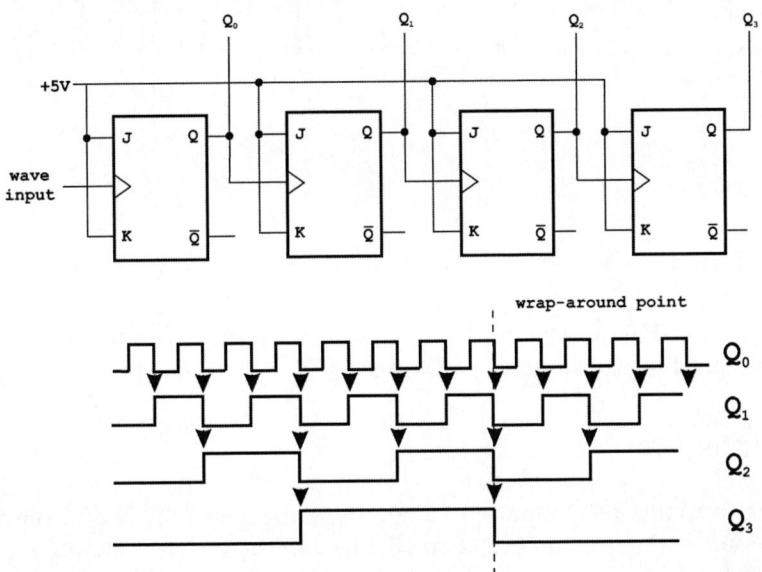

Figure 6-19 Synchronous Four-Bit Up Counter

In Figure 6-19 note that the first flip-flop (the one with the Q_0 output) has a positive-edge triggered clock input. The lowest-order flip-flop toggles with each rising edge of the clock signal (not shown in the illustration). The second flip-flop to the right toggles with every falling edge of the signal from its neighbor to the left. And so on to the last flip-flop in the chain. The arrows in the waveform portion of Figure 6-19 show that each succeeding output bit is toggled by the transition from high to low of its lower-ordered neighbor. Also note the dashed line that marks the point where all found counters are transitioning from high-to-low. At this point all four counters wrap around to zero and a new count begins.

Observe that the not-Q output line transitions opposite to the Q output. That is, when the Q output line goes high, not-Q goes low, and viceversa. So if the pulse into each successive flip-flop originated in the not-Q line, instead of the Q line, then the resulting circuit would be a synchronous counter that transitions on the positive edge (low-to-high) instead of in the negative edge, as is the case with the counter in Figure 6-19. Furthermore, the not-Q line provides a set of negated outputs in reference to the Q lines; therefore, it is possible to come up with a circuit that serves both as an up- and down-counter according to the selected set of outputs. Such a circuit is shown in Figure 6-20.

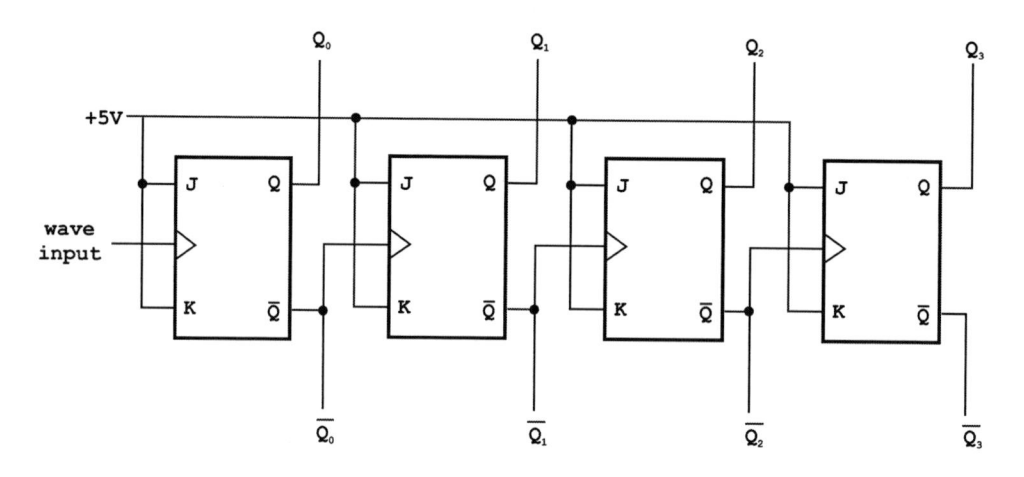

Figure 6-20 Synchronous 4-bit Up- and Down-counter

In the counter of Figure 6-20 the Q outputs generate the up-count series while the not-Q outputs produce the down-count series.

6.3.6 Counter ICs

Counters are available as standard TTL components. The 7493 is an asynchronous 4-bit ripple counter that counts from 0 to 15. The 7490 is another version of the ripple counter, called a *decade counter*, since the count output is in the range 0 to 9. The 74193 is a 4-bit synchronous up/down counter in the range 0 to 15. Figure 6-21 is a pin diagram of the 74193.

Circuit Components

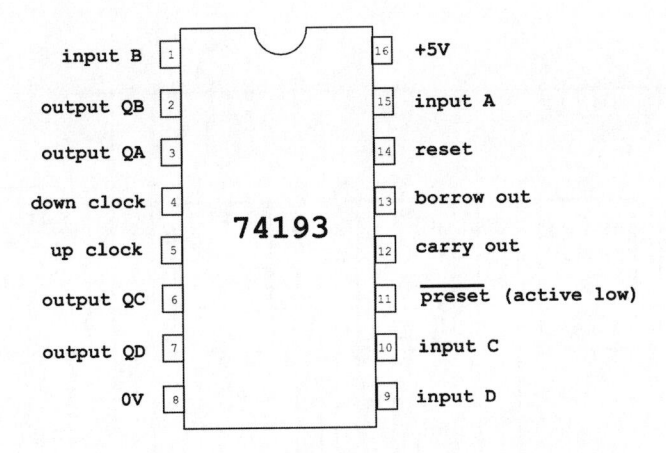

```
           input B  ┤1      16├  +5V
         output QB  ┤2      15├  input A
         output QA  ┤3      14├  reset
        down clock  ┤4  74193 13├  borrow out
          up clock  ┤5      12├  carry out
         output QC  ┤6      11├  preset (active low)
         output QD  ┤7      10├  input C
                0V  ┤8       9├  input D
```

Figure 6-21 74193 Asynchronous Up/Down Counter Pin Diagram

The 74193 is a synchronous counter, so its output changes precisely at each clock pulse. This is convenient since it allows connecting its output to other logic gates and avoids the glitches associated with ripple counters. Note from Figure 6-21 that the 74194 has separate clock inputs for counting up and counting down. The count increases as the up clock input becomes high (on the rising-edge). The count decreases as the down clock input becomes high (on the rising-edge). In both cases the other clock input should be high. For normal operation the Preset input should be high and the Reset input low. When the Reset input is high it resets the count to zero, that is, lines QA to QD are low. The counter can be preset by placing any desired binary number on inputs A to D and making the Preset input low. These inputs may be left unconnected if not required.

Several 74193 counters can be chained by wiring a common Reset line, connecting the carry to the up clock line of the next counter and the borrow to the down line.

6.3.7 Shift Registers

In chapter 4 we discussed logical operations that shift and rotate the operand bits. These manipulations are useful in inspecting individual bits, in performing fast multiplications, in implementing time delays, and in converting parallel input to serial output, and vice versa. The hardware implementation of shift-and-rotate operations are called *shift counters* or *shift registers*.

Shift counters are often based on the D-type flip-flop. Actually, several D-type flip-flops can be chained together so that the D output of one goes into the D input of the next one. If all the flip-flops are driven by the same clock signal, then the effect would be to shift the bits from one flip-flop into the next one at each rising clock pulse.

A common implementation of a shift counter is called a *parallel-in/serial-out shift register*, as the one in Figure 6-22.

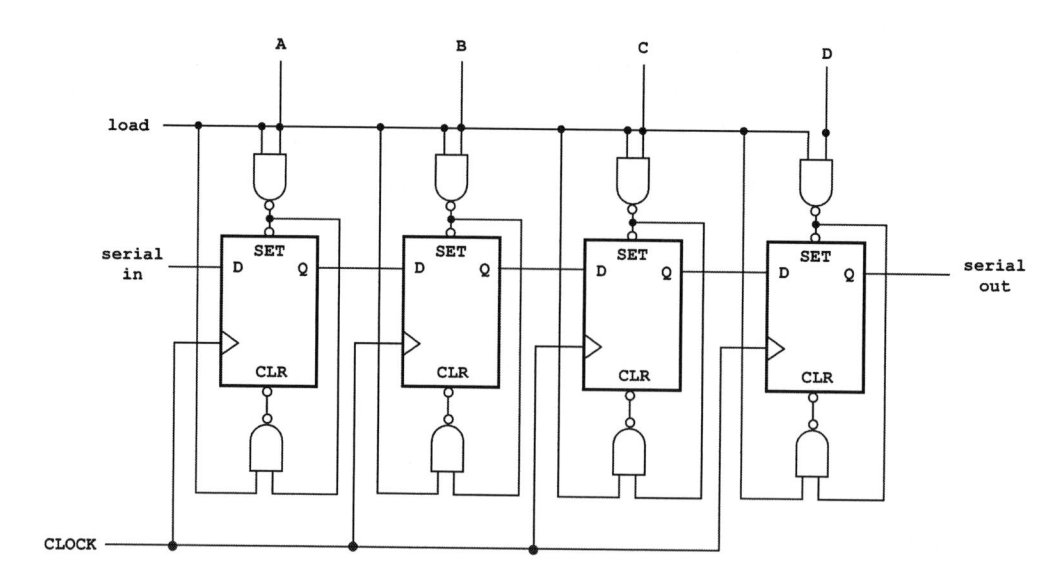

Figure 6-22 Four-Bit Parallel-In/Serial-Out Shift Counter

The circuit in Figure 6-22 shows four flip-flops connected so that the output of one feeds into the input of the next one. Also, a set of NAND gates allow parallel data input. When the load signal is set high the flip-flops in the shift register are loaded simultaneously with the logic values at the inputs A, B, C, and D. The 74165 IC is an 8-bit parallel-in/serial-out shift register with asynchronous parallel load and two OR-gated clock inputs. Figure 6-23 is a pin diagram of the 74165 IC.

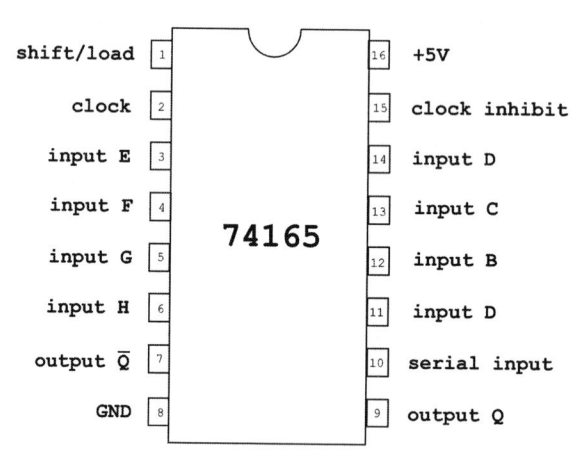

Figure 6-23 Pin Diagram of IC 74165

The serial input line in the diagram of Figure 6-22 and in the 74165 IC in Figure 6-23 allows cascading multiple chips.

Parallel-in/serial-out shift registers find common use in the implementation of serial ports. In serial communications data is sent one bit at a time over a single wire.

Circuit Components

In order to accomplish this, data is first loaded into a parallel-in/serial-out shift register. The individual bits are then shifted out one at a time. The frequency of the driving clock in this case corresponds to the baud rate being used. To receive the data on a serial communications line a second type of shift register is used. In this case the operation is serial-in/parallel-out. The circuit that accomplishes this is based on D-type flip-flops in which the Q outputs are connected to the D input lines.

The 74164 IC is one such device. In actual serial ports, the transmitting and receiving shift registers are contained in a single device called a *UART* (universal asynchronous receiver/transmitter).

6.4 Multiplexers and Demultiplexers

There are many situations in digital electronics where different signals must be sent out on a single output line, or several signals must be received in a single input line. The digital circuits that perform these operations are called multiplexers and demultiplexers. Multiplexers and demultiplexers are TTL analogs of the many-to-one and one-to-many mechanical switches.

6.4.1 Multiplexers

Multiplexing (also called *muxing*) is a way of combining data of two or more input channels into a single output channel. The hardware multiplexer, also called a *mux*, combines several electrical signals into a single one. In other words, the multiplexer performs a many-into-one function while the demultiplexer performs one-into-many. Sometimes multiplexers and demultiplexers are combined into a single device, which is still referred to as a "multiplexer." Figure 6-24 shows the schematics diagram of a multiplexer.

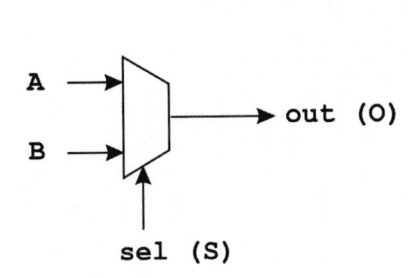

A	B	S	O
0	0	0	B (0)
0	0	1	A (0)
0	1	0	B (1)
0	1	1	A (0)
1	0	0	B (0)
1	0	1	A (1)
1	1	0	B (1)
1	1	1	A (1)

Figure 6-24 Multiplexer Schematics

The truth table in Figure 6-24 describes the multiplexer operation. The line labeled "sel" in the illustration is the selector line. If the selector is low (S = 0) then input line, B, is mirrored in the output line O. Otherwise, input line A is vectored to the output.

The Boolean expression for the multiplexer in Figure 6-24 is:

$$O = (A \wedge S) \vee (B \wedge \neg S)$$

Often, a multiplexer circuit is preceded by a decoder circuit so that input can be compressed into fewer lines. For example, a four-to-one multiplexer receives a binary value in the range 0 to 3 (00 to 11) on two input lines and sets high one of four output lines accordingly. Figure 6-25 shows the circuit diagram for such a device.

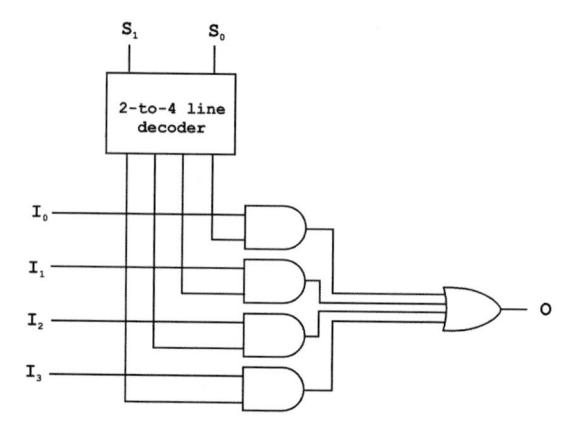

Figure 6-25 Two-Bit to One-of-Four Multiplexer

In the circuit of Figure 6-25 there are four input lines. Which one of these four lines is copied to the multiplexer output depends on the binary value in the two S lines at the top of the illustration. The 2-to-4 line decoder converts this value into one of four selector lines, which are in one of these four states:

```
LLLL   LLHL   LHLL   HLLL
```

Whichever line is high from the decoder output selects the corresponding input line. By analogy to the circuit in Figure 6-25, an 8-input multiplexer has eight data inputs and three binary selection inputs, which are converted into one-of-eight selection lines by the decoder. By the same token, a 16-input multiplexer requires four binary digits in the decoder input, which are converted into one-of-sixteen selection lines.

Alternatively, the decoder circuit can be eliminated by using multiple AND gates and negating the input signals, as shown in Figure 6-26.

In Figure 6-26 assume that the input bits are both low, that is, S1 = 0 and S2 = 0. The first-level NOR gates change the L signals to H. The two high signals go into the first multiple AND gate, as shown by the solid lines in the illustration. This determines that the first input line (I0) is copied to the circuit output. In fact, the four inverters at the top of the illustration perform the function of the two-to-four decoder in the circuit of Figure 6-25.

Circuit Components

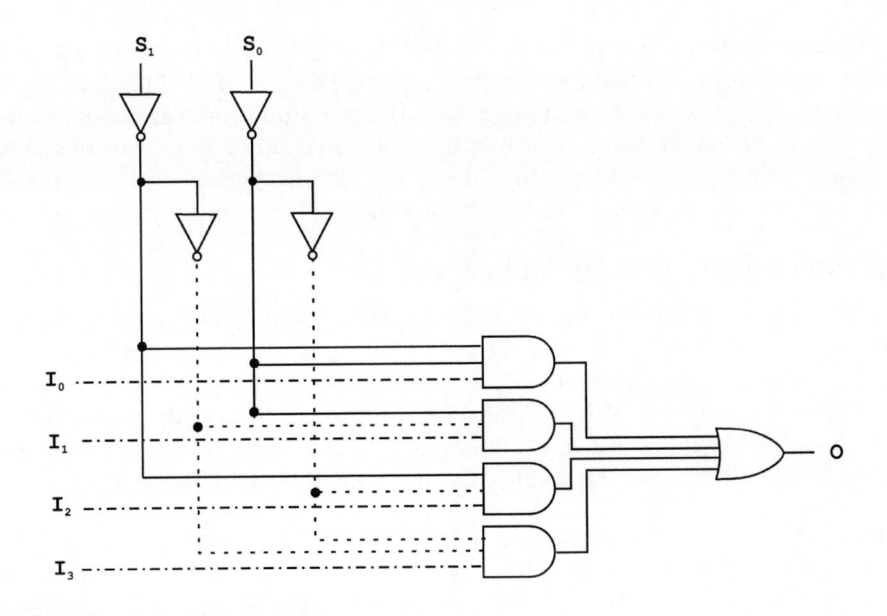

Figure 6-26 Multiplexer with Multiple AND Gates

6.4.2 Demultiplexers

A demultiplexer takes one data input and a number of selection inputs, and returns multiple outputs. So while the multiplexer performs a many-into-one operation, the demultiplexer performs a one-into-many. For example, a 4-output demultiplexer has one data input line, two selection inputs, and four data output lines. Figure 6-27 shows such a circuit.

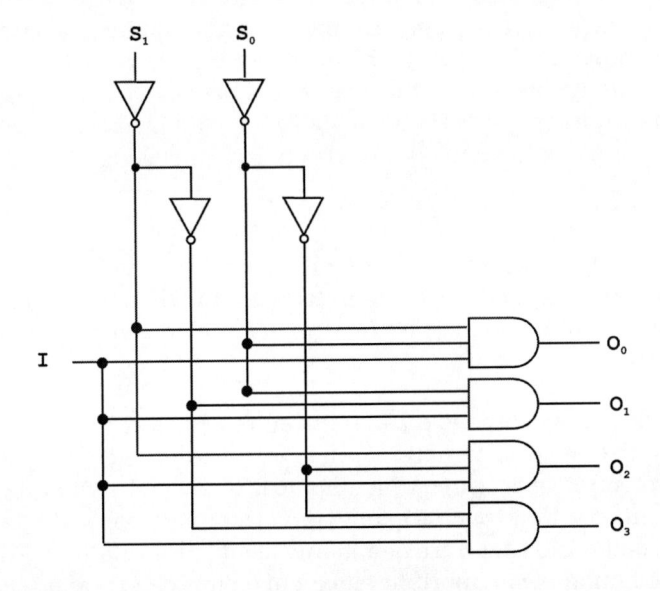

Figure 6-27 Two-Bit into Four-of-One Demultiplexer

Demultiplexers can be made to act as decoders by holding the input line high. For example, the circuit in Figure 6-27 performs as a binary to four-line decoder if the I line is held high. The binary bit patterns on the two input lines are converted into a single output in one of the four output lines. Thus, if there were four devices, each one connected to one of the output lines, the demultiplexer circuit would select which one is enabled according to the binary value of the input.

6.4.3 Multiplexer and Demultiplexer ICs

Several ICs are available that perform multiplexing and demultiplexing operations. For example, the 74138 is a 3-line to 8-line decoder and demultiplexer. With this IC any of eight inputs can be selected by placing the corresponding 3-bit number on the device's three address lines. The 74151 is a 1-of-8 data selector/multiplexer. This device routes data from eight sources to a single output line. Here again, a 3-bit selector is used to determine which of the eight inputs is routed to the output.

An important use of the multiplexer ICs is to encode row and column addresses into the address lines of dynamic RAM, although more often tristate buffers such as the 74541 are used. Another important use of multiplexers is in implementing dual-port memories for video displays.

6.5 Input Devices

Electronic devices, including computers and microcontrollers, often receive the data and commands required for their operation. In computer technology the most common input device is the keyboard, which allows entering text data as well as keystroke orders. Alternate computer input devices are the mouse, trackballs, light pens, graphical tablets, scanners, speech recognition devices, optical character recognition devices, and many others. Although these devices are not excluded from use in microcontroller-based systems, a more typical case is that microcontroller input devices are much simpler and limited. In this section we discuss the two most commonly used devices for microcontrollers: the *switch* and the *keypad*. Keep in mind that specialized systems often use special input devices; for example, a radio receiver could be the input device for a radio-controller microcontroller system.

6.5.1 Switches

The electrical switch is a device for changing current flow in a circuit. Although mechanical switches find use in fields such as railroads and fluid flow control, here we refer to switches used in controlling electrical power or electronic telecommunications.

In abstract terms the switch is often referred to as a "gate", in the same sense as the logic gates discussed in Chapter 5. In this sense an electronic device can be viewed as a system of logic gates. The simplest electrical switch has two components, called contacts, that touch to *make* the circuit and separate to *break* the circuit. The terms make and break are commonly used in this context. The selection of material for the contacts is important since corrosion can form an insulating layer that prevents the switch from performing its function. One possible solution is plating the contacts with noble metals, such as gold or silver.

Circuit Components

In a switch, the actuator is the part that applies the operating force to the contacts. Common switch types are rocker, toggle, push-button, DIP, rotary, tactile, slide, keylock, snap-action, thumbwheel, and several others. Figure 6-28 shows several switches commonly found in microcontroller circuit boards.

Figure 6-28 Electrical Switches

In switches, contacts are "closed" when there is no space between them, thus allowing electricity to flow. When the contacts are separated by a space, they are "open." In this case no electricity flows through the switch.

Switches are classified according to the various contact arrangements. In the normally open switch the contacts are separated until some force causes them to close. In the normally closed switch the contacts are held together until some force separates them. Some switches can be selected to operate as either normally open or normally closed. The term *pole* is used in reference to a single set of contacts on a switch. The term *throw* refers to the positions that a switch can adopt. Figure 6-29, on the following page, shows some common switch designs and their electrical symbols.

A multi-throw switch can have two possible transient behaviors as it transits from one position to the other one. One possibility is that the new contact is made before the old one is broken. This *make-before-break* action ensures that the line is never an open circuit. Alternatively, there is a *break-before-make* action, where the old contact is broken before the new one is made. This mode of switch operations ensures that the two fixed contacts are never shorted. Both designs are in common use.

Diagram	Electronic abbreviation	Description
	SPST	Single pole, single throw. On-off switch such as a household light switch.
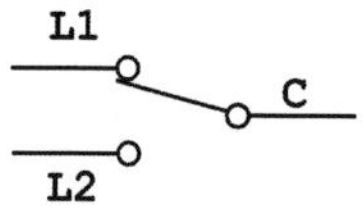 L1 / C / L2	SPDT SPCO	Single pole, double throw. Single pole, changeover. Changeover switch. C is connected to either L1 or L2.
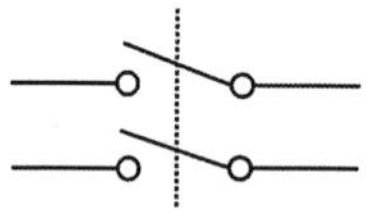	DPST	Double pole, single throw. Equivalent to two SPST switches operated by the same mechanism.
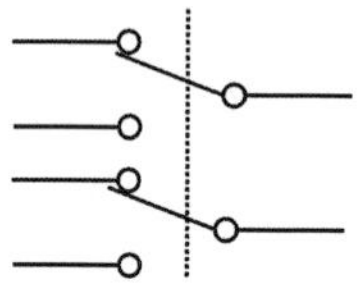	DPDT	Double pole, double throw. Equivalent to two SPDT switches operated by the same mechanism.

Figure 6-29 Switch Symbols and Types

A *biased switch* is one in which the actuator is automatically returned to a certain position, usually by the action of a spring. A *push-button switch* is a type of biased switch, of which the most common type is a *push-to-make* switch. In this case, the contact makes when the button is pressed and breaks when it is released. A push-to-break switch, on the other hand, breaks contact when the button is pressed. Many other special function switches are available; for example, tilt switches, such as the mercury switch, in which contact is made by a blob of mercury inside a glass bulb as the switch is tilted. Other specialized switches are activated by vibration, pressure, fluid level (as in the float switch), linear or rotary movement, the turning of a key, a radio signal, or a magnetic field.

6.5.2 Switch Contact Bounce

Switch contact bounce is a common problem of electrical switches. Switch contacts are metal surfaces that are forced into contact by an actuator. Due to momentum and elasticity, the striking action of the contacts causes a rapidly pulsating electrical current instead of a clean transition from zero to full current. Parasitic inductance and capacitance in the circuit can further modify the waveform resulting in a series of sinusoidal oscillations.

Switch bounce sometimes causes problems in logic circuits that are not designed to cope with oscillating voltages, particularly in sequential digital logic circuits. Several methods of switch debouncing have been developed. These can be divided into timing-based schemes and hysteresis-based schemes. Timing-based techniques are based on adding sufficient delays so as to prevent the bounce from being detected. The main advantage of using timing to control bouncing is that it does not require any special switch design. Alternatively, it is possible to use hysteresis to separate the positions where the make and break actions are detected. We discussed hysterisis in the context of Schmitt trigger inverters, which are actually switches, in Section 5.6.8.

The actual hardware circuits used in switch debouncing belong to three common types: RS flip-flops, CMOS gate debouncers, and integrated RC circuit debouncers. The debouncing action of the RS flip-flop is obvious from its operation, that is, when the key is in a position in which neither contact is touched (*key bouncing*) the inputs are pulled low by the pull-down resistors. In this case, the key appears as being pressed. Before being pressed, the key is touching the set input and appears as an RS flip-flop, which was covered in Section 6.1.2.

Alternatively, switch debouncing can be accomplished by means of CMOS buffer circuit with high input impedance. One such circuit is the 4050 hex buffer IC, with eight input and eight output gates. When the switch is pressed, the input line of the 4050 chip is gounded, and output is forced low. The output voltage, by means of an internal resistor, is also kept low when the switch is bouncing. The effect is that the switch action is debounced.

Finally, switch debouncing can be implemented by means of a simple resistor-capacitor circuit. The circuit action is based on the rate at which the capacitor recharges once the ground connection is broken by the switch. As long as the capacitor voltage is below the threshold level of the logic zero value, the output signal continues to appear as logic zero.

6.5.3 Keypads

In the context of microcontroller-based circuits, a *keypad* (also called a numeric keypad) is a set of pushbutton switches sometimes labeled with digits, mathematical symbols, or letters of the alphabet. For example, a calculator keypad contains the decimal (occasionally hexadecimal) digits, the decimal point, and keys for the mathematical features of the calculator. Although in theory the computer keyboard is a keypad, the keypad is usually limited to a smaller arrangement of buttons or to part of a computer keyboard consisting mainly of numeric keys.

By convention, the keys on calculator-style keypads and keypads on computer keyboards are arranged such that the keys 123 are on the bottom row. On the other hand, telephone keypads have the 123 keys on the top row.

Keypads are usually implemented as pushbutton switches located in a row and column matrix. The location of any key on the keypad can be based on two coordinates: the row and column position for that key. Therefore only eight outputs are required from the keypad: one for each row and one for each column. Determining which switch on a keypad has been activated can be done either by polling or by means of an interrupt routine. In the polling approach the controller checks the status of each switch in a loop. A more efficient approach is to implement and interrupt-driven routine that notifies the processor of a keystroke.

Keypads, like the switches that they incorporate, require debouncing. The three methods of switch debouncing described in Section 6.5.2 apply to keypads.

6.6 Output Devices

As is the case with input devices, electronic systems, including computers and microcontrollers, must provide data output in a human-readable form. Here again, computer technology uses many different types of output devices, including video displays, printers, plotters, film recorders, projectors, sound systems, and even holographic devices. Although these output devices cannot be excluded from use in microcontroller systems, they use simpler and limited output means. In this section we discuss two common output devices used in microcontroller-based circuits: the seven-segment LED and the liquid crystal display. Simple devices, such as LEDs and buzzers, are sometimes used as output devices. LEDs were covered in Chapter 5. Buzzers are such simple components that their operation does not require a detailed explanation.

6.6.1 Seven-segment LED

Digital devices often need to output a numeric value. Although individual LEDs can be combined to represent binary, decimal, or hexadecimal digits, a far more convenient device consists of seven built-in LEDs which can be combined to represent all ten decimal digits and even the six letters of the hex character set. Such a circuit is furnished in a single IC, called a *seven-segment LED*, that is common in clocks, watches, calculators, and household appliances.

Seven-segment displays have been in use since the first generation of calculators came to market. The scheme consists of placing lighted bars in a figure-eight pattern. By selecting which bars are lighted, all the digits and some letters of the alphabet can be represented. In addition, seven-segment LEDs are usually capable of displaying one or two decimal points. Figure 6-30 shows the layout of a seven-segment LED and the combinations to generate the decimal and hex digit sets.

Note in Figure 6-30 that two of the letters (b and d) of the hexadecimal set are displayed in lower-case while the others are in upper-case. This is a limitation of the seven-segment LED since an upper-case letter "D" would coincide with the digit "0", and an upper-case letter "B" with the digit "8."

Some seven-segment LED displays are slanted to make the digits appear in italics. It is used in clock displays where the two digits are inverted so that the decimal points appear like a colon between the digits. In addition seven-segment displays are packaged in several different ways. Sometimes several digits are combined in a single IC. Another packaging is in the form of a 14-pin DIP.

Circuit Components

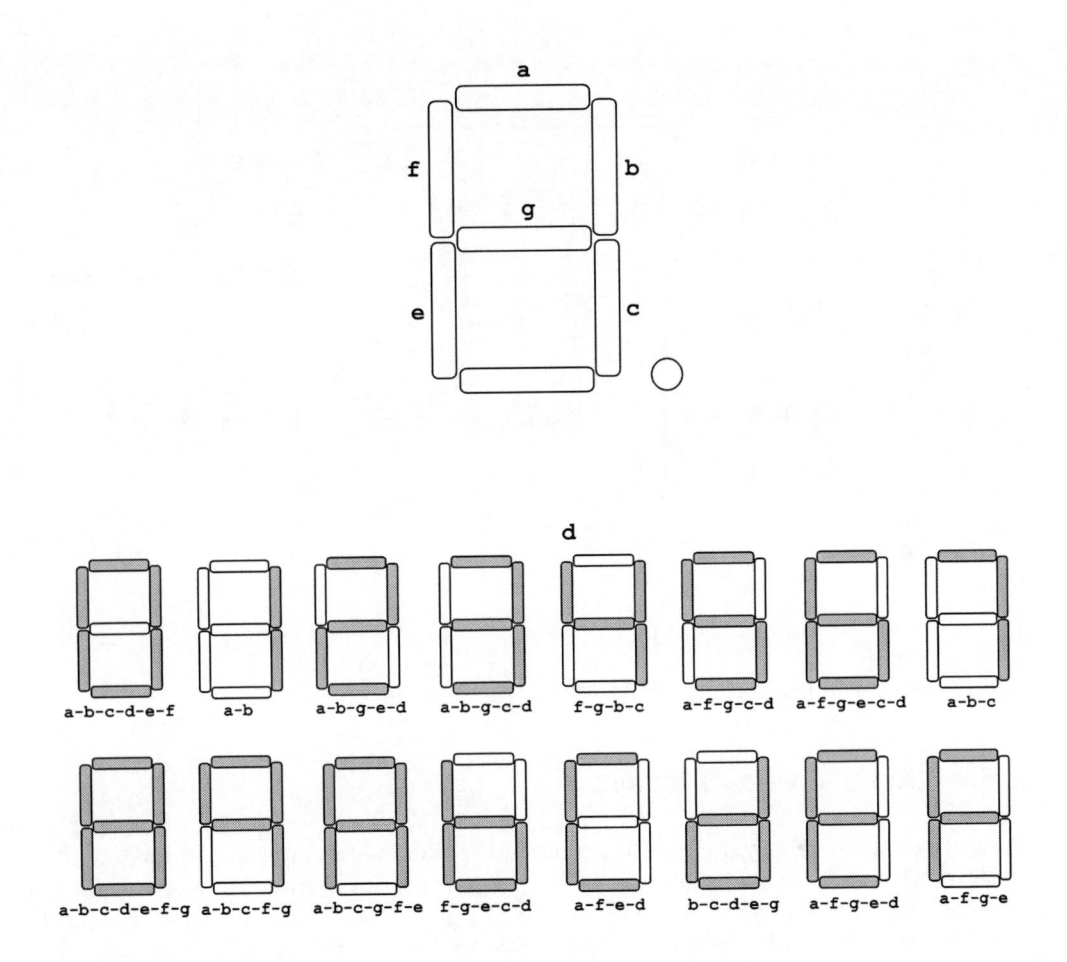

Figure 6-30 Seven-Segment LED Layout and Digit Patterns

Seven-segment displays are also furnished using display technologies other than LEDs. Many line-powered devices and home appliances, such as clocks and microwave ovens, use fluorescent seven-segment displays. Battery-powered devices, such as watches and miniature digital instruments, use seven-segment liquid crystal displays. Liquid crystal technologies are covered in sections that follow.

The LEDs in a seven-segment display are interconnected. The two interconnection modes are to wire together the cathodes of all individual LEDs, or to do so with the anodes. In one case the device is said to have a common-cathode and in the other one a common-anode. This circuit scheme simplifies the wiring and reduces the number of connections, since only one line is necessary for controlling each LED. There is no intrinsic advantage to either system since each one is suited to different applications. Figure 6-31, on the following page, shows the pin diagram for a common-cathode seven-segment LED in a DIP package.

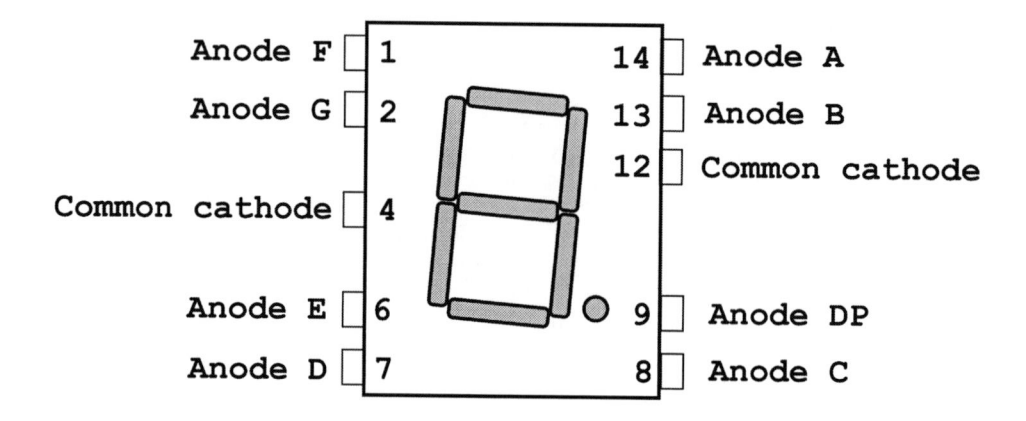

```
        Anode F  │ 1          14 │ Anode A
        Anode G  │ 2          13 │ Anode B
                              12 │ Common cathode
Common cathode   │ 4

        Anode E  │ 6           9 │ Anode DP
        Anode D  │ 7           8 │ Anode C
```

Figure 6-31 Pin Diagram for a Common Cathode Seven-Segment LED

6.6.2 Liquid Crystal Displays

A *liquid crystal display* (LCD) is a pixilated output device capable of displaying ASCII characters and dot-based graphics. LCDs can be color or monochrome according to their construction. One of the advantages of LCD displays is their very small consumption of electrical power, making them suitable for battery-powered devices. In operation the liquid crystal display consists of two pieces of polarized glass with perpendicular axes of polarity. Sandwiched between the polarizers is a layer of nematic crystals, as shown schematically in Figure 6-32.

In the top image of Figure 6-32 light cannot pass through the system since the liquid crystal layer preserves the original angle of vibration of the light which cannot pass through the polarizer. In the lower image the various molecular layers of the liquid crystal are twisted approximately 90 degrees. This twisting of the liquid crystal also changes the light's pane of vibration. So when light reaches the second polarized filter it vibrates at the same angle as the final molecule layer of the liquid crystal and can pass through the polarizer. Note that the electrical current applied to the crystals has the effect of straightening the various molecular layers. When the current is released, the various molecular layers resume their twisted form. By varying the amount of twist in the liquid crystals the amount of light that passes through can be controlled.

Circuit Components

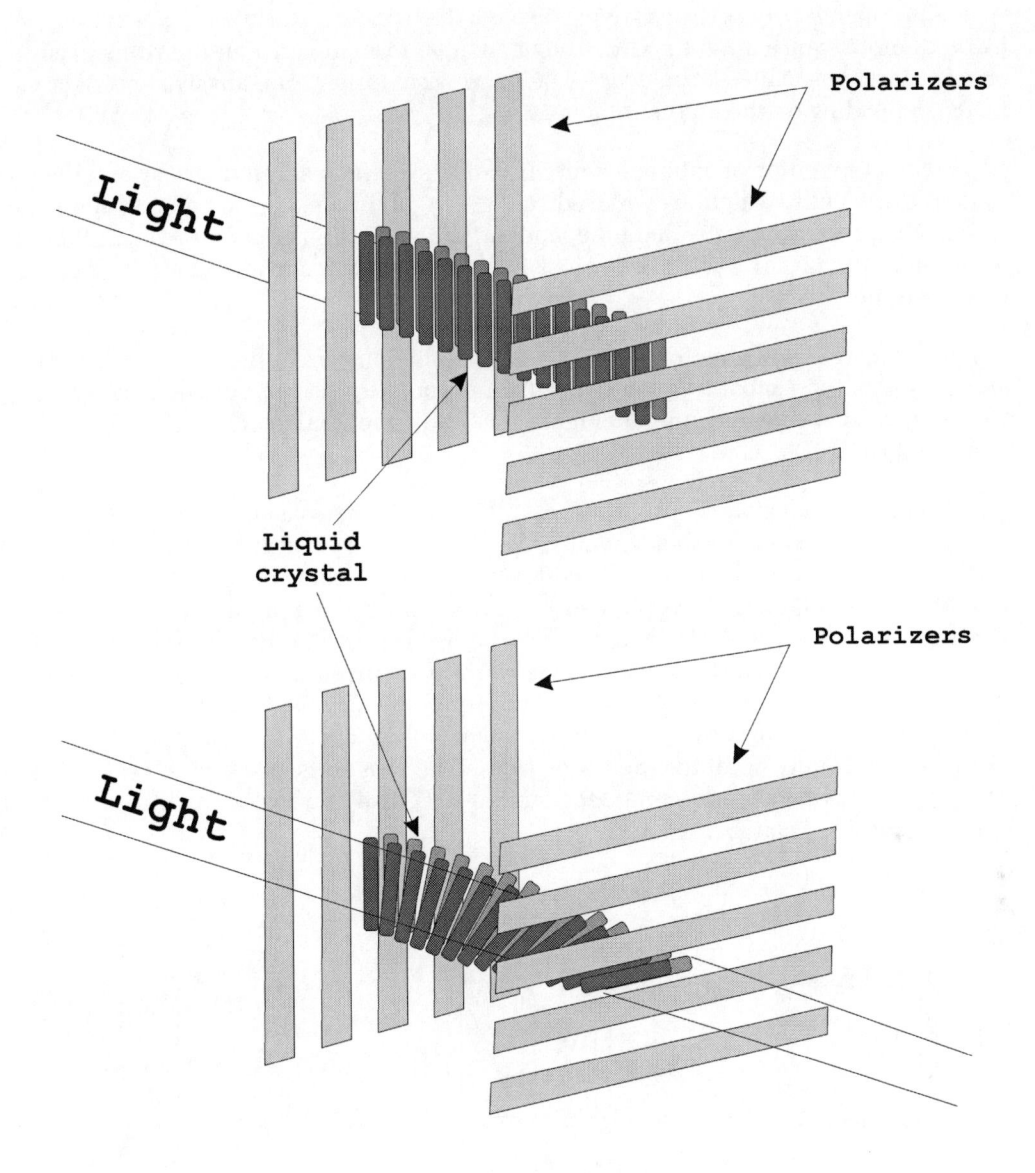

Figure 6-32 Schematic Representation of a LCD Display

6.6.3 LCD Technologies

Depending on the positioning of the light source LCDs can be either transmissive or reflective. A *transmissive LCD* is illuminated from the back and viewed from the front. This type is common in applications that require high levels of illumination, as is the case with computer displays and television sets. *Reflective LCDs*, on the other hand, are illuminated by an external source. This type finds use in digital watches and calculators. Reflective technology produces a darker black color than the transmissive type, since light is forced to pass twice through the liquid crystal layer. Since reflective LCDs do not require a light source they consume less power than the transmissive ones. A third type, called *transflective LCDs*, work as either transmissive or reflective LCDs, depending on the ambient light.

LCDs can be color or monochrome. In color systems each individual pixel consists of three cells, which are colored red, green, and blue. These cells, sometimes called *subpixels*, are controlled independently to yield thousands (or even millions) of possible colors for each pixel. Most LCDs used in microcontroller systems are monochrome.

According to display technology, LCDs are alphanumeric or dot-addressable. The alphanumeric type, most frequently used in microcontroller applications, uses a matrix composed of linear segments. Figure 6-33 shows several possible electrode configurations of LCDs.

The first two electrode configurations in Figure 6-33 are based on linear segments similar to the ones in seven-segment LEDs. Segmented electrodes are suitable for simple alphanumeric displays as are often required in small digital devices such as watches or calculators. To display entire character sets or graphics, a dot-addressable matrix of electrodes is necessary. This setup is shown in the rightmost image in Figure 6-34. However, such power comes at a price, since the more addressable elements in the display, the greater the number of connections and the more complex the driver logic required to operate the system. Note that the 5 x 7 matrix display in Figure 6-34 actually contains eight dot rows. The reason is that the lowest row is used for displaying the cursor. Most popular LCD displays for microcontroller circuits use the 5 x 7 matrix format.

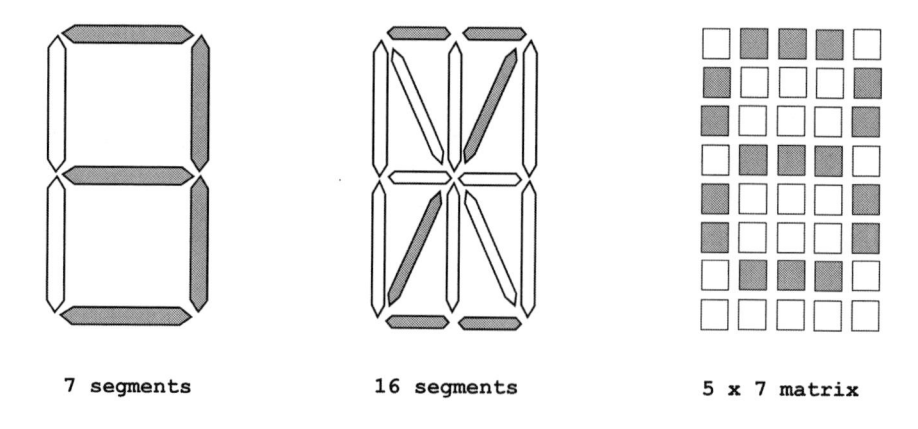

Figure 6-33 Electrode Configurations in LCD Displays

One way of reducing the number of electrical connections in an LCD is by means of a method called *passive matrix display*. Here the pixels to be lighted are determined by the crossing points between the row and the column selector electrodes. For example, in the 5 x 7 matrix display in Figure 6-34, the pixel at the center of the character is selected by picking row number 4 and column number 3. The name passive matrix originates in the fact that each pixel must retain its state between refreshes. As the number of pixels to be refreshed increases so does the time required for the refresh cycle. As a consequence of their design, passive matrix displays usually have slow response times and poor contrast.

In high-resolution and color LCDs an *active matrix display* is used. In this design a grid of thin-film transistors is added to the polarizing and color filters. Each pixel contains its own dedicated transistor and each row line and column line is addressed individually. During the refresh cycle each pixel row is activated sequentially. Active matrix displays are brighter and sharper and have quicker response time than passive matrix. Active matrix displays are also known as thin-film-transistor or *TFT displays*.

Chapter 7

The Microchip PIC

A microcontroller is a type of microprocessor furnished in a single integrated circuit and needing a minimum of support chips. Its principal nature is self-sufficiency and low cost. It is not intended to be used as a computing device in the conventional sense; that is, a microcontroller is not designed to be a data processing machine, but rather an intelligent core for a specialized dedicated system.

Microcontrollers are embedded in many control, monitoring, and processing systems. Some are general-purpose devices but most microcontrollers are used in specialized systems such as washing machines, telephones, microwave ovens, automobiles, and weapons of many kinds. A microcontroller usually includes a central processor, input and output ports, memory for program and data storage, an internal clock, and one or more peripheral devices such as timers, counters, analog-to-digital converters, serial communication facilities, and watchdog circuits.

More than two dozen companies in the world manufacture and market microcontrollers. They range from 8- to 32-bit devices. Those at the low end are intended for very simple circuits and provide limited functions and program space, while those at the high end have many of the features associated with microprocessors. The most popular ones include several from Intel (such as the 8051), Zilog (derivatives of their famous Z-80 microprocessor), Motorola (such as the 68HC05), Atmel (the AVR), Parallax (the BASIC Stamp), and Microchip. Some of the latter ones are the main topic of this book.

7.0 The PICMicro Microcontroller

PIC is a family of microcontrollers made by Microchip Technology. The original one was the PIC1650 developed by General Instruments. This device was called PIC for *"Programmable Intelligent Computer"* although it is now associated with *"Programmable Interface Controller."* Microchip does not use PIC as an acronym. Instead they prefer the brand name PICmicro. Popular wisdom relates that PIC is a registered brand in Germany and Microchip is unable to use it internationally.

The original PIC was built to be used with General Instruments' CP1600 processor, which had poor I/O performance. The PIC was designed to take over the I/O tasks for the CPU, thus improving performance. In 1985, the PIC was upgraded with EPROM to produce a programmable controller. Today, a huge variety of PICs are available with many different on-board peripherals and program memories ranging from a few hundred words to 32K.

PICs use an instruction set that varies in length from about 35 instructions for the low-end PICs to more than 70 for the high-end devices. The *accumulator*, which is known as the *work register* in PIC documentation, is part of many instructions since the PIC contains no other internal registers accessible to the programmer. The PICs are programmable in their native Assembly Language, which is straightforward and not difficult to learn. In addition, C language and BASIC compilers have been developed for the PIC. Open-source Pascal, JAL, and Forth compilers are also available for PIC programming.

One of the reasons for the success of the PIC is the support provided by Microchip. This includes a professional-quality development environment called MPLAB which can be downloaded free from the company's website (). The MPLAB package includes an assembler, a linker, a debugger, and a simulator. Microchip also sells a low-cost in-circuit debugger called MPLAB ICD 2. Other development products intended for the professional market are available from Microchip. The Microchip website furnishes hundreds of free support documents, including data sheets, application notes, and sample code.

In addition to the documents and products in the Microchip website, the PIC microcontrollers have gained the support of many hobbyists, enthusiasts, and entrepreneurs who develop code and support products and publish their results on the Internet. This community of PIC users is a treasure trove of information and know-how easily accessible to the beginner and useful even to the professional. One such Internet resource is an open-source collection of PIC tools named GPUTILS, which is distributed under the GNU General Public License. GPUTILS includes an assembler and a linker. The software works on Linux, Mac OS, OS/2, and Windows. Another product named GPSIM is an Open Source simulator featuring PIC hardware modules.

7.0.1 Programming the PIC

Programming a PIC microcontroller requires the following tools and components:

1. An Assembler or high-level language compiler. The software package usually includes a debugger, simulator, and other support programs.
2. A computer (usually a PC) in which to run the development software.
3. A hardware device called a programmer that connects to the computer through the serial, parallel, or USB line. The PIC is inserted in the programmer and "blown" by downloading the executable code generated by the development system. The hardware programmer usually includes the support software.
4. A cable or connector for connecting the programmer to the computer.
5. A PIC microcontroller.

Figure 7-1 USB PIC Programmer by MicroPro

PIC Programmers

The development system (assembler or compiler) and the programmer driver are the software components. The computer, programmer, and connectors are the hardware elements. Figure 7-1 shows a commercial programmer that connects to the USB port of a PC. The one in the illustration is made by MicroPro.

Many other programmers are available on the market. Microchip offers several high-end models with *in circuit serial programming* (ICSP) and *low voltage programming* (LVP) capabilities. These devices allow the PIC to be programmed in the target circuit. Some PICs can write to their own program memory. This makes possible the use of so-called *bootloaders*, which are small resident programs that allow loading user software over the RS-232 or USB lines. Programmer/debugger combinations are also offered by Microchip and other vendors.

Development Boards

A *development board* is a demonstration circuit that usually contains an array of connected and connectable components. Their main purpose is as a learning and experimental tool. Like programmers, PIC development boards come in a wide range of prices and levels of complexity. Most boards target a specific PIC microcontroller or a PIC family of related devices. Lacking a development board the other option is to build the circuits oneself, a time-consuming but valuable experience. Figure 7-2 (in the following page) shows the LAB-X1 development board for the 16F87x PIC family.

The LAX-X1 board, as well as several other models, is a product of microEngineering Labs, Inc. Some of the sample programs developed for this book were tested on a LAB-X1 board. Development boards from Microchip and other vendors are also available.

Figure 7-2 LAB-X1 Development Board

7.0.2 Prototyping the PIC Circuit

Very few of us are satisfied with writing a PIC program and assuming that it works correctly. Testing software is a simple matter if there happens to be a development board at hand, if the board is compatible with the PIC, and if it provides the hardware that we need to test. But often one of these elements is missing and it becomes necessary to build the circuit for which the program was designed. Here again, there are several options. These range from having the circuit built for us by a professional engineering firm, to using a *breadboard* to prototype the circuit ourselves.

Breadboarding a prototype circuit is one of the options. A breadboard is a reusable, solderless device that allows building a prototype circuit, usually for temporary use. Breadboards have strips down one or both sides that are used as power rails. One strip carries the circuit's positive voltage and the other one is wired to the ground of the power supply. Wire jumper kits provide connectors of different lengths and colors for making the circuit connections on the breadboard. For complex circuits several breadboards can be easily interconnected. Figure 7-3 shows two interconnected breadboards used to test one of the programs developed for this book.

The Microchip PIC 133

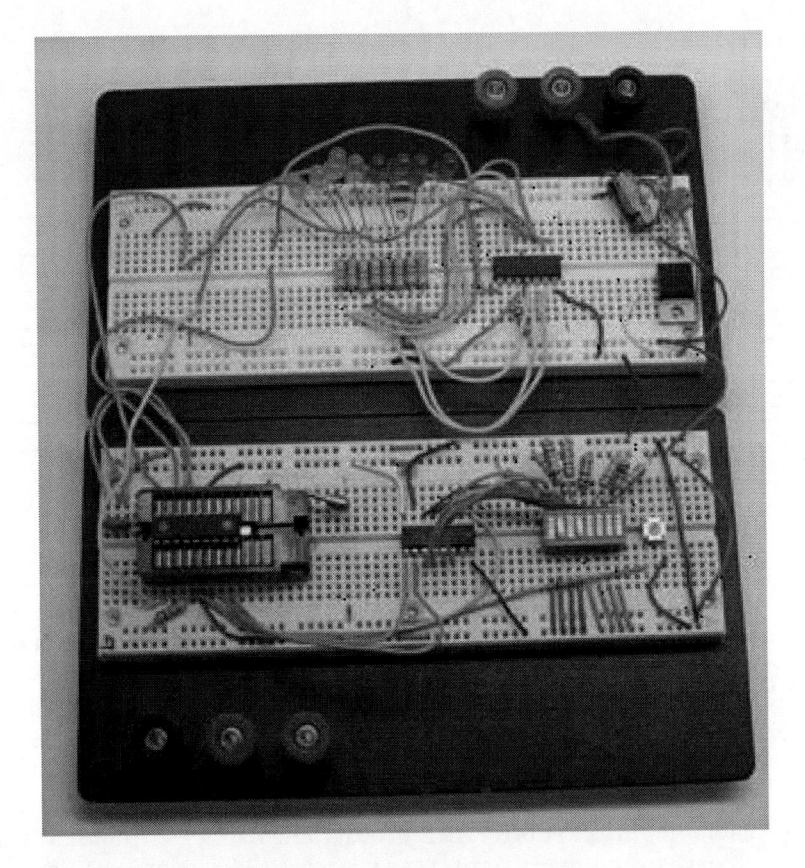

Figure 7-3 Circuits in Two Interconnected Breadboards

Once a circuit and the software have been tested, there are several available technologies for building a more permanent prototype. These include wire wrap, stripboards, and several other circuit board building tools and techniques, including prototyping boards specially designed for PIC circuits.

Finally, one can build a semi-professional quality *printed circuit board* (called a *PCB*) and solder the components to it. A PCB is used to mechanically support the electronic components and provides conductive pathways, called *traces*, that implement the circuit. The components are soldered to the PCB board using either surface mount or through-the-board technology. The PCB board is made of a non-conductive material and the conductive pathways are etched out of copper sheets laminated on one or both sides of the board. Once the board has been populated with electronic components it becomes a *printed circuit assembly*, or PCA. Industrial quality PCB boards are suited to high-volume production. The circuits of the development board in Figure 7-2 are on a commercial PCB.

Building one's own PCB is quite possible and requires few tools and resources. Appendix B describes one technique that has been used successfully. Figure 7-4, in the followng page, shows a drawing of both sides of a simple PCB board.

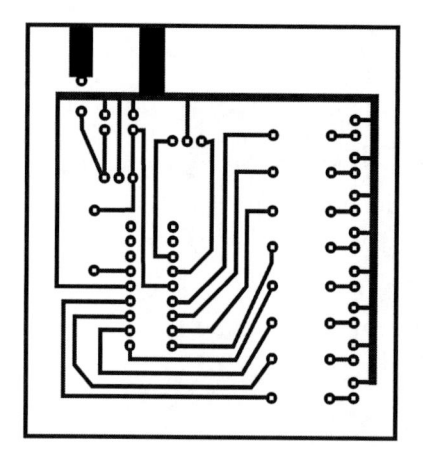

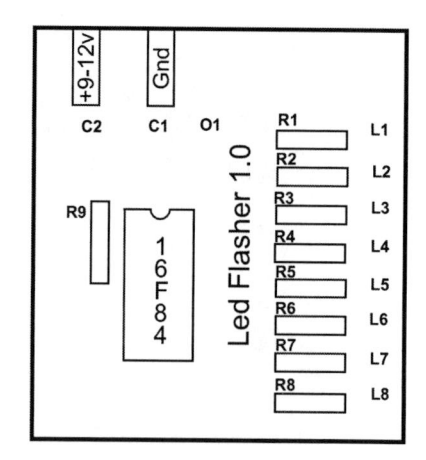

Figure 7-4 Drawing for Etching a PCB Board

The PCB in Figure 7-4 is intended for a copper-plated single-sided blank. The left-side image shows the actual circuit that is etched on the copper side of the board. The text and diagrams on the right-hand image are engraved (usually by silk screening) on the back side of the board and serve as a guide for welding the components. Refer to Appendix B for details on designing and building PCBs at the amateur level.

Several firms on the Internet offer PCB prototyping services from the circuit diagrams. In some cases the advertised turnaround time is a couple of days. One of these companies furnishes software tools for drawing the PCB in a format that they can use directly in manufacturing the prototypes. Googling "PCB prototypes" produces many hits.

7.1 PIC Architecture

PIC controllers are roughly classified by Microchip into three groups: baseline, mid-range, and high-performance. Within each of the groups the PICs are classified based on the first two digits of the PIC's family type. However, the subclassification is not very strict, since there is some overlap. For this reason we find PICs with 16X designations that belong to the baseline family and others that belong to the mid-range group. In the following subsections we describe the basic characteristics of the various subgroups of the three major PIC families with 8-bit architectures.

7.1.1 Baseline PIC Family

This group includes members of the PIC10, PIC12, and PIC16 families. The devices in the Baseline group have 12-bit program words and are supplied in 6- to 28-pin packages. The microcontrollers in the baseline group are described as being suited for battery-operated applications since they have low power requirements. The typical member of the Baseline group has a low pin count, flash program memory, and low power requirements. The following types are in the Baseline group.

PIC10 Devices

The PIC10 devices are low-cost, 8-bit, flash-based CMOS microcontrollers. They use 33 single-word, single-cycle instructions (except for program branches, which take two cycles). The instructions are 12-bits wide. The PIC10 devices feature power-on reset, an internal oscillator mode that saves having to use ports for an external oscillator. They have a power-saving SLEEP mode, a Watchdog Timer, and optional code protection.

The recommended applications of the PIC10 family range from personal care appliances and security systems to low-power remote transmitters and receivers. The PICs of this family have a small footprint and are manufactured in formats suitable for both through-hole and surface mount technologies. Table 7.1 summarizes the characteristics of PIC10 devices.

Table 7.1
PIC10F Devices

	10F200	10F202	10F204	10F206
Clock:				
Maximum Frequency of Operation (MHz)	4	4	4	4
Memory:				
Flash Program Memory	256	512	256	512
Data Memory (bytes)	16	24	16	24
Peripherals:				
Timer Module(s)	TMR0	TMR0	TMR0	TMR0
Wake-up from Sleep	Yes	Yes	Yes	Yes
Comparators	0	0	1	1
Features:				
I/O Pins	3	3	3	3
Input Only Pins	1	1	1	1
Internal Pull-ups	Yes	Yes	Yes	Yes
In-Circuit Serial Programming	Yes	Yes	Yes	Yes
Instructions	33	33	33	33
Packages:				
6-pin SOT-23				
8-pin PDIP				

Two other PICs of this series are the 10F220 and the 10F222. These versions include four I/O pins and two analog-to-digital converter channels. Program memory is 256 words on the 10F220 and 512 in the 10F222. Data memory is 16 bytes on the F220 and 23 in the F222.

PIC12 Devices

The PIC12C5XX family are 8-bit, fully static, EEPROM/EPROM/ROM-based CMOS microcontrollers. They use RISC architecture and have 33 single-word, single-cycle instructions (except for program branches, which take two cycles). Like the PIC10 family, the PIC12C5XX chips have power-on reset, device reset, and internal timer. Four oscillator options can be selected, including a port-saving internal oscillator and a low-power oscillator. These devices can operate in SLEEP mode and have Watchdog Timer and code-protection features.

Table 7.2
PIC 12Cxxx and 12CExxx Devices

	12C508(A) 12C509A 12CR509A	12C518	12CE519	12C671 12C672	12CE674
Clock:					
Maximum Frequency of Operation (MHz)	4	4	4	10	10
Memory:					
EPROM Program Memory	512/1024/1024 x12	512x12	1024x12	1024/2048/1024x12	2048x14
RAM Data Memory (bytes)	25/41/41	25	41	128	128
Peripherals:					
EEPROM Data Memory (bytes)	—	16	16	0/0/16	16
Timer Module(s)	TMR0	TMR0	TMR0	TMR0	TMR0
A/D Converter (8-bit) Channels	—	—	—	4	4
Features:					
Wake-up from SLEEP on pin change	Yes	Yes	Yes	Yes	Yes
Interrupt Sources	—	—	—	4	4
I/O Pins	5	5	5	5	5
Input Pins	1	1	1	1	1
Internal Pull-ups	Yes/Yes/No	Yes	Yes	Yes	Yes
In-Circuit Serial Programming	Yes/No	Yes	Yes	Yes	Yes
Number of Instructions	33	33	33	35	35
Packages	8-pin DIP SOIC	8-pin DIP JW, SOIC	8-pin DIP JW. SOIC	8-pin DIP SOIC	8-pin DIP JW

The PIC12C5XX devices are recommended for applications including personal care appliances, security systems, and low-power remote transmitters and receivers. The internal EEPROM memory makes possible the storage of user-defined codes and passwords as well as appliance setting and receiver frequencies. The various packages allow through-hole or surface mounting technologies. Table 7.2 lists the characteristics of some selected members of this PIC family.

Two other members of the PIC12 family are the 12F510 and the 16F506. In most respects these devices are similar to the other members of the PIC12 family previously described, except that the 12F510 and 16F506 both have flash program memory. Table 7.3 lists the most important features of these two PICs.

Table 7.3
PIC12F510 and 12F506

	16F506	12F510
Clock:		
Maximum Frequency of Operation (MHz)	20	8
Memory:		
Flash Program Memory	1024	1024
Data Memory (bytes)	67	38
Peripherals:		
Timer Module(s)	TMR0	TMR0
Wake-up from Sleep on Pin Change	Yes	Yes
Features:		
I/O Pins	11	5
Input Only Pin	1	1
Internal Pull-ups	Yes	Yes
In-Circuit Serial Programming	Yes	Yes
Number of Instructions	33	33
Packages	14-pin PDIP, SOIC, TSSOP	8-pin PDIP, SOIC, MSOP

Two other members of the PIC12F are the 12F629 and 12F675. The only difference between these two devices is that the 12F675 has a 10-bit analog-to-digital converter while the 629 has no A/D converter. Table 7.4 lists some important features of both PICs.

Table 7.4
PIC12F629 and 12F675

	12F629	12F675
Clock:		
Maximum Frequency of Operation (MHz)	20	20
Memory:		
Flash Program Memory	1024	1024
Data Memory (SRAM bytes)	64	64
Peripherals:		
Timers 8/16 bits	1/1	1/1
Wake-up from Sleep on Pin Change	Yes	Yes
Features:		
I/O Pins	6	6
Analog comparator module	Yes	Yes
Analog-to-digital converter	No	10-bit
In-Circuit Serial Programming	Yes	Yes
Enhanced Timer1 module	Yes	Yes
Interrupt capability	Yes	Yes
Number of Instructions	35	35
Relative addressing	Yes	Yes
Packages	8-pin PDIP, SOIC, DFN-S	8-pin PDIP, SOIC, DFN-S

Several members of the PIC12 family; the 12F635, 12F636, 12F639, and 12F683, are equipped with special power-management features (called nano-watt technology). These devices were especially designed for systems that require extended battery life.

PIC14 Devices

The single member of this family is the PIC14000. The 14000 is built with CMOS technology; this makes it fully static and gives the PIC an industrial temperature range. The 14000 is recommended for battery chargers, power supply controllers, power management system controllers, HVAC controllers, and for sensing and data acquisition applications. Table 7.5 lists the most important characteristics of this PIC.

Table 7.5
PIC14000

Clock:	
Maximum Frequency of Operation (MHz)	20
Memory:	
Flash Program Memory	4096
Data Memory (SRAM bytes)	192
Peripherals:	
Timers (16 bits with capture)	1
Wake-up from Sleep on Pin Change	Yes
Features:	
I/O Pins	22
Analog-to-digital converter	2 channels
On-chip temperature sensor	1
On-chip comparator modules	2
In-Circuit Serial Programming	Yes
Interrupt capability:	
Internal	6 sources
External	5 sources
I2C-compatible serial port	1
Number of Instructions	35
Relative addressing	Yes
Packages	22-pin PDIP, SOIC, SSOP, Windowed CERDIOP

7.1.2 Mid-range PIC Family

The mid-range PIC family includes members of the PIC12 and PIC16 groups. According to Microchip, the mid-range PICs all have 14-bit program words with either flash or OTP program memory. Those with flash program memory have EEPROM data memory and support interrupts. Some members of the mid-range group have USB, I2C, LCD, USART, and A/D converters. Implementations range from 8 to 64 pins. In the following subsections the basic characteristics of some mid-range PICs are listed.

PIC16 Devices

This is by far the most extensive PIC family. Currently, over 80 versions of the PIC16 are listed in production by Microchip. The remainder of this book is devoted to programming two of these PICs: the 16F84 and the 16F877. Here we listed a few of the most prominent members of the PIC16 family and their most important features. The Microchip website has more detailed information on these devices.

Table 7.6
PIC16 Devices

	16C432	16C58	16C770	16F54	16F84A	16F946
Clock:						
Maximum Frequency MHz	20	40	20	20	20	20
Memory:						
Program memory type	OTP	OTP	OTP	Flash	Flash	Flash
K-bytes	3.5	3	3.5	0.75	1.75	14
K-words	2	2	2	0.5	1	8
Data EEPROM	0	0	0	0	64	256
Peripherals:						
I/O channels	12	12	16	12	13	53
ADC channels	0	0	6	0	0	8
Comparators	0	0	0	0	0	2
Timers	1/8-bit	1/8-bit, 1/16-bit	2/8-bit	1/8-bit	1/8-bit	2/8-bit, 1/16-bit
Watchdog timer	Yes	Yes	Yes	Yes	Yes	Yes
Features:						
ICSP	Yes	No	Yes	No	Yes	Yes
ICD	No	No	No	No	0	1
Pin count	20	18	20	18	18	64
Communications	-	-	MPC/SPI	-	-	AUSART
Packages	20/CERDIP, 20/SSOP, 208mil	18/CERDIP, 18/PDIP, 18/SOIC 300mil	20/CERDIP, 20/PDIP, 20/SOIC 300mil	18/PDIP, 18/SOIC 300mil	18/PDIp, 18/SOIC 300mil	64/TQFP

7.1.3 High-Performance PIC Family

The high-performance PICs belong to the PIC18 group. They have 16-bit program words, flash program memory, a linear memory space of up to two Mbytes, and protocol-based communications facilities. They all support internal and external interrupts and have a much larger instruction set than members of the baseline and mid-range families.

PIC18 Devices

The PIC18 family is also a large one, with over 70 different variations currently in production. The PIC18 family uses 16-bit program words and are furnished in 18 to 80 pin packages. Microchip describes the PICs in this family as high-performance with integrated A/D converters. They have 32-level stacks and support interrupts. The instruction set is much larger and starts at 79 instructions. The PICs in this family have flash program memory, a linear memory space of up to 2 Mbytes, 8-by-8 bit hardware multiplier, and communications peripherals and protocols. Table 7.7 lists some members of the PIC18 family.

Table 7.7
PIC18 Devices

	18F222	18F2455	18F2580	18F4580	18F8622
Clock:					
Maximum Frequency MHz	40	48	40	40	40
Memory:					
Program memory type	flash	flash	flash	flash	flash
K-bytes	4	24	32	32	64
K-words	2	12	16	16	321
Data EEPROM	256	256	256	256	1024
Peripherals:					
I/O channels	25	23	25	36	70
ADC channels	10	10	8	11	16
Comparators	2	2	0	2	2
Timers	1/8-bit 3/16-bit	1/8-bit 3/16-bit	1/8-bit 3/16-bit	1/8-bit 3/16-bit	2/8-bit 3/16-bit
Watchdog timer	Yes	Yes	Yes	Yes	Yes
Features:					
EUSART	Yes	Yes	Yes	Yes	2
ICSP	Yes	Yes	Yes	Yes	Yes
ICD	1	3	3	3	3
Pin count	28	28	28	44	80
Communications	MPC/SPI	MPC/SPI/USB	MPC/SPI	MPC/SPI	2-MPC/SPI
Packages	28/PDIP, 28/SOIC 300mil	28/SOIC 28/PDIP 300mil	28/QFN 20/PDIP 300mil 44/TQFP	40/PDIP 44/QFN	80/TQFP

Chapter 8

Mid-range PIC Architecture

In Chapter 7 we encountered the three major PIC families of 8-bit devices. In the remainder of this book we focus on the mid-range family. Our reason for concentrating our attention on this group is that it is the mid-range PICs that have achieved greater success and popularity.

In addition, as the PIC architecture increases in complexity and power, so does the size, intricacy, and cost of the devices. For many purposes an 80-pin PIC with 64Kbytes of program memory, 1K EERPOM, 70 I/O ports, 16 A/D channels, is more complex than necessary. In fact, some high-end PICs appear to be closer to microprocessors than to microcontrollers. Furthermore, the programming complexity of these high-end PICs is also much greater than their mid-range counterparts because their instruction set has double the number of instructions and the assembly language itself is more difficult to learn and follow. Finally, the circuits in which we typically find the high-end devices are more advanced and elaborate and their design requires greater engineering skills. For these reasons, and for the natural space limitations of a single volume, we do not discuss the high-performance family or 8-bit PICs nor any of the 16-bit products.

It can be argued that the baseline PICs do find extensive use and are quite practical for many applications. Although this is true, the baseline PICs are quite similar in architecture and programming to their mid-range relatives. In most cases the difference between a baseline and mid-range device is that the low-end one lacks some features or has less program space or storage. So someone familiar with the mid-range devices can easily port their knowledge to any of the simpler baseline products.

Our conclusion has been to limit the coverage to the mid-range family of PICs. Within this family we have concentrated our attention on the two most used, documented, and popular PICs: the 16F84 (also 16F84A) and the 16F877. The F84 sets the lower limit of complexity and sophistications and the F877 the higher limit.

8.0 Processor Architecture and Design

PIC microcontrollers are unique in many ways. We start by mentioning several general characteristics of the PIC: Harvard architecture, RISC processor design, single-word instructions, machine and data memory configuration, and characteristic instruction formats.

8.0.1 Harvard Architecture

The PIC microcontrollers do not use the conventional von Neumann architecture but a different hardware design often referred to as *Harvard architecture*. Originally, Harvard architecture referred to a computer design in which data and instruction used different signal paths and storage areas. In other words, data and instructions are not located in the same memory area but in separate ones. One consequence of the traditional von Neumann architecture is that the processor can either read or write instructions or data but cannot do both at the same time, since both instructions and data use the same signal lines. In a machine with a Harvard architecture, on the other hand, the processor can read and write instructions and data to and from memory at the same time. This results in a faster, albeit more complex, machine. Figure 8-1 shows the program and data memory space in a mid-range PIC.

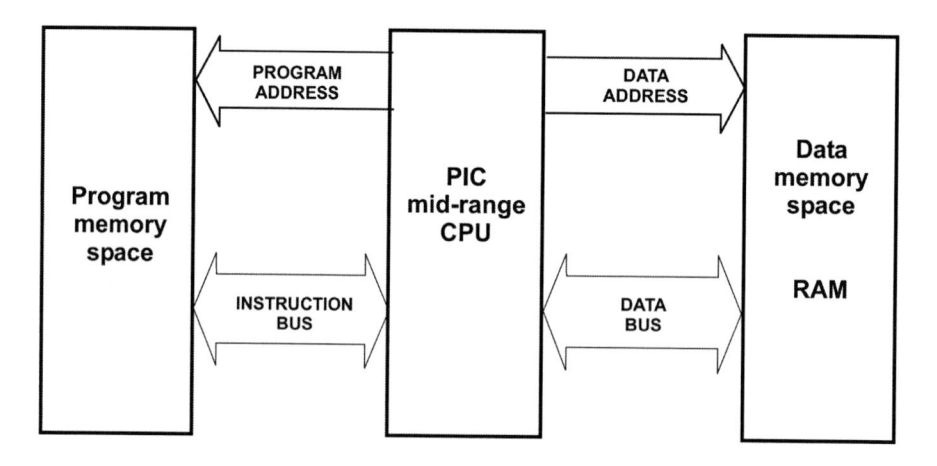

Figure 8-1 Mid-range PIC Memory (Harvard Architecture)

The most recent arguments in favor of the Harvard architecture are based on the access speed to main memory. Making a CPU faster while memory accesses remain at the same speed represents little total gain, especially if many memory accesses are required. This situation is often referred to as the von Neumann bottleneck and machines that suffer from it are said to be *memory bound*.

Several generations of microcontrollers, including the Microchip PICs, have been based on the Harvard architecture. These processors have separate storage for program and data and a reduced instruction set. The midrange PICs, in particular, have

8-bit data words but either 12-, 14-, or 16-bit program instructions. Since the instruction size is much wider than the data size, an instruction can contain a full-size data constant.

8.0.2 RISC CPU Design

The *CISC* (Complete Instruction Set Computer) design is based on each low-level instruction performing several operations. For example, one Intel 80x86 opcode can decrement a counter register, determine the state of a processor flag, and execute a jump instruction if the flag is set or cleared. Another CISC instruction moves a number of bytes of data contained in a counter register from an area pointed at by a source register, into another area pointed at by a destination register. Any popular Intel CISC CPU contains about 120 primitive operations in its instruction set. The original design idea of the CISC architecture was to provide high-level instructions in order to facilitate the implementation of high-level languages. Supposedly, this would be achieved through complex instruction sets, multiple addressing modes, and primitive operations that performed multiple functions.

However, some argued that the CISC architecture did not result in better performance. Furthermore, the more complex the instruction set resulted in greater decoding time. At the same time, implementing large instruction sets required more silicon space and considerably more design effort. Some CISC processors developed in the 1960s and 70s are the IBM System/360, the PDP-11, the Motorola 68000 family, and Intel 80x86 CPUs.

In contrast, a *RISC* (Reduced Instruction Set Computer) machine contains fewer instructions and each instruction performs more elementary operations. Consequences of this are a smaller silicon area, faster execution, and reduced program size with fewer accesses to main memory. The PIC designers have followed the RISC route. Other CPUs with RISC design are the MIPS, the IBM Power PC, and the DEC Alpha.

8.0.3 Single-word Instructions

One of the consequences of the PIC's Harvard architecture is that the instructions can be wider than the 8-bit data size. Since the device has separate buses for instructions and data, it is possible for instructions to be sized differently than data items. Being able to vary the number of bits in each instruction opcode makes possible the optimization of program memory and the use of single-word instructions that can be fetched in one bus cycle.

In the mid-range PICs each instruction is 14-bits wide and every fetch operation brings into the execution unit one complete operation code. Since each instruction takes up one 14-bit word, the number of words of program memory in a device exactly equals the number of program instructions that can be stored. In a von Neumann machine, instruction storage and fetching becomes a much more complicated issue. Since von Neumann instructions can span multiple bytes, there is no assurance that each program memory location contains the first opcode of a multi-byte instruction.

As in conventional processors, the PIC architecture has a two-stage instruction pipeline; however, since the fetch of the current instruction and the execution of the previous one can overlap in time, one complete instruction is fetched and executed at every machine cycle. This is known as *instruction pipelining*. Since each instruction is 14-bits wide and the program memory bus is also 14-bits wide, each instruction contains all the necessary information, so it can be executed without any additional fetching. The one exception is when an instruction modifies the contents of the Program Counter. In this case, a new instruction must be fetched, requiring an additional machine cycle.

The PIC clocking system is designed so that an instruction is fetched, decoded, and executed every four clock cycles. In this manner, a PIC equipped with a 4MHz oscillator clock beats at a rate of 0.25 µs. Since each instruction executes at every four clock cycles, each instruction takes 1 µs.

8.0.4 Instruction Format

All members of the mid-range family of PICs have 14-bit instructions and a set of 35 instructions. The format for the instructions follows three different patterns: byte-oriented, bit-oriented, and literal and control instructions. Figure 8-2 shows the bitmaps for the three types.

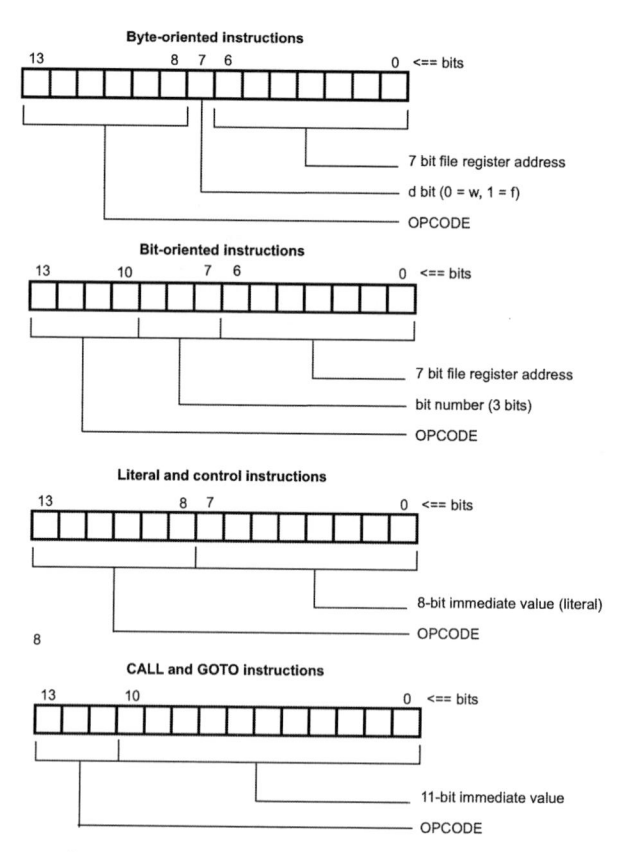

Figure 8-2 Mid-Range Instruction Formats

Note that the opcode field has variable number of bits in the PIC instruction set. This scheme allows implementing 35 different instructions while using a minimum of the 14 available opcode bits. Also note that instructions that reference a file register do so in a 7-bit field. The numerical range of seven bits is 128 values. For this reason, the mid-range PICs that address more than 128 data memory locations must resort to *banking techniques*. In this case, a bit or bit field in the STATUS register serves to select the bank currently addressed.

A similar situation arises when addressing program memory with an 11-bit field. Eleven bits allow 2048 addresses, so if a PIC is to have more than 2K program memory it is necessary to adopt a paging scheme in which a special function register is used to select the memory page where the instruction is located. Paging is required only in devices that exceed the 2K program space limit that can be encoded in 11 bits.

8.0.5 Mid-Range Device Versions

The device names used by Microchip use different encodings to represent different versions of the various devices. For example, the first letter following the family affiliation designator represents the memory type of the device, as follows:

1. The letter C, as in PIC16Cxxx, refers to devices with EPROM type memory.
2. The letters CR, as in PIC16CRxxx, refer to devices with ROM type memory.
3. The letter F, as in PIC16Fxxx, refers to devices with flash memory.

The letter L immediately following the affiliation designator refers to devices with an extended voltage range. For example, the PIC16LFxxx designation corresponds to devices with extended voltage range.

8.1 The Mid-range Core Features

Core features refer to the device oscillator, reset mechanism, CPU architecture and operation, Arithmetic-Logic Unit, memory organization, interrupts, and instruction set. We have already referred to the architecture and general features of the CPU. Memory organization is discussed in a separate section later in this chapter. The remaining topics are covered in the following subsections.

8.1.1 Oscillator

Mid-range PICs require an external device to produce the clock cycles required for its operation. The PIC executes an instruction every four clock cycles, so the *oscillator* speed determines the device performance.

Mid-range PICs support up to eight different oscillator modes. For example, in the 16F877, any of the eight modes can be used, while in the 16F84 only four oscillator modes are available. The oscillator mode is selected at device programming time and cannot be changed at runtime. The configuration bits, which are non-volatile flags set during device programming, determine which oscillator mode is used by the program, among the following:

1. LP Low Frequency Crystal
2. XT Crystal Resonator
3. HS High Speed Crystal Resonator
4. RC External Resistor/Capacitor
5. EXTRC External Resistor/Capacitor
6. EXTRC External Resistor/Capacitor with CLKOUT
7. INTRC Internal 4 MHz Resistor/Capacitor
8. INTRC Internal 4 MHz Resistor/Capacitor with CLKOUT

The resistor/capacitor oscillator option is the least expensive to implement, but also the least accurate one. This option is used only in systems where clock accuracy and consistency are not issues. The low-power frequency crystal option is the one with lowest power consumption and can be used in systems where the power consumption element is important.

The first three oscillator modes (LP, XT, and HS) allow selecting different frequency ranges. The *HS option* has the highest frequency range and consumes the most power. The *XT option* is based on a standard crystal resonator and has a mid-range power consumption. The *LP option* has low gain and consumes the least power of the three crystal modes. The general rule is to use the oscillator with the lowest possible gain that still meets the circuit requirements. The RC mode with EXTRC and CLKOUT features has the same functionality as the straight RC oscillator option.

The XT option (crystal resonator) can be purchased in a ceramic package. This device, called a *ceramic resonator*, contains three pins. The ones on the extremes are connected to the corresponding oscillator input lines on the PIC, labeled OSC1 and OSC2. The center pin is connected to ground. Figure 8-3 shows the circuit diagram for an oscillator and a crystal resonator.

Figure 8-3 Circuit Diagram for Oscillators

Mid-range PIC Architecture

Alternatively, the oscillator function is provided by an integrated circuit (such as the ICS502) that can generate several different clock frequencies. Some circuits, especially in PIC demonstration boards, contain jumper pins that allow selecting among several clock rates.

8.1.2 System Reset

The reset mechanism places the PIC in a known condition. The reset mechanism is used to gain control of a runaway or hung-up program, as a forced interrupt in program execution, or to make the device ready at program load time. The processor's !MCLR pin produces the reset action when it reads logic zero. The exclamation sign preceding the pin's name (or a line over it) indicates that the action is active-low. To prevent accidental resets the !MCLR pin must be connected to the positive voltage supply through a 5K or 10K resistor. When a resistor serves to place a logic one on a line it is called a *pull-up resistor*.

The mid-range PICs are capable of several reset actions:

1. Reset during power on (POR).
2. !MCLR reset during normal operation.
3. Reset during SLEEP mode.
4. Watchdog timer reset (WDT).
5. Brown-out reset (BOR).
6. Parity error reset.

The first two reset sources in the preceding list are the most common. POR reset serves to bring all PIC registers to an initial state, including the program counter register. The second source of reset action takes place when the !MCLR line is intentionally brought down, usually by the action of a push-button reset switch. This switch is useful during program development since it provides a way of forcefully restarting execution. Figure 8-4 shows a typical wiring of the !MCLR line to provide a reset action.

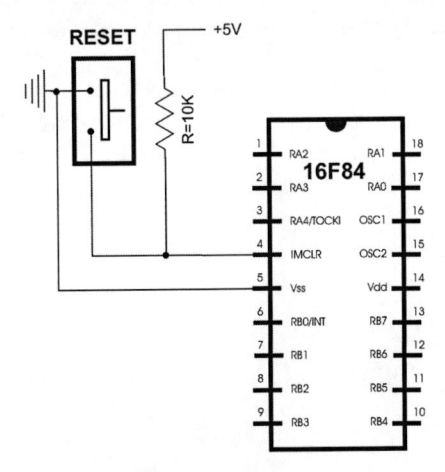

Figure 8-4 Typical Wiring of the Reset Switch

The second one is a product of purposefully bringing-in a logical zero to the MCLR pin during normal operation of the microcontroller. This second one is often used in program development.

User RAM memory is not affected by a reset. The GPRs (general purpose register) are in an unknown state during power-up and are not changed by reset. SFR registers, on the other hand, are reset to an initial state. The initialization conditions for each of the SFRs are found in the device data sheet. The most important of these is the *program count* (PC) which is reset to zero. This action directs execution to the first instruction and effectively restarts the program.

During power-up the processor itself initiates a reset and the power supply voltage increases from 1.2 to 1.8V. Several bits in various registers are related to the reset action, but these are not available in all mid-range devices. For example, some high-end devices in the mid-range group, such as the 16F87x, contain two reset-related bits in the PCON register. One of them (named !POR) determines the power-on reset status. The other one (named !BOR) informs about the brown-out reset status. However, the PCON register does not exist in the 16F84 or 16F84A.

8.1.3 Interrupts

The *interrupt* mechanism provides a way of having the microcontroller respond to events as they occur, rather than having to poll devices in order to determine their state. Thus, the interrupt works like a "tap on the shoulder" on the microcontroller, calling its attention to an event that requires an action or device that needs servicing. After responding to or ignoring the interrupt, the CPU resumes processing where it left off.

In computer technologies the interrupt mechanism is a complicated hardware/software system that often includes *programmable interrupt controller* ICs. Processors and microprocessors usually support hardware and software interrupts and maskable and non-maskable interrupts; interrupts originate in practically any device connected to the system. In the PICs, the interrupt mechanism is much simpler and varies considerably even among members of the same PIC family.

All PICs of the mid-range family to some degree support interrupts. The interrupt source usually originates in one of the hardware modules, although some sources generate more than one interrupt. The following are interrupt sources in the mid-range family, although not all are supported by every PIC.

- INT Pin Interrupt (external interrupt)
- TMR0 Overflow Interrupt
- PORTB Change Interrupt
- Comparator Change Interrupt
- Parallel Slave Port Interrupt
- USART Interrupts
- Receive and Transmit Interrupt

- A/D Conversion Complete Interrupt
- LCD Interrupt
- Data EEPROM Write Complete Interrupt
- Timer Overflow Interrupt
- CCP Interrupt
- SSP Interrupt

Several SFRs are related to the interrupt systems. The INTCON register provides interrupt enabling and control and the PIE1, PIE2, PIR1, and PIR2 registers have specific device-related functions. Programming interrupts is discussed in the context of the corresponding operations later in this book.

8.2 Mid-Range CPU and Instruction Set

In a digital system, the *central processing unit* (CPU) is the component that executes the program instructions and processes data. It provides the fundamental functionality of a digital system and is responsible for its programmability. In the PIC architecture, the CPU is the part of the device which fetches and executes the instructions contained in a program.

The *arithmetic-logic unit* (ALU) is the CPU element that performs arithmetic, bitwise, and logical operations. It also controls the bits in the STATUS register as they are changed by the execution of the various program instructions. For example, if the result of executing an instruction is zero, the ALU sets the zero bit in the STATUS register.

8.2.1 Mid-Range Instruction Set

The mid-range PIC instruction set consists of 35 instructions, divided into three general groups:

1. Byte-oriented and byte-wise file register operations
2. Bit-oriented and bit-wise file register operations
3. Literal and control instructions

Table 8.1 lists and briefly describes each instruction in the mid-range set.

Table 8.1

Mid-range PIC Instruction Set

MNEMONIC	OPERAND	DESCRIPTION	CYCLES	BITS AFFECTED
		BYTE-ORIENTED OPERATIONS:		
ADDWF	f,d	Add w and f	1	C,DC,Z
ANDWF	f,d	AND w with f	1	Z
CLRF	f	Clear f	1	Z
CLRW	-	Clear w	1	Z
COMF	f,d	Complement f	1	Z
DECF	f,d	Decrement f	1	Z

(continues)

Table 8.1

Mid-range PIC Instruction Set (continued)

MNEMONIC	OPERAND	DESCRIPTION	CYCLES	BITS AFFECTED
\multicolumn{5}{c}{BYTE-ORIENTED OPERATIONS}				
DECFSZ	f,d	Decrement, skip if 0	1(2)	-
INCF	f,d	Increment f	1	Z
INCFSZ	f,d	Increment, skip if 0	1(2)	-
IORWF	f,d	Inclusive OR w and f	1	Z
MOVF	f,d	Move f	1	Z
MOVWF	f	Move w to f	1	-
NOP	-	No operation	1	-
RLF	f,d	Rotate left through carry	1	C
RRF	f,d	Rotate right through carry	1	C
SUBWF	f,d	Subtract w from f	1	C,DC,Z
SWAPF	f,d	Swap nibbles in f	1	-
XORWF				
\multicolumn{5}{c}{BIT-ORIENTED OPERATIONS}				
BCF	f,b	Bit clear in f	1	-
BSF	f,b	Bit set in f	1	-
BTFSC	f,b	Bit test, skip if clear	1	-
BTFSS	f,b	Bit test, skip if set	1	-
\multicolumn{5}{c}{LITERAL AND CONTROL OPERATIONS}				
ADDLW	k	Add literal and w	1	C,DC,Z
ANDLW	k	AND literal and w	1	Z
CALL	k	Call procedure	2	-
CLRWDT	-	Clear watchdog timer	1	TO,PD
GOTO	k	Go to address	2	-
IORLW	k	Inclusive OR literal with w	1	Z
MOVLW	k	Move literal to w	1	-
RETFIE	-	Return from interrupt	2	-
RETLW	k	Return literal in w	2	-
RETURN	-	Return from procedure	2	-
SLEEP	-	Go into SLEEP mode	1	TO,PD
SUBLW	k	Subtract literal and w	1	C,DC,Z
XORLW	k	Exclusive OR literal with w	1	Z

Legend:
  f = file register
  d = destination: 0 = w register
                    1 = file register
  b = bit position
  k = 8-bit constant

8.2.2 STATUS and OPTION Registers

The *STATUS register* is one of the SFRs in the mid-range PICs. The bits in this register reflect the arithmetic status of the ALU, the RESET status, and the bits that select which memory bank is currently being accessed. Because the bank selection bits are in the STATUS register it must be present and at the same relative position in every bank. Figure 8-5 is a bitmap of the STATUS register.

bits:	7	6	5	4	3	2	1	0
	IRP	RP-1	RP-0	TO	PD	Z	DC	C

```
bit 7 IRP: Register Bank Select bit (used for indirect
           addressing)
           1 = Bank 2, 3 (0x100 - 0x1ff)
           0 = Bank 0, 1 (0x000 - 0xff)
           For devices with only Bank0 and Bank1 the
           IRP bit is reserved, always maintain this
           bit clear.
bit 6:5 RP1:RP0:
           Register Bank Select bits (used for direct
           addressing)
           11 = Bank 3 (0x180 - 0x1ff)
           10 = Bank 2 (0x100 - 0xx17f)
           01 = Bank 1 (0x80 - 0xff)
           00 = Bank 0 (0x00 - 0x7f)
           Each bank is 128 bytes. For devices with only
           Bank0 and Bank1 the IRP bit is reserved,
           always maintain this bit clear.
bit 4 TO:  Time-out bit
           1 = After power-up, CLRWDT instruction, or
               SLEEP instruction
           0 = A WDT time-out occurred
bit 3 PD:  Power-down bit
           1 = After power-up or by the CLRWDT instruction
           0 = By execution of the SLEEP instruction
bit2 Z:    Zero bit
           1 = The result of an operation is zero
           0 = The result of an operation is not zero
bit 1 DC:  Digit carry/borrow bit for ADDWF, ADDLW, SUBLW,
           and SUBWF instructions. For borrow the polarity
           is reversed.
           1 = A carry-out from the 4th bit of the result
           0 = No carry-out from the 4th bit of the result
bit 0 C:   Carry/borrow bit for ADDWF, ADDLW, SUBLW, and
           SUBWF instructions
           1 = A carry-out from the most significant bit
           0 = No carry-out from the most significant bit
```

Figure 8-5 STATUS Register Bitmap

The STATUS register can be the destination for any instruction. If it is, and the Z, DC, or C bits are affected, then the write operation to these bits is disabled. In addition, the TO and PD bits are not writable.

Some instructions may have an unexpected action on the STATUS register bits, for example, the instruction

```
Clrf    STATUS
```

clears the upper 3 bits, sets the Z bit, and leaves all other bits unchanged. For this reason, it is recommended that only instructions that do not change the Z, C, and DC bits be used to alter the STATUS register. The only ones that qualify are BCF, BSF, SWAPF, and MOVWF.

The *OPTION register* is actually named the *OPTION_REG* to avoid name clash with the option instruction. The OPTION_REG register contains several bits related to interrupts, the internal timers, and the watchdog timer. Figure 8-6 is a bitmap of the OPTION_REG register.

bits:	7	6	5	4	3	2	1	0
	RPBU	INTEDG	T0CS	T0SE	PSA	PS2	PS1	PS0

bit 7 **RBPU:**   PORTB Pull-up Enable bit
                          1 = PORTB pull-ups are disabled
                          0 = PORTB pull-ups are enabled by individual port latch values

bit 6 **INTEDG:**   Interrupt Edge Select bit
                          1 = Interrupt on rising edge of INT pin
                          0 = Interrupt on falling edge of INT pin

bit 5 **T0CS:**   TMR0 Clock Source Select bit
                          1 = Transition on T0CKI pin
                          0 = Internal instruction cycle clock (CLKOUT)

bit 4 **T0SE:**   TMR0 Source Edge Select bit
                          1 = Increment on high-to-low transition on T0CKI pin
                          0 = Increment on low-to-high transition on T0CKI pin

bit 3 **PSA:**   Prescaler Assignment bit
                          1 = Prescaler is assigned to the WDT
                          0 = Prescaler is assigned to the Timer0

bit 2-0 **PS2:PS0:** Prescaler Rate Select bits

Figure 8-6 Bitmap of the OPTION_REG Register

8.3 EEPROM Data Storage

EEPROM (pronounced double-e PROM or *e*-squared PROM) stands for electrically-erasable programmable read-only memory. EEPROM is used in computers and digital devices as non-volatile storage. EEPROM is not RAM, since RAM is volatile and EEPROM retains its data after power is removed. EEPROM is found in USB flash drives and in the non-volatile storage of several microcontrollers, including many PICs.

One advantage of EEPROM is that it can be erased and written electrically, without removing the chip. The predecessor technology, named *EPROM*, required that the chip be removed from the circuit and placed under ultraviolet light. EEPROM simplifies the erasing and re-writing process.

EEPROM data memory refers to both on-board EEPROM memory and to EEPROM memory ICs as separate circuit components. In general, EEPROM elements are classified according to their electrical interfaces into serial and parallel. Most EEPROM memories used in PICs are serial EEPROMs, also called *SEEPROMs*. The typical use of serial EEPROM on-board memory and EEPROM on ICs is in the storage of passwords, codes, configuration settings, and other information to be remembered after the system is turned off. For example, a PIC-based security system can use EEPROM memory to store the system password. Since EEPROM can be written, the user can change this password and the new one will also be remembered.

8.3.1 EEPROM in Mid-Range PICs

The mid-range PICs are equipped with EEPROM memory in three possible sizes: 64 bytes, 128 bytes, and 256 bytes. EEPROM memory allows read and write operations. This memory is not mapped into the processor's data or program area, but in a separate block that is addressed through some SFRs. The registers related to EEPROM operations are:

1. EECON1
2. EECON2 (not a physically implemented register)
3. EEDATA
4. EEADR

EECON1 contains the control bits, and EECON2 is used to initiate the EEPROM read and write operations. The 8-bit data item to be written must first be stored in the EEDATA register, while the address of the location in EEPROM memory is stored in the EEADR register. The EEPROM address space always starts at 0x00 and extends linearly to maximum in the device.

When a write operation is performed, the contents of the EEPROM location are automatically erased. The EEPROM memory used in PICs is rated for high erase/write cycles. EEPROM programming is the topic of Chapter 15.

8.4 Data Memory Organization

The structure and organization of data memory in the PIC hardware also has some unique and interesting features. The programmer accustomed to the flat, addressable memory space of the von Neumann computer with its multiple machine registers may require some time in order to gain familiarity with the PIC's data formats.

8.4.1 The w Register

PICs have only one addressable register called the *work register* or the *w register*. The CISC programmer who is used to having multiple general purpose registers into which data can be moved and later retrieved has to become used to a single machine register that takes part in practically every instruction. Add to this the lack of an addressable stack into which data can be pushed and popped, and you see that PIC programming is a different paradigm.

8.4.2 The Data Registers

PIC's data memory consists of *registers*, also called *file registers*. These behave more like conventional variables, and can be addressed directly and indirectly. All data registers are 8-bits. Data registers come in two types: *general purpose registers* (GPRs) and *special function registers* (SFRs).

Memory Banks

The PIC instruction format devotes seven bits to the address field (see Figure 8-2, Section 8.0.4). A 7-bit address allows access to only 128 memory locations. Since many PICs of the mid-range family have more than 128 bytes of data memory, an addressing scheme based on memory banks must be implemented. The memory banking mechanism adopted by the PICs is effective, although not very user-friendly.

The number of banks vary according to the amount of available RAM, always in multiples of 128-bytes. All mid-range PICs have banked memory. Banking is accomplished through the special bank-select bits in the STATUS register (see Figure 8-5). Not all banking bits are implemented in all devices. For example, the 16F84/16F84A contain two memory banks; therefore, bank shifting requires a single bank-select bit (RP0). In this case the RP1 bit is not implemented. In devices with more than two memory banks bank selection is as shown in Table 8.2.

Table 8.2
Mid-Range Bank Selection Options in Direct Addressing

BANK ACCESSED	STATUS REGISTER BITS (RP1:RP0)
0	0 : 0
1	0 : 1
2	1 : 0
3	1 : 1

Figure 8-7 shows how banked memory is accessed in direct addressing. The illustration refers to a mid-range PIC with four banks, as is the case with the 16F87x.

Mid-range PIC Architecture

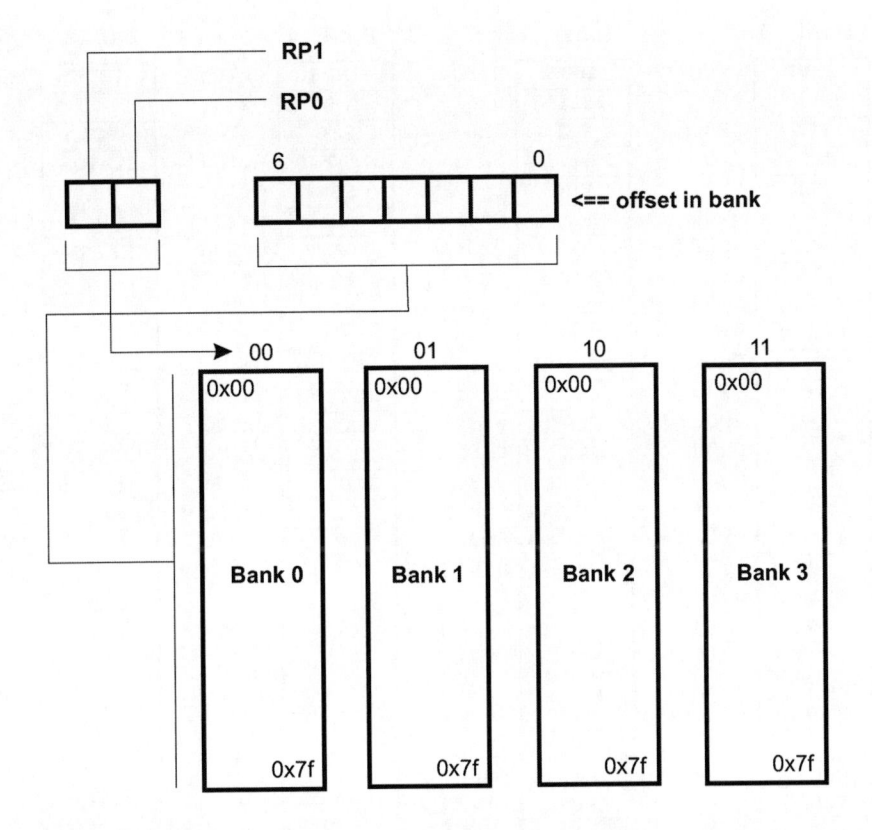

Figure 8-7 Memory Access in Direct Addressing

The SFRs

The special function registers are defined by the device architecture and have reserved names. For example, the TMR0 register is part of the system timer, the STATUS register holds several processor flags, and the INTCON register is used in controlling interrupts. Some SFRs can be written and read and others are read-only. Some reserved and not-implemented SFR bits always read as zero. Two SFR registers, which are used in indirect addressing, have special characteristics: one of them (the indirect address register) is not a physical register, and the other one (the FSR register) is used to initialize the indirect pointer. The SFR are allocated starting at the lowest RAM address (address 0).

Figure 8-8 (in the following page) is a map of the register file in the 16F87x family. Note in Figure 8-8 that the general purpose registers do not start at the same address offset in each bank. However, there is a common area that extends from 0x70 to 0x7f that is accessible no matter which bank is selected. In applications that require frequent bank switching, this 16-byte area is very valuable real-estate since user variables created in it are accessible no matter which bank is currently selected. GPRs created outside this common area are only accessible when the corresponding bank is selected.

Bank 0		Bank 1		Bank 2		Bank 3	
INDF	0x00	**INDF**	0x80	**INDF**	0x100	**INDF**	0x180
TMR0	0x01	OPTION*	0x81	TMR0	0x101	OPTION*	0x181
PCL	0x02	**PCL**	0x82	**PCL**	0x102	**PCL**	0x182
STATUS	0x03	**STATUS**	0x83	**STATUS**	0x103	**STATUS**	0x183
FSR	0x04	**FSR**	0x84	**FSR**	0x104	**FSR**	0x184
PORTA	0x05	TRISA	0x85		0x105		0x185
PORTB	0x06	TRISB	0x86	PORTB	0x106	TRISB	0x186
PORTC	0x07	TRISC	0x87		0x107		0x187
PORTD	0x08	TRISD	0x88		0x108		0x188
PORTE	0x09	TRISE	0x89		0x109		0x189
PCLATH	0x0a	**PCLATH**	0x8a	**PCLATH**	0x10a	**PCLATH**	0x18a
INTCON	0x0b	**INTCON**	0x8b	**INTCON**	0x10b	**INTCON**	0x18b
PIR1	0x0c	PIE1	0x8c	EEDATA	0x10c	EECON1	0x18c
PIR2	0x0d	PIE2	0x8d	EEADR	0x10d	EECON2	0x18d
TMR1L	0x0e	PCON	0x8e	EEDATH	0x10e	Reserved	0x18e
TMR1H	0x0f		0x8f	EEADRH	0x10f	Reserved	0x18f
T1CON	0x10		0x90		0x110		0x190
TMR2	0x11	SSPCON2	0x91				
T2CON	0x12	PR2	0x92				
SSPBUF	0x13	SSPADD	0x93				
SSPCON	0x14	SSPTAT	0x94				
CCPR1L	0x15		0x95				
CCPR1H	0x16		0x96				
CCP1CON	0x17		0x97				
RCSTA	0x18	TXSTA	0x98	General Purpose Registers		General Purpose Registers	
TXREG	0x19	SPBRG	0x99				
RCREG	0x1a		0x9a				
CCPR2L	0x1b		0x9b				
CCPR2H	0x1c		0x9c				
CCP2CON	0x1d		0x9d				
ADRESH	0x1e	ADRESL	0x9e				
ADCON0	0x1f	ADCON1	0x9f				
	0x20		0xA0				
General Purpose Registers		General Purpose Registers					
			0xef		0x16f		0x1ef
Common area 0x70-0x7f		Common area 0x70-0x7f	0xf0	Common area 0x70-0x7f	0x170	Common area 0x70-0x7f	0x1f0
	0x7f		0xff		0x17f		0x1ff

* Actual name is OPTION_REG

Figure 8-8 16F87x File Register Map

The registers in boldface in Figure 8-8 are accessible from any bank. These registers, such as STATUS and the indirect addressing registers FSR and INDF, are bank-independent. Also, some registers are mirrored in more than one bank. For example, the PORTB register is accessible in bank 0 and in bank 2, and the TRISB register in bank 1 and bank 3. The *mirrored registers* are designed to simplify data access and minimize bank changes in applications.

Other members of the mid-range PIC group, such as the 16F84 and 16F84A, have a different memory footprint. Figure 8-9 is a bitmap of the 16F84A.

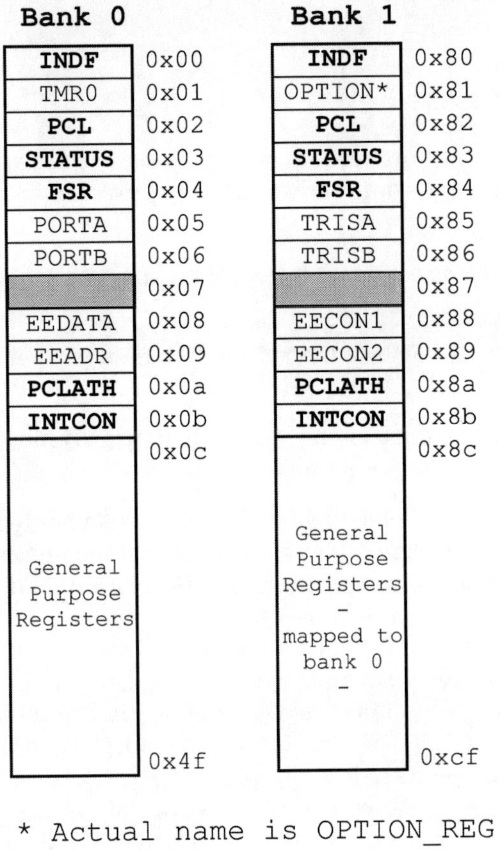

Bank 0		Bank 1	
INDF	0x00	INDF	0x80
TMR0	0x01	OPTION*	0x81
PCL	0x02	PCL	0x82
STATUS	0x03	STATUS	0x83
FSR	0x04	FSR	0x84
PORTA	0x05	TRISA	0x85
PORTB	0x06	TRISB	0x86
▨	0x07	▨	0x87
EEDATA	0x08	EECON1	0x88
EEADR	0x09	EECON2	0x89
PCLATH	0x0a	PCLATH	0x8a
INTCON	0x0b	INTCON	0x8b
	0x0c		0x8c
General Purpose Registers		General Purpose Registers - mapped to bank 0 -	
	0x4f		0xcf

* Actual name is OPTION_REG

Figure 8-9 16F84A File Register Map

Here again, the general purpose registers do not start at the same address offset in each bank. Also note that all GPRs are mapped to bank 0. In the 16F84A, this means that user-defined registers created in bank 0 are accessible no matter which bank is currently selected.

The GPRs

General purpose registers are created and named by the programmer and must be allocated in the reserved memory space. In the 16F84A all GPRs are mapped to the same memory area, no matter in which bank they are defined. The GPR memory space actually extends from 0x0c to 0x4f (68 bytes). A different situation exists in the 16F87x PICs, in which only 16 bytes of GPR space is mirrored in all three banks. This is the memory referred to as the *common area* in Figure 8-8. In the 16F87x the total available GPR space is as follows:

```
BANK 0         BANK 1         BANK2          BANK3
96 bytes       80 bytes       96 bytes       96 bytes

Total = 368 bytes
```

8.4.3 Indirect Addressing

The instruction set of most processors, including the PICs, provides a mechanism for accessing memory operands indirectly. *Indirect addressing* is based on the following capabilities:

1. The address of a memory operand is loaded into a register. This register is called the *pointer*.

2. The *pointer register* is then used to indirectly access the memory location at the address it "points to."

3. The value in the pointer register can be modified (usually incremented or decremented) so as to allow access to other memory operands.

In the PIC architecture indirect addressing is implemented using two registers: INDF and FSR. The *INDF register*, always located at memory address 0x00 and mirrored in all banks, is not a physical register, in the sense that it cannot be directly accessed by code. The *FSR register* is the pointer register that is initialized to the address of a memory operand. Once a memory address is placed in FSR, any action on the INDF register takes place at the memory location pointed at by FSR. For example, if the FSR register is initialized to memory address 0x20, then clearing the INDF register has the effect of clearing the memory location at address 0x20. In other words, the action on the INDF register actually takes place at the address contained in the FSR register. Now, if FSR (the pointer register) is incremented and INDF is again cleared, the memory location at address 0x21 is cleared. Indirect addressing is covered in detail in the programming chapters.

8.5 Mid-range I/O and Peripheral Modules

Mid-range devices contain special modules to implement peripheral and I/O functions. The more complex the device the more peripheral modules are likely to be present. For example, a simple mid-range PIC like the 16F84A contains few peripheral modules, specifically, EEPROM data memory, I/O ports, and a timer module. The 16F87x PICs, on the other hand, in addition to I/O ports, EEPROM, and three individual timers, have a parallel slave port, a WPM (capture and compare) module, an MSSP (master synchronous serial port) module, a USART (universal asynchronous/synchronous receiver and transmitter) module, and an A/D (analog-to-digital converter) module.

Other members of the mid-range family have additional or different peripheral and I/O modules. In the following sections, we briefly describe the architecture of the most common peripheral modules. The programming details are covered elsewhere in the book.

Implementation of many different functions in a device with a small footprint requires multiplexing many of the PIC's access connections. Figure 8-10 shows the pinout of the 16F84A and the 16F877 and the multiple functions of most pins in both devices.

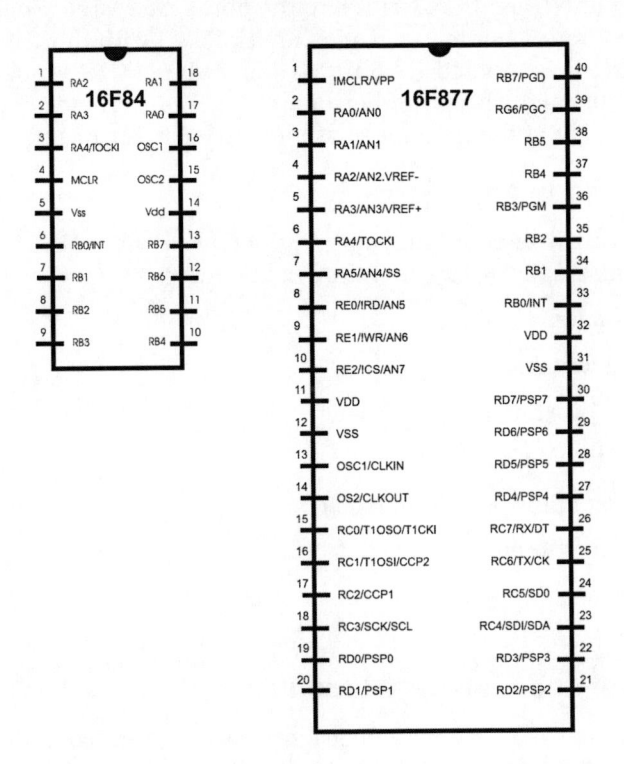

Figure 8-10 16F84A and 16F877 Pin Diagrams

8.5.1 I/O Ports

Ports provide PICs access to the outside world and are mapped to physical pins on the device. In some mid-range PICs (see Figure 8-10) some port pins for I/O ports are multiplexed with alternate functions of peripheral modules. When a peripheral module is enabled, that pin ceases to be a general purpose I/O.

Port pins can be configured either as input or output, that is, general ports are bidirectional. Each port has a corresponding TRIS register which determines if a port is designated as input or output. A value of 1 in the port's TRIS register makes the port an input and a value of 0 makes the mapped port an output. Typically, input ports are used in communicating with input devices, such as switches, keypads, and input data lines from hardware devices. Output ports are used in communicating with output devices, such as LEDs, seven-segment displays, LCDs (liquid-crystal displays), and data output lines to hardware devices.

Although port pins are bitmapped, they are read and written as a unit. For example, the PORTA register holds the status of the eight pins possibly mapped to Port-A, while writing to PORTA writes to the port latches. Write operations to ports are actually read-modify-write operations. In other words, the port pins are first read, then the value is modified, and then written to the port's data latch. Some of the port pins are multiplexed; for example, pin RA4 is multiplexed with the Timer0 module clock input; therefore, it is labeled RA4/T0CKI pin. Other PORTA pins are multiplexed with analog inputs and with other peripheral functions. The device data sheets contain information about the functions assigned to each device pin.

8.5.2 Timer Modules

Timer modules are available in all mid-range devices. The TIMER0 module is present in all PICs of this family. It has the following features:

1. 8-bit timer/counter
2. Readable and writable
3. 8-bit software programmable prescaler
4. Internal or external clock select
5. Interrupt on overflow from FFh to 00h
6. Edge select for external clock

Chapter 12 is devoted entirely to the architecture and programming of timers and counters.

8.5.3 Capture-and-Compare Module

Some mid-range devices contain one or more *capture-and-compare modules*, designated as Capture/Compare/PWM modules. In Figure 8-10 you can see that one of the functions multiplexed onto pin 17 of the 16F877 is labeled CCP1 (capture-and-compare module number 1). The CCP2 module is multiplexed onto pin number 16. The principal function of the capture-and-compare modules is to enhance timer operations. Each module contains the following elements:

- A 16-bit register which can operate as:

    a 16-bit capture register or a 16-bit compare register

- A PWM Master/Slave Duty Cycle register

When more than one capture-and-compare module is implemented in a single device, they are all identical in operation. In the 16F877, the two available modules are designated as CCP1 and CCP2 respectively. In each module a Capture/Compare/PWM Register1 (CCPR1) is comprised of two 8-bit registers: *CCPR1L* (low byte) and *CCPR1H* (high byte). The CCP1CON register controls the operation of CCP1.

The CCP modules find use in recording events, measuring time periods, counting, generating pulses and periodic waveforms, and voltage averaging, among others. However, since these applications are not commonly found in the simple PIC circuits covered in this book we make no further reference to this topic.

8.5.4 Master Synchronous Serial Port (MSSP) Module

Some mid-range PICs come equipped with hardware modules to implement serial protocols, including SPI and I2C. The module that provides these interfaces is named the *Master Synchronous Serial Port*, or *MSSP*. The MSSP module can operate in either the slave or the master mode, as well as in a *free-bus mode*, also called the multi-master function.

The MSSP module is useful for communicating with other peripheral or microcontroller devices. The peripheral devices can be serial EEPROMs, shift registers, display drivers, A/D converters, etc. The MSSP module is discussed in Chapter 8, in the context of EEPROM data memory programming.

8.5.5 USART Module

The *Universal Synchronous Asynchronous Receiver Transmitter* (*USART*) module in the 16F87x family is also known as a Serial Communications Interface, or SCI. The USART module is used in communicating with devices and systems that support RS-232 communications, including computers and terminals. It can be configured as an asynchronous full duplex device, as a synchronous half-duplex master, or as a synchronous half-duplex slave. In the synchronous mode, the USART is useful in communicating with analog-to-digital and digital-to-analog integrated circuits or for accessing serial EEPROMS. The USART is discussed extensively in Chapter 14 and, in the context of programming serial EEPROMS, in Chapter 15.

8.5.6 A/D Module

Until recently, A/D conversions required the use of dedicated devices, usually in the form of an integrated circuit component. Mid-range PICs now come with on-board A/D hardware. One of the advantages of using on-board A/D converters is saving interface lines. Interfacing with a hardware IC usually requires three to four lines. A similar function can be implemented with on-board A/C hardware with a single line. Since I/O lines are often needed in PIC circuits, the advantage of on board A/C hardware is significant.

Mid-range PICs equipped with A/D converters have either 8- or 10-bit resolution and can receive analog input in 2 to 16 different channels. For example, the 16F877 contains eight analog input channels at a 10-bit resolution. An A/D converter uses a sample-and-hold capacitor to store the analog charge and performs a successive approximation algorithm to produce the digital result. When the converter resolution is 10 bits these are stored in two 8-bit registers, one of them having only four significant bits.

The *A/D module* has high- and low-voltage reference inputs which are selected by software. The module can operate while the processor is in the SLEEP mode, but only if the A/D clock pulse is derived from its internal RC oscillator. The A/D converter module is discussed in detail in Chapter 16.

Chapter 9

PIC Programming: Tools and Techniques

PIC microcontrollers can be programmed in high-level languages or in their native machine language. Machine language programming is facilitated by the use of an assembler program, and thus becomes assembly language programming. Although assembly language is the most used and popular way of PIC programming, there is an ongoing debate regarding the use of high-level languages.

The major argument in favor of high-level languages is their ease of use and their faster learning curve. The advantages of assembly language, on the other hand, are better control and greater efficiency. It is true that arguments that favor high-level languages find some justification in the computer world, but these reasons are not as valid regarding PIC programming. In the first place, the PIC programmer cannot avoid complications and technical details by resorting to a high-level language, since PIC programs relate closely to hardware devices and to electronic circuits. These devices and circuits must be understood at their most essential level if they are to be controlled and operated by software. For example, consider a PIC program to provide some sort of thermostatic control. In this case, the programmer must become familiar with temperature sensors, analog-to-digital conversions, motor controls, and so on. This is true whether the program is written in a low- or a high-level language.

Another reason for using assembly language in PIC programming is that the language itself is quite simple. The mid-range PICs have 35 instructions in their instruction sets and many of them are quite similar; so learning assembly language for PIC programming takes no great effort. Additionally, assembly language development tools are free from Microchip, while most high-level languages must be purchased from their developers. For these reasons we have excluded all high-level languages from consideration in this book.

9.0 Microchip's MPLAB

The PIC assembly language development system provided by Microchip is named *MPLAB*. The package is furnished as an *IDE (integrated development environment)* and can be downloaded from the company's web site at www.microchip.com.

One limitation of the MPLAB package is that it is furnished only for the PC. If you are a Mac, UNIX, or Linux user you cannot use MPLAB. Development packages for other operating systems are available on the Web.

The MPLAB IDE is intended for software development of *embedded systems*. An embedded system is designed for a specific purpose, in contrast with a computer system which is a general purpose machine. The embedded system is designed to perform specific and predefined tasks; for example, control a microwave oven, control a TV receiver, or operate a model railroad. The software of a general-purpose computer can be easily changed. You may, at will, run a word processor, a web browser, or a database management system on your computer. The software in an embedded system is usually fixed and cannot be easily changed; for this reason it is called "*firmware*."

9.0.1 Embedded Systems

At the heart of an embedded system is a microcontroller (such as a PIC), sometimes several of them. These devices are programmed to perform one, or at most a few, tasks. In the most typical case, an embedded system also includes one or more "peripheral" circuits which are operated by dedicated ICs or by functionality contained in the microcontroller itself. The term "embedded system" refers to the fact that the device is often found inside another one; for instance, the control circuit is embedded in a microwave oven. Furthermore, embedded systems do not have (in most cases) general purpose devices such as hard disk drives, video controllers, printers, and network cards.

A typical embedded system is a control for a microwave oven. In this case, the controller includes a timer to clock various operations, a temperature sensor to provide information about the oven's operation, a motor to rotate the oven's tray, a sensor to detect when the oven door is open, and a set of pushbutton switches to select the operational options. A program running on the embedded microcontroller reads the commands and parameters input through the keyboard, programs the timer and the rotating table, detects the state of the door, and turns the heating element on and off as required by the user's selection. Many other daily devices including automobiles, digital cameras, cell phones, and home appliances use embedded systems and many of them are PIC-based.

The development process of an embedded system consists of the following steps:

1. Define the system specifications. This step includes listing the functions that the system is to perform and determining the tests that are to validate their operations.
2. Select the system components according to the specifications. This step includes locating the microcontroller that best suits the system as well as the other hardware components.
3. Design the system hardware. This step includes drawing the circuit diagrams.
4. Implement a prototype of the system hardware by means of breadboards, wire boards, or any other changeable implementation technology.

5. Develop, load, and test the software. Loading software into a PIC is referred to as "*burning*" or "*blowing*" the PIC.
6. Implement the final system and test hardware and software.

9.1 Integrated Development Environment

The MPLAB development system consists of a system of programs that run on a PC. This software package is designed to help develop, edit, test, and debug PIC code.

Installing the MPLAB package is straightforward and simple. The package includes the following components:

1. *MPLAB editor*. This tool allows creating and editing the assembly language source code. It behaves like any Windows editor and contains the standard editor functions, including cut-and-paste, search-and-replace, and undo and redo functions.
2. *MPLAB assembler*. The assembler reads the source file produced in the editor and generates either absolute or relocatable code. *Absolute code* executes directly in the PIC. *Relocatable code* can be linked with other separately assembled modules or with libraries.
3. *MPLAB linker*. This component combines modules generated by the assembler with libraries or other object files, into a single executable file in .hex format.
4. *MPLAB debuggers*. Several debuggers are compatible with the MPLAB development system. Debuggers are used to single-step through the code, breakpoint at critical places in the program, and watch variables and registers as the program executes. In addition to being a powerful tool for detecting and fixing program errors, debuggers provide an internal view of the processor; this is a valuable learning tool.
5. *MPLAB In-circuit emulators*. These are development tools that allow performing basic debugging functions while the processor is installed in the circuit.

Figure 9-1 (in the following page) is a screen image of the MPLAB program. The application on the editor window is one of the programs developed later in this book.

9.1.1 Installing MPLAB

In normal installation, the MPLAB executable is placed in the following path:

```
C:\Program Files\Microchip\MPASM Suite
```

Once the development environment is installed, the software is executed by clicking the MPLAB IDE icon. It is usually a good idea to drag and drop the icon onto the desktop so that the program can be easily activated.

With the MPLAB software installed, it may be a good idea to check that the applications were placed in the correct paths and folders. Failure to do so produces assembly-time failure errors with cryptic messages. To check the correct path for the software, open the **Project** menu and select the **Select Language Toolsuite** command. Figure 9-2 shows the command screen.

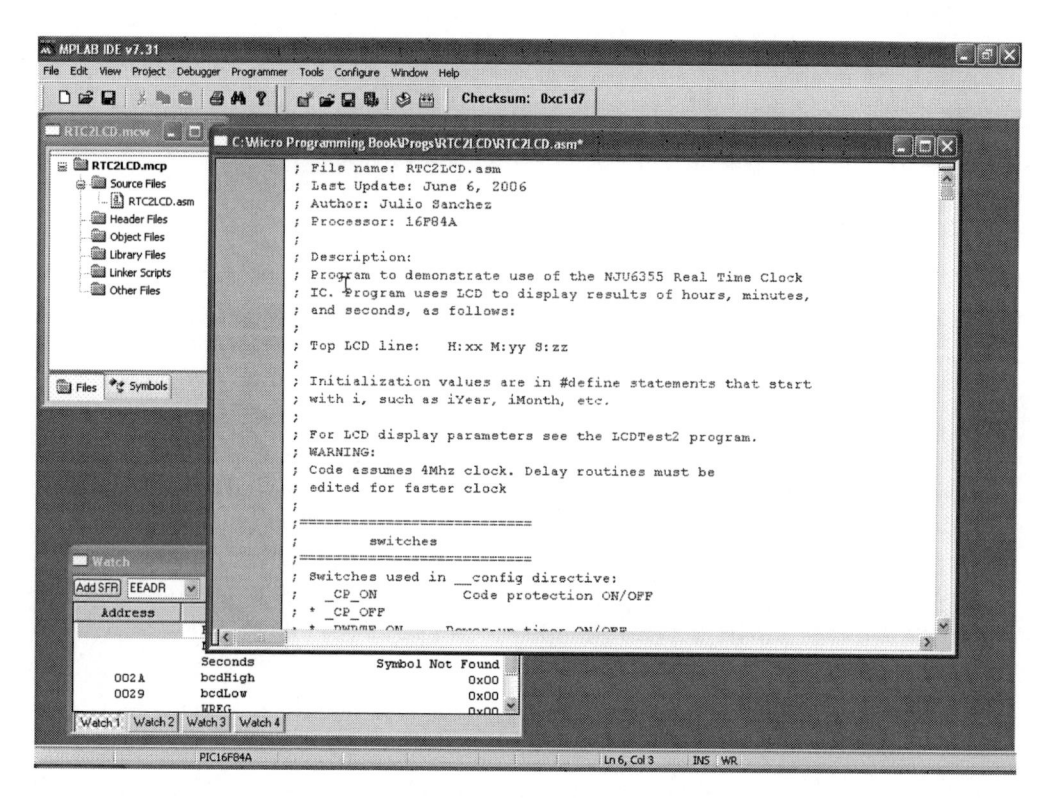

Figure 9-1 Screen Image of the MPLAB IDE

In the toolsuite window make sure that the file location coincides with the actual installation path for the software. If in doubt, use the **<Browse>** button to navigate through the installation directories until the executable program is located. In this case, **mpasmwin.exe**. Follow the same process for all the executables in the **Toolsuite Contents** window.

Figure 9-2 MPLAB Select Language Toolsuite Screen

PIC Programming: Tools and Techniques

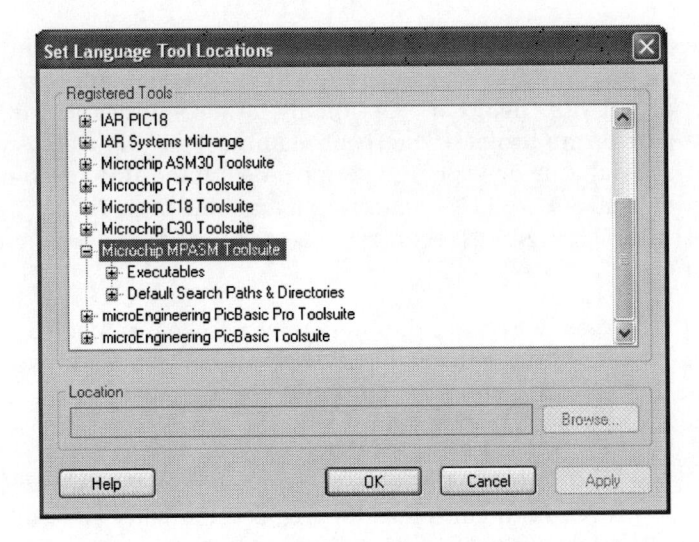

Figure 9-3 MPLAB Set Language Tools Locations Screen

A more detailed control over the location of the various individual tools is provided by the **Set Language Tools Location** command, also in the **Project** menu. This command allows setting the installation path not only to the major suites, but also to the individual tools. Figure 9-3 shows the display screen of this command.

9.1.2 Creating the Project

In MPLAB, a project is a group of files generated or recognized by the IDE. Figure 9-4 shows the structure of an assembly language project.

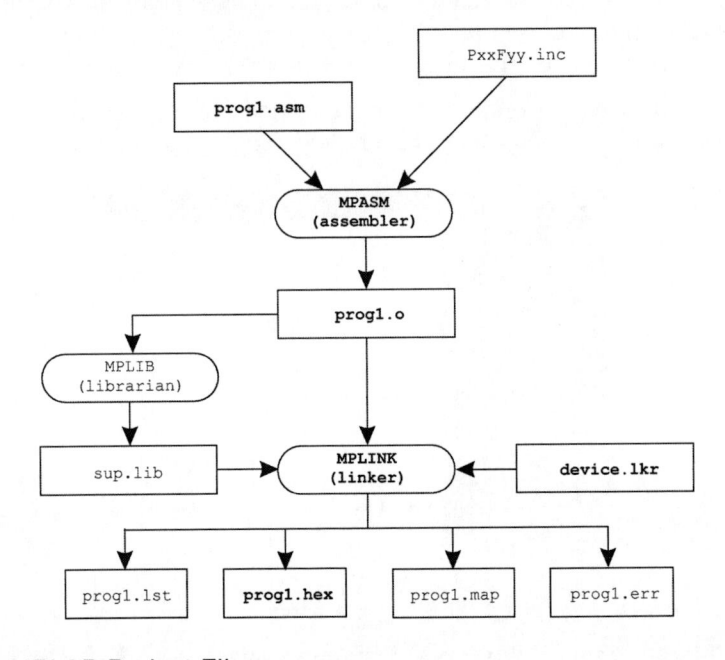

Figure 9-4 MPLAB Project Files

Figure 9-4 shows an assembly language source file (**prog1.asm**) and an optional processor-specific **include** file which are used by the assembler program (MPASM) to produce an object file (**prog1.o**). Optionally, other sources and other **include** files may form part of the project. The resulting object file, as well as one or more optional libraries, and a device-specific script file (**device.lkr**) are then fed to the linker program (MPLINK). MPLINK generates a machine code file (**prog1.hex**) and several support files with listings, error reports, and map files. The **.hex** file is used to blow the PIC.

In addition to the files in Figure 9-4, others may also be produced by the development environment according to the selected tools and options. For example, the assembler or the linker can generate a file with the extension **.cod** that contains symbols and references used in debugging.

Projects can be created using the **<New>** command in the **Project** menu. The programmer then proceeds to configure the project manually and add to it the required files. An alternative option, much to be preferred when learning the environment, is using the **<Project Wizard>** command in the **Project** menu. The wizard prompts you for all the decisions and options that are required, as follows:

1. Device selection. Here the programmer selects the PIC hardware for the project, for example 16F84A.

2. Select language toolsuite. This screen is the same one shown in Figure 9-2. Its purpose is to make sure that the proper development tools and paths are active.

3. Next, the wizard prompts the user for a project name and directory. It is possible to create a new directory at this time.

4. In the next step, the user is given the option of adding existing files to the project and renaming these files if necessary. This can be a useful option, since most projects reuse a template, an include file, or other preexisting resources.

5. Finally, the wizard displays a summary of the project parameters. When the user clicks on the **<Finish>** button, the project is created and programming can begin.

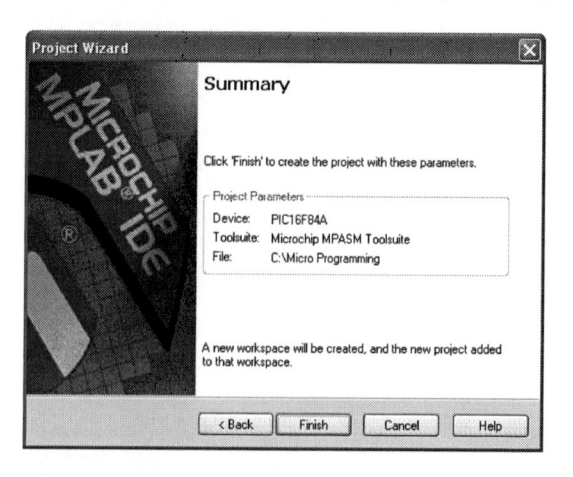

Figure 9-5 Final Screen of the Project Creation Wizard

9.1.3 Project Build Options

The **<Build Options: Project>** command in the **Project** menu allows the user to customize the development environment. Of the tabs available on the **Build Options** screen, the **MPASM Assembler** is probably the most used. The screen is shown in Figure 9-6.

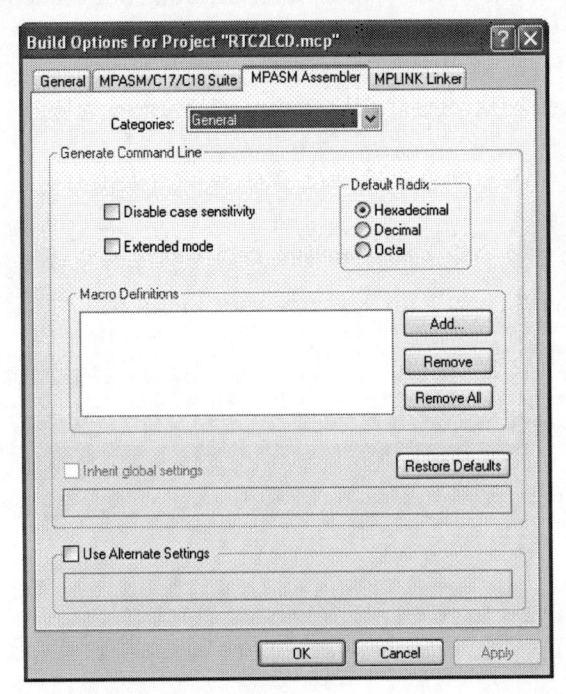

Figure 9-6 MPASM Assembler Tab in the Build Options Screen

The **MPASM Assembler** tab allows performing the following customizations:

1. Disable/enable case sensitivity. Normally the assembler is case-sensitive. Enabling this option turns all variables and labels to upper case.

2. Select the default radix. Numbers without formatting codes are assumed to be hex, decimal, or octal according to the selected option.

3. The **Macro Definition** window allows adding macro directives. *Macros* are discussed later in this chapter.

4. The **Use Alternate Settings** text box is provided for command line commands in non-GUI environments.

5. The **Restore Defaults** box turns off all custom configurations.

9.1.4 Building the Project

Once all the options have been selected, the installation checked, and the assembly language source file written or imported, the development environment builds the project. Building consists of calling the assembler, the linker, and any other support program in order to generate the files shown in Figure 9-4 and any others that result from a particular project or IDE configuration.

The *build process* is initiated by selecting the **<Build All>** command in the **Project** menu. Once the building concludes, a screen labeled **Output** is displayed showing the results of the build operation. If the build succeeded, the last line of the **Output** screen shows this result. Figure 9-7 shows the output screen after a successful build.

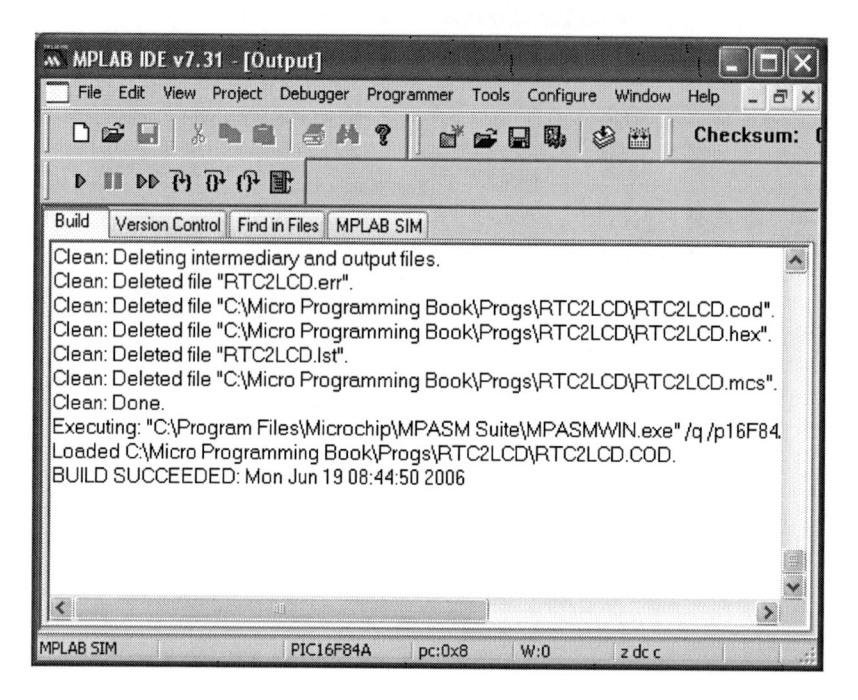

Figure 9-7 Output Window showing the Build Command Result

9.2 Simulators and Debuggers

In the context of MPLAB documentation the term *debugger* is reserved for hardware debuggers while the software versions are called *simulators*. Although this distinction is not always enforceable, we will abide by this terminology (whenever possible) in order to avoid confusion. The reader should note that there are MPLAB functions in which the IDE considers a simulator as a debugger.

The MPLAB standard simulator is called MPLAB SIM. SIM is part of the Integrated Development Environment and can be selected at any time. The hardware debuggers currently offered by Microchip are named ICD 2, ICE 2000, and ICE 4000. A simulator, as the term implies, allows simulating the execution of a program one instruction at a time and viewing file registers and symbols defined in the code. Debuggers, on the other hand, allow executing a program one step at a time or to a predefined breakpoint while the PIC is installed in the target system. This makes

PIC Programming: Tools and Techniques

possible realtime viewing of the processor's internals, and also the state of circuit components.

In the sections that follow we present an overview of PIC simulators and debuggers and their use.

9.2.1 MPLAB SIM

Microchip documentation describes the SIM program as a *discrete-event simulator*. SIM is part of the MPLAB IDE and is selected by clicking on the **<Select Tool>** command in the **Debugger** menu. The command offers several options, one of them being MPLAB SIM. Once the SIM program is selected, a special debug toolbar is displayed. The toolbar and its functions is shown in Figure 9-8.

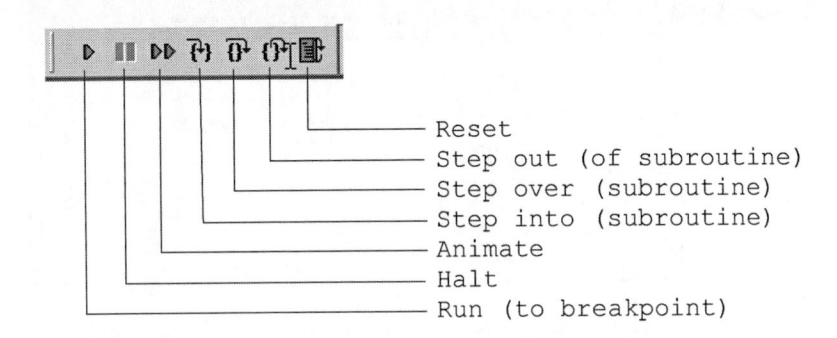

```
                    Reset
                    Step out (of subroutine)
                    Step over (subroutine)
                    Step into (subroutine)
                    Animate
                    Halt
                    Run (to breakpoint)
```

Figure 9-8 SIM Toolbar

In order for the simulator to work the program must first be successfully built. The most commonly used simulator methods are single-stepping through the code and breakpoints. A *breakpoint* is a mark at a program line at which the simulator stops and waits for user actions.

Breakpoints provide a way of inspecting program results at a particular place in the code. *Single-stepping* is executing the program one instruction at a time. The three buttons labeled **<Step...>** are used in single-stepping. The first one allows breaking out of a subroutine or procedure. The second one is for bypassing a procedure or subroutine while in step mode. The third one single steps into whatever line follows.

Breakpoints are set by double-clicking at the desired line while using the editor. The same action removes an existing breakpoint. Lines in which breakpoints have been placed are marked, on the left document margin, by a letter "B" enclosed in a red circle. Right-clicking while the cursor is on the program editor screen provides a context menu with several simulator-related commands. These include commands to set and clear breakpoints, to run to the cursor, and to set the program counter to the code location at the cursor.

The **View** menu contains several commands that provide useful features during program simulation and debugging. These include commands to program memory, file registers, EEPROM, and special function registers. One command in particular, named **<Watch>**, provides a way of inspecting the contents of FSRs and GPRs on the same screen. The **<Watch>** command displays a program window that contains references to all file registers used by the program. The user then selects which registers to view and these are shown in the **Watch** window. The **Watch** window is shown in Figure 9-9.

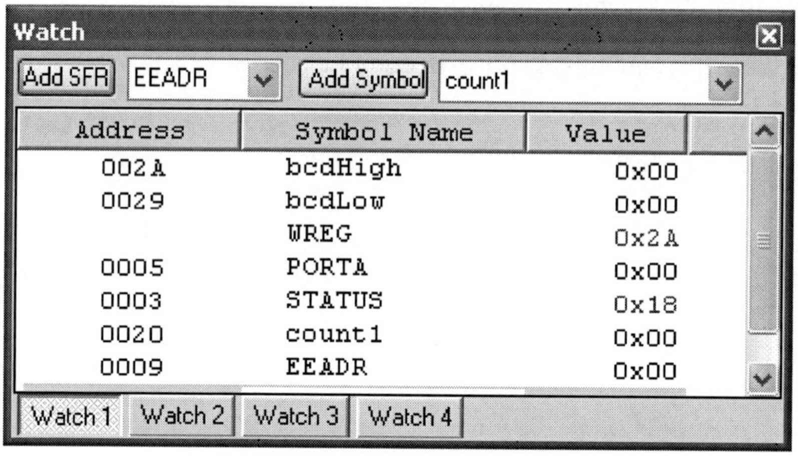

Figure 9-9 Use of Watch Window in MPLAB SIM

When the program is in the single-step mode or breakpoint modes, the contents of the various registers can be observed in the **Watch** window. Those that have changed since the last step or breakpoint are displayed in red. The user can click the corresponding arrows on the **Watch** window to display all the symbols or registers. The **<Add Symbol>** or **<Add FSR>** button is then used to display the item on the **Watch** screen. Four different **Watch** windows can be enabled, labeled **Watch 1** to **Watch 4** at the bottom of the screen in Figure 9-9.

Another valuable tool available from the View menu is the one labeled <Simulator Trace>. The Simulator Trace window provides a view of the machine instruction combined with a window that displays the source code. The Simulator Trace window is shown in Figure 9-10.

9.2.2 MPLAB Hardware Debuggers

A more powerful and versatile debugging tool is a hardware or in-circuit debugger. Hardware debuggers allow tracing, breakpointing, and single-stepping through code while the PIC is installed in the target circuit. The typical in-circuit debugger requires several hardware components, as shown in Figure 9-11.

PIC Programming: Tools and Techniques

Trace									
Line	Addr	Op	Label	Instruction	SA	SD	DA	DD	Cycles
0	0000	2808		GOTO 0x8	----	--	----	--	000000000000
1	0001	0000		NOP	----	--	----	--	000000000001
2	0008	3000	main	MOVLW 0	W	--	W	00	000000000002
3	0009	1683		BSF 0x3, 0x5	0003	18	0003	33	000000000003
4	000A	0085		MOVWF 0x5	----	--	0085	00	000000000004
5	000B	0086		MOVWF 0x6	----	--	0086	00	000000000005
6	000C	1283		BCF 0x3, 0x5	0083	38	0003	18	000000000006
7	000D	3000		MOVLW 0	W	--	W	00	000000000007
8	000E	0085		MOVWF 0x5	----	--	0005	00	000000000008
9	000F	0086		MOVWF 0x6	----	--	0006	00	000000000009

C:\MICRO PROGRAMMING BOOK\PROGS\RTC2LCD\RTC2LCD.ASM

```
138         org     0       ; start at address
139         goto    main
140 ; Space for interrupt handlers
141         org     0x08
142
143 main:
144         movlw   b'00000000' ; All lines to output
145         Bank1
146         movwf   TRISA     ; in port A
147         movwf   TRISB     ; and port B
148         Bank0
149         movlw   b'00000000' ; All outputs ports low
150         movwf   PORTA
151         movwf   PORTB
```

Figure 9-10 MPLAB SIM Simulator Trace Window

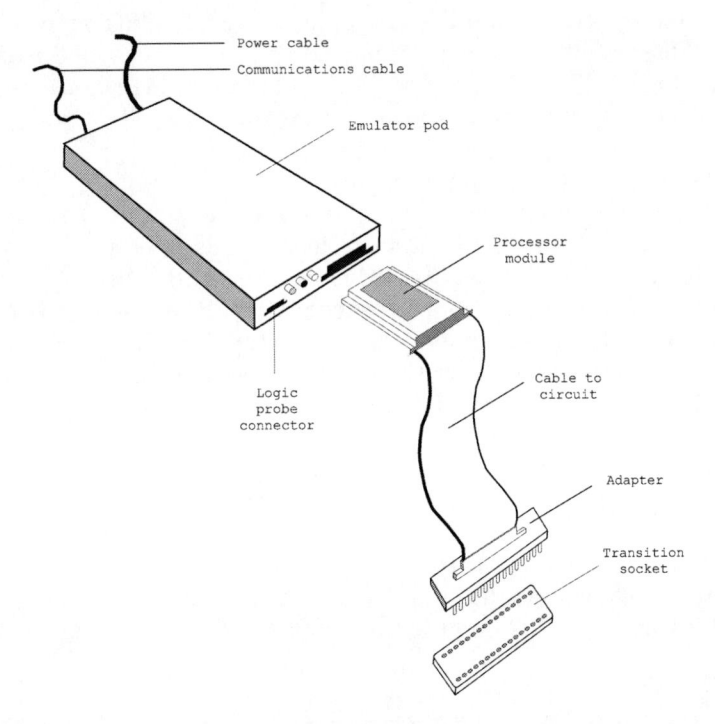

Figure 9-11 Components of a Typical Hardware Debugger

The emulator pod with power supply and communications cable provides the basic communications and functionality of the debugger. The communications line between the PC and the debugger can be an RS-232, a USB, or a parallel port line. The processor module fits into a slot at the front of the pod module. The processor is device-specific and provides these functions to the debugger. A flex cable connects the processor module to an interchangeable device adapter that allows connecting to the several PICs supported by the system. The transition socket allows connecting the device adapter to the target hardware. A separate socket allows connecting logic probes to the debugger.

Microchip provides two models of their in-circuit hardware debuggers, which they call *In-Circuit Emulators*, or *ICEs*. The ICE 2000 is designed to work with most PICs of the mid-range and lower series, while the ICE 4000 is for the PIC18x high-end family of PICs. Recently Microchip has released an in-circuit debugger designated as ICD 2 that offers many of the features of their full-fledged in-circuit emulators at a much reduced price. One of the disadvantages of the ICD 2 system is that it requires the exclusive use of some hardware and software resources in the target. Furthermore, the ICD requires that the system be fully functional. The ICEs, on the other hand, provide memory and clocks so that the processor can run code even if it is not connected to the application board.

9.2.3 A "Quick-and-Dirty" Debugger

The functionality of an actual hardware debugger can be replaced with a little ingenuity and a few lines of code. Most PICs are equipped with EEPROM memory. Programmers (covered in the following section) have the ability to read all the data stored in the PIC, including EEPROM. These two facts can be combined to obtain run-time information without resorting to a hardware debugger.

Suppose a defective application is suspected of not finding the expected value in a PIC port. The developer can write a few lines of code to store the port value on an EEPROM memory cell. An endless loop following this operation ensures that the stored value is not changed. Now the PIC is inserted in the circuit and the application executed. When the endless loop is reached, the PIC is removed from the circuit and placed back in the programmer. The value stored in EEPROM can now be inspected so as to determine the run-time state of the machine. In many cases, this simple trick is less complicated and time consuming than setting up a hardware debugger, even if such a device is available.

9.3 Programmers

In the context of microcontroller technology, a *programmer* is a device that allows transferring the program onto the chip. The process is called "burning" a PIC, or more commonly "blowing" a PIC. Most programmers have three components:

1. A software package that runs on the PC
2. A cable connecting the PC to the programmer
3. A programmer device

PIC Programming: Tools and Techniques

Dozens of PIC programmers are available on the Internet. When Microchip released the programming specifications of the PIC to the public without requiring a nondisclosure agreement, they originated a cottage industry. The commercial programmers on the Internet range from a "no parts" PIC programmer that has been around since 1998, to sophisticated devices costing hundreds of dollars and providing many additional features and refinements. For the average new PIC user, a nice USB programmer with a ZIF (zero insertion-force) socket and the required software can be purchased for about $50.00. Build-it-yourself versions are available for about half this amount.

An alternative programmer is made possible by the fact that some of the newer flash-based PICs can write to their own program memory. This allows placing a small bootloader program in PIC memory which loads an application over the RS-232 or USB lines.

Figure 9-12 is a screen capture of the driver software for a popular programmer from MicroPro.

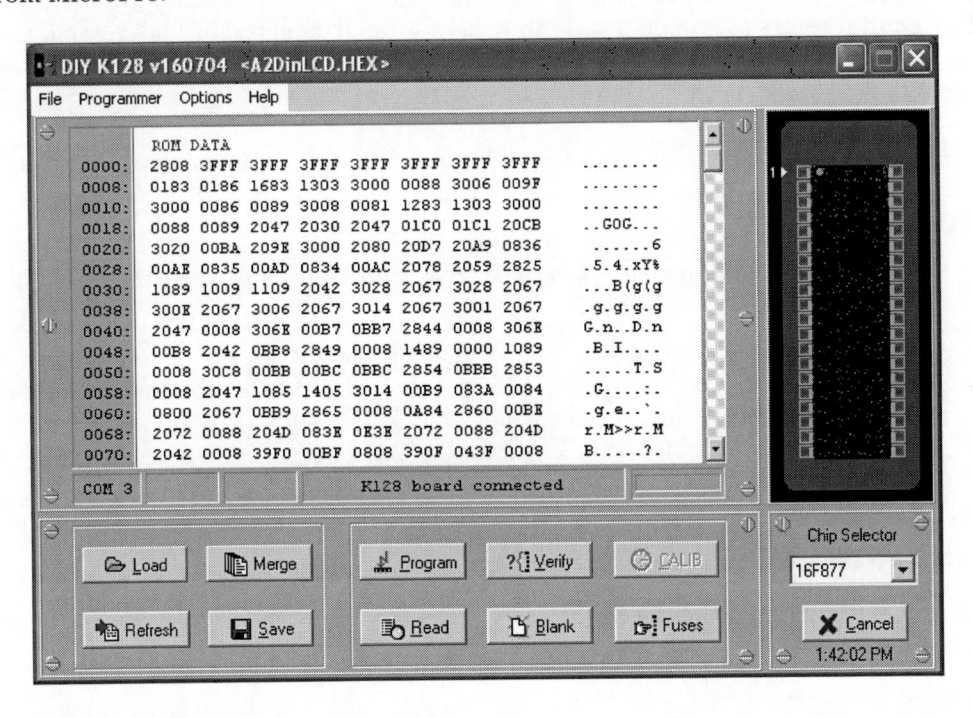

Figure 9-12 Control Program for the DIY MicroPro Programmer

9.4 Engineering PIC Software

The program developer's main challenge is writing code that performs the task at hand. In this context this means writing a PIC assembly language program that assembles without errors (usually after some effort) and makes the circuit perform as intended. We have already reviewed the IDE (integrated development environment) and the various hardware components and software tools. We now focus on the various elements that are used in developing the program itself.

9.4.1 Using Program Comments

One of the first realizations of beginning programmers is how quickly we forget the reasoning and logic that went into our code. It is common that a few weeks, even a few hours, after we coded a routine we find that what was obvious then is now undecipherable and that the ideas that were clear in our minds a short time ago now evade our understanding. The only solution is to write good program comments that explain, not the elementary, but the trains of thought behind our code.

In PIC assembly language, the comment symbol is the *semicolon* (;). The presence of a semicolon indicates to the assembler that everything that follows, to the end of the line, must be ignored. Using comments judiciously and with good taste is the mark of the expert software engineer. Programs with few, cryptic, or confusing comments fall into the category of "spaghetti code." In programming lingo, "spaghetti code" refers to a coding style that cannot be deciphered or understood. One of the worst offenses that can be said about one's programming style is that it is spaghetti code.

How we use comments to explain our code or even to embellish it is a matter of personal preference. However, there are certain common sense rules that should always be considered:

1. Do not use program comments to explain the programming language or reflect on the obvious.
2. Abstain from humor in comments. Comedy has a place in the world but it is not in programs. By the same token, stay away from vulgarity, racial or sexist remarks, and anything that could be offensive. You can never anticipate who will read your code.
3. Write short, clear, readable comments that explain how the program works. Decorate or embellish your code using comments according to your tastes.

Program Header

Every program should have a commented header that contains the following information:

1. Program name
2. Programmer's or software company's name
3. Copyright notice, if pertinent
4. Target device or hardware
5. Development environment
6. Development dates
7. Program description

Some of these elements allow various levels of detail. For example, the target device can be a simple reference to the PIC for which the program is written, a more-or-less detailed description of the target system, or a reference to a circuit diagram or board drawing. The development environment can also be described briefly or in detail. The date element can be a single entry that lists the first or the last program change, or a detailed description of all program changes, tests, and updates. The program description can be a short sentence or a mini-manual on using the application. In any case, the level of detail and the contents of each category are determined by the programmer's style and the complexity and purpose of the application.

The following lines show the header of one of the programs developed for this book:

```
; File name: RTC2LCD.asm
; Last Update: June 6, 2006
; Author: Julio Sanchez
; Processor: 16F84A
;
; Description:
; Program to demonstrate use of the NJU6355 Real Time Clock
; IC. Program uses LCD to display results in hours, minutes,
; and seconds, as follows:
;
; Top LCD line:    H:xx M:yy S:zz
;
; Initialization values are in #define statements that start
; with i, such as iYear, iMonth, etc.
;
; For LCD display parameters see the LCDTest2 program.
;
; WARNING:
; Code assumes 4Mhz clock. Delay routines must be
; edited for faster clock
```

Commented Banners

Often, we need to scroll through the code in search of a particular line or routine. Having banners that signal critical places in the program facilitates this search. Banners are created using comments and a framing symbol, as in the following code fragment:

```
;================================
;   first text string procedure
;================================
storeMS1:
; Procedure to store in PIC RAM buffer the message
; contained in the code area labeled msg1
; ON ENTRY:
;         variable pic_ad holds address of text buffer
;         in PIC RAM
;         w register hold offset into storage area
;         msg1 is routine that returns the string characters
;         and a zero terminator
;         index is local variable that hold offset into
;         text table. This variable is also used for
;         temporary storage of offset into buffer
```

Sometimes, the programmer needs to emphasize a program area with a large banner that extends from margin to margin, as follows:

```
;=================================================================
;=================================================================
;              L O C A L     P R O C E D U R E S
;=================================================================
;=================================================================
;=========================
; init LCD for 4-bit mode
;=========================
initLCD:
; Initialization for Densitron LCD module as follows:
;    4-bit interface
;    2 display lines of 20 characters each
;    cursor on
;    left-to-right increment
;    cursor shift right
;    no display shift
```

Commented Bitmaps

It is also possible to use comments to signal the function of bit fields and individual bits of an operand, as in the following code fragment:

```
; OPTION_REG bitmap
;    7  6  5  4  3  2  1  0 <= OPTION bits
;    |  |  |  |  |  |__|__|_____ PS2-PS0 (prescaler bits)
;    |  |  |  |  |               Values for Timer0
;    |  |  |  |  |                000 = 1:2      001 = 1:4
;    |  |  |  |  |                010 = 1:8      011 = 1:16
;    |  |  |  |  |                100 = 1:32     101 = 1:64
;    |  |  |  |  |                110 = 1:128  *111 = 1:256
;    |  |  |  |  |_____ PSA (prescaler assign)
;    |  |  |  |                   *1 = to WDT
;    |  |  |  |                    0 = to Timer0
;    |  |  |  |_____ TOSE (Timer0 edge select)
;    |  |  |                      *0 = increment on low-to-high
;    |  |  |                       1 = increment in high-to-low
;    |  |  |_____ TOCS (TMR0 clock source)
;    |  |                         *0 = internal clock
;    |  |                          1 = RA4/TOCKI bit source
;    |  |_____ INTEDG (Edge select)
;    |                             *0 = falling edge
;    |_____ RBPU (Pullup enable)
;                                  *0 = enabled
;                                   1 = disabled
; * indicates options selected
      movlw        b'00001000'  ; Value installed
      movwf        OPTION_REG
```

Clearly commented bitmaps, banners, and many other code embellishments do not add to the quality and functionality of the code. It is quite possible to write very sober and functional programs without using them. The decision of how to comment and embellish programs is one of style.

9.4.2 Defining Data Elements

Most programs require the use of general purpose file registers. These registers are allocated to memory addresses reserved for this purpose in the PIC architecture, as shown in Figure 8-8 and Figure 8-9. Since the areas at these memory locations are already reserved for use as GPRs, the program can access the location either by address or by assigning to that address a name. The **equ** (equate) directory performs this function, as follows:

```
var1 equ   0x0c      ; Name var1 is assigned to location 0x0c
```

Actually, the name (in this case *var1*) becomes an alias for the memory address to which it is linked. From this point on, program code can access the memory cell at address 0x0c as follows:

```
    movf    var1,w   ; Contents of var1 to w
```

or:

```
    movf    0x0c,w   ; Same variable to w
```

In addition to the **equ** directive, PIC assembly language recognizes the C-like **#define** directive, so the name assignation could be done as follows:

```
#define var1 0x0c
```

Although most of the time-named variables are to be preferred to hard-coding addresses, there are times when we need to access an internal element of some multi-byte structure. In these cases, the hard-coded form could be convenient, although not absolutely necessary.

The cblock Directive

Another way of defining memory data is by using one of the data directives available in PIC assembly language. Although there are several of these, perhaps the most useful is the **cblock** directive. The **cblock** directive specifies an address for the first item and other items listed are allocated from this first address. The group ends with the endc directive. The following code fragment shows the use of the **cblock/endc** directives.

```
; Reserve 20 bytes for string buffer
    cblock    0x20
    strData
    endc

; Reserve three bytes for ASCII digits
    cblock    0x34
    asc100
    asc10
    asc1
    endc
```

In reality, the **cblock** directive defines a group of constants which are assigned consecutive addresses in RAM. In the previous code fragment the allocation of 20 bytes for the buffer named **strData** is illusory since no memory is actually reserved. The illusion works because the second **cblock** starts at address 0x34 which is 20 bytes after **strData**, and also because the programmer abstains from allocating other variables in the buffer space.

9.4.3 Banking Techniques

Having to deal with memory banks is one of the aggravations of PIC programming. Banks are numbered starting with bank 0. All PICs of the mid-range family have at least two banks, so bank shifting operations are virtually unavoidable. The issue is more how to switch bank designation since there are several possible techniques.

Bank selection is by means of bit RP0 and RP1 in the STATUS register. In mid-range PICs with four banks, the various combinations are as shown in Table 9.1.

Table 9.1
STATUS Register Bank Selection Bits

RP1	RP0	BANK	ADDRESS RANGE
1	1	*Bank 3	0x180 - 0x1ff
1	0	*Bank 2	0x100 - 0xx17f
0	1	Bank 1	0x80 - 0xff
0	0	Bank 0	0x00 - 0x7f

* RP1 bit is not used in devices with two banks

The most direct way to select the current bank is by clearing or setting the corresponding bits in the STATUS register. For example, to select bank 2 in a 4-bank device you could code:

```
    bsf     STATUS,6    ; Set bit 6 in STATUS register
    bcf     STATUS,5    ; Clear bit 5
```

The banksel Directive

Alternatively the application can use the **banksel** directive which selects the bank in which a particular register is located. For example, to select the bank in which the **ADCON1** register is located code could be as follows:

```
    banksel    ADCON1
```

The **banksel** directive also works with registers defined by the user (GPRs).

Bank Selection Macros

An alternative way of performing bank selection is by coding the corresponding bank select macros. A *macro* is an assembler structure that allows defining a series of instructions inserted in the code every time the macro is referenced. The PIC macro language defines the following format:

```
    label macro [arg1, arg2... argn]
        .
        .
        .
    endm
```

The ellipses are placeholders for the PIC instructions, assembler directives, macro directives, and macro calls. Macros are usually defined at the beginning of the program since forward references to macros are not allowed. The optional arguments passed to the macro (*arg1*, *arg2*, etc) are assigned values when the macro is

PIC Programming: Tools and Techniques

invoked. For example, the following macros make the corresponding bank selections in a mid-range PIC with four banks.

```
; Macros to select the register banks
Bank0           MACRO                   ; Select RAM bank 0
        bcf     STATUS,RP0
        bcf     STATUS,RP1
        ENDM

Bank1           MACRO                   ; Select RAM bank 1
        bsf     STATUS,RP0
        bcf     STATUS,RP1
        ENDM

Bank2           MACRO                   ; Select RAM bank 2
        bcf     STATUS,RP0
        bsf     STATUS,RP1
        ENDM

Bank3           MACRO                   ; Select RAM bank 3
        bsf     STATUS,RP0
        bsf     STATUS,RP1
        ENDM
```

Once the bank switching macros have been defined, the application can change banks simply by calling the macro name; for example, if we know that the **ADCON1** register is in bank 1 we can select the bank by calling:

```
    Bank1
```

At this point in the code the macro expansion inserts the corresponding operations to make the switch.

Which method to use when switching banks is a matter of personal preference and program constraints. Setting and clearing the RP1/RP0 bits is simple enough, but can be error-prone. Using the **banksel** directive is convenient since we do not need to know in which bank the item is located. The objection to using **banksel** is that some unnecessary bank changes may take place. For example, if the program is already in bank 1 and the **banksel** directive appears with a register file in that same bank, bank switching is generated.

The use of bank selection macros seems like a suitable method for most conditions. One advantage of the macro approach is that programs for different PICs can have their own banking macros. This way code can be easily ported to a different architecture.

Deprecated Banking Instructions

Several instructions in the mid-range instruction set have been deprecated and are no longer recommended by Microchip. These instructions are **tris** and **option**. Microchip's reason for not recommending these instructions is to maintain compatibility with future mid-range products. From a programmer's viewpoint, it is difficult to see why using these instructions may be undesirable. In the unlikely case that code using **tris** or **option** is ported to a future device that does not support them, it will be easy enough to modify.

The **tris** and **option** instructions are convenient since they allow loading the contents of the w register to the OPTION, TRISA, and TRISB registers directly, without bank concerns. For example, the following code fragment sets port line 1 to input and all others to output:

```
movlw    b'00000010'    ; Line 1 is input
tris     PORTA
```

We continue to use the deprecated instructions in programs in which there is no concern about future consequences. In programs in which portability is an issue, we use the banking macros discussed previously.

9.4.4 Processor and Configuration Controls

PIC programs must define the processor to be used by the development software. The processor directive **assembler** (and also the **list** directive) allows defining the PIC type. For example, a program for the 16F877 would contain the following line:

```
processor 16f877
```

Configuration Bits

The PIC microcontrollers contain a special register called the *configuration register*. The bits in this register allow customizing certain processor features. These bits are mapped to program memory location 0x2007. This memory location can be accessed only during the programming mode, so the bits cannot be changed during normal program operation. The configuration bits cannot be read at runtime.

Microchip recommends that the configuration bits be set by means of the **__config** directive. The bits are mapped as follows:

```
CP1:CP0: Code Protection bits
    11 = Code protection off
    10 = See device data sheet
    01 = See device data sheet
    00 = All memory is code protected
```

Some devices use different numbers of bits to determine the level of code protection. Some use a single bit. In this case, the encoding is as follows:

```
1 = Code protection off
0 = Code protection on
```

DP: Data EEPROM Memory Code Protection bit

```
1 = Code protection off
0 = Data EEPROM Memory is code protected
```

BODEN: Brown-Out Reset Enable bit

```
1 = BOR enabled
0 = BOR disabled
```

PIC Programming: Tools and Techniques

Enabling *Brown-out Reset* automatically enables PWRT (the *Power-up Timer*) regardless of the value of bit *PWRTE*. The Power-up Timer must be enabled any time that the Brown-out Reset is enabled.

```
PWRTE: Power-up Timer Enable bit
    1 = PWRT disabled
    0 = PWRT enabled
See note about the BODEN bit.
MCLRE: MCLR Pin Function Select bit
    1 = Pin's function is MCLR
    0 = Pin's function is as a digital I/O.
        MCLR is internally tied to VDD.
WDTE: Watchdog Timer Enable bit
    1 = WDT enabled
    0 = WDT disabled
FOSC1:FOSC0: Oscillator Selection bits
    11 = RC oscillator
    10 = HS oscillator
    01 = XT oscillator
    00 = LP oscillator
FOSC2:FOSC0: Oscillator Selection bits
    111 = EXTRC oscillator, with CLKOUT
    110 = EXTRC oscillator
    101 = INTRC oscillator, with CLKOUT
    100 = INTRC oscillator
    011 = Reserved
    010 = HS oscillator
    001 = XT oscillator
    000 = LP oscillator
```

The **__config** directive is used to embed configuration data in the source file. Alternatively, the configuration bits can be set at the time the PIC is blown. The following code fragment shows setting the configuration bits for a 16F877 PIC:

```
; Switches used in __config directive:
;   _CP_ON            Code protection ON/OFF
; * _CP_OFF
; * _PWRTE_ON         Power-up timer ON/OFF
;   _PWRTE_OFF
;   _BODEN_ON         Brown-out reset enable ON/OFF
; * _BODEN_OFF
; * _PWRTE_ON         Power-up timer enable ON/OFF
;   _PWRTE_OFF
;   _WDT_ON           Watchdog timer ON/OFF
; * _WDT_OFF
;   _LPV_ON           Low voltage IC programming enable ON/OFF
; * _LPV_OFF
;   _CPD_ON           Data EE memory code protection ON/OFF
; * _CPD_OFF
; OSCILLATOR CONFIGURATIONS:
;   _LP_OSC           Low power crystal osccillator
;   _XT_OSC           External parallel resonator/crystal ocillator
; * _HS_OSC           High speed crystal resonator
;   _RC_OSC           Resistor/capacitor oscillator
;    |                  (simplest, 20% error)
;    |_____ * indicates setup values presently selected
       __CONFIG _CP_OFF & _WDT_OFF & _BODEN_OFF & _PWRTE_ON & _HS_OSC &
_WDT_OFF & _LVP_OFF & _CPD_OFF
```

9.4.5 Naming Conventions

The programmer must decide on the conventions to be followed for program labels and variable (register) names. The MPLAB assembler is case sensitive by default, so PORTB and portb can refer to different registers.

Using the **equ** or **#define** directives, the programmer can define all of the registers (SFRs and GPRs) used by an application. A safer approach is to import an **include** file (**.inc** extension) furnished in the MPALB package for each different PIC. The **include** files have the names of all SFRs and bits used by a particular device. The following code fragment is a listing of the MPLAB **include** file for the 16f84a:

```
        LIST
; P16F84A.INC  Standard Header File, Version 2.00
; Microchip Technology, Inc.
        NOLIST
; This header file defines configurations, registers, and other
; useful bits of information for the PIC16F84 microcontroller.
; These names are taken to match the data sheets as closely as
; possible.
; Note that the processor must be selected before this file is
; included.  The processor is selected by using:
;       1. Command line switch:
;               C:\ MPASM MYFILE.ASM /PIC16F84A
;       2. LIST directive in the source file
;               LIST    P=PIC16F84A
;       3. Processor Type entry in the MPASM full-screen interface
;==================================================================
;
;       Revision History
;
;==================================================================

;Rev:   Date:    Reason:

;1.00   2/15/99 Initial Release

;==================================================================
;
;       Verify Processor
;
;==================================================================

        IFNDEF __16F84A
           MESSG "Processor-header file mismatch.  Verify selected
 processor."
        ENDIF

;==================================================================
;
;       Register Definitions
;
;==================================================================

W                       EQU     H'0000'
F                       EQU     H'0001'

;-- Register Files------------------------
```

```
INDF            EQU     H'0000'
TMR0            EQU     H'0001'
PCL             EQU     H'0002'
STATUS          EQU     H'0003'
FSR             EQU     H'0004'
PORTA           EQU     H'0005'
PORTB           EQU     H'0006'
EEDATA          EQU     H'0008'
EEADR           EQU     H'0009'
PCLATH          EQU     H'000A'
INTCON          EQU     H'000B'

OPTION_REG      EQU     H'0081'
TRISA           EQU     H'0085'
TRISB           EQU     H'0086'
EECON1          EQU     H'0088'
EE
Z               EQU     H'0002'
DC              EQU     H'0001'
C               EQU     H'0000'

;---- INTCON Bits ----------------------------------

GIE             EQU     H'0007'
EEIE            EQU     H'0006'
T0IE            EQU     H'0005'
INTE            EQU     H'0004'
RBIE            EQU     H'0003'
T0IF            EQU     H'0002'
INTF            EQU     H'0001'
RBIF            EQU     H'0000'

;---- OPTION_REG Bits-------------------------------

NOT_RBPU        EQU     H'0007'
INTEDG          EQU     H'0006'
T0CS            EQU     H'0005'
T0SE            EQU     H'0004'
PSA             EQU     H'0003'
PS2             EQU     H'0002'
PS1             EQU     H'0001'
PS0             EQU     H'0000'

;---- EECON1 Bits ----------------------------------

EEIF            EQU     H'0004'
WRERR           EQU     H'0003'
WREN            EQU     H'0002'
WR              EQU     H'0001'
RD              EQU     H'0000'

;==================================================================
;
;       RAM Definition
;
;==================================================================

        __MAXRAM H'CF'
        __BADRAM H'07', H'50'-H'7F', H'87'
```

```
;================================================================
;
;       Configuration Bits
;
;================================================================
_CP_ON              EQU     H'000F'
_CP_OFF             EQU     H'3FFF'
_PWRTE_ON           EQU     H'3FF7'
_PWRTE_OFF          EQU     H'3FFF'
_WDT_ON             EQU     H'3FFF'
_WDT_OFF            EQU     H'3FFB'
_LP_OSC             EQU     H'3FFC'
_XT_OSC             EQU     H'3FFD'
_HS_OSC             EQU     H'3FFE'
_RC_OSC             EQU     H'3FFF'
```

Names in the **include** file are defined in all-capital letters. It is probably a good idea to adhere to this style instead of creating alternate names in lower case. The C-like **#include** directive is used to refer the **.inc** files at assembly time, for example:

```
#include <p16f84a.inc>
```

9.4.6 Errorlevel Directive

This directive allows controlling the warning and error messages produced at assembly and link times. One particular type of warning can be disturbing: those that refer to bank changes. Applications often turn off bank change related warning with the following line:

```
    errorlevel -302
```

9.5 Pseudo Instructions

Sometimes a code listing contains instructions that are not part of the standard set for the particular device. The reason this happens is that MPLAB includes a set of pseudo-instructions for 12- and 14-bit devices. Table 9.2 lists these pseudo-instructions and their standard equivalents:

Table 9.2
PIC Pseudo Instructions

MNEMONIC	DESCRIPTION	EQUIVALENT OPERATION(S)	STATUS BIT CHANGED
ADDCF f,d	Add Carry to File Register	BTFSC 3,0	Z
		INCF f,d	
ADDDCF f,d	Add Digit Carry to File Register	BTFSC 3,1	Z
		INCF f,d	
B k	Branch	GOTO k	-
BC k	Branch on Carry	BTFSC 3,0	-
		GOTO k	

(continues)

PIC Programming: Tools and Techniques

Table 9.2
PIC Pseudo Instructions

MNEMONIC	DESCRIPTION	EQUIVALENT OPERATION(S)	STATUS BIT CHANGED
BDC k	Branch on Digit Carry	BTFSC GOTO k	3,1 -
BNC k	Branch on No Carry	BTFSS 3,0 GOTO k	-
BNDC k	Branch on No Digit Carry	BTFSS GOTO k	3,1 -
BNZ k	Branch on No Zero	BTFSS GOTO k, 2	3,2 -
BZ k	Branch on Zero	BTFSC 3,2 GOTO k	-
CLRC	Clear Carry	BCF 3,0	-
CLRDC	Clear Digit Carry	BCF 3,1	-
CLRZ	Clear Zero	BCF 3,2	-
LCALL k	Long Call	BCF/BSF 0x0a,3 BCF/BSF 0x0a,4 CALL k	
LGOTO k	Long GOTO	BCF/BSF 0x0a,3 BCF/BSF 0x0a,4 GOTO k	
MOVFW f	Move File to W	MOVF f,0	Z
NEGF f,d	Negate File	COMF f,1 INCF f,d	-
SETC	Set Carry	BSF 3,0	-
SETDC	Set Digit Carry	BSF 3,1	-
SETZ	Set Zero	BSF 3,2	-
SKPC	Skip on Carry	BTFSS 3,0	-
SKPDC	Skip on Digit Carry	BTFSS 3,1	-
SKPNC	Skip on No Carry	BTFSC 3,0	-
SKPNDC	Skip on No Digit Carry	BTFSC 3,1	-
SKPNZ	Skip on Non Zero	BTFSC 3,2	-
SKPZ	Skip on Zero	BTFSS 3,2	-
SUBCF f,d	Subtract Carry from File	BTFSC 3,0 DECF f,d	Z
SUBDCF f,d	Subtract Digit Carry from File	BTFSC 3,1 DECF f,d	Z
TSTF f	Test File	MOVF f,1	Z

We have listed the PIC pseudo-instructions to provide a reference. In our programming we prefer to stay away from using them since they tend to make code less readable. Microchip recommends not using the pseudo-instructions.

Chapter 10

Programming Essentials: Input and Output

In this chapter, we discuss the simplest circuits and programming operations. Using a PIC to control an LED or read a switch is as elementary as it gets. However, neither of these operations is trivial, since there is more to it than a few lines of code. Other input/output devices that are also considered are seven-segment LED displays and multiple switches, sometimes called *toggle switches*. A bank of multiple LEDs can also function as a *binary output device*.

10.0 16F84A Programming Template

We have found that program development can be simplified considerably by using *code templates*. A code template is a program devoid of functionality that serves to implement the most common and typical features of an application. The template not only saves the effort of redoing the same tasks, but reminds the programmer of program elements that could otherwise be forgotten. A professional developer will have collected many different templates over the years for different types of applications on various processors. The following template is for the 16F84A PIC:

```
;==============================================================
; File name:
; Date:
; Author:
; Processor:
; Reference circuit:
;==============================================================
; Copyright notice:
;==============================================================
; Program Description:
;
;===========================
; configuration switches
;===========================
; Switches used in __config directive:
```

```
;       _CP_ON          Code protection ON/OFF
; *     _CP_OFF
; *     _PWRTE_ON       Power-up timer ON/OFF
;       _PWRTE_OFF
;       _WDT_ON         Watchdog timer ON/OFF
; *     _WDT_OFF
;       _LP_OSC         Low power crystal oscillator
; *     _XT_OSC         External parallel resonator
;       _HS_OSC         High speed crystal resonator (8 to 10 MHz)
;       _RC_OSC         Resistor/capacitor oscillator
;                       (simplest, 20% error)
;               |
;               |_____ * indicates setup values

;=========================
; setup and configuration
;=========================
        processor 16f84A
        include   <p16f84A.inc>
        __config  _XT_OSC & _WDT_OFF & _PWRTE_ON & _CP_OFF


;=============================================================
;                   constant definitions
;=============================================================
;=============================================================
;                   PIC register equates
;=============================================================
;=============================================================
;                   variables in PIC RAM
;=============================================================
        cblock    0x0c
        endc


;=============================================================
;                           program
;=============================================================
                org             0         ; start at address
                goto    main
; Space for interrupt handlers
                org             0x08

main:

;=============================================================
        end                     ; END OF PROGRAM
;=============================================================
```

In addition to the *template file* the program developer should keep at hand the necessary *include files*. In this case, **p16f84a.inc**.

10.1 Introducing the 16F84A

The circuits and programs in this chapter use the 16F84A, probably the most popular of all mid-range PIC microcontrollers. Although we have discussed the mid-range architecture, we start with a review of this processor in order to establish a base for the material that follows.

10.1.1 Template Circuit for 16F84A

Like the programmer uses a programming template for developing 16F84A code, the circuit designer uses a *template circuit*. This circuit contains the components that most 16F84A boards require. The elements include a diagram of the PIC itself with the pin-out, as well as the wiring of the standard components, including the power source, ground, the *reset pin* (MCLR), and the most commonly used oscillator. Figure 10-1 shows a circuit template for the 16F84A.

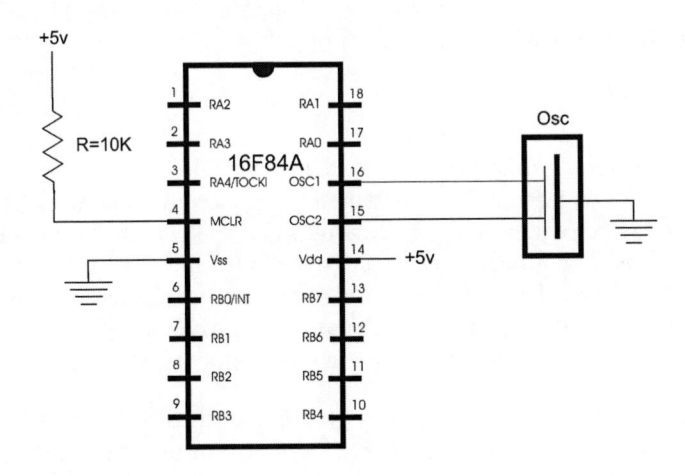

Figure 10-1 16F84A Circuit Template

The circuit template in Figure 10-1 does not suit every possible circuit. Even the simplest components must sometimes be configured differently; for example, the reset line could be wired to a pushbutton switch, or a different oscillator may be used. In any case, it is always easier to make modifications to an existing base than to start from scratch every time.

10.1.2 Power Supplies

Every PIC-based circuit board requires a +5V power source. A possible source of power is one or more batteries. There is an enormous selection of battery types, sizes, and qualities. The most common ones for use in experimental circuits are listed in Table 10.1.

Table 10.1
Common Dry Cell Alkaline Battery Types

DESIGNATION	VOLTAGE MM.	LENGTH MM.	DIAMETER
D	1.5	61.5	34.2
C	1.5	50	26.2
AA	1.5	50	14.2
AAA	1.5	44.5	10.4
AAAA	1.5	42.5	8.3

All of the batteries in Table 10.1 produce 1.5V. A PIC with a supply voltage of 2 to 6 volts uses two to four batteries. Note that in selecting the battery power source for a PIC-based circuit, other elements beside the microcontroller itself must be considered, such as the oscillator. Holders for several interconnected batteries are available at electronic supply sources.

Alternatively, the power supply can be a transformer with 120VAC input and 3 to 12VDC called *AC/DC adapters*. The most useful type for the experimenter are the ones with an ON/OFF switch and several selectable *output voltages*. Color-coded alligator clips at the output wires are convenient.

Voltage Regulator

A useful device for a typical PIC-based power source is a *voltage regulator IC*. The 7805 voltage regulator is ubiquitous in most PIC-based boards with AC/DC adapter sources. The IC is a three-pin device whose purpose is to ensure a stable voltage source which does not exceed the device rating. The 7805 is rated for 5V and produces this output from any input source in the range 8 to 35V. Since the excess voltage is dissipated as heat the 7805 is equipped with a metallic plate intended for attaching a *heat sink*. The heat sink is not required in a typical PIC application but it is a good idea to maintain the supply voltage closer to the device minimum rather than its maximum.

The voltage regulator circuit requires two capacitors: one electrolytic and the other one not. Figure 10-2 shows a power source circuit using the 7805.

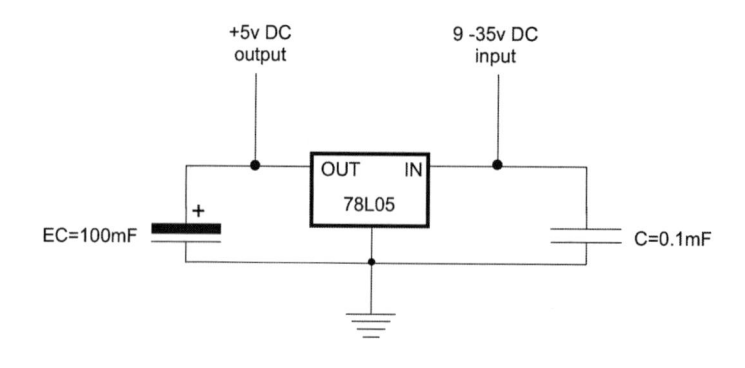

Figure 10-2 Voltage Stabilizer Circuit

10.1.3 Comparisons in PIC Programming

The power and usefulness of programs is due, in great measure, to their decision-making ability, and decisions are based on comparison. In a comparison code, it is able to make decisions based on the relative values of two operands. For example, compare the values a and b. If a is greater than b execute a certain code routine. If b is greater than a, execute another one. If both operands have the same value then proceed to a third code branch.

CISC and even some *RISC* microprocessors contain a *compare* operator in their instruction set. However, the compare can be substituted, with some inconvenience, by a *subtraction*. Since there is no compare operation in the PIC instruction set, we have to simulate the comparison by subtracting the w register from a literal value or from a file register. The **sublw** and **subwf** instructions can be used. After the subtraction takes place, code can make decisions based on the state of the zero and the carry flags. For example, the following code fragment compares the value in the two registers, labeled OP1 and OP2 respectively, and directs execution to three possible routines:

```
; Declare variables at 2 memory locations
OP1       equ       0x0c      ; First operand
OP2       equ       0x0d      ; second operand
.
.
.
main:
          movlw     0x30      ; First operand
          movwf     OP1       ; to OP1 register
          movlw     0x50      ; Second operand
          movwf     OP2       ; To OP2 register
          movf      OP2,w     ; OP2 to w register (not really
                              ; necessary)
          subwf     OP1,w     ; Subtract w (OP2) from OP1
          btfsc     STATUS,2  ; 2 is zero bit. Test zero flag.
                              ; Skip next instruction if Z bit = 0,
                              ; that is if both numbers are not the
                              ; same
          goto      ops_are_eq      ; OP2 = w routine
; At this point the zero flag is not set. Therefore the two operands
; are not equal
; Now test the carry flag for OP1 < OP2, in this case C = 1
          btfss     STATUS,0 ; 0 is carry bit. Test carry flag
                              ; and skip next instruction if
                              ; C bit = 1
          goto      op2big    ; OP2 > w routine
; Processing for the case OP1 > OP2
          nop
          goto      done
```

```
ops_are_eq:
;    Processing for the case OP1 = OP21
        nop
        nop
        goto    done
op2big:
;    Processing for the case OP1 < OP2
        nop
        nop
done:
        goto    done
        end
```

The Infamous PIC Carry Flag

In PIC programming, the effects on the carry flag are different in addition than in subtraction. During addition (**addwf** and **addlw**) the carry flag indicates a carry-out of the most significant bit of the result. In this case, $C = 1$ if there was a carry out, and $C = 0$ otherwise. However, in subtraction the carry flag is described in the Microchip documentation as behaving as an *inverted borrow*. This means that when two numbers are subtracted and the result is too big to fit in the destination operand, then the carry flag is clear. What this amounts to is that in PIC subtraction (**sublw** and **subwf** operations) the carry bit is set if there is no carry-out of the high-order bit. This unusual behavior is shown in the preceding code fragment.

10.2 Simple Circuits and Programs

In the following sections we describe very simple PIC-based circuits that can be assembled with few components on a breadboard. The corresponding programs exercise the circuit components. The beginner should not skip building these circuits and coding the programs since they demonstrate essential hardware and software elements.

As a learning experience, it is a good idea to *reverse engineer* the code in these sample programs. With the processor's instruction set at hand, listed in Appendix C, proceed to follow the code one instruction at a time until you can understand every processing detail.

10.2.1 A Single LED Circuit

One of the simplest circuits consists of a single LED lamp wired to Port-B, line 0, of a 16F84A PIC, as shown in Figure 10-3.

The power source for the circuit in Figure 10-3 is not shown in the diagram. Typically, a battery source or an AC/DC converter and a voltage stabilizer circuit as in the one in Figure 10-2 are used.

A program to turn on the LED on Port-B, line 0, requires a few but essential processing operations. Code must perform the following operations:

Programming Essentials: Input and Output 195

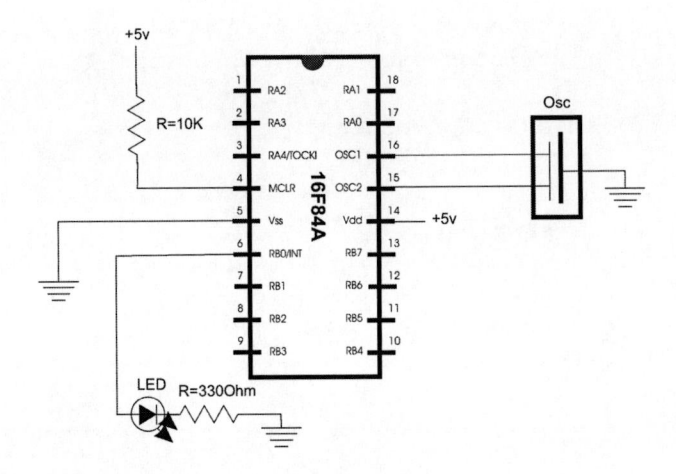

Figure 10-3 Simple LED Circuit

1. Define and select processor (in this case 16F84A).
2. Link-in the corresponding **include** file (**p16f84A.inc**).
3. Select the oscillator type (here external resonator, _XT type).
4. Direct execution to the main label.
5. Initialize Port-B for output.
6. Set line 0 in Port-B high.

The entire program is as follows:

```
; File: LEDOn.asm
; Date: June 1, 2006
; Author: Julio Sanchez
; Processor: 16F84A
;
; Description:
; Turn on LED wired to Port-B, line 0
;============================
;         switches
;============================
; Switches used in __config directive:
;   _CP_ON          Code protection ON/OFF
; * _CP_OFF
; * _PWRTE_ON       Power-up timer ON/OFF
;   _PWRTE_OFF
;   _WDT_ON         Watchdog timer ON/OFF
; * _WDT_OFF
;   _LP_OSC         Low power crystal occilator
; * _XT_OSC         External parallel resonator/crystal oscillator

;   _HS_OSC         High speed crystal resonator (8 to 10 MHz)
```

```
;                       Resonator: Murate Erie CSA8.00MG = 8 MHz
;       _RC_OSC         Resistor/capacitor oscillator
;       |
;       |_____ * indicates setup values

            processor  16f84A
            include    <p16f84A.inc>
            __config   _XT_OSC & _WDT_OFF & _PWRTE_ON & _CP_OFF
;==========================================================
;                    variables in PIC RAM
;==========================================================
; None used
;==========================================================
;                    m a i n   p r o g r a m
;==========================================================
            org     0               ; start at address 0
            goto    main
;=============================
; space for interrupt handler
;=============================
            org             0x04
;=============================
;       main program
;=============================
main:
; Initialize all line in Port-B for output
            movlw   B'00000000'     ; w = 00000000 binary
            tris    PORTB           ; Set up Port-B for output
; Turn on line 0 in Port-B. All others remain off
            movlw   B'00000001'
                    ;    ----|
                    ;    |   |____ Line 0 ON
                    ;    |_____ All others off
            movwf   PORTB
; Endless loop intentionally hangs up program
wait:
            goto    wait

            end
```

The preceding program, named LEDOn, can be found in the book's online software.

LED Flasher Program

A different program makes the LED in the circuit in Figure 10-3 flash on and off. All that is necessary is a delay loop using a file register counter. The logic turns on the LED and counts down to zero. Then it turns the LED off and counts down again.

Programming Essentials: Input and Output

The counter routine demonstrates the creation of a *procedure* in PIC programming. In fact, a procedure is nothing more than a routine called by a *label* at its entry point and terminated with a **return** statement. The procedure is executed by a **call** statement to its initial label, as follows:

```
         call    delay              ; Call to procedure
           .
           .
           .
; Elsewhere in the program
delay:
         ; procedure instructions go here
         return                     ; End of procedure
```

The simplest delay loop consists of wasting processor time. Since each instruction takes four clock cycles, the delay can be calculated by multiplying the number of instructions in the loop by the device's clock speed divided by four. The details of delay loops are discussed in Chapter 12, on timers and counters. Here we just present a *double-counter loop* without entering into timing details.

The timer loop requires two counters, since the maximum value that can be stored in a register file is 255 and a delay of 255 machine cycles is very short. In this example, we get around this limitation by creating *double counters*: an inner loop counts down 200 cycles and an outer loop repeats the inner loop 200 times. The result is that the routine repeats 200 multiplied by 200 times, or 40,000 iterations, which is sufficient for the purpose at hand. Code is as follows:

```
delay:
         movlw   .200      ; w = 200 decimal
         movwf   j         ; j = w
jloop:
         movwf   k         ; k = w
kloop:
         decfsz  k,f       ; k = k-1, skip next if zero
         goto    kloop
         decfsz  j,f       ; j = j-1, skip next if zero
         goto    jloop
         return
```

Code assumes that two variables were created in the processor's GPR space, as follows:

```
; Declare variables at 2 memory locations
j        equ              0x0c
k        equ              0x0d
```

The listing for the entire LEDFlash program, contained in the book's online software, is as follows:

```
;   File: LEDFlash.asm
;   Date: June 2, 2006
;   Author: Julio Sanchez
;   Processor: 16F84A
;
;   Description:
;   Turn on and off LED wired to Port-B, line 0
;============================
;          switches
;============================
;   Switches used in __config directive:
;     _CP_ON        Code protection ON/OFF
; *   _CP_OFF
; *   _PWRTE_ON     Power-up timer ON/OFF
;     _PWRTE_OFF
;     _WDT_ON       Watchdog timer ON/OFF
; *   _WDT_OFF
;     _LP_OSC       Low power crystal occilator
; *   _XT_OSC       External parallel resonator/crystal oscillator
;
;     _HS_OSC       High speed crystal resonator (8 to 10 MHz)
;                   Resonator: Murate Erie CSA8.00MG = 8 MHz
;     _RC_OSC       Resistor/capacitor oscillator
;    |
;    |_____ * indicates setup values

        processor 16f84A
        include   <p16f84A.inc>
        __config  _XT_OSC & _WDT_OFF & _PWRTE_ON & _CP_OFF
;=======================================================
;             variables in PIC RAM
;=======================================================
; Declare variables at 2 memory locations
j       equ           0x0c
k       equ           0x0d
;=======================================================
;             m a i n    p r o g r a m
;=======================================================
        org     0          ; start at address 0
        goto    main
;============================
; space for interrupt handler
;============================
        org           0x04
;============================
;       main program
;============================
main:
```

```
; Initialize all line in Port-B for output
        movlw   B'00000000'         ; w = 00000000 binary
        tris    PORTB               ; Set up Port-B for output
;
; Program loop to turn LED on and off
LEDonoff:
; Turn on line 0 in Port-B. All others remain off
        movlw   B'00000001'         ; LED ON
        movwf   PORTB
        call    delay               ; Local delay routine
; Turn off line 0 in Port-B.
        movlw   B'00000000'         ; LED OFF
        movwf   PORTB
        call    delay
        goto    LEDonoff
;=================================
;           delay subroutine
;=================================
delay:
        movlw   .200                ; w = 200 decimal
        movwf   j                   ; j = w
jloop:
        movwf   k                   ; k = w
kloop:
        decfsz  k,f                 ; k = k-1, skip next if zero
        goto    kloop
        decfsz  j,f                 ; j = j-1, skip next if zero
        goto    jloop
        return

        End
```

10.2.2 LED/Pushbutton Circuit

A slightly more complex circuit contains a pushbutton switch. In this case, the program monitors the state of the pushbutton and lights the LED accordingly. Figure 10-4 (in the following page) shows one possible wiring for the LED/pushbutton circuit.

If a switch reports a zero bit when active, it is described as *active-low*. A switch that reports a one-bit when pressed is said to be *active-high*. The pushbutton switch on the preceding figure is active-low. In the same manner, an output device can be wired so that it is turned on with a logic 0 and off with logic 1 on the port pin. A device turned on by the port current is said to be a *source current device*. When the device is turned on when the port reports logic 0 the line is said to *sink the current*. PICs and other CMOS devices operate better sinking than sourcing current. Table 10.2 shows the maximum sink and source currents for the 16F84 ports.

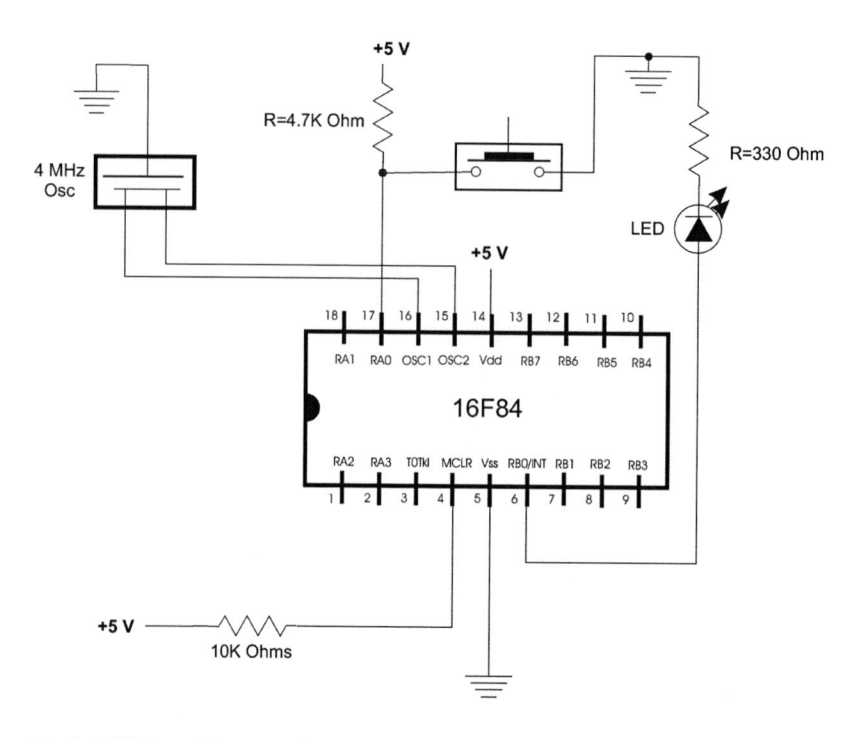

Figure 10-4 LED/pushbutton Experimental Circuit

Table 10.2

Sink and Source Current for 16F84 Ports

SOURCE	ANY I/O PIN	PORT A	PORT-B
sink current	25 mA	80 mA	150 mA
source current	20 mA	50 mA	100 mA

The 4.7K Ohm resistor in the circuit of Figure 10-4 keeps RA0 high until the switch is pressed. This switch action determines that RA0 reads binary one when the switch is released and binary zero (low) when the switch is pressed (active).

To test if the switch in the circuit of Figure 10-4 is closed, the application can read RA0. If the value in the port is 1, then the switch is open (released). If 0, then the switch is closed. The following program, named LEDandPb, exercises the circuit in Figure 10-4:

```
; File: LEDandPb.asm
; Date: June 2, 2006
; Author: Julio Sanchez
; Processor: 16F84A
;
; Description:
; Circuit with LED wired to RB0 and pushbutton switch,
; active low, wired to RA0. Pushbutton action turns LED
; OFF when pressed and ON when released.
```

```
;============================
;          switches
;============================
; Switches used in __config directive:
;     _CP_ON         Code protection ON/OFF
; *   _CP_OFF
; *   _PWRTE_ON      Power-up timer ON/OFF
;     _PWRTE_OFF
;     _WDT_ON        Watchdog timer ON/OFF
; *   _WDT_OFF
;     _LP_OSC        Low power crystal occilator
; *   _XT_OSC        External parallel resonator/crystal oscillator

;     _HS_OSC        High speed crystal resonator (8 to 10 MHz)
;                    Resonator: Murate Erie CSA8.00MG = 8 MHz
;     _RC_OSC        Resistor/capacitor oscillator (simplest, 20%
error)
; |
; |_____ * indicates setup values

        processor 16f84A
        include    <p16f84A.inc>
        __config   _XT_OSC & _WDT_OFF & _PWRTE_ON & _CP_OFF
;========================================================
;                variables in PIC RAM
;========================================================
; Not used in this program
;========================================================
;                m a i n   p r o g r a m
;========================================================
        org     0              ; start at address 0
        goto    main
;============================
; space for interrupt handler
;============================
        org             0x04
;============================
;     main program
;============================
main:
; Initialize all lines in Port-B for output
        movlw   B'00000000'    ; w = 00000000 binary
        tris    PORTB          ; Set up Port-B for output
; Initialize Port-A, line 0, for input
        movlw   B'00000001'    ; w = 00000001 binary
        tris    PORTA          ; Set up RA0 for input
; Program loop to test state of pushbutton switch
;==============================
```

```
;       read PB switch state
;===============================
LEDctrl:
; Push button switch on demo board is wired to Port-A bit 0
; Switch logic is active low
        btfss   PORTA,0         ; Test. Skip next line if
                                ; bit is set
        goto    turnOFF         ; Turn LED off routine
; At this point Port-A bit 0 is not set
; Switch is pressed (active low action)
; Turn ON line 0 in Port-B
        bsf     PORTB,0         ; RB0 high
        goto    LEDctrl
turnOFF:
; Routine to turn OFF LED
        bcf     PORTB,0         ; RB0 low
        goto    LEDctrl

                End
```

10.2.3 Multiple LED Circuit

The following circuit allows a few more programming complications since it contains a battery of eight LEDs, all wired to Port-B.

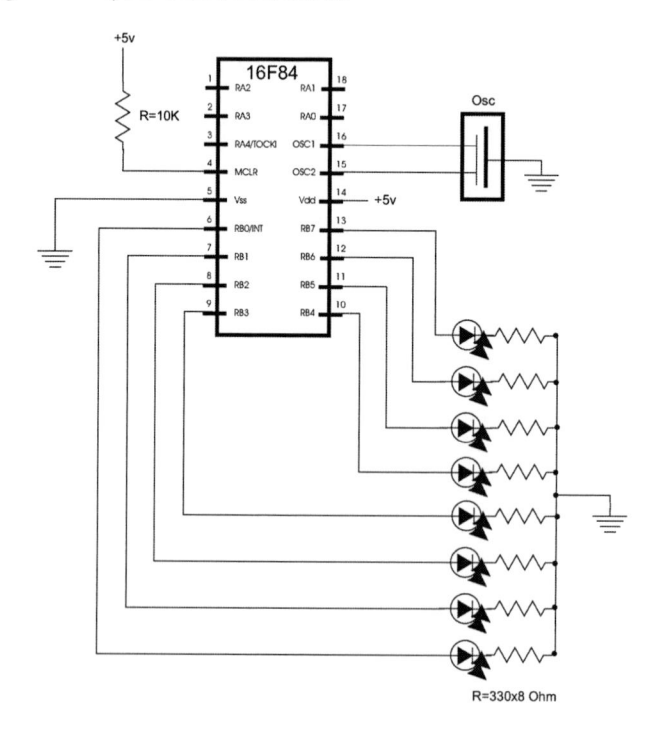

Figure 10-5 Multiple LED Circuit

The circuit in Figure 10-5 can be programmed to do different functions. For example, the eight LEDs can be visualized as representing an 8-bit binary number and the circuit can be programmed to count in binary from 0 to 255. Since the eight LEDs are all wired to Port-B, the binary count can be directly echoed on the port. The following program, named LEDCount, performs this operation:

```
; File: LEDCount.asm
; Date: June 3, 2006
; Author: Julio Sanchez
; Processor: 16F84A
; Description:
; Circuit with eight LEDs wired to RB0 to RB7.
; Program displays a binary count from 0 to 255 on
; LEDs.
;===========================
;         switches
;===========================
; Switches used in __config directive:
;   _CP_ON          Code protection ON/OFF
; * _CP_OFF
; * _PWRTE_ON       Power-up timer ON/OFF
;   _PWRTE_OFF
;   _WDT_ON         Watchdog timer ON/OFF
; * _WDT_OFF
;   _LP_OSC         Low power crystal occilator
; * _XT_OSC         External parallel resonator/crystal oscillator
;
;   _HS_OSC         High speed crystal resonator (8 to 10 MHz)
;                   Resonator: Murate Erie CSA8.00MG = 8 MHz
;   _RC_OSC         Resistor/capacitor oscillator
; |
; |_____ * indicates setup values
        processor 16f84A
        include   <p16f84A.inc>
        __config  _XT_OSC & _WDT_OFF & _PWRTE_ON & _CP_OFF
;======================================================
;              variables in PIC RAM
;======================================================
; Declare variables at 2 memory locations
j       equ             0x0c
k       equ             0x0d
;======================================================
;              m a i n   p r o g r a m
;======================================================
        org     0           ; start at address 0
        goto    main
;=============================
; space for interrupt handler
;=============================
```

```
                org             0x04
;===============================
;       main program
;===============================
main:
; Initialize all lines in Port-B for output
                movlw   B'00000000'     ; w = 00000000 binary
                tris    PORTB           ; Set up Port-B for output
; Set Port-B bit 0 ON
                movlw   B'00000000'     ; w := 0 binary
                movwf   PORTB           ; Port-B itself := w
; Clear the carry bit
                bcf     STATUS,C
mloop:
                incf    PORTB,f         ; Add 1 to register value
                call    delay
                goto    mloop
;================================
;       delay sub-routine
;================================
delay:
                movlw   .200            ; w = 200 decimal
                movwf   j               ; j = w
jloop:
                movwf   k               ; k = w
kloop:
                decfsz  k,f             ; k = k-1, skip next if zero
                goto    kloop
                decfsz  j,f             ; j = j-1, skip next if zero
                goto    jloop
                return

                end
```

10.3 Programming the Seven-segment LED

A 7-segment display can be connected to output ports on the PIC and used to display numbers and some digits. The circuit in Figure 10-6 shows one possible wiring scheme.

As the name indicates, the seven-segment display has seven linear LEDs that allow forming all the decimal and hex digits and some symbols and letters. Once the mapping of the individual bars of the display to the PIC ports has been established, digits and letters are shown by selecting which port lines are set and which are not. For example, in the seven-segment LED of Figure 10-5, the digit 2 is displayed by setting segments a, b, g, e, and d. In this particular wiring, these segments correspond to Port-B lines 0, 1, 6, 4, and 5.

Programming Essentials: Input and Output

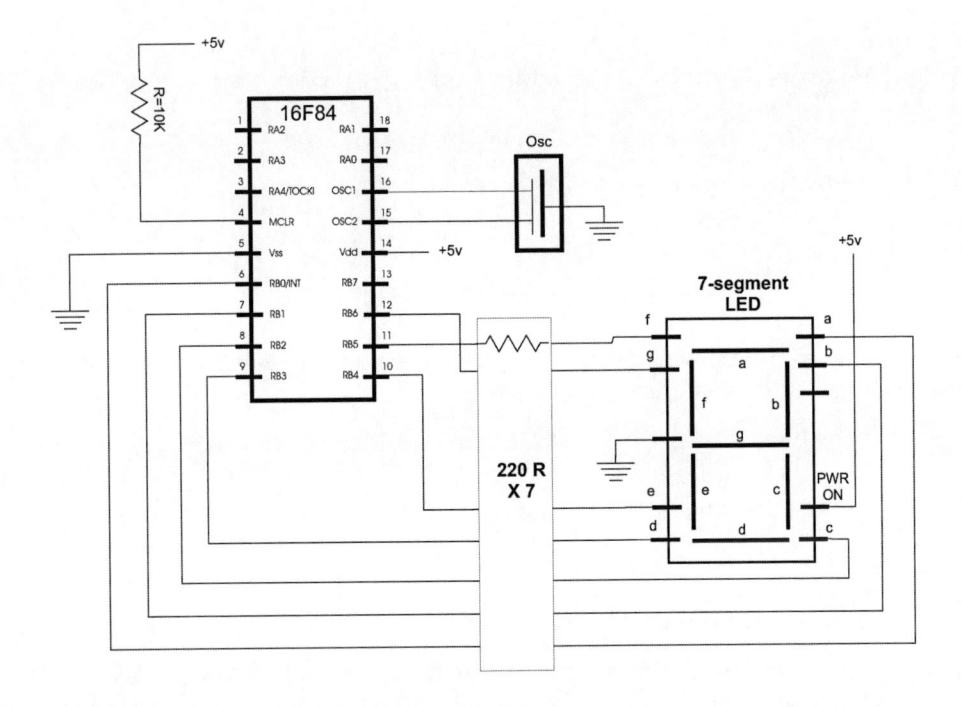

Figure 10-6 Seven-segment LED Circuit

As the name indicates, the seven-segment display has seven linear LEDs that allow forming all the decimal and hex digits and some symbols and letters. Once the mapping of the individual bars of the display to the PIC ports has been established, digits and letters are shown by selecting which port lines are set and which are not. For example, in the seven-segment LED of Figure 10-6, the digit 2 is displayed by setting segments a, b, g, e, and d. In this particular wiring, these segments correspond to Port-B lines 0, 1, 6, 4, and 5.

Conversion of the individual digits to port display codes is easily accomplished by means of a *lookup table*. The processing depends on three special features of PIC assembly language:

- The *program counter* file register (labeled PC and located at offset 0x02) holds the address in memory of the current instruction. Since each PIC instruction takes up a single byte (except for those that modify the PC), one can jump to consecutive entries in a table by adding an integer value to the program counter.

- The **addwf** instruction is used to add a value in the w register to the program counter.

- The **retlw** instruction returns to the caller a literal value stored in the w register. In the case of **retlw** the literal value is the instruction operand.

If the lookup table is located at a subroutine called getcode, then the processing can be implemented as follows:

```
getcode:
        addwf    PC,f      ; Add value in w register to program
counter
        retlw    0x3f      ; code for number 0
        retlw    0x06      ; code for number 1
        retlw    0x5b      ; code for number 2
        ...
        retlw    0x6f      ; code for number 9
```

The calling routine places in the w register the numeric value whose code is desired, and then calls the table lookup, as follows:

```
        movlw    0x03      ; Code for number 3 desired
        call     getcode
        movwf    PORTB     ; Display 3 in 7-segment display
```

10.4 A Demonstration Board

A *demonstration board*, also known as a demo board, is a useful tool in mastering PIC programming. Many are available commercially; like programmers, there is a cottage industry of PIC demo boards on the internet. Constructing your own demo boards and circuits is not difficult. The components can be placed on a breadboard, or wire-wrapped onto a special circuit board, or a printed circuit board can be homemade, or ordered through the internet. These options have been previously discussed and Appendix B contains instructions on how to build your own PCBs.

Figure 10-7 shows a simple 16F84A-based demo board with a seven-segment LED, buzzer, pushbutton switch, and a bank of four toggle switches.

10.4.1 PCB Images for Demo Board

Some PCBs contain circuit etchings on both sides. In this case two circuit board images are required. In addition, most boards contain a top-side image of the components, company logos, model numbers, and other information. Commercially, this image is silk-screened onto the board.

The homemade board (see Appendix B) usually contains a single etched image and a top-side image with informational text and graphics. Both images can be created with a conventional drawing program, such as Corel Draw, Adobe Illustrator, or Windows Paint, or with a specialized application, several of which are available free and for purchase on the Web. Figure 10-7 shows the images used for making the PCB for the circuit in Figure 10-8.

Note that the top-side (text) image has been mirrored on the horizontal plane. This is necessary so that the text and graphics coincide with the circuit etchings once the images are transferred to the board. The process for making your own PCBs is described in Appendix B.

Programming Essentials: Input and Output

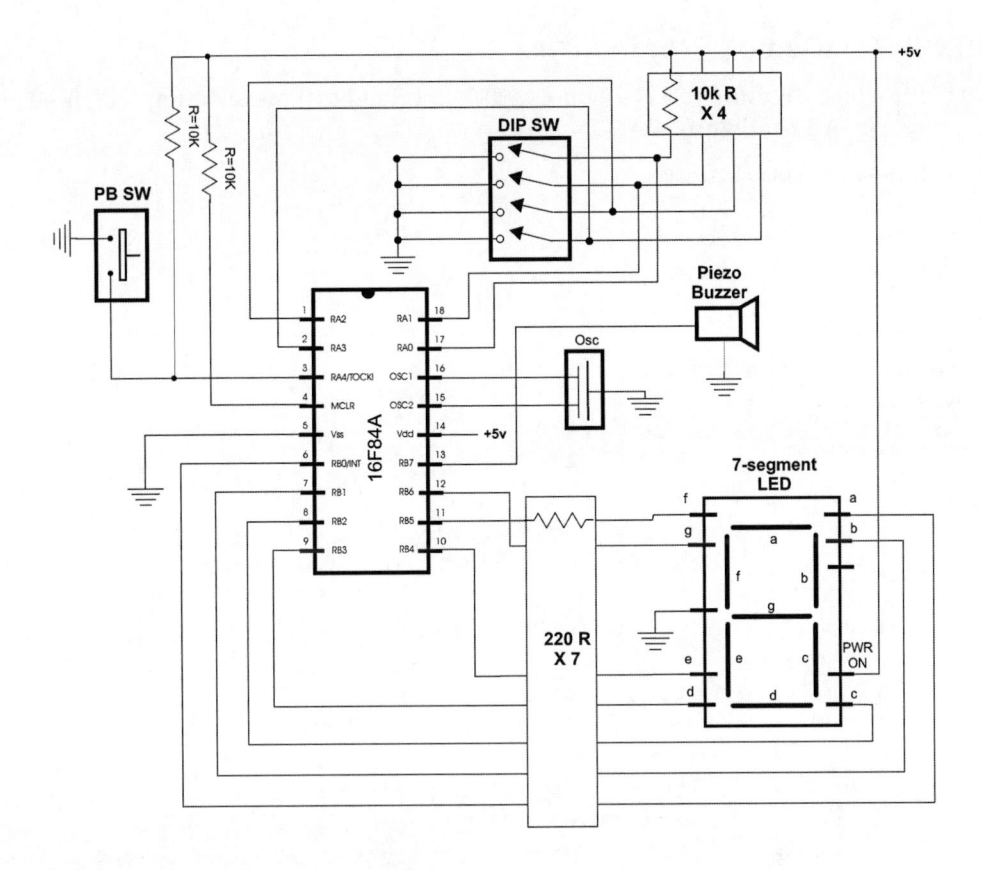

Figure 10-7 PIC 16F87A Demo Board

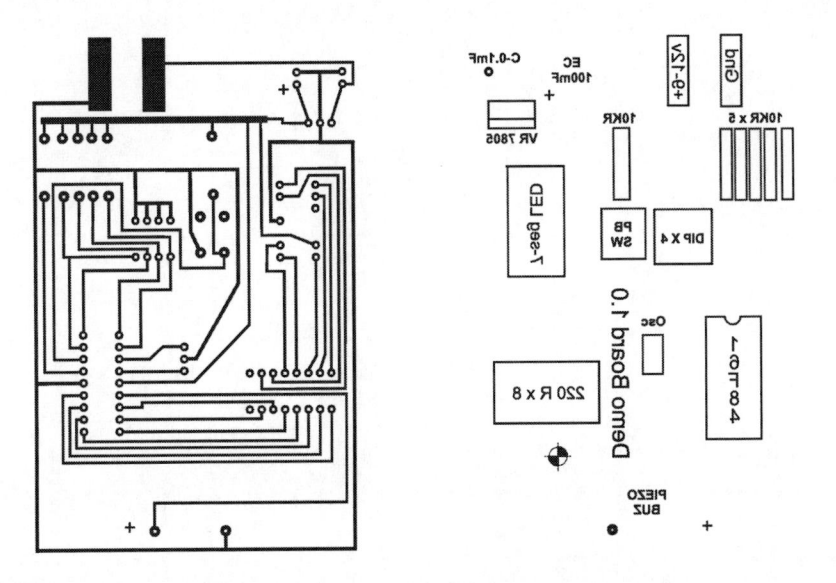

Figure 10-8 Bottom- and Top-side images of a PCB.

10.4.2 TestDemo1 Program

The following program exercises some of the experiments that can be implemented on the demo boards in Figure 10-7:

```
; File: TestDemo1.asm
; Date: June 2, 2006
; Author: Julio Sanchez
; Processor: 16F84A
;
; Description:
; Program to exercise the demonstration circuit and board
; number 1
;============================
;       switches
;============================
; Switches used in __config directive:
;   _CP_ON         Code protection ON/OFF
; * _CP_OFF
; * _PWRTE_ON      Power-up timer ON/OFF
;   _PWRTE_OFF
;   _WDT_ON        Watchdog timer ON/OFF
; * _WDT_OFF
;   _LP_OSC        Low power crystal oscillator
; * _XT_OSC        External parallel resonator/crystal
;                  oscillator
;   _HS_OSC        High speed crystal resonator (8 to 10 MHz)
;                  Resonator: Murate Erie CSA8.00MG = 8 MHz
;   _RC_OSC        Resistor/capacitor oscillator
; |
; |_____ * indicates setup values

        processor 16f84A
        include   <p16f84A.inc>
        __config  _XT_OSC & _WDT_OFF & _PWRTE_ON & _CP_OFF
;========================================================
;               variables in PIC RAM
;========================================================
        cblock    0x0c      ; Start of block
        count1              ; Counter # 1
        j                   ; counter J
        k                   ; counter K
        endc
;========================================================
;                   P R O G R A M
;========================================================
        org       0         ; start at address 0
        goto      main
;
```

Programming Essentials: Input and Output

```
; Space for interrupt handlers
        org             0x08
main:
; Port A (5 lob) for input
        movlw    B'00011111'        ; w := 00001111 binary
        tris     PORTA              ; Port-A (lines 0 to 4) to input
; Port-Bit (8 lines) for output
        movlw    B'00000000'        ; w := 00000000 binary
        tris     PORTB              ; Port-B to output
;==============================
; Pushbutton switch processing
;==============================
pbutton:
; Push button switch on demo board is wired to RA4
; Switch logic is active low
        btfss    PORTA,4            ; Test and skip if bit is set
        goto     buzzit             ; Buz if switch ON
; At this point Port-A bit 4 is set (switch is off)
        call     buzoff             ; Buzzer off
        goto     readdip            ; Read DIP switches
buzzit:
        call     buzon              ; Turn on buzzer
        goto     pbutton
;==============================
;      DIP switch processing
;==============================
; Read all bits of Port-A
readdip:
        movf     PORTA,w            ; Port A bits to w
; If board uses active low then all switch bits must be negated
; This is done by XORing with 1-bits
        xorlw    b'11111111'        ; Invert all bits in w
; Eliminate all 4 high order bits
        andlw    b'00001111'        ; And with mask
; Get digit into w
        call     segment            ; get digit code
        movwf    PORTB              ; Display digit
        call     delay              ; Give time
; Update digit and loop counter
        goto     pbutton

;*****************************
;   7-segment table of hex codes
;*****************************
segment:
        addwf    PCL,f              ; PCL is program counter latch
        retlw    0x3f               ; 0 code
```

```
        retlw     0x06        ; 1
        retlw     0x5b        ; 2
        retlw     0x4f        ; 3
        retlw     0x66        ; 4
        retlw     0x6d        ; 5
        retlw     0x7d        ; 6
        retlw     0x07        ; 7
        retlw     0x7f        ; 8
        retlw     0x6f        ; 9
        retlw     0x77        ; A
        retlw     0x7c        ; B
        retlw     0x39        ; C
        retlw     0x5b        ; D
        retlw     0x79        ; E
        retlw     0x71        ; F
        retlw     0x7f        ; Just in case all on
;****************************
;    piezo buzzer ON
;****************************
; Routine to turn on piezo buzzer on Port-B bit 7
buzon:
        bsf       PORTB,7     ; Tune on bit 7, Port-B
        return
;****************************
;    piezo buzzer OFF
;****************************
; Routine to turn off piezo buzzer on Port-B bit 7
buzoff:
        bcf       PORTB,7     ; Bit 7 Port-B clear
        return
;================================
;        delay subroutine
;================================
delay:
        movlw     .200        ; w = 200 decimal
        movwf     j           ; j = w
jloop:
        movwf     k           ; k = w
kloop:
        decfsz    k,f         ; k = k-1, skip next if zero
        goto      kloop
        decfsz    j,f         ; j = j-1, skip next if zero
        goto      jloop
        return
        end
```

Chapter 11

Interrupts

An interrupt is an asynchronous signal calling for processor attention. Interrupts can originate in hardware or in software. The interrupt mechanism is a way to avoid wasting processor time, since without interrupts code has to poll hardware devices in ineffective, closed loops. With interrupts, the processor can continue to do its work since the interrupt mechanism ensures that the CPU receives a signal whenever an event occurs that requires its attention. PIC microcontrollers provide varying levels of support for interrupts. We focus on interrupts on the 16F84.

11.0 Interrupts on the 16F84

Four different sources of interrupts are available in the 16F84. These are discussed in Section 11.1. One instruction (**RETFIE** for *return-from-interrupt*) is specifically related to interrupt processing. Its purpose is to return to the program counter the address of the instruction that follows the location in code where the interrupt took place. It does so by loading into the program counter register the 13-bit address saved at the top of the stack. In addition, **RETFIE** sets the *Global Interrupt Enable* bit in the INTCON register (discussed in Section 11.0.1) automatically re-enabling interrupts.

In addition to the **RETFIE** instruction, two PIC hardware elements relate directly to interrupts: the OPTION register and the INTCON register. Both registers are readable and writeable and contain bits that allow setting up, controlling, and detecting the various interrupts. INTCON records individual interrupt requests in flag bits. It also contains the individual and global interrupt enable bits. The OPTION register has several bits that must be accessed in order to initialize interrupts.

11.0.1 The Interrupt Control Register

INTCON (the Interrupt Control Register) is a readable and writeable register located at offset 0x08 in bank 0. The INTCON register contains two classes of bits: bits to enable and disable the various interrupt sources, and flag bits that allow detecting the occurrence of the various interrupts. The bits to enable and disable interrupts have names that end with the letter E, while the interrupt flag bit names end with the letter F. They are known collectively as the *INTCON E* and *INTCON F* bits. Figure 11-1 is a bitmap of the INTCON Register.

```
    bit 7                                              bit 0
   +------+------+------+------+------+------+------+------+
   | GIE  | EEIE | TOIE | INTE | RBIE | TOIF | INTF | RBIF |
   +------+------+------+------+------+------+------+------+
```

bit 7 GIE: Global Interrupt Enable bit
     1 = Enables all unmasked interrupts
     0 = Disables all interrupts
bit 6 EEIE: EE Write Complete Interrupt Enable bit
     1 = Enables the EE Write Complete interrupts
     0 = Disables the EE Write Complete interrupt
bit 5 T0IE: TMR0 Overflow Interrupt Enable bit
     1 = Enables the TMR0 interrupt
     0 = Disables the TMR0 interrupt
bit 4 INTE: RB0 Interrupt Enable bit
     1 = Enables the RB0 external interrupt
     0 = Disables the RB0 external interrupt
bit 3 RBIE: Port Change Interrupt Enable bit
     1 = Enables the RB port change interrupt
     0 = Disables the RB port change interrupt
bit 2 T0IF: TIMER0 Overflow Interrupt Flag bit
     1 = TMR0 register has overflowed
     0 = TMR0 register did not overflow
bit 1 INTF: RB0 External Interrupt Flag bit
     1 = The RB0/INT external interrupt occurred
     0 = The RB0/INT external interrupt did not occur
bit 0 RBIF: RB0-RB3 Port Change Interrupt Flag bit
     1 = At least one of the RB7:RB4 pins changed state
     0 = None of the RB7:RB4 pins have changed state

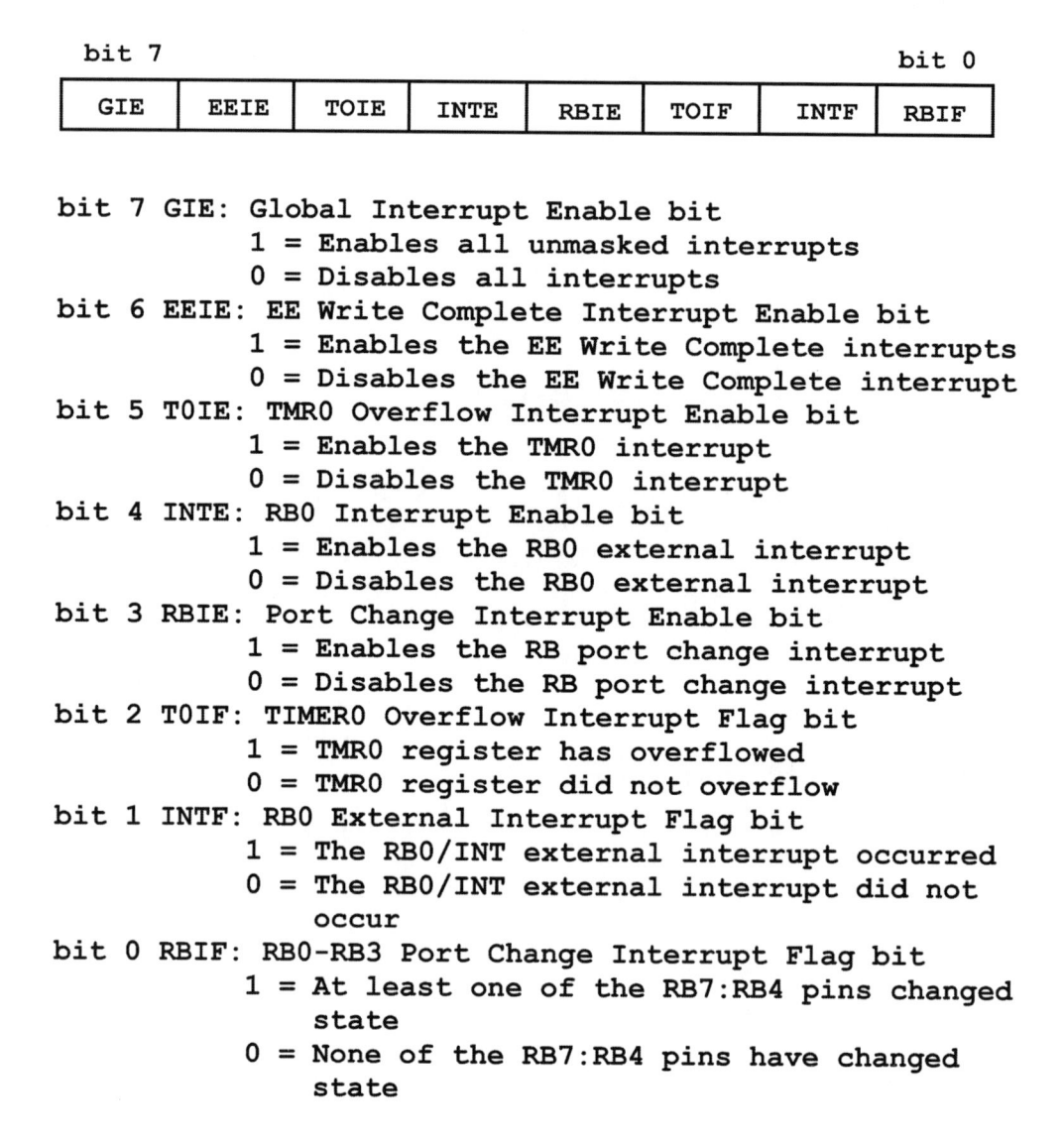

Figure 11-1 INTCON Register Bitmap

11.0.2 The OPTION Register

The *OPTION Register* is a readable and writeable register that contains controls for configuring the prescaler bits and assigning them to either TIMER0 or the Watchdog Timer, for selecting the increment mode on the RA4/TOCKI pin, the TIMER0 source clock, the rising or falling edge in the RB0 interrupt, and for enabling and disabling the internal Port-B's pull-up resistors. The OPTION register is located in Bank1, at address 0x81. Although this register is not directly related to interrupts, several of its bits are related to the various interrupts. Figure 11-2 is a bitmap of the OPTION register.

Interrupts

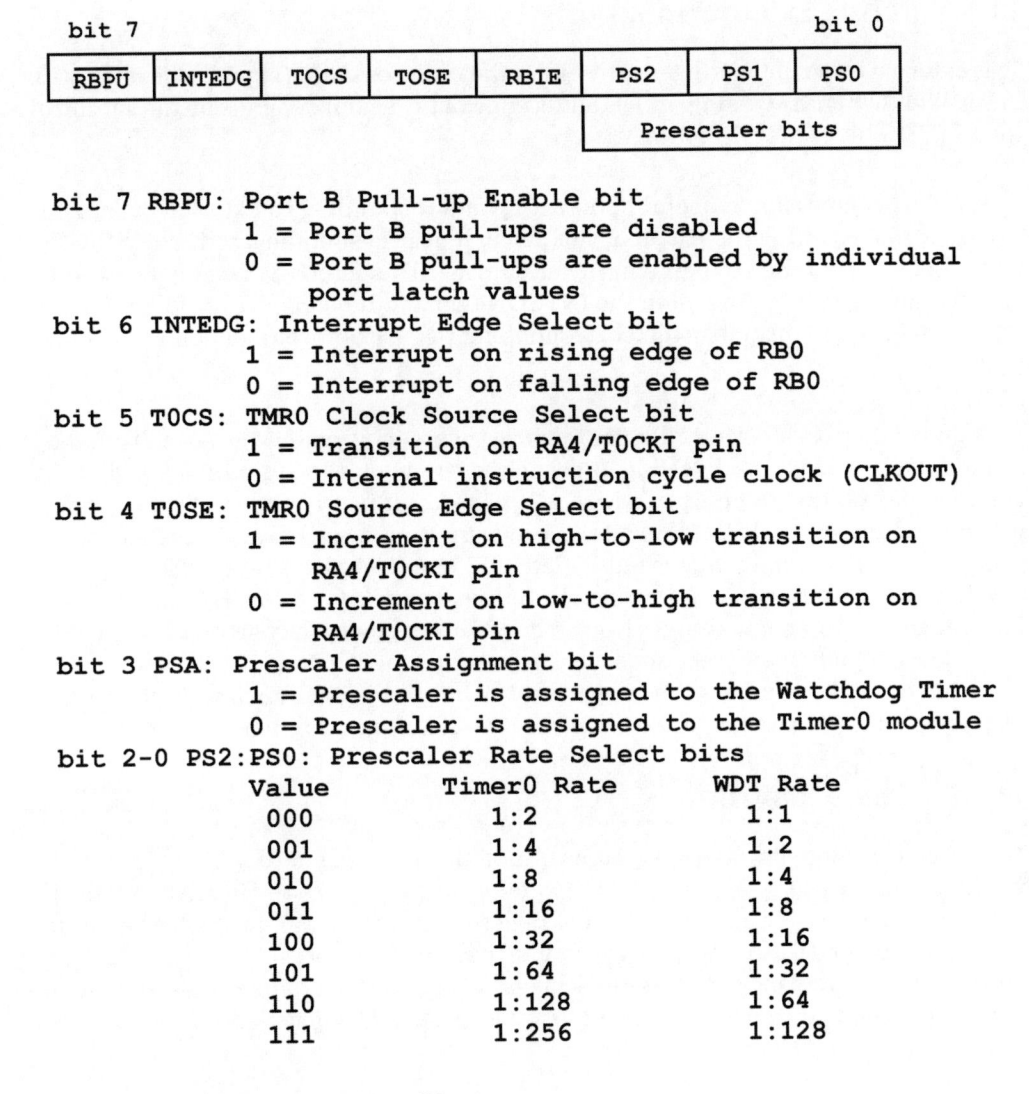

bit 7							bit 0
$\overline{RBPU}$	INTEDG	T0CS	T0SE	RBIE	PS2	PS1	PS0
					Prescaler bits		

```
bit 7 RBPU: Port B Pull-up Enable bit
            1 = Port B pull-ups are disabled
            0 = Port B pull-ups are enabled by individual
                port latch values
bit 6 INTEDG: Interrupt Edge Select bit
            1 = Interrupt on rising edge of RB0
            0 = Interrupt on falling edge of RB0
bit 5 T0CS: TMR0 Clock Source Select bit
            1 = Transition on RA4/T0CKI pin
            0 = Internal instruction cycle clock (CLKOUT)
bit 4 T0SE: TMR0 Source Edge Select bit
            1 = Increment on high-to-low transition on
                RA4/T0CKI pin
            0 = Increment on low-to-high transition on
                RA4/T0CKI pin
bit 3 PSA: Prescaler Assignment bit
            1 = Prescaler is assigned to the Watchdog Timer
            0 = Prescaler is assigned to the Timer0 module
bit 2-0 PS2:PS0: Prescaler Rate Select bits
                Value       Timer0 Rate     WDT Rate
                000         1:2             1:1
                001         1:4             1:2
                010         1:8             1:4
                011         1:16            1:8
                100         1:32            1:16
                101         1:64            1:32
                110         1:128           1:64
                111         1:256           1:128
```

Figure 11-2 OPTION Register Bitmap

11.1 Interrupt Sources

The 16F8X supports four different sources of interrupt:

1. External interrupt detected by line 0 of Port-B

2. Interrupts that originate in the timer (TMR0 overflow interrupt)

3. Interrupts that originate in changes of lines RB7 to RB4 in Port-B

4. EEPROM complete data write interrupt

11.1.1 Port-B External Interrupt

This external interrupt is triggered by either the rising or falling signal edge on port-B, line 0. Whether it is the rising or the falling edge of the signal depends on the setting of the INTEDG bit of the OPTION register.

The Port-B interrupt is useful in detecting and responding to external events; for example, in measuring the frequency of a signal or in responding with some PIC action to a change in the state of a hardware device. This interrupt can be disabled by clearing the corresponding bit in the INTCON register. If enabled, once the interrupt takes place, code must clear the corresponding flag bit before re-enabling the interrupt.

Suppose there is a circuit that contains an emergency switch that is activated by some critical event. One possible approach is to check the state of the switch by continuously polling the port to which it is wired. But in a complex program it may be difficult to ensure that the switch polling routine is called with sufficient frequency so that an emergency event is detected immediately. A more effective solution is to connect the emergency switch to line number 0 of Port-B and set up the Port-B external interrupt source. Now, whenever the emergency switch is activated, the program immediately responds via the interrupt mechanism. Furthermore, once the interrupt code has been developed and debugged it continues to function correctly no matter what changes are made to the rest of the program.

11.1.2 Timer0 Interrupt

The 16F84 is equipped with a special timer module, named Timer0, which serves both as a timer and as a counter. The Timer0 module, which is discussed in greater detail in Chapter 12, consists of an 8-bit readable register operated by an internal or external clock and attached to an 8-bit programmable *prescaler*. The prescaler is used to delay the timer by dividing the previous clock signal. The timer0 module can be set up to interrupt on overflow. In this case, an interrupt is generated whenever the counter goes from 0xff to 0x00.

The Timer0 counter interrupt can be used to measure events and to respond to elapsed periods. For example, the timer is used to measure events by determining the number of timer interrupts that have taken place since an event occurred. The timer of each interrupt is determined from the processor clock speed and the prescaler set up. The event time is calculated by multiplying the time of each interrupt by the number of interrupts that have occurred. In this case, the interrupt routine increments a counter register that is accessible to code anywhere in the program; so the actual count can be reset from inside or outside the service routine.

In responding to an elapsed period, the Timer0 interrupt service routine not only keeps track of the time elapsed since the event, but also tests for a certain counter value that represents the desired time limit. Once the timer counter reaches this pre-set limit, the service routine responds directly with the required action.

One powerful and common application of a Timer0 interrupt is in implementing serial communications. In this case, the timer interrupt is set up to take place at the baud rate at which the serial line is polled for data or at which individual data bits are sent. The sample program LapseTmrInt, developed in Chapter 12, demonstrates this use of the timer interrupt.

11.1.3 Port-B Line Change Interrupt

The third 16F84 interrupt source relates to a change in the values stored in Port-B lines 4 to 7. When this interrupt is enabled, any change in status in any of the four Port-B pins labeled RB7, RB6, RB5, and RB4 can trigger an interrupt. The interrupt is set up to take place when their status changes from logic one to logic zero, or vice versa. For this interrupt to take place, Port-B pins 4 to 7 must be defined as input. Otherwise, the interrupt does not take place.

The Port-B line-change interrupt provides a mechanism for monitoring up to four different interrupt sources, typically originating in hardware devices. When the interrupt is enabled, the current state of the Port-B lines is constantly compared to the old values. If there is a change in state in any of the four lines the interrupt is generated.

Implementation of the line change interrupt is not without complications. The characteristics of the external signal are necessary to develop code that correctly handles the various possible sources. Two pieces of information that are necessary in this case are:

1. The signal's rising and falling edges
2. The pulse width of the interrupt trigger

The signal's rising and falling edges determine the service routine's entry point. For example, if the device is an active-low pushbutton switch, an interrupt typically is desired on the signal's falling edge, that is, when it goes from high-to-low.

Knowledge about the signal's width determines the processing required by the service routine. This is due to the fact that both the rising and the falling edge of the signal can trigger the interrupt. So, if the triggering signal has a small pulse width compared to the time of execution of the interrupt handler, then the interrupt line has returned to the inactive state before the service routine completes and a possible false interrupt on the signal's falling edge is not possible. On the other hand, if the pulse width of the interrupt signal is large and the service routine completes before the signal returns to the inactive state, then the signal's falling edge can trigger a false interrupt. Figure 11-3 (in the following page) shows both situations.

In the context of Figure 11-3, the period between the edge that triggers the interrupt and the termination of the interrupt handler is called the *mismatch period*. The mismatch period terminates when the service routine completes and the corresponding interrupt is re-enabled. If this happens after the interrupt signal is reset, no possible false interrupt takes place and no special provision is required in the handler. In fact, the interrupt handler runs correctly as long as the service routine takes longer to execute than the interrupt frequency. However, if the handler terminates before the signal returns to its original state, then the handler must make special provisions to handle a possible false interrupt. In order to do this, the handler must first determine if the interrupt took place on the rising or the falling signal edge, which can be done by examining the corresponding port-B line. For example, if the interrupt is to take place on the rising edge only, and the line is low, then it can be ignored since it takes place on the falling edge.

```
               CASE 1: relatively small pulse width
   Signal              ___
               _____|   |_____     ___
                                                    |___|
                       ▲         ▲         ▲
   Raising edge _____|                   |_____  Service routine complete
   triggers interrupt                              Interrupt flag clear
                                                    No possible false interrupt
                                 |
                        Interrupt handler
                            progress


               CASE 2: relatively large pulse width
   Signal                _____
               _____|                        |_____
                       ▲       ▲       ▲       ▲
                                               |_____ Falling edge can trigger
   Raising edge _____|                              false interrupt
   triggers interrupt
                                                       Service routine complete
                                                       Interrupt flag cleared
                               |
                        Interrupt handler
                          in progress
```

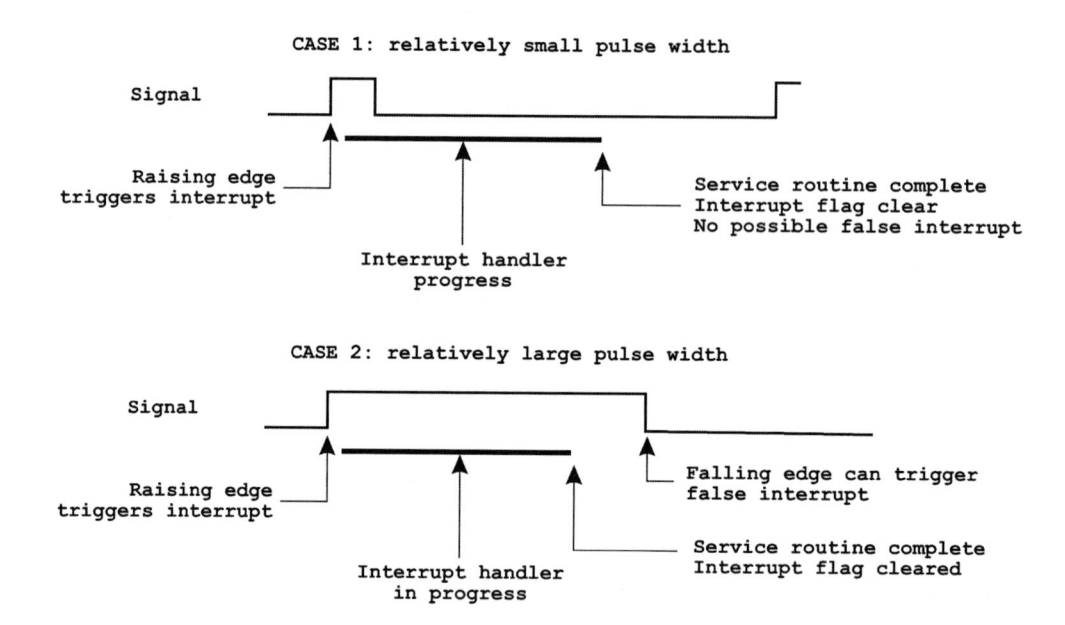

Figure 11-3 Signal Pulse Width and Interrupt Latency

When an interrupt can take place on either the rising or the falling edge of the triggering signal, the interrupt source must have a minimum pulse width in order to ensure that both edges are detected. In this case, the minimum pulse width is the maximum time from the edge that triggered the interrupt to the moment when the interrupt flag is cleared. Otherwise, the interrupt is lost since the interrupt mechanism is disabled at the time it takes place.

The preceding discussion leads directly to the possibility of an interrupt taking place while the service routine of a previous interrupt is still in progress. These are called *reentrant* or *nested interrupts*. Several things must happen to allow reentrant interrupts. One of them is that interrupts must be re-enabled before the handler terminates. In addition, the service routine must be able to create different instances of the variables in use, usually allocated in the stack. The lack of a program-accessible stack and the PIC interrupt mechanism itself forces the conclusion that reentrant interrupts should not be attempted in PIC programs.

Multiple External Interrupts

One of the practical applications of the port-B line-change interrupt is in handling several different interrupt sources; for example, a circuit containing four push-button switches that activate four different circuit responses. If the switches are wired to the corresponding pins in Port-B (RB4 to RB7) and the line-change interrupt is enabled, the interrupt takes place when any one of the four switches changes level, that is, when any one of the interrupt lines go from high to low or from low to high. The interrupt handler software can easily determine which of the switches changed state and if the change took place on the signal's rising or falling edge. The corresponding software routines then handle each case.

Later in this chapter we develop a sample program that uses the Port-B line-change interrupt to respond to action on four pushbutton switches.

11.1.4 EEPROM Data Write Interrupt

The origin of this interrupt relates to the relative slowness of the EEPROM data write operation, which is of 10 ms. The interrupt serves no other function than to allow the microcontroller to continue execution while the data write operation is in progress. The interrupt service routine informs the microcontroller when writing has ended through the EEIF bit located in the EECON1 register. The use of this interrupt is considered in Chapter 15, in the context of EEPROM data memory access and programming.

11.2 Interrupt Handlers

The *interrupt handler*, also called the interrupt service routine or the *ISR*, is the code that receives control upon occurrence of the interrupt. Most of the programming that goes into the service routine is specific to the application; however, there are certain housekeeping operations that should be included. The following list describes the structure of an interrupt service routine for the mid-range PICs:

1. Preserve the value in the w register.
2. Preserve the value of the STATUS register.
3. Execute the application-specific operations.
4. Restore the value of the STATUS register at the time of the interrupt.
5. Restore the value of the w register at the time of the interrupt.
6. Issue the RETFIE instruction to end the interrupt handler.

In the PIC 16F84, the interrupt service routine must be located at offset 0x004 in code memory. A simple **org** directive takes care of ensuring this location, as in the following code fragment:

```
        org     0x000           ; Beginning of code area
        goto    start           ; Jump to program start
        org     0x004           ; Start of Service routine
          .
          .             ; SERVICE ROUTINE GOES HERE
          .
```

```
        retfie              ; End of ISR
start:                      ; Program starts here
```

Alternatively, code can place a jump at offset 0x004 and locate the Service Routine elsewhere in the code. In this case, it is important to remember not to *call* the Service Routine, but to access it with a **goto** instruction. The reason is that the **call** opcode places a return address in the stack, which then polls for the **retfie** instruction.

11.2.1 Context Saving Operations

The only value automatically preserved by the interrupt mechanism is PC (the Program Counter), which is stored in the stack. Applications often need to restore the processor to the same state as when the interrupt took place, so the first operation of most interrupt handlers is saving the processor's context. This usually includes the w and the STATUS registers and occasionally others used by the specific implementation.

Saving w and STATUS Registers

Saving the w and the STATUS registers requires using register variables, but the process requires special care. Saving the w register is simple enough: its value at the start of the Service Routine is stored in a local variable from which it is restored at termination. But saving the STATUS register cannot be done with the **MOVF** instruction, since this instruction changes the zero flag. The solution is to use the **SWAPF** instruction which does not affect any of the flags. Of course, **SWAPF** inverts the nibbles in the operand, so it must be repeated so as to restore the original state. The following code fragment assumes that file register variables named old_w and old_status were previously created.

```
save_cntx:
        movwf    old_w              ; Save w register
        swapf    STATUS,w ; STATUS to w
        movwf    old_status         ; Save STATUS
;
; Interrupt handler operations go here
;
        swapf    old_status,w  ; Saved status to w
        movfw    STATUS        ; To STATUS register
; At this point all operations that change the
; STATUS register must be avoided, but swapf does not.
        swapf    old_w,f  ; Swap file register in itself
        swapf    old_w,w  ; reswap back to w
        retfie
```

11.3 Interrupt Programming

In the sections that follow, we discuss programming interrupts that originate in Port-B, line 0, and those that originate in changes of port-B lines RB4 to RB7. Interrupts that relate to the Timer0 overflow or to EEPROM data write operations are cov-

Interrupts

ered in the chapter on Serial Communications and the one on EEPROM Data Operations, respectively.

11.3.1 Programming the External Interrupt

Port-B, line 0, is referred to as the *External Interrupt* source. The name is not the most adequate since other interrupts can also have external sources. One of the important uses of this interrupt source is to wake the processor from the SLEEP mode. This allows developing applications that can run on a small power source (such as batteries) since the program uses almost no power until some action associated with the interrupt source wakes up the PIC. A sample program using the RB0 interrupt is developed later in this chapter. Our first sample program is a simple demonstration of the installation and action of the interrupt. The program is based on the circuit in Figure 11-4.

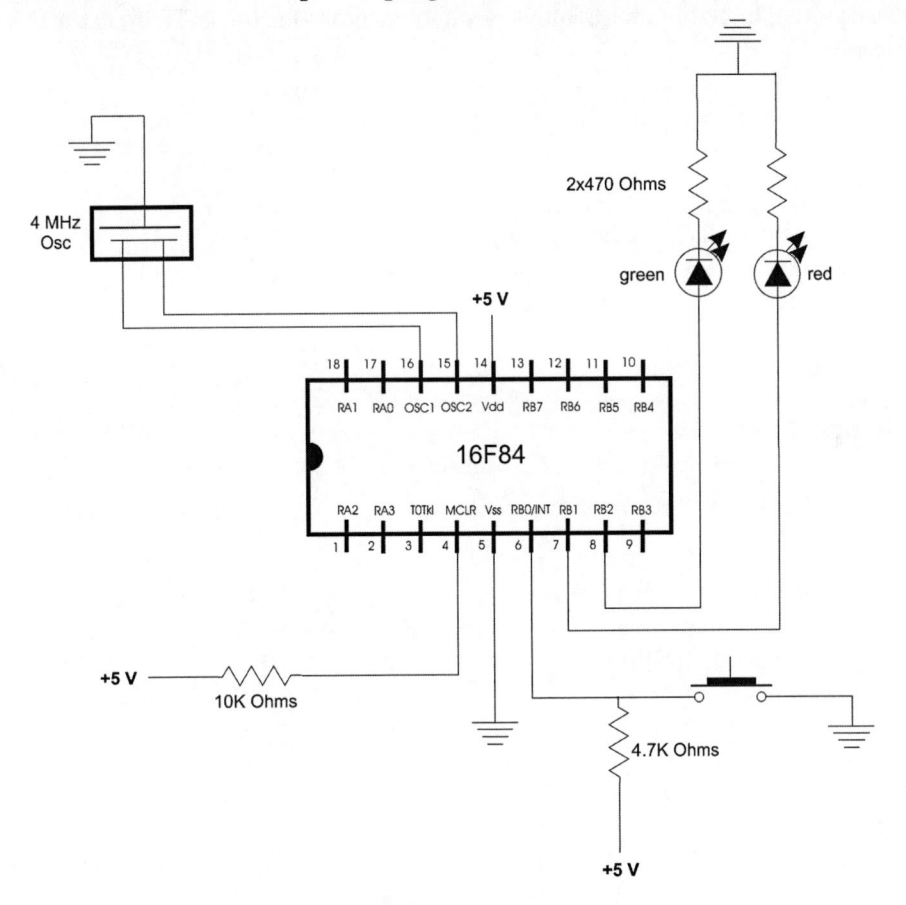

Figure 11-4 Circuit for RB0 Interrupt Demonstration

In the circuit of Figure 11-4, a pushbutton switch is wired to the RB0 port. It is this switch which produces the interrupt when pressed. A red LED is wired to port RB1 and a green LED to port RB2. The main program flashes the green LED on and off at a rate of approximately one-half second. The red LED is toggled on and off when the pushbutton switch is pressed. The switch contains a 4.7K Ohm resistor that keeps the port high until the contact is made and sent to ground. This makes the switch active low and the interrupt is programmed on the falling edge of the signal, which takes place when the contact is made.

RB0 Interrupt Initialization

In order to initialize the RB0 interrupt, the following operations must take place:
1. Port-B, line 0, must be initialized for input.
2. The interrupt source must be set to take place either on the falling or the rising edge of the signal.
3. The external interrupt flag (INTF in the INTCON Register) must be initially cleared.
4. Global interrupts must be enabled by setting the GIE bit in the INTCON Register.
5. The External Interrupt on RB0 must be enabled by setting the INTE bit in the INTCON Register.

The following code fragment, from the program RB0Int in the book's online software package, performs these operations:

```
;===============================
;      interrupt handler
;===============================
        org             0x04
        goto            IntServ
;===============================
;      main program
;===============================
main:
; Set up interrupt on falling edge
; by clearing OPTION register bit 6
        movlw   b'10111111'
        option
        movlw   b'11111111'     ; Set Port-A for input
        tris    porta           ; (not necessary for this program)
        movlw   b'00000001'     ; Port-B bit 0 is input
        tris    portb           ; all others are output
        clrf    portb           ; All Port-B to 0
; Initially turn on LED
        bsf     portb,0         ; Set line 0 bit
;===============================
;      setup interrupts
;===============================
; Clear external interrupt flag (intf = bit 1)
        bcf     INTCON,intf     ; Clear flag
; Enable global interrupts (gie = bit 7)
; Enable RB0 interrupt (inte = bit 4)
        bsf     INTCON,gie      ; Enable global int (bit 7)
        bsf     INTCON,inte     ; Enable RB0 int (bit 4)
;===============================
;        flash LED
```

```
;============================
; Program flashes LED wired to Port-B, line 2
lights:
        movlw   b'00000010'         ; Mask with bit 1 set
        xorwf   portb,f             ; Complement bit 1
        call    long_delay          ; Local delay routine
        call    long_delay
        call    long_delay
        goto    lights
```

RB0 Interrupt Service Routine

The Service Routine for the RB0 interrupt depends on the specific application. Nevertheless, the following processing steps should be considered:

1. Determine if the source is an RB0 interrupt.
2. Clear the RB0 interrupt flag (INTF bit) in the INTCON Register.
3. Save the context. Which registers and variables need to be saved depends on the specific application.
4. Perform the interrupt action.
5. Restore the context.
6. Return from the interrupt with the **retfie** instruction.

In addition, the interrupt handler may have to perform operations that are specific to the application. For example, debounce a switch or initialize local variables. The following Interrupt Service routine is from the program RB0Int in the book's on-line software:

```
;==========================================================
;               Interrupt Service Routine
;==========================================================
; Service routine receives control when there is
; action on pushbutton switch wired to port-B, line 0
IntServ:
; First test if source is an RB0 interrupt
        btfss   INTCON,INTF         ; INTF flag is RB0 interrupt
        goto    notRB0              ; Go if not RB0 origin
; Save context
        movwf   old_w               ; Save w register
        swapf   STATUS,w            ; STATUS to w
        movwf   old_status          ; Save STATUS
;=========================
;    interrupt action
;=========================
; Debounce switch
;       Logic:
;       Debounce algorithm consists in waiting until the
; same level is repeated on a number of samplings of the
```

```
; switch. At this point the RB0 line is clear since the
; interrupt takes place on the falling edge. The routine
; waits until the low value is read several times.
        movlw   D'10'       ; Number of repetitions
        movwf   count2      ; To counter
wait:
; Check to see that port-B bit 0 is still 0
; If not, wait until it changes
        btfsc   portb,0     ; Is bit set?
        goto    exitISR     ; Go if bit not 0
; At this point RB0 bit is clear
        decfsz  count2,f    ; Count this iteration
        goto    wait                ; Continue if not zero
; Interrupt action consists of toggling bit 2 of
; port-B to turn LED on and off
        movlw   b'00000100' ; Xoring with a 1-bit produces
                            ; the complement
        xorwf   portb,f     ; Complement bit 2, port-B
;=========================
;       exit ISR
;=========================
exitISR:
; Restore context
        swapf   old_status,w ; Saved status to w
        movfw   STATUS       ; To STATUS register
        swapf   old_w,f      ; Swap file register in itself
        swapf   old_w,w      ; re-swap back to w
notRB0:
; Reset interrupt
        bcf     INTCON,intf  ; Clear INTCON bit 1
        retfie
```

Note that the interrupt handler listed previously contains a debouncing routine that cleans the switch's signal. In this particular implementation the detection of a signal of the wrong value determines that the interrupt is aborted. For the particular switch used in the test circuit this approach seemed to work better. Alternatively, the routine can be designed so that if a wrong edge is detected, execution continues in the wait loop. In any case, the entire complication of software debouncing can be avoided by debouncing the switch in hardware.

11.3.2 Wakeup from SLEEP Using the RB0 Interrupt

The PIC microcontroller *sleep mode* provides a useful mechanism for saving power. It is particularly useful in battery-operated devices.

The sleep mode is activated by executing the **SLEEP** instruction; it suspends all normal operations and switches of the clock oscillator.

Interrupts

The sleep mode is suitable for applications that are not required to run continuously. For example, a device that records temperature at daybreak can be designed so that a light-sensitive switch generates an interrupt that turns the device on each morning. Once the data is recorded, the device goes into the sleep mode until the next daybreak.

Several events can make the device wake up from the sleep mode:

1. A device reset on the !MCLR pin
2. Watchdog timer wake-up signal, if WDT is enabled
3. Interrupt on RB0 line
4. Port change interrupt on RB4 to RB7 lines
5. EEPROM write complete interrupt

In the sleep mode, the device is placed on a power-down state that generates the lowest power consumption. The system clock is turned off in the sleep mode so signals that depend on the clock cannot be used to terminate the sleep. If enabled, the Watchdog Timer is cleared by the sleep instruction but keeps running. The PD bit in the STATUS register is also cleared and the TO bit is set. The ports maintain the status they had before the **SLEEP** instruction was executed.

The TO and PD bits in the STATUS register can be used to determine the cause of wake-up, since the TO bit is cleared if a Watchdog Timer wake-up took place. The corresponding interrupt enable bit must be set for the device to wake-up up due to an interrupt. Wake-up takes place regardless of the state of the *General Interrupt Enable (GIE)* bit. If the bit is clear, the device continues execution at the instruction following **SLEEP**. Otherwise, the device executes the instruction after the **SLEEP** instruction and then branches to the interrupt address. If the execution of the instruction following **SLEEP** is undesirable, the program should contain a **NOP** instruction after the **SLEEP** instruction.

The SleepDemo Program

The program named SleepDemo in the book's online software package is a trivial demonstration of using the RB0 interrupt to wake the processor from sleep mode. The program can be tested using the circuit in Figure 11-4. SleepDemo flashes the green LED at ½ second intervals during 20 iterations and then goes into sleep mode. Pressing the pushbutton switch on line RB0 generates an interrupt that wakes the processor from sleep mode. The following code fragment shows the coding of the main loop in the program:

```
;============================
;     flash LED 20 times
;============================
wakeUp:
; Program flashes LED wired to port-B, line 2
; 20 times before entering the sleep state
        movlw   D'20'     ; Number of iterations
        movwf   count2    ; To counter
```

```
lights:
        movlw    b'00000010' ; Mask with bit 1 set
        xorwf    portb,f  ; Complement bit 1
        call     long_delay
        call     long_delay
        call     long_delay
        decfsz   count2   ; Decrement counter
        goto     lights
; 20 iterations have taken place
        clrwdt            ; Clear WDT
        sleep
        nop               ; Recommended!
        goto     wakeUp   ; Resume execution
```

In the SleepDemo program the Interrupt Service Routine does nothing. Its coding is as follows:

```
;===========================================================
;               Interrupt Service Routine
;===========================================================
; The interrupt service routine performs no operation
IntServ:
        bcf      INTCON,INTF    ; Clear flag
        retfie
```

The initialization of the RB0 interrupt is identical to the one in the RB0Int program previously listed.

11.3.3 Port-B Bits 4-7 Status Change Interrupt

In the PIC 16F84 microcontroller, a change of input signal on Port-B, lines 4 to 7, generates an interrupt. This interrupt sets the RBIF bit in the INTCON Register to indicate that at least one of the ports have changed value. The port change takes place when the port's previous value changes from logic one to logic zero or vice versa. In order for port pins to recognize this interrupt, they must have been defined as input. If any one of the port pins (4 to 7) is defined as output the interrupt takes place. The status change of the ports is in reference to the last time port-B was read.

The principal application of this interrupt source is in detecting several different interrupt sources. Its principal disadvantage is that it forces the declaration of four port-B lines as input, although during processing not all lines need be recognized as interrupt sources. The conclusion is that applications that only need a single external interrupt source should use the RB0 interrupt described in previous sections. Only applications that require more than one external interrupt should use the Port-B lines 4 to 7 interrupt on change source.

Interrupts

Since the interrupt takes place on any status change (high-to-low or low-to-high) the service routine executes on both signal edges. If interrupt processing is required on only one edge, that is, either when the port goes high or low, then the filtering must be performed in software. The circuit in Figure 11-5 allows testing the Port-B Status Change Interrupt.

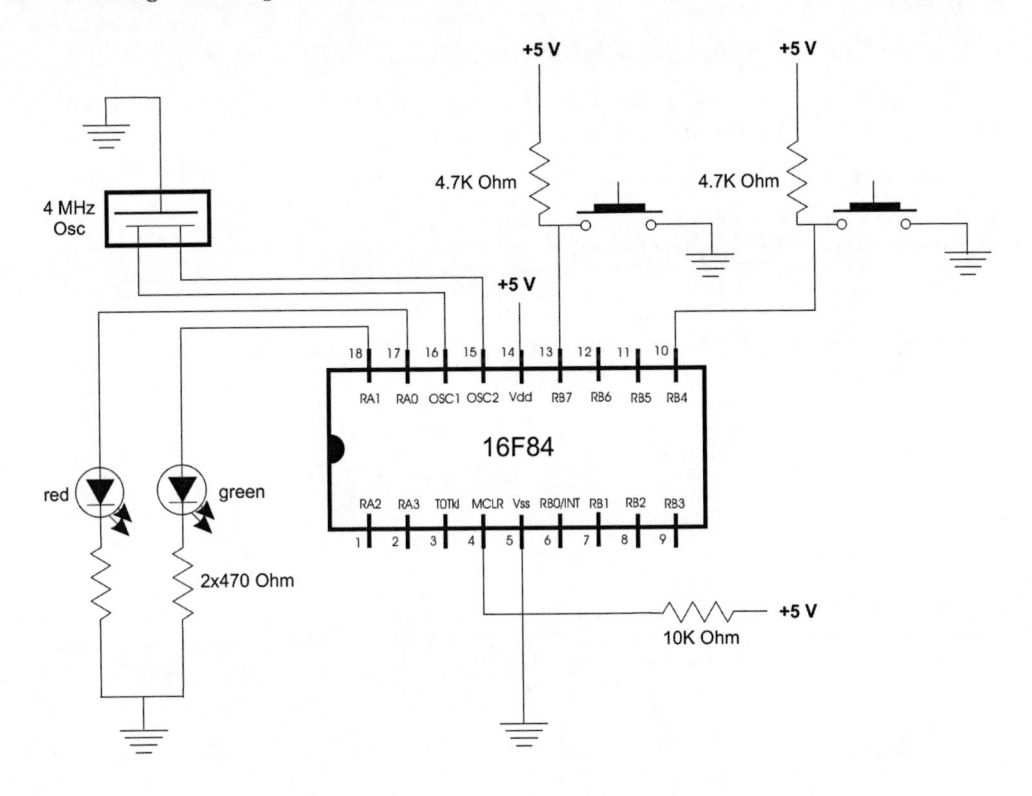

Figure 11-5 Circuit for Testing the Port-B Status Change Interrupt

In the circuit of Figure 11-5, a pushbutton switch is wired to the RB7 port and another one to RB4. Both of these switches produce the interrupt when pressed. A red LED is wired to port RA1 and a green LED to port RA0. The red and green LEDs are toggled on and off when the corresponding pushbutton switches are pressed. The switches contain a 4.7K Ohm resistor that keeps the port high until the contact is made and sent to ground. This makes both switches active low and the interrupt is programmed on the falling edge of the signal.

RB4-7 Interrupt Initialization

In order to initialize the RB4-7 change interrupt the following operations must take place:

1. Port-B lines 4 to 7 must be initialized for input.
2. The interrupt source must be set to take place either on the falling or the rising edge of the signal.

3. The RB port change interrupt flag (RBIF in the INTCON Register) must be initially cleared.
4. Global interrupts must be enabled by setting the GIE bit in the INTCON Register.
5. The RB port change interrupt must be enabled by setting the RBIE bit in the INTCON Register.
6. Internal pull-ups on port-B should be disabled in the OPTION register.

The following code fragment from the program RB4to7Int in the book's online software package shows the required processing:

```
;==============================
;          main program
;==============================
main:
; Disable port-B internal pull-ups
; Interrupts on falling edge of pushbutton action
          Movlw               b'10111111'
          option
; Wiring:
;          7  6  5  4  3  2  1  0   <= port-B
;          |        |_____ red pushbutton
;          |_____ black pushbutton
;
;          7  6  5  4  3  2  1  0   <= Port-A
;                            |  |_____ red LED
;                            |_____ green LED
;
          movlw     b'00000000'     ; Set Port-A for ouput
          tris      porta
          movlw     b'11110000'     ; Port-B bit 0-3 are output
                                    ; bits 4-7 are input
          tris      portb           ; all others are output
          clrf      portb           ; All port-B to 0
          movlw     b'00000000'     ; Zero to w
          movwf     bitsB47         ; Store in local variable
; Initially turn on LEDs
          bsf       porta,0         ; Set LEDs on line 0
          bsf       porta,1         ; and on line 1
;==============================
;         setup interrupts
;==============================
; Clear external interrupt flag (intf = bit 1)
          bcf       INTCON,rbif     ; Clear flag
; Enable global interrupts (gie = bit 7)
; Enable RB0 interrupt (inte = bit 4)
          bsf       INTCON,gie      ; Enable global int (bit 7)
          bsf       INTCON,rbie     ; Enable RB0 int (bit 3)
```

RB4-7 Change Interrupt Service Routine

The Service Routine for the RB4-7 change interrupt depends on the specific application. Nevertheless, the following processing steps should be considered:

1. Determine if the source is an RB4-7 change interrupt.
2. Clear the RBIF interrupt flag in the INTCON Register.
3. Save the context. Which registers and variables need to be saved depends on the specific application.
4. Perform the interrupt action.
5. Restore the context.
6. Return from the interrupt with the retfie instruction.

In addition, the interrupt handler may have to perform operations that are specific to the application; for example, debounce a switch or initialize local variables. The following Interrupt Service routine is from the program RB4to7Int in the book's online software:

```
;===========================================================
;                Interrupt Service Routine
;===========================================================
; Service routine receives control whenever any of
; port-B lines 4 to 7 change state
IntServ:
; First test: make sure source is an RB4-7 interrupt
        btfss   INTCON,rbif     ; RBIF flag is interrupt
        goto    notRBIF         ; Go if not RBIF origin
; Save context
        movwf   old_w           ; Save w register
        swapf   STATUS,w        ; STATUS to w
        movwf   old_status      ; Save STATUS
;=========================
;    interrupt action
;=========================
; The interrupt occurs when any of port-B bits 4 to 7
; have changed status.
        movf    portb,w         ; Read port-B bits
        movwf   temp            ; Save reading
        xorwf   bitsB47,f       ; Xor with old bits,
                                ; result in f
; Test each meaningful bit (4 and 7 in this example)
        btfsc   bitsB47,4       ; Test bit 4
        goto    bit4Chng ; Routine for changed bit 4
; At this point bit 4 did not change
        btfsc   bitsB47,7       ; Test bit 7
        goto    bit7Chng ; Routine for changed bit 7
; Invalid port line change. Exit
        goto    pbRelease
```

```
;========================
; bit 4 change routine
;========================
; Check for signal falling edge, ignore if not
bit4Chng:
        btfsc   portb,4         ; Is bit 4 high
        goto    pbRelease       ; Bit is high. Ignore
; Toggling bit 1 of Port-A turns LED on and off
        movlw   b'00000010'     ; Xoring with a 1-bit produces
                                ; the complement
        xorwf   porta,f         ; Complement bit 1, Port-A
        goto    pbRelease
;========================
; bit 7 change routine
;========================
; Check for signal falling edge, ignore if not
bit7Chng:
        btfsc   portb,7         ; Is bit 7 high
        goto    exitISR         ; Bit is high. Ignore
; Toggling bit 0 of Port-A turns LED on and off
        movlw   b'00000001'     ; Xoring with a 1-bit produces
                                ; the complement
        xorwf   porta,f         ; Complement bit 1, Port-A
;
pbRelease:
        call    delay           ; Debounce switch
        movf    portb,w  ; Read port-B into w
        andlw   b'10010000'  ; Eliminate unused bits
        btfsc   STATUS,z  ; Check for zero
        goto    pbRelease       ; Wait
; At this point all port-B pushbuttons are released
;========================
;         exit ISR
;========================
exitISR:
; Store new value of port-B
        movf    temp,w          ; This port-B value to w
        movwf   bitsB47         ; Store
; Restore context
        swapf   old_status,w    ; Saved status to w
        movfw   STATUS   ; To STATUS register
        swapf   old_w,f  ; Swap file register in itself
        swapf   old_w,w  ; re-swap back to w
; Reset,interrupt
notRBIF:
        bcf     INTCON,rbif     ; Clear INTCON bit 0
        retfie
```

Processing by the interrupt service routine is straightforward. The code first determines which line caused the interrupt and takes the corresponding action in each case. In either case, the handler waits until all pushbuttons have been released before returning from the interrupt. This serves to debounce the switches.

11.4 Sample Programs

The following programs demonstrate the programming discussed in this chapter.

11.4.1 The RB0Int Program

```
; File: RB0Int.ASM
; Date: April 22, 2006
; Author: Julio Sanchez
; Processor: 16F84A
;
; Description:
; Program to test interrupt on port RB0
; A pushbutton switch is connected to port RB0.
; The pushbutton toggles a LED on port-B, line 2
; Another LED on port-B, line 1, flashes on and off
; at 1/2 second intervals
;============================
;         switches
;============================
; Switches used in __config directive:
;   _CP_ON         Code protection ON/OFF
; * _CP_OFF
; * _PWRTE_ON     Power-up timer ON/OFF
;   _PWRTE_OFF
;   _WDT_ON       Watchdog timer ON/OFF
; * _WDT_OFF
;   _LP_OSC       Low power crystal occilator
; * _XT_OSC       External parallel resonator/crystal oscillator

;   _HS_OSC       High speed crystal resonator (8 to 10 MHz)
;                 Resonator: Murate Erie CSA8.00MG = 8 MHz
;   _RC_OSC       Resistor/capacitor oscillator (simplest, 20%
error)
; |
; |_____ * indicates setup values

;==========================
; setup and configuration
;==========================
         processor 16f84A
         include   <p16f84A.inc>
         __config  _XT_OSC & _WDT_OFF & _PWRTE_ON & _CP_OFF
;=========================================================
```

```
;                 variables in PIC RAM
;=========================================================
; Local variables
        cblock   0x0d                ; Start of block
        J                            ; counter J
        K                            ; counter K
        count1                       ; Auxiliary counter
        count2                       ; ISR counter
        old_w                        ; Context saving
        old_STATUS  ; Idem
        endc

;=========================================================
;                 m a i n   p r o g r a m
;=========================================================
        org      0                   ; start at address 0
        goto     main
;
;=============================
;       interrupt handler
;=============================
        org              0x04
        goto     IntServ
;=============================
;       main program
;=============================
main:
; Set up interrupt on falling edge
; by clearing OPTION register bit 6
        movlw    b'10111111'
        option
        movlw    b'11111111'         ; Set port a for input
        tris     PORTA
        movlw    b'00000001'         ; Port-B bit 0 is input
        tris     PORTB               ; all others are output
        clrf     PORTB               ; All port-B to 0
; Initially turn on LED
        bsf      PORTB,0             ; Set line 0 bit
;=============================
;       setup interrupts
;=============================
; Clear external interrupt flag (INTF = bit 1)
        bcf      INTCON,INTF         ; Clear flag
; Enable global interrupts (GIE = bit 7)
; Enable RB0 interrupt (INTE = bit 4)
        bsf      INTCON,GIE          ; Enable global int (bit 7)
        bsf      INTCON,INTE         ; Enable RB0 int (bit 4)
;=============================
```

```
;               flash LED
;==============================
; Program flashes LED wired to Port-B, line 2
lights:
        movlw   b'00000010'     ; Mask with bit 1 set
        xorwf   PORTB,f         ; Complement bit 1
        call    long_delay
        call    long_delay
        call    long_delay
        goto    lights
;==========================================================
;               Interrupt Service Routine
;==========================================================
; Service routine receives control when there is
; action on pushbutton switch wired to Port-B, line 0
IntServ:
; First test if source is an RB0 interrupt
        btfss   INTCON,INTF     ; INTF flag is RB0 interrupt
        goto    notRB0          ; Go if not RB0 origin
; Save context
        movwf   old_w           ; Save w register
        swapf   STATUS,w ;      ; STATUS to w
        movwf   old_STATUS      ; Save STATUS
; Make sure that interrupt occurred on the falling edge
; of the signal. If not, abort handler
        btfsc   PORTB,0         ; Is bit set?
        goto    exitISR         ; Go if clear
;=========================
;   interrupt action
;=========================
; Debounce switch
;       Logic:
;       Debounce algorithm consists in waiting until the
; same level is repeated on a number of samplings of the
; switch. At this point the RB0 line is clear since the
; interrupt takes place on the falling edge. An initial
; short delay makes sure that spikes are ignored.
        movlw   D'10'           ; Number of repetitions
        movwf   count2          ; To counter
wait:
; Check to see that port-B bit 0 is still 0
; If not, wait until it changes
        btfsc   PORTB,0         ; Is bit set?
        goto    exitISR         ; Go if bit not 0
; At this point RB0 bit is clear
        decfsz  count2,f ;      Count this iteration
        goto    wait            ; Continue if not zero
; Interrupt action consists of toggling bit 2 of
```

```
; port-B to turn LED on and off
        movlw   b'00000100'         ; Xoring with a 1-bit produces
                                    ; the complement
        xorwf   PORTB,f             ; Complement bit 2, port-B
;=========================
;           exit ISR
;=========================
exitISR:
; Restore context
        swapf   old_STATUS,w        ; Saved STATUS to w
        movfw   STATUS              ; To STATUS register
        swapf   old_w,f             ; Swap file register in itself
        swapf   old_w,w             ; re-swap back to w
; Reset,interrupt
notRB0:
        bcf     INTCON,INTF         ; Clear INTCON bit 1
        retfie
;======================
;   Procedure to delay
;     10 machine cycles
;======================
delay:
        movlw   D'4'                ; Repeat 12 machine cycles
        movwf   count1              ; Store value in counter
repeat
        decfsz  count1,f            ; Decrement counter
        goto    repeat              ; Continue if not 0
        return
;=============================
;   long delay sub-routine
;       (for debugging)
;=============================
long_delay
        movlw   D'200'      ; w = 200 decimal
        movwf   J           ; J = w
jloop:  movwf   K           ; K = w
kloop:  decfsz  K,f         ; K = K-1, skip next if zero
        goto    kloop
        decfsz  J,f         ; J = J-1, skip next if zero
        goto    jloop
        return
        end
```

11.4.2 The SleepDemo Program

```
; File: SleepDemo
; Date: April 25, 2006
```

Interrupts

```
; Author: Julio Sanchez
; Processor: 16F84A
;
; Description:
; Program to use the External Interrupt on port RB0
; to terminate the power-down state caused by the
; SLEEP instruction. A pushbutton switch is connected to
; port RB0. The pushbutton generates the interrupt that
; ends the SLEEP conditions.
; Demonstration:
; A LED on port-B, line 1, flashes on and off at 1/2
; second intervals for 20 iterations. At that time the
; program enters the SLEEP condition. Pressing the
; pushbutton switch on line RB0 generates the interrupt
; that ends the SLEEP.
;===========================
;          switches
;===========================
; Switches used in __config directive:
;   _CP_ON         Code protection ON/OFF
; * _CP_OFF
; * _PWRTE_ON      Power-up timer ON/OFF
;   _PWRTE_OFF
;   _WDT_ON        Watchdog timer ON/OFF
; * _WDT_OFF
;   _LP_OSC        Low power crystal occilator
; * _XT_OSC        External parallel resonator/crystal oscillator

;   _HS_OSC        High speed crystal resonator (8 to 10 MHz)
;                  Resonator: Murate Erie CSA8.00MG = 8 MHz
;   _RC_OSC        Resistor/capacitor oscillator (simplest, 20%
error)
; |
; |_____ * indicates setup values

;=========================
; setup and configuration
;=========================
        processor 16f84A
        include   <p16f84A.inc>
        __config  _XT_OSC & _WDT_OFF & _PWRTE_ON & _CP_OFF


;========================================================
;              variables in PIC RAM
;========================================================
; Local variables
        cblock    0x0d              ; Start of block
        J                           ; counter J
```

```
            K               ; counter K
            count1          ; Auxiliary counter
            count2          ; Second auxiliary counter
            old_w           ; Context saving
            old_STATUS  ; Idem
            endc

;===========================================================
;                m a i n   p r o g r a m
;===========================================================
            org     0           ; start at address 0
            goto    main
;
;==============================
;       interrupt handler
;==============================
            org             0x04
            goto    IntServ
;==============================
;       main program
;==============================
main:
; Set up interrupt on falling edge
; by clearing OPTION register bit 6
            movlw   b'10111111'
            option
            movlw   b'11111111'     ; Set port a for input
            tris    PORTA
            movlw   b'00000001'     ; Port-B bit 0 is input
            tris    PORTB           ; all others are output
            clrf    PORTB           ; All port-B to 0
;==============================
;       setup interrupts
;==============================
; Clear external interrupt flag (INTF = bit 1)
            bcf     INTCON,INTF     ; Clear flag
; Enable global interrupts (GIE = bit 7)
; Enable RB0 interrupt (INTE = bit 4)
            bsf     INTCON,GIE      ; Enable global int (bit 7)
            bsf     INTCON,INTE     ; Enable RB0 int (bit 4)
;==============================
;   flash LED 20 times
;==============================
wakeUp:
; Program flashes LED wired to port-B, line 2
; 20 times before entering the sleep state
            movlw   D'20'           ; Number of iterations
            movwf   count2          ; To counter
```

Interrupts

```
lights:
        movlw   b'00000010'         ; Mask with bit 1 set
        xorwf   PORTB,f             ; Complement bit 1
        call    long_delay
        call    long_delay
        call    long_delay
        decfsz  count2,f ; Decrement counter
        goto    lights
; 20 iterations have taken place
        sleep
        nop                         ; Recommended!
        goto    wakeUp              ; Resume execution

;============================================================
;              Interrupt Service Routine
;============================================================
; The interrupt service routine performs no operation
IntServ:
        bcf     INTCON,INTF         ; Clear flag
        retfie

;=============================
;   long delay sub-routine
;=============================
long_delay
        movlw   D'200'      ; w = 200 decimal
        movwf   J           ; J = w
jloop:
        movwf   K           ; K = w
kloop:
        decfsz  K,f         ; K = K-1, skip next if zero
        goto    kloop
        decfsz  J,f         ; J = J-1, skip next if zero
        goto    jloop
        return

        end
```

11.4.3 The RB4to7Int Program

```
; File: RB4to7Int.ASM
; Date: April 26, 2006
; Author: Julio Sanchez
; Processor: 16F84A
;
; Description:
; Program to test the port-B, bits 4 to 7, STATUS
```

```
; change interrupt. Pushbutton switches are connected
; to port-B lines 4 and 7. A red LED is wired to port
; RA1 and a green LED to port RA0. The pushbuttons
; generate interrupts that toggle a LEDs on and off.
;===========================
;          switches
;===========================
; Switches used in __config directive:
;   _CP_ON           Code protection ON/OFF
; * _CP_OFF
; * _PWRTE_ON        Power-up timer ON/OFF
;   _PWRTE_OFF
;   _WDT_ON          Watchdog timer ON/OFF
; * _WDT_OFF
;   _LP_OSC          Low power crystal occilator
; * _XT_OSC          External parallel resonator/crystal oscillator

;   _HS_OSC          High speed crystal resonator (8 to 10 MHz)
;                    Resonator: Murate Erie CSA8.00MG = 8 MHz
;   _RC_OSC          Resistor/capacitor oscillator (simplest, 20%
error)
; |
; |_____ * indicates setup values

;=========================
; set up and configuration
;=========================
        processor 16f84A
        include   <p16f84A.inc>
        __config  _XT_OSC & _WDT_OFF & _PWRTE_ON & _CP_OFF
;======================================================
;               variables in PIC RAM
;======================================================
; Local variables
        cblock    0x0d      ; Start of block
        J                   ; counter J
        K                   ; counter K
        count1              ; Auxiliary counter
        count2              ; ISR counter
        old_w               ; Context saving
        old_STATUS ; Idem
        bitsB47             ; Storage for previous value
                            ; in port-B bits 4-7
        temp                ; Temporary storage
        endc

;======================================================
;                main    program
```

Interrupts

```
;===========================================================
        org     0               ; start at address 0
        goto    main
;
;==============================
;       interrupt handler
;==============================
        org                 0x04
        goto    IntServ
;==============================
;       main program
;==============================
main:
; Disable port-B internal pullups
; Interrupts on falling edge of pushbutton action
        movlw   b'10111111'
        option
; Wiring:
;       7   6   5   4   3   2   1   0   <= port-B
;       |           |_____ red pushbutton
;       |_____ black pushbutton
;
;       7   6   5   4   3   2   1   0   <= Port-A
;                           |   |_____ red LED
;                           |_____ green LED
;
        movlw   b'00000000'     ; Set Port-A for ouput
        tris    PORTA
        movlw   b'11110000'     ; Port-B bit 0-3 are output
                                ; bits 4-7 are input
        tris    PORTB           ; all others are output
        clrf    PORTB           ; All Port-B to 0
        movlw   b'00000000'     ; Zero to w
        movwf   bitsB47         ; Store in local variable
; Initially turn on LEDs
        bsf     PORTA,0         ; Set LEDs on line 0
        bsf     PORTA,1         ; and on line 1
;==============================
;       set up interrupts
;==============================
; Clear external interrupt flag (intf = bit 1)
        bcf     INTCON,RBIF     ; Clear flag
; Enable global interrupts (GIE = bit 7)
; Enable RB0 interrupt (inte = bit 4)
        bsf     INTCON,GIE      ; Enable global int (bit 7)
        bsf     INTCON,RBIE     ; Enable RB0 int (bit 3)
;==============================
;           flash LED
```

```
;==============================
; Main program does nothing. All action takes place in
; Interrupt Service Routine
lights:
        nop
        goto    lights
;============================================================
;               Interrupt Service Routine
;============================================================
; Service routine receives control whenever any of
; port-B lines 4 to 7 change state
IntServ:
; First test: make sure source is an RB4-7 interrupt
        btfss   INTCON,RBIF     ; RBIF flag is interrupt
        goto    notRBIF         ; Go if not RBIF origin
; Save context
        movwf   old_w           ; Save w register
        swapf   STATUS,w ; STATUS to w
        movwf   old_STATUS      ; Save STATUS
;=========================
;   interrupt action
;=========================
; The interrupt occurs when any of Port-B bits 4 to 7
; have changed STATUS.
        movf    PORTB,w         ; Read Port-B bits
        movwf   temp            ; Save reading
        xorwf   bitsB47,f       ; Xor with old bits,
                                ; result in f
; Test each meaningful bit (4 and 7 in this example)
        btfsc   bitsB47,4       ; Test bit 4
        goto    bit4Chng ; Routine for changed bit 4
; At this point bit 4 did not change
        btfsc   bitsB47,7       ; Test bit 7
        goto    bit7Chng ; Routine for changed bit 7
; Invalid port line change. Exit
        goto    pbRelease
;========================
; bit 4 change routine
;========================
; Check for signal falling edge, ignore if not
bit4Chng:
        btfsc   PORTB,4         ; Is bit 4 high
        goto    pbRelease       ; Bit is high. Ignore
; Toggling bit 1 of Port-A turns LED on and off
        movlw   b'00000010'     ; Xoring with a 1-bit produces
                                ; the complement
        xorwf   PORTA,f         ; Complement bit 1, Port-A
        goto    pbRelease
```

```
;========================
; bit 7 change routine
;========================
; Check for signal falling edge, ignore if not
bit7Chng:
        btfsc   PORTB,7         ; Is bit 7 high
        goto    exitISR         ; Bit is high. Ignore
; Toggling bit 0 of Port-A turns LED on and off
        movlw   b'00000001'     ; Xoring with a 1-bit produces
                                ; the complement
        xorwf   PORTA,f         ; Complement bit 1, Port-A
;
pbRelease:
        call    delay           ; Debounce switch
        movf    PORTB,w         ; Read port-B into w
        andlw   b'10010000'     ; Eliminate unused bits
        btfsc   STATUS,Z        ; Check for zero
        goto    pbRelease       ; Wait
; At this point all port-B pushbuttons are released
;========================
;       exit ISR
;========================
exitISR:
; Store new value of port-B
        movf    temp,w          ; This port-B value to w
        movwf   bitsB47         ; Store
; Restore context
        swapf   old_STATUS,w    ; Saved STATUS to w
        movfw   STATUS          ; To STATUS register
        swapf   old_w,f         ; Swap file register in itself
        swapf   old_w,w         ; re-swap back to w
; Reset,interrupt
notRBIF:
        bcf     INTCON,RBIF     ; Clear INTCON bit 0
        retfie

;========================
;   Procedure to delay
;     10 machine cycles
;========================
delay:
        movlw   D'6'            ; Repeat 18 machine cycles
        movwf   count1          ; Store value in counter
repeat:
        decfsz  count1,f        ; Decrement counter
        goto    repeat          ; Continue if not 0
        return
;==============================
```

```
;       long delay sub-routine
;          (for debugging)
;=============================
long_delay
        movlw   D'200'   ; w = 200 decimal
        movwf   J        ; J = w
jloop:
        movwf   K        ; K = w
kloop:
        decfsz  K,f      ; K = K-1, skip next if zero
        goto    kloop
        decfsz  J,f      ; J = J-1, skip next if zero
        goto    jloop
        return

        end
```

Chapter 12

Timers and Counters

This chapter is about using the built-in timing and counting circuits on the 16F84. It relates to Chapter 11 since timing and counting operations can be set up to generate interrupts. The material also serves as background for Chapter 14, on serial communications, since these require precise pulses that are usually obtained through the timers.

12.0 The 16F84 Timer0 Module

One of the timers on the 16F84 PIC is known as the *Timer0 module*, the *free-running timer*, the *timer/counter*, or as *TMR0*. Timer0 is an internal 8-bit register that increments automatically with every PIC instruction cycle until the count overflows timer capacity. This takes place when the timer count goes from 0xff to 0x00. At that time, the timer restarts the count. The timer has the following characteristics:

1. A timer register that is readable and writeable by software
2. Can be powered by an external or internal clock
3. Timing edge for external clock can be selected
4. 8-bit software programmable prescaler
5. Interrupt capability
6. Can be used as a timer or as a counter

12.0.1 Timer0 Operation

Timer operation can be assigned to the internal clock or to the PIC's RA4/TOCKI pin. Bit 5 of the OPTION register (labeled TOCS) performs this selection. If TOCS is set, then the timer is linked to the RA4/TOCKI pin. In this mode, the timer is used as a counter. If TOCS is reset, then the timer uses the PIC's instruction cycle clock signal. If an external source is selected by setting the TOCS bit, then bit 4 of the OPTION register (labeled TOSE) allows selecting whether the timer increments on the high-to-low or low-to-high transition of the signal on the RA4/TOCKI pin. As shown in Figure 12-1, bits 6 and 7 of the OPTION register are not used in configuring the Timer0 module.

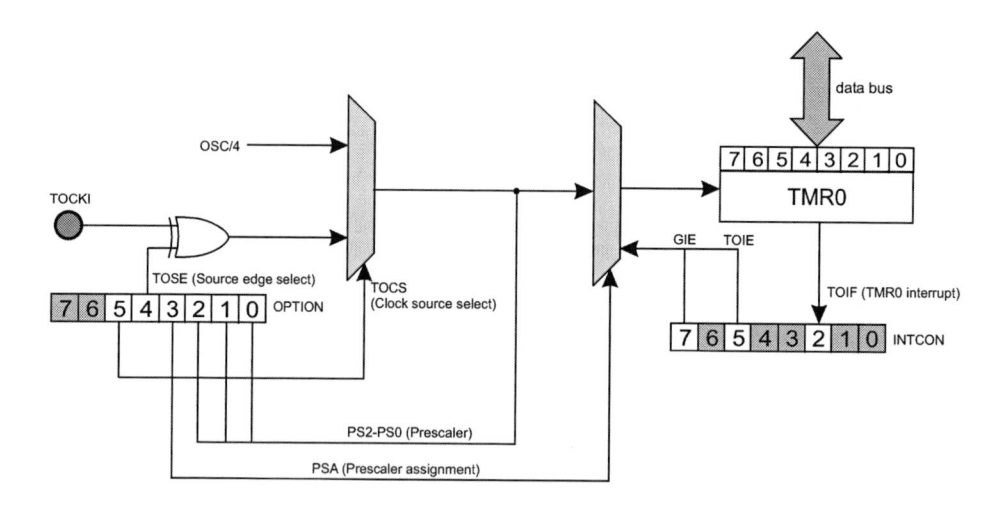

Figure 12-1 Timer0 Block Diagram

When used as a timer, Timer0 can be visualized as a register that increments with every instruction cycle at ¼ the clock rate, without using the prescaler. In a PIC equipped with a 4 Mhz oscillator the timer register increments at a rate of one pulse per millisecond. Since there are eight bits in the counter register, the value stored is in the range 0 to 255 decimal. When the counter overflows the register is reset. Figure 12-1 is a simplified block diagram of the Timer0 hardware.

Timer0 Interrupt

Software can read the timer register directly or set up the timer to generate an interrupt at every transition from 0xff to 0x00. The timer register can be accessed in bank 0, offset 0x01. The timer interrupt is enabled by setting bit 5 (labeled TOIE) of the INTCON register. In this case the *Global Interrupt Enable* bit (labeled *GIE*) of INTCON register must also be set. Once the timer interrupt is enabled, the *Timer Interrupt Flag*, assigned to bit 2 of the INTCON Register and labled TOIF, is set on every overflow of the timer register. At that time an interrupt takes place. The TOIF bit (Timer0 flag) must be cleared by the interrupt handler so that the timer interrupt can take place again. Later in this chapter we develop a sample program that uses Timer0 as an interrupt source.

Timer0 Prescaler

The counter prescaler consists of the three low-order bits in the OPTION register. These bits allow selecting eight possible values that serve as a divisor for the counter rate. When the prescaler is disabled, the counter rate is one-fourth the processor's clock speed. If the prescaler is set to the maximum value (255) then one of 255 clock signals actually reach the timer. Table 12-1 shows the prescaler settings and their action on the rate of the Timer0 module and the Watchdog Timer, covered later in this chapter.

Table 12.1
Prescaled Bits Selected Rates

BIT VALUE	TMR0 RATE	WDT RATE
000	1:2	1:1
001	1:4	1:2
010	1:8	1:4
011	1:16	1:8
100	1:32	1:16
101	1:64	1:32
110	1:128	1:64
111	1:256	1:128

The prescaler can be assigned to either Timer0 or the Watchdog Timer, but not to both. If bit 3 of the OPTION register is set, then the prescaler is assigned to the Watchdog Timer if it is clear it is assigned to the Timer0 module.

12.1 Delays Using Timer0

The simplest application of the Timer0 module is as an instruction cycle counter in implementing delay loops. Applications in which the Timer0 register is polled directly are said to use a *free running* timer. There are two advantages of using free running timers over conventional delay loops: the prescaler provides a way of slowing down the count, and the delay is independent of the number of machine cycles in the loop body. In most cases, it is easier to implement an accurate time delay using the Timer0 module than by counting instruction cycles.

Calculating the time taken by each counter iteration consists of dividing the clock speed by four. For example, a PIC running on a 4 Mhz oscillator clock increments the counter every 1 Mhz. If the prescaler is not used, the counter register is incremented at a rate of 1 µs; the timer beats at a rate of 1,000,000 times per second. If the prescaler is set to the maximum divisor value (256) then each increment of the timer takes place at a rate of 1,000,000/256 µs, which is approximately 3.906 ms. Since this is the slowest possible rate of the timer in a machine running at 4 Mhz, it is often necessary to employ supplementary counters in order to achieve larger delays.

The fact that the timer register (Tmr0) is both readable and writeable makes possible some interesting timing techniques. For example, an application can set the Timer register to an initial value and then count until a predetermined limit is reached. For example, if the difference between the limit and the initial value is 100, then the routine counts 100 times the timer rate per beat. In another example, if a routine allows the timer to start from zero and count unrestrictedly, then when the count reaches the maximum value (0xff) the routine would have introduced a delay of 256 times the timer beat rate, as is the case in the previous example, in which a maximum value was used in the prescaler and the timer ran at a rate of 1,000,000 beats per second. Applying the prescaler, each timer beat takes place at a rate of 1,000,000/256, or approximately 3,906 timer beats per second. If we develop a routine that delays execution until the maximum value has been reached in the counter register, then the delay can be calculated by dividing the number of beats per second

(3,906) by the number of counts in the delay loop. In this case, 3,906/256 results in a delay of approximately 15.26 iterations of the delay routine per second.

A general formula for calculating the number of timer beats per second is as follows:

$$T = \frac{C}{4PR}$$

where T is the number of clock beats per second, C is the system clock speed in Hz, P is the value stored in the prescaler, and R is the number of iterations counted in the TMR0 register. The range of both P and R in this formula is 1 to 256. Also, note that the reciprocal of T ($1/T$) gives the time delay, in seconds, per iteration of the delay routine.

12.1.1 Long Delay Loops

In the previous section we saw that even when using the largest possible prescaler and counting the maximum number of timer beats, the longest possible timer delay in a 4 Mhz system is approximately 1/15th of a second. Also consider that applications must sometimes devote the prescaler to the Watchdog Timer, which impedes its use in Timer0. Without the prescaler, the maximum delay is of approximately 3,906 timer beats per second. Applications that measure time in seconds or in minutes must find ways for keeping count of large numbers of repetitions of the timer beat.

In implementing counters for larger delays we have to be careful not to introduce round-off errors. For instance, in the previous example a timer cycles at the rate of 15.26 times per second. The closest integer to 15.26 is 15, so if we now set up a seconds counter that counts 15 iterations, the counter would introduce an error of approximately 2 percent.

Considering that in the previous example each iteration of the timer contains 256 individual beats, there are 3,906.25 individual timer beats per second at the maximum prescaled rate. So if we were to implement a counter to keep track of individual prescaled beats, instead of timer iterations, the count would proceed from 0 to 3,906 instead of from 0 to 15. Approximating 3,906.25 to the closest integer, 3,906, introduces a much smaller round-off error than approximating 15.26 with 15.

Finally, in this same example, we could eliminate the prescaler so that the timer beats at the clock rate, that is, at 1,000,000 beats per second. In this option, a counter that counts from 0 to 1,000,000 would have no intrinsic error due to round off.

Which solution is more adequate depends on the accuracy required by the application and the complexity tolerated. A timer counter in the range of 0 to 15 can be implemented in a single 8-bit register. A counter in the range 0 to 3,906 requires two bytes. One to count from 0 to 1,000,000 requires three bytes. Since arithmetic operations in the 16F84 are 8-bits, manipulating multiple-register counters requires more complicated processing.

How Accurate the Delay?

The actual implementation of a delay routine based on multi-byte counters presents some difficulties. If the timer register (TMR0) is used to keep track of timer beats, then detecting the end of the count poses a subtle problem. Intuitively, our program could detect timer overflow by reading the TMR0 and testing the zero flag in the status register. Since the **movf** instruction affects the zero flag, one could be tempted to code:

```
wait:
        movf            tmr0,w    ; Timer value into w
        btfss           status,z  ; Was it zero?
        goto            wait
; If this point is reached TMR0 has overflowed
```

But there is a problem: the timer ticks as each instruction executes. Since the **goto** instruction takes two machine cycles, it is possible that the timer overflows while the **goto** instruction is in progress; therefore the overflow condition would not be detected. One possible solution found in the Microchip documentation is to check for less than a nominal value by testing the carry flag, as follows:

```
wait1:
        movlw           0x03                ; 3 to w
        subwf           tmr0,w    ; Subtract w - TMR0
        btfsc           status,c  ; Test carry
        goto            wait1
```

One adjustment that is sometimes necessary in free running timers arises from the fact that when the TMR0 register is written, the count is inhibited for the following two instruction cycles. Software compensates for the skip by writing an adjusted value to the timer register. If the prescaler is assigned to Timer0, then a write operation to the timer register determines that the timer does not increment for four clock cycles.

The Black-Ammerman Method

A more elegant and accurate solution has been described by Roman Black in a Web article titled *Zero-Error One Second Timer*. Black credits Bob Ammerman with the suggestion of using Bresenham's algorithm for creating accurate PIC timer periods. In the Black-Ammerman method, the counter works in the background, either by being polled or interrupt-driven, so the program can continue executing while the counter runs. In both cases, the timer-count value is stored in a 3-byte register decremented by the software.

In their interrupt-driven version, **TMR0** generates an interrupt whenever the counter register overflows, that is, every 256th timer beat (assuming no prescaler). The interrupt handler routine decrements the mid-order register that holds the 3-byte timer count. This is appropriate since every unit in the mid-order register represents 256 units of the low-order counter, which in this case is the TMR0 register. If the mid-order register underflows when decremented, then the high-order one is decremented. If the high-order one underflows, then the count has reached zero and the delay ends. Since the counter is interrupt-driven, the processor continues to do other work in the foreground.

An even more ingenious option proposed by Black is a background counter that does not rely on interrupts. This is accomplished by introducing a 1:2 delay in the timer by means of the prescaler. Since now the timer beats at one-half the instruction rate, 128 timer cycles are required for one complete iteration at the full instruction rate. By testing the high-order bit of the timer counter, the routine detects when the count reaches 128. At that time the mid-range and high-range counter variables are updated (as in the non-interrupt version of the software described in the previous paragraph). The high-order bit of the timer is then cleared, but the low-order bits are not changed. This allows the timer counter not to lose step in the count, which remains valid until the next time the high-order bit is again set. During the period between the updating of the 3-byte counter and the next polling of the timer register, the program continues to perform other tasks.

Many of the details of the Black-Ammerman method are missing in our description. The reader should refer to the Internet article for a thorough coverage of this algorithm.

12.2 Timer0 as a Counter

In Section 12.0.1, we saw that Timer0 operation can be assigned to the PIC's RA4/TOCKI pin by setting bit 5 of the OPTION register (labeled TOCS). This mode is referred to as the *counter mode*. When the timer is set up to work as a counter, then bit 4 of the OPTION register (labeled TOSE) allows selecting whether the counter increments on the high-to-low or low-to-high transition of the signal.

When an external clock input is present in the RA4/TOCKI pin, it must meet certain requirements. Used for Timer0, these requirements ensure that the external source can be synchronized with the internal phase clock. When no prescaler is used, the external clock input must be high and low for at least twice the internal clock rate. In addition, there must be a resistor-capacitor induced delay of 20 ns on both the high and the low cycles.

When a prescaler is used, the external clock input must be high and low for at least four times the rate of the internal clock rate. In addition, there must be a resistor-capacitor induced delay of 40 ns on both the high and the low cycles.

Once the counter mode is enabled, any pulse on pin RA4/TOCKI is automatically counted in the TMR0 register. The mechanism can be compared to an automatic interrupt since no program action is required to keep track of the number of pulses. The routine can be coded so that when the timer count overflows, an interrupt is generated. The interrupt handler then increments a supplementary counter so that events that exceed 256 pulses are recorded. The program named Tmr0Counter developed later in this chapter and contained in the book's online software is an example of using the counter function of the Timer0 module.

12.3 Timer0 Programming

Software routines that use the Timer0 module range in complexity from simple approximate delay loops to configurable, interrupt-driven counters that must meet very high timing accuracy requirements. When the time period to be measured does not exceed the one obtained with the prescaler and the timer register count, then the coding is straightforward and the processing is uncomplicated. But often this is not the case. The following elements should be examined before attempting to design and code a Timer0-based routine:

1. What is the required accuracy of the timer delay?
2. Can the prescaler be used or is the prescaler devoted to the Watchdog Timer?
3. Does the program suspend execution while the delay is in progress, or does the application continue executing in the foreground?
4. Can the timer be interrupt-driven or must it be polled?
5. Will the delay be the same on all calls to the timer routine, or must the routine provide delays of different magnitude?
6. How long must the delay last?

In this section we explore several timer routines of different complexity and requirements. The first one uses the Timer0 module as a counter, as described in Section 12.2. Later, we develop a simple delay loop that uses the Timer0 register instead of an instruction count. We conclude with an interrupt-driven timer routine that can be changed to implement different delays.

12.3.1 Programming a Counter

The 16F84 can be programmed so that port RA4/TOCKI is used to count events or pulses by initializing the Timer0 module as a counter. Without interrupts, the process requires the following preparatory steps:

1. Port-A, line 4, (RA4/TOCKI) is defined for input.
2. The Timer0 register (TMR0) is cleared.
3. The Watchdog Timer internal register is cleared by means of the clrwdt instruction.
4. The OPTION register bits PSA and PS0:PS2 are initialized if the prescaler is to be used.
5. The OPTION register bit TOSE is set so as to increment the count on the high-to-low transition of the port pin if the port source is active low. Otherwise the bit is cleared.
6. The OPTION register bit TOCS is set to select action on the RA4/TOCKI pin.

Once the timer is set up as a counter, any pulse received on the RA4/TOCKI pin that meets the restrictions mentioned in Section 12.2 is counted in the TMR0 register. Software can read and write to the TMR0 register, located at address 0x01 in bank 0, in order to obtain or change the event count. If the timer interrupt is enabled when the timer is defined as a counter, the interrupt takes place every time the counter overflows, that is, when the count cycles from 0xff to 0x00.

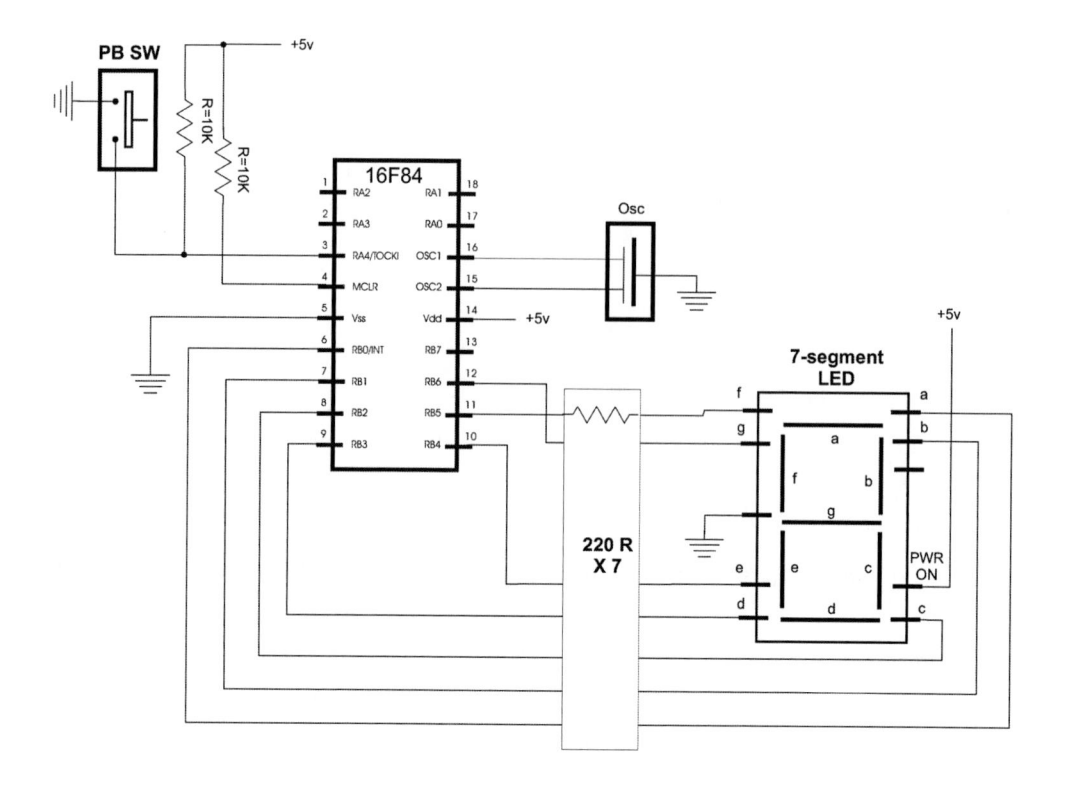

Figure 12-2 Test Circuit for Timer/Counter Program

A Timer/Counter Test Circuit

The circuit shown in Figure 12-2 contains a pushbutton switch wired to port RA4/TOCKI and a seven-segment LED display wired to Port-B lines 0 to 6.

The Tmr0Counter Program

The program named Tmr0Counter in the book's online software package uses the circuit in Figure 12-2 to demonstrate the programming of the Timer0 module in the counter mode. The program detects and counts action on the pushbutton switch wired to port RA4/TOCKI. The value of the count in hex digits ranging 0x00 to 0x0f is displayed in the seven-segment LED connected to Port-B.

The following code fragment shows the program's initialization routine to set up the ports and the timer:

```
main:
; Clear the Watchdog Timer and reset prescaler
        clrwdt
; Set up the OPTION regiser bit map
        movlw           b'10111000'
;       7  6  5  4  3  2  1  0 <= OPTION bits
;       |  |  |  |  |  |__|__|____ PS2-PS0 (prescaler bits)
;       |  |  |  |  |              Values for Timer0
```

```
;         |   |   |   |   |           *000 = 1:2    001 = 1:4
;         |   |   |   |   |            010 = 1:8    011 = 1:16
;         |   |   |   |   |            100 = 1:32   101 = 1:64
;         |   |   |   |   |            110 = 1:128 *111 = 1:256
;         |   |   |   |   |_____ PSA (prescaler assign)
;         |   |   |   |               *1 = to WDT
;         |   |   |   |                0 = to Timer0
;         |   |   |   |_____ TOSE (Timer0 edge select)
;         |   |   |                    0 = increment on low-to-high
;         |   |   |                   *1 = increment in high-to-low
;         |   |   |_____ TOCS (TMR0 clock source)
;         |   |                        0 = internal clock
;         |   |                       *1 = RA4/TOCKI bit source
;         |   |_____ INTEDG (Edge select)
;         |                           *0 = falling edge
;         |_____ RBPU (Pullup enable)
;                                      0 = enabled
;                                     *1 = disabled
          option
; Set up ports
          movlw     0x00        ; Set Port-B to output
          tris      portb
          clrf      portb       ; All Port-B to 0
; Port-A. Five low-order lines set for input
          movlw     B'00011111' ; w = 00011111 binary
          tris      porta       ; Port-A (lines 0 to 4) to input
```

Once the hardware is initialized, program operation consists of reading the value stored in TMR0, scaling this value to the display range 0 to 15, and displaying it on the seven-segment LED. Processing is as follows:

```
;==================================
; Check value in TMR0 and display
;==================================
; Every press of the pushbutton switch connected to line
; RA4/TOCKI adds one to the value in the TMR0 register.
; Loop checks this value, adjusts to the range 0 to 15
; and displays the result in the seven-segment LED on
; Port-B
;
checkTmr0:
          movf               mr0,w    ; Timer register to w
; Eliminate four high order bits
          andlw              b'00001111' ; Mask off high bits
; At this point the w register contains a 4-bit value
; in the range 0 to 0xf. Use this value (in w) to
; obtain seven-segment display code
;
          call               segment
```

```
                movwf       portb           ; Display switch bits
                goto        checkTmr0       ; Endless loop
;
;================================
;   routine to return 7-segment
;           codes
;================================
segment:
                addwf       PCL,f           ; PCL is program counter latch
                retlw       0x3f            ; 0 code
                retlw       0x06            ; 1
                retlw       0x5b            ; 2
                retlw       0x4f            ; 3
                retlw       0x66            ; 4
                retlw       0x6d            ; 5
                retlw       0x7d            ; 6
                retlw       0x07            ; 7
                retlw       0x7f            ; 8
                retlw       0x6f            ; 9
                retlw       0x77            ; A
                retlw       0x7c            ; B
                retlw       0x39            ; C
                retlw       0x5b            ; D
                retlw       0x79            ; E
                retlw       0x71            ; F
                retlw       0x7f            ; Just in case all on
```

The programming of seven-segment LEDs was discussed in Chapter 10.

12.3.2 Timer0 as a Simple Delay Timer

Perhaps the simplest use of the Timer0 module is to implement a delay loop. In this application, the Timer0 module is initialized to use the internal clock by setting the TOSE bit of the OPTION register. If the prescaler is to be used, as is most likely, the PSA bit is cleared and the desired prescaling is entered in bits PS2 to PS0 of the OPTION register. The circuit in Figure 12-3 allows testing several timer-related programs developed in this chapter.

The program named Timer0, in the book's online software package, uses a timer-based delay loop to flash in sequence eight LEDs that display the binary values from 0x00 to 0xff. The delay routine executes in the foreground, so that processing is suspended while the count is in progress. The initialization requires clearing the TOCS bit in the OPTION register to select the internal clock. The prescaler is assigned to Timer0 by clearing the PSA bit and bits PS2 to PS0 are set to assign a 1:256 prescale to the timer. The following code fragment shows the processing.

Timers and Counters

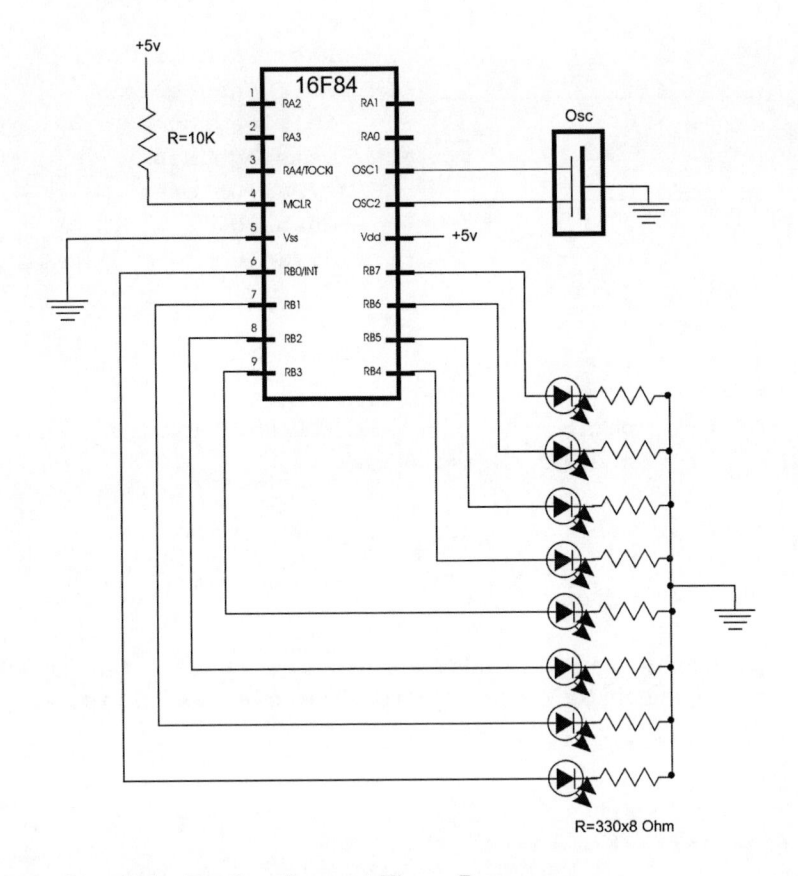

Figure 12-3 Circuit for Testing Several Timer Programs

```
main:
; Clear the Watchdog Timer and reset prescaler
        clrwdt
; Set up the OPTION register
        movlw               b'11010111'
;    7   6   5   4   3   2   1   0  <= OPTION bits
;    |   |   |   |   |   |__|__|_____ PS2-PS0 (prescaler bits)
;    |   |   |   |   |                Values for Timer0
;    |   |   |   |   |                000 = 1:2     001 = 1:4
;    |   |   |   |   |                010 = 1:8     011 = 1:16
;    |   |   |   |   |                100 = 1:32    101 = 1:64
;    |   |   |   |   |                110 = 1:128  *111 = 1:256
;    |   |   |   |   |_____  PSA (prescaler assign)
;    |   |   |   |                    1 = to WDT
;    |   |   |   |                   *0 = to Timer0
;    |   |   |   |_____  TOSE (Timer0 edge select)
;    |   |   |                        0 = increment on low-to-high
;    |   |   |                       *1 = increment in high-to-low
;    |   |   |_____  TOCS (TMR0 clock source)
;    |   |                           *0 = internal clock
```

```
;       |   |                           1 = RA4/TOCKI bit source
;       |   |_____    INTEDG (Edge select)
;       |                                   0 = falling edge
;       |                                  *1 = raising edge
;       |_____    RBPU (Pullup enable)
;                                           0 = enabled
;                                          *1 = disabled
            option
; Set up ports
            movlw       0x00            ; Set Port-B to output
            tris        portb
            clrf        portb           ; All Port-B to 0
; Port-A is not used in this program
mloop:
            incf        portb,f  ; Add 1 to register value
            call        TM0delay
            goto        mloop
```

The delay procedure named TM0delay provides the necessary time lapse between successive increments in the count displayed. The code is as follows:

```
;*****************************
;       delay sub-routine
;           uses Timer0
;*****************************
TM0delay:
; Initialize the timer register
            clrf            tmr0        ; Clear SFR for Timer0
; Routine tests the value in the TMR0 register by
; subtracting 0xff from the value in TMR0. The zero flag
; is set if TMR0 = 0xff
cycle:
            movf        tmr0,w          ; Timer to w
; w has TMR0 register value
            sublw       0xff            ; Subtract max value
; Zero flag is set if value in TMR0 = 0xff
            btfss       status,z ; Test for zero
            goto        cycle           ; Repeat
            Return
```

12.3.3 Measured Time Lapse

A variable time-lapse routine that can be edited or adjusted to produce delays within a specific time range is a useful tool in any programmer's library. In previous sections, we developed delay routines that do so by counting timer pulses. This same idea can be used to develop a routine that produces accurate delays within a range.

The routine can be implemented to varying degrees of sophistication. One extreme would be a procedure that receives the desired time lapse as a parameter. Another option would be a procedure that reads the desired time lapse from program constants. In the program named LapseTimer contained in the book's online software we develop a procedure in which the calling code passes the desired time delay in three variables containing the number of machine cycles necessary for the desired wait period. By using machine cycles instead of time units (such as microseconds or milliseconds), the procedure becomes easily adaptable to devices running at different clock speeds. Since each instruction requires four clock cycles, the device's clock speed in Hz is divided by four in order to determine the number of machine cycles per time unit.

For example, a processor equipped with an 8 Mhz clock executes at a rate of 8,000,000/4 machine cycles per second; that is, 2,000,000 instruction cycles per second. To produce a one-quarter second delay requires a wait period of 2,000,000/4 or 500,000 instruction cycles. By the same token, the 16F84 running at 4 Mhz executes 1,000,000 instructions per second. In this case a one-quarter second delay would require waiting 250,000 instruction cycles.

The program titled lapseTimer in the book's online software package uses Timer0 to produce a *variable-lapse delay*. The delay is calculated based on the number of machine cycles necessary for the desired wait period, as described in the preceding paragraph. The program uses the Black-Ammerman methods, which require a prescaler of 1:2 so that each timer iteration takes place at one-half the clock rate. The program initializes the OPTION register and the ports as follows:

```
main:
; Clear the Watchdog Timer and reset prescaler
        clrf    tmr0
        clrwdt
; Set up the OPTION regiser bit map
        movlw   b'11010000'
;       7  6  5  4  3  2  1  0 <= OPTION bits
;       |  |  |  |  |  |__|__|_____ PS2-PS0 (prescaler bits)
;       |  |  |  |  |              Values for Timer0
;       |  |  |  |  |              *000 = 1:2    001 = 1:4
;       |  |  |  |  |               010 = 1:8    011 = 1:16
;       |  |  |  |  |               100 = 1:32   101 = 1:64
;       |  |  |  |  |               110 = 1:128 *111 = 1:256
;       |  |  |  |  |_____ PSA (prescaler assign)
;       |  |  |  |                  1 = to WDT
;       |  |  |  |                 *0 = to Timer0
;       |  |  |  |_____ TOSE (Timer0 edge select)
;       |  |  |                     0 = increment on low-to-high
;       |  |  |                    *1 = increment in high-to-low
;       |  |  |_____ TOCS (TMR0 clock source)
;       |  |                       *0 = internal clock
;       |  |                        1 = RA4/TOCKI bit source
;       |  |_____ INTEDG (Edge select)
```

```
;         |                               *0 = falling edge
;         |_____   RBPU (Pullup enable)
;                                          0 = enabled
;                                         *1 = disabled
        option
; Set up ports
        movlw   0x00            ; Set Port-B to output
        tris    portb
        clrf    portb           ; All Port-B to 0
; Port-A is not used in this program
```

The LapseTimer program is designed to produce a one-half second delay on a 16F84 running at 4 Mhz; therefore, the delay requires 500,000 clock beats. The value is converted to hexadecimal and stored in a 3-byte counter, as follows:

$$500,000 = 0x07a120 \text{ or}$$
$$countL = 0x20$$
$$countM = 0xa1$$
$$countH = 0x07$$

The variables countL, countM, and countH are defined locally and initialized by a procedure named onehalfSec, as follows:

```
; Procedure to initialize local variables for a
; delay of one-half second on a 16F84 at 4 Mhz.
; Timer is set up for 500,000 clock beats as
; follows: 500,000 = 0x07 0xa1 0x20
; 500,000 = 0x07 0xa1 0x20
;           __  __  __
;           |   |   |___ countL)
;           |   |_____ countM
;           |_____ countH
onehalfSec:
        movlw   0x07
        movwf   countH
        movlw   0xa1
        movwf   countM
        movlw   0x20
        movwf   countL
        return
```

The delay routine uses the Timer0 register to provide the low-order level of the count. Since the counter counts up from zero to ensure that the initial low-level delay count is correct—the value 128 - ($xx/2$) must be calculated, where xx is the value in the original countL register. The program performs the division by 2 by shifting bits to the right by one position. The resulting value is subtracted from 128 and the result stored in TMR0, as follows:

Timers and Counters

```
; First calculate xx/2 by bit shifting
        bcf             status,c ; Clear carry flag
        rrf             countL,f ; Divide by 2
; now subtract 128 - (xx/2)
        movf            countL,w ; w holds low-order byte
        sublw           d'128'
; Now w has adjusted result. Store in TMR0
        movwf           tmr0
```

The delay routine detects timer overflow by testing bit 7 of the TMR0 register. If the bit is set, then 256 time cycles have elapsed and the mid-order counter register is decremented. If the mid-order register underflows when it is decremented, then the high-order register is decremented. If it underflows, the counter has gone to zero and the delay routine ends. Processing is as follows:

```
cycle:
        btfss    tmr0,7        ; Is bit 7 set?
        goto     cycle         ; Wait if not set
;
; At this point TMR0 bit 7 is set
; Clear the bit
        bcf      tmr0,7        ; All other bits are preserved
; Subtract 256 from beat counter by decrementing the
; mid-order byte
;
        decfsz   countM,f
        goto     cycle         ; Continue if mid-byte not zero
; At this point the mid-order byte has overflowed.
; High-order byte must be decremented.
        decfsz   countH,f
        goto     cycle
; At this point the time cycle has elapsed
        return
```

The circuit in Figure 12-3 can be used to test the lapseTimer program.

Interrupt-driven Timer

Interrupt-driven timers and counters have several advantages over polled routines: first, the time lapse counting takes place in the background so that the application can continue to do other work in the foreground. Another advantage of an interrupt-driven counter is that the prescaler is unnecessary and can be used for the Watchdog Timer. Developing a timer routine that is interrupt-driven presents no additional challenges over the conventional interrupt-driven examples covered in Chapter 11. The initialization consists of configuring the OPTION and the INTCON register bits for the task at hand. In the particular case of an interrupt-driven timer, the following are necessary:

1. The external interrupt flag (INTF in the INTCON Register) must be initially cleared.
2. Global interrupts must be enabled by setting the GIE bit in the INTCON Register.
3. The Timer0 overflow interrupt must be enabled by setting the T0IE bit in the INTCON register.

In this example program, named LapseTmrInt, the prescaler is not used with the timer, so the initialization code sets the PSA bit in the OPTION register and the prescaler is assigned to the Watchdog Timer. The following code fragment is from the LapseTmrInt program:

```
main:
; Clear the Watchdog Timer and reset prescaler
        clrf     tmr0
        clrwdt
; Set up the OPTION register bit map
        movlw              b'11011000'
;    7  6  5  4  3  2  1  0 <= OPTION bits
;    |  |  |  |  |  |__|__|_____ PS2-PS0 (prescaler bits)
;    |  |  |  |  |              Values for Timer0
;    |  |  |  |  |              000 = 1:2    001 = 1:4
;    |  |  |  |  |              010 = 1:8    011 = 1:16
;    |  |  |  |  |              100 = 1:32   101 = 1:64
;    |  |  |  |  |              110 = 1:128 *111 = 1:256
;    |  |  |  |  |_____ PSA (prescaler assign)
;    |  |  |  |                 *1 = to WDT
;    |  |  |  |                  0 = to Timer0
;    |  |  |  |_____ T0SE (Timer0 edge select)
;    |  |  |                     0 = increment on low-to-high
;    |  |  |                    *1 = increment in high-to-low
;    |  |  |_____ T0CS (TMR0 clock source)
;    |  |                       *0 = internal clock
;    |  |                        1 = RA4/TOCKI bit source
;    |  |_____ INTEDG (Edge select)
;    |                          *0 = falling edge
;    |_____ RBPU (Pullup enable)
;                                0 = enabled
;                               *1 = disabled
        option
; Set up ports
        movlw    0x00              ; Set Port-B to output
        tris     portb
        clrf     portb             ; All Port-B to 0
; Port-A is not used in this program
;============================
;     set up interrupts
;============================
; Clear external interrupt flag (intf = bit 1)
        bcf              INTCON,intf     ; Clear flag
```

```
; Enable global interrupts (gie = bit 7)
; Enable RB0 interrupt (inte = bit 4)
        bsf     INTCON,gie      ; Enable global int (bit 7)
        bsf     INTCON,toie     ; Enable TMR0 overflow
                                ; interrupt
```

As in the program LapseTimer, developed previously in this chapter, the timer operates by decrementing a 3-byte counter that holds the number of timer beats required for the programmed delay. In the case of the LapseTmrInt program, the routine that initializes the register variables for a one-half second delay also correctly adjusts the initial value loaded into the TMR0 register. The code is as follows:

```
;===============================
;  set register variables for
;       one-half second delay
;===============================
; Procedure to initialize local variables for a delay of
; one-half second on a 16F84 at 4 Mhz. Timer is set up for a
; 500,000 clock beats as follows: 500,000 = 0x07 0xa1 0x20
; 500,000 = 0x07 0xa1 0x20
;             __  __  __
;             |   |   |___ countL)
;             |   |_____ countM
;             |_____ countH
onehalfSec:
        movlw   0x07
        movwf   countH
        movlw   0xa1
        movwf   countM
        movlw   0x20
        movwf   countL
; The TMR0 register provides the low-order level of
; the count. Since the counter counts up from zero,
; in order to ensure that the initial low-level delay
; count is correct, the value 256 - xx must be calculated
; where xx is the value in the original countL variable.
        movf    countL,w ; w holds low-order byte
        sublw   d'256'
; Now w has adjusted result. Store in TMR0
        movwf   tmr0
        return
```

The interrupt service routine in the LapseTmrInt program receives control when the TMR0 register underflows, that is, when the count goes from 0xff to 0x00. The service routine then proceeds to decrement the mid-range counter register and adjust, if necessary, the high-order counter. If the count goes to zero, the handler toggles the LED on Port-B, line 0, and re-initializes the counter variables by calling the onehalfSec procedure described previously. The interrupt handler is coded as follows:

```
;=========================================================
;           Interrupt Service Routine
;=========================================================
; Service routine receives control when there the timer
; register TMR0 overflows, that is, when 256 timer beats
; have elapsed
IntServ:
; First test if source is a Timer0 interrupt
        btfss     INTCON,toif      ; TOIF is Timer0 interrupt
        goto      notTOIF          ; Go if not RB0 origin
; If so clear the timer interrupt flag so that count continues
        bcf       INTCON,toif      ; Clear interrupt flag
; Save context
        movwf     old_w            ; Save w register
        swapf     STATUS,w ; STATUS to w
        movwf     old_status       ; Save STATUS
;=========================
;    interrupt action
;=========================
; Subtract 256 from beat counter by decrementing the
; mid-order byte
        decfsz    countM,f
        goto      exitISR          ; Continue if mid-byte not
                                   ; zero
; At this point the mid-order byte has overflowed.
; High-order byte must be decremented.
        decfsz    countH,f
        goto      exitISR
; At this point count has expired so the programmed time
; has elapsed. Service routine turns the LED on line 0,
; Port-B on and off at every conclusion of the count.
; This is done by XORing a mask with a one-bit at the
; Port-B line 0 position
        movlw     b'00000001'      ; XORing with a 1-bit produces
                                   ; the complement
        xorwf     portb,f          ; Complement bit 2, Port-B
; Reset one-half second counter
        call      onehalfSec
;=========================
;       exit ISR
;=========================
exitISR:
; Restore context
        swapf     old_status,w  ; Saved status to w
        movfw     STATUS   ; To STATUS register
        swapf     old_w,f  ; Swap file register in itself
        swapf     old_w,w  ; re-swap back to w
; Return from interrupt
```

```
notTOIF:
        retfie
```

One of the initial operations of the service routine is to clear the TOIF bit in the INTCON register. This action re-enables the timer interrupt and prevents counting cycles to be lost. Since the interrupt is generated every 256 beats of the timer, there is no risk that by enabling the timer interrupt flag a re-entrant interrupt will take place.

The interrupt-based timer program named LapseTmrInt can be tested on the same circuit shown in Figure 12-3.

12.4 The Watchdog Timer

The 16F84 contains an independent timer with its own clock source called the *Watchdog Timer*, or WDT. The Watchdog Timer provides a way for the processor to recover from a software error that impedes program continuation, such as an *endless loop*. The Watchdog Timer is not designed to recover from hardware faults, such as a brown-out.

The Watchdog Timer hardware is independent of the PIC's internal clock. Its time-out period lasts approximately 18ms to 2.3s, depending on whether the prescaler is used and on its setting. It is not very accurate due to its sensitivity to temperature. According to Microchip's documentation, under worst-case conditions, its time-out period can take up to several seconds. The following program elements relate to Watchdog Timer operation:

1. Configuration bit 2, labeled WDTE, enables and disables the Watchdog Timer during system configuration. The WDT cannot be set or reset at runtime. It is enabled and disabled during programming.

2. The PSA bit in the OPTION register selects whether the prescaler is assigned to the Watchdog Timer or to the Timer0 module.

3. Bits PS2 to PS0 in the OPTION register allow assigning eight rates to the Watchdog Timer, from 1:1 to 1:128.

4. Bit 4 of the STATUS register, named the TO bit, is cleared when a time-out condition occurred that originated in the WDT.

5. The power-down bit (PD) in the STATUS register is set after the execution of the clrwdt instruction.

6. The clrwdt instruction clears the Watchdog Timer. It also clears the prescaler count (if the prescaler is assigned to the Watchdog Timer) and sets STATUS bits TO and PD.

The WDT provides a *recovery mechanism* for software errors. When the WDT times-out, the TO flag in the STATUS register is cleared and the program counter is reset to 0000 so that the program restarts. Applications can prevent the reset by issuing the **clrwdt** instruction before the time-out period ends. When **clrwdt** executes the WDT time-out period restarts.

12.4.1 Watchdog Timer Programming

Not much information is available regarding the details of operation of the Watchdog Timer in the 16F84. Using the WDT in applications is not just a simple matter of restarting the counter with the **clrwdt** instruction. The timer is designed to detect software errors that can hang up a program, but how it detects these errors and which conditions trigger the WDT operation are not clear from the information provided by Microchip. For example, an application that contains a long delay loop may find that the Watchdog Timer forces an untimely break out of the loop. The Watchdog Timer provides a powerful error-recovery mechanism, but its use requires careful consideration of program conditions that could make the timer malfunction.

12.5 Sample Programs

The following programs demonstrate the programming discussed in this chapter.

12.5.1 The Tmr0Counter program

```
; File name: Tmr0Counter.asm
; Date: April 30, 2006
; Author: Julio Sanchez
; Processor: 16F84A
;
; Reference: SevenSeg Circuit and Board
;
; Description:
; Test program for the Timer0 counter. The program counts
; the number of presses of the pushbutton switch on port
; RA4/TOCKI and displays the count on a seven segment LED.
; Switch is wired active low.
;
; Switches used in __config directive:
;   _CP_ON         Code protection ON/OFF
; * _CP_OFF
; * _PWRTE_ON      Power-up timer ON/OFF
;   _PWRTE_OFF
;   _WDT_ON        Watchdog Timer ON/OFF
; * _WDT_OFF
;   _LP_OSC        Low power crystal occilator
; * _XT_OSC        External parallel resonator/crystal oscillator
;
;   _HS_OSC        High speed crystal resonator (8 to 10 MHz)
;                  Resonator: Murate Erie CSA8.00MG = 8 MHz
;   _RC_OSC        Resistor/capacitor oscillator
; |
; |_____ * indicates set up values
;
;==========================
; set up and configuration
;==========================
```

```
            processor 16f84A
            include    <p16f84A.inc>
            __config   _XT_OSC & _WDT_OFF & _PWRTE_ON & _CP_OFF
;
;=========================================================
;                    constant definitions
;               (per circuit wiring diagram)
;=========================================================
#define Pb_sw  4 ; Port-A line 4 to push button switch
;
;===========================
;       local variables
;===========================
            cblock    0x0c        ; Start of block
            J                     ; counter J
            K                     ; counter K
            endc
;=============================================================
;                          program
;=============================================================
            org       0           ; start at address 0
            goto      main
;
; Space for interrupt handlers
            org       0x08

main:
; Clear the timer and the watchdog
            clrf      TMR0
            clrwdt
; Set up the OPTION register bit map
            movlw     b'10111000'
;    7  6  5  4  3  2  1  0 <= OPTION bits
;    |  |  |  |  |  |__|__|____ PS2-PS0 (prescaler bits)
;    |  |  |  |  |              Values for Timer0
;    |  |  |  |  |              *000 = 1:2    001 = 1:4
;    |  |  |  |  |               010 = 1:8    011 = 1:16
;    |  |  |  |  |               100 = 1:32   101 = 1:64
;    |  |  |  |  |               110 = 1:128 *111 = 1:256
;    |  |  |  |  |_____ PSA (prescaler assign)
;    |  |  |  |                 *1 = to WDT
;    |  |  |  |                  0 = to Timer0
;    |  |  |  |_____ T0SE (Timer0 edge select)
;    |  |  |                     0 = increment on low-to-high
;    |  |  |                    *1 = increment in high-to-low
;    |  |  |_____ T0CS (TMR0 clock source)
;    |  |                        0 = internal clock
;    |  |                       *1 = RA4/T0CKI bit source
```

```
;       |   |_____ INTEDG (Edge select)
;       |                              *0 = falling edge
;       |_____ RBPU (Pullup enable)
;                                       0 = enabled
;                                      *1 = disabled
            option
; Set up ports
            movlw    0x00         ; Set Port-B to output
            tris     PORTB
            clrf     PORTB        ; All Port-B to 0
; Port-A. Five low-order lines set for for input
            movlw    B'00011111'  ; w = 00011111 binary
            tris     PORTA        ; Port-A (lines 0 to 4) to
                                  ; input
;=================================
; Check value in TMR0 and display
;=================================
; Every press of the pushbutton switch connected to line
; RA4/TOCKI adds one to the value in the TMR0 register.
; Loop checks this value, adjusts to the range 0 to 15
; and displays the result in the seven-segment LED on
; Port-B
checkTmr0:
            movf     TMR0,w       ; Timer register to w
; Eliminate four high order bits
            andlw    b'00001111'  ; Mask off high bits
; At this point the w register contains a 4-bit value
; in the range 0 to 0xf. Use this value (in w) to
; obtain seven-segment display code
            call     segment
            movwf    PORTB        ; Display switch bits
            goto     checkTmr0    ; Endless loop
;
;=================================
;   routine to returns 7-segment
;               codes
;=================================
segment:
            addwf    PCL,f        ; PCL is program counter latch
            retlw    0x3f         ; 0 code
            retlw    0x06         ; 1
            retlw    0x5b         ; 2
            retlw    0x4f         ; 3
            retlw    0x66         ; 4
            retlw    0x6d         ; 5
            retlw    0x7d         ; 6
            retlw    0x07         ; 7
            retlw    0x7f         ; 8
```

```
            retlw    0x6f    ; 9
            retlw    0x77    ; A
            retlw    0x7c    ; B
            retlw    0x39    ; C
            retlw    0x5b    ; D
            retlw    0x79    ; E
            retlw    0x71    ; F
            retlw    0x7f    ; Just in case all on
            end
```

12.5.2 The Timer0 Program

```
; File: Timer0.ASM
; Date: April 27, 2006
; Author: Julio Sanchez
; Processor: a6F84A
;
; Description:
; Program to demonstrate programming of the 16F84A
; TIMER0 module. Program flashes eight LEDs in sequence
; counting from 0 to 0xff. Timer0 is used to delay
; the count.
;===========================
;        switches
;===========================
; Switches used in __config directive:
;   _CP_ON         Code protection ON/OFF
; * _CP_OFF
; * _PWRTE_ON     Power-up timer ON/OFF
;   _PWRTE_OFF
;   _WDT_ON        Watchdog Timer ON/OFF
; * _WDT_OFF
;   _LP_OSC        Low power crystal occilator
; * _XT_OSC        External parallel resonator/crystal oscillator

;   _HS_OSC        High speed crystal resonator (8 to 10 MHz)
;                  Resonator: Murate Erie CSA8.00MG = 8 MHz
;   _RC_OSC        Resistor/capacitor oscillator (simplest, 20%
error)
; |
; |_____ * indicates set up values

         processor 16f84A
         include   <p16f84A.inc>
         __config  _XT_OSC & _WDT_OFF & _PWRTE_ON & _CP_OFF
;=========================================================
;              variables in PIC RAM
;=========================================================
```

```
; None in this application
;
;===========================================================
;               m a i n    p r o g r a m
;===========================================================
          org      0              ; start at address 0
          goto     main
;
;============================
;       interrupt handler
;============================
          org              0x08
;============================
;       main program
;============================
main:
; Clear the Watchdog Timer and reset prescaler
          clrwdt
; Set up the OPTION register bit map
          movlw    b'11010111'
;   7  6  5  4  3  2  1  0 <= OPTION bits
;   |  |  |  |  |  |__|__|_____ PS2-PS0 (prescaler bits)
;   |  |  |  |  |               Values for Timer0
;   |  |  |  |  |               000 = 1:2    001 = 1:4
;   |  |  |  |  |               010 = 1:8    011 = 1:16
;   |  |  |  |  |               100 = 1:32   101 = 1:64
;   |  |  |  |  |               110 = 1:128 *111 = 1:256
;   |  |  |  |  |_____ PSA (prescaler assign)
;   |  |  |  |                   1 = to WDT
;   |  |  |  |                  *0 = to Timer0
;   |  |  |  |_____ TOSE (Timer0 edge select)
;   |  |  |                      0 = increment on low-to-high
;   |  |  |                     *1 = increment in high-to-low
;   |  |  |_____ TOCS (TMR0 clock source)
;   |  |                        *0 = internal clock
;   |  |                         1 = RA4/TOCKI bit source
;   |  |_____ INTEDG (Edge select)
;   |                           *0 = falling edge
;   |_____ RBPU (Pullup enable)
;                                0 = enabled
;                               *1 = disabled
          option
; Set up ports
          movlw    0x00
          tris     PORTB           ; Set Port-B to output
          clrf     PORTB           ; All Port-B to 0
; Port-A is not used in this program
mloop:
```

```
            incf    PORTB,f         ; Add 1 to register value
            call    TM0delay
            goto    mloop
;*****************************
;       delay sub-routine
;          uses Timer0
;*****************************
TM0delay:
; Initialize the timer register
            clrf    TMR0            ; Clear SFR for Timer0
; Routine tests the value in the TMR0 register by
; subtracting 0xff from the value in TMR0. The zero flag
; is set if TMR0 = 0xff
cycle:
            movf    TMR0,w          ; Timer to w
; w has TMR0 register value
            sublw   0xff            ; Subtract max value
; Zero flag is set if value in TMR0 = 0xff
            btfss   STATUS,Z ; Test for zero
            goto    cycle           ; Repeat
            return

            end
```

12.5.3 The LapseTimer Program

```
; File: LapseTimer.ASM
; Date: May 1, 2006
; Author: Julio Sanchez
; Processor: 16F84A
;
; Description:
; Using Timer0 to produce a variable-lapse delay.
; The delay is calculated based on the number of machine
; cycles necessary for the desired wait period. For
; example, a machine running at a 4 Mhz clock rate
; executes 1,000,000 instructions per second. In this
; case a 1/2 second delay requires 500,000 instructions.
; The wait period is passed to the delay routine in three
; program registers which hold the high-, middle-, and
; low-order bytes of the counter.
;===========================
;          switches
;===========================
; Switches used in __config directive:
;    _CP_ON          Code protection ON/OFF
; *  _CP_OFF
; *  _PWRTE_ON       Power-up timer ON/OFF
```

```
;       _PWRTE_OFF
;       _WDT_ON         Watchdog Timer ON/OFF
; *     _WDT_OFF
;       _LP_OSC         Low power crystal occilator
; *     _XT_OSC         External parallel resonator/crystal oscillator

;       _HS_OSC         High speed crystal resonator (8 to 10 MHz)
;                       Resonator: Murate Erie CSA8.00MG = 8 MHz
;       _RC_OSC         Resistor/capacitor oscillator
;    |
;    |_____ * indicates set up values

        processor 16f84A
        include   <p16f84A.inc>
        __config  _XT_OSC & _WDT_OFF & _PWRTE_ON & _CP_OFF
;=========================================================
;               variables in PIC RAM
;=========================================================
; Local variables
        cblock  0x0d    ; Start of block
                        ; 3-byte auxiliary counter for delay.
        countH          ; High-order byte
        countM          ; Medium-order byte
        countL          ; Low-order byte
        endc
;=========================================================
;               m a i n   p r o g r a m
;=========================================================
        org     0       ; start at address 0
        goto    main
;
;=============================
;       interrupt handler
;=============================
        org     0x04
;       goto    IntServ
;=============================
;       main program
;=============================
main:
; Clear the Watchdog Timer and reset prescaler
        clrf    TMR0
        clrwdt
; Set up the OPTION register bit map
        movlw   b'11010000'
;   7  6  5  4  3  2  1  0 <= OPTION bits
;   |  |  |  |  |  |__|__|_____ PS2-PS0 (prescaler bits)
;   |  |  |  |  |               Values for Timer0
```

```
;       |   |   |   |   |                   *000 = 1:2    001 = 1:4
;       |   |   |   |   |                    010 = 1:8    011 = 1:16
;       |   |   |   |   |                    100 = 1:32   101 = 1:64
;       |   |   |   |   |                    110 = 1:128 *111 = 1:256
;       |   |   |   |   |_____ PSA (prescaler assign)
;       |   |   |   |                         1 = to WDT
;       |   |   |   |                        *0 = to Timer0
;       |   |   |   |_____ TOSE (Timer0 edge select)
;       |   |   |                             0 = increment on low-to-high
;       |   |   |                            *1 = increment in high-to-low
;       |   |   |_____ TOCS (TMR0 clock source)
;       |   |                                *0 = internal clock
;       |   |                                 1 = RA4/TOCKI bit source
;       |   |_____ INTEDG (Edge select)
;       |                                    *0 = falling edge
;       |_____ RBPU (Pullup enable)
;                                             0 = enabled
;                                            *1 = disabled
        option
; Set up ports
        movlw   0x00            ; Set Port-B to output
        tris    PORTB
        clrf    PORTB           ; All Port-B to 0
; Port-A is not used in this program
;============================
;    display loop
;============================
mloop:
; Turn on LED
        bsf     PORTB,0
; Initialize counters and delay
        call    onehalfSec
        call    TM0delay
; Turn off LED
        bcf     PORTB,0
; Re-initialize counter and delay
        call    onehalfSec
        call    TM0delay
        goto    mloop
;===================================
;   variable-lapse delay procedure
;          using Timer0
;===================================
; ON ENTRY:
;         Variables countL, countM, and countH hold
;         the low-, middle-, and high-order bytes
;         of the delay period, in timer units
; Routine logic:
```

```
; The prescaler is assigned to Timer0 and set up so
; that the timer runs at 1:2 rate. This means that
; every time the counter reaches 128 (0x80) a total
; of 256 machine cycles have elapsed. The value 0x80
; is detected by testing bit 7 of the counter
; register.
; Note:
;     The Timer0 register provides the low-order level
; of the count. Since the counter counts up from zero,
; in order to ensure that the initial low-level delay
; count is correct the value 128 - (xx/2) must be calculated
; where xx is the value in the original countL register.
; First calculate xx/2 by bit shifting
TM0delay:
        bcf     STATUS,C ; Clear carry flag
        rrf     countL,f ; Divide by 2
; now subtract 128 - (xx/2)
        movf    countL,w ; w holds low-order byte
        sublw   d'128'
; Now w has adjusted result. Store in TMR0
        movwf   TMR0
; Routine tests timer overflow by testing bit 7 of
; the TMR0 register.
cycle:
        btfss   TMR0,7              ; Is bit 7 set?
        goto    cycle               ; Wait if not set
; At this point TMR0 bit 7 is set
; Clear the bit
        bcf     TMR0,7              ; All other bits are preserved
; Subtract 256 from beat counter by decrementing the
; mid-order byte
        decfsz  countM,f
        goto    cycle               ; Continue if mid-byte not
zero
; At this point the mid-order byte has overflowed.
; High-order byte must be decremented.
        decfsz  countH,f
        goto    cycle
; At this point the time cycle has elapsed
        return
;==============================
;   set register variables for
;      one-half second delay
;==============================
; Procedure to initialize local variables for a
; delay of one-half second on a 16F84 at 4 Mhz.
; Timer is set up for 500,000 clock beats as
; follows: 500,000 = 0x07 0xa1 0x20
```

```
;  500,000 = 0x07 0xa1 0x20
;             ——   ——   ——
;             |    |    |___ countL)
;             |    |_____ countM
;             |_____ countH
onehalfSec:
        movlw   0x07
        movwf   countH
        movlw   0xa1
        movwf   countM
        movlw   0x20
        movwf   countL
        return

        end
```

12.5.4 The LapseTmrInt Program

```
; File: LapseTmrInt.ASM
; Date: May 1, 2006
; Author: Julio Sanchez
; Processor: 16F84A
;
; Description:
; Interrupt-driven version of the LapseTimer program.
; Using Timer0 to produce a variable-lapse delay.
; The delay is calculated based on the number of machine
; cycles necessary for the desired wait period. For
; example, a machine running at a 4 Mhz clock rate
; executes 1,000,000 instructions per second. In this
; case a 1/2 second delay requires 500,000 instructions.
; The wait period is passed to the delay routine in three
; register variables which hold the high-, middle-, and
; low-order bytes of the counter.
;===========================
;       switches
;===========================
; Switches used in __config directive:
;   _CP_ON          Code protection ON/OFF
; * _CP_OFF
; * _PWRTE_ON       Power-up timer ON/OFF
;   _PWRTE_OFF
;   _WDT_ON         Watchdog Timer ON/OFF
; * _WDT_OFF
;   _LP_OSC         Low power crystal occilator
; * _XT_OSC         External parallel resonator/crystal oscillator

;   _HS_OSC         High speed crystal resonator (8 to 10 MHz)
```

```
;                       Resonator: Murate Erie CSA8.00MG = 8 MHz
;       _RC_OSC         Resistor/capacitor oscillator
;       |
;       |_____ * indicates set up values

        processor 16f84A
        include    <p16f84A.inc>
        __config   _XT_OSC & _WDT_OFF & _PWRTE_ON & _CP_OFF

;=========================================================
;                  variables in PIC RAM
;=========================================================
; Local variables
        cblock     0x0d       ; Start of block
                              ; 3-byte auxiliary counter for delay.
        countH                ; High-order byte
        countM                ; Medium-order byte
        countL                ; Low-order byte
        old_w                 ; Context saving
        old_STATUS            ; Idem
        endc
;=========================================================
;                  m a i n    p r o g r a m
;=========================================================
        org        0          ; start at address 0
        goto       main
;
;=============================
;     interrupt handler
;=============================
        org        0x04
        goto       IntServ
;=============================
;     main program
;=============================
main:
; Clear the Watchdog Timer and reset prescaler
        clrf       TMR0
        clrwdt
; Set up the OPTION register bit map
        movlw      b'11011000'
;       7 6 5 4 3 2 1 0 <= OPTION bits
;       | | | | | |_|_|_____ PS2-PS0 (prescaler bits)
;       | | | | |            Values for Timer0
;       | | | | |            000 = 1:2    001 = 1:4
;       | | | | |            010 = 1:8    011 = 1:16
;       | | | | |            100 = 1:32   101 = 1:64
;       | | | | |            110 = 1:128 *111 = 1:256
```

```
;        |   |   |   |   |_____  PSA (prescaler assign)
;        |   |   |   |                     *1 = to WDT
;        |   |   |   |                      0 = to Timer0
;        |   |   |   |_____   TOSE (Timer0 edge select)
;        |   |   |                          0 = increment on low-to-high
;        |   |   |                         *1 = increment in high-to-low
;        |   |   |_____   TOCS (TMR0 clock source)
;        |   |                             *0 = internal clock
;        |   |                              1 = RA4/TOCKI bit source
;        |   |_____   INTEDG (Edge select)
;        |                                 *0 = falling edge
;        |_____   RBPU (Pullup enable)
;                                           0 = enabled
;                                          *1 = disabled
         option
; Set up ports
         movlw   0x00              ; Set Port-B to output
         tris    PORTB
         clrf    PORTB             ; All Port-B to 0
; Port-A is not used in this program
;=============================
;       set up interrupts
;=============================
; Clear external interrupt flag (INTF = bit 1)
         bcf     INTCON,INTF       ; Clear flag
; Enable global interrupts (GIE = bit 7)
; Enable RB0 interrupt (inte = bit 4)
         bsf     INTCON,GIE        ; Enable global int (bit 7)
         bsf     INTCON,T0IE       ; Enable TMR0 overflow
                                   ; interrupt
; Init count
         call    onehalfSec
;=============================
;      do-nothing loop
;=============================
; All work is performed by the interrupt handler
mloop:
         goto    mloop
;==============================
;  set register variables for
;      one-half second delay
;==============================
; Procedure to initialize local variables for a
; delay of one-half second on a 16F84 at 4 Mhz.
; Timer is set up for 500,000 clock beats as
; follows: 500,000 = 0x07 0xa1 0x20
; 500,000 = 0x07 0xa1 0x20
;                    —  —  —
```

```
;                |        |        |___ countL)
;                |        |_____ countM
;                |_____ countH
onehalfSec:
         movlw    0x07
         movwf    countH
         movlw    0xa1
         movwf    countM
         movlw    0x20
         movwf    countL
; The Timer0 register provides the low-order level
; of the count. Since the counter counts up from zero,
; in order to ensure that the initial low-level delay
; count is correct the value 256 - xx must be calculated
; where xx is the value in the original countL register.
         movf     countL,w ; w holds low-order byte
         sublw    d'255'
; Now w has adjusted result. Store in TMR0
         movwf    TMR0
         return
;===========================================================
;              Interrupt Service Routine
;===========================================================
; Service routine receives control when the timer
; register TMR0 overflows, that is, when 256 timer beats
; have elapsed
IntServ:
; First test if source is a Timer0 interrupt
         btfss    INTCON,T0IF     ; T0IF is Timer0 interrupt
         goto     notTOIF    ; Go if not RB0 origin
; If so clear the timer interrupt flag so that count continues
         bcf              INTCON,T0IF     ; Clear interrupt flag
; Save context
         movwf    old_w              ; Save w register
         swapf    STATUS,w ; STATUS to w
         movwf    old_STATUS         ; Save STATUS
;==========================
;    interrupt action
;==========================
; Subtract 256 from beat counter by decrementing the
; mid-order byte
         decfsz   countM,f
         goto     exitISR            ; Continue if mid-byte not zero
; At this point the mid-order byte has overflowed.
; High-order byte must be decremented.
         decfsz   countH,f
         goto     exitISR
```

```
; At this point count has expired so the programmed time
; has elapsed. Service routine turns the LED on line 0,
; Port-B on and off at every conclusion of the count.
; This is done by xoring a mask with a one-bit at the
; Port-B line 0 position
        movlw   b'00000001'     ; Xoring with a 1-bit produces
                                ; the complement
        xorwf   PORTB,f         ; Complement bit 2, Port-B
; Reset one-half second counter
        call    onehalfSec
;=========================
;       exit ISR
;=========================
exitISR:
; Restore context
        swapf   old_STATUS,w  ; Saved STATUS to w
        movfw   STATUS    ; To STATUS register
        swapf   old_w,f   ; Swap file register in itself
        swapf   old_w,w   ; re-swap back to w
; Reset,interrupt
notTOIF:
        retfie

        end
```

Chapter 13

LCD Interfacing and Programming

This chapter is about programming liquid *crystal displays* and interfacing the *LCD* with the PIC 16f84 microcontroller. LCDs are one of the most used devices for alphanumeric output in microcontroller-based circuits. Their advantages are their reduced size and cost and the convenience of mounting the LCD directly on the circuit board.

LCDs are classified according to their interface into serial and parallel. *Serial LCDs* require less I/O resources but execute slower than their parallel counterparts. In addition, they are considerably more expensive. In this chapter we discuss parallel-driven LCD devices based on the Hitachi HD44780 character-based controller, which is by far the most popular controller for PIC-driven LCDs. Serial interface with LCD devices is discussed in Chapter 13.

13.0 LCD Features and Architecture

The HD44780 is a *dot-matrix* liquid crystal display controller and driver. The device displays ASCII alphanumeric characters, Japanese kana characters, and some symbols. A single HD44780 can display up to two 28-character lines. An available extension diver makes possible addressing up to 80 characters.

The HD44780U contains a 9,920 bit character-generator ROM that produces a total of 240 characters: 208 characters with a 5 × 8 dot resolution and 32 characters at a 5 × 10 dot resolution. The device is capable of storing 64 x 8-bit character data in its character generator RAM. This corresponds to eight custom characters in 5 x 8-dot resolution or four characters in 5 x 10-dot resolution.

The controller is programmable to three different dy cycles: 1/8 for one line of 5 × 8 dots with cursor, 1/11 for one line of 5 × 10 dots with cursor, and 1/16 for two lines of 5 × 8 dots with cursor. The built-in commands include clearing the display, homing the cursor, turning the display on and off, turning the cursor on and off, setting display characters to blink, shifting the cursor and the display left-to-right or right-to-left, and reading and writing data to the character generator and to display data ROM.

13.0.1 LCD Functions and Components

The following hardware elements form part of the HD44780 controller: two internal registers labeled the data register and the instruction register, a busy flag, an address counter, a RAM area of display data (DDRAM), a character generator ROM, a character generator RAM, a timing generation circuit, a liquid crystal display driver circuit, and a cursor and blink control circuit. The controller itself is often referred to as the *MPU* in the Hitachi literature.

Internal Registers

The HD44780 contains an *IR* (*instruction register*) and a DR (*data register*). The IR is used to store instruction codes, such as those to clear the display, define an address, or store a bit-map in character generator RAM. The IR is written only from the controller.

The *data register, DR*, is used to temporarily store data to be written into DDRAM or CGRAM as well as temporarily store data read from DDRAM or CGRAM. Data placed in the data register is automatically written into DDRAM or CGRAM by an internal operation.

Busy Flag

When *BF* (the *busy flag*) is 1, the HD44780U is in the internal operation mode, and the next instruction not accepted. The busy flag is mapped to data bit 7. Software must ensure that the busy flag is reset (BF = 0) before the next instruction is entered.

Address Counter

AC (the *address counter*) stores the current address used in operations that access DDRAM or CGRAM. When an instruction contains address information, the address is stored in the address counter. The RAM area accessed—DDRAM or CGRAM—is also determined by the instruction that stores the address in the AC.

The AC is automatically incremented or decremented after each instruction that writes or reads DDRAM or CGRAM data. The variations and options in operations that change the AC are described later in this chapter.

Display Data RAM (DDRAM)

DDRAM (the *display data RAM area*) is used to store the 8-bit bitmaps that represent the display characters and graphics. Display data is represented in 8-bit character codes. When equipped with the extension, its capacity is 80 x 8 bits, or 80 characters. The area not used for storing display character can be used by software for storing any other 8-bit data. The mapping of DDRAM locations to the LCD display is discussed in Section 13.1.4.

Character Generator ROM (CGROM)

The character generator is a ROM that has the bitmaps for 208 characters in 5 x 8 dot resolution or 32 characters in 5 x 10 dot resolution. Figure 13-1 shows the standard character set in the HD44780.

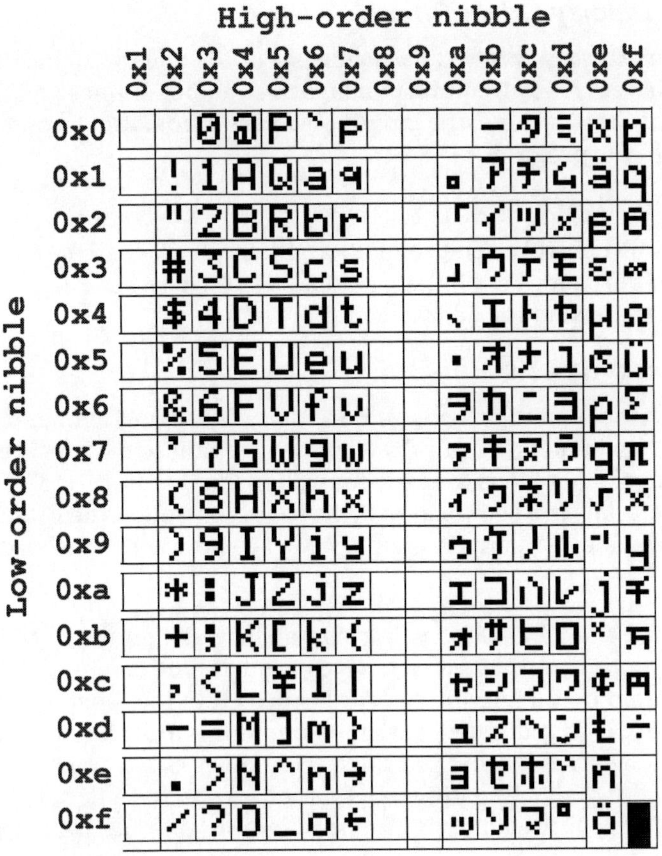

LCD Interfacing and Programming

Figure 13-1 HD44780 Character Set

With a few exceptions, the characters in the range 0x20 to 0x7f correspond to those of the ASCII character set. The remaining characters are Japanese kana characters and special symbols. The characters in the range 0x0 to 0x1f, ASCII control characters, do not function as such in the HD44780. Sending a backspace (0x08), a bell (0x07), or a carriage return (0x0d) code to the controller has no effect.

Character Generator RAM (CGRAM)

CGRAM (the *character generator RAM*) allows the creation of customized characters by defining the corresponding 5 x 8 bitmaps. Eight custom characters can be stored in the 5 x 8 dot resolution and four in the 5 x 10 resolution. The creation and use of custom characters is addressed later in this chapter.

Timing Generation Circuit

This circuit produces the timing signals for the operation of internal components circuits such as DDRAM, CGROM, and CGRAM. The *timing generation circuit* is not accessible to the program.

Liquid Crystal Display Driver Circuit

The *liquid crystal display driver circuit* consists of 16 common signal drivers and 40 segment signal drivers. The circuit responds to the number of lines and the character font selected. Once this is done, the circuit performs automatically and is not otherwise accessible to the program.

Cursor/Blink Control Circuit

The cursor and blink control circuit generates both the cursor and the character blinking. The cursor or the character blinking is applied to the character located in the data RAM address referenced in the address counter (AC).

13.0.2 Connectivity and Pin-Out

LCDs are powerful yet complex devices. Fortunately, the programmer does not have to deal with all the complexities of LCD displays since these devices are usually furnished in a module that includes the LCD controller. Furthermore, most LCDs used in microcontroller circuits are equipped with the same controller, the Hitachi HD44780. This controller provides a relatively simple interface between a microcontroller and the LCD.

But the fact that the HD44780 has become almost ubiquitous in LCD controller technology does not mean that these devices are without complications. The first difficulty confronted by the circuit designer is selecting the most appropriate LCD for the application among dozens (perhaps hundreds) of available configurations, each one with its own resolution, interface technology, size, graphics options, pin patterns, and other individual features. In this sense, it may be better to experiment with a simple LCD in a breadboard circuit before attempting a final circuit with hardware.

Two common connectors used with the 44780-based LCDs have either 14 pins in a single row, each pin spaced 0.100" apart, or two rows of eight pins each, also spaced 0.100" apart. In both cases, the pins are labeled in the LCD board. The two common connectors are shown in Figure 13-2.

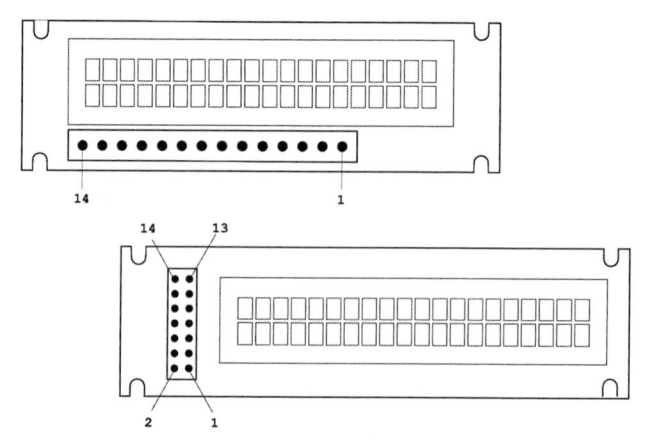

Figure 13-2 Typical HD44780 Connector Pin-Outs

In LCDs with a backlight option sometimes the connectors have two extra pins, usually numbered 15 and 16. Pin number 15 is connected to a 5V source for the backlight and pin number 16 to ground. Typical LCD wiring is shown in Table 13.1.

Table 13.1

Hitachi HD44780 LCD Controller Pin-Out (80 characters or less)

PIN NUMBER	SYMBOL	DESCRIPTION
1	Vss	Ground
2	Vcc	Vcc (Power supply +5V)
3	Vee	Contrast control
4	RS	Set/reset 0 = instruction input 1 = data input
5	R/W	R/W (read/write select) 0 = write to LCD 1 = read LCD data
6	E	Enable. Clock signal to initiate data transfer
7	DB0	Data bus line 0
8	DB1	Data bus line 1
9	DB2	Data bus line 2
10	DB3	Data bus line 3
11	DB4	Data bus line 4
12	DB5	Data bus line 5
13	DB6	Data bus line 6
14	DB7	Data bus line 7

The pin-out in Table 13.1 refers to controllers that address no more than 80 characters. In addition, some LCDs with LED backlighting contain two additional pins, usually numbered 15 and 16. In these cases, pin number 15 is a +5 VDC source for the backlight and pin 16 is the backlight ground.

From the pin-out in Table 13.1, it is evident that the interface to the LCD uses eight parallel lines (lines 7 to 14). However, it is also possible to drive the LCD using just four lines, saving connections on limited circuits.

The reader should beware that LCDs are often furnished in custom boards that may or may not have other auxiliary components. These boards are often wired differently from the examples shown in Figure 13-2. In all cases, the device's documentation and the corresponding data sheets should provide the appropriate wiring information.

13.1 Interfacing with the HD44780

The Hitachi 44780 controller allows parallel interfacing using 4- or 8-bit data paths. In the 4-bit mode, each data byte must be divided into a high-order and a low-order nibble and are transmitted sequentially, the high-nibble first. In the 8-bit parallel mode, each data byte is transmitted from the PIC to the controller as a unit. The advantage of using the 4-bit mode is greater economy of I/O lines on the PIC side. The disadvantages are slightly more complicated programming and minimally slower execution speed. Our first example and circuit uses the 8-bit data mode so as to avoid complications. Once the main processing routines are developed, make the necessary modifications so as to make possible the 4-bit data mode.

In addition to the *data transmission mode*, there are other circuit options to be considered. Two control lines between the microcontroller and the HD44780-driven LCD are necessary in all cases: one to the RS line to select between data and instruction input modes, and another one to the E line to provide the pulse that initiates the data transfer. The *R/W control line*, which selects between the read and the write mode of the LCD controller, can be connected or grounded. If the R/W line is not connected to a microcontroller port, then the HD44780 operates only in the write data mode and all read operations are unavailable.

13.1.1 Busy Flag or Timed Delay Options

Since many applications do not read text data from controller memory, the write-only mode is often an attractive option, especially considering that microcontroller I/O ports are often in short supply and that this option saves one port for other duties. However, there is a less apparent drawback to not being able to read LCD data, which is that the application is not able to monitor the **busy** flag. This flag, which indicates that the controller has concluded its operation, is mapped to bit 7. Since testing the BF requires reading this bit, not connecting the R/W line has the effect that applications cannot use the busy flag and must rely on timing routines to ensure that each operation completes before the next one begins. The timing requirements for each instruction are listed in the rightmost column in Table 13.3. The subjects of timing and delay routines are discussed in detail later in this chapter.

For the circuit designer, to read or not to read controller data is a decision with several tradeoffs. Using time delay routines to ensure that each controller operation has concluded is a viable option that saves one interface line. On the other hand, code that relies on timing routines is externally dependent on the clocks and timer hardware. If code that relies on timing routines is ported to another circuit with a different microcontroller, clocks, or timer hardware, the delays may change and the routines could fail. Furthermore, the use of delay routines often is not efficient, since controller operations can terminate before the timed delay has expired.

On the other hand, code that reads the busy flag to determine the termination of a controller operation is not without dangers. If the controller or the circuit fails, then the program can hang up in an endless loop, waiting for the busy flag to clear. To be absolutely safe, the code would have to contain an external wait loop when testing the busy flag, so that if the external loop expires, then the processing can assume that there is a hardware problem and break out of the flag test loop. The programmer must decide whether this safety mechanism for reading the busy flag is necessary since its implementation requires a somewhat complicated exception response.

In the code samples developed in this chapter, we implement both ways of ensuring operation completion. The code also furnishes a software switch that allows selecting the preferred option.

13.1.2 Contrast Control

In addition to the control lines that require processor interface, the HD44780 contains other control lines. One such line is used for the LCD contrast. The *contrast control line* (usually labeled *Vee*) is connected to pin number 3 (see Table 13.1). The actual implementation of the contrast control function varies according to the manufacturer. In general, for an LCD with a normal temperature range, the contrast control line is wired as shown in Figure 13-3.

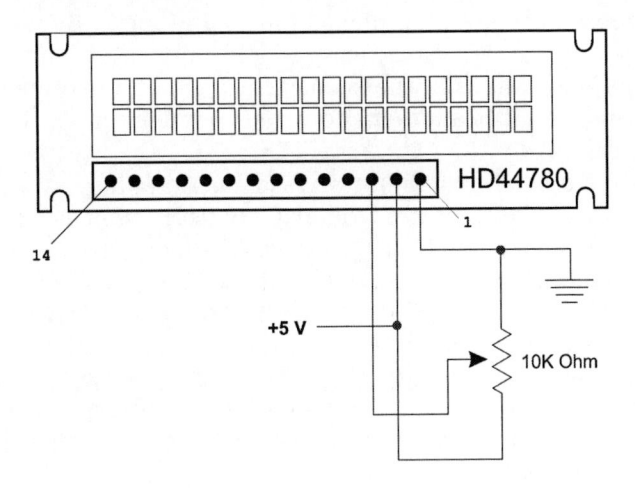

Figure 13-3 Typical Contrast Adjustment Circuit

13.1.3 Display Backlight

Some LCDs are equipped with a LED backlight so as to make the displayed characters more visible. In different LCDs, backlight is implemented in different ways. Some manufacturers wire the backlight directly to the LCD power supply, while others provide additional pins that allow turning the backlight on or off independently of the LCD display. Backlit displays with 14 pins belong to the first type, while those with 16 pins have independent backlight control. If the backlight pins are adjacent to the other display pins, then they are numbered 15 and 16. In this case pin number 15 is wired, through a current limiting resistor, to the +5V source and pin 16 to ground. Sometimes the current-limiting resistor is built into the display. This information is available in the device's data sheet.

Note that some 4-line displays use pins 15 and 16 for other purposes. In these systems, backlight control, if available, is provided by separate pins.

13.1.4 Display Memory Mapping

The Hitachi HD44780 is a memory-mapped system in which characters are displayed by storing their ASCII codes in the corresponding memory address associated with each digit-display area. The area of controller RAM mapped to character-display memory has a capacity of 80 characters. This area is known as *display data RAM* or *DDRAM*.

In order to save circuitry, the common lines of the controller outputs to the liquid crystal display hardware are multiplexed. In this context, the duty ratio of a system is the number of multiplexed common lines. The most common duty ratio is 1/16, although 1/8 and 1/11 are found in some systems. Since the duty ratio measures the number of multiplexed lines, it also determines the display mapping. For example, in a single-line-by-16 character display with a 1/16 duty ratio the first eight characters are mapped to one set of consecutive memory addresses and the second eight characters to another set of addresses. The reason is that in every display line, sixteen common access lines are multiplexed, instead of eight. By the same token, a two-line-by-sixteen character display with a 1/16 duty ratio requires 16 common lines. In this case, the address of the second lines is not a continuation of the address of the first line, but is in another address set not contiguous to the first one.

For example, in a typical two-line-by-sixteen character display, the addresses of the 16 characters in the first line are from 0x00 to 0x0F, while the addresses of the characters in the second line are from 0x40 to 0x4F. Since there are 80 memory locations in the controller's DDRAM, each line contains storage for a total of 40 characters. The range of the entire first line is from 0x00 to 0x27 (40 characters total) but of these, only 16 are actually displayed. The same applies to the second line of 16 characters. In this case, the storage area is in the range 0x28 to 0x4f, but only 16 characters are displayed. In the single-line-by-sixteen character display mentioned first the addresses of the first eight characters would be a set from 0x00 to 0x07 and the addresses of the second eight characters in the line are from 0x40 to 0x47. Table 13.2 lists the memory address mapping of some common LCD configurations.

Table 13.2

7-bit DDRAM Address Mapping for Common LCDs

CHARACTERS/ROW	LINE NUMBER	CHARACTER NUMBER	FIRST IN GROUP	NEXT IN GROUP	LAST IN GROUP
8/1	1	1	0x00	0x01	0x07
8/2	1	1	0x00	0x01	0x07
	2	1	0x40	0x41	0x47
16/1	1	1	0x00	0x01	0x07
	1	9	0x40	0x41	0x47
16/2	1	1	0x00	0x01	0x0f
	2	1	0x40	0x41	0x4f
20/2	1	1	0x00	0x01	0x13
	2	1	0x40	0x41	0x53
24/2	1	1	0x00	0x01	0x17
	2	1	0x40	0x41	0x57
16/4	1	1	0x00	0X01	0x0f
	2	1	0x40	0x41	0x4f
	3	1	0x10	0x11	0x1f
	4	1	0x50	0x51	0x5f
20/4	1	1	0x00	0x01	0x13
	2	1	0x40	0x41	0x53
	3	1	0x14	0x15	0x27
	4	1	0x54	0x55	0x67

LCD Interfacing and Programming

Note that systems that exceed a total of 80 characters require two or more HD44780 controllers. Although the information provided in Table 13.3 corresponds to the mapping in most LCDs, it is a good idea to consult the data sheet of the specific hardware in order to corroborate the address mapping in a particular device.

Table 13.3 contains the seven low-order bits of DDRAM addresses. HD44780 commands to set the DDRAM address for read or write operations require that the high-order bit (bit number 7) be set. Therefore, to write to DDRAM memory address 0x07, code uses the value 0x87, and to write to DDRAM address 0x43, code uses 0xc3 as the instruction operand.

13.2 HD44780 Instruction Set

The HD44780 instruction set includes operators to initialize the system and set operational modes, clear the display, manipulate the cursor, set, reset, and control automatic display address shift, set and reset the interface parameters, poll the busy flag, read and write to CGRAM and DDRAM memory.

13.2.1 Instruction Set Overview

Pin number 4 in Table 13.1 selects two modes of operation on the HD44780 controller: instruction and data input. When the instruction mode is enabled (RS pin is set low) the controller receives commands that set up the hardware and determine its configuration and mode of operation. These commands are part of the HD44780 instruction set shown in Table 13.3.

Table 13.3
HD44780 Instruction Set

INSTRUCTION	RS	R/W	B7	B6	B5	B4	B3	B2	B1	B0	TIME
Clear Display	0	0	0	0	0	0	0	0	0	1	1.64
Return home	0	0	0	0	0	0	0	0	1	#	1.64
Entry mode set	0	0	0	0	0	0	0	1	I/D	S	37
Display/Cursor ON/OFF	0	0	0	0	0	0	1	D	C	B	37
Cursor/display shift	0	0	0	0	0	1	S/C	R/L	#	#	37
Function set	0	0	0	0	1	DL	N	F	#	#	37
Set CGRAM address	0	0	0	1	---------- address ------------------						37
Set DDRAM address	0	0	1	------------------ address ------------------							37
Read busy flag and Address register	0	1	BF	---------------- address ------------------							0
Write data	1	0	-------------------------- data --------------------								37
Read data	0	1	-------------------------- data --------------------								37

Note: Bits labeled # have no effect.

Clearing the Display

Clearing the display clears the display with blanks by writing the code 0x20 into all DDRAM addresses. It also returns the cursor to the *home position* (top-left display corner) and sets address 0 in the DDRAM address counter. After this command executes, the display disappears and the cursor goes to the left edge of the display.

Return home

Return home returns the cursor to home position at the upper left position of the first character line. It sets DDRAM address 0 in the address counter and sets the display to its default status if it was shifted. DDRAM contents remain unchanged.

Entry mode set

Entry mode set sets the direction of cursor movement and the display shift mode. If B1 (I/D) bit is set, cursor handling is set to the increment mode, that is, left-to-right. If this bit is clear, then cursor movement is set to the decrement mode, that is, right-to-left.

If B0 (S) bit is set, *display shift* is enabled. In the display shift mode, it appears as if the display moves instead of the cursor; otherwise display shift is disabled. Operations that read or write to CGRAM and operations that read DDRAM do not shift the display.

Display and Cursor ON/OFF

If B2 (D) bit is set, display is turned on. Otherwise, it is turned off. When the display is turned off data in DDRAM is not changed.

If B1 (C) bit is set, the cursor is turned on. Otherwise, it is turned off. Operations that change the current address in the DDRAM Address register, like those to automatically increment or decrement the address, are not affected by turning off the cursor. The cursor is displayed at the eighth line in the 5 x 8 character matrix.

If B0 (B) bit is set, the character at the current cursor position blinks. Otherwise, the character does not blink. Note that character blinking and cursor are independent operations and that both can be set to work simultaneously.

Cursor/display shift

Cursor/display shift moves the cursor or shifts the display according to the selected mode. The operation does not change the DDRAM content. Since the cursor position always coincides with the value in the Address register, the instruction provides software with a mechanism for making DDRAM corrections or to retrieve display data at specific DDRAM locations. Table 13.4 lists the four available options:

Table 13.4

Cursor/Display Shift Options

BITS S/C	R/L	OPERATION
0	0	Cursor position is shifted left. Address counter is decremented by one.
0	1	Cursor position is shifted right. Address counter is incremented by one.
1	0	Cursor and display are shifted left.
1	1	Cursor and display are shifted right.

Function set

Function set sets the parallel interface data length, the number of display lines, and the character font. If B4 (DL) bit is set, then the interface is set to eight bits. Otherwise it is set to four bits. If B3 (N) bit is zero, the display is initialized for 1/8 or 1/11 duty cycle. When the N bit is set, the display is set to 1/16 duty cycle. Displays with multiple lines typically use the 1/16 duty cycle. The 1/16 duty cycle on a one-line display appears as if it were a two-line display, that is, the line consists of two separate address groups (see Table 13.2).

If B2 (F) bit is set then the display resolution is 5 x 10 pixels. Otherwise the resolution is 5 x 8 pixels. This bit is not significant when the 1/16 duty cycle is selected; that is, when the N bit is set.

The function set instruction should be issued during controller initialization. No other instruction can be executed before this one, except for changing the interface data length.

Set CGRAM address

Set CGRAM address sets the *CGRAM (character generator RAM)* address to which data is sent or received after this operation. The CGRAM address is a six-bit field in the range 0 to 64 decimal. Once a value is entered in the CGRAM Address register, data can be read or written from CGRAM.

Set DDRAM address

Set DDRAM address sets the *DDRAM (display data RAM)* address to which data is sent or received after this operation. The DDRAM address is a seven-bit field in the range 0 to 127 decimal. Once a value is entered in the DDRAM Address register, data can be read or written from CGRAM. DDRAM address mapping is discussed in Section 13.1.4.

Read busy flag and Address register

Read busy flag and Address register reads the busy flag to determine if an internal operation is in progress and reads the address counter content. The value in the Address register is reported in bits 0 to 6. Bit 7 (BF) is the busy flag bit. This bit is read only. The address counter is incremented or decremented by 1 (according to the mode set) after the execution of a data write or read instruction.

Write data

Write data writes eight data bits to CGRAM or DDRAM. Before data is written to either controller RAM area, software must first issue a set DDRAM address or set CGRAM address instruction (described previously). These two instructions not only set the next valid address in the Address register, but also select either CGRAM or DDRAM for writing operations. What other actions take place as data is written to the controller depends on the settings selected by the *entry mode set* instruction. If the direction of cursor movement or data shift is in the increment mode, then the data write operation adds one to the value in the Address register. If the cursor movement is enabled, then the cursor is moved accordingly after data write takes place. If the display shift mode is active, then the displayed characters are shifted either right or left.

Read data

Read data reads eight data bits to CGRAM or DDRAM. Before data is read from either controller RAM area, software must first issue a set DDRAM address or set CGRAM address instruction. These instructions not only set the next valid address in the Address register, but also select either CGRAM or DDRAM for writing operations. Failing to set the corresponding RAM area results in reading invalid data.

What other actions take place as data is read from the controller RAM depends on the settings selected by the entry mode set instruction. If the direction of cursor movement or data shift is in the increment mode, then the data read operation adds one to the value in the Address register. However, display is no shifted by a read operation even if the display shift is active.

The cursor shift instruction has the effect of changing the content of the Address register. So if a cursor shift precedes a data read instruction, there is not need to reset the address by means of an address set command.

13.2.2 A 16F84 8-bit Data Mode Circuit

The first circuit presented in this chapter is experimental. Its purpose is to exercises LCD display functions in the simplest forms. Therefore, the circuit uses 8-bit *parallel data transmission interfacing* with a 16F84 microcontroller. The circuit is shown in Figure 13-4.

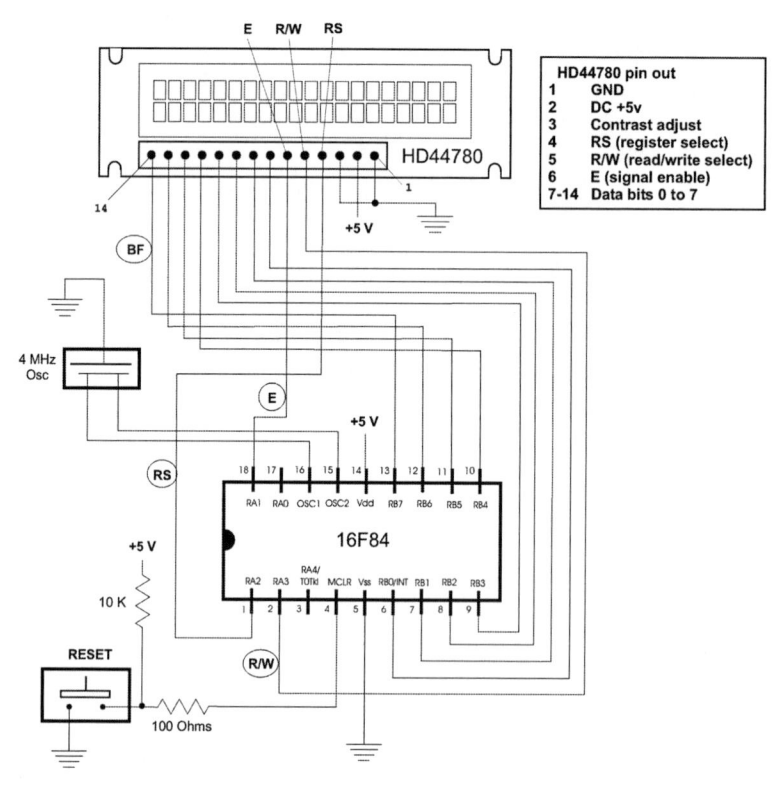

Figure 13-4 16F84 to LCD 8-bit Mode Circuit

In the circuit of Figure 13-4, three control lines are wired between the microcontroller and the LCD. The line designations are shown inside ovals. The R/W line is not necessary, since it is possible to devise a system that does not read LCD data. In spite of this, the R/W line is not included since it allows reading the busy flag in synchronizing operations. Table 13.5 shows the control and data connections for the circuit in Figure 13-4.

Table 13.5
Connections for 16F84/LCD 8-bit Data Mode Circuit

16F84 PIN	PORTBIT	LCD PIN	LINE NAME	FUNCTION
1	A2	4	RS	Select instruction/data register
2	A3	5	R/W	Read/write select
18	A1	6	E	Enable signal
13	B7	14	before	Busy flag.
6-13	B0-B7	7-14	Data	Data lines

13.3 LCD Programming

LCD programming is usually device-specific. Before attempting to write code, the programmer should become familiar with the circuit wiring diagram, the set up parameters, and the specific hardware requirements. It is risky to make assumptions that a specific device conforms exactly to the HD44780 interface since often a style sheet contains specifications that are not in strict conformance with the standard. In addition to the PIC set up and initialization functions, code to display a simple text message on the LCD screen consists of the following display-related functions:

1. Define the required constants, variables, and buffers.
2. Set up and initialize ports used by the LCD.
3. Initialize the LCD to circuit and software specifications.
4. Store text in PIC text buffer.
5. Select DDRAM start address on LCD.
6. Display text by transferring characters in PIC text buffer to LCD DDRAM.

If the LCD display consists of multiple lines, then the previous steps 4, 5, and 6 are repeated for each line. LCD initialization and display operations vary according to whether the interface is 4- or 8-bits and whether the code uses delay loops or busy flag monitoring to synchronize operations. All of these variations are considered in the examples in this chapter.

13.3.1 Defining Constants and Variables

In any program, defining and documenting constants and fixed parameters should be done centrally, rather than hard-coded through the code. Centralizing the elements that are variable under different circumstances makes it possible to adapt code to circuit and hardware changes.

Two common ways are available for defining constants: the C-like **#define** directive and the **equ** (equate) directive. In most cases, it is a matter of personal preference which is used, but a general guideline is to use the **#define** statement to create literal constants; that is, constants that are not associated with program registers or variables. The **equ** directive is then used to define registers, flags, and local variables.

According to this scheme, an LCD display driver program could use **#define** statements to create literals that are related to the wiring diagram or the specific LCD values obtained from the data sheet, such as the DDRAM addresses for each display line, as in the following code fragment:

```
;========================================================
;                constant definitions
;   for PIC-to-LCD pin wiring and LCD line addresses
;========================================================
#define E_line 1            ; |
#define RS_line 2           ; | - from wiring diagram
#define RW_line 3           ; |
; LCD line addresses (from LCD data sheet)
#define LCD_1 0x80          ; First LCD line constant
#define LCD_2 0xc0          ; Second LCD line constant
```

By the same token, the values associated with PIC register addresses and bit flags are defined using equ, as follows:

```
;========================================================
;                 PIC register equates
;========================================================
porta    equ        0x05
portb    equ        0x06
fsr      equ        0x04
status   equ        0x03
indf     equ        0x00
z        equ        2
```

One advantage of this scheme is that constants are easier to locate, since they are grouped by device. Those for the LCD are in **#define** directives area and those for the PIC hardware in an area of **equ** directives.

There are also drawbacks to this approach, since symbols created in **#define** directive are not available for viewing in the MPLAB debuggers. However, if the use of the **#define** directive is restriced to literal constants, then their viewing during a debugging session is not essential.

MPLAB also supports the **constant** directive for creating a constant symbol. Its use is identical to the **equ** directive but the latter is more commonly found in code.

Using MPLAB Data Directives

Often a program needs to define a block of sequential symbols and assign to each one a corresponding name. In the PIC 16f84, the address space allocated to general purpose registers allocated by the user is of 68 bytes, starting at address 0x0c. One possible way of allocating user-defined registers is to use the **equ** directive to assign addresses in the PIC SRAM space:

```
        Var1    equ     0x0c
        Var2    equ     0x0d
        Var3    equ     0x0e
        Buf1    equ     0x0f    ; 10-byte buffer space
        Var4    equ     0x19    ; Next variable
```

Although this method is functional, it depends on the programmer calculating the location of each variable in the PIC's available SRAM space. Alternatively, MPLAP provides a **cblock** directive that allows defining a group of consecutive sequential symbols while referring only to the address of the first element in the group. If no address is entered in **cblock**, then the assembler assigns the address. This address is one higher than the final address in the previous **cblock**. Each **cblock** ends with the **endc** directive. The following code fragment showing the use of the **cblock** directive is from one of the sample programs for this chapter.

```
;========================================================
;                variables in PIC RAM
;========================================================
; Reserve 16 bytes for string buffer
        cblock  0x0c
                strData
        endc
; Leave 16 bytes and continue with local variables
        cblock  0x1d    ; Start of block
                count1          ; Counter # 1
                count2          ; Counter # 2
                count3          ; Counter # 3
                pic_ad          ; Storage for start of text area
                                ; (labeled strData) in PIC RAM
                J               ; counter J
                K               ; counter K
                index           ; Index into text table
        endc
```

Note in the preceding code fragment, the allocation for the 16-byte buffer space named strData is ensured by entering the corresponding start address in the second **cblock**. The PIC microcontrollers do not contain a directive for reserving memory areas inside **cblock**, although the **res** directive can be used to reserve memory for individual variables.

13.3.2 LCD Initialization

LCD initialization depends on the specific hardware in use and on the circuit wiring. Information about the specific LCD can be obtained from the device's data sheet. Sometimes, the data sheet includes examples of initialization values for different conditions and even code listings. The information is usually sufficient to ensure correct initialization.

A word of warning: the popular LCD literature available online often contains initialization "myths" for specific components requiring that a certain mystery code be used for no documented reason, or that a certain function be repeated a given number of times. The programmer should make sure that the code is rational and that every operation is actually required and documented.

Before the LCD initialization commands are used it is necessary to set the communications lines correctly. The E line should be low, the RS line should be low for command, and the R/W line (if connected) should be low for *write mode*. After the lines are set accordingly, there should be a 125ms delay. Note that at this point, the LCD busy flag is not yet reliable. The following code fragment shows the processing:

```
        bcf     porta,E_line        ; E line low
        bcf     porta,RS_line       ; RS line low for command
        bcf     porta,RW_line       ; Write mode
        call    delay_125           ; delay 125 microseconds
```

The procedure delay_125 in the previous code fragment is described later in this chapter.

Function Set Command

Function set is the first initialization command sent to the LCD. The command determines whether the display font consists of 5 x 10 or 5 x 7 pixels. The latter is by far the more common. It determines the duty cycle, which is typically 1/8 or 1/11 for single-line displays and 1/16 for multiple lines. The interface width is also determined in the Function Set command. It is 4-bits or 8-bits. The following code fragment shows the commented code for the Function Set command:

```
;*********************|
;    Function Set     |
;*********************|
        movlw   0x38        ; 0 0 1 1 1 0 0 0 (FUNCTION SET)
                            ;         | | | |__ font select:
                            ;         | | |      1 = 5x10 in 1/8 or 1/11
                            ;         | | |      0 = 1/16 dc
                            ;         | | |___ Duty cycle select
                            ;         | |        0 = 1/8 or 1/11
                            ;         | |        1 = 1/16 (multiple ines)
                            ;         | |___ Interface width
                            ;         |        0 = 4 bits
                            ;         |        1 = 8 bits
                            ;         |___ FUNCTION SET COMMAND
```

LCD Interfacing and Programming

```
        movwf   portb
        call    pulseE   ;pulse E line to force LCD command
```

In the preceding code fragment, the LCD is initialized to multiple lines, 5 x 7 font, and 8-bit interface, as in the program LCDTest1 found in the book's online software package.

The procedure named pulseE sets the E line bit off and on to force command recognition by the LCD. The procedure is listed and described later in the chapter.

Display Off

Some initialization routines in LCD documentation and data sheets require that the display be turned off following the Function Set command. If so, the Display Off command is executed as follows:

```
;***********************|
;       Display Off     |
;***********************|
        movlw   0x08    ; 0 0 0 0 1 0 0 0 (DISPLAY ON/OFF)
                        ;         | | | |___ Blink character at
                        ;         | | | |    Cursor
                        ;         | | |      1 = on, 0 = off
                        ;         | | |___ Curson on/off
                        ;         | |        1 = on, 0 = off
                        ;         | |___ Display on/off
                        ;         |          1 = on, 0 = off
                        ;         |___ COMMAND BIT
        movwf   portb
        call    pulseE  ; pulse E line to force LCD command
```

Display and Cursor On

Whether or not the display is turned off, it must be turned on first. Also code must select if the cursor is on or off, and whether the character at the cursor position is to blink. The following command sets the cursor and the display on and the character blink off:

```
;***********************|
; Display and Cursor On |
;***********************|
        movlw   0x0e    ; 0 0 0 0 1 1 1 0 (DISPLAY ON/OFF)
                        ;         | | | |___ Blink character at
                        ;         | | | |    cursor
                        ;         | | |      1 = on, 0 = off
                        ;         | | |___ Curson on/off
                        ;         | |        1 = on, 0 = off
                        ;         | |___ Display on/off
                        ;         |          1 = on, 0 = off
                        ;         |___ COMMAND BIT
```

```
            movwf     portb
            call      pulseE      ; pulse E line to force LCD command
```

Set Entry Mode

The Entry Mode Command sets the direction of cursor movement or *display shift mode*. Normally, the display is set to the *increment mode* when writing in the Western European languages. The Entry Mode command controls *display shift*. If enabled, the displayed characters appear to scroll. This mode is used to simulate an electronic billboard effect by storing more than one line of characters in DDRAM and then scrolling the characters left-to-right. The following code sets entry mode to increment mode and no shift:

```
;**********************|
;   Set Entry Mode     |
;**********************|
            movlw     0x06        ; 0 0 0 0 0 1 1 0 (ENTRY MODE SET)
                                  ;           | | |___ display shift
                                  ;           | |        1 = shift
                                  ;           | |        0 = no shift
                                  ;           | |____ cursor increment
                                  ;           |            mode
                                  ;           |        1 = left-to-right
                                  ;           |        0 = right-to-left
                                  ;           |____ COMMAND BIT
            movwf     portb       ;00000110
            call      pulseE
```

Operations that read or write to CGRAM and operations that read DDRAM do not shift the display.

Cursor and Display Shift

These commands determine whether the cursor or the display shift according to the selected mode. Shifting the cursor or the display provides a software mechanism for making DDRAM corrections or for retrieving display data at specific DDRAM locations. The four available options appear in Table 13.4 previously in this chapter. The following instructions set the cursor to *shift right* and disable *display shift*:

```
;**********************|
; Cursor/Display Shift |
;**********************|
            movlw     0x14        ; 0 0 0 1 0 1 0 0 (CURSOR/DISPLAY
                                  ;         | | | | |     SHIFT)
                                  ;         | | | |_|___ don't care
                                  ;         | |_|__ cursor/display shift
                                  ;         |         00 = cursor shift left
                                  ;         |         01 = cursor shift right
                                  ;         |         10 = cursor and display
                                  ;         |                 shifted left
```

LCD Interfacing and Programming

```
                        ;        |       11 = cursor and display
                        ;        |            shifted right
                        ;        |___ COMMAND BIT
        movwf   portb   ;0001 1111
        call    pulseE
```

Clear Display

The final initialization command is usually one to clear the display. It is entered as follows:

```
;*********************|
;    Clear Display    |
;*********************|
        movlw   0x01    ; 0 0 0 0 0 0 0 1 (CLEAR DISPLAY)
                        ;                |___ COMMAND BIT
        movwf   portb   ;0000 0001
        call    pulseE
        call    delay_5 ;delay 5 milliseconds after init
```

Note that the last command is followed by a 5ms delay. The delay procedure delay_5 is listed and described later in this chapter.

13.3.3 Auxiliary Operations

Several support routines are required for effective text display in LCD devices. These include time delay routines for timed access, a routine to pulse the E line in order to force the LCD to execute a command or to read or write text data, routines to read the busy flag when this is the method used for processor/LCD synchronization, and routines to merge data with port bits so as to preserve the status of port lines not being addressed by code.

Time Delay Routine

There are several ways of producing time delays in PIC microcontroller. The Bibliography lists a title by David Benson devoted almost entirely to timing and counting routines. The present concern is quite simple: to develop a software routine that ensures the time delay that must take place in LCD programming, as shown in Table 13.3.

One mechanism for producing time delays in PIC programming is by means of the TIMER0 module, a built-in 8-bit timer counter. Once enabled, Port-A pin 4, labeled the TOCKI bit and associated with file register 01 (TMR0), is used to time processor operations. In the particular case of LCD timing routines, using the TIMER0 module seems somewhat of an overkill, in addition to the fact that it requires the use of a Port-A line which is often required for other purposes.

Alternatively, timing routines that serve the purpose at hand can be developed using simple delay loops. In this case, no port line is sacrificed and coding is considerably simplified. These routines are generically labeled *software timers*, in contrast with the hardware timers that depend on the PIC timer/counter device described previously. Software timers provide the necessary delay by means of program loops;

that is, by wasting time. The length of delay provided by the routine depends on the execution time of each instruction and on the number of repeated instructions.

Instructions on the PIC 16f84 consume four clock cycles. If the processor clock is running at 4 MHz, then one fourth of 4 MHz is the execution time for each instruction, which is 1 µs. So if each instructions requires 1 µs, repeating 1000 instructions produces a delay of 1 ms. The following routines provide convenient delays for LCD interfacing:

```
;=======================
;   Procedure to delay
;   125 microseconds
;=======================
delay_125mics:
          movlw    D'42'              ; Repeat 42 machine cycles
          movwf    count1             ; Store value in counter
repeat:
          decfsz   count1,f           ; Decrement counter (1 cycle)
          goto     repeat             ; Continue if not 0 (2 cycles)
                                      ; 42 * 3 = 126
          return                      ; End of delay
;=======================
;   Procedure to delay
;   5 milliseconds
;=======================
delay_5ms:
          movlw    D'41'              ; Counter = 41
          movwf    count2             ; Store in variable
delay:
          call     delay_125mics      ; Delay 41 microseconds
          decfsz   count2,f           ; 41 times 125 = 5125 ms.
                                      ; or approximately 5 ms
          goto     delay
          return                      ; End of delay
```

Actually, the delay loop of the procedure named delay_5ms is not exactly the product of 41 iterations times 125 µs, since the instruction to decrement the counter and the **goto** to the label delay are also inside the loop. Three instruction cycles must be added to those consumed by the delay_125mics procedure. This results in a total of 41 * 3 or 123 instruction cycles that must be added to the 5,125 consumed by delay_125mics. In fact, there are several other minor delays by the instructions to initialize the counters that are not included in the calculation. In reality, the delay loops required for LCD interfacing need not be exact, as long as they are not shorter than the recommended minimums.

For calculating software delays in the 16f84, the instruction execution time is determined by an external clock either in the form of an oscillator crystal, a resonator, or an RC oscillator furnished in the circuit. The PIC 16f84A is available in various processor speeds, from 4MHz to 20MHz. These speeds describe the maximum ca-

LCD Interfacing and Programming

pacity of the PIC hardware. The actual instruction speed is determined by the clocking device, so a 20 MHZ 16f84A using a 4 MHz oscillator effectively runs at 4 MHz.

Pulsing the E Line

The LCD hardware does not recognize data as it is placed in the input lines. When the various control and data pins of the LCD are connected to ports in the PIC and data is placed in the port bits, no action takes place in the LCD controller. In order for the controller to respond to commands or to perform read or write operations, it must be activated by pulsating (or strobing) the E line. The pulsing or strobing mechanism requires that the E line be kept low and then raised momentarily. The LCD checks the state of its lines on the raising edge of the E line. Once the command has completed, the E line is brought low again. The following code fragment pulses the E line in the manner described.

```
;========================
;     pulse E line
;========================
pulseE
        bsf     porta,E_line    ; pulse E line
        bcf     porta,E_line
        call    delay_125mics   ; delay 125 microseconds
        return
```

Note that the listed routine includes a 125μs delay following the pulsing operation. This delay is not part of the pulse function but is required by most LCD hardware. Some pulse functions in the popular PIC literature include a *no operation opcode* (**nop**) between the commands to set and clear the E line. In most cases this short delay does not hurt, but some LCDs require a minimum time lapse during the pulse and will not function correctly if the **nop** is inserted in the code.

Reading the Busy Flag

Synchronization between LCD commands and between data access operations is based on time delay loops or on reading the LCD busy flag. The busy flag, which is in the same pin as the bit 7 data line, is read clear when the LCD is ready to receive the next command, read, or write operation and is set if the device is not ready. By reading the state of the busy flag, code can accomplish more effective synchronization than with time delay loops. The sample program named LCDTest2, in the book's online software package, performs LCD display using the busy flag method. The following procedure shows busy flag synchronization:

```
;========================
; busy flag test routine
;========================
; Procedure to test the HD44780 busy flag
; Execution returns when flag is clear
busyTest:
        movlw   b'11111111'     ; All lines to input
        tris    portb           ;         in port B
        bcf     porta,RS_line   ; RS line low for control
```

```
            bsf     porta,RW_line       ; Read mode
            bsf     porta,E_line        ; E line high
            movf    portb,w             ; Read port B into W
                                        ; Port B bit 7 is busy flag
            bcf     porta,E_line        ; E line low
            andlw   0x80                ; Test bit 7, high is busy
            btfss   status,z            ; Test zero bit in STATUS
            goto    busyTest            ; Repeat if set
; At this point busy flag is clear
; Reset R/W line and port B to output
            bcf     porta,RW_line       ; Clear R/W line
            movlw   b'00000000'         ; All lines to output
            tris    portb                       ; in port B
            return
```

Note that testing the busy flag requires setting the LCD in read mode, which in turn requires implementing a connection between a PIC port and the R/W line. Also that the listed procedure contains no safety mechanism for detecting a hardware error condition in which the busy flag never clears. If such were the case, the program would hang in a forever loop. To detect and recover from this error the routine would have to include an external timing loop or some other means of recovering a possible hardware error.

Bit Merging Operations

Often, PIC/LCD circuits do not use all of the lines in an individual port. In this case the routines that manipulate PIC/LCD port access should not change the settings of other port bits. This situation is not exclusive to LCD interfacing; the discussion that follows has many other applications in PIC programming.

A processing routine can change one or more port lines without affecting the remaining ones. For example, an application that uses a 4-bit interface between the PIC and the LCD typically leaves four unused lines in the access port, or uses some of these lines for interface connections. In this case, the programming problem can be described as merging bits of the data byte to be written to the port and some existing port bits. One operand is the access port value and the other one is the new value to write to this port. If the operation at hand uses the four high-order port bits, then its four low-order bits must be preserved. The logic required is simple: AND the corresponding operands with masks that clear the unneeded bits and preserve the significant ones, then OR the two operands. The following procedure shows the required processing:

```
;==================
;   merge bits
;==================
; Routine to merge the 4 high-order bits of the
; value to send with the contents of port B
; so as to preserve the 4 low-bits in port B
; Logic:
;       AND value with 1111 0000 mask
```

LCD Interfacing and Programming

```
;       AND port B with 0000 1111 mask
;       At this point low nibble in value and high
;       nibble in port B are all 0 bits:
;                  value = vvvv 0000
;                  port B = 0000 bbbb
;       OR value and port B resulting in:
;                  vvvv bbbb
; ON ENTRY:
;       w contain value bits
; ON EXIT:
;       w contains merged bits
merge4:
        andlw   b'11110000'      ; ANDing with 0 clears the
                                 ; bit. ANDing with 1 preserves
                                 ; the original value
        movwf   store2           ; Save result in variable
        movf    portb,w          ; port B to w register
        andlw   b'00001111'      ; Clear high nibble in port b
                                 ; and preserve low nibble
        iorwf   store2,w         ; OR two operands in w
        return
```

Note that this particular example refers to merging two operand nibbles. The code can be adapted to merge other size bit-fields by modifying the corresponding masks. For example, the following routine merges the high-order bit of one operand with the seven low-order bits of the second one:

```
; Routine to merge the high-order bit of the first operand with
; the seven low-order bits of the second operand
; ON ENTRY:
;          w contains value bits of first operand
;          port b is the second operand
merge1:
        andlw   b'10000000'      ; ANDing with 0 clears the
                                 ; bit. ANDing with 1 preserves
                                 ; the original value
        movwf   store2           ; Save result in variable
        movf    portb,w          ; port B to w register
        andlw   b'01111111'      ; Clear high-order bit in
                                 ; port b and preserve the
                                 ; seven low order bits
        iorwf   store2,w         ; OR two operands in w
        return
```

Popular PIC literature describes routines to merge bit fields by assuming certain conditions in the destination operand, then testing the first operand bit to determine if the assumed condition should be preserved or changed. This type of operation is sometimes called "bit flipping," for example:

```
flipBit7:
; Code fragment to test the high-order bit in the variable named
; oprnd1 and preserve its status in the register variable portb
        bcf     portb,7         ; Assume oprnd1 bit is reset
        btfsc   oprnd1,7        ; Test operand bit and skip if
                                ; clear (assumption valid)
        bsf     portb,7         ; Set bit if necessary
        return
```

The logic in bit-flipping routines contains one critical flaw: if the assumed condition is false then the second operand is changed improperly, even if for only a few microseconds. However, the incorrect value can produce errors in execution if it is used by another device during this period. Since there is no such objection to the merge routines based on masking, the programmer should always prefer them.

13.3.4 Text Data Storage and Display

Text display operations require some way of generating the ASCII characters that are to be stored in DDRAM memory. Although the PIC Assembler contains several operators to generate ASCII data in program memory, there is no convenient way of storing a string in the General Purpose register area. Even if this was possible, SRAM is typically in short supply and text strings gobble up considerable data space.

Several possible approaches are available. The most suitable one depends on the total string length to be generated or stored, whether the strings are reused in the code, and other program-related circumstances. In this sense, short text-strings can be produced character-by-character and sent sequentially to DDRAM memory by placing the characters in the corresponding port and pulsing the E line.

The following code fragment consecutively displays the characters in the word "Hello." Code assumes that the command to set the Address register has been entered previously:

```
; Generate characters and send directly to DDRAM
        movlw   'H'             ; ASCII for H in w
        movwf   portb           ; Store code in port B
        call    pulseE          ; Pulse E line
        movlw   'e'             ; Continues
        movwf   portb
        call    pulseE
        movlw   'l'
        movwf   portb
        call    pulseE
        movlw   'l'
        movwf   portb
        call    pulseE
        movlw   'o'
        movwf   portb
        call    pulseE
        call    delay_5
```

LCD Interfacing and Programming

Note in the preceding fragment, the code assumes that the LCD has been initialized to automatically increment the Address register left-to-right. For this reason, the Address register is bumped to the next address with each port access.

Generating and Storing a Text String

An alternative approach suitable for generating and displaying longer strings consists of storing the string data in a local variable (sometimes called a *buffer*) and then transferring the characters, one by one, from the buffer to DDRAM. This kind of processing has the advantage of allowing the reuse of the same string and the disadvantage of using up scarce data memory. The logic for one possible routine consists of first generating and storing in PIC RAM the character string, then retrieving the characters from the PIC RAM buffer and displaying them. The character generation and storage logic is shown in Figure 13-5.

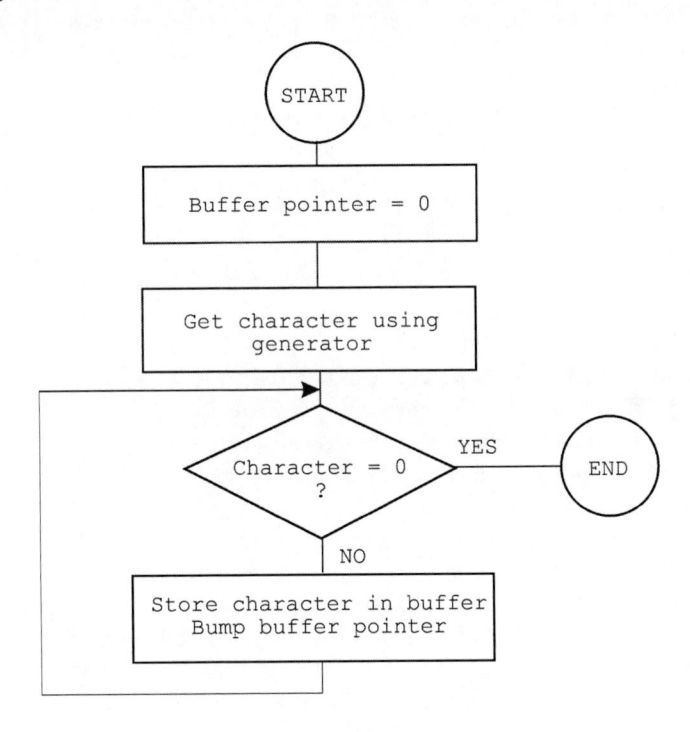

Figure 13-5 Flowchart for String Generation Logic

The processing is demonstrated in the following procedure.

```
;=================================
;   first text string procedure
;=================================
storeMN:
; Procedure to store in PIC RAM buffer the message
; contained in the code area labeled msg1
; ON ENTRY:
;          variable pic_ad holds address of text buffer
;          in PIC RAM
```

```
;               w register hold offset into storage area
;               msg1 is routine that returns the string characters
;               and a zero terminator
;               index is local variable that hold offset into
;               text table. This variable is also used for
;               temporary storage of offset into buffer
; ON EXIT:
;               Text message stored in buffer
;
; Store offset into text buffer (passed in the w register)
; in temporary variable
            movwf   index           ; Store w in index
; Store base address of text buffer in fsr
            movf    pic_ad,w ; first display RAM address to W
            addwf   index,w ; Add offset to address
            movwf   fsr     ; W to FSR
; Initialize index for text string access
            movlw   0       ; Start at 0
            movwf   index   ; Store index in variable
; w still = 0
get_msg_char:
            call    msg1    ; Get character from table
; Test for zero terminator
            andlw   0x0ff
            btfsc   status,z ; Test zero flag
            goto    endstr1  ; End of string
; ASSERT: valid string character in w
;           store character in text buffer (by fsr)
            movwf   indf    ; store in buffer by fsr
            incf    fsr,f   ; increment buffer pointer
; Restore table character counter from variable
            movf    index,w ; Get value into w
            addlw   1       ; Bump to next character
            movwf   index   ; Store table index in variable
            goto    get_msg_char    ; Continue
endstr1:
            return
; Routine for returning message stored in program area
msg1:
            addwf   PCL,f           ; Access table
            retlw   'M'
            retlw   'i'
            retlw   'n'
            retlw   'n'
            retlw   'e'
            retlw   's'
            retlw   'o'
            retlw   't'
```

LCD Interfacing and Programming

```
        retlw   'a'
        retlw   0                       ; terminator character
```

The auxiliary procedure named msg1, listed in the preceding code fragment, performs the character-generator function by producing each of the ASCII characters in the message string. Since a **retlw** instruction is necessary for each character, one instruction space in program memory is used for each character generated, plus a final binary zero for the string terminator.

Displaying the Text String

Once the string is stored in a local buffer, it is displayed by moving each ASCII code from the buffer into LCD DDRAM. Here again, we assume that the LCD has previously been set to the auto increment mode and that the Address register has been properly initialized with the corresponding DDRAM address. The following procedure demonstrates initialization of the DDRAM Address register to the value defined in the constant named LCD_1:

```
;========================
; Set Address register
;    to LCD line 1
;========================
; ON ENTRY:
;       Address of LCD line 1 in constant LCD_1
line1:
        bcf     porta,E_line    ; E line low
        bcf     porta,RS_line   ; RS line low, set up for
                                ; control
        call    delay_125       ; delay 125 microseconds
; Set to second display line
        Movlw   LCD_1           ; Address and command bit
        movwf   portb
        call    pulseE          ; Pulse and delay
; Set RS line for data
        bsf     porta,RS_line   ; Set up for data
        call    delay_125mics   ; Delay
        return
```

Once the Address register has been set up, the display operation consists of transferring characters from the PIC RAM buffer into LCD DDRAM. The following procedure can be used for this:

```
;==============================
;   LCD display procedure
;==============================
; Sends 16 characters from PIC buffer, with address stored
; in variable pic_ad, to LCD line previously selected
display16:
; Set up for data
        bcf     porta,E_line    ; E line low
```

```
                bsf     porta,RS_line       ; RS line low for control
                call    delay_125           ; Delay
; Set up counter for 16 characters
                movlw   D'16'               ; Counter = 16
                movwf   count3
; Get display address from local variable pic_ad
                movf    pic_ad,w ; First display RAM address to W
                movwf   fsr                 ; W to FSR
getchar:
                movf    indf,w   ; get character from display RAM
                                 ; location pointed to by file select
                                 ; register
                movwf   portb
                call    pulseE   ;send data to display
; Test for 16 characters displayed
                decfsz  count3,f ; Decrement counter
                goto    nextchar ; Skipped if done
                return
nextchar:
                incf    fsr,f               ; Bump pointer
                goto    getchar
```

Note the procedure display16, previously listed, assumes that the address of the local buffer is stored in a variable name pic_ad. This allows reusing the procedure to display text stored at other locations in PIC RAM.

The previously listed procedures demonstrate just one of many possible variations on this technique. Another approach is to store the characters directly in DDRAM memory as they are produced by the message-returning routine, thus avoiding the display procedure entirely. In this last case, the programming saves some data memory space at the expense of having to generate the message characters each time they are needed. Which approach is the most suitable one depends on the application.

13.3.5 Data Compression Techniques

Circuits based on the parallel data transfer of eight data bits require eight devoted port lines. Assuming that three other lines are required for LCD commands and interfacing (RS, E, and R/W lines), then 11 PIC-to-LCD lines are needed, leaving two free port lines at the most, on an 16f84 microcontroller. Not many useful devices can make do with just two port lines. Several possible solutions allow compressing the data transfer function. The most obvious one is to use the 4-bit data transfer mode to free four port lines. Other solutions are based on dedicating logic components to the LCD function.

4-bit Data Transfer Mode

One possible solution is to use the capability of the Hitachi 44780 controller that allows a parallel interface using just four data paths instead of eight. The objections are that programming in 4-bit mode is slightly more convoluted and there is a very minor performance penalty. In 4-bit mode, data must be sent one nibble at a time, so execu-

LCD Interfacing and Programming

tion is slower. Since the delay is required only after the second nibble, the execution time penalty for 4-bit transfers is not very large.

Many of the previously developed routines for 8-bit data mode can be reused without modification in the 4-bit mode. Others require minor changes, and there is one specific display procedure that must be developed ad hoc. The first required change is in the LCD initialization since bit 4 in the Function Set command must be clear for a 4-bit interface. The remaining initialization commands should require no further change, although it is a good idea to consult the data sheet for the LCD hardware in use.

Displaying data using a 4-bit interface consists of sending the high-order nibble followed by the low-order nibble, through the LCD 4-high-order data lines, usually labeled DB5 to DB7. The pulsing of line E follows the last nibble sent. It is usually the case in the 16f84 PIC that circuit wiring in the 4-bit mode uses four of five lines in Port-A, or four of eight lines in port B. Software must provide a way of reading and writing to the appropriate port lines, the ones used in the data transfer, without altering the value stored in the port bits dedicated to other uses. Bit merging routines, discussed in Section 13.3, are quite suitable for the purpose at hand.

The following procedures are designed to send the two nibbles of a data byte through the four high-order lines in port B. The auxiliary procedure named merge4 performs the bit-merging operation while the procedure named send8 does the actual write operation:

```
;=========================
;   send 2 nibbles in
;      4-bit mode
;=========================
; Procedure to send two 4-bit values to port B lines
; 7, 6, 5, and 4. High-order nibble is sent first
; ON ENTRY:
;         w register holds 8-bit value to send
send8:
        movwf      store1              ; Save original value
        call       merge4              ; Merge with port B
; Now w has merged byte
        movwf      portb
        call       pulseE              ; w to port B
                                       ; Send data to LCD
; High nibble is sent
        movf       store1,w            ; Recover byte into w
        swapf      store1,w            ; Swap nibbles in w
        call       merge4
        movwf      portb
        call       pulseE              ; Send data to LCD
        call       delay_125
        return
;==================
;   merge bits
```

```
;=================
; Routine to merge the 4 high-order bits of the
; value to send with the contents of port B
; so as to preserve the 4 low-bits in port B
; Logic:
;       AND value with 1111 0000 mask
;       AND port B with 0000 1111 mask
;       Now low nibble in value and high nibble in
;       port B are all 0 bits:
;               value = vvvv 0000
;               port B = 0000 bbbb
;       OR value and port B resulting in:
;               vvvv bbbb
; ON ENTRY:
;       w contain value bits
; ON EXIT:
;       w contains merged bits
merge4:
        andlw   b'11110000'     ; ANDing with 0 clears the
                                ; bit. ANDing with 1 preserves
                                ; the original value
        movwf   store2          ; Save result in variable
        movf    portb,w         ; port B to w register
        andlw   b'00001111'     ; Clear high nibble in port b
                                ; and preserve low nibble
        iorwf   store2,w        ; OR two operands in w
        return
```

The program named LCDTest3 in the book's online software package is a demonstration using the 4-bit interface mode. Figure 13-6 shows a PIC/LCD circuit that is wired for the 4-bit data transfer mode.

Note in the circuit of Figure 13-6 that a total of six port lines remain unused. Two of these lines are in Port-A and four in Port-B.

Master/Slave Systems

To this point we have assumed that driving the LCD is one of the functions performed by the PIC microcontroller, which also executes the other circuit functions. In practice, such a scheme is rarely viable for two reasons: the number of interface lines required and the amount of PIC code space used up by the LCD driver routines. A more efficient approach is to dedicate a PIC exclusively to controlling the LCD hardware, while one or more other microcontrollers perform the main circuit functions. In this scheme, the PIC devoted to the LCD function is referred to as a *slave* while the one that sends the display commands is called the *master*.

LCD Interfacing and Programming

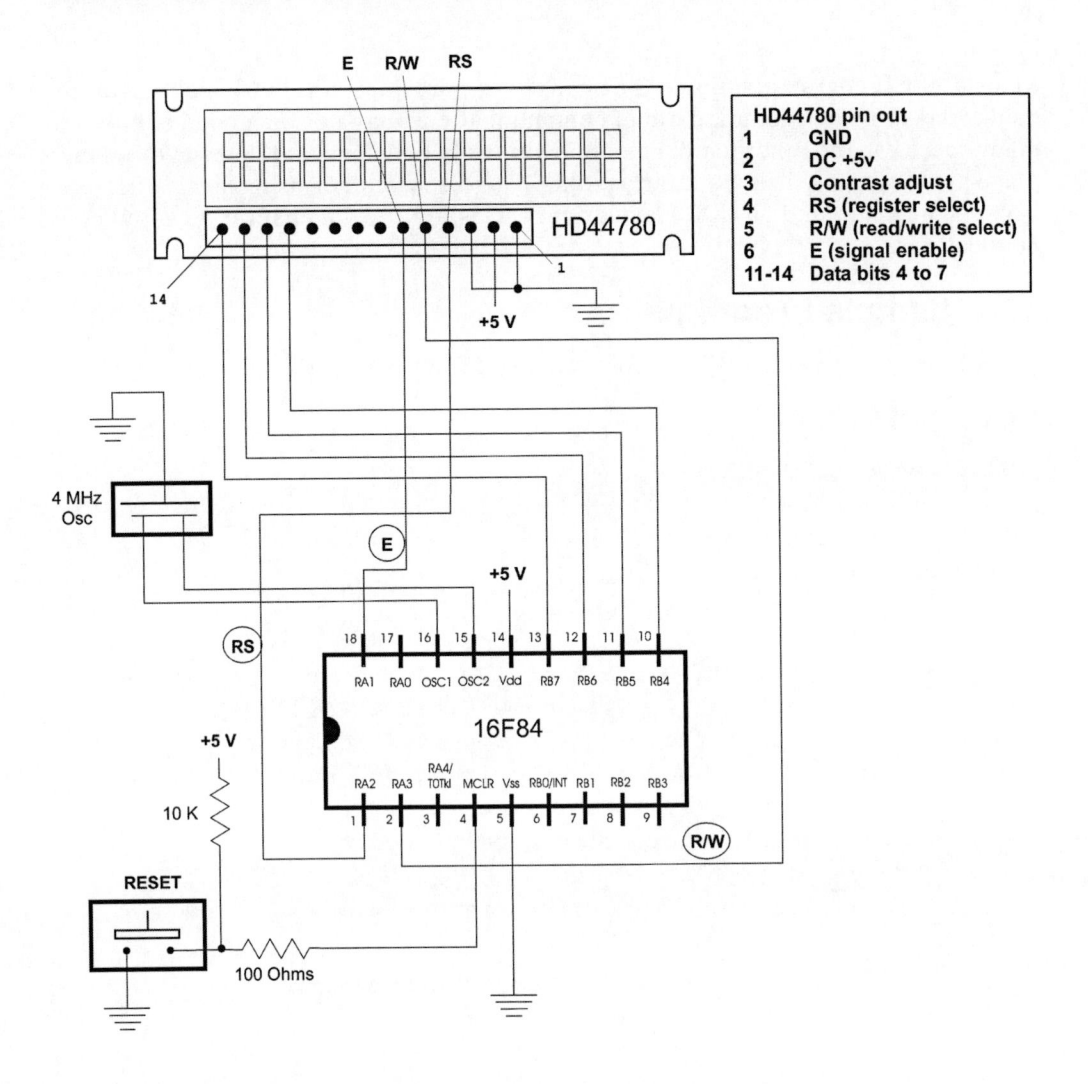

HD44780 pin out
1 GND
2 DC +5v
3 Contrast adjust
4 RS (register select)
5 R/W (read/write select)
6 E (signal enable)
11-14 Data bits 4 to 7

Figure 13-6 PIC/LCD Circuit for 4-bit Data Mode

When sufficient numbers of interface lines are available, the connection between master and slave can be simplified by using a *parallel interface*. For example, if four port lines are used to interconnect the two PICs, then 16 different command codes can be sent to the slave. The slave reads the communications lines much like it would read a multiple toggle switch. A simple protocol can be devised so that the slave uses these same interface lines to provide feedback to the master. For example, the slave sets all four lines low to indicate that it is ready for the next command, and sets them high to indicate that command execution is in progress and that no new commands can be received. The master, in turn, reads the communications lines to determine when it can send another command to the slave.

But using parallel communications between master and slave can be a self-defeating proposition, since it requires at least seven interface lines to be able to send ASCII characters. Since the scarcity of port lines is the original reason for using a master/slave set up, parallel communications may not be a good solution in many cases. On the other hand, communications between master and slave can take place serially, using a single interface line. The discussion of using *serial interface* between a master and an LCD slave driver PIC is left for the chapter on serial communications.

13.4 Sample Programs

The following section lists the sample programs discussed in this chapter.

13.4.1 LCDTest1

```
; File name: LCDTest1.asm
; Date: April 13, 2006
; Author: Julio Sanchez
; Processor: 16F84A
;
; Description:
; Program to exercise 8-bit PIC-to-LCD interface.
; Code assumes that LCD is driven by Hitachi HD44780
; controller and that the display supports two lines
; each one with 16 characters. The wiring and base
; address of each display line is stored in #define
; statements. These statements can be edited to
; accommodate a different set-up.
; Program uses delay loops for interface timing.
; WARNING:
; Code assumes 4Mhz clock. Delay routines must be
; edited for faster clock

; Displays: Minnesota State, Mankato
;
;===========================
;         switches
;===========================
; Switches used in __config directive:
;    _CP_ON         Code protection ON/OFF
; *  _CP_OFF
; *  _PWRTE_ON      Power-up timer ON/OFF
;    _PWRTE_OFF
;    _WDT_ON        Watchdog timer ON/OFF
; *  _WDT_OFF
;    _LP_OSC        Low power crystal occilator
; *  _XT_OSC        External parallel resonator/crystal oscillator

;    _HS_OSC        High speed crystal resonator (8 to 10 MHz)
;                   Resonator: Murate Erie CSA8.00MG = 8 MHz
```

LCD Interfacing and Programming

```
;    _RC_OSC          Resistor/capacitor oscillator (simplest, 20%
error)
;    |
;    |_____ * indicates set up values

;==========================
; set up and configuration
;==========================
        processor 16f84A
        include   <p16f84A.inc>
        __config  _XT_OSC & _WDT_OFF & _PWRTE_ON & _CP_OFF

;========================================================
;                constant definitions
;   for PIC-to-LCD pin wiring and LCD line addresses
;========================================================
#define E_line  1           ; |
#define RS_line 2           ; | - from wiring diagram
#define RW_line 3           ; |
; LCD line addresses (from LCD data sheet)
#define LCD_1 0x80          ; First LCD line constant
#define LCD_2 0xc0          ; Second LCD line constant
; Note: The constants that define the LCD display line
;       addresses have the high-order bit set in
;       order to facilitate the controller command
;
;========================================================
;                variables in PIC RAM
;========================================================
; Reserve 16 bytes for string buffer
        cblock    0x0c
        strData
        endc
; Leave 16 bytes and Continue with local variables
        cblock    0x1d              ; Start of block
        count1              ; Counter # 1
        count2              ; Counter # 2
        count3              ; Counter # 3
        pic_ad              ; Storage for start of text area
                            ; (labeled strData) in PIC RAM
        J                   ; counter J
        K                   ; counter K
        index               ; Index into text table (also used
                            ; for auxiliary storage)
        endc

;================================================================
;                           program
```

```
;===============================================================
        org     0               ; start at address
        goto    main
; Space for interrupt handlers
        org             0x08

main:
        movlw   b'00000000'     ; All lines to output
        tris    PORTA           ; in Port-A
        tris    PORTB           ; and port B
        movlw   b'00000000'     ; All outputs ports low
        movwf   PORTA
        movwf   PORTB
; Wait and initialize HD44780
        call    delay_5ms       ; Allow LCD time to initialize
                                ; itself
        call    initLCD         ; Then do forced
                                ; initialization
        call    delay_5ms       ; Wait.
; Store base address of text buffer in PIC RAM
        movlw   0x0c            ; Start address of text buffer
        movwf   pic_ad          ; to local variable
;======================
;   first LCD line
;======================
; Store 16 blanks in PIC RAM, starting at address stored
; in variable pic_ad
        call    blank16
; Call procedure to store ASCII characters for message
; in text buffer
        movlw   d'3'            ; Offset into buffer
        call    storeMN
; Set DDRAM address to start of first line
        call    line1
; Call procedure to display 16 characters in LCD
        call    display16
;========================
;   second LCD line
;========================
        call    delay_125mcs    ; Wait for termination
        call    blank16         ; Blank buffer
; Call procedure to store ASCII characters for message
; in text buffer
        movlw   d'1'            ; Offset into buffer
        call    storeUniv
        call    line2           ; DDRAM address of LCD line 2
        call    display16
;========================
```

```
;               done!
;========================
loopHere:
          goto     loopHere   ;done

;****************************************************************
;                  INITIALIZE LCD PROCEDURE
;****************************************************************
initLCD
; Initialization for Densitron LCD module as follows:
;    8-bit interface
;    2 display lines of 16 characters each
;    cursor on
;    left-to-right increment
;    cursor shift right
;    no display shift
;*********************|
;     COMMAND MODE    |
;*********************|
          bcf      PORTA,E_line      ; E line low
          bcf      PORTA,RS_line     ; RS line low for command
          bcf      PORTA,RW_line     ; Write mode
          call     delay_125mcs              ;delay 125
microseconds
;*********************|
;     FUNCTION SET    |
;*********************|
          movlw    0x38     ; 0 0 1 1 1 0 0 0 (FUNCTION SET)
                            ;         | | | |__ font select:
                            ;         | | |    1 = 5x10 in 1/8 or 1/11
                            ;         | | |    0 = 1/16 dc
                            ;         | | |___ Duty cycle select
                            ;         | |      0 = 1/8 or 1/11
                            ;         | |      1 = 1/16
                            ;         | |___ Interface width
                            ;         |      0 = 4 bits
                            ;         |      1 = 8 bits
                            ;         |___ FUNCTION SET COMMAND
          movwf    PORTB    ;0011 1000
          call     pulseE   ;pulseE and delay

;*********************|
;     DISPLAY OFF     |
;*********************|
          movlw    0x08     ; 0 0 0 0 1 0 0 0 (DISPLAY ON/OFF)
                            ;             | | | |___ Blink character
                            ;             | | |     1 = on, 0 = off
                            ;             | | |___ Cursor on/off
```

```
                           ;          | |       1 = on, 0 = off
                           ;          | |____ Display on/off
                           ;          |         1 = on, 0 = off
                           ;          |____ COMMAND BIT
            movwf   PORTB
            call    pulseE    ;pulseE and delay

;*********************|
; DISPLAY AND CURSOR ON |
;*********************|
            movlw   0x0e      ; 0 0 0 0 1 1 1 0 (DISPLAY ON/OFF)
                           ;           | | | |___ Blink character
                           ;           | | |       1 = on, 0 = off
                           ;           | | |___ Cursor on/off
                           ;           | |         1 = on, 0 = off
                           ;           | |____ Display on/off
                           ;           |          1 = on, 0 = off
                           ;           |____ COMMAND BIT
            movwf   PORTB
            call    pulseE    ;pulseE and delay

;*********************|
;    ENTRY MODE SET    |
;*********************|
            movlw   0x06      ; 0 0 0 0 0 1 1 0 (ENTRY MODE SET)
                           ;             | | |___ display shift
                           ;             | |       1 = shift
                           ;             | |       0 = no shift
                           ;             | |____ increment mode
                           ;             |         1 = left-to-right
                           ;             |         0 = right-to-left
                           ;             |___ COMMAND BIT
            movwf   PORTB    ;00000110
            call    pulseE

;*********************|
; CURSOR/DISPLAY SHIFT  |
;*********************|
            movlw   0x14      ; 0 0 0 1 0 1 0 0 (CURSOR/DISPLAY
                           ;         | | | | |   SHIFT)
                           ;         | | | |_|___ don't care
                           ;         | |_|__ cursor/display shift
                           ;         |          00 = cursor shift left
                           ;         |          01 = cursor shift right
                           ;         |          10 = cursor and display
                           ;         |                 shifted left
                           ;         |          11 = cursor and display
```

```
                              ;      |         shifted right
                              ;      |___ COMMAND BIT
        movwf    PORTB    ;0001 1111
        call     pulseE

;*********************|
;   CLEAR DISPLAY     |
;*********************|
        movlw    0x01     ; 0 0 0 0 0 0 0 1 (CLEAR DISPLAY)
                          ;                 |___ COMMAND BIT
        movwf    PORTB    ;0000 0001
;
        call     pulseE
        call     delay_5ms      ;delay 5 milliseconds after
init
        return
;****************************************************************
;                 DELAY AND PULSE PROCEDURES
;****************************************************************
;======================
;  Procedure to delay
;   42 microseconds
;======================
delay_125mcs
        movlw    D'42'          ; Repeat 42 machine cycles
        movwf    count1         ; Store value in counter
repeat
        decfsz   count1,f       ; Decrement counter
        goto     repeat         ; Continue if not 0
        return                  ; End of delay
;======================
;  Procedure to delay
;   5 milliseconds
;======================
delay_5ms
        movlw    D'41'          ; Counter = 41
        movwf    count2         ; Store in variable
delay
        call     delay_125mcs   ; Delay
        decfsz   count2,f       ; 40 times = 5 milliseconds
        goto     delay
        return                  ; End of delay
;======================
;     pulse E line
;======================
pulseE
        bsf      PORTA,E_line   ;pulse E line
        bcf      PORTA,E_line
```

```
                call    delay_125mcs    ;delay 125 microseconds
        return

;==============================
;   long delay sub-routine
;       (for debugging)
;==============================
long_delay
        movlw   D'200'          ; w = 200 decimal
        movwf   J               ; J = w
jloop:
        movwf   K               ; K = w
kloop:
        decfsz  K,f             ; K = K-1, skip next if zero
        goto    kloop
        decfsz  J,f             ; J = J-1, skip next if zero
        goto    jloop
        return
;==============================
;   LCD display procedure
;==============================
; Sends 16 characters from PIC buffer with address stored
; in variable pic_ad to LCD line previously selected
display16:
; Set up for data
        bcf     PORTA,E_line    ; E line low
        bsf     PORTA,RS_line   ; RS line low for control
        call    delay_125mcs    ; Delay
; Set up counter for 16 characters
        movlw   D'16'           ; Counter = 16
        movwf   count3
; Get display address from local variable pic_ad
        movf    pic_ad,w ; First display RAM address to W
        movwf   FSR     ; W to FSR
getchar:
        movf    INDF,w  ; get character from display RAM
                        ; location pointed to by file select
                        ; register
        movwf   PORTB
        call    pulseE  ;send data to display
; Test for 16 characters displayed
        decfsz  count3,f        ; Decrement counter
        goto    nextchar ; Skipped if done
        return
nextchar:
        incf    FSR,f   ; Bump pointer
        goto    getchar
;==========================
```

LCD Interfacing and Programming

```
;         blank buffer
;=========================
; Procedure to store 16 blank characters in PIC RAM
; buffer starting at address stored in the variable
; pic_ad
blank16:
        movlw   D'16'           ; Set up counter
        movwf   count1
        movf    pic_ad,w ; First PIC RAM address
        movwf   FSR             ; Indexed addressing
        movlw   0x20            ; ASCII space character
storeit:
        movwf   INDF            ; Store blank character in PIC RAM
                                ; buffer using FSR register
        decfsz  count1,f        ; Done?
        goto    incfsr          ; no
        return                  ; yes
incfsr:
        incf    FSR,f           ; Bump FSR to next buffer space
        goto    storeit

;=========================
; Set Address register
;     to LCD line 1
;=========================
; ON ENTRY:
;         Address of LCD line 1 in constant LCD_1
line1:
        bcf     PORTA,E_line    ; E line low
        bcf     PORTA,RS_line   ; RS line low, set up for control
        call    delay_125mcs    ; delay 125 microseconds
; Set to second display line
        movlw   LCD_1           ; Address and command bit
        movwf   PORTB
        call    pulseE          ; Pulse and delay
; Set RS line for data
        bsf     PORTA,RS_line   ; Set up for data
        call    delay_125mcs    ; Delay
        return
;=========================
; Set Address register
;     to LCD line 2
;=========================
; ON ENTRY:
;         Address of LCD line 2 in constant LCD_2
```

```
line2:
        bcf     PORTA,E_line    ; E line low
        bcf     PORTA,RS_line   ; RS line low, set up for
                                ; control
        call    delay_125mcs    ; delay
; Set to second display line
        movlw   LCD_2           ; Address with high-bit set
        movwf   PORTB
        call    pulseE          ; Pulse and delay
; Set RS line for data
        bsf     PORTA,RS_line   ; RS = 1 for data
        call    delay_125mcs    ; delay
        return

;===============================
;   first text string procedure
;===============================
storeMN:
; Procedure to store in PIC RAM buffer the message
; contained in the code area labeled msg1
; ON ENTRY:
;           variable pic_ad holds address of text buffer
;           in PIC RAM
;           w register hold offset into storage area
;           msg1 is routine that returns the string characters
;           and a zero terminator
;           index is local variable that hold offset into
;           text table. This variable is also used for
;           temporary storage of offset into buffer
; ON EXIT:
;           Text message stored in buffer
;
; Store offset into text buffer (passed in the w register)
; in temporary variable
        movwf   index           ; Store w in index
; Store base address of text buffer in FSR
        movf    pic_ad,w ; first display RAM address to W
        addwf   index,w         ; Add offset to address
        movwf   FSR             ; W to FSR
; Initialize index for text string access
        movlw   0               ; Start at 0
        movwf   index           ; Store index in variable
; w still = 0
get_msg_char:
        call    msg1            ; Get character from table
; Test for zero terminator
        andlw   0x0ff
        btfsc   STATUS,Z ; Test zero flag
```

LCD Interfacing and Programming

```
            goto      endstr1           ; End of string
; ASSERT: valid string character in w
;         store character in text buffer (by FSR)
            movwf     INDF              ; store in buffer by FSR
            incf      FSR,f             ; increment buffer pointer
; Restore table character counter from variable
            movf      index,w           ; Get value into w
            addlw     1                 ; Bump to next character
            movwf     index             ; Store table index in variable
            goto      get_msg_char      ; Continue
endstr1:
            return

; Routine for returning message stored in program area
msg1:
            addwf     PCL,f             ; Access table
            retlw     'M'
            retlw     'i'
            retlw     'n'
            retlw     'n'
            retlw     'e'
            retlw     's'
            retlw     'o'
            retlw     't'
            retlw     'a'
            retlw     0


;==================================
;    second text string procedure
;==================================
storeUniv:
; Processing identical to procedure StoreMSU
            movwf     index             ; Store w in index
; Store base address of text buffer in FSR
            movf      pic_ad,0 ; first display RAM address to W
            addwf     index,0           ; Add offset to address
            movwf     FSR               ; W to FSR
; Initialize index for text string access
            movlw     0                 ; Start at 0
            movwf     index             ; Store index in variable
; w still = 0
get_msg_char2:
            call      msg2              ; Get character from table
; Test for zero terminator
            andlw     0x0ff
            btfsc     STATUS,Z ; Test zero flag
```

```
            goto    endstr2             ; End of string
; ASSERT: valid string character in w
;         store character in text buffer (by FSR)
            movwf   INDF                ; Store in buffer by FSR
            incf    FSR,f               ; Increment buffer pointer
; Restore table character counter from variable
            movf    index,w             ; Get value into w
            addlw   1                   ; Bump to next character
            movwf   index               ; Store table index in
variable
            goto    get_msg_char2       ; Continue
endstr2:
            return

; Routine for returning message stored in program area
msg2:
            addwf   PCL,f               ; Access table
            retlw   'S'
            retlw   't'
            retlw   'a'
            retlw   't'
            retlw   'e'
            retlw   ','
            retlw   0x20
            retlw   'M'
            retlw   'a'
            retlw   'n'
            retlw   'k'
            retlw   'a'
            retlw   't'
            retlw   'o'
            retlw   0

            end
```

13.4.2 LCDTest2 Program

```
; File name: LCDTest2.asm
; Date: April 16, 2006
; Author: Julio Sanchez
; Processor: 16F84A
;
; Description:
; Program to exercises 8-bit PIC-to-LCD interface.
; Code assumes that LCD is driven by Hitachi HD44780
; controller and that the display supports two lines
; each one with 16 characters. The wiring and base
```

LCD Interfacing and Programming

```
; address of each display line is stored in #define
; statements. These statements can be edited to
; accommodate a different set-up.
; Program uses the busy flag to synchronize processor
; access, although delay loops are still required in
; some cases.
; Displays: Minnesota State, Mankato
; WARNING:
; Code assumes 4Mhz clock. Delay routines must be
; edited for faster clock
;
; Displays: Minnesota State, Mankato
;
;===========================
;         switches
;===========================
; Switches used in __config directive:
;   _CP_ON          Code protection ON/OFF
; * _CP_OFF
; * _PWRTE_ON       Power-up timer ON/OFF
;   _PWRTE_OFF
;   _WDT_ON         Watchdog timer ON/OFF
; * _WDT_OFF
;   _LP_OSC         Low power crystal occilator
; * _XT_OSC         External parallel resonator/crystal oscillator

;   _HS_OSC         High speed crystal resonator (8 to 10 MHz)
;                   Resonator: Murate Erie CSA8.00MG = 8 MHz
;   _RC_OSC         Resistor/capacitor oscillator
; |                 (simplest, 20% error)
; |
; |_____  * indicates set up values presently selected

;=========================
; set up and configuration
;=========================
        processor 16f84A
        include   <p16f84A.inc>
        __config  _XT_OSC & _WDT_OFF & _PWRTE_ON & _CP_OFF


;========================================================
;                 constant definitions
;  for PIC-to-LCD pin wiring and LCD line addresses
;========================================================
#define E_line 1            ;|
#define RS_line 2           ;| - from wiring diagram
#define RW_line 3           ;|
; LCD line addresses (from LCD data sheet)
```

```
#define LCD_1 0x80        ; First LCD line constant
#define LCD_2 0xc0        ; Second LCD line constant
; Note: The constants that define the LCD display line
;       addresses have the high-order bit set in
;       order to facilitate the controller command
;
;=========================================================
;                 variables in PIC RAM
;=========================================================
; Reserve 16 bytes for string buffer
        cblock    0x0c
        strData
        endc
; Leave 16 bytes and Continue with local variables
        cblock    0x1d              ; Start of block
        count1                      ; Counter # 1
        count2                      ; Counter # 2
        count3                      ; Counter # 3
        pic_ad                      ; Storage for start of text area
                                    ; (labeled strData) in PIC RAM
        J                           ; counter J
        K                           ; counter K
        index                       ; Index into text table (also used
                                    ; for auxiliary storage)
        endc

;=========================================================
;                          program
;=========================================================
        org       0                 ; start at address
        goto      main
; Space for interrupt handlers
        org       0x08

main:
        movlw     b'00000000'       ; All lines to output
        tris      PORTA             ; in Port-A
        tris      PORTB             ; and port B
        movlw     b'00000000'       ; All outputs ports low
        movwf     PORTA
        movwf     PORTB
; Wait and initialize HD44780
        call      delay_5           ; Allow LCD time to initialize
                                    ; itself
        call      initLCD           ; Then do forced
initialization
; Store base address of text buffer in PIC RAM
        movlw     0x0c              ; Start address for buffer
```

LCD Interfacing and Programming

```
            movwf    pic_ad           ; to local variable
;======================
;    first LCD line
;======================
; Store 16 blanks in PIC RAM, starting at address stored
; in variable pic_ad
            call     blank16
; Call procedure to store ASCII characters for message
; in text buffer
            movlw    d'3'             ; Offset into buffer
            call     storeMSU
; Set DDRAM address to start of first line
            call     line1
; Call procedure to display 16 characters in LCD
            call     display16
;========================
;    second LCD line
;========================
            call     busyTest ; Wait for termination
            call     blank16           ; Blank buffer
; Call procedure to store ASCII characters for message
; in text buffer
            movlw    d'1'             ; Offset into buffer
            call     storeUniv
            call     line2            ; DDRAM address of LCD line 2
            call     display16
;======================
;       done!
;======================
loopHere:
            goto     loopHere    ;done

;*************************************************************
;                    INITIALIZE LCD PROCEDURE
;*************************************************************
initLCD:
;*********************|
;     COMMAND MODE    |
;*********************|
            bcf                PORTA,E_line    ;E line low
            bcf                PORTA,RS_line   ;RS line low
            call     delay_125                 ;delay 125 microseconds
;*********************|
;     FUNCTION SET    |
;*********************|
            movlw    0x38     ; 0 0 1 1 1 0 0 0 (FUNCTION SET)
                              ;         | | | |__ font select:
```

```
                        ;       | | |       1 = 5x10 in 1/8 or 1/11
                        ;       | | |       0 = 1/16 dc
                        ;       | | |___ Duty cycle select
                        ;       | |         1 = 1/8 or 1/11
                        ;       | |         0 = 1/16
                        ;       | |___ Interface width
                        ;       |         0 = 4 bits
                        ;       |         1 = 8 bits
                        ;       |___ FUNCTION SET COMMAND
        movwf   PORTB   ;0011 1000
        call    pulseE  ;pulseE and delay

;**********************|
;     DISPLAY OFF      |
;**********************|
        movlw   0x08    ; 0 0 0 0 1 0 0 0 (DISPLAY ON/OFF)
                        ;         | | | |___ Blink character
                        ;         | | |       1 = on, 0 = off
                        ;         | | |___ Cursor on/off
                        ;         | |         1 = on, 0 = off
                        ;         | |___ Display on/off
                        ;         |           1 = on, 0 = off
                        ;         |___ COMMAND BIT
        movwf   PORTB
        call    pulseE  ;pulseE and delay

;**********************|
; DISPLAY AND CURSOR ON |
;**********************|
        movlw   0x0e    ; 0 0 0 0 1 1 1 0 (DISPLAY ON/OFF)
                        ;         | | | |___ Blink character
                        ;         | | |       1 = on, 0 = off
                        ;         | | |___ Cursor on/off
                        ;         | |         1 = on, 0 = off
                        ;         | |___ Display on/off
                        ;         |           1 = on, 0 = off
                        ;         |___ COMMAND BIT
        movwf   PORTB
        call    pulseE  ;pulseE and delay

;**********************|
;    ENTRY MODE SET    |
;**********************|
        movlw   0x06    ; 0 0 0 0 0 1 1 0 (ENTRY MODE SET)
                        ;           | | |___ display shift
                        ;           | |       1 = shift
                        ;           | |       0 = no shift
                        ;           | |___ increment mode
```

LCD Interfacing and Programming

```
                              ;            |        1 = left-to-right
                              ;            |        0 = right-to-left
                              ;            |___ COMMAND BIT
        movwf    PORTB        ;00000110
        call     pulseE

;*********************|
;   CURSOR/DISPLAY SHIFT   |
;*********************|
        movlw    0x14         ; 0 0 0 1 0 1 0 0 (CURSOR/DISPLAY
                              ;         | | | | |       SHIFT)
                              ;         | | | |_|___ don't care
                              ;         |  |_|__ cursor/display shift
                              ;         |         00 = cursor shift left
                              ;         |         01 = cursor shift right
                              ;         |         10 = cursor and display
                              ;         |              shifted left
                              ;         |         11 = cursor and display
                              ;         |              shifted right
                              ;         |___ COMMAND BIT
        movwf    PORTB        ;0001 1111
        call     pulseE

;*********************|
;     CLEAR DISPLAY        |
;*********************|
        movlw    0x01         ; 0 0 0 0 0 0 0 1 (CLEAR DISPLAY)
                              ;                 |___ COMMAND BIT
        movwf    PORTB        ;0000 0001
;
        call     pulseE
        call     busyTest   ; Test for busy
        return

;========================
; busy flag test routine
;========================
; Procedure to test the HD44780 busy flag
; Execution returns when flag is clear
busyTest:
        movlw    b'11111111'       ; All lines to input
        tris     PORTB             ; in port B
        bcf      PORTA,RS_line     ; RS line low for control
        bsf      PORTA,RW_line     ; Read mode
        bsf      PORTA,E_line      ; E line high
        movf     PORTB,w           ; Read port B into W
; Port B bit 7 is busy flag
        bcf      PORTA,E_line      ; E line low
```

```
            andlw    0x80                 ; Test bit 7, high is busy
            btfss    STATUS,Z ; Test zero bit in STATUS
            goto     busyTest ; Repeat if set
; At this point busy flag is clear
; Reset R/W line and port B to output
            bcf      PORTA,RW_line        ; Clear R/W line
            movlw    b'00000000'          ; All lines to output
            tris     PORTB                ; in port B
            return

;=======================
;   Procedure to delay
;   42 microseconds
;=======================
delay_125:
            movlw    D'42'                ; Repeat 42 machine cycles
            movwf    count1               ; Store value in counter
repeat:
            decfsz   count1,f             ; Decrement counter
            goto     repeat               ; Continue if not 0
            return                        ; End of delay

;=======================
;   Procedure to delay
;   5 milliseconds
;=======================
delay_5:
            movlw    D'41'                ; Counter = 41
            movwf    count2               ; Store in variable
delay:
            call     delay_125            ; Delay
            decfsz   count2,f             ; 40 times = 5 milliseconds
            goto     delay
            return                        ; End of delay
;=======================
;     pulse E line
;=======================
pulseE
            bsf      PORTA,E_line         ; Pulse E line
            nop                           ; Delay
            bcf      PORTA,E_line
            call     delay_5              ; Wait
            return

;=============================
;   long delay sub-routine
;      (for debugging)
;=============================
```

LCD Interfacing and Programming

```
long_delay
        movlw   D'200'          ; w = 200 decimal
        movwf   J               ; J = w
jloop:
        movwf   K               ; K = w
kloop:
        decfsz  K,f             ; K = K-1, skip next if zero
        goto    kloop
        decfsz  J,f             ; J = J-1, skip next if zero
        goto    jloop
        return
;=============================
;   LCD display procedure
;=============================
; Sends 16 characters from PIC buffer with address stored
; in variable pic_ad to LCD line previously selected
display16
        call    busyTest        ; Make sure not busy
; Set up for data
        bcf     PORTA,E_line    ; E line low
        bsf     PORTA,RS_line   ; RS line high for data
; Set up counter for 16 characters
        movlw   D'16'           ; Counter = 16
        movwf   count3
; Get display address from local variable pic_ad
        movf    pic_ad,w        ; First display RAM address to W
        movwf   FSR             ; W to FSR
getchar
        movf    INDF,w          ; get character from display RAM
                                ; location pointed to by file select
                                ; register
        movwf   PORTB
        call    pulseE          ;send data to display
; Test for 16 characters displayed
        decfsz  count3,f        ; Decrement counter
        goto    nextchar        ; Skipped if done
        return
nextchar:
        incf    FSR,f           ; Bump pointer
        goto    getchar
;=========================
;       blank buffer
;=========================
; Procedure to store 16 blank characters in PIC RAM
; buffer starting at address stored in the variable
; pic_ad
blank16:
        movlw   D'16'           ; Set up counter
```

```
            movwf   count1
            movf    pic_ad,w        ; First PIC RAM address
            movwf   FSR             ; Indexed addressing
            movlw   0x20            ; ASCII space character
storeit:
            movwf   INDF            ; Store blank character in PIC RAM
                                    ; buffer using FSR register
            decfsz  count1,f        ; Done?
            goto    incfsr          ; no
            return                  ; yes
incfsr
            incf    FSR,f           ; Bump FSR to next buffer
                                    ; space
            goto    storeit
;========================
; Set Address register
;    to LCD line 1
;========================
; ON ENTRY:
;       Address of LCD line 1 in constant LCD_1
line1:
            bcf     PORTA,E_line    ; E line low
            bcf     PORTA,RS_line   ; RS line low, set up for control
            call    busyTest ; busy?
; Set to second display line
            movlw   LCD_1           ; Address and command bit
            movwf   PORTB
            call    pulseE          ; Pulse and delay
; Set RS line for data
            bsf     PORTA,RS_line   ; Set up for data
            call    busyTest ; Busy?
            return
;========================
; Set Address register
;    to LCD line 2
;========================
; ON ENTRY:
;       Address of LCD line 2 in constant LCD_2
line2:
            bcf     PORTA,E_line    ; E line low
            bcf     PORTA,RS_line   ; RS line low, set up for control
            call    busyTest        ; Busy?
; Set to second display line
            movlw   LCD_2           ; Address with high-bit set
            movwf   PORTB
```

LCD Interfacing and Programming

```
            call    pulseE          ; Pulse and delay
; Set RS line for data
            bsf     PORTA,RS_line   ; RS = 1 for data
            call    busyTest ; Busy?
            return

;===============================
;  first text string procedure
;===============================
storeMSU:
; Procedure to store in PIC RAM buffer the message
; contained in the code area labeled msg1
; ON ENTRY:
;           variable pic_ad holds address of text buffer
;           in PIC RAM
;           w register hold offset into storage area
;           msg1 is routine that returns the string characters
;           and a zero terminator
;           index is local variable that hold offset into
;           text table. This variable is also used for
;           temporary storage of offset into buffer
; ON EXIT:
;           Text message stored in buffer
;
; Store offset into text buffer (passed in the w register)
; in temporary variable
            movwf   index           ; Store w in index
; Store base address of text buffer in FSR
            movf    pic_ad,w ; first display RAM address to W
            addwf   index,w         ; Add offset to address
            movwf   FSR             ; W to FSR
; Initialize index for text string access
            movlw   0               ; Start at 0
            movwf   index           ; Store index in variable
; w still = 0
get_msg_char:
            call    msg1            ; Get character from table
; Test for zero terminator
            andlw   0x0ff
            btfsc   STATUS,Z ; Test zero flag
            goto    endstr1         ; End of string
; ASSERT: valid string character in w
;           store character in text buffer (by FSR)
            movwf   INDF            ; store in buffer by FSR
            incf    FSR,f           ; increment buffer pointer
; Restore table character counter from variable
            movf    index,w         ; Get value into w
            addlw   1               ; Bump to next character
```

```
                movwf   index                   ; Store table index in
variable
                goto    get_msg_char            ; Continue
endstr1:
                return

; Routine for returning message stored in program area
msg1:
                addwf   PCL,f                   ; Access table
                retlw   'M'
                retlw   'i'
                retlw   'n'
                retlw   'n'
                retlw   'e'
                retlw   's'
                retlw   'o'
                retlw   't'
                retlw   'a'
                retlw   0


;==================================
;    second text string procedure
;==================================
storeUniv:
; Processing identical to procedure StoreMSU
                movwf   index                   ; Store w in index
; Store base address of text buffer in FSR
                movf    pic_ad,0 ; first display RAM address to W
                addwf   index,0                 ; Add offset to address
                movwf   FSR                     ; W to FSR
; Initialize index for text string access
                movlw   0                       ; Start at 0
                movwf   index                   ; Store index in variable
; w still = 0
get_msg_char2:
                call    msg2                    ; Get character from table
; Test for zero terminator
                andlw   0x0ff
                btfsc   STATUS,Z ; Test zero flag
                goto    endstr2                 ; End of string
; ASSERT: valid string character in w
;         store character in text buffer (by FSR)
                movwf   INDF                    ; Store in buffer by FSR
                incf    FSR,f                   ; Increment buffer pointer
; Restore table character counter from variable
                movf    index,w                 ; Get value into w
                addlw   1                       ; Bump to next character
                movwf   index                   ; Store table index in
```

LCD Interfacing and Programming

```
                                    ; variable
        goto    get_msg_char2       ; Continue
endstr2:
        return

; Routine for returning message stored in program area
msg2:
        addwf   PCL,f               ; Access table
        retlw   'S'
        retlw   't'
        retlw   'a'
        retlw   't'
        retlw   'e'
        retlw   ','
        retlw   0x20
        retlw   'M'
        retlw   'a'
        retlw   'n'
        retlw   'k'
        retlw   'a'
        retlw   't'
        retlw   'o'
        retlw   0

        End
```

13.4.3 LCDTest3 Program

```
; File name: LCDTest3.asm
; Date: April 16, 2006
; Author: Julio Sanchez
; Processor: 16F84A
;
; Description:
; Program to exercise 4-bit PIC-to-LCD interface.
; Code assumes that LCD is driven by Hitachi HD44780
; controller and that the display supports two lines
; each one with 16 characters. The wiring and base
; address of each display line is stored in #define
; statements. These statements can be edited to
; accommodate a different set-up.
; Program uses delay loops for interface timing.
; WARNING:
; Code assumes 4Mhz clock. Delay routines must be
; edited for faster clock
;
; Displays: Minnesota State, Mankato
;
```

```
;===========================
;           switches
;===========================
; Switches used in __config directive:
;     _CP_ON          Code protection ON/OFF
; *   _CP_OFF
; *   _PWRTE_ON       Power-up timer ON/OFF
;     _PWRTE_OFF
;     _WDT_ON         Watchdog timer ON/OFF
; *   _WDT_OFF
;     _LP_OSC         Low power crystal occilator
; *   _XT_OSC         External parallel resonator/crystal oscillator
;
;     _HS_OSC         High speed crystal resonator (8 to 10 MHz)
;                     Resonator: Murate Erie CSA8.00MG = 8 MHz
;     _RC_OSC         Resistor/capacitor oscillator
; |                   (simplest, 20% error)
; |
; |_____ * indicates set up values presently selected
;
;=========================
; set up and configuration
;=========================
        processor 16f84A
        include    <p16f84A.inc>
        __config   _XT_OSC & _WDT_OFF & _PWRTE_ON & _CP_OFF

;=========================================================
;                  constant definitions
;   for PIC-to-LCD pin wiring and LCD line addresses
;=========================================================
#define E_line  1         ; |
#define RS_line 2         ; | — from wiring diagram
#define RW_line 3         ; |
; LCD line addresses (from LCD data sheet)
#define LCD_1 0x80        ; First LCD line constant
#define LCD_2 0xc0        ; Second LCD line constant
; Note: The constants that define the LCD display line
;       addresses have the high-order bit set in
;       order to facilitate the controller command
;
;=========================================================
;                variables in PIC RAM
;=========================================================
; Reserve 16 bytes for string buffer
        cblock  0x0c
        strData
        endc
```

LCD Interfacing and Programming

```
; Leave 16 bytes and Continue with local variables
        cblock   0x1d                ; Start of block
        count1                       ; Counter # 1
        count2                       ; Counter # 2
        count3                       ; Counter # 3
        pic_ad                       ; Storage for start of text area
                                     ; (labeled strData) in PIC RAM
        J                            ; counter J
        K                            ; counter K
        index                        ; Index into text table (also used
                                     ; for auxiliary storage)
        store1                       ; Local temporary storage
        store2                       ; Storage # 2
        endc

;============================================================
;                       program
;============================================================
        org      0                   ; start at address
        goto     main
; Space for interrupt handlers
        org      0x08

main:
        movlw    b'00000000'         ; All lines to output
        tris     PORTA               ; in Port-A
        tris     PORTB               ; and port B
        movlw    b'00000000'         ; All outputs ports low
        movwf    PORTA
        movwf    PORTB
; Wait and initialize HD44780
        call     delay_5             ; Allow LCD time to initialize
                                     ; itself
        call     delay_5
        call     initLCD             ; Then do forced
initialization
        call     delay_5             ; Wait again
; Store base address of text buffer in PIC RAM
        movlw    0x0c                ; Start address for buffer
        movwf    pic_ad              ; to local variable
;======================
;   first LCD line
;======================
; Store 16 blanks in PIC RAM, starting at address stored
; in variable pic_ad
        call     blank16
; Call procedure to store ASCII characters for message
; in text buffer
```

```
                movlw   d'3'                ; Offset into buffer
                call    storeMSU
; Set DDRAM address to start of first line
                call    line1
; Call procedure to display 16 characters in LCD
                call    display16
;========================
;    second LCD line
;========================
                call    delay_5             ; Wait for termination
                call    blank16             ; Blank buffer
; Call procedure to store ASCII characters for message
; in text buffer
                movlw   d'1'                ; Offset into buffer
                call    storeUniv
                call    line2               ; DDRAM address of LCD line 2
                call    display16
;======================
;      done!
;======================
loopHere:
                goto    loopHere    ;done


;===============================================================
;                   initialize LCD for 4-bit mode
;===============================================================
initLCD:
; Initialization for Densitron LCD module as follows:
;       4-bit interface
;    2 display lines of 16 characters each
;    cursor on
;    left-to-right increment
;    cursor shift right
;    no display shift
;======================|
;    set command mode  |
;======================|
                bcf     PORTA,E_line        ; E line low
                bcf     PORTA,RS_line       ; RS line low
                bcf     PORTA,RW_line       ; Write mode
                call    delay_125                   ; delay 125
microseconds
;*********************|
;      FUNCTION SET   |
;*********************|
                movlw   0x28       ; 0 0 1 0 1 0 0 0 (FUNCTION SET)
                           ;         | | | |__ font select:
                           ;         | | |    1 = 5x10 in 1/8 or 1/11
```

```
                                ;       | | |     0 = 1/16 dc
                                ;       | | |___ Duty cycle select
                                ;       | |       0 = 1/8 or 1/11
                                ;       | |       1 = 1/16
                                ;       | |___ Interface width
                                ;       |       0 = 4 bits
                                ;       |       1 = 8 bits
                                ;       |___ FUNCTION SET COMMAND
        call    send8           ; 4-bit send routine
; Set 4-bit mode command must be repeated
        movlw   0x28
        call    send8

;*********************|
; DISPLAY AND CURSOR ON |
;*********************|
        movlw   0x0e            ; 0 0 0 0 1 1 1 0 (DISPLAY ON/OFF)
                                ;         | | | |___ Blink character
                                ;         | | |       1 = on, 0 = off
                                ;         | | |___ Cursor on/off
                                ;         | |       1 = on, 0 = off
                                ;         | |___ Display on/off
                                ;         |       1 = on, 0 = off
                                ;         |___ COMMAND BIT
        call    send8
;*********************|
;   set entry mode    |
;*********************|
        movlw   0x06            ; 0 0 0 0 0 1 1 0 (ENTRY MODE SET)
                                ;           | | |___ display shift
                                ;           | |       1 = shift
                                ;           | |       0 = no shift
                                ;           | |___ increment mode
                                ;           |       1 = left-to-right
                                ;           |       0 = right-to-left
                                ;           |___ COMMAND BIT
        call    send8

;*********************|
; cursor/display shift |
;*********************|
        movlw   0x14            ; 0 0 0 1 0 1 0 0 (CURSOR/DISPLAY
                                ;       | | | | |   SHIFT)
                                ;       | | | |_|___ don't care
                                ;       | |_|__ cursor/display shift
                                ;       |         00 = cursor shift left
                                ;       |         01 = cursor shift right
                                ;       |         10 = cursor and display
```

```
                             ;         |          shifted left
                             ;         |     11 = cursor and display
                             ;         |          shifted right
                             ;         |___ COMMAND BIT
           call    send8
;********************|
;    clear display      |
;********************|
           movlw   0x01      ; 0 0 0 0 0 0 0 1 (CLEAR DISPLAY)
                             ;                 |___ COMMAND BIT
           call    send8
; Per documentation
           call    delay_5   ; Test for busy
           return


;======================
;   Procedure to delay
;     42 microseconds
;======================
delay_125
           movlw   D'42'          ; Repeat 42 machine cycles
           movwf   count1         ; Store value in counter
repeat
           decfsz  count1,f       ; Decrement counter
           goto    repeat         ; Continue if not 0
           return                 ; End of delay


;======================
;   Procedure to delay
;     5 milliseconds
;======================
delay_5:
           movlw   D'41'          ; Counter = 41
           movwf   count2         ; Store in variable
delay:
           call    delay_125      ; Delay
           decfsz  count2,f       ; 40 times = 5 milliseconds
           goto    delay
           return                 ; End of delay
;========================
;      pulse E line
;========================
pulseE
           bsf     PORTA,E_line   ; Pulse E line
           nop
           bcf     PORTA,E_line
           return
```

LCD Interfacing and Programming

```
;==============================
;   long delay sub-routine
;       (for debugging)
;==============================
long_delay:
        movlw   D'200'          ; w = 200 decimal
        movwf   J               ; J = w
jloop:
        movwf   K               ; K = w
kloop:
        decfsz  K,f             ; K = K-1, skip next if zero
        goto    kloop
        decfsz  J,f             ; J = J-1, skip next if zero
        goto    jloop
        return
;==============================
;   LCD display procedure
;==============================
; Sends 16 characters from PIC buffer with address stored
; in variable pic_ad to LCD line previously selected
display16
        call    delay_5         ; Make sure not busy
; Set up for data
        bcf     PORTA,E_line    ; E line low
        bsf     PORTA,RS_line   ; RS line high for data
; Set up counter for 16 characters
        movlw   D'16'           ; Counter = 16
        movwf   count3
; Get display address from local variable pic_ad
        movf    pic_ad,w ; First display RAM address to W
        movwf   FSR             ; W to FSR
getchar:
        movf    INDF,w   ; get character from display RAM
                         ; location pointed to by file select
                         ; register
        call    send8    ; 4-bit interface routine
; Test for 16 characters displayed
        decfsz  count3,f        ; Decrement counter
        goto    nextchar        ; Skipped if done
        return
nextchar:
        incf    FSR,f           ; Bump pointer
        goto    getchar


;=======================
;   send 2 nibbles in
;       4-bit mode
;=======================
```

```
; Procedure to send two 4-bit values to port B lines
; 7, 6, 5, and 4. High-order nibble is sent first
; ON ENTRY:
;          w register holds 8-bit value to send
send8:
        movwf   store1              ; Save original value
        call    merge4              ; Merge with port B
; Now w has merged byte
        movwf   PORTB               ; w to port B
        call    pulseE              ; Send data to LCD
; High nibble is sent
        movf    store1,w            ; Recover byte into w
        swapf   store1,w            ; Swap nibbles in w
        call    merge4
        movwf   PORTB
        call    pulseE              ; Send data to LCD
        call    delay_125
        return
;==================
;   merge bits
;==================
; Routine to merge the 4 high-order bits of the
; value to send with the contents of port B
; so as to preserve the 4 low-bits in port B
; Logic:
;       AND value with 1111 0000 mask
;       AND port B with 0000 1111 mask
;       Now low nibble in value and high nibble in
;       port B are all 0 bits:
;            value = vvvv 0000
;            port B = 0000 bbbb
;       OR value and port B resulting in:
;                    vvvv bbbb
; ON ENTRY:
;       w contain value bits
; ON EXIT:
;       w contains merged bits
merge4:
        andlw   b'11110000'         ; ANDing with 0 clears the
                                    ; bit. ANDing with 1 preserves
                                    ; the original value
        movwf   store2              ; Save result in variable
        movf    PORTB,w             ; port B to w register
        andlw   b'00001111'         ; Clear high nibble in port b
                                    ; and preserve low nibble
        iorwf   store2,w ; OR two operands in w
        return
```

LCD Interfacing and Programming

```
;=========================
;     blank buffer
;=========================
; Procedure to store 16 blank characters in PIC RAM
; buffer starting at address stored in the variable
; pic_ad
blank16
        movlw    D'16'       ; Set up counter
        movwf    count1
        movf     pic_ad,w   ; First PIC RAM address
        movwf    FSR         ; Indexed addressing
        movlw    0x20        ; ASCII space character
storeit
        movwf    INDF        ; Store blank character in PIC RAM
                             ; buffer using FSR register
        decfsz   count1,f             ; Done?
        goto     incfsr      ; no
        return               ; yes
        incfsr
        incf     FSR,f       ; Bump FSR to next buffer space
        goto     storeit

;=========================
; Set Address register
;    to LCD line 1
;=========================
; ON ENTRY:
;         Address of LCD line 1 in constant LCD_1
line1:
        bcf      PORTA,E_line    ; E line low
        bcf      PORTA,RS_line   ; RS line low, set up for
control
        call     delay_5         ; busy?
; Set to second display line
        movlw    LCD_1           ; Address and command bit
        call     send8           ; 4-bit routine
; Set RS line for data
        bsf      PORTA,RS_line   ; Set up for data
        call     delay_5         ; Busy?
        return
;=========================
; Set Address register
;    to LCD line 2
;=========================
; ON ENTRY:
;         Address of LCD line 2 in constant LCD_2
line2:
        bcf      PORTA,E_line    ; E line low
```

```
            bcf     PORTA,RS_line   ; RS line low, set up for
                                    ; control
            call    delay_5         ; Busy?
; Set to second display line
            movlw   LCD_2           ; Address with high-bit set
            call    send8
; Set RS line for data
            bsf     PORTA,RS_line   ; RS = 1 for data
            call    delay_5         ; Busy?
            return

;================================
;   first text string procedure
;================================
storeMSU:
; Procedure to store in PIC RAM buffer the message
; contained in the code area labeled msg1
; ON ENTRY:
;         variable pic_ad holds address of text buffer
;         in PIC RAM
;         w register hold offset into storage area
;         msg1 is routine that returns the string characters
;         andiy a zero terminator
;         index is local variable that hold offset into
;         text table. This variable is also used for
;         temporary storage of offset into buffer
; ON EXIT:
;         Text message stored in buffer
;
; Store offset into text buffer (passed in the w register)
; in temporary variable
            movwf   index           ; Store w in index
; Store base address of text buffer in FSR
            movf    pic_ad,w ; first display RAM address to W
            addwf   index,w         ; Add offset to address
            movwf   FSR             ; W to FSR
; Initialize index for text string access
            movlw   0               ; Start at 0
            movwf   index           ; Store index in variable
; w still = 0
get_msg_char:
            call    msg1            ; Get character from table
; Test for zero terminator
            andlw   0x0ff
            btfsc   STATUS,Z ; Test zero flag
            goto    endstr1         ; End of string
; ASSERT: valid string character in w
;           store character in text buffer (by FSR)
```

LCD Interfacing and Programming

```
            movwf    INDF              ; store in buffer by FSR
            incf     FSR,f             ; increment buffer pointer
; Restore table character counter from variable
            movf     index,w           ; Get value into w
            addlw    1                 ; Bump to next character
            movwf    index             ; Store table index in
variable
            goto     get_msg_char      ; Continue
endstr1:
            return
; Routine for returning message stored in program area
msg1:
            addwf    PCL,f             ; Access table
            retlw    'M'
            retlw    'i'
            retlw    'n'
            retlw    'n'
            retlw    'e'
            retlw    's'
            retlw    'o'
            retlw    't'
            retlw    'a'
            retlw    0


;==================================
;    second text string procedure
;==================================
storeUniv:
; Processing identical to procedure StoreMSU
            movwf    index             ; Store w in index
; Store base address of text buffer in FSR
            movf     pic_ad,0 ; first display RAM address to W
            addwf    index,0           ; Add offset to address
            movwf    FSR               ; W to FSR
; Initialize index for text string access
            movlw    0                 ; Start at 0
            movwf    index             ; Store index in variable
; w still = 0
get_msg_char2:
            call     msg2              ; Get character from table
; Test for zero terminator
            andlw    0x0ff
            btfsc    STATUS,Z ; Test zero flag
            goto     endstr2           ; End of string
; ASSERT: valid string character in w
;        store character in text buffer (by FSR)
            movwf    INDF              ; Store in buffer by FSR
            incf     FSR,f             ; Increment buffer pointer
```

```
; Restore table character counter from variable
        movf    index,w             ; Get value into w
        addlw   1                   ; Bump to next character
        movwf   index               ; Store table index in
                                    ; variable
        goto    get_msg_char2       ; Continue
endstr2:
        return

; Routine for returning message stored in program area
msg2:
        addwf   PCL,f               ; Access table
        retlw   'S'
        retlw   't'
        retlw   'a'
        retlw   't'
        retlw   'e'
        retlw   ','
        retlw   0x20
        retlw   'M'
        retlw   'a'
        retlw   'n'
        retlw   'k'
        retlw   'a'
        retlw   't'
        retlw   'o'
        retlw   0

        end
```

Chapter 14

Communications

In this chapter we focus on digital communications techniques used in PIC interfacing with I/O devices, integrated circuits, and with other forms of programmable logic. Communications, in general, refers to the exchange of information following rules, sometimes called a *protocol*. Digital and computer communications come in two flavors: serial and parallel. Serial communications take place when the data is sent one bit at a time over the communications channel. In parallel communications all the bits that compose a single symbol or character are sent simultaneously.

Common wisdom regards serial communications as slower than parallel communications but with modern-day technologies this is often not the case, since serial techniques often match or even excel parallel methods in speed and performance. Computer networks such as Ethernet and fiber optic links are able to achieve high performance even though they use serial bit streams. The preference for serial over parallel communications is often more related to hardware, since parallel transmissions require more communication lines than serial transmissions.

14.0 PIC Communications Overview

Many communications standards were created with other interface and hardware requirements in mind and are not ideally suited for PIC applications. For example, RS-232-C, a serial protocol developed over 35 years ago, originated in an age of teletypewriters and modems. The voltage levels and circuit requirements of RS-232-C are not suited for PIC hardware. The more modern USB standard is more suited to PIC interfacing, but adopting a standard; RS-232-C, EIA-485, USB, or any other conventions, requires adhering to special configurations in hardware and the use of ad hoc software protocols. This compliance with a standard comes at a price of added hardware components and increased software complexity.

When PIC-based circuits must interface with other systems or devices that follow these standards, then there is no alternative but to design circuits and write programs that comply with the standards. On the other hand, when the communications take place in dedicated circuits, which do not interface with devices or systems that follow standard communications protocols, then pure PIC communications techniques and hardware are often simpler and more effective. In other words, adhering to a communications protocol usually implies an additional cost in software and hardware complexity. Here are two examples: a PIC-based circuit that interfaces with a PC through the RS-232-C port would be a case where compliance with RS-232-C is required. Another case would be a PIC-based circuit that sends serial data to an onboard LCD display. In this case, the circuit and the software need not comply with any communications standards or protocols. Programmers often refer to techniques that use serial communications without the presence of specialized hardware, such as UART or USART chips, as *bit-banging*.

In the following sections, we discuss serial and parallel communications at their most essential level. In the general literature, communications concerns often focus on transmission speeds, system performance, and minimum processing time. Typically, PIC applications do not transfer large data files or communicate interactively on the Internet or in networks. In a typical PIC application, communication functions are used to upload stored data to a PC, sometimes called *data-logging*, or to receive small data sets or commands from a host machine. In this context there are no major concerns regarding super-fast transmission rates or maximum performance.

14.1 Serial Data Transmission

Serial communications take place by transmitting and receiving data in a stream of consecutive electrical pulses that represent data bits and control codes. The *Electronic Industries Association* (EIA) has sponsored the development of several standards for serial communications, such as RS-232-C, RS-422, RS-423, RS 449, EIA232E, and EIA232F, among others. In this designation the characters RS stand for the words *Recommended Standard*. The oldest, simplest to implement, and most-used serial communications standard is the *RS-232-C voltage level convention*. In the following sections we present the essential concepts of the RS-232-C standard. Most of the material also applies to the various updates of the standard. Later in the chapter we briefly discuss the EIA485 Standard.

14.1.1 Asynchronous Serial Transmission

The information in a serial bit stream is contained in a time-dependent waveform, that is, each bit code (data, control, or error) is transmitted for a fixed time period, known as the *baud period*. The word baud was chosen to honor the French scientist and inventor *Jean Maurice Emile Baudot* who studied various serial encodings in the late 19th century.

The serial bit streams used in data transmission follow a very simple encoding: one bit is transmitted during each baud period. A binary 1 bit is represented with a negative voltage level and a binary 0 bit by a positive voltage. The line condition during the logic 1 transmission is called a *marking state*, and the one for a logic 0 a *spacing state*. The baud rate is equal to the number of bits per second being transmitted or received. Note that the voltage levels that represent a one and a zero bit in RS232 are somewhat counter-intuitive, since one would expect a logic 1 to be represented with a positive voltage, and not a negative one.

One possible approach to sending information bit-by-bit is based on the transmitter and receiver clocks being synchronized at the same frequency. That is, both receiver and transmitter operate at the same baud rate. Note that the expression "synchronized at the same frequency" implies not only that their clocks have the same speed, but that the high and the low portions of the waveform coincide.

In typical asynchronous serial communications bits are transmitted as separate groups, usually 7 to 10 bits long. Each group is called a *character*. The name character relates to the fact that in alphanumeric transmissions each bit group represents one numeric or alphabetic symbol. In reality, the term "character" is also applied to control codes, error codes, and other non-alphanumeric encodings.

Each character is sent in a frame consisting of a start bit, followed by a set of character bits, followed (optionally) by a parity bit, and finalized by one or more stop bits. The serial line is normally held *marking*, that is, at a logic 1 state. The change from logic high to logic low, signaled by the start bit, tells the receiver that a *frame* follows. The receiver reads the number of character bits expected according to the adopted protocol until a logic high, represented by one or more stop bits, marks the end of the frame.

Figure 14-1 shows the different elements in a serial communications bit stream. The term *asynchronous* reflects the fact that the time period separating characters is variable. The transmitter holds the line to logic high (marking state) until it is ready to send. The start bit (*spacing state*) is used to signal the start of a new character. The start bit is also used by the receiver to synchronize with the transmitter. The logic high and low regions of the signal wave occur at the same time. This compensates for drifts and small errors in the baud rate.

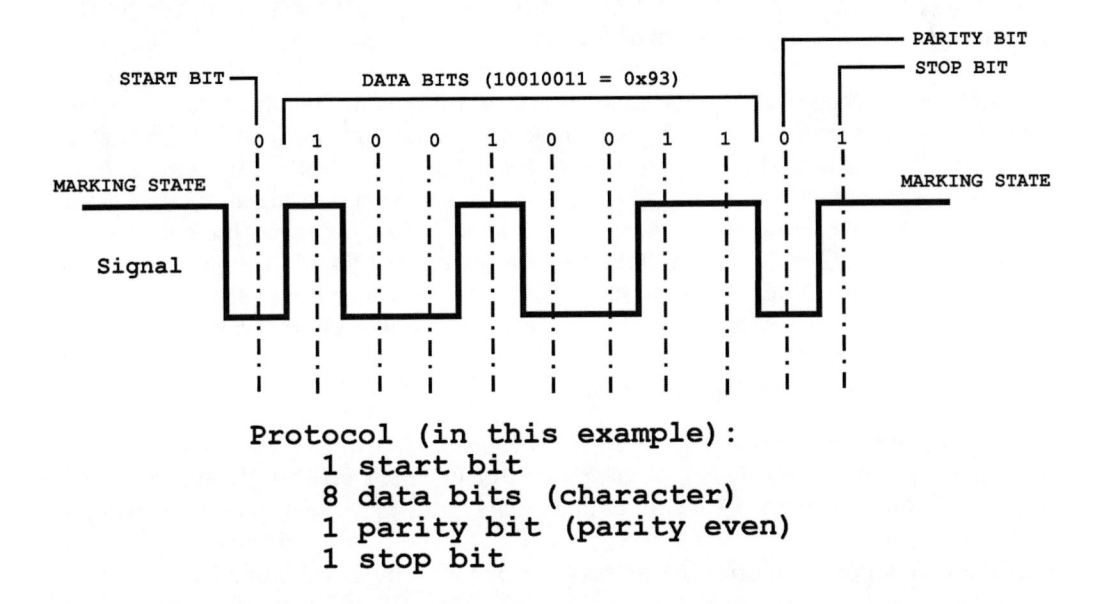

```
Protocol (in this example):
    1 start bit
    8 data bits (character)
    1 parity bit (parity even)
    1 stop bit
```

Figure 14-1 Serial Communications Bit Stream

This form of transmitting serial data is called *asynchronous* because the receiver resynchronizes itself to the transmitter using the start bit of each frame. The lack of synchronization does not refer to the bits within each frame, which must be in fact "synchronized," but to the fact that characters need not come at a fixed time interval.

14.1.2 Synchronous Serial Transmission

An alternative approach to asynchronous serial data transmission is one in which the characters are sent in blocks with no framing bits surrounding them. In asynchronous communications, each character is framed by a start and a stop signal so that the receiver can know exactly where the character bits are located. In *synchronous communications*, the sender and receiver are synchronized with a clock or a signal that is part of the data stream.

In theory, synchronous communications implies that characters are sent out at a constant rate, in step with a clock signal. This scheme assumes that a separate line (or wire) is used for the clock signal, although, in some variations, the clock signal is contained in the transmitted characters. Alternatively, a clock line can be used to synchronize the moment in time at which the receiver reads the data line. In either case, it is this contained clock or command signal that identifies a synchronous transmission.

Most legacy PC communications systems are asynchronous, although the EIA232F standard supports both synchronous and asynchronous methods. The most common chip used in PC communications is the *UART* (*Universal Asynchronous Receiver and Transmitter*). An alternative chip called the USRT is used for synchronous communications and the USART (*Universal Synchronous/Asynchronous Receiver and Transmitter)* supports both.

Synchronous communications can be block- or bit-based. The block-based modes are also called character-based. In this mode, characters are grouped in blocks with each block having a starting flag, similar to the start bit used in asynchronous communications. Once the receiver and the transmitter are synchronized, the transmitter inserts two or more control characters known as *synchronous idle characters*, or SYNs. Then the block is sent and the receiver places the data in a memory storage area for later processing. Bit-oriented methods, on the other hand, are used for the transmission of binary data that is not tied to any particular character set.

14.1.3 PIC Serial Communications

Serial communications are often used in PIC programming, mostly due to the scarcity of available port lines. For example, an application in which a 16F84 PIC needs to read data in parallel from eight DIP switches and display the result, also in parallel, in eight LEDs, requires a total of 16 available port lines. But the 16F84 only has 13 lines, 8 in Port-B and 5 in Port-A; therefore, the application would not be feasible.

One possible solution is to find some way of reading the DIP switches serially; this requires three lines at most. Alternatively, the output data to the LEDs could be transmitted serially, thus reducing the total lines required from 16 for parallel transmission, to six, or even less for serial transmission.

PIC communications can be designed both asynchronously and synchronously. Asynchronous modes are used when the same or compatible clock signals are available to both receiver and transmitter. For example, two PICs both running at the same clock rate can transmit and receive data using a single communications line, plus a common ground. PIC-to-PIC asynchronous data transmission mode is demonstrated later in this chapter with both circuit and code.

Asynchronous communications can be implemented by incorporating a dedicated IC, such as a UART or USART chip, in the circuit. PCs usually have one of these ICs, or functionally equivalent ones, in their implementation of the serial port. Some PICs include one or more serial circuits, which sometimes include a USART module. For example, the 16F877 PIC has two serial communication modules. One of them is the *Master Asynchronous Serial Port*, or MSSP. The other one is a USART. Later in this chapter we present serial communications programming examples using the USART module in the 16F877 PIC. Programs using the MSSP module are found in the chapter on EEPROM programming.

When communications take place between a PIC and a device that does not contain a clock, or whose clock runs at a different speed than the PIC's, then synchronous communications is used. For example, a circuit can be designed using a shift register IC, such as the 74HC164, that performs an 8-bit serial-in, parallel-out function. In the previous example, it is possible to reduce the number of transmission lines by connecting the eight LEDs to the output ports of the 74HC164. But the 74HC164 contains no internal clock that runs at the speed of the 16F84. Thus, communications between the PIC and the shift register IC (74HC164 in this case) requires a clock or command signal transmitted through a separate line; that is, a synchronous serial transmission. In this chapter we present circuits and sample code showing synchronous communications between a PIC and one or more shift register ICs.

14.1.4 The RS-232-C Standard

RS-232-C was developed jointly by the Electronic Industries Association (EIA), the Bell Telephone System, and modem and computer manufacturers. The standard has achieved such widespread acceptance that its name is often used as a synonym for the serial port. EIA232F, published in 1997, is the latest update of RS-232-C. Today, RS-232-C is gradually being replaced by USB for local communications. USB is faster, has lower voltage levels, and uses smaller connectors that are easier to wire. USB has software support in most PC operating systems. On the other hand, USB is a more complex standard, requiring more complex software. Furthermore, serial ports are used to directly control hardware devices, such as relays and lamps, since the RS-232-C control lines can be easily manipulated by software. This is not feasible with USB.

In the following sections we describe the essential terminology and communications principles of RS-232-C.

Essential Concepts

The RS-232-C convention specifies that, with respect to ground, a voltage more negative than -3 V is interpreted as a 1 bit and a voltage more positive than +3 V as a 0 bit. Serial communications, according to RS-232-C, require that transmitter and receiver agree on a communications protocol. The following terminology refers to the RS-232-C communications protocol:

- **Baud period:** The rate of transmission measured in bits per second, also called the baud rate. In serial protocols, the transmitter and the receiver clocks must be synchronized to the same baud period.
- **Marking state:** The time period during which no data is transmitted. During the marking period, the transmitter holds the line at a steady high voltage, indicating logic 0.
- **Spacing state:** The time period during which data is transmitted. During the spacing period, the transmitter holds the line at a steady low voltage, indicating logic 1.
- **Start bit:** The transition that indicates that data transmission is about to start. The voltage low state that occurs during the start bit is called the spacing state.
- **Character bits:** The data stream composed of 5, 6, 7, or 8 bits that encode the character transmitted. The least significant bit is the first one transmitted.
- **Parity bit:** An optional bit, transmitted following the character bits, used in checking for transmission errors. If even parity is chosen, the transmitter sets or clears the parity bit so as to make the sum of the character's 1 bits and the parity bit an even number. In odd parity, the sum of 1 bits is an odd number. If parity is not correct, the receiver sets an error flag in a special register.
- **Stop bits:** One or more logic high bits inserted in the stream following the character bits or the parity bit, if there is one. The stop bit or bits ensure that the receiver has enough time to get ready for the next character.
- **DTE (Data Terminal Equipment):** The device at the far end of the connection. It is usually a computer or terminal. The DTE uses a male DB-25 connector, and utilizes 22 of the 25 available pins. DB-9 connectors with 9 pins are also used.
- **DCE (Data Circuit-terminating Equipment):** Refers to the modem or other terminal of the telephone line interface. DCE has a female DB-25 connector, and utilizes the same 22 pins as the DTE for signals and ground. DB-9 connectors are also used.
- **Half-duplex:** A system that allows serial communications in both directions, but only one direction at a time. Half-duplex communications are reminiscent of radio communications where one user says the word "Over" to indicate the end of transmission. In other words, half-duplex is similar to a one-lane road in which with traffic controllers at each end can direct flow in either direction, but only in one direction at a time.
- **Full-duplex:** A full-duplex system allows communication in both directions simultaneously. A full-duplex system is reminiscent of a two-lane highway in which traffic can flow in both directions at once.

The Serial Bit Stream

In the RS-232-C protocol, the transmission/reception parameters are selected from a range of standard values. The following are the most common ones:

Baud rate: 50, 110, 300, 600, 1200, 2400, 4800, 9600, and 19200

Data bits: 5, 6, 7, or 8.

Parity bit: Odd, even, or no parity.

Stop bits: 1, 1.5, or 2.

RS-232-C defines *DTE (Data Terminal Equipment)* and *DCE (Data Circuit-terminating Equipment)*, sometimes called *Data Communications Equipment*. According to the standard, the DTE designation includes both terminals and computers and DCE refers to modems, transducers, and other devices. The serial port in a computer is defined as a DTE device.

Parity Testing

In RS-232 communications, a bit called a parity bit may optionally be transmitted along with the data. A parity bit provides a simple, but not too reliable, error test to detect data corruption that takes place during transmission. Parity can be even, odd, or none. Even or odd parity refers to the number of 1-bits in each data byte. The parity bit immediately follows the data bits.

If even parity is selected, the parity bit is transmitted with a value of 0 if the number of high bits is even. For example, the binary value:

```
0110 0011
```

contains a total of 4 one-bits; therefore, the parity bit is 0. By the same token, if even parity is selected, then the binary value

```
0101 0001
```

requires that the parity bit be 1. One way of describing the parity bit is to say that the bit is set to indicate a parity error; therefore, it serves as a parity error detector. Another description is that the parity coincides with the number of one-bits in the data, plus the parity bit. Thus, when even parity is selected the parity bit is added to the number of one-bits in the data to produce an even number.

Odd parity is the opposite of even parity. If odd parity were selected then the parity bit in the last example would be 0. Given odd or even parity, the sender counts the number of 1-bits and sets or clears the parity bit accordingly. The receiver, knowing that the parity is odd or even, can do likewise to determine if the number of 1-bits received matches the required parity setting.

Parity error checking is very primitive. In the first place, the parity error does not identify the bit or bits that cause the error. Furthermore, if an even number of bits are incorrect, then the parity bit would not show the error. On the other hand, over a long transmission, the parity check is likely to detect garbled data.

Connectors and Wiring

The RS-232-C standard requires specific hardware connectors with either 25 or 9 pins. The 25-pin connector is called a *D-shell connector*, or DB-25. The connector with 9 pins is called the *9-pin D-shell connector* or DB-9. In addition, the RJ-45 connector (the name stands for *Registered-Jack 45*) is used for twisted-pair cables. RJ-45 use in RS-232-C serial interface is regulated by the EIA/TIA-561 standard. A common application of RJ-45 connectors is in *Ethernet* cables. Figure 14-2 shows the male DB-25, DB-9, and the female RJ-45 connectors.

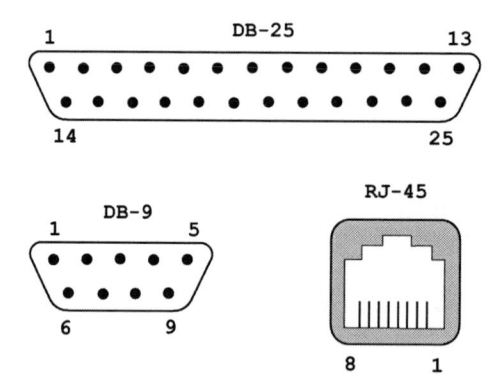

Figure 14-2 DB-25, DB-9, and RJ-45 Connectors

The function assigned to each pin varies in the common connectors. Table 14.1 lists the assignation of the RS-232-C lines in the different hardware. The cable linking DTE and DCE devices is a parallel straight-through cable with no cross-over or self-connects.

Table 14.1
Definition of Common RS-232-C Lines

DB-25	CONNECTOR DB-9	RJ-45	FUNCTION	CODE NAME	DIRECTION
1		4	Ground	G	
2	3	6	Transmit data	TXD	Output
3	2	5	Receive data	RXD	Input
4	7	8	Request to send	RTS	Output
5	8	7	Clear to send	CTS	Input
6	6		Data set ready	DSR	Input
7	5		Chassis ground	G	
8	1	2	Carrier detect	CD	
20	4	3	Data terminal ready	DTR	Output
22	9	1	Ring indicator	RI	Input

The Null Modem

The RS-232-C standards describe the way a computer communicates with a peripheral device, such as a modem. In this case, the DTE and DCE lines serve as a communications control. In this context, DTE means data terminal equipment, such as a computer, and DCE is the abbreviation of data communication equipment, such as modems. Often, communications must take place in an environment that does not include a modem; for example, computers communicating with each other or with other devices such as a PIC-based board. In these cases, the use of the DTE/DCE communication lines in flow control is not well defined. The common RS-232-C control and data signals appear in Table 14.2.

Communications

Table 14.2

Definition of Common RS-232-C Lines

SIGNAL NAME	DIRECTION	PURPOSE
CONTROL SIGNALS		
Request to Send	DTE -> DCE	DTE wishes to send
Clear to Send	DTE <- DCE	Response to Request to Send
Data Set Ready	DTE <- DCE	DCE ready to operate
Data Terminal Ready	DTE -> DCE	DTE ready to operate
Ring Indicator	DTE <- DCE	DTE receiving telephone ringing signal
Carrier Detect	DTE <- DCE	DTE receiving a carrier signal
DATA SIGNALS		
Transmitted Data	DTE -> DCE	Data generated by DTE
Received Data	DTE <- DCE	Data generated by DCE

The term *null modem* refers to situations in which serial communications take place without the presence of a modem. In this case, the connection between the communicating devices, usually a cable, is wired in such a way so as to allow data transmission without a modem.

In Table 14.1 two pins are used in flow control: *RTS (request to send)* and *CTS (clear to send)*. In conventional RS232 communication (as is the case when a computer communicates with a modem), the RTS signal is an output and DCE an input. Before a character is sent, the sender sets the RTS line high to ask the DTE's permission. Until the DTE grants permission, no data is sent. The DTE grants its permission by setting the CTS line high. If the DCE cannot receive new data it keeps the CTS signal low. This interface, which provides a simple mechanism for flow control in a single direction, is called a *handshake*.

In full duplex transmission the handshake must take place in both directions, that is, both devices must be able to signal their status. The *DTR (data terminal ready)* and *DSR (data set ready)* signals can be used for a second level of flow control. Finally, the *CD (carrier detect)* signal serves as an indication of the state of a modem.

The Null Modem Cable

Implementing handshaking without a modem requires that we take into account that two communicating devices can expect to find certain signals on given lines. For example, a device checks the CTS signal for a high value before sending data. If the CTS signal never goes high, transmission does not take place. When a cable is wired so that two devices can communicate without one of them being a modem, the cable is said to be a *null modem*.

One simple approach is to completely eliminate handshaking. In this case, cable wiring interconnects the transmit and the receive lines and the ground wire. The remaining pins are left unconnected, as shown in the null modem cable in Figure 14-3 (in the following page).

```
        DB-9
      (female)
    ●5 ●4 ●3 ●2 ●1
     ●9 ●8 ●7 ●6
```

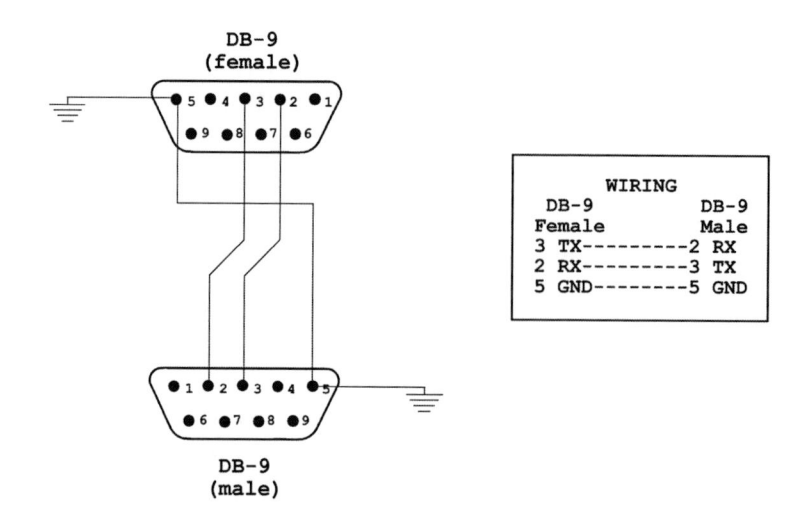

```
              WIRING
     DB-9              DB-9
    Female             Male
    3 TX---------2 RX
    2 RX---------3 TX
    5 GND--------5 GND
```

```
     ●1 ●2 ●3 ●4 ●5
      ●6 ●7 ●8 ●9
        DB-9
       (male)
```

Figure 14-3 Null Modem with No Handshaking

The three-wire null modem cable can be used to interface devices that do not use modem control signals. However, if one of the devices checks one of the handshake lines, such as RTS/CTS, then the three-wire modem cable fails. To solve this problem, a modem cable can be designed so that the handshake signals are interconnected. For example, DTS to DSR and vice versa. Not knowing which handshake signals are to be used, manufacturers of standard modem cables usually interconnect all handshake lines, as shown in Figure 14-4.

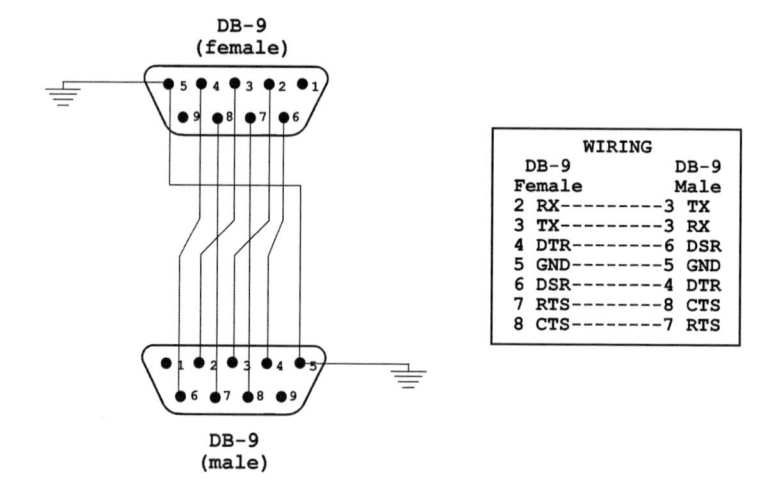

```
        DB-9
      (female)
    ●5 ●4 ●3 ●2 ●1
     ●9 ●8 ●7 ●6
```

```
              WIRING
     DB-9              DB-9
    Female             Male
    2 RX---------3 TX
    3 TX---------3 RX
    4 DTR--------6 DSR
    5 GND--------5 GND
    6 DSR--------4 DTR
    7 RTS--------8 CTS
    8 CTS--------7 RTS
```

```
     ●1 ●2 ●3 ●4 ●5
      ●6 ●7 ●8 ●9
        DB-9
       (male)
```

Figure 14-4 Null Model With Full Handshaking

Some variations of the full-handshake null modem connect the DTR to the CD line at each end. Pin number 1 (CD) in both male and female connectors is dummied-out to pin number 4 (CDR).

A conventional, straight-through serial cable can be converted to null modem by means of a commercial null modem adapter that crosses over the corresponding signal lines. A continuity test is used to determine whether a serial cable is wired as null modem or not. If it is null modem, pin number 2 on one end would show continuity with number 3 pin on the other end.

A circuit tester is used to diagnose serial cables. The tester, which is plugged into the port connector, contains a LED for each of the communications lines. When the corresponding LED lights up the line is active. LED colors indicate positive or negative voltages, with green usually indicating positive and red negative. The light pattern is used to identify different handshakes. Figure 14-5 shows a DB-25 mini tester.

Figure 14-5 DB-25 RS232 Line Tester

14.1.5 The EIA-485 Standard

EIA-485 provides a two-wire, half-duplex serial connection standard, also known as RS-485. This convention provides a multipoint connection with differential signaling. The connection can be made full-duplex by using four wires. In this standard, data is conveyed by voltage differences. One polarity represents logic 1 and the reverse one logic 0. The standard requires that the difference of potential be at least 0.2 volts, but any voltage between +12 and -7 volts allows correct operation.

EIA-485 does not specify a data transmission protocol, making possible the implementation of simple, inexpensive local networks and communications links. Its data transmission speeds can reach 35 Mbits/s at distances of up to 10 m, and 100 kbit/s at distances up to 1200 m. The use of a twisted wire pair and the differential balanced line allows spanning distances of up to 4000 ft.

EIA-485 is often used with common UARTs and USARTs to implement low-speed data communications that require minimal hardware. It is also found in programmable logic controllers that are used with proprietary data communications systems. In factories and other electrically charged environments, the differential feature of EIA-485 makes it resistant to electromagnetic interference from motors and other equipment. The standard also finds use in large sound systems, such as those found in theaters and music events. EIA-485 does not specify any connector.

EIA-485 in PIC-based Systems

In PIC-based systems, EIA-485 is often used to provide strong serial signals that can travel up to 4000 ft at high baud rates in noisy electrical environments. Only two wires are needed to carry the EIA-485 signals. These are usually labeled the A and B lines. Once the A/B data line is established, up to 32 devices can be connected to it. The system is referred to as an EIA-485 network.

Implementing the EIA-485 network requires some way of converting the 485 signal levels to the TTL-levels in the PIC circuit. This is accomplished by means of a dedicated IC, such as the *Texas Instruments Differential Bus Transceiver* chip named the SN75176. The chip actually converts 485 signals to RS-232-C TTL-level signals. This allows devices that traditionally communicate over RS-232-C serial connections to communicate over a two-wire EIA-485 network. Figure 14-6 shows the pin diagram of the SN75176.

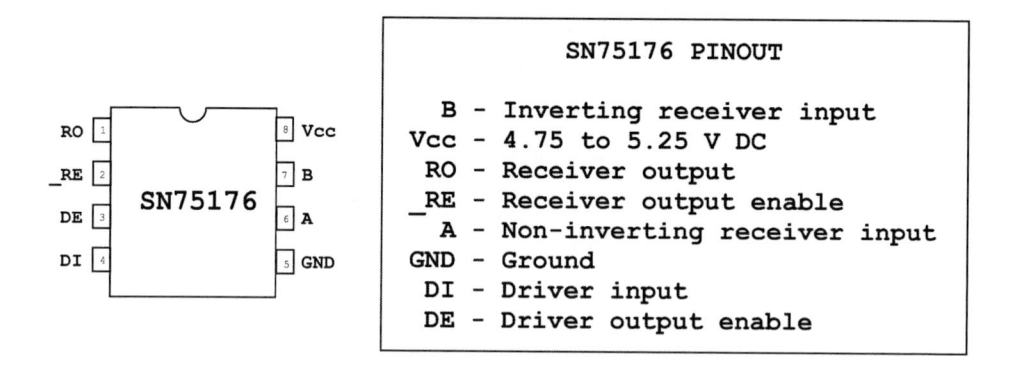

```
                         SN75176 PINOUT

                    B  - Inverting receiver input
  RO [1]    [8] Vcc Vcc - 4.75 to 5.25 V DC
 _RE [2]    [7] B   RO - Receiver output
  DE [3] SN75176 [6] A _RE - Receiver output enable
  DI [4]    [5] GND  A  - Non-inverting receiver input
                    GND - Ground
                    DI - Driver input
                    DE - Driver output enable
```

Figure 14-6 Pin Out of the SN75176 IC

In addition to the SN75176, an EIA-485 circuit requires a 485 chip such as the MAX485. In PIC-based systems, the EIA-485 is sometimes used to communicate with multiple devices in a chain. It uses the same 8-bit asynchronous serial communications format as was described previously for RS-232-C.

14.2 Parallel Data Transmission

Parallel communications is the process of sending several bits of data simultaneously over individual data lines. In the computer environment, parallel communications are often associated with a popular printer interface developed by Centronics and sometimes called the Centronics or *printer interface*. Originally, the Centronics interface was designed for one-way communications. Later, it was made bi-directional, allowing its use in high-speed data transfers. The Centronics or parallel printer interface is now considered a legacy port.

In PIC-based systems, parallel communications often refer to the general principle rather than to the specific Centronics implementation. For example, wiring an 8-line toggle switch to the eight pins of the 16F85 Port-B line provides parallel communications between the switch and the PIC.

PIC circuits that use parallel data transfers offer many advantages. In the first place, parallel transmission is fast and the software is simple to develop. The hardware implementation is straightforward and does not require many additional components. Examples are connecting a multiple toggle switch to each of the lines of a PIC input port, or each of the pins of a seven-segment LED to the various pins of a PIC output port. The disadvantages of parallel systems are the distance limitations and the cost in system resources. Furthermore, parallel data transfers do not work well for data transmission over long distances. Many of the circuits and programs covered in previous chapters use parallel data transmission techniques. Since PIC-based systems rarely communicate with parallel printers or use the Centronics standard for data transfer, no further discussion of the Centronics standard is justifiable in this context.

14.2.1 PIC Parallel Slave Port (PSP)

Some PICs are equipped with an 8-bit *Parallel Slave Port* module (PSP). At present, the PSP is multiplexed onto Port D and is found in PICs of the mid-range family, such as the 16F877. The PSP is also called the *microprocessor port*.

The PSP module provides an interface mechanism with one or more microprocessors. The parallel slave port has an operating speed of 200 ns with a clock rate of 20 MHz, as well as several on-chip peripheral functions for implementing real world interfaces.

In PICs equipped with the PSP, the parallel slave port functions are assigned to Port D, with some Port E bits providing control signals. To initialize PSP mode, data direction bits in the TRISE register that correspond to RD, WR, and CS (TRISE<2:0>) are configured as inputs and the control bit PSPMODE (TRISE) is set. When the PSP mode is active, Port D is asynchronously readable and writable through the chip Select (RE2/CS), Read (RE0/RD), and Write (RE1/WR) control inputs.

At this time, not many general-purpose applications for the PSP port have been documented, outside of its use as a multi-microprocessor interface. For this reason we have excluded PSP programming from this context.

14.3 PIC "Free-style" Serial Programming

This section is about PIC serial programming and circuit design that does not follow any specific communications protocol. In this sense, we have used the expression "free-style" as opposed to circuits and programs constrained by the requirements of a standard or convention. Many self-contained PIC circuits that do not interface with standardized components can benefit from not having to follow any specific standard. Later in this chapter, and in other chapters in the book, we present examples of PIC circuits and programs that follow established communications protocols. The titles of the corresponding sections refer to the specific standards or protocols; for example, the section titled *PIC RS-232-C Serial Programming* found in this chapter.

The advantages of so-called "free-style" circuit design and programming are greater in ease in development and the use of fewer hardware components. When designer and programmer are not constrained by the specifications of a standard, the circuit can be implemented with a minimal number of hardware components. By the same token, software is simpler and easier to develop.

The following examples of free-style communications systems are presented in the sections that follow:

1. A PIC to PIC communications circuit and program. Two programs are required: one for the receiver PIC and one for the sender.
2. Serial-to-parallel and parallel-to-serial circuit and program. Circuit uses 74HC164 and 74HC165 ICs.

14.3.1 PIC-to-PIC Serial Communications

Perhaps the most obvious and straightforward mode of PIC serial communications is one that takes place between two PICs. In this case, one PIC acts as a sender, or master, and the other one as a receiver or slave, although it is also possible for sender and receiver to exchange roles. Consider a circuit in which one PIC polls the state of a bank of switches and then sends the result serially to a second PIC that controls a bank of LEDs to be lighted according to the switch settings. The reason for this circuit is that some PICs may not have a sufficient number of ports to monitor eight switches and control eight LEDs.

PIC-to-PIC Serial Communications Circuits

Actually, the system required for one PIC reading data and serially sending the result to another PIC that outputs the data can be visualized as two separate circuits. One circuit is used to read the state of the eight DIP switches and to send the data serially to another PIC circuit that displays the results. Figure 14-7 shows the two PIC-based circuits.

Structurally, the circuits in Figure 14-7 are quite similar to ones described previously in this book. The bottom circuit contains eight DIP switches wired to ports RB0 to RB7. A pushbutton switch is wired to port RA2 and a LED to port RA3. The serial output is through port RA1. The circuit at the top of Figure 14-7 has eight LEDs wired to ports RB0 to RB7. There is a pushbutton on port RA2 and a LED on port RA3. Input into the circuit is through port RA0. In the remainder of this description we refer to the bottom circuit as the *sender circuit and PIC* and the one on the top as the *receiver circuit and PIC*.

The pushbuttons are necessary so that sender and receiver are synchronized. In operation, the receiver circuit is first activated by pressing the switch labeled "receive ready." The LED on the top circuit lights to indicate the ready state. The sender circuit has a LED labeled "ready" that indicates its state. The user presses the switch labeled "send ready" in the sender circuit. At that time, the program in the sender reads the state of the DIP switches and sends the data out, one bit at a time, through the line labeled "serial out" in the diagram. The receiver reads the eight bits in its "serial in" line and lights the LEDs accordingly.

Communications

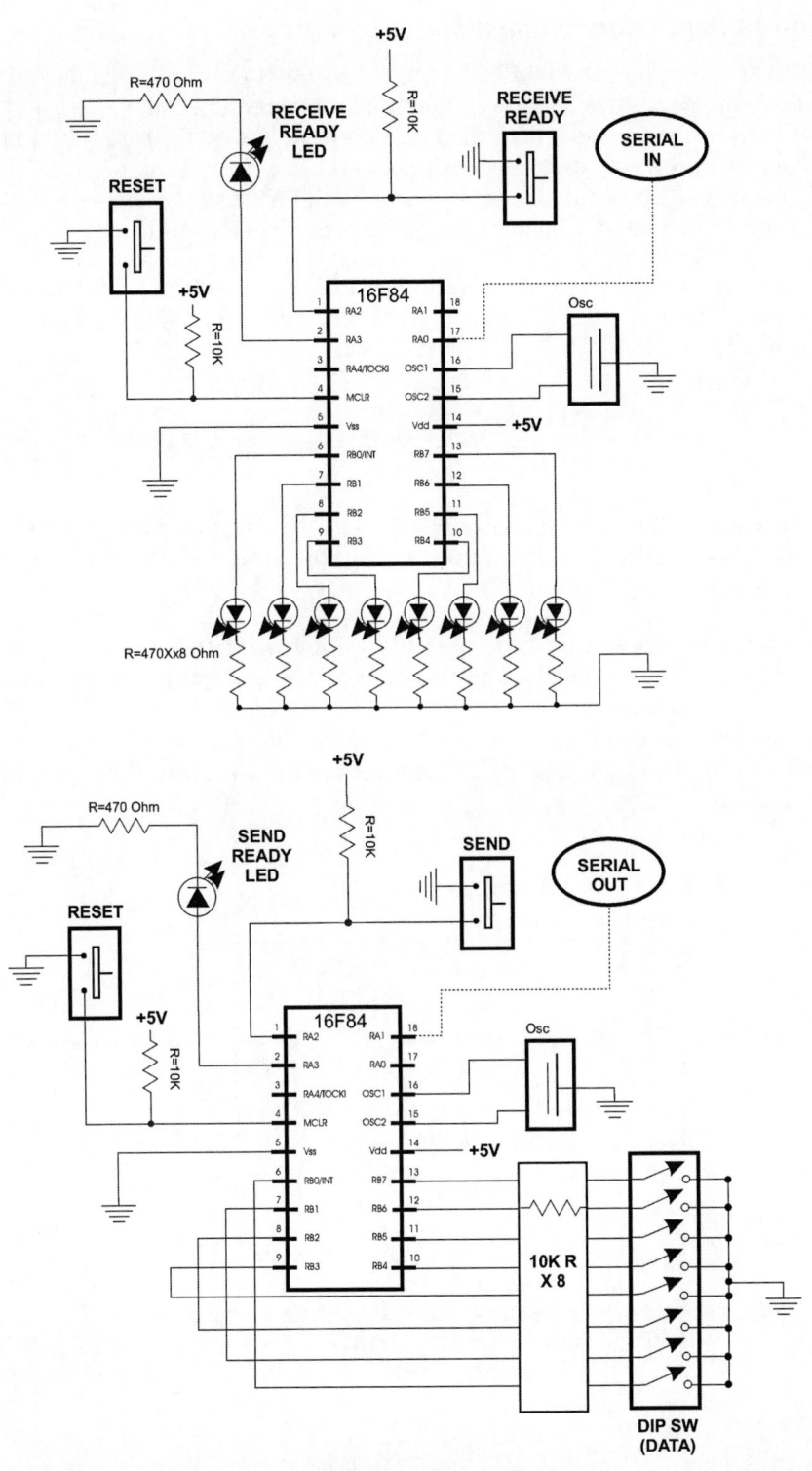

Figure 14-7 PIC-to-PIC Serial Communications Circuits

PIC-to-PIC Serial Communications Programs

The software consists of two different programs, one to run in the sender PIC and one in the receiver PIC. Asynchronous communications require that sender and receiver operate at the same data speed. Both devices need not run at the same clock speed, but both must synchronize data transmission and reception at the same clock rate. Since the easiest way to accomplish this is to have both PICs use the same oscillator at the same speed, we make this assumption in the programs that follow.

The instruction time and clock rate of a PIC are one-fourth of its clock speed. Thus, a PIC with a 4Mhz clock runs at 1,000,000 cycles per second, and the default timer speed is:

$$\frac{1,000,000}{256} = 3,906.25 \, \mu s. \, per \, bit$$

approximately 3,906μs per clock cycle. Although 3,906μs is not a standard baud rate, the present application is self-contained, therefore there is no need to conform to RS232-C or any other protocol.

Since it seems more intuitive to associate a high voltage with a logic 1 and a low voltage with a logic 0, we will adopt this convention in the present application. Nevertheless, we will borrow the character structure from the RS-232-C convention, that is, information will contain a start bit, a series of eight data bits, and a stop bit. No parity is implemented. Figure 14-8 shows the bit structure for one character in our application.

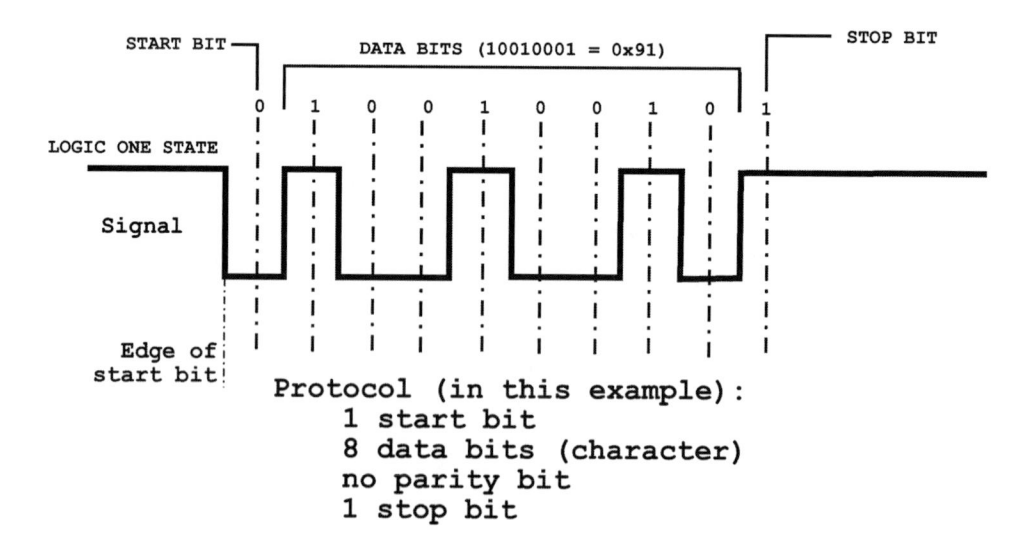

Figure 14-8 Data Structure for PIC-to-PIC Application

Communications

The sender program, named SerialSnd, performs the following initialization operations:

1. Line RA2 is initialized for input since the pushbutton switch is located on this line. Lines RB0 to RB7 are also input, since they are connected to the DIP switch array.
2. The prescaler is assigned to the Watchdog Timer so that channel TMR0 runs at full processor speed.
3. Interrupts are disabled.

Initialization code is as follows:

```
; Port-A, bit 2 is input. All others are output
        movlw    b'00000100'      ; Port-A bit 2 is input
                                  ; all others are output
        tris     porta
; Port-B is all input
        movlw    b'11111111'
        tris     portb
        bsf      porta,1          ;Marking bit
; Prepare to set prescaler
        clrf     tmr0
        clrwdt
; Setup OPTION register for full timer speed
        movlw              b'11011000'
;    1  1  0  1  1  0  0  0 <= OPTION bits
;    |  |  |  |  |  |__|__|____ PS2-PS0 (prescaler bits)
;    |  |  |  |  |              Values for Timer0
;    |  |  |  |  |             *000 = 1:2    001 = 1:4
;    |  |  |  |  |              010 = 1:8    011 = 1:16
;    |  |  |  |  |              100 = 1:32   101 = 1:64
;    |  |  |  |  |              110 = 1:128  111 = 1:256
;    |  |  |  |  |_____ PSA (prescaler assign)
;    |  |  |  |                *1 = to WDT
;    |  |  |  |                 0 = to Timer0
;    |  |  |  |_____ TOSE (Timer0 edge select)
;    |  |  |                    0 = increment on low-to-high
;    |  |  |                   *1 = increment in high-to-low
;    |  |  |_____ TOCS (TMR0 clock source)
;    |  |                      *0 = internal clock
;    |  |                       1 = RA4/TOCKI bit source
;    |  |_____ INTEDG (Edge select)
;    |                          0 = falling edge
;    |                         *1 = raising edge
;    |_____ RBPU pullups
;                               0 = enabled
;                              *1 = disabled
        option
; Disable interrupts
        bcf      intcon,5        ; Timer0 overflow disabled
```

```
            bcf     intcon,7    ; Global interupts disabled
```

Once initialized, the program performs the following functions:

1. The SEND READY LED is turned on.
2. Code monitors the SEND pushbutton switch.
3. Once the switch is pressed, the program turns off the SEND READY LED.
4. The state of the DIP switches is obtained by reading RB0 to RB7.
5. The byte from Port-B is sent through the serial line.

The following code fragment shows the procedure to send serial data.

```
;============================================================
;                procedure to send serial data
;============================================================
; ON ENTRY:
;         local variable dataReg holds 8-bit value to be
;         transmitted through port labeled serialLN
; OPERATION:
;         1. The timer at register TMR0 is set to run at
;            maximum clock speed, that is, 256 clock beats.
;            The timer overflow flag in the INTCON register
;            is set when the timer cycles from 0xff to 0x00.
;         2. Each bit (start, data, and stop bits) is sent
;            at a rate of 256 timer beats. That is, each bit is
;            held high or low for one full timer cycle (256
;            clock beats).
;         3. The procedure tests the timer overflow flag
;            (tmrOVF) to determine when the timer cycle has
;            ended, that is when 256 clock beats have passed.
;
sendData:
         movlw   0x08            ; Setup shift counter
         movwf   bitCount
;========================
;    send START bit
;========================
; Set line low then hold for 256 timer clock beats.
         bcf     PORTA,serialLN  ; Send start bit
; First reset timer
         clrf    TMR0            ; Reset timer counter
         bcf     INTCON,tmrOVF   ; Reset TMR0 overflow flag
; Wait for 256 timer clock beats
startBit:
         btfss   INTCON,tmrOVF   ; timer overflow?
         goto    startBit        ; Wait until set
; At this point timer has cycled. Start bit has ended
         bcf     INTCON,tmrOVF   ; Clear overflow flag
```

```
;=========================
;     send 8 DATA bits
;=========================
; Eight data bits are sent through the serial line
; starting with the high-order bit. The data byte is
; stored in the register named dataReg. The bits are
; rotated left to the carry flag. Code assumes the bit
; is zero and sets the serial line low. Then the carry
; flag is tested. If the carry is set the serial line
; is changed to high. The line is kept low or high for
; 256 timer beats.
send8:
        rlf     dataReg,f       ; Bit into carry flag
        bcf     PORTA,serialLN  ; 0 to serial line
; Code can assume the bit is a zero and set the line
; low since, if low is the wrong state, it will only
; remain for two timer beats. The receiver will not
; check the line for data until 128 timer beats have
; elapsed, so the error will be harmless. In any case,
; there is no assurance that the previous line state is
; the correct one, so leaving the line in its previous
; state could also be wrong.
        btfsc   STATUS,c        ; Test carry flag
        bsf     PORTA,serialLN  ; Bit is set. Fix error.
bitWait:
        btfss   INTCON,tmrOVF   ; Timer cycled?
        goto    bitWait         ; Not yet
; At this point timer has cycled.
; Test for end of byte, if not, send next bit
        bcf     INTCON,tmrOVF   ; Clear overflow flag
        decfsz  bitCount,f      ; Last bit?
        goto    send8           ; not yet
;=========================
;     hold MARKING state
;=========================
; All 8 data bits have been sent. The serial line must
; now be held high (MARKING) for one clock cycle
        bsf     PORTA,serialLN  ; Marking state
markWait:
        btfss   INTCON,tmrOVF   ; Done?
        goto    markWait        ; not yet
;=========================
;     end of transmission
;=========================
        return
```

The code comments explain the routine's operation.

The receiving program, named SerialRcv, runs in the receiver PIC. In this case, the serial line is RA0. Input from the sender program is received through this line. The program performs the following initialization operations:

1. Lines RA0 and RA2 are initialized for input since the pushbutton switch is located on RA2 and RA0 is the serial input line. Lines RB0 to RB7 are output since they are wired to the eight LEDs.
2. The prescaler is assigned to the Watchdog Timer so that channel TMR0 runs at full processor speed.
3. Interrupts are disabled.

Once initialized, code performs the following functions:

1. The SEND READY LED is turned on.
2. Code monitors the RECEIVE READY pushbutton switch.
3. Once the switch is pressed, the program turns on the RECEIVE READY LED.
4. Code then monitors the serial line for the first low that indicates the leading edge of the start bit.
5. Once the start bit is detected, code waits for 128 clock cycles to locate the center of the start bit. This synchronizes the receiver with the sender and accommodates small timing errors.
6. The eight data bits are then received and stored.
7. After waiting for the stop bit, code turns off the RECEIVE READY LED and sets the eight LEDs according to the data received through the serial line.

The following code fragment is the procedure rcvData from the SerialRcv program:

```
;================================================================
;                procedure to receive serial data
;================================================================
; ON ENTRY:
;        local variable dataReg is used to store 8-bit value
;        received through port (labeled serialLN)
; OPERATION:
;        1. The timer at register TMR0 is set to run at
;           maximum clock speed, that is, 256 clock beats.
;           The timer overflow flag in the INTCON register
;           is set when the timer cycles from 0xff to 0x00.
;        2. When the START signal is received, the code
;           waits for 128 timer beats so as to read data in
;           the middle of the send period.
;        3. Each bit (start, data, and stop bits) is read
;           at intervals of 256 timer beats.
;        4. The procedure tests the timer overflow flag
;           (tmrOVF) to determine when the timer cycle has
```

```
;               ended, that is when 256 clock beats have passed.
;================================================================
rcvData:
        clrf    TMR0            ; Reset timer
        movlw   0x08            ; Initialize bit counter
        movwf   bitCount
;=========================
;   wait for START bit
;=========================
startWait:
        btfsc   PORTA,0 ; Is port A0 low?
        goto    startWait               ; No. Wait for mark
;=========================
;   offset 128 clock beats
;=========================
; At this point the receiver has found the falling
; edge of the start bit. It must now wait 128 timer
; beats to synchronize in the middle of the sender's
; data rate, as follows:
;             |<========= falling edge of START bit
;             |
;             |-----|<====== 128 clock beats offset
;    ----------.    |    .-------
;             |     |    <== SIGNAL
;             -----------
;             |<---256--->|
;
        movlw   0x80            ; 128 clock beats offset
        movwf   TMR0            ; to TMR0 counter
        bcf     INTCON,tmrOVF   ; Clear overflow flag
offsetWait:
        btfss   INTCON,tmrOVF   ; Timer overflow?
        goto    offsetWait              ; Wait until
        btfsc   PORTA,0         ; Test start bit for error
        goto    offsetWait              ; Recycle if a false
start
;=========================
;   receive data
;=========================
        clrf    TMR0            ; Restart timer
        bcf     INTCON,tmrOVF   ; Clear overflow flag
; Wait for 256 timer cycles for first/next data bit
bitWait:
        btfss   INTCON,tmrOVF   ; Timer cycle end?
        goto    bitWait         ; Keep waiting
; Timer has counter 256 beats
        bcf     INTCON,tmrOVF   ; Reset overflow flag
        movf    PORTA,w         ; Read Port-A into w
```

```
             movwf    temp          ; Store value read
             rrf      temp,f        ; Rotate bit 0 into carry flag
             rlf      rcvReg,f ;    Rotate carry into rcvReg bit 0
             decfsz   bitCount,f    ; 8 bits received
             goto     bitWait       ; Next bit
; Wait for one time cycle at end of reception
markWait:
             btfss    INTCON,tmrOVF ; Timer overflow flag
             goto     markWait      ; keep waiting
;========================
;    end of reception
;========================
             return
```

Neither the SerialRcv nor the SerialSnd programs contain any handshake signal. The programs rely on the user turning-on the receiver before the send function is activated. If this is not the case, the programs fail to communicate. But looking at the circuit diagram in Figure 14-7, we notice that there are available ports in both receiver and sender circuits. The circuit designer could interconnect two ports, one in the receiver and one in the sender, so as to provide a handshake signal.

For example, lines RA4 in both circuits can be interconnected. Then Port-A, line 4, in the sender circuit is defined as input and the same line as output in the receiver. The receiver could then set the handshake line high to indicate that it is ready to receive. The sender monitors this same port and does not start the transmission of each character until it reads that the handshake line is high. In this manner, the receiver can suspend transmission at any time and prevent data from being lost. At the same time, the "receiver ready" and "send ready" LEDs can be eliminated.

14.3.2 Program Using Shift Register ICs

The problem of handling multiple input and output lines, which was resolved in the previous example by using two PICs, can also be tackled by means of special-purpose integrated circuits. The term *shift register* refers to the fact that register input and output are connected in a way that data is shifted-down a set of flip-flops when the circuits are activated. Many variations of shift registers ICs are available, the most popular ones being serial-in to serial-out, parallel-in to parallel-out, serial-in to parallel-out, and parallel-in to serial-out. In shift register terminology the *in* and *out* terms refer to the function in the registers themselves, and are not related to the functions that these elements perform in a particular circuit. Figure 14-9 shows an input/output circuit using shift registers.

The circuit in Figure 14-9 shows the use of a parallel-to-serial IC (74HC165) that reads the state of eight input switches, and a serial-to-parallel IC (74HC164) that outputs data to eight LEDs. Without the shift register ICs, the circuit would require sixteen ports, more than those available in the 16F84. Using the shift registers, only six PIC ports are required, leaving eight ports available on the PIC. The demonstration program for the circuit in Figure 14-9 is named Serial6465.

Communications

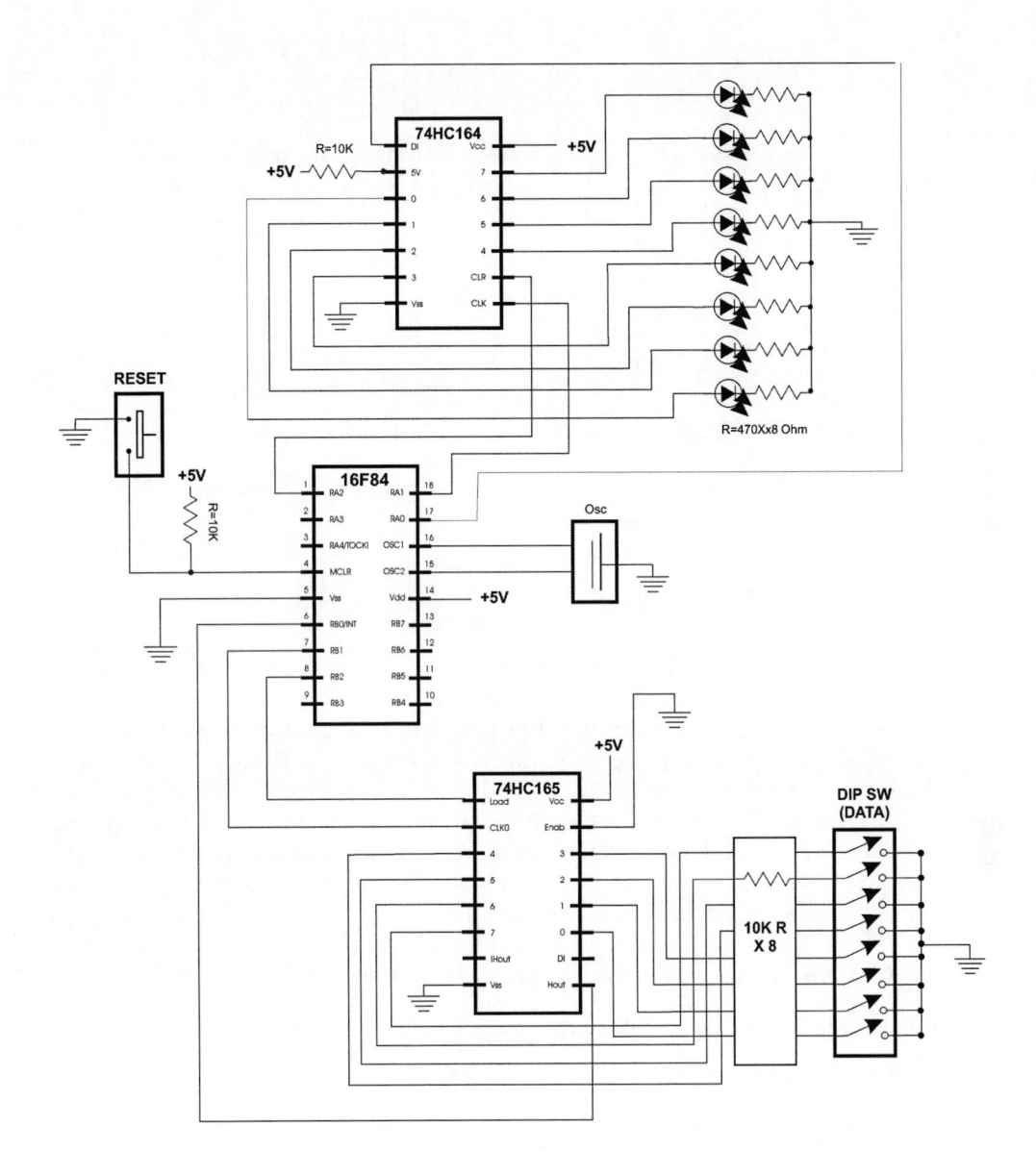

Figure 14-9 Input/output Circuit using Shift Registers

The 74HC165 Parallel-to-Serial Shift Register

The 74HC165 (sometimes called the 165) is a parallel-in, serial-out high-speed 8-bit shift register. Shift registers are discussed in Section 6.4.7. Figure 14-10 (in the following page) shows the pin-out of the 74HC165.

```
                    ┌──────┐
    shift/load  │1         16│  +5V
         clock  │2         15│  clock inhibit
            D4  │3         14│  D3
            D5  │4 74HC165 13│  D2
            D6  │5         12│  D1
            D7  │6         11│  D0
 serial output  │7         10│  serial input
           GND  │8          9│  serial output
                    └──────┘
```

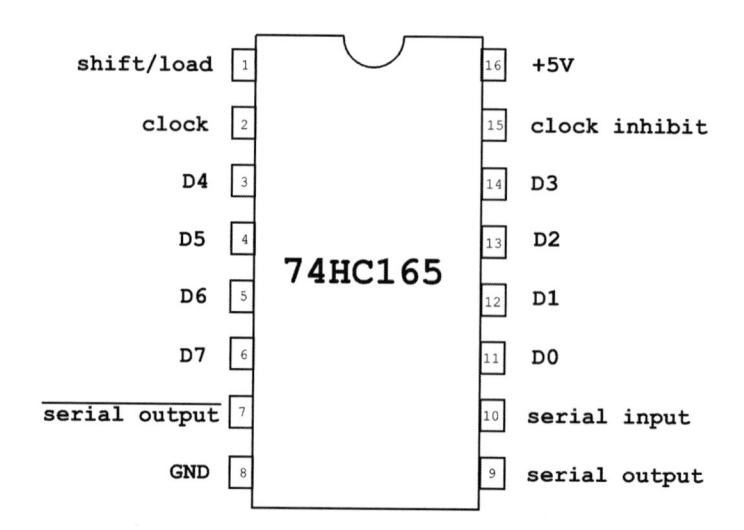

Figure 14-10 74HC165 Pin-Out

In the 165, pins 3 to 6 and 11 to 14 (labeled D0 to D7) are used as parallel data input lines. Normally these pins are connected to input sources, such as switches or other two-state devices. Serial output takes place through pin number 9, labeled serial output Q. An inverted output is available at pin number 7. The *shift/load control line*, at pin number 1, is used to latch the data into the 165 shift registers. For example, assume that the 165's input lines are connected to sources that can change state in time. These highs and lows are not recorded internally in the 165 until the shift/load line is pulsed. When this line is pulsed, line values are said to be *latched*. After the data lines are latched, the 165 clock-line is pulsed in order to sequentially shift-out each of the eight bits stored internally. Shifting takes place with the most significant bit first. The actual operations are as follows:

1. A local data storage register is cleared and a local counter is initialized for 8 data bits.
2. The 165 shift/load line is pulsed to reset the shift register.
3. The status of the serial output line (165 pin number 9) can now be read to determine the value of the bit shifted out.
4. The bit is stored in a data register and the bit counter is decremented. If the last bit was read the routine ends.
5. If not, the clock line is pulsed to shift-out the next bit. Execution continues at step number 3.

The wiring of the 165 normally requires at least three interface lines with the PIC. One line connects to the 165 serial output (pin number 9), another one to the clock line (pin number 2), and a third one to the shift/load line (pin number 1). The eight 165 data lines are normally wired to the input source.

Communications

The following code fragment lists a procedure to interface a 16F84 PIC with a 74HC165 parallel-to-serial shift register:

```
;===============================================================
;          constant definitions from wiring diagram
;===============================================================
#define clk65LN 1    ;| — 74HC165 lines
#define loadLN 2     ;|
   .
   .
   .
;===============================================================
;     74HC165 procedure to read parallel data and send
;                     serially to PIC
;===============================================================
; OPERATION:
;       1. Eight DIP switches are connected to the input
;          ports of an 74HC165 IC. Its output line Hout,
;          and its control lines CLK and load are connected
;          to the PIC's Port-B lines 0, 1, and 2
;          respectively.
;       2. Procedure sets a counter (bitCount) for 8
;          iterations and clears a data holding register
;          (dataReg).
;       3. Port-B bits are read into w. Only the lsb of
;          Port-B is relevant. Value is stored in a working
;          register and the meaningful bit is rotated into
;          the carry flag, then the carry flag bit is
;          then shifted into the data register.
;       4. The iteration counter is decremented. If this
;          is the last iteration the routine ends. Otherwise
;          the bitwise read-and-write operation is repeated.

in165:
         clrf     dataReg            ; Clear data register
         movlw    0x08               ; Initialize counter
         movwf    bitCount
         bcf      PORTB,loadLN       ; Reset shift register
         bsf      PORTB,loadLN
nextBit:
         movf     PORTB,w            ; Read Port-B (only LOB is
                                     ; meaningful in this routine)
         movwf    workReg            ; Store value in local
                                     ;register
         rrf      workReg,f          ; Rotate LOB bit into carry
                                     ; flag
         rlf      dataReg,f          ; Carry flag into dataReg
         decfsz   bitCount,f         ; Decrement bit counter
         goto     shiftBits          ; Continue if not zero
```

```
        Return                          ; done
shiftBits:
        bsf       PORTB,clk65LN         ; Pulse clock
        bcf       PORTB,clk65LN
        goto      nextBit               ; Continue
```

The procedure in165 is in the program Serial6465 listed at the end of this chapter.

74HC164 Serial-to-Parallel Shift Register

The circuit in Figure 14-9 also uses a 74HC164 serial-to-parallel shift register for output to the eight LEDs. Figure 14-11 shows the pin-out of the 74HC164 IC.

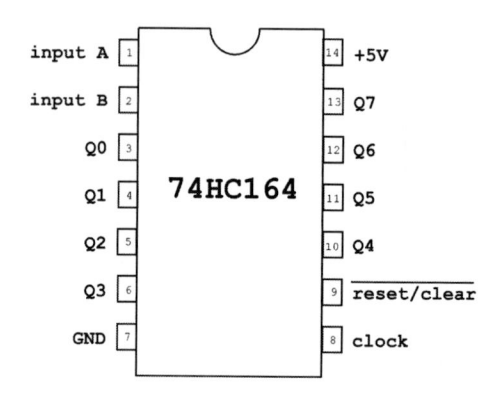

Figure 14-11 74HC164 Pin Out

Serial input into the 164 is through the input A line (pin number 1). Parallel output is through the lines labeled Q0 to Q7. The reset/clear line (on pin 9) and the clock line (on pin 8) provide the control functions. The operations are as follows:

1. A local data storage register holds the 8-bit value that serves as data input. A local counter is initialized for 8 data bits.
2. The 164 shift register is cleared by pulsing the reset/clear line.
3. The first/next bit of the data operand is placed on the input line.
4. Bit is shifted-in by pulsing the 164 clock line.
5. Bit counter is decremented. If it goes to zero the routine ends.
6. Otherwise, the bits in the source operand are shifted and execution continues at step number 3.

The following code fragment lists a procedure to interface a 16F84 PIC with a 74HC164 serial-to-parallel shift register:

```
;=========================================================
;       constant definitions from wiring diagram
;=========================================================
#define clockLN 1              ; |
```

```
#define clearLN 2          ; | - 74HC164 lines
#define dataLN 0           ; |
...
;================================================================
;           74HC164 procedure to send serial data
;================================================================
; ON ENTRY:
;       local variable dataReg holds 8-bit value to be
;       transmitted through port labeled serialLN
; OPERATION:
;       1. A local counter (bitCount) is initialized to
;          8 bits
;       2. Code assumes that the first bit is zero by
;          setting the data line low. Then the high-order
;          bit in the data register (dataReg) is tested.
;          If set, the data line is changed to high.
;       3. Bits are shifted in by pulsing the 74HC164
;          clock line (CLK).
;       4. Data bits are then shifted left and the bit
;          counter is tested. If all 8 bits have been sent
;          the procedure returns.
out164:
; Clear 74HC164 shift register
        bcf     PORTA,clearLN    ; 74HC164 CLR clear low
        bsf     PORTA,clearLN    ; then high again
; Init counter
        movlw   0x08             ; Initialize bit counter
        movwf   bitCount
sendBit:
        bcf     PORTA,dataLN     ; Set data line low (assume)
; Using this assumption is possible because the bit is not
; shifted in until the clock line is pulsed.
        btfsc   dataReg,highBit  ; test number bit 7
        bsf     PORTA,dataLN     ; Change assumption if set
;=========================
;    pulse clock line
;=========================
; Bits are shifted in by pulsing the 74HC164 CLK line
        bsf     PORTA,clockLN    ; CLK high
        bcf     PORTA,clockLN    ; CLK low
;=========================
; Rotate data bits left
;=========================
        rlf     dataReg,f        ; Shift left data bits
        decfsz  bitCount,f       ; Decrement bit counter
        goto    sendBit          ; Repeat if not 8 bits
;=========================
;    end of transmission
```

```
;========================
            return
```

It is important to note that serial communications that use shift register ICs are described as synchronous. Synchronous serial transmission requires that the sender and receiver use the same clock signal or that the sender provide signal or pulse so as to indicate to the receiver when to read the next data element from the line. In the circuits discussed in this section the shift/load, reset/clear, and clock lines provide this synchronous interface between the PIC and the shift register IC.

The program named Serial6465, in the book's on line software, is a demonstration of PIC-to-shift register interfacing.

14.4 PIC Protocol-based Serial Programming

In the preceding sections we discussed circuits and developed software using PIC serial communications that did not conform to any particular protocol or standard. This style is adequate for stand-alone applications and circuits. On the other hand, PIC-based circuits sometimes communicate with systems that conform to a specific communications standard, for example, with a PC through its RS-232-C serial port. In this case, the PIC software and hardware must conform with the protocol, at least to an operational minimum that ensures satisfactory interfacing with the protocol-based system.

In the context of protocol-based programming, two situations are possible: either the PIC in use supports the communications standard or protocol or it does not. In the case of the smaller PICs, such as the 16F84, the software emulates communications protocols since hardware provides no support. The more complex PICs, on the other hand, often contain hardware modules that provide a functionality equivalent to that required by the various standards. In this sense, mid-range and high-range PICs often include hardware support for one or more communication standards and conventions. For instance, the 16F87X PIC family includes an *MSSP* (*Master Synchronous Serial Port*) module and a *USART* (*Universal Synchronous/asynchronous Receiver and Transmitter*) module.

In the sections that follow we develop circuits and programs for cases in which the on-board PIC does not contain hardware support for the standard and for cases in which it does. Examples with PICs that do not provide hardware support for serial communications use the 16F84. Examples with PICs that provide hardware serial communications support use the 16F877, which contains an MSSP and a USART module. The 16F877 circuits and applications in the present chapter use the processor's USART module. The 16F877 MSSP module is demonstrated in the chapter on EEPROM programming.

14.4.1 RS-232-C Communications on the 16F84

The *UART* (*Universal Asynchronous Receiver/Transmitter*) controller is a serial communications IC found in computers and other data communication devices. In the PC, the UART was originally National Semiconductor INS8250. With the introductions of the PC AT, IBM changed its serial IC to the NC16450, an improved 8250. Later PCs adopted the NS16550A UART as their serial communications controllers. Other vendors, including Intel and Western Digital, furnish clones of the NS16550A and other UARTs.

The UART-based serial port implementation and circuitry in the PC is compliant with RS/EIA232. For a PIC-based circuit to communicate with a PC's serial port it must either implement in hardware or emulate in software the RS232 signals and protocol. One possibility is to include a UART or UART-like IC in the circuit. But this option is not simple to implement since RS-232-C requires voltage levels that are not TTL-compatible.

For PIC-based systems without a UART module, a viable approach is to emulate UART functions in software, at least those required for interfacing with the PC hardware. This is quite feasible due to the availability of dedicated ICs that provide RS-232-C-compatible signals and voltage levels in systems in which a ±12 volt source is not available. These chips, sometimes called RS-232-C Drivers/Receivers or Transceivers, are especially useful in interfacing UART and USART-based systems with PIC-based hardware.

The RS-232-C Transceiver IC

RS-232-C interface ICs are available from several vendors, although the ones from Dallas Semiconductors' Maxim line are probably the most popular. These chips, sometimes called RS-232-C driver/receivers, have in common the use of so-called charge-pump DC/DC converters that generate, from the +5 volt TTL power source, the polarities and voltage levels required by RS-232-C.

One of the most popular implementations of the RS-232-C transceiver used in PIC-based systems is the MAX232 and its upgrade, the MAX202. One improvement in the MAX202 is to provide some degree of human-body *electrostatic discharge protection* (ESD), a desirable feature in experimenter boards. Other versions are the MAX233 and MAX203, which do not require external capacitors. Other RS-232-C transceiver ICs with various additional features, such as automatic shutdown, are available. Figure 14-12 is a pin-out of the MAX232 and 203 ICs.

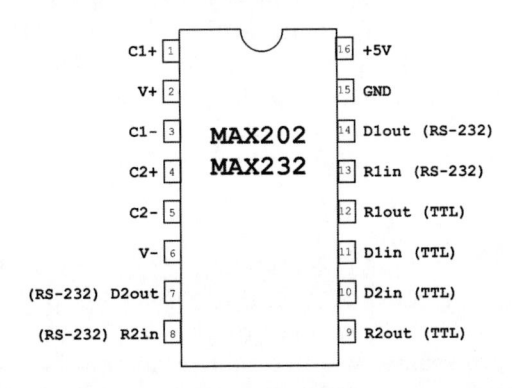

```
              C1+  [1]          [16] +5V
               V+  [2]          [15] GND
               C1- [3]  MAX202  [14] D1out (RS-232)
               C2+ [4]  MAX232  [13] R1in  (RS-232)
               C2- [5]          [12] R1out (TTL)
               V-  [6]          [11] D1in  (TTL)
   (RS-232) D2out [7]          [10] D2in  (TTL)
   (RS-232)  R2in [8]           [9] R2out (TTL)
```

Figure 14-12 MAX202 and MAX232 Transceiver Pin Out

Note that the MAX232 and MAX202 consist of two drivers and two receivers per chip. Lines 14 and 7 (labeled D1out and D2out) provide RS-232-C output. Lines 13 and 8 (labeled R1in and R2in) are RS-232-C input. Lines 10 and 11 (labeled D1in and D2in) are TTL (or CMOS) inputs. Lines 9 and 12 (labeled R2out and R1out) are TTL output. In this designation the letter R stands for receiver and the letter D for driver. The digit 1 indicates the first driver/receiver set and the digit 2 the second one. The lines labeled D are wired to capacitors.

A circuit using the transceiver ICs is simple and easy to build. If a single communication line is required, then the TTL input line can be wired to pin 10 (D2in) and the TTL output to pin 9 (R2out). The RS-232-C input is wired to pin 8 (R2in) and the output to pin 7 (D2out). Later in this section, we present a circuit that uses the MAX202 with a 16F84 PIC.

PIC to PC Communications

Often, a PIC-based circuit has to communicate with a device that conforms to a standard communications protocol. One of the most common cases is a PIC board that interfaces with a computer, usually a PC or Mac with an RS-232-C port. For example, a PIC board is placed somewhere to collect information, such as temperature, pressure, and humidity. Before the internal storage capacity of the PIC board is exhausted, it is connected to a laptop PC and the data is downloaded from the PIC board to the computer. Once this is done, then the local PIC memory is cleared so that new data can be collected and stored. This application, called a *data logger*, requires some way of transferring data from the PIC-based board to the PC. The RS-232-C line is often available on the PC end and the required interface hardware and programming is uncomplicated.

On the PC end, the communications software can be off-the-shelf applications or especially developed programs. If the purpose is simply to download data to the PC or send simple commands to the PIC board, then a standard utility is used. For example, the Windows program named *Hyper Terminal* allows sending and receiving files and commands at various baud rates and RS-232-C communications parameters. *Hyper Terminal* is included with most Windows versions or can be downloaded free from the developer's website.

The PIC board must have a system that conforms to the communications protocol of the device, in this case, the PC. In order to use the PC's serial port, PIC hardware and software must be able to generate required signal levels, baud rate, and other RS-232-C communications parameters. Hardware interfacing is implemented by using a transceiver chip, such as the MAX232 or 202 previously described. If the PIC contains a UART or USART module, then the communications software is easy to develop. This case is explored later in this chapter.

An RS-232-C TTY Board

The terms "teletype" and "teletypewriter" refer to an obsolete electro-mechanical typewriter that was used to send and receive information through a simple communication channel. In a modern sense, TTY refers to a simple style of communications where the same device sends and receives text messages interactively. The current board is actually a TTY receiver since it does not contain a keyboard that allows sending data. Figure 14-13 shows the circuit diagram for an 16F84-based PC-to-PIC serial communications board.

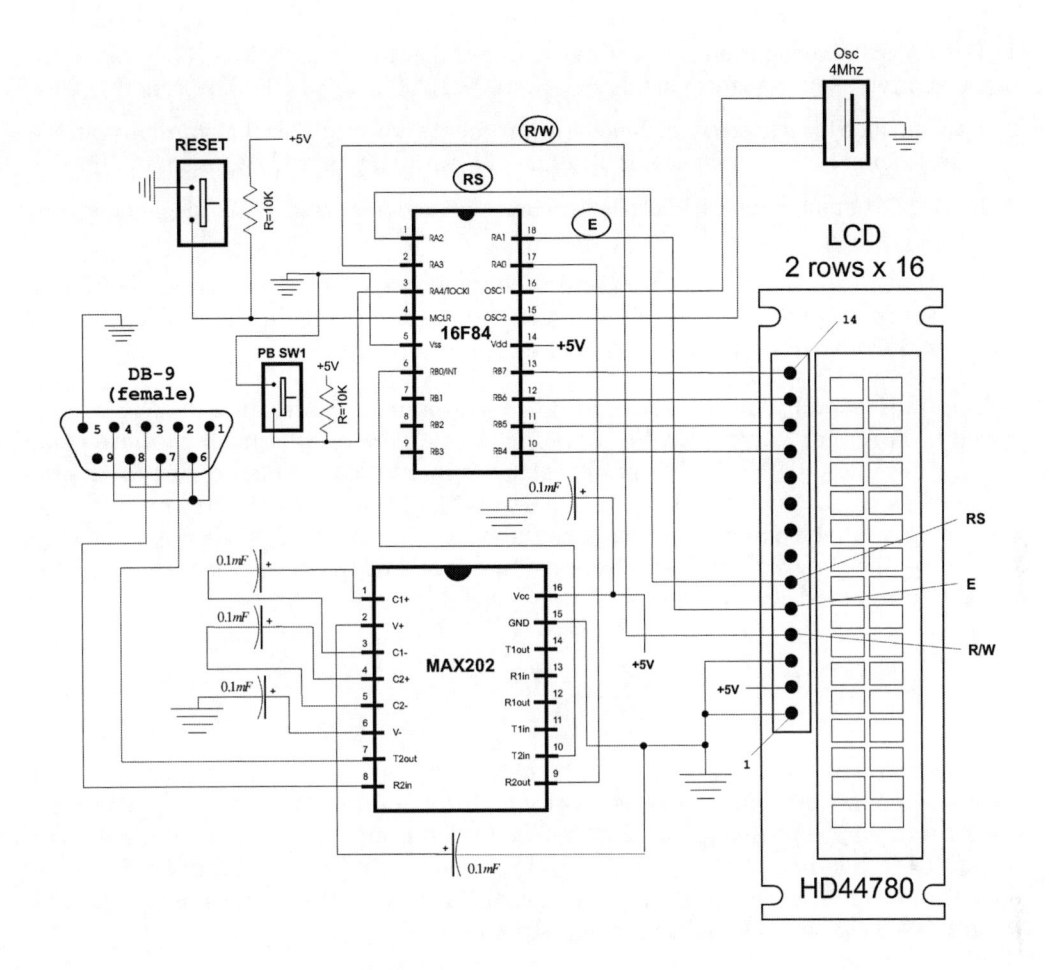

Figure 14-13 PC-to-PIC Serial Communications Circuit

The circuit in Figure 14-13 contains previously discussed components. The LCD is wired in 4-bit mode, with control lines for RS (reset), E (pulse), and R/W (read/write). The MAX202 provides the TTL-to-RS-232-C conversion and vice versa. The physical connection between the PC and the PIC board is by means of a DB-9 connector and a standard null modem cable. The cable is not shown in the circuit diagram.

A 16F84A UART Emulation

The 16F84A PIC contains no built-in facilities for RS-232-C communications. Therefore, a 16F84A application that communicates through the serial port using the RS-232-C protocol must emulate the protocol in software. The programs previously developed for PIC-to-PIC communications, discussed in Section 14.3.1, serve as a base for the UART emulation application. The major differences between a "free style" PIC communications program and one that complies for RS-232-C are the following:

1. Data must be transmitted and received at one of the standard RS-232-C baud rates. The most often-used baud rates in this case are: 600, 1,200, 2,400, 4,800, 9,600, and 19,200.
2. Data must be formatted according to the protocol's conventions; that is, a start bit, 5, 6, 7, or 8 data bits, the presence or absence of a parity bit, and 1, 1½, or 2 stop bits.
3. RS-232-C communication data is transmitted and received with the least-significant-bit first.

The first problem (transmitting and receiving at a standard baud rate) often requires an approximation. The PIC's instructions execute at the rate of its internal clock, which also determines the rate of its timer module.

The time taken by each counter iteration is obtained by dividing the PIC's clock speed by four. For example, a PIC running on a 4 Mhz oscillator clock increments the counter every 1 Mhz. The counter register is incremented at a rate of 1μs (assuming no prescaler). If we were to use the unmodified timer rate to measure bit time, the result would be a baud rate of approximately 3,906. Since 3,906 is not a standard baud rate, the timer is adjusted to approximate one of the standard RS-232-C baud rates. For example, at 4,800 baud the time per bit is:

$$\frac{1}{4,800} = 208.33 \mu s.$$

Since the timer of a PIC with a 4 Mhz clock runs at 1 μs per timer iteration, then we could count up from 0 to 208 iterations of the counter in order to approximate the bit time of 208 μs needed at 4,800 baud. In addition, we would have to calculate one-half the bit time since synchronization requires offsetting the timer from the edge to the center of the start bit (see Section 14.3.1). In this case, to delay approximately 104 μs we would count up from 0 to 104.

But counting up is inconvenient with the PIC timer/counter since the signal is produced when the counter reaches its maximum. A better solution is to preset the Timer counter (TMR0) to a calculated value such that the desired time lapse occurs when the Timer register reaches 255. So the actual delays for 4,800 baud are as follows:

```
DELAY        CALCULATION      TMR0 PRESET
208 µs       255 - 208        48
104 µs       255 - 104        151
```

Once we have obtained the clock rate for a standard baud rate, it is easy to obtain slower standard rates by slowing down the clock with the prescaler. For example, if the prescaler is assigned to the timer/counter register with a bit value of 000, then the counter rate is one-half the unscaled rate. This would produce a baud rate of 2,400 baud. By the same token, assigning a 1:4 prescaler to the timer produces a baud rate of 1,200 baud using the same preset values previously calculated. Faster baud rates are easily calculated by the same method.

Formatting the data transmission according to the RS-232-C protocol presents no major problem. In fact, the communications programs previously listed in this chapter use a start bit to commence character transmission, followed by eight data bits, and one stop bit to end it, with no parity bit. This same format is compatible with RS-232-C.

The third compatibility issue refers to the bit order in RS-232-C, which requires that the low-order bit be transmitted first. In previous applications, we have sent the high-order bit first by rotating the bits left inside the holding register and testing the carry flag. In the RS-232-C routine, the bits are rotated right into the carry flag and then the carry flag is rotated into the storage variable.

The demonstration program for the circuit in Figure 14-13, named TTYUsart, uses a 2-line by 16 character LCD to display the characters received from the PC through the serial line. The program initially sends the test string "Ready-" to the PC to test the data transmission routine and to let the PC user know that the PIC board is ready to receive. The program operates at 2,400 baud, one start bit, eight data bits, no parity, and one stop bit. The communications program on the PC must be set to these parameters.

An LCD Scrolling Routine

LCDs have limited capacity for data display. A 2-line by 16 character LCD fills the screen when 32 characters are displayed. For some applications it is convenient to have a procedure that takes some reasonable action when the LCD screen is full. One approach is to detect when the last character in the second LCD line is displayed, then move the second line to the first line, clear the second line, and continue displaying at the start of the second line. This is the standard screen handling for a computer program.

An LCD screen scroll routine can be called as each character is displayed. For the scroll to work, the program must keep track of the currently selected LCD line (variable LCDline can be 0 for line 1, and 1 for line 2), of the number of characters displayed on that line (variable LCDcount), and of the total capacity of the line (constant LCDlimit). Given this information, the logic for an LCD line scrolling routine can be as follows:

1. Add current character to LCDcount. If LCDcount is equal to LCD limit then the end of a line was reached. If not, exit routine.

2. If line end reached is for line 1, set current display address to start of line 2. Reset variable LCDcount. Exit routine.

3. If line end reached is for line 2, then copy the characters displayed in line 2 to line 1. Clear line 2. Reset the display address to the start of line 2. Reset LCDline variable to line 2. Reset variable LCD count. Exit routine.

Of these operations, copying the characters from the second line to the first one can be the most troublesome. One possibility is to read the data from the LCD directly. This approach requires that the connection between the PIC and the LCD includes the R/W line. Another option is to create a buffer in RAM and copy each character displayed to this area. In the case of an LCD with 16 characters per line the buffer requires a capacity of 16 bytes. Since the line input is "remembered" in the buffer, the program scrolls a line by copying the contents of the buffer to the other line. This alternative does not require reading the LCD and saves implementing the R/W line.

Storing the characters received in a local buffer first requires reserving a 16-byte area (the buffer) in PIC RAM. There are several ways of accomplishing this. A simple one is using the **cblock** directive, as shown in the following code fragment:

```
;============================================================
;              buffer and variables in PIC RAM
;============================================================
; Create a 16-byte storage area
        cblock   0x0c      ; Start of first data block
        lineBuf                ; buffer for text storage
        endc
; Leave 16 bytes and continue with local variables;
        cblock   0x1c      ; Second data block
        count1             ; Counter # 1
        count2             ; Counter # 2
. . . other variables can go here
        endc
```

In reality, the buffer is most likely accessed by indirect addressing, so a buffer name (lineBuf in this case) is not really necessary. This is due to the fact that PIC assembly language does not contain a directive for finding the address of a variable. So the buffer address has to be hard-coded or defined in a constant. But, in any case, having a buffer name does not cost storage capacity and it may help make the code clearer.

In our design, the scrolling routine depends on finding the characters in the ending line stored in the RAM area mentioned in the preceding paragraph. The buffer locations are accessed directly by referencing the address. For example, the first byte in lineBuf is stored at addres 0x0c, the second one at 0xod, and so on. A more effective way of using a buffer is by creating and keeping a *buffer pointer variable* that has the current offset from the start of the buffer. The buffer pointer is then added to the buffer's base address in order to access the current buffer location. Indirect addressing using the FSR and the INDF registers simplify the process, as shown in the following code fragment:

```
; Store character in local line buffer using indirect
; addressing. Byte to store is in rcvData variable.
; 16-byte buffer named lineBuf starts at address 0x0c
; Register variable bufPtr holds offset into buffer
        movlw    0x0c           ; Buffer base address
        addwf    bufPtr,w       ; Add pointer in w
        movwf    FSR            ; Value to index register
        movf     rcvData,w      ; Character into w
        movwf    INDF           ; Store w in [FSR]
        incf     bufPtr,f       ; Bump pointer
```

The manipulation requires loading the base address of the buffer (0x0c in this case) in the w register, adding the value stored in the buffer pointer variable (bufPtr), and storing the sum in the FSR register. The character is then loaded into the w register and moved into the INDF register, which has the effect of storing it in the address pointed at by FSR. Conventionally, brackets are used to indicate indirect addressing, so [FSR] means the memory location referenced by the FSR register.

Once the line characters are stored locally, all that is left is the design of a line scrolling routine following the processing steps previously listed. The following procedure performs the necessary operations:

```
;===========================
;    scroll LCD line 2
;===========================
; Procedure to count the number of characters displayed on
; each LCD line. If the number reaches the value in the
; constant LCDlimit, then display is scrolled to the second
; LCD line. If at the end of the second line, then the
; second line is scrolled to the first line and display
; continues at the start of the second line
; reset to the first line.
LCDscroll:
        incf     LCDcount,f              ; Bump counter
; Test for line limit
        movf     LCDcount,w
        sublw    LCDlimit               ; Count minus limit
        btfss    STATUS,z               ; Is count - limit = 0
        goto     scrollExit             ; Go if not at end of line
; At this point the end of the LCD line was reached
; Test if this is also the end of the second line
        movf     LCDline,w
        sublw    0x01                   ; Is it line 1?
        btfsc    STATUS,z               ; Is LCDline minus 1 = 0?
        goto     line2End               ; Go if end of second line
; At this point it is the end of the top LCD line
        call     line2                  ; Scroll to second line
        clrf     LCDcount               ; Reset counter
        incf     LCDline,f              ; Bump line counter
        goto     scrollExit
; End of second LCD line
line2End:
; Scroll second line to first line. Characters to be
; scrolled are stored in buffer starting at address 0x0c.
; 16 characters are to be moved
; First clear LCD
        call     initLCD
        call     delay_5                ; Make sure not busy
; Set up for data
```

```
            bcf      PORTA,E_line     ; E line low
            bsf      PORTA,RS_line    ; RS line high for data
; Set up counter for 16 characters
            movlw    D'16'            ; Counter = 16
            movwf    count2
; Get address of storage buffer
            movlw    0x0c
            movwf    FSR              ; W to FSR
getchar:
            movf     INDF,w   ; get character from display RAM
                              ; location pointed to by file select
                              ; register
            call     send8    ; 4-bit interface routine
; Test for 16 characters displayed
            decfsz   count2,f ; Decrement counter
            goto     nextchar ; Skipped if done
; At this point scroll operation has concluded
            clrf     LCDcount ; Clear counters
; Stay at line 2
            clrf     LCDline
            incf     LCDline,f
            call     line2    ; Set for second line
scrollExit:
            return
nextchar:
            incf     FSR,f    ; Bump pointer
            goto     getchar
;===========================
;    clear line buffer
;===========================
; Use indirect addressing to store 16 blanks in the
; buffer located at 0x0c
blankBuf:
            Bank0
            movlw    0x0c     ; Pointer to RAM
            movwf    FSR      ; To index register
blank16:
            clrf     INDF     ; Clear memory pointed at by FSR
            incf     FSR,f    ; Bump pointer
            btfss    FSR,4    ; 000x0000 when bit 4 is set
                              ; count reached 16
            goto     blank16
            return
;========================
; Set address register
;    to LCD line 1
;========================
; ON ENTRY:
```

```
;            Address of LCD line 1 in constant LCD_1
line1:
        bcf     PORTA,E_line        ; E line low
        bcf     PORTA,RS_line       ; RS line low, set up for control
        call    delay_5             ; busy?
; Set to second display line
        movlw   LCD_1               ; Address and command bit
        call    send8               ; 4-bit routine
; Set RS line for data
        bsf     PORTA,RS_line       ; Setup for data
        call    delay_5             ; Busy?
; Clear buffer and pointer
        call    blankBuf
        clrf    bufPtr              ; Pointer
        return
;========================
; Set address register
;     to LCD line 2
;========================
; ON ENTRY:
;            Address of LCD line 2 in constant LCD_2
line2:
        bcf     PORTA,E_line        ; E line low
        bcf     PORTA,RS_line       ; RS line low, setup for control
        call    delay_5             ; Busy?
; Set to second display line
        movlw   LCD_2               ; Address with high-bit set
        call    send8
; Set RS line for data
        bsf     PORTA,RS_line       ; RS = 1 for data
        call    delay_5             ; Busy?
; Clear buffer and pointer
        call    blankBuf
        clrf    bufPtr              ; Pointer
        return
```

The entire program, named TTYUsart, is found in the book's online software package.

14.4.2 RS-232-C Communications on the 16F87x

The second alternative for protocol-compliant communications is using a PIC that provides hardware support for the standard. The 16F84, our workhorse in this book, contains no such facilities. However, other midrange PICs do provide hardware support to one or several serial communications protocols.

For the examples that follow, we have selected what is perhaps the second most popular PIC of the midrange family (after the 16F84): the 16F87x. The architecture and basic programming facilities of the 16F87x PIC family were discussed in Chapter 8. At this time, we should recall that 16F87x includes the PIC 16F873, 16F874, 16F876, and 16F877. For our sample programs we have selected the 16F877 since it is the most powerful one of the group. The 16F877 has an operating frequency of up to 20Mhz, 8K of flash program memory, 368 bytes of data memory, 256 bytes of EEPROM, 5 input/output ports, and contains two modules for serial communications: a Master Synchronous Serial Port and a Universal Synchronous/Asynchronous Receiver and Transmitter. We focus on the USART module and leave the MSSP for the chapter on EEPROM programming.

The 16F87x USART Module

The Universal Synchronous Asynchronous Receiver Transmitter (USART) module in the 16F87X family is also known as a *Serial Communications Interface*, or *SCI*. The USART module is useful in communicating with devices and systems that support RS-232-C communications, including computers and terminals. It can be configured as an asynchronous full-duplex device, as a synchronous half-duplex master, or as a synchronous half-duplex slave. In the synchronous mode, the USART module is used mostly in communicating with analog-to-digital and digital-to-analog integrated circuits or for accessing serial EEPROMS. Both of these functions are discussed in later chapters.

Five registers relate to USART operation in the 16F877: RCSTA, TXREG, RCREG, TXSTA, and SPBRG. The first three are located in bank 0 and the second two in bank 1. TXSTA is the Transmit Status and Control register and the RCSTA the Receive Status and Control register. Figure 14-14 shows the bitmap for the TXSTA register located at address 0x98 in bank 1.

The RCSTA register contains control and status bits for the receive function. The register is found at address 0x18 in bank 0. Figure 14-15 (in the following page) is a bitmap of the RCSTA register.

The USART Baud Rate Generator

In the USART emulation programs for the 16F84 we were forced to approximate the RS-232-C baud rate with the system clock. The USART module in the 16F87X PICs contains its own baud rate generator, but it is also dependent on the system clock.

Setting the baud rate in the USART module consists of manipulating the *Baud Rate Generator* (BRG) unit. The BRG is a dedicated 8-bit generator that supports both the asynchronous and synchronous modes. The SPBRG is an 8-bit register that controls the rate of a dedicated timer. In the asynchronous mode, the bit labeled BRGH in the TXSTA register (see Figure 14-14) also relates to the baud rate since it allows setting either slow-speed or high-speed baud rate. The baud-rate-speed-select bit is inactive in the synchronous mode.

```
 bit 7                                                    bit 0
┌──────┬──────┬──────┬──────┬──────┬──────┬──────┬──────┐
│ CSRC │ TX9  │ TXEN │ SYNC │      │ BRGH │ TRMT │ TX9D │
└──────┴──────┴──────┴──────┴──────┴──────┴──────┴──────┘
```

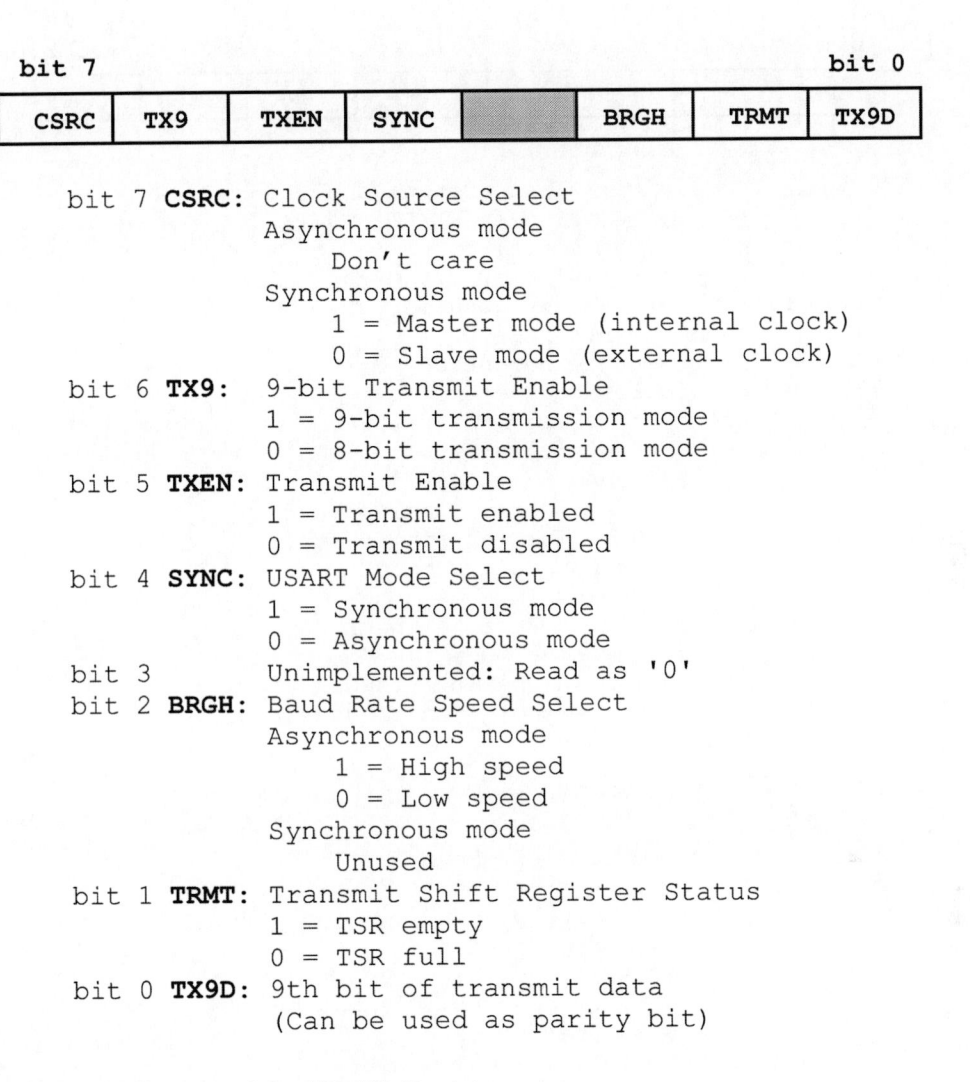

```
bit 7 CSRC: Clock Source Select
            Asynchronous mode
                Don't care
            Synchronous mode
                1 = Master mode (internal clock)
                0 = Slave mode (external clock)
bit 6 TX9:  9-bit Transmit Enable
            1 = 9-bit transmission mode
            0 = 8-bit transmission mode
bit 5 TXEN: Transmit Enable
            1 = Transmit enabled
            0 = Transmit disabled
bit 4 SYNC: USART Mode Select
            1 = Synchronous mode
            0 = Asynchronous mode
bit 3       Unimplemented: Read as '0'
bit 2 BRGH: Baud Rate Speed Select
            Asynchronous mode
                1 = High speed
                0 = Low speed
            Synchronous mode
                Unused
bit 1 TRMT: Transmit Shift Register Status
            1 = TSR empty
            0 = TSR full
bit 0 TX9D: 9th bit of transmit data
            (Can be used as parity bit)
```

Figure 14-14 Bitmap of the TXSTA Register

The formula for computing the baud rate takes into account the *system oscillator speed* (Fosc), the setting of the *Baud-Rate-Speed-Select* bit (BRGH), which is set for the high-speed mode and cleared for slow-speed, and also the setting of the SYNC bit in TXSTA register, which selects either asynchronous or synchronous mode. The formula is as follows:

$$ABR = \frac{Fosc}{S(x+1)}$$

where *ABR* represent the Asynchronous Baud Rate, x is the value in the SPRGB register (range 0 to 255), *S* is 64 in the high-speed mode (BRGH bit is 1) and 16 in the slow speed mode (BRGH bit is 0). Solving the formula in terms of the value to be placed in the SPRGB register we get:

bit 7							bit 0
SPEN	RX9	SREN	CREN		FERR	OERR	RX9D

```
bit 7 SPEN:  Serial Port Enable
             1 = Serial port enabled
                 (Configures RX/DT and TX/CK pins
                  as serial pins)
             0 = Serial port disabled
bit 6 RX9:   9-bit Receive Enable
             1 = 9-bit reception
             0 = 8-bit reception
bit 5 SREN:  Single Receive Enable
             Asynchronous mode
                 Don't care
             Synchronous master mode
                 1 = Enables single receive
                 0 = Disables single receive
             Synchronous slave mode
                 Unused in this mode
bit 4 CREN:  Continuous Receive Enable
             Asynchronous mode
                 1 = Enables continuous receive
                 0 = Disables continuous receive
             Synchronous mode
                 1 = Enables continuous receive until CREN
                     bit is cleared
                 0 = Disables continuous receive
bit 3        Unimplemented: Read as '0'
bit 2 FERR:  Framing Error bit
             1 = Framing error
             0 = No framing error
bit 1 OERR:  Overrun Error bit
             1 = Overrun error (cleared by CREN bit)
             0 = No overrun error
bit 0 RX9D:  9th bit of received data
             (can be used for parity bit)
```

Figure 14-15 Bitmap of the RCSTA Register

$$ABR = \frac{Fosc}{S(x+1)}$$

For example, to calculate the setting of the SPRGB register for 9,600 baud, with a 16Mhz oscillator, at the high-speed rate ($S = 64$) the equation becomes:

$$x = \left(\frac{16{,}000{,}000}{9{,}600 \cdot 64}\right) - 1 = 25.042 \approx 25$$

Communications

In this case, the value to store in the SPRGB register is 25. The actual baud rate can now be calculated using the first equation, as follows:

$$ABR = \frac{16{,}000{,}000}{64 \cdot (25+1)} = 9615.38$$

The percent error in the baud rate can be estimated by dividing the difference between the desired and the actual baud rate by the desired baud rate. The percent error is 0.16.

16F87x USART Asynchronous Transmitter

The USART in the 16F87x PICs uses a *non-return-to-zero format*, consisting of one start bit, eight or nine data bits, no parity, and one stop bit. In compliance with RS-232-C the USART transmits and receives the least significant bit first. Transmitter and receiver units are functionally independent but use the same data format and baud rate.

Although parity is not directly supported by the hardware, it can be implemented in software by using the ninth data bit. Figure 14-16 shows the 16F87x registers related to asynchronous transmission.

REGISTER NAME	7	6	5	4	3	2	1	0	bits
TXSTA		TX9	TXEN	SYNC		BRGH	TRMT	TX9D	
RCSTA	SPEN								
TXREG	TX7	TX6	TX5	TX4	TX3	TX2	TX1	TX0	
PIR1				TXIF					
PIE1				TXIE					
SPBRG	(Baud Rate Generator)								
INTCON	GIE	PEIE							

Figure 14-16 16F87x Registers used in Asynchronous Transmission

The transmitter function also uses the *Transmit Shift register* (TSR), which is not mapped in memory and is thus not accessible to code. TSR obtains its data from the read/*write transmit buffer*, named *TXREG*, which is loaded in software after the stop bit is received. Then TXREG transfers the data to TSR and becomes empty. At this time the TXIF flag bit is set. An interrupt related to the TXIF bit is enabled/disabled by setting/clearing the TXIE enable bit in the PIE1 register. However, the TXIF flag bit is set regardless of the state of the TXIE enable bit. The TXIF flag is reset automatically when new data is loaded into TXREG.

While the TXIF flag indicates the status of TXREG, the TRMT bit, in TXSTA, reflects the status of TSR. TRMT is set when TSR is empty. This is a read-only bit. No interrupts are linked to the TRMT bit, so the program has to poll this bit to determine if TSR is empty. Transmission is enabled by setting the TXEN bit in TXSTA. The actual transmission does not occur until TXREG is loaded with data and the *baud rate generator* (BRG) has produced a clock beat. Alternatively, transmission can be started by loading TXREG and then setting the TXEN enable bit.

When transmission starts, the (not accessible) TSR register usally is empty. Thereafter, transferring data to TXREG results in a transfer to TSR, which then produces an empty TXREG. This mechanism makes possible the *back-to-back transfer*. Clearing the TXEN enable bit during transmission aborts the transmission. This action also resets the transmitter and sets the TX/CK pin high.

16F87x USART Asynchronous Receiver

When Asynchronous mode is selected by setting the SYNC bit in TXSTA, then reception can be enabled by setting the CREN bit

In the RCSTA register. Figure 14-17 shows the registers related to asynchronous reception.

REGISTER NAME	7	6	5	4	3	2	1	0	bits
TXSTA				SYNC		BRGH			
RCSTA	SPEN	RX9		CREN		FERR	OERR	RX9D	
RCREG	RX7	RX6	RX5	RX4	RX3	RX2	RX1	RX0	
PIR1				RCIF					
PIE1				RCIE					
SPBRG	(Baud Rate Generator)							
INTCON	GIE	PEIE							

Figure 14-17 Registers used in Asynchronous Reception

The main operational register is the *RSR (Receive Shift Register)*, which, like TSR, is not accessible to application software. As soon as the stop bit is detected in the RX/TX pin, the received data in RSR is transferred to RCREG if it is empty. In this case, the RCIF flag bit is set. The interrupt linked to the RCIF flag is enabled or disabled by means of the RCIE in the PIE1 register. The RCIF flag bit is read-only and can be cleared only by hardware; this happens when the RCREG register has been read and is empty.

RCREG is double-buffered, meaning that it is possible for two bytes of data to be started simultaneously while a third byte begins shifting to RSR. If the stop bit is detected while RCREG is not empty, then the *overrun error bit* (OERR) is set in RCSTA. RCREG operates in *first-in-first-out order*. When it is read twice the two bytes are retrieved in this order.

The *overrun error bit* (OERR) inhibits transfer from RSR into RCREG; therefore, it is important to clear this bit once the error is detected. The *framing error bit* (FERR) in the RCSTA register is set if a stop bit is not detected.

The following steps are followed in initializing and executing asynchronous reception:

1. The SPBRG register is set up for the selected baud rate.
2. Asynchronous reception is enabled by clearing the SYNC bit in the TXSTA register and setting the SPEN bit in the RCSTA register.
3. To enable the receive data interrupt, the RCIE, GIE, and PEIE bits must be set.
4. Reception is activated by setting the CREN bit in RCSTA.
5. When reception has concluded, the RCIF bit in the PIE1 register is set. At that time, an interrupt is generated if the RCIE bit was set.
6. Received data is retrieved by reading RCREG.
7. If any error occurred the CREN bit must be cleared.

PIC-to-PC RS-232-C Communications Circuit

To demonstrate serial communications with the RS-232-C protocol we developed a circuit consisting of a 4-by-4 keypad and a 2-line by 20-character LCD display. Characters typed on the keypad are converted to ASCII codes for the hexadecimal digit set, that is, the numeral digits and the letters A through F. When a key is pressed, the corresponding ASCII code is displayed in the LCD and transmitted through the serial port to a PC application. Characters received through the serial line are displayed on the LCD. Figure 14-18 (in the following page) is a wiring diagram of the circuit.

The program SerComLCD demonstrates the circuit in Figure 14-18:

16F877 PIC Initialization Code

The following code fragment shows the initialization of the UART module in the 16F877 PIC for 2400 baud, 8 bits, no parity, and one stop bit. No interrupts are used in this example.

```
;===================================================================
;                USART initialization procedure
;===================================================================
; Initialize serial port for 2400 baud, 8 bits, no parity,
; 1 stop
InitSerial:
        Bank1                  ; Macro to select bank1
; Bits 6 and 7 of Port C are multiplexed as TX/CK and RX/DT
```

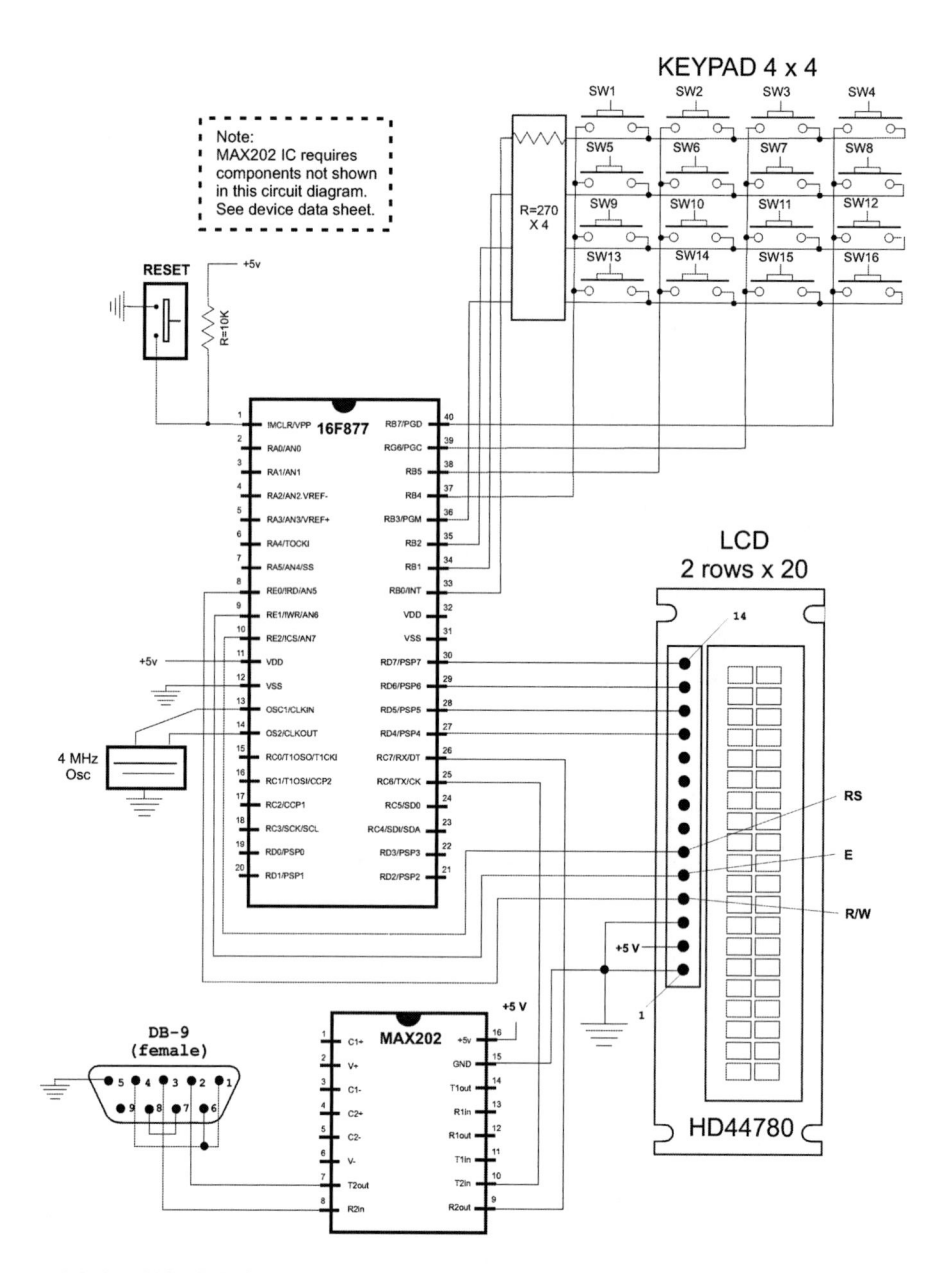

Figure 14-18 USART Communications Circuit with PIC 16F877

```
; for USART operation. These bits must be set to input in the
; TRISC register
        movlw   b'11000000'     ; Bits for TX and RX
        iorwf   TRISC,f         ; OR into Trisc register
; The asynchronous baud rate is calculated as follows:
;                   Fosc
;           ABR = --------
;                  S*(x+1)
```

```
; Where x is the value in the SPBRG register and S is 64 if the
; high baud rate select bit (BRGH) in the TXSTA control register
; is clear, and 16 if the BRGH bit is set. For setting to 9600
; baud using a 4Mhs oscillator at a high-speed baud rate the
; formula is:
;           4,000,000       4,000,000
;           ----------      ---------   = 9,615 baud (0.16% error)
;           16*(25+1)          416
;
; At slow speed (BRGH = 0)
;           4,000,000       4,000,000
;           ---------       ---------   = 2,403.85 (0.16% error)
;           64*(25+1)         1,664
;
        movlw   spbrgVal        ; Value in spbrgVal = 25
        movwf   SPBRG           ; Place in baud rate generator
;
; TXSTA (Transmit Status and Control Register) bit map:
;   7  6  5  4  3  2  1  0   <== bits
;   |  |  |  |  |  |  |  |_____ TX9D 9nth data bit on
;   |  |  |  |  |  |  |              ? (used for parity)
;   |  |  |  |  |  |  |_____ TRMT Transmit Shift Register
;   |  |  |  |  |  |                 1 = TSR empty
;   |  |  |  |  |  |               * 0 = TSR full
;   |  |  |  |  |  |_____ BRGH High Speed Baud Rate
;   |  |  |  |  |                    (Asynchronous mode only)
;   |  |  |  |  |                    1 = high speed (* 4)
;   |  |  |  |  |                  * 0 = low speed
;   |  |  |  |  |_____ NOT USED
;   |  |  |  |_____ SYNC USART Mode Select
;   |  |  |                          1 = syncrhonous mode
;   |  |  |                        * 0 = asynchronous mode
;   |  |  |_____ TXEN Transmit Enable
;   |  |                           * 1 = transmit enabled
;   |  |                             0 = transmit disabled
;   |  |_____ TX9 Enable 9-bit Transmit
;   |                                1 = 9-bit transmission mode
;   |                              * 0 = 8-bit mode
;   |_____ CSRC Clock Source Select
;                                    Not used in asynchronous mode
;                                    Synchronous mode:
;                                      1 = Master Mode (internal clock)
;                                    * 0 = Slave mode (external clock)
; Setup value: 0010 0000 = 0x20
        movlw   0x20            ; Enable transmission and high
baud rate
        movwf   TXSTA
        Bank0                   ; Bank 0
```

```
;    RCSTA (Receive Status and Control Register) bit map:
;       7   6   5   4   3   2   1   0   <== bits
;       |   |   |   |   |   |   |   |_____ RX9D 9th data bit received
;       |   |   |   |   |   |   |           ? (can be parity bit)
;       |   |   |   |   |   |   |_____ OERR Overrun errror
;       |   |   |   |   |   |               ? 1 = error (cleared by software)
;       |   |   |   |   |   |_____ FERR Framing Error
;       |   |   |   |   |                   ? 1 = error
;       |   |   |   |   |_____ NOT USED
;       |   |   |   |_____ CREN Continuous Receive Enable
;       |   |   |                               Asynchronous mode:
;       |   |   |                           *   1 = Enable continuous receive
;       |   |   |                               0 = Disables continuous receive
;       |   |   |                               Synchronous mode:
;       |   |   |                               1 = Enables until CREN cleared
;       |   |   |                               0 = Disables continuous receive
;       |   |   |_____ SREN Single Receive Enable
;       |   |                               ? Asynchronous mode = don't care
;       |   |                                   Synchronous master mode:
;       |   |                                   1 = Enable single receive
;       |   |                                   0 = Disable single receive
;       |   |_____ RX9 9th-bit Receive Enable
;       |                                   1 = 9-bit reception
;       |                                 * 0 = 8-bit reception
;       |_____ SPEN Serial Port Enable
;                                         * 1 = RX/DT and TX/CK are serial pins
;
;                                           0 = Serial port disabled
; Setup value: 1001 0000 = 0x90
            movlw    0x90        ; Enable serial port and continuous
                                 ; reception
            movwf    RCSTA
;
            clrf     errorFlags  ; Clear local error flags register
            Return
```

USART Receive and Transmit Routines

The transmit data routine is quite simple. Code checks the TXIF bit in PIR1. If the bit is set, data is transmitted by storing the data byte in TXREG. The following procedure performs the required operations.

```
;===============================
;       transmit data
;===============================
; Test for Transmit Register Empty and transmit data in w
SerialSend:
        Bank0                          ; Select bank 0
busyWait:
```

Communications

```
        btfss   PIR1,TXIF       ; check if transmitter busy
        goto    busyWait ; wait until transmitter is not busy
        movwf   TXREG           ; and transmit the data
        return
```

Receiving data is more complicated than transmitting it. One of the reasons is that code must test for and handle several possible errors that can occur during reception. The following code fragment shows the local variables and processing required for simple data reception.

```
;=========================================================
;                   variables in PIC RAM
;=========================================================
; Local variables
        cblock  0x20                    ; Start of block
            .
            .
            .
; Communications variables
        newData                 ; not 0 if new data received
        ascVal
        errorFlags
        endc


;=============================================================
;                   USART receive data procedure
;=============================================================
; Procedure to test line for data received and return value
; in w. Overrun and framing errors are detected and
; remembered in the variable errorFlags, as follows:
;       7  6  5  4  3  2  1  0    <== errorFlags
;       — not used —   |  |___ overrun error
;                      |_____ framing error
SerialRcv:
        clrf    newData ; Clear new data received register
        Bank0           ; Select bank 0
; Bit 5 (RCIF) of the PIR1 Register is clear if the USART
; receive buffer is empty. If so, no data has been received
        btfss   PIR1,RCIF       ; Check for received data
        return                  ; Exit if no data
; At this point data has been received. First eliminate
; possible errors: overrun and framing.
; Bit 1 (OERR) of the RCSTA register detects overrun
; Bit 2 (FERR( of the RCSTA register detects framing error
        btfsc   RCSTA,OERR      ; Test for overrun error
        goto    OverErr ; Error handler
        btfsc   RCSTA,FERR      ; Test for framing error
        goto    FrameErr ; Error handler
; At this point no error was detected
```

```
; Received data is in the USART RCREG register
        movf    RCREG,w    ; get received data
        bsf     newData,7  ; Set bit 7 to indicate new data
; Clear error flags
        clrf    errorFlags
        return
;===========================
;    error handlers
;===========================
; Overrun error detected
OverErr:
        bsf     errorFlags,0   ; Bit 0 is overrun error
; Reset system
errExit:
        bcf     RCSTA,CREN     ; Clear continuous receive bit
        bsf     RCSTA,CREN     ; Set to re-enable reception
        return
; Error. FERR framing error bit is set
FrameErr:
        bsf     errorFlags,1   ; Bit 1 is framing error
        movf    RCREG,W        ; Read and throw away bad data
        goto    errExit
```

The procedures listed previously are from the program SerComLCD in the book's online software. The applicable circuit is shown in Figure 14-18.

The USART Receive Interrupt

Polled routines for serial communications are adequate when the application does little else but check transmission lines. If the application has other tasks to perform, polled routines can waste processing time and even lose data. In this sense, the send function is usually less critical. An application can typically determine when to send data and have available all the data when the send operation activates. This is often not the case in receiving data, especially in applications that execute full-duplex.

A practical solution is to use interrupts for receiving characters through the serial line. The 60F87x includes facilities for implementing interrupt routines by both the send and the receive functions. To enable interrupts for the USART receive operation the following preparatory steps are necessary:

1. Peripheral and global interrupts must be enabled by setting bits 6 and 7 of the INTCON register.
2. The receive interrupt must be enabled by setting the RCIF bit in the PIE1 register.

The handler for the serial reception interrupt usually performs the following functions:

1. The context is saved. This includes, but is not limited to, the status register, the w register, the PCLATH register, and the FSR register.
2. Code tests for received data by checking the RCIF bit in the PIR1 register. If this bit is clear the interrupt did not originate in received data.
3. Code can also check if the *interrupt enable bit* (RCIE) is set in the RCIE register. If not enabled the interrupt is related to serial data.
4. The handler usually checks two possible errors during reception: overflow and framing error. The first one by checking the OERR bit and the second one by checking the FERR bit, both in the RCSTA register. If reception errors have taken place, the handler takes appropriate action.
5. If no error is detected then the received data can be retrieved from the RCREG.
6. On exit the interrupt handler restores the context and issues the retfie instruction.

The following code fragment lists the variables and processing routine for an interrupt handler for serial data reception:

```
;========================================================
;                variables in PIC RAM
;========================================================
; Local variables
        cblock    0x20              ; Start of block
            .
            .
            .
; Communications variables
        errorFlags
; Temporary storage used by interrupt handler
        tempW
        tempStatus
        tempPclath
        tempFsr
        endc


;================================================================
;================================================================
;           interrupt handler for received characters
;================================================================
;================================================================
IntServ:
        movwf     tempW       ; Save W
        movf      STATUS,W ; Store STATUS in W
        clrf      STATUS    ; Select bank0
        movwf     tempStatus          ; Save STATUS
        movf      PCLATH,W ; Store PCLATH in W
        movwf     tempPclath          ; Save PCLATH
        clrf      PCLATH    ; Select program memory page 0
        movf      FSR,W     ; Store FSR in W
```

```
                movwf     tempFsr     ; Save FSR value
; Test for received data interrupt
                Bank0                 ; select bank0
;   7  6  5  4  3  2  1  0  <= PIR1
;            |_____ (RCIF) USART receive interrupt
;                                     flag
                Btfsc     PIR1,RCIF   ; Test bit 5
                bsf       STATUS,RP0  ; Bank 1 if RCIF set
;   7  6  5  4  3  2  1  0  <= PIE1
;            |_____ (RCIE) Receive interrupt enable
;                                     bit
                btfss     PIE1,RCIE   ; Test if interrupt is enabled
                goto      IntExit ; Go if not enabled
;==============================
;      received data
;==============================
; Routine to handler received data. Overrun and framing
; errors are detected and remembered in the variable
; errorFlags, as follows:
;       7  6  5  4  3  2  1  0   <== errorFlags
;       — not used —  |  |___ overrun error
;                        |_____ framing error
                Bank0                 ; Select bank 0
; Test for overrun and framing errors.
; Bit 1 (OERR) of the RCSTA register detects overrun
; Bit 2 (FERR) of the RCSTA register detects framing error
                btfsc     RCSTA,OERR  ; Test for overrun error
                goto      OverErr ; Error handler
                btfsc     RCSTA,FERR  ; Test for framing error
                goto      FrameErr ; Error handler
; At this point no error was detected
; Received data is in the USART RCREG register
                movf      RCREG,w  ; Received data into w
; Clear error flags
                clrf      errorFlags
                goto      IntExit
;==========================
;      error handlers
;==========================
; Errors are returned as bits in the errorFlags register
;   7  6  5  4  3  2  1  0  <= errorFlags
;   -- not used --  |  |____ overrun error
;                      |_____ framing error
; Error responses to be made by main code
OverErr:
                bsf       errorFlags,0    ; Bit 0 is overrun error
; Reset system
                bcf       RCSTA,CREN      ; Clear continuous receive bit
```

Communications

```
            bsf     RCSTA,CREN      ; Set to re-enable reception
            goto    IntExit
FrameErr:
            bsf     errorFlags,1; Bit 1 is framing error
            movf    RCREG,W  ; Read and throw away bad data
;==============================
;    interrupt handler exit
;==============================
IntExit:
            Bank0
            movf    tempFsr,w       ; Recover FSR value
            movwf   FSR             ; Restore in register
            movf    tempPclath,w    ; Recover PCLATH value
            movwf   PCLATH          ; Restore in register
            movf    tempStatus,W    ; Recover STATUS
            movwf   STATUS          ; Restore in register
            swapf   tempW,F         ; Swap file register in itself
            swapf   tempW,W         ; Restore in register
            retfie
```

The program SerIntLCD in the book's online software is an interrupt-driven demonstration for the circuit in Figure 14-18.

14.5 Sample Programs

The sample programs listed in the following sections refer to the programming discussed in this chapter.

14.5.1 SerialSnd Program

```
; File name: SerialSnd.asm
; Date: May 5, 2006
; Author: Julio Sanchez
; Processor: 16F84A
;
; Description:
; Two programs to exercise serial communications between
; two PIC 16F84A both running at 4Mhs. One program sends
; data through a single line and the other one receives
; it. This program is the sender.
;
; Circuit:
;       Port A1 is the serial transmission line
;       Port A2 is an active-low pushbutton switch that
;               serves to initiate communications.
;       Port A3 is a LED that is ON when the program is
;               ready to send data. Once data starts
;               being sent the LED is turned OFF.
;       Port-B0-B7 is a 8 x toggle switch that provides
```

```
;                   the data byte to be sent
;           A pushbutton switch is in the 16F84 RESET line
;                   and serves to restart the program
;
; Communications parameters:
;           Timer channel TMR0 is used for synchronizing data
;           transmission. The timer runs at the maximum rate of
;           256 cycles per iteration. In a 4Mhz system the
;           timer rate is 1Mhz, thus the bit rate is
;                           1,000,000/256
;           which is approximately 3,906 microseconds per bit.
;
;==========================
;         switches
;==========================
; Switches used in __config directive:
;     _CP_ON          Code protection ON/OFF
; *   _CP_OFF
; *   _PWRTE_ON       Power-up timer ON/OFF
;     _PWRTE_OFF
;     _WDT_ON         Watchdog timer ON/OFF
; *   _WDT_OFF
;     _LP_OSC         Low power crystal occilator
; *   _XT_OSC         External parallel resonator/crystal oscillator

;     _HS_OSC         High speed crystal resonator (8 to 10 MHz)
;                     Resonator: Murate Erie CSA8.00MG = 8 MHz
;     _RC_OSC         Resistor/capacitor oscillator
; |                   (simplest, 20% error)
; |
; |_____ * indicates setup values presently selected

;==========================
; setup and configuration
;==========================
        processor 16f84A
        include    <p16f84A.inc>
        __config  _XT_OSC & _WDT_OFF & _PWRTE_ON & _CP_OFF
;=============================================================
;                        M A C R O S
;=============================================================
; Macros to select the register banks
Bank0   MACRO                       ; Select RAM bank 0
        bcf     STATUS,RP0
        ENDM

Bank1   MACRO                       ; Select RAM bank 1
        bsf     STATUS,RP0
```

Communications

```
        ENDM
;==========================================================
;           constant definitions for pin wiring
;==========================================================
#define readySW   2           ; |
#define readyLED  3           ; | — from wiring diagram
#define serialLN  1           ; |
;==========================================================
;              PIC register flag equates
;==========================================================
c         equ     0           ; Carry flag
tmrOVF    equ     2           ; Timer overflow bit
;==========================================================
;              variables in PIC RAM
;==========================================================
        cblock  0x0d    ; Start of block
        bitCount ; Counter for 8 bits
        dataReg         ; Data to send
        endc


;==========================================================
;                       program
;==========================================================
        org     0          ; start at address
        goto    main
; Space for interrupt handlers
        org     0x04

main:
; Port-A, bit 2 is input. Rest is output
        Bank1
        movlw   b'00000100'         ; Port-A bit 2 is input
                                    ; all others are output
        movwf   TRISA
; Port-B is all input
        movlw   b'11111111'
        movwf   TRISB
        Bank0
        bsf     PORTA,1             ;Marking bit
; Prepare to set prescaler
        clrf    TMR0
        clrwdt
; Setup OPTION register for full timer speed
        movlw   b'11011000'
;    1  1  0  1  1  0  0  0 <= OPTION bits
;    |  |  |  |  |  |__|__|____ PS2-PS0 (prescaler bits)
;    |  |  |  |  |              Values for Timer0
;    |  |  |  |  |                *000 = 1:2    001 = 1:4
```

```
;       |   |   |   |   |                010 = 1:8    011 = 1:16
;       |   |   |   |   |                100 = 1:32   101 = 1:64
;       |   |   |   |   |                110 = 1:128  111 = 1:256
;       |   |   |   |   |_____ PSA (prescaler assign)
;       |   |   |   |                    *1 = to WDT
;       |   |   |   |                    0  = to Timer0
;       |   |   |   |_____  TOSE (Timer0 edge select)
;       |   |   |                        0  = increment on low-to-high
;       |   |   |                        *1 = increment in high-to-low
;       |   |   |_____   TOCS (TMR0 clock source)
;       |   |                            *0 = internal clock
;       |   |                            1  = RA4/TOCKI bit source
;       |   |_____    INTEDG (Edge select)
;       |                                0  = falling edge
;       |                                *1 = raising edge
;       |_____  RBPU pullups
;                                        0  = enabled
;                                        *1 = disabled
        option
; Disable interrupts
        bcf     INTCON,5          ; Timer0 overflow disabled
        bcf     INTCON,7          ; Global interupts disabled
; Turn on ready LED
        bsf     PORTA,3           ; LED on
;===========================
;    wait for READY switch
;        to be pressed
;===========================
ready2send:
        btfsc   PORTA,readySW
        goto    ready2send
;===========================
;     send serial data
;===========================
; At this point program proceeds to send data through
; the serial port line
; Turn off LED
        bcf     PORTA,readyLED
; Read switches and store in local variable
        movf    PORTB,w
        movwf   dataReg
;===========================
;    call serial output
;         procedure
;===========================
        call    sendData ; call serial output procedure
;===========================
;        wait forever
```

```
;===========================
endloop:
        goto    endloop

;================================================================
;                procedure to send serial data
;================================================================
; ON ENTRY:
;         local variable dataReg holds 8-bit value to be
;         transmitted through port labeled serialLN
; OPERATION:
;       1. The timer at register TMR0 is set to run at
;          maximum clock speed, that is, 256 clock beats.
;          The timer overflow flag in the INTCON register
;          is set when the timer cycles from 0xff to 0x00.
;       2. Each bit (start, data, and stop bits) is sent
;          at a rate of 256 timer beats. That is, each bit is
;          held high or low for one full timer cycle (256
;          clock beats).
;       3. The procedure tests the timer overflow flag
;          (tmrOVF) to determine when the timer cycle has
;          ended, that is when 256 clock beats have passed.
;
sendData:
        movlw   0x08            ; Setup shift counter
        movwf   bitCount
;========================
;    send START bit
;========================
; Set line low then hold for 256 timer clock beats.
        bcf     PORTA,serialLN  ; Send start bit
; First reset timer
        clrf    TMR0            ; Reset timer counter
        bcf     INTCON,tmrOVF   ; Reset TMR0 overflow flag
; Wait for 256 timer clock beats
startBit:
        btfss   INTCON,tmrOVF   ; timer overflow?
        goto    startBit ; Wait until set
; At this point timer has cycled. Start bit has ended
        bcf     INTCON,tmrOVF   ; Clear overflow flag
;========================
;    send 8 DATA bits
;========================
; Eight data bits are sent through the serial line
; starting with the high-order bit. The data byte is
; stored in the register named dataReg. The bits are
; rotated left to the carry flag. Code assumes the bit
; is zero and sets the serial line low. Then the carry
```

```
; flag is tested. If the carry is set the serial line
; is changed to high. The line is kept low or high for
; 256 timer beats.
send8:
        rlf     dataReg,f          ; bit into carry flag
        bcf     PORTA,serialLN ; 0 to serial line
; Code can assume the bit is a zero and set the line
; low since, if low is the wrong state, it will only
; remain for two timer beats. The receiver will not
; check the line for data until 128 timer beats have
; elapsed, so the error will be harmless. In any case,
; there is no assurance that the previous line state is
; the correct one, so leaving the line in its previous
; state could also be wrong.
        btfsc   STATUS,c           ; test carry flag
        bsf     PORTA,serialLN     ; bit is set. Fix error.
bitWait:
        btfss   INTCON,tmrOVF      ; Timer cycled?
        goto    bitWait            ; not yet
; At this point timer has cycled.
; Test for end of byte, if not, send next bit
        bcf     INTCON,tmrOVF      ; clear overflow flag
        decfsz  bitCount,f         ; Last bit?
        goto    send8              ; not yet
;=========================
;    hold MARKING state
;=========================
; All 8 data bits have been sent. The serial line must
; now be held high (MARKING) for one clock cycle
        bsf             PORTA,serialLN    ; Marking state
markWait:
        btfss   INTCON,tmrOVF      ; Done?
        goto    markWait ; not yet
;=========================
;    end of transmission
;=========================
        return                     ; done


;===========================================================
;                        end of program
;===========================================================
        end
```

14.5.2 SerialRcv Program

```
; File name: SerialRcv.asm
; Date: May 6, 2006
; Author: Julio Sanchez
```

```
; Processor: 16F84A
;
; Description:
; Two programs to exercise serial communications between
; two PIC 16F84A both running at 4Mhs. One program sends
; data through a single line and the other one receives
; it. This program is the receiver.
;
; Circuit:
;        Port A0 is the serial transmission line
;        Port A2 is an active-low pushbutton switch that
;               serves to initiate communications.
;        Port A3 is a LED that is ON when the program is
;               ready to receive data. Once data starts
;               being received the LED is turned OFF.
;        Port-B0-B7 are 8 LEDs that display the data bits
;               that have been received.
;        A pushbutton switch is in the 16F84 RESET line
;               and serves to restart the program
;
; Communications parameters:
;        Timer channel TMR0 is used for synchronizing data
;        transmission. The timer runs at the maximum rate of
;        256 cycles per iteration. In a 4Mhz system the
;        timer rate is 1Mhz, thus the bit rate is
;                    1,000,000/256
;        which is approximately 3,906 microseconds per bit.
;
;        Upon receiving the START bit, the program waits for
;        one half a clock cycle (128 timer beats) to
;        synchronize with the sender.
;===========================
;          switches
;===========================
; Switches used in __config directive:
;    _CP_ON         Code protection ON/OFF
; *  _CP_OFF
; *  _PWRTE_ON      Power-up timer ON/OFF
;    _PWRTE_OFF
;    _WDT_ON        Watchdog timer ON/OFF
; *  _WDT_OFF
;    _LP_OSC        Low power crystal occilator
; *  _XT_OSC        External parallel resonator/crystal oscillator

;    _HS_OSC        High speed crystal resonator (8 to 10 MHz)
;                   Resonator: Murate Erie CSA8.00MG = 8 MHz
;    _RC_OSC        Resistor/capacitor oscillator
; |                 (simplest, 20% error)
```

```
;   |
;   |_____  * indicates setup values presently selected

;==========================
; setup and configuration
;==========================
        processor 16f84A
        include   <p16f84A.inc>
        __config  _XT_OSC & _WDT_OFF & _PWRTE_ON & _CP_OFF
;================================================================
;                       M A C R O S
;================================================================
; Macros to select the register banks
Bank0   MACRO                   ; Select RAM bank 0
        bcf       STATUS,RP0
        ENDM

Bank1   MACRO                   ; Select RAM bank 1
        bsf       STATUS,RP0
        ENDM
;================================================================
;       constant definitions for pin wiring
;================================================================
#define readySW  2       ;|
#define readyLED 3       ;| - from wiring diagram
#define serialLN 0       ;|
;================================================================
;              PIC register and flag equates
;================================================================
c         equ     0       ; Carry flag
tmrOVF    equ     2       ; Timer overflow bit
;
;================================================================
;              variables in PIC RAM
;================================================================
        cblock    0x0c    ; Start of block
        bitCount          ; Counter for 8 bits
        rcvReg            ; Data to send
        temp
        endc
;================================================================
;                       program
;================================================================
        org       0       ; start at address
        goto      main
; Space for interrupt handlers
        org       0x04
```

Communications

```
main:
          Bank1
; Port-A bits 0 and 2 are input. All others are output
          movlw     b'00000101'  ; Port-A setup
          movwf     TRISA
; Port-B is all output
          movlw     b'00000000'  ; Port-B setup
          MOVWF     TRISB
          Bank0
; Turn off all Port-B LEDs
          clrf      PORTB
; And receiver register
          clrf      rcvReg
; Prepare to set prescaler
          clrf      TMR0
          clrwdt
; Setup OPTION register for full timer speed
          movlw     b'11011000'
;   1  1  0  1  1  0  0  0  <= OPTION bits
;   |  |  |  |  |  |__|__|_____ PS2-PS0 (prescaler bits)
;   |  |  |  |  |              Values for Timer0
;   |  |  |  |  |              *000 = 1:2    001 = 1:4
;   |  |  |  |  |               010 = 1:8    011 = 1:16
;   |  |  |  |  |               100 = 1:32   101 = 1:64
;   |  |  |  |  |               110 = 1:128  111 = 1:256
;   |  |  |  |  |_____ PSA (prescaler assign)
;   |  |  |  |                 *1 = to WDT
;   |  |  |  |                  0 = to Timer0
;   |  |  |  |_____ TOSE (Timer0 edge select)
;   |  |  |                     0 = increment on low-to-high
;   |  |  |                    *1 = increment in high-to-low
;   |  |  |_____ TOCS (TMR0 clock source)
;   |  |                       *0 = internal clock
;   |  |                        1 = RA4/TOCKI bit source
;   |  |_____ INTEDG (Edge select)
;   |                           0 = falling edge
;   |                          *1 = raising edge
;   |_____ RBPU pullups
;                               0 = enabled
;                              *1 = disabled
          option
; Disable interrupts
          bcf       INTCON,5   ; Timer0 overflow disabled
          bcf       INTCON,7   ; Global interrupts disabled
;===========================
;  wait for READY switch
;      to be pressed
;===========================
```

```
ready2rcv:
         btfsc    PORTA,readySW    ; Test switch
         goto     ready2rcv        ; loop
; Turn ON the ready-to-receive LED
         bsf      PORTA,readyLED
;===========================
;       receiving
;===========================
         call     rcvData          ; Call serial input procedure
;===========================
;       data received
;===========================
; Turn ready to receive LED off
         bcf      PORTA,readyLED
; Display received data
         movf     rcvReg,w         ; Byte received to w
         movwf    PORTB            ; display in Port-B
;===========================
;       wait forever
;===========================
endloop:
         goto     endloop
;================================================================
;                procedure to receive serial data
;================================================================
; ON ENTRY:
;         local variable dataReg is used to store 8-bit value
;         received through port (labeled serialLN)
; OPERATION:
;       1. The timer at register TMR0 is set to run at
;          maximum clock speed, that is, 256 clock beats.
;          The timer overflow flag in the INTCON register
;          is set when the timer cycles from 0xff to 0x00.
;       2. When the START signal is received, the code
;          waits for 128 timer beats so as to read data in
;          the middle of the send period.
;       3. Each bit (start, data, and stop bits) is read
;          at intervals of 256 timer beats.
;       4. The procedure tests the timer overflow flag
;          (tmrOVF) to determine when the timer cycle has
;          ended, that is when 256 clock beats have passed.

;================================================================
rcvData:
         clrf     TMR0             ; Reset timer
         movlw    0x08             ; Initialize bit counter
         movwf    bitCount
;===========================
```

Communications

```
;    wait for START bit
;=========================
startWait:
        btfsc   PORTA,0             ; Is port A0 low?
        goto    startWait           ; No. Wait for mark
;=========================
;   offset 128 clock beats
;=========================
; At this point the receiver has found the falling
; edge of the start bit. It must now wait 128 timer
; beats to synchronize in the middle of the sender's
; data rate, as follows:
;              |<========= falling edge of START bit
;              |
;              |-----|<====== 128 clock beats offset
;    ---------  |         .---------
;              |         |  <== SIGNAL
;                  -----------
;              |<-- 256--->|
;
        movlw   0x80                ; 128 clock beats offset
        movwf   TMR0                ; to TMR0 counter
        bcf     INTCON,tmrOVF       ; Clear overflow flag
offsetWait:
        btfss   INTCON,tmrOVF       ; timer overflow?
        goto    offsetWait          ; Wait until
        btfsc   PORTA,0             ; Test start bit for error
        goto    offsetWait          ; Recycle if a false start
;=========================
;        receive data
;=========================
        clrf    TMR0                ; Restart timer
        bcf              INTCON,tmrOVF   ; Clear overflow flag
; Wait for 256 timer cycles for first/next data bit
bitWait:
        btfss   INTCON,tmrOVF       ; Timer cycle end?
        goto    bitWait             ; Keep waiting
; Timer has counter 256 beats
        bcf     INTCON,tmrOVF       ; Reset overflow flag
        movf    PORTA,w             ; Read Port-A into w
        movwf   temp                ; Store value read
        rrf     temp,f              ; Rotate bit 0 into carry flag
        rlf     rcvReg,f            ; Rotate carry into rcvReg 0
        decfsz  bitCount,f          ; 8 bits received
        goto    bitWait             ; Next bit
; Wait for one time cycle at end of reception
markWait:
        btfss   INTCON,tmrOVF       ; timer overflow flag
```

```
                goto    markWait ; keep waiting
;========================
;   end of reception
;========================
        return


;============================================================
;                      end of program
;============================================================
        end
```

14.5.3 Serial6465 Program

```
; File name: Serial6465.asm
; Last update: May 7, 2006
; Author: Julio Sanchez
; Processor: 16F84A
;
; Description:
; Program to exercise serial communications using a
; PIC 16F84A and two shift registers: a 74HC164, and a
; 74HC165. The 74HC165 inputs 8 lines from a DIP switch
; and transmits settings to PIC through a serial line.
; PIC sends data serially to an 74HC164 which is wired
; to 8 LEDs that display the received data. A total of
; 6 PIC lines are used in interfacing 8 input switches
; to 8 output LEDs.
; Circuit:
;          * Port A0 is the serial transmission line which
;            comes from the 74HC165.
;        * Port A1 is wired to the 74HC164 CLOCK pin
;        * Port A2 is wired to the 74HC164 CLEAR pin
;        * 74HC164 output pins 0 to 7 are wired to LEDs.
;        * Port B0 is wired to the 74HC165 Hout line
;        * Port B1 is wired to the 74HC165 CLK line
;        * Port B2 is wired to the 74HC165 load line
;        * A pushbutton switch is in the 16F84 RESET line
;          and serves to restart the program
; Communications protocol:
;          Communication between PIC and the 74HC164 and
;          74HC165 is synchronous since the shift registers
;          clock lines serve to shift in and out the data
;          bits.
;
;===========================
;        switches
;===========================
; Switches used in __config directive:
;   _CP_ON        Code protection ON/OFF
```

```
;   *   _CP_OFF
;   *   _PWRTE_ON        Power-up timer ON/OFF
;       _PWRTE_OFF
;       _WDT_ON          Watchdog timer ON/OFF
;   *   _WDT_OFF
;       _LP_OSC          Low power crystal occilator
;   *   _XT_OSC          External parallel resonator/crystal oscillator
;
;       _HS_OSC          High speed crystal resonator (8 to 10 MHz)
;                        Resonator: Murate Erie CSA8.00MG = 8 MHz
;       _RC_OSC          Resistor/capacitor oscillator
;   |                    (simplest, 20% error)
;   |
;   |_____ * indicates setup values presently selected

;=========================
; setup and configuration
;=========================
        processor 16f84A
        include    <p16f84A.inc>
        __config   _XT_OSC & _WDT_OFF & _PWRTE_ON & _CP_OFF


;=============================================================
;                       M A C R O S
;=============================================================
; Macros to select the register banks
Bank0   MACRO                   ; Select RAM bank 0
        bcf                     STATUS,RP0
        ENDM

Bank1   MACRO                   ; Select RAM bank 1
        bsf                     STATUS,RP0
        ENDM
; Note: in the case of the 16F84A the bank select macros
;       do not make the code more efficient, however, they
;       do serve to clarify the bank selection operations.
;========================================================
;       constant definitions from wiring diagram
;========================================================
#define clockLN 1               ; |
#define clearLN 2               ; | - 74HC164 lines
#define dataLN 0  ; |
;
#define clk65LN 1               ; | - 74HC165 lines
#define loadLN 2                ; |
;========================================================
;               PIC register and flag equates
;========================================================
```

```
highBit  equ         7              ; High order bit
;==========================================================
;                variables in PIC RAM
;==========================================================
        cblock   0x0d      ; Start of block
        bitCount ; Counter for 8 bits
        dataReg           ; Data to send
        workReg           ; Work register for bit shifts
        endc
;==========================================================
;                        program
;==========================================================
        org      0         ; start at address
        goto     main
; Space for interrupt handlers
        org      0x04

main:
; Port-A is all output
        Bank1
        movlw    b'00000000'
        movwf    TRISA
; Port-B line 0 is input, all others are output
        movlw    b'00000001'
        movwf    TRISB
        Bank0
; Make sure Port-A line 2 (clear line) is high
        movlw    b'00000100'
        movwf    PORTA
;========================
; read input from 165 IC
;========================
        call     in165         ; Local procedure
; dataReg contains input
;==========================
;    call serial output
;        procedure
;==========================
        call     out164        ; Call serial output procedure
;==========================
;       wait forever
;==========================
endloop:
        goto     endloop
;==============================================================
;           74HC164 procedure to send serial data
;==============================================================
; ON ENTRY:
```

```
;                 local variable dataReg holds 8-bit value to be
;                 transmitted through port labeled serialLN
; OPERATION:
;           1. A local counter (bitCount) is initialized to
;              8 bits
;           2. Code assumes that the first bit is zero by
;              setting the data line low. Then the high-order
;              bit in the data register (dataReg) is tested.
;              If set, the data line is changed to high.
;           3. Bits are shifted in by pulsing the 74HC164
;              clock line (CLK).
;           4. Data bits are then shifted left and the bit
;              counter is tested. If all 8 bits have been sent
;              the procedure returns.
out164:
; Clear 74HC164 shift register
          bcf        PORTA,clearLN     ; 74HC164 CLR clear low
          bsf        PORTA,clearLN     ; then high again
; Init counter
          movlw      0x08              ; Initialize bit counter
          movwf      bitCount
sendBit:
          bcf        PORTA,dataLN      ; Set data line low (assume)
; Using this assumption is possible because the bit is not
; shifted in until the clock line is pulsed.
          btfsc      dataReg,highBit   ; test number bit 7
          bsf        PORTA,dataLN      ; Change assumption if set
;=========================
;    pulse clock line
;=========================
; Bits are shifted in by pulsing the 74HC164 CLK line
          bsf        PORTA,clockLN     ; CLK high
          bcf        PORTA,clockLN     ; CLK low
;=========================
; Rotate data bits left
;=========================
          rlf        dataReg,f         ; Shift left data bits
          decfsz     bitCount,f        ; Decrement bit counter
          goto       sendBit           ; Repeat if not 8 bits
;=========================
;    end of transmission
;=========================
          return


;================================================================
;       74HC165 procedure to read parallel data and send
;                      serially to PIC
;================================================================
```

```
;   OPERATION:
;           1.  Eight DIP switches are connected to the input
;               ports of an 74HC165 IC. Its output line Hout,
;               and its control lines CLK and load are connected
;               to the PIC's Port-B lines 0, 1, and 2
;               respectively
;           2.  Procedure sets a counter (bitCount) for 8
;               iterations and clears a data holding register
;               (dataReg).
;           3.  Port-B bits are read into w. Only the lsb of
;               Port-B is relevant. Value is stored in a working
;               register and the meaningful bit is rotated into
;               the carry flag, then the carry flag bit is
;               then shifted into the data register.
;           4.  The iteration counter is decremented. If this
;               is the last iteration the routine ends. Otherwise
;               the bitwise read-and-write operation is repeated.

in165:
        clrf    dataReg             ; Clear data register
        movlw   0x08                ; Initialize counter
        movwf   bitCount
        bcf     PORTB,loadLN        ; Reset shift register
        bsf     PORTB,loadLN
nextBit:
        movf    PORTB,w  ; Read Port-B (only LOB is
                         ; meaningful in this routine)
        movwf   workReg             ; Store value in local
                                    ; register
        rrf     workReg,f           ; Rotate LOB bit into carry
                                    ; flag
        rlf     dataReg,f           ; Carry flag into dataReg
        decfsz  bitCount,f          ; Decrement bit counter
        goto    shiftBits           ; Continue if not zero
        return                      ; done
shiftBits:
        bsf     PORTB,clk65LN       ; Pulse clock
        bcf     PORTB,clk65LN
        goto    nextBit             ; Continue
;===========================================================
;                        end of program
;===========================================================
        end
```

14.5.4 TTYUsart Program

```
; File name: TTYUsart.asm
; Last update: May, 2006
```

```
; Author: Julio Sanchez
; Processor: 16F84A
;
; Description:
; Program to emulate USART operation in PIC code. Uses
; PIC-to-LCD interface. Display has 2 lines, each with
; 16 characters.
; Program operation:
; Characters received from the RS232 line are displayed on
; the LCD. LCD lines scroll automatically. A pushbutton
; activates the send operation by transmitting the text
; string: Ready- which is also displayed on the LCD.
;
; Program communications and LCD parameters are stored in
; #define statements. These statements can be edited to
; accommodate a different set-up. Program uses delay loops
; for interface timing.
;
; WARNING:
; Code assumes 4Mhz clock. Delay routines must be
; edited for faster clock
;
; BAUD RATE CALCULATIONS:
; A 4Mhz clock oscillator has a clock frequency of 1 Mhz:
; Since the baud rate is the number of clock cycles per
; second, for a 4Mhz clock it is:
;                    1
; bit time = ———  sec. = 208.33 microseconds
;                  4,800
; Calculating one half the baud rate allows resetting the
; clock from the edge to the center of a time pulse:
;
;          |<======== falling edge of start bit
;          |          |<======== center of bit time
;       >|          |< one-half baud rate
;          |          |
;_____.          |          ._____.
;          |_____|          |_____
;              208/2 = 104
; The PIC clock counts up from 0 to 255. So to implement
; a 104 microsecond delay we must start counting at
; clock beat:
;                255 - 104 = 151
; plus one microsecond for movlw instruction used to
; initialize the clock:
;                151 + 1 = 152
; For one full baud rate delay:
;                255 - 208 = 47 + 1 = 48
```

```
; The following two constants are stored in #define
; statements:
;                  halfBaud = 152
;                  fullBaud = 48
; Setting the prescaler to TMR0 reduces the baud rate
; to one-half. Other prescaler values will reduce the
; baud rate accordingly.
;
; Wiring diagram:
;    RB4-RB7 ===> LCD data lines 4 to 7 (output)
;    RB0    ======> MAX202 T2in line (output)
;    RA0    ======> MAX202 R2out line (input)
;    RA1    ======> LCD E line (output)
;    RA2    ======> LCD RS line (output)
;    RA3    ======> LCD R/W line (output - not used)
;    RA4    ======> Pushbutton switch 1
;                  (input - active low)
;
;==========================
;         switches
;==========================
; Switches used in __config directive:
;    _CP_ON         Code protection ON/OFF
; *  _CP_OFF
; *  _PWRTE_ON      Power-up timer ON/OFF
;    _PWRTE_OFF
;    _WDT_ON        Watchdog timer ON/OFF
; *  _WDT_OFF
;    _LP_OSC        Low power crystal oscillator
; *  _XT_OSC        External parallel resonator/crystal oscillator

;    _HS_OSC        High speed crystal resonator (8 to 10 MHz)
;                   Resonator: Murate Erie CSA8.00MG = 8 MHz
;    _RC_OSC        Resistor/capacitor oscillator
;    |              (simplest, 20% error)
;    |
;    |_____ * indicates setup values presently selected

;==========================
; setup and configuration
;==========================
        processor 16f84A
        include   <p16f84A.inc>
        __config  _XT_OSC & _WDT_OFF & _PWRTE_ON & _CP_OFF

;================================================================
;                    M A C R O S
;================================================================
```

```
; Macros to select the register banks
Bank0     MACRO             ; Select RAM bank 0
          bcf     STATUS,RP0
          ENDM

Bank1     MACRO             ; Select RAM bank 1
          bsf     STATUS,RP0
          ENDM
;================================================================
;                   constant definitions
;       for PIC-to-LCD pin wiring and LCD line addresses
;================================================================
#define E_line 1            ; |
#define RS_line 2           ; | — from wiring diagram
#define RW_line 3           ; |
; LCD line addresses (from LCD data sheet)
#define LCD_1 0x80          ; First LCD line constant
#define LCD_2 0xc0          ; Second LCD line constant
#define LCDlimit .16; Number of characters per line
; 4800 baud clock countdown values
; Code reduces rate to 2400 baud by entering a minimal
; presclaer to TRM0
#define halfBaud .152       ; For one-half bit time
#define fullBaud .48        ; For one full bit time
;
; Note: The constants that define the LCD display line
;       addresses have the high-order bit set in
;       order to facilitate the controller command
;
;=========================================================
;              buffer and variables in PIC RAM
;=========================================================
; Create a 16-byte storage area
          cblock   0x0c     ; Start of first data block
          lineBuf           ; buffer for text storage
          endc
; Leave 16 bytes and Continue with local variables
          cblock   0x1c     ; Second data block
          count1            ; Counter # 1
          count2            ; Counter # 2
          J                 ; counter J
          K                 ; counter K
          store1            ; Local temporary storage
          store2            ; Storage # 2
; For LCDscroll procedure
          LCDcount ; Counter for characters per line
          LCDline           ; Current display line (0 or 1)
          bufPtr            ; Buffer pointer
```

```
; Variables for serial communications
         tempData ; Temporary storage for bit manipulations
         rcvData           ; Final storage for received character
         bitCount ; Bit counter
         sendData ; Character to send
         endc


;============================================================
;              m a i n   p r o g r a m
;============================================================
         org            0              ; start at address
         goto    main
; Space for interrupt handlers
         org            0x08
main:
         Bank1
         movlw   b'00010001'     ; Port-A lines I/O setup
                                 ; RA0 = RS232 input (R2out)
                                 ; RA4 = Pushbutton SW # 1
         movwf   TRISA
         movlw   b'00000000' ; Port-B lines as follow:
;      RB4-RB7 ===> LCD data lines 4 to 7 (output)
;      RB0 =======> MAX202 T2in line (output)
         movwf   TRISB
         Bank0
; Clear bits in Port-A output lines
         bcf     PORTA,1
         bcf     PORTA,2
         bcf     PORTA,3
         movlw   b'00000000'     ; All outputs ports low
         movwf   PORTB
; Wait and initialize HD44780
         call    delay_5         ; Allow LCD time to initialize
                                 ; itself
         call    delay_5
         call    initLCD         ; Then do forced
initialization
         call    delay_5         ; Wait again
; Set Port-B, line 0 high so start bit is detected
         bsf     PORTB,0
;=============================
;   wait for start command
;=============================
; Program waits until pushbutton number 1 is pressed
; to continue execution. Pushbutton 1 is active low
; and wired to RA4
pb1Wait:
         btfsc   PORTA,4         ; Test Port-A, line 4
```

Communications

```
            goto    pb1Wait             ; Loop if not clear
;=============================
;  display and send "Ready-"
;=============================
; Set LCD base address
            call    line1
; Initialize system for UART emulation at 2400 baud
            call    initTTY
; Display on LCD and test serial transmission by sending
; the string "Ready-"
            movlw   'R'
            movwf   sendData ; Store in send register
            call    send8               ; Local LCD display procedure
            call    sendTTY             ; Local send procedure
            movlw   'e'
            movwf   sendData ; Store in send register
            call    send8               ; Local LCD display procedure
            call    sendTTY             ; Local send procedure
            movlw   'a'
            movwf   sendData ; Store in send register
            call    send8               ; Local LCD display procedure
            call    sendTTY             ; Local send procedure
            movlw   'd'
            movwf   sendData ; Store in send register
            call    send8               ; Local LCD display procedure
            call    sendTTY             ; Local send procedure
            movlw   'y'
            movwf   sendData ; Store in send register
            call    send8               ; Local LCD display procedure
            call    sendTTY             ; Local send procedure
            movlw   '-'
            movwf   sendData ; Store in send register
            call    send8               ; Local LCD display procedure
            call    sendTTY             ; Local send procedure
; Init  character counter and line counter variables for
; LCD line scroll procedure
            movlw   0x06                ; 6 characters already displayed
            movwf   LCDcount
            clrf    LCDline             ; LCD line counter
;=============================
;     monitor RS232 line
;=============================
nextChar:
            call    rcvTTY              ; Receive character
; Store character in local line buffer using indirect
; addressing
; 16-byte buffer named lineBuf starts at address 0x0c
```

```
; Register variable bufPtr holds offset into buffer
        movlw   0x0c            ; Buffer base address
        addwf   bufPtr,w        ; Add pointer in w
        movwf   FSR             ; Value to index register
        movf    rcvData,w       ; Character into w
        movwf   INDF            ; Store w in [FSR]
        incf    bufPtr,f        ; Bump pointer
; Send character (still in w)
        call    send8           ; Display it
        call    LCDscroll       ; Scroll display lines
        goto    nextChar        ; Continue

;================================================================
;                 initialize LCD for 4-bit mode
;================================================================
initLCD:
; Initialization for Densitron LCD module as follows:
;       4-bit interface
;   2 display lines of 16 characters each
;   cursor on
;   left-to-right increment
;   cursor shift right
;   no display shift
;=====================|
;   set command mode  |
;=====================|
        bcf     PORTA,E_line    ; E line low
        bcf     PORTA,RS_line   ; RS line low
        bcf     PORTA,RW_line   ; Write mode
        call    delay_125       ; delay 125 microseconds
;**********************|
;      FUNCTION SET    |
;**********************|
        movlw   0x28    ; 0 0 1 0 1 0 0 0 (FUNCTION SET)
                        ;       | | | |__ font select:
                        ;       | | |     1 = 5x10 in 1/8 or 1/11
                        ;       | | |     0 = 1/16 dc
                        ;       | | |___ Duty cycle select
                        ;       | |       0 = 1/8 or 1/11
                        ;       | |       1 = 1/16
                        ;       | |___ Interface width
                        ;       |       0 = 4 bits
                        ;       |       1 = 8 bits
                        ;       |___ FUNCTION SET COMMAND
        call    send8   ; 4-bit send routine

; Set 4-bit mode command must be repeated
        movlw   0x28
```

```
              call    send8
;**********************|
; DISPLAY AND CURSOR ON |
;**********************|
        movlw   0x0e    ; 0 0 0 0 1 1 1 0 (DISPLAY ON/OFF)
                        ;         | | | |___ Blink character
                        ;         | | |        1 = on, 0 = off
                        ;         | | |___ Cursor on/off
                        ;         | |        1 = on, 0 = off
                        ;         | |____ Display on/off
                        ;         |         1 = on, 0 = off
                        ;         |____ COMMAND BIT
        call    send8
;**********************|
;   set entry mode     |
;**********************|
        movlw   0x06    ; 0 0 0 0 0 1 1 0 (ENTRY MODE SET)
                        ;           | | |___ display shift
                        ;           | |       1 = shift
                        ;           | |       0 = no shift
                        ;           | |____ increment mode
                        ;           |        1 = left-to-right
                        ;           |        0 = right-to-left
                        ;           |____ COMMAND BIT
        call    send8

;**********************|
; cursor/display shift |
;**********************|
        movlw   0x14    ; 0 0 0 1 0 1 0 0 (CURSOR/DISPLAY
                        ;       | | | | |   SHIFT)
                        ;       | | | |_|___ don't care
                        ;       | |_|__ cursor/display shift
                        ;       |         00 = cursor shift left
                        ;       |         01 = cursor shift right
                        ;       |         10 = cursor and display
                        ;       |              shifted left
                        ;       |         11 = cursor and display
                        ;       |              shifted right
                        ;       |____ COMMAND BIT
        call    send8
;**********************|
;    clear display     |
;**********************|
        movlw   0x01    ; 0 0 0 0 0 0 0 1 (CLEAR DISPLAY)
                        ;                 |___ COMMAND BIT
        call    send8
```

```
; Per documentation
        call    delay_5         ; Test for busy
        return


;======================
;   Procedure to delay
;    42 microseconds
;======================
delay_125:
        movlw   D'42'           ; Repeat 42 machine cycles
        movwf   count1          ; Store value in counter
repeat:
        decfsz  count1,f        ; Decrement counter
        goto    repeat          ; Continue if not 0
        return                  ; End of delay


;======================
;   Procedure to delay
;    5 milliseconds
;======================
delay_5:
        movlw   D'41'           ; Counter = 41
        movwf   count2          ; Store in variable
delay:
        call    delay_125       ; Delay
        decfsz  count2,f        ; 40 times = 5 milliseconds
        goto    delay
        return                  ; End of delay
;======================
;       pulse E line
;======================
pulseE
        bsf     PORTA,E_line    ; Pulse E line
        nop
        bcf     PORTA,E_line
        return


;============================
;   long delay sub-routine
;       (for debugging)
;============================
long_delay
                movlw   D'200'  ; w = 200 decimal
                movwf   J       ; J = w
jloop:
        movwf   K               ; K = w
kloop:
        decfsz  K,f             ; K = K-1, skip next if zero
```

```
                        goto     kloop
                        decfsz   J,f           ; J = J-1, skip next if zero
                        goto     jloop
                        return
;========================
;   send 2 nibbles in
;      4-bit mode
;========================
; Procedure to send two 4-bit values to Port-B lines
; 7, 6, 5, and 4. High-order nibble is sent first
; ON ENTRY:
;         w register holds 8-bit value to send
send8:
            movwf    store1            ; Save original value
            call     merge4            ; Merge with Port-B
; Now w has merged byte
            movwf    PORTB             ; w to Port-B
            call     pulseE            ; Send data to LCD
; High nibble is sent
            movf     store1,w ; Recover byte into w
            swapf    store1,w ; Swap nibbles in w
            call     merge4
            movwf    PORTB
            call     pulseE            ; Send data to LCD
            call     delay_125
            return
;==================
;   merge bits
;==================
; Routine to merge the 4 high-order bits of the
; value to send with the contents of Port-B
; so as to preserve the 4 low-bits in Port-B
; Logic:
;       AND value with 1111 0000 mask
;       AND Port-B with 0000 1111 mask
;       Now low nibble in value and high nibble in
;       PortB are all 0 bits:
;                 value = vvvv 0000
;       PortB = 0000 bbbb
;       OR value and Port-B resulting in:
;                 vvvv bbbb
; ON ENTRY:
;       w contains value bits
; ON EXIT:
;       w contains merged bits
merge4:
            andlw    b'11110000'       ; ANDing with 0 clears the
                                       ; bit. ANDing with 1 preserves
```

```
                    movwf   store2          ; the original value
                                            ; Save result in variable
                    movf    PORTB,w         ; Port-B to w register
                    andlw   b'00001111'     ; Clear high nibble in Port-B
                                            ; and preserve low nibble
                    iorwf   store2,w ; OR two operands in w
                    return
;========================
; Set address register
;     to LCD line 1
;========================
; ON ENTRY:
;           Address of LCD line 1 in constant LCD_1
line1:
            bcf     PORTA,E_line    ; E line low
            bcf     PORTA,RS_line   ; RS line low, set up for
                                    ; control
            call    delay_5         ; busy?
; Set to second display line
            movlw   LCD_1           ; Address and command bit
            call    send8           ; 4-bit routine
; Set RS line for data
            bsf     PORTA,RS_line   ; Setup for data
            call    delay_5         ; Busy?
; Clear buffer and pointer
            call    blankBuf
            clrf    bufPtr          ; Clear
            return
;========================
; Set address register
;     to LCD line 2
;========================
; ON ENTRY:
;           Address of LCD line 2 in constant LCD_2
line2:
            bcf     PORTA,E_line    ; E line low
            bcf     PORTA,RS_line   ; RS line low, setup for
control
            call    delay_5         ; Busy?
; Set to second display line
            movlw   LCD_2           ; Address with high-bit set
            call    send8
; Set RS line for data
            bsf     PORTA,RS_line   ; RS = 1 for data
            call    delay_5         ; Busy?
; Clear buffer and pointer
            call    blankBuf
            clrf    bufPtr
```

Communications

```
        return

;===========================
;   scroll LCD line 2
;===========================
; Procedure to count the number of characters displayed on
; each LCD line. If the number reaches the value in the
; constant LCDlimit, then display is scrolled to the second
; LCD line. If at the end of the second line, then the
; second line is scrolled to the first line and display
; continues at the start of the second line
; reset to the first line.
LCDscroll:
        incf    LCDcount,f              ; Bump counter
; Test for line limit
        movf    LCDcount,w
        sublw   LCDlimit                ; Count minus limit
        btfss   STATUS,Z                ; Is count - limit = 0
        goto    scrollExit              ; Go if not at end of line
; At this point the end of the LCD line was reached
; Test if this is also the end of the second line
        movf    LCDline,w
        sublw   0x01                    ; Is it line 1?
        btfsc   STATUS,Z                ; Is LCDline minus 1 = 0?
        goto    line2End                ; Go if end of second line
; At this point it is the end of the top LCD line
        call    line2                   ; Scroll to second line
        clrf    LCDcount                ; Reset counter
        incf    LCDline,f               ; Bump line counter
        goto    scrollExit
; End of second LCD line
line2End:
; Scroll second line to first line. Characters to be
; scrolled are stored in buffer starting at address 0x0c.
; 16 characters are to be moved
; First clear LCD
        call    initLCD
        call    delay_5                 ; Make sure not busy
; Set up for data
        bcf     PORTA,E_line            ; E line low
        bsf     PORTA,RS_line           ; RS line high for data
; Set up counter for 16 characters
        movlw   D'16'                   ; Counter = 16
        movwf   count2
; Get address of storage buffer
        movlw   0x0c
        movwf   FSR                     ; W to FSR
getchar:
```

```
                movf    INDF,w      ; get character from display RAM
                                    ; location pointed to by file select
                                    ; register
                call    send8       ; 4-bit interface routine
; Test for 16 characters displayed
                decfsz  count2,f            ; Decrement counter
                goto    nextchar    ; Skipped if done
; At this point scroll operation has concluded
                clrf    LCDcount    ; Clear counters
; Stay at line 2
                clrf    LCDline
                incf    LCDline,f
                call    line2       ; Set for second line
scrollExit:
                return
nextchar:
                incf    FSR,f           ; Bump pointer
                goto    getchar

;=============================
;    clear line buffer
;=============================
; Use indirect addressing to store 16 blanks in the
; buffer located at 0x0c
blankBuf:
                Bank0
                movlw   0x0c        ; Pointer to RAM
                movwf   FSR         ; To index register
blank16:
                clrf    INDF        ; Clear memory pointed at by FSR
                incf    FSR,f       ; Bump pointer
                btfss   FSR,4       ; 000x0000 when bit 4 is set
                                    ; count reached 16
                goto    blank16
                return


;================================================================
;                       initialize for TTY
;================================================================
; Procedure to initialize RS232 reception
; Assumes:
;                       2400 baud
;                       8 data bits
;                       no parity
;                       one stop bit
initTTY:
; First initialize receiver to RS-232-C line parameters
; Disable global and peripheral interrupts
```

```
;       7  6  5  4  3  2  1  0   <= INTCON bitmap
;       |  ?  |  ?  ?  ?  ?  ?   (? = unrelated bits)
;       |     |_____  Timer0 interrupt on overflow
;       |_____ Global interrupts
            bcf     INTCON,5         ; Disable TMR0 interrupts
            bcf     INTCON,7         ; Disable global interrupts
            clrf    TMR0             ; Reset timer
            clrwdt                   ; Clear WDT for prescaler
assign
            Bank1
; Set up the OPTION register bit map
;       7  6  5  4  3  2  1  0  <= OPTION bits
;       1  1  0  1  1  0  0  0  <= setup
;       |  |  |  |  |  |__|__|_____ PS2-PS0 (prescaler bits)
;       |  |  |  |  |               Values for Timer0
;       |  |  |  |  |               *000 = 1:2    001 = 1:4
;       |  |  |  |  |               010 = 1:8    011 = 1:16
;       |  |  |  |  |               100 = 1:32   101 = 1:64
;       |  |  |  |  |               110 = 1:128  111 = 1:256
;       |  |  |  |  |_____ PSA (prescaler assign)
;       |  |  |  |                  1 = to WDT
;       |  |  |  |                  *0 = to Timer0
;       |  |  |  |_____ TOSE (Timer0 edge select)
;       |  |  |                      0 = increment on low-to-high
;       |  |  |                      *1 = increment in high-to-low
;       |  |  |_____ TOCS (TMR0 clock source)
;       |  |                         *0 = internal clock
;       |  |                         1 = RA4/TOCKI bit source
;       |  |_____ INTEDG (Edge select)
;       |                            0 = falling edge
;       |                            *1 = raising edge
;       |_____ RBPU (Pullup enable)
;                                    0 = enabled
;                                    *1 = disabled
            movlw   b'11010000'      ; set up timer/counter
            movwf   OPTION_REG
            Bank0
            return
;============================================================
;                    receive character
;============================================================
; Receive a single character through the serial port.
; Assumes: 4800 baud, 8 data bits, no parity, 1 stop bit.
; Receiving line is Port-A, line 0
rcvTTY:
            movlw   0x08             ; Counter for 8 bits
            movwf   bitCount
; The start of character transmission is signaled by
```

```
; the sender by setting the line low
startBit:
        btfsc    PORTA,0            ; Test for low on line
        goto     startBit ; Go if not low
;=========================
;   offset to data bit
;=========================
; At this point the receiver has found the falling
; edge of the start bit. It must now wait one and
; one-half the baud rate to synchronize in the center
; of the sender's first data bit, as follows:
;      |<========= falling edge of START bit
;      |      |<========== center of start bit
;      |      |            |<====== center of data bit
;      |-----|-----|
;_____                _____              _____
;      |      |       |            |              <== SIGNAL
;      -----------             ----------
;      |<-- 208 -->|h| <====== ms. for 4800 baud
;
; Clock start count for one-half bit  = 255 - 104 = 151
; Clock start count for one full bit  =  255 - 208 = 47
; One clock cycle is added for the movwf instruction:
;     clkHalf = 152 (for one-half bit countdown)
;     clkFull = 48 (for one full bit countdown)
        movlw    halfBaud           ; Skip one-half bit
        movwf    TMR0      ; Initialize tmr0 and start count
        bcf      INTCON,2           ; Clear overflow flag
;=============================
;          start bit
;=============================
wait1:
        btfss    INTCON,2           ; Timer count overflow?
        goto     wait1              ; No, keep waiting
; At this point we are at the center of the start bit
        btfsc    PORTA,0            ; Check to see it is still low
        goto     startBit ; No, it is high. False start
; At this point the clock is at the center of the start
; bit. The first data bit must be read one full baud
; period later
        movlw    fullBaud           ; One full bit delay
        movwf    TMR0               ; Start timer
        bcf      INTCON,2           ; clear tmr0 overflow flag
wait2:
        btfss    INTCON,2           ; End of one full baud period?
        goto     wait2              ; Wait if not end of period
; Timer is now at the center of the first/next data bit
; Timer must be reset immediately so that code will not
```

Communications

```
; lose synchronization with sender
        movlw   fullBaud ; Skip to next data bit
        movwf   TMR0             ; Restart timer
        bcf     INTCON,2         ; Reset overflow flag
; Now the data bit can be read and and stored
        movf    PORTA,w          ; Read Port-B
        movwf   tempData ; Store in temporary variable
        rrf     tempData,f       ; Rotate bit 0 into carry flag
        rrf     rcvData,f        ; Rotate carry flag into
                                 ; storage register high-order
                                 ; bit
        decfsz  bitCount,f       ; End of data?
        goto    wait2    ; Continue until 8 bits received
;===========================
;            stop bit
;===========================
stopWait:
        btfss   INTCON,2         ; Test time
        goto    stopWait ; Wait
        return                   ; Exit


;=============================================================
;                        send character
;=============================================================
; Procedure to send one character through the RS232 line.
; Assumes: 2400 baud, 8 data bits, no parity, one stop bit
; Sending line is Port-B, line 0
; ON ENTRY:
;         variable sendData holds character to send
sendTTY:
        movlw   0x08             ; Init bit counter
        movwf   bitCount
        bcf     PORTB,0          ; Low for start bit
        movlw   fullBaud ; For one baud space
        movwf   TMR0             ; Start timer
        bcf     INTCON,2         ; Clear timer flag
start2snd:
        btfss   INTCON,2         ; Full baud done?
        goto    start2snd        ; No
        movlw   fullBaud ; Reset for one full bit period
        movwf   TMR0             ; Start timer
        bcf     INTCON,2         ; Clear flag
; At this point the start bit has been sent
; Data follows
sendOut:
        rrf     sendData,f       ; Rotate bit into carry
        bcf     PORTB,0          ; Assume data bit is 0
        btfsc   STATUS,c         ; Test if carry set
```

```
                bsf       PORTB,0   ; Change bit to 1 if clear
; Hold bit for 1 baud period
timeBit:
                btfss     INTCON,2            ; Wait for baud period to end
                goto      timeBit             ; Loop if not yet
                movlw     fullBaud            ; Reset timer
                movwf     TMR0                ; Start timer
                bcf       INTCON,2            ; Clear flag
; Test for last bit
                decfsz    bitCount,f          ; Count this bit
                goto      sendOut             ; Continue if not last bit
; Done. Send stop bit
                bsf       PORTB,0             ; High for stop bit
stopBit:
                btfss     INTCON,2            ; Timer done?
                goto      stopBit             ; No
; Set Port-B line 0 high back again
                bsf                 PORTB,0
                call      delay_5             ; And hold
                return

                End
```

14.5.5 SerComLCD Program

```
; File name: SerComLCD.asm
; Last revision: May 14, 2006
; Author: Julio Sanchez
; Processor: 16F877
;
; Description:
; Decode 4 x 4 keypad, display scan code in LCD, and send
; ASCII character through the serial port. Also receive
; data through serial port and display on LCD. LCD lines
; are scrolled by program.
; Default serial line setting:
;             2400 baud
;             no parity
;             1 stop bit
;             8 character bits
;
; Program uses 4-bit PIC-to-LCD interface.
; Code assumes that LCD is driven by Hitachi HD44780
; controller and PIC 16F977. Display supports two lines
; each one with 20 characters. The length, wiring and base
; address of each display line is stored in #define
; statements. These statements can be edited to accommodate
; a different set-up.
```

```
; Keypad switch wiring (values are scan codes):
;       -- KEYPAD --
;       0   1   2   3   <= port B0 |
;       4   5   6   7   <= port B1 |-- ROWS = OUTPUTS
;       8   9   A   B   <= port B2 |
;       C   D   E   F   <= port B3 |
;       |   |   |   |
;       |   |   |   |_____ port B4 |
;       |   |   |_____ port B5 |-- COLUMNS = INPUTS
;       |   |_____ port B6 |
;       |_____ port B7 |
;
; Operations:
; 1. Key press action generates a scan code in the range
;    0x0 to 0xf.
; 2. Scan code is converted to an ASCII digit and displayed
;    on the LCD. LCD lines are scrolled as end-of-line is
;    reached.
; 3. Characters typed on the keypad are also transmitted
;    through the serial port.
; 4. Serial port is polled for received characters. These
;    are displayed on the LCD.
;
; WARNING:
; Code assumes 4Mhz clock. Delay routines must be
; edited for faster clock. Clock speed also determines
; values for baud rate setting (see spbrgVal constant).
;
;===========================
;      16F877 switches
;===========================
; Switches used in __config directive:
;    _CP_ON           Code protection ON/OFF
; *  _CP_OFF
; *  _PWRTE_ON        Power-up timer ON/OFF
;    _PWRTE_OFF
;    _BODEN_ON        Brown-out reset enable ON/OFF
; *  _BODEN_OFF
; *  _PWRTE_ON        Power-up timer enable ON/OFF
;    _PWRTE_OFF
;    _WDT_ON          Watchdog timer ON/OFF
; *  _WDT_OFF
;    _LPV_ON          Low voltage IC programming enable ON/OFF
; *  _LPV_OFF
;    _CPD_ON          Data EE memory code protection ON/OFF
; *  _CPD_OFF
; OSCILLATOR CONFIGURATIONS:
;    _LP_OSC          Low power crystal oscillator
```

```
;       _XT_OSC         External parallel resonator/crystal oscillator
; *     _HS_OSC         High speed crystal resonator
;       _RC_OSC         Resistor/capacitor oscillator
; |                     (simplest, 20% error)
; |
; |_____ * indicates setup values presently selected

        processor       16f877          ; Define processor
        #include <p16f877.inc>
        __CONFIG _CP_OFF & _WDT_OFF & _BODEN_OFF & _PWRTE_ON &
_HS_OSC & _WDT_OFF & _LVP_OFF & _CPD_OFF

; __CONFIG directive is used to embed configuration data
; within the source file. The labels following the directive
; are located in the corresponding .inc file.
;==============================================================
;                       M A C R O S
;==============================================================
; Macros to select the register banks
Bank0   MACRO           ; Select RAM bank 0
        bcf     STATUS,RP0
        bcf     STATUS,RP1
        ENDM

Bank1   MACRO           ; Select RAM bank 1
        bsf     STATUS,RP0
        bcf     STATUS,RP1
        ENDM

Bank2   MACRO           ; Select RAM bank 2
        bcf     STATUS,RP0
        bsf     STATUS,RP1
        ENDM

Bank3   MACRO           ; Select RAM bank 3
        bsf     STATUS,RP0
        bsf     STATUS,RP1
        ENDM
;========================================================
;               constant definitions
;   for PIC-to-LCD pin wiring and LCD line addresses
;========================================================
#define E_line 1                ;|
#define RS_line 0               ;| - from wiring diagram
#define RW_line 2               ;|
; LCD line addresses (from LCD data sheet)
#define LCD_1 0x80              ; First LCD line constant
#define LCD_2 0xc0              ; Second LCD line constant
```

```
#define LCDlimit  .20; Number of characters per line
#define spbrgVal  .25; For 2400 baud on 4Mhz clock
; Note: The constants that define the LCD display
;       line addresses have the high-order bit set
;       so as to meet the requirements of controller
;       commands.
;
;==========================================================
;               variables in PIC RAM
;==========================================================
; Local variables
        cblock  0x20            ; Start of block
        count1                  ; Counter # 1
        count2                  ; Counter # 2
        count3                  ; Counter # 3
        J                       ; counter J
        K                       ; counter K
        store1                  ; Local storage
        store2
; For LCDscroll procedure
        LCDcount ; Counter for characters per line
        LCDline                 ; Current display line (0 or 1)
; Keypad processing variables
        keyMask                 ; For keypad processing
        rowMask                 ; For masking-off key rows
        rowCode                 ; Row addend for calculating scan code
        rowCount ; Counter for key rows (0 to 3)
        scanCode ; Final key code
        newScan                 ; 0 if no new scan code detected
; Communications variables
        newData                 ; not 0 if new data received
        ascVal
        errorFlags
        endc


;==============================================================
;                       P R O G R A M
;==============================================================
                org             0       ; start at address
                goto    main
; Space for interrupt handlers
        org     0x08
main:
; Wiring:
;       LCD data to Port D, lines 0 to 7
;       E line -> port E, 1
;       RW line -> port E, 2
;       RS line -> port E, 0
```

```
; Set PORTE D and E for output
; Data memory bank selection bits:
; RP1:RP0           Bank
;  0:0               0        Ports A,B,C,D, and E
;  0:1               1        Tris A,B,C,D, and E
;  1:0               2
;  1:1               3
; First, initialize Port-B by clearing latches
        clrf    STATUS
        clrf    PORTB
; Select bank 1 to tris Port D for output
        bcf     STATUS,RP1        ; Clear banks 2/3 selector
        bsf     STATUS,RP0        ; Select bank 1 for tris
                                  ; registers
; Tris Port D for output. Port D lines 4 to 7 are wired
; to LCD data lines. Port D lines 0 to 4 are wired to LEDs.
        movlw   B'00000000'
        movwf   TRISD   ; and Port D
; By default Port-A lines are analog. To configure them
; as digital code must set bits 1 and 2 of the ADCON1
; register (in bank 1)
        movlw   0x06    ; binary 0000 0110 is code to
                        ; make all Port-A lines digital
        movwf   ADCON1
; Port-B, lines are wired to keypad switches, as follows:
;    7 6 5 4 3 2 1 0
;    | | | | |_|_|_|_____ switch rows (output)
;    |_|_|_|_____ switch columns (input)
; rows must be defined as output and columns as input
        movlw   b'11110000'
        movwf   TRISB
; Tris port E for output
        movlw   B'00000000'
        movwf   TRISE             ; Tris port E
; Enable Port-B pullups for switches in OPTION register
;    7 6 5 4 3 2 1 0 <= OPTION bits
;    | | | | | |_|_|_____ PS2-PS0 (prescaler bits)
;    | | | | |            Values for Timer0
;    | | | | |            000 = 1:2    001 = 1:4
;    | | | | |            010 = 1:8    011 = 1:16
;    | | | | |            100 = 1:32   101 = 1:64
;    | | | | |            110 = 1:128 *111 = 1:256
;    | | | | |_____ PSA (prescaler assign)
;    | | | |             *1 = to WDT
;    | | | |              0 = to Timer0
;    | | | |_____ TOSE (Timer0 edge select)
;    | | |               *0 = increment on low-to-high
;    | | |                1 = increment in high-to-low
```

```
;         |   |   |_____ T0CS (TMR0 clock source)
;         |   |                             *0 = internal clock
;         |   |                              1 = RA4/T0CKI bit source
;         |   |_____ INTEDG (Edge select)
;         |                                 *0 = falling edge
;         |_____ RBPU (Pullup enable)
;                                           *0 = enabled
;                                            1 = disabled
        movlw   b'00001000'
        movwf   OPTION_REG
; Back to bank 0
        bcf             STATUS,RP0
; Initialize serial port for 9600 baud, 8 bits, no parity
; 1 stop
        call    InitSerial
; Test serial transmission by sending "RDY-"
        movlw   'R'
        call    SerialSend
        movlw   'D'
        call    SerialSend
        movlw   'Y'
        call    SerialSend
        movlw   '-'
        call    SerialSend
        movlw   0x20
        call    SerialSend
; Clear all output lines
        movlw   b'00000000'
        movwf   PORTD
        movwf   PORTE
; Wait and initialize HD44780
        call    delay_5    ; Allow LCD time to initialize itself
        call    initLCD    ; Then do forced initialization
        call    delay_5    ; (Wait probably not necessary)
; Clear character counter and line counter variables
        clrf    LCDcount
        clrf    LCDline
; Set display address to start of second LCD line
        call    line1
;===============================================================
;                        scan keypad
;===============================================================
; Keypad switch wiring:
;         x   x   x   x   <= port B0 |
;         x   x   x   x   <= port B1 |-- ROWS = OUTPUTS
;         x   x   x   x   <= port B2 |
;         x   x   x   x   <= port B3 |
;         |   |   |   |
```

```
;              |    |    |    |_____ port B4 |
;              |    |    |_____ port B5  |-- COLUMNS = INPUTS
;              |    |_____ port B6  |
;              |_____ port B7  |
; Switches are connected to Port-B lines
; Clear scan code register
        clrf    scanCode
;===========================
;   scan keypad and display
;===========================
keyScan:
; Port-B, lines are wired to pushbutton switches, as follows:
;    7 6 5 4 3 2 1 0
;    | | | | |_|_|_|_____ switch rows (output)
;    |_|_|_|_____ switch columns (input)
; Keypad processing:
; switch rows are successively grounded (row = 0)
; Then column values are tested. If a column returns 0
; in a 0 row, that switch is down.
; Initialize row code addend
        clrf    rowCode         ; First row is code 0
        clrf    newScan         ; No new scan code detected
; Initialize row count
        movlw   D'4'            ; Four rows
        movwf   rowCount        ; Register variable
        movlw   b'11111110'     ; All set but LOB
        movwf   rowMask
keyLoop:
; Initialize row eliminator mask:
; The row mask is ANDed with the key mask to successively
; mask-off each row, for example:
;
;                    |----- row 3
;                    ||---- row 2
;                    |||--- row 1
;                    ||||-- row 0
;             0000 1111 <= key mask
;       AND   1111 1101 <= mask for row 1
;             ------
;             0000 1101 <= row 1 is masked off
;
; The row mask, which is initally 1111 1110, is rotated left
; through the carry in order to mask off the next row
        movlw   b'00001111'     ; Mask off all lines
        movwf   keyMask         ; To local register
; Set row mask for current row
        movf    rowMask,w       ; Mask to w
        andwf   keyMask,f       ; Update key mask
```

```
        movf    keyMask,w        ; Key mask to w
        movwf   PORTB            ; Mask-off Port-B lines
; Read Port-B lines 4 to 7 (columns are input)
        btfss   PORTB,4
        call    col0             ; Key column procedures
        btfss   PORTB,5
        call    col1
        btfss   PORTB,6
        call    col2
        btfss   PORTB,7
        call    col3
; Index to next row by adding 4 to row code
        movf    rowCode,w        ; Code to w
        addlw   D'4'
        movwf   rowCode
;=========================
;       shift row mask
;=========================
; Set the carry flag
        bsf     STATUS,C
        rlf     rowMask,f        ; Rotate mask bits in storage
;=========================
;       end of keypad?
;=========================
; Test for last key row (maximum count is 4)
        decfsz  rowCount,f       ; Decrement counter
        goto    keyLoop
;===========================================================
;===========================================================
;           display, send, and receive data
;===========================================================
;===========================================================
; At this point all keys have been tested.
; Variable newScan = 0 if no new scan code detected, else
; variable scanCode holds scan code
        movf    newScan,f        ; Copy onto intsef (sets Z
                                 ; flag)
        btfsc   STATUS,Z ; Is it zero
        goto    receive
; At this point a new scan code is detected
        movf    scanCode,w       ; To w
; If scan code is in the range 0 to 9, that is, a decimal
; digit, then ASCII conversion consists of adding 0x30.
; If the scan code represents one of the hex letters
; (0xa to 0xf) then ASCII conversion requires adding
; 0x37
        sublw   0x09             ; 9 - w
; if w from 0 to 9 then 9 - w = positive (C flag = 1)
```

```
; if w = 0xa then 9 - 10 = -1 (C flag = 0)
; if w = 0xc then 9 - 12 = -2 (C flag = 0)
          btfss     STATUS,C ; Test carry flag
          goto      hexLetter ; Carry clear, must be a letter
; At this point scan code is a decimal digit in the
; range 0 to 9. Convert to ASCII by adding 0x30
          movf      scanCode,w       ; Recover scan code
          addlw     0x30             ; Convert to ASCII
          goto      displayDig
hexLetter:
          movf      scanCode,w       ; Recover scan code
          addlw     0x37             ; Convert to ASCII
displayDig:
; Store so it can be sent
          movwf     ascVal
          call      send8            ; Display routine
          call      LCDscroll
          call      long_delay       ; Debounce
; Recover ASCII
          movf      ascVal,w
          call      SerialSend
          goto      scanExit
;==========================
;    receive serial data
;==========================
receive:
; Call serial receive procedure
          call      SerialRcv
; HOB of newData register is set if new data
; received
          btfss     newData,7
          goto      scanExit
; At this point new data was received
          call      send8            ; Display in LCD
          call      LCDscroll        ; Scroll at end of line
scanExit:
          goto      keyScan          ; Continue
;==========================
;    calculate scan code
;==========================
; The column position is added to the row code (stored
; in rowCode register). Sum is the scan code
col0:
          movf      rowCode,w        ; Row code to w
          addlw     0x00             ; Add 0 (clearly not
necessary)
          movwf     scanCode ; Final value
          incf      newScan,f                    ; New scan code
```

```
                return
col1:
        movf    rowCode,w       ; Row code to w
        addlw   0x01            ; Add 1
        movwf   scanCode
        incf    newScan,f
        return

col2:
        movf    rowCode,w       ; Row code to w
        addlw   0x02            ; Add 2
        movwf   scanCode
        incf    newScan,f
        return

col3:
        movf    rowCode,w       ; Row code to w
        addlw   0x03            ; Add 3
        movwf   scanCode
        incf    newScan,f
        return
;================================================================
;================================================================
;               L O C A L     P R O C E D U R E S
;================================================================
;================================================================
;==========================
; init LCD for 4-bit mode
;==========================
initLCD:
; Initialization for Densitron LCD module as follows:
;       4-bit interface
;   2 display lines of 16 characters each
;   cursor on
;   left-to-right increment
;   cursor shift right
;   no display shift
;=====================|
;   set command mode  |
;=====================|
        bcf     PORTE,E_line    ; E line low
        bcf     PORTE,RS_line   ; RS line low
        bcf     PORTE,RW_line   ; Write mode
        call    delay_125                       ; delay 125 microseconds
;*********************|
;     FUNCTION SET    |
```

```
;**********************|
         movlw    0x28      ; 0 0 1 0 1 0 0 0 (FUNCTION SET)
                            ;         | | | |__ font select:
                            ;         | | |     1 = 5x10 in 1/8 or 1/11
                            ;         | | |     0 = 1/16 dc
                            ;         | | |___ Duty cycle select
                            ;         | |      0 = 1/8 or 1/11
                            ;         | |      1 = 1/16
                            ;         | |___ Interface width
                            ;         |      0 = 4 bits
                            ;         |      1 = 8 bits
                            ;         |___ FUNCTION SET COMMAND
         call     send8     ; 4-bit send routine

; Set 4-bit mode command must be repeated
         movlw    0x28
         call     send8

;**********************|
; DISPLAY AND CURSOR ON |
;**********************|
         movlw    0x0e      ; 0 0 0 0 1 1 1 0 (DISPLAY ON/OFF)
                            ;         | | | |___ Blink character
                            ;         | | |      1 = on, 0 = off
                            ;         | | |___ Cursor on/off
                            ;         | |      1 = on, 0 = off
                            ;         | |____ Display on/off
                            ;         |       1 = on, 0 = off
                            ;         |____ COMMAND BIT
         call     send8
;**********************|
;   set entry mode      |
;**********************|
         movlw    0x06      ; 0 0 0 0 0 1 1 0 (ENTRY MODE SET)
                            ;           | | |___ display shift
                            ;           | |      1 = shift
                            ;           | |      0 = no shift
                            ;           | |____ increment mode
                            ;           |       1 = left-to-right
                            ;           |       0 = right-to-left
                            ;           |___ COMMAND BIT
         call     send8

;**********************|
; cursor/display shift  |
;**********************|
         movlw    0x14      ; 0 0 0 1 0 1 0 0 (CURSOR/DISPLAY
SHIFT)
```

```
                         ;              |   |   |   |_|___ don't care
                         ;              |   |_|__ cursor/display shift
                         ;              |           00 = cursor shift left
                         ;              |           01 = cursor shift right
                         ;              |           10 = cursor and display
                         ;              |                     shifted left
                         ;              |           11 = cursor and display
                         ;              |                     shifted right
                         ;              |___ COMMAND BIT
        call    send8
;*********************|
;   clear display     |
;*********************|
        movlw   0x01       ; 0 0 0 0 0 0 0 1 (CLEAR DISPLAY)
                           ;                 |___ COMMAND BIT
        call    send8
; Per documentation
        call    delay_5    ; Test for busy
        return

;======================
;   Procedure to delay
;     42 microseconds
;======================
delay_125:
        movlw   D'42'              ; Repeat 42 machine cycles
        movwf   count1             ; Store value in counter
repeat
        decfsz  count1,f           ; Decrement counter
        goto    repeat             ; Continue if not 0
        return                     ; End of delay

;======================
;   Procedure to delay
;     5 milliseconds
;======================
delay_5:
        movlw   D'42'              ; Counter = 41
        movwf   count2             ; Store in variable
delay
        call    delay_125          ; Delay
        decfsz  count2,f           ; 40 times = 5 milliseconds
        goto    delay
        return                     ; End of delay
;========================
;     pulse E line
;========================
pulseE
```

```
            bsf        PORTE,E_line     ; Pulse E line
            nop
            bcf        PORTE,E_line
            return

;=============================
;    long delay sub-routine
;=============================
long_delay
                       movlw      D'200'     ; w delay count
                       movwf      J          ; J = w
jloop:      movwf      K                     ; K = w
kloop:
            decfsz     K,f                   ; K = K-1, skip next if zero
            goto       kloop
            decfsz     J,f                   ; J = J-1, skip next if zero
            goto       jloop
            return

;========================
;    send 2 nibbles in
;      4-bit mode
;========================
; Procedure to send two 4-bit values to Port-B lines
; 7, 6, 5, and 4. High-order nibble is sent first
; ON ENTRY:
;          w register holds 8-bit value to send
send8:
            movwf      store1          ; Save original value
            call       merge4          ; Merge with Port-B
; Now w has merged byte
            movwf      PORTD
            call       pulseE          ; Send data to LCD
; High nibble is sent
            movf       store1,w ; Recover byte into w
            swapf      store1,w ; Swap nibbles in w
            call       merge4
            movwf      PORTD
            call       pulseE          ; Send data to LCD
            call       delay_125
            return
;==========================
;       merge bits
;==========================
; Routine to merge the 4 high-order bits of the
; value to send with the contents of Port-B
; so as to preserve the 4 low-bits in Port-B
; Logic:
```

```
;       AND value with 1111 0000 mask
;       AND Port-B with 0000 1111 mask
;       Now low nibble in value and high nibble in
;       Port-B are all 0 bits:
;           value  = vvvv 0000
;           Port-B = 0000 bbbb
;       OR value and Port-B resulting in:
;                  vvvv bbbb
; ON ENTRY:
;       w contain value bits
; ON EXIT:
;       w contains merged bits
merge4:
        andlw   b'11110000'     ; ANDing with 0 clears the
                                ; bit. ANDing with 1 preserves
                                ; the original value
        movwf   store2          ; Save result in variable
        movf    PORTD,w         ; Port-B to w register
        andlw   b'00001111'     ; Clear high nibble in Port-B
                                ; and preserve low nibble
        iorwf   store2,w ; OR two operands in w
        return

;===========================
;   Set address register
;       to LCD line 2
;===========================
; ON ENTRY:
;           Address of LCD line 2 in constant LCD_2
line2:
        bcf     PORTE,E_line    ; E line low
        bcf     PORTE,RS_line   ; RS line low, setup for
control
        call    delay_5         ; Busy?
; Set to second display line
        movlw   LCD_2           ; Address with high-bit set
        call    send8
; Set RS line for data
        bsf     PORTE,RS_line   ; RS = 1 for data
        call    delay_5         ; Busy?
        return

;===========================
;   Set address register
;       to LCD line 1
;===========================
; ON ENTRY:
;           Address of LCD line 1 in constant LCD_1
```

```
line1:
        bcf     PORTE,E_line        ; E line low
        bcf     PORTE,RS_line       ; RS line low, set up for
control
        call    delay_5             ; busy?
; Set to second display line
        movlw   LCD_1               ; Address and command bit
        call    send8               ; 4-bit routine
; Set RS line for data
        bsf     PORTE,RS_line       ; Setup for data
        call    delay_5             ; Busy?
        return

;==========================
;   scroll to LCD line 2
;==========================
; Procedure to count the number of characters displayed on
; each LCD line. If the number reaches the value in the
; constant LCDlimit, then display is scrolled to the second
; LCD line. If at the end of the second line, then LCD is
; reset to the first line.
LCDscroll:
        incf    LCDcount,f          ; Bump counter
; Test for line limit
        movf    LCDcount,w
        sublw   LCDlimit ; Count minus limit
        btfss   STATUS,Z ; Is count - limit = 0
        goto    scrollExit          ; Go if not at end of line
; At this point the end of the LCD line was reached
; Test if this is also the end of the second line
        movf    LCDline,w
        sublw   0x01                ; Is it line 1?
        btfsc   STATUS,Z ; Is LCDline minus 1 = 0?
        goto    line2End ; Go if end of second line
; At this point it is the end of the top LCD line
        call    line2               ; Scroll to second line
        clrf    LCDcount ; Reset counter
        incf    LCDline,f           ; Bump line counter
        goto    scrollExit
; End of second LCD line
line2End:
        call    initLCD             ; Reset
        clrf    LCDcount ; Clear counters
        clrf    LCDline
        call    line1               ; Display to first line
scrollExit:
        return
```

```
;=================================================================
;              communications procedures
;=================================================================
; Initizalize serial port for 2400 baud, 8 bits, no parity,
; 1 stop
InitSerial:
        Bank1                   ; Macro to select bank1
; Bits 6 and 7 of Port C are multiplexed as TX/CK and RX/DT
; for USART operation. These bits must be set to input in the
; TRISC register
        movlw    b'11000000'    ; Bits for TX and RX
        iorwf    TRISC,f        ; OR into Trisc register
; The asynchronous baud rate is calculated as follows:
;                     Fosc
;            ABR = ---------
;                    S*(x+1)
; Where x is value in the SPBRG register and S is 64 if the high
; baud rate select bit (BRGH) in the TXSTA control register is
; clear, and 16 if the BRGH bit is set. For setting to 9600 baud
; using a 4Mhs oscillator at a high-speed baud rate the formula
; is:
;            4,000,000     4,000,000
;            ---------  =  ---------  = 9,615 baud (0.16% error)
;            16*(25+1)        416
;
; At slow speed (BRGH = 0)
;            4,000,000     4,000,000
;            ---------  =  ---------  = 2,403.85 (0.16% error)
;            64*(25+1)       1,664
;
        movlw    spbrgVal       ; Value in spbrgVal = 25
        movwf    SPBRG          ; Place in baud rate generator
; TXSTA (Transmit Status and Control Register) bit map:
;   7  6  5  4  3  2  1  0   <== bits
;   |  |  |  |  |  |  |  |_____ TX9D 9nth data bit on
;   |  |  |  |  |  |  |          ? (used for parity)
;   |  |  |  |  |  |  |_____ TRMT Transmit Shift Register
;   |  |  |  |  |  |             1 = TSR empty
;   |  |  |  |  |  |           * 0 = TSR full
;   |  |  |  |  |  |_____ BRGH High Speed Baud Rate
;   |  |  |  |  |                (Asynchronous mode only)
;   |  |  |  |  |                1 = high speed (* 4)
;   |  |  |  |  |              * 0 = low speed
;   |  |  |  |  |_____ NOT USED
;   |  |  |  |_____ SYNC USART Mode Select
;   |  |  |                      1 = synchronous mode
;   |  |  |                    * 0 = asynchronous mode
;   |  |  |_____ TXEN Transmit Enable
```

```
;         |   |                        *  1 = transmit enabled
;         |   |                           0 = transmit disabled
;         |   |_____ TX9 Enable 9-bit Transmit
;         |                           1 = 9-bit transmission mode
;         |                        *  0 = 8-bit mode
;         |_____ CSRC Clock Source Select
;                                     Not used in asynchronous mode
;                                     Synchronous mode:
;                                        1 = Master Mode (internal clock)
;                                     *  0 = Slave mode (external clock)
; Setup value: 0010 0000 = 0x20
          movlw     0x20      ; Enable transmission and high baud
                              ; rate
          movwf     TXSTA
          Bank0               ; Bank 0
; RCSTA (Receive Status and Control Register) bit map:
;    7  6  5  4  3  2  1  0   <== bits
;    |  |  |  |  |  |  |  |_____ RX9D 9th data bit received
;    |  |  |  |  |  |  |       ? (can be parity bit)
;    |  |  |  |  |  |  |_____ OERR Overrun errror
;    |  |  |  |  |  |          ? 1 = error (cleared by software)
;    |  |  |  |  |  |_____ FERR Framing Error
;    |  |  |  |  |             ? 1 = error
;    |  |  |  |  |_____ NOT USED
;    |  |  |  |_____ CREN Continuous Receive Enable
;    |  |  |                         Asynchronous mode:
;    |  |  |                      *  1 = Enable continuous receive
;    |  |  |                         0 = Disables continuous receive
;    |  |  |                         Synchronous mode:
;    |  |  |                         1 = Enables until CREN cleared
;    |  |  |                         0 = Disables continuous receive
;    |  |  |_____ SREN Single Receive Enable
;    |  |                         ? Asynchronous mode = don't care
;    |  |                            Synchronous master mode:
;    |  |                               1 = Enable single receive
;    |  |                               0 = Disable single receive
;    |  |_____ RX9 9th-bit Receive Enable
;    |                               1 = 9-bit reception
;    |                             * 0 = 8-bit reception
;    |_____ SPEN Serial Port Enable
;                                 * 1 = RX/DT and TX/CK are serial pins

;                                   0 = Serial port disabled
; Setup value: 1001 0000 = 0x90
          movlw     0x90           ; Enable serial port and
                                   ; continuous reception
          movwf     RCSTA
;
```

Communications

```
            clrf    errorFlags          ; Clear local error flags
                                        ; register
            return

;===============================
;       transmit data
;===============================
; Test for Transmit Register Empty and transmit data in w
SerialSend:
            Bank0                       ; Select bank 0
            btfss   PIR1,TXIF           ; check if transmitter busy
            goto    $-1                 ; wait until transmitter is
                                        ; not busy
            movwf   TXREG               ; and transmit the data
            return
;===============================
;       receive data
;===============================
; Procedure to test line for data received and return value
; in w. Overrun and framing errors are detected and
; remembered in the variable errorFlags, as follows:
;       7  6  5  4  3  2  1  0    <== errorFlags
;       — not used —   |  |___ overrun error
;                      |_____ framing error
SerialRcv:
            clrf    newData    ; Clear new data received register
            Bank0              ; Select bank 0
; Bit 5 (RCIF) of the PIR1 Register is clear if the USART
; receive buffer is empty. If so, no data has been received
            btfss   PIR1,RCIF          ; Check for received data
            return                     ; Exit if no data
; At this point data has been received. First eliminate
; possible errors: overrun and framing.
; Bit 1 (OERR) of the RCSTA register detects overrun
; Bit 2 (FERR) of the RCSTA register detects framing error
            btfsc   RCSTA,OERR         ; Test for overrun error
            goto    OverErr            ; Error handler
            btfsc   RCSTA,FERR         ; Test for framing error
            goto    FrameErr ; Error handler
; At this point no error was detected
; Received data is in the USART RCREG register
            movf    RCREG,w            ; get received data
            bsf     newData,7  ; Set bit 7 to indicate new data
; Clear error flags
            clrf    errorFlags
            return
;==========================
;     error handlers
```

```
;=========================
OverErr:
        bsf     errorFlags,0        ; Bit 0 is overrun error
; Reset system
        bcf     RCSTA,CREN          ; Clear continuous receive bit
        bsf     RCSTA,CREN          ; Set to re-enable reception
        return
;error because FERR framing error bit is set
;can do special error handling here - this code simply clears
; and continues
FrameErr:
        bsf     errorFlags,1        ; Bit 1 is framing error
        movf    RCREG,W             ; Read and throw away bad data
        return

        end
```

14.5.6 SerIntLCD Program

```
; File name: SerIntLCD.asm
; Last revision: May 14, 2006
; Author: Julio Sanchez
; Processor: 16F877
;
; Interrupt-driven version of the SerComLCD program
;
; Description:
; Decode 4 x 4 keypad, display scan code in LCD, and send
; ASCII character through the serial port. Also receive
; data through serial port and display on LCD. LCD lines
; are scrolled by program.
; Default serial line setting:
;               2400 baud
;               no parity
;               1 stop bit
;               8 character bits
;
; Program to uses 4-bit PIC-to-LCD interface.
; Code assumes that LCD is driven by Hitachi HD44780
; controller and PIC 16F977. Display supports two lines
; each one with 20 characters. The length, wiring and base
; address of each display line is stored in #define
; statements. These statements can be edited to accommodate
; a different setup.
; Keypad switch wiring (values are scan codes):
;         -- KEYPAD -
;         0   1   2   3   <= port B0 |
;         4   5   6   7   <= port B1 |-- ROWS = OUTPUTS
```

```
;           8    9    A    B    <= port B2 |
;           C    D    E    F    <= port B3 |
;           |    |    |    |
;           |    |    |    |_____ port B4 |
;           |    |    |_____ port B5 |-- COLUMNS = INPUTS
;           |    |_____ port B6 |
;           |_____ port B7 |
;
; Operations:
; 1. Key press action generates a scan code in the range
;    0x0 to 0xf.
; 2. Scan code is converted to an ASCII digit and displayed
;    on the LCD. LCD lines are scrolled as end-of-line is
;    reached.
; 3. Characters typed on the keypad are also transmitted
;    through the serial port.
; 4. Received characters generate an interrupt. The interrupt
;    handler displays received characters on the LCD.
;
; WARNING:
; Code assumes 4Mhz clock. Delay routines must be
; edited for faster clock. Clock speed also determines
; values for baud rate setting (see spbrgVal constant).
;
;===========================
;      16F877 switches
;===========================
; Switches used in __config directive:
;   _CP_ON           Code protection ON/OFF
; * _CP_OFF
; * _PWRTE_ON        Power-up timer ON/OFF
;   _PWRTE_OFF
;   _BODEN_ON        Brown-out reset enable ON/OFF
; * _BODEN_OFF
; * _PWRTE_ON        Power-up timer enable ON/OFF
;   _PWRTE_OFF
;   _WDT_ON          Watchdog timer ON/OFF
; * _WDT_OFF
;   _LPV_ON          Low voltage IC programming enable ON/OFF
; * _LPV_OFF
;   _CPD_ON          Data EE memory code protection ON/OFF
; * _CPD_OFF
; OSCILLATOR CONFIGURATIONS:
;   _LP_OSC          Low power crystal oscillator
;   _XT_OSC          External parallel resonator/crystal oscillator
;
; * _HS_OSC          High speed crystal resonator
;   _RC_OSC          Resistor/capacitor oscillator
```

```
;   |                    (simplest, 20% error)
;   |
;   |_____  *  indicates setup values presently selected

        processor       16f877          ; Define processor
        #include <p16f877.inc>
        __CONFIG _CP_OFF & _WDT_OFF & _BODEN_OFF & _PWRTE_ON &
_HS_OSC & _WDT_OFF & _LVP_OFF & _CPD_OFF

; __CONFIG directive is used to embed configuration data
; within the source file. The labels following the directive
; are located in the corresponding .inc file.
;=============================================================
;                       M A C R O S
;=============================================================
; Macros to select the register banks
Bank0   MACRO                   ; Select RAM bank 0
        bcf     STATUS,RP0
        bcf     STATUS,RP1
        ENDM

Bank1   MACRO                   ; Select RAM bank 1
        bsf     STATUS,RP0
        bcf     STATUS,RP1
        ENDM

Bank2   MACRO                   ; Select RAM bank 2
        bcf     STATUS,RP0
        bsf     STATUS,RP1
        ENDM

Bank3   MACRO                   ; Select RAM bank 3
        bsf     STATUS,RP0
        bsf     STATUS,RP1
        ENDM
;=========================================================
;                   constant definitions
;   for PIC-to-LCD pin wiring and LCD line addresses
;=========================================================
#define E_line 1            ; |
#define RS_line 0           ; | - from wiring diagram
#define RW_line 2           ; |
; LCD line addresses (from LCD data sheet)
#define LCD_1 0x80          ; First LCD line constant
#define LCD_2 0xc0          ; Second LCD line constant
#define LCDlimit .20; Number of characters per line
#define spbrgVal .25; For 2400 baud on 4Mhz clock
; Note: The constants that define the LCD display
```

```
;               line addresses have the high-order bit set
;               so as to meet the requirements of controller
;               commands.
;
;===========================================================
;                   variables in PIC RAM
;===========================================================
; Local variables
            cblock  0x20    ; Start of block
            count1          ; Counter # 1
            count2          ; Counter # 2
            count3          ; Counter # 3
            J               ; counter J
            K               ; counter K
            store1          ; Local storage
            store2
; For LCDscroll procedure
            LCDcount ; Counter for characters per line
            LCDline         ; Current display line (0 or 1)
; Keypad processing variables
            keyMask         ; For keypad processing
            rowMask         ; For masking-off key rows
            rowCode         ; Row addend for calculating scan code
            rowCount ; Counter for key rows (0 to 3)
            scanCode ; Final key code
            newScan         ; 0 if no new scan code detected
; Communications variables
            ascVal
            errorFlags
; Temporary storage used by interrupt handler
            tempW
            tempStatus
            tempPclath
            tempFsr
            endc


;===========================================================
;                       P R O G R A M
;===========================================================
            org     0               ; start at address
            goto    main
; Space for interrupt handlers
            org             0x04
InterruptCode:
            goto    IntServ         ; Interrupt service routine

;===========================================================
;                       main program
```

```
;============================================================
main:
; Wiring:
;       LCD data to Port D, lines 0 to 7
;       E line -> port E, 1
;       RW line -> port E, 2
;       RS line -> port E, 0
; Set PORTE D and E for output
; Data memory bank selection bits:
; RP1:RP0           Bank
;   0:0              0      Ports A,B,C,D, and E
;   0:1              1      Tris A,B,C,D, and E
;   1:0              2
;   1:1              3
; First, initialize Port-B by clearing latches
        clrf    STATUS
        clrf    PORTB
; Select bank 1 to tris Port D for output
        Bank1
; Tris Port D for output. Port D lines 4 to 7 are wired
; to LCD data lines. Port D lines 0 to 4 are wired to LEDs.
        movlw   B'00000000'
        movwf   TRISD               ; and Port D
; By default Port-A lines are analog. To configure them
; as digital code must set bits 1 and 2 of the ADCON1
; register (in bank 1)
        movlw   0x06                ; binary 0000 0110 is code to
                                    ; make all Port-A lines digial
        movwf   ADCON1
; Port-B, lines are wired to keypad switches, as follows:
;   7 6 5 4 3 2 1 0
;   | | | | |_|_|_|_____ switch rows (output)
;   |_|_|_|_____ switch columns (input)
; rows must be defined as output and columns as input
        movlw   b'11110000'
        movwf   TRISB
; Tris port E for output
        movlw   B'00000000'
        movwf   TRISE               ; Tris port E
; Enable Port-B pullups for switches in OPTION register
;   7 6 5 4 3 2 1 0 <= OPTION bits
;   | | | | | |_|_|_____ PS2-PS0 (prescaler bits)
;   | | | | |            Values for Timer0
;   | | | | |            000 = 1:2    001 = 1:4
;   | | | | |            010 = 1:8    011 = 1:16
;   | | | | |            100 = 1:32   101 = 1:64
;   | | | | |            110 = 1:128 *111 = 1:256
;   | | | | |_____ PSA (prescaler assign)
```

```
;        |   |   |   |                    *1 = to WDT
;        |   |   |   |                     0 = to Timer0
;        |   |   |   |_____ TOSE (Timer0 edge select)
;        |   |   |                        *0 = increment on low-to-high
;        |   |   |                         1 = increment in high-to-low
;        |   |   |_____ TOCS (TMR0 clock source)
;        |   |                            *0 = internal clock
;        |   |                             1 = RA4/TOCKI bit source
;        |   |_____ INTEDG (Edge select)
;        |                                *0 = falling edge
;        |_____ RBPU (Pullup enable)
;                                         *0 = enabled
;                                          1 = disabled
         movlw   b'00001000'
         movwf   OPTION_REG
; Back to bank 0
         Bank0
; Initialize serial port for 9600 baud, 8 bits, no parity
; 1 stop
         call    InitSerial
; Test serial transmission by sending "RDY-"
         movlw   'R'
         call    SerialSend
         movlw   'D'
         call    SerialSend
         movlw   'Y'
         call    SerialSend
         movlw   '-'
         call    SerialSend
         movlw   0x20
         call    SerialSend
; Clear all output lines
         movlw   b'00000000'
         movwf   PORTD
         movwf   PORTE
; Wait and initialize HD44780
         call    delay_5   ; Allow LCD time to initialize itself
         call    initLCD   ; Then do forced initialization
         call    delay_5   ; (Wait probably not necessary)
; Clear character counter and line counter variables
         clrf    LCDcount
         clrf    LCDline
; Set display address to start of second LCD line
         call    line1
;===============================================================
;                           scan keypad
;===============================================================
; Keypad switch wiring:
```

```
;            x   x   x   x    <= Port B0 |
;            x   x   x   x    <= Port B1 |-- ROWS = OUTPUTS
;            x   x   x   x    <= port B2 |
;            x   x   x   x    <= port B3 |
;            |   |   |   |
;            |   |   |   |____ port B4 |
;            |   |   |_____ port B5 |-- COLUMNS = INPUTS
;            |   |_____ port B6 |
;            |_____ port B7 |
; Switches are connected to Port-B lines
; Clear scan code register
         clrf     scanCode
;===========================
;  scan keypad and display
;===========================
keyScan:
; Port-B, lines are wired to pushbutton switches, as follows:
;   7 6 5 4 3 2 1 0
;   | | | | |_|_|_|_____ switch rows (output)
;   |_|_||_____ switch columns (input)
; Keypad processing:
; switch rows are successively grounded (row = 0)
; Then column values are tested. If a column returns 0
; in a 0 row, that switch is down.
; Initialize row code addend
         clrf     rowCode         ; First row is code 0
         clrf     newScan         ; No new scan code detected
; Initialize row count
         movlw    D'4'            ; Four rows
         movwf    rowCount        ; Register variable
         movlw    b'11111110'     ; All set but LOB
         movwf    rowMask
keyLoop:
; Initialize row eliminator mask:
; The row mask is ANDed with the key mask to successively
; mask-off each row, for example:
;
;                     |----- row 3
;                     ||---- row 2
;                     |||--- row 1
;                     ||||-- row 0
;              0000 1111 <= key mask
;       AND    1111 1101 <= mask for row 1
;              ---------
;              0000 1101 <= row 1 is masked off
;
; The row mask, which is initially 1111 1110, is rotated left
; through the carry in order to mask off the next row
```

```
                movlw   b'00001111'         ; Mask off all lines
                movwf   keyMask             ; To local register
; Set row mask for current row
                movf    rowMask,w           ; Mask to w
                andwf   keyMask,f           ; Update key mask
                movf    keyMask,w           ; Key mask to w
                movwf   PORTB               ; Mask-off Port-B lines
; Read Port-B lines 4 to 7 (columns are input)
                btfss   PORTB,4
                call    col0                ; Key column procedures
                btfss   PORTB,5
                call    col1
                btfss   PORTB,6
                call    col2
                btfss   PORTB,7
                call    col3
; Index to next row by adding 4 to row code
                movf    rowCode,w           ; Code to w
                addlw   D'4'
                movwf   rowCode
;=========================
;       shift row mask
;=========================
; Set the carry flag
                bsf     STATUS,C
                rlf     rowMask,f           ; Rotate mask bits in storage
;=========================
;       end of keypad?
;=========================
; Test for last key row (maximum count is 4)
                decfsz  rowCount,f          ; Decrement counter
                goto    keyLoop
;=================================================================
;=================================================================
;                       display and send data
;=================================================================
;=================================================================
; At this point all keys have been tested.
; Variable newScan = 0 if no new scan code detected, else
; variable scanCode holds scan code
                movf    newScan,f           ; Copy onto itself
                btfsc   STATUS,Z ; Is it zero
                goto    ScanExit
; At this point a new scan code is detected
                movf    scanCode,w          ; To w
; If scan code is in the range 0 to 9, that is, a decimal
; digit, then ASCII conversion consists of adding 0x30.
; If the scan code represents one of the hex letters
```

```
; (0xa to 0xf) then ASCII conversion requires adding
; 0x37
        sublw   0x09              ; 9 - w
; if w from 0 to 9 then 9 - w = positive (C flag = 1)
; if w = 0xa then 9 - 10 = -1 (C flag = 0)
; if w = 0xc then 9 - 12 = -2 (C flag = 0)
        btfss   STATUS,C ; Test carry flag
        goto    hexLetter         ; Carry clear, must be a letter
; At this point scan code is a decimal digit in the
; range 0 to 9. Convert to ASCII by adding 0x30
        movf    scanCode,w        ; Recover scan code
        addlw   0x30              ; Convert to ASCII
        goto    displayDig
hexLetter:
        movf    scanCode,w        ; Recover scan code
        addlw   0x37              ; Convert to ASCII
displayDig:
; Store so it can be sent
        movwf   ascVal
        call    send8             ; Display routine
        call    LCDscroll
        call    long_delay        ; Debounce
; Recover ASCII
        movf    ascVal,w
        call    SerialSend
ScanExit:
        goto    keyScan           ; Continue
;===========================
;   calculate scan code
;===========================
; The column position is added to the row code (stored
; in rowCode register). Sum is the scan code
col0:
        movf    rowCode,w         ; Row code to w
        addlw   0x00              ; Add 0
        movwf   scanCode          ; Final value
        incf    newScan,f         ; New scan code
        return
col1:
        movf    rowCode,w         ; Row code to w
        addlw   0x01              ; Add 1
        movwf   scanCode
        incf    newScan,f
        return

col2:
        movf    rowCode,w         ; Row code to w
```

```
                addlw       0x02                ; Add 2
                movwf       scanCode
                incf        newScan,f
                return

col3:
                movf        rowCode,w           ; Row code to w
                addlw       0x03                ; Add 3
                movwf       scanCode
                incf        newScan,f
                return
;================================================================
;================================================================
;                  L O C A L    P R O C E D U R E S
;================================================================
;================================================================
;==========================
; init LCD for 4-bit mode
;==========================
initLCD:
; Initialization for Densitron LCD module as follows:
;    4-bit interface
;    2 display lines of 16 characters each
;    cursor on
;    left-to-right increment
;    cursor shift right
;    no display shift
;======================|
;   set command mode   |
;======================|
                bcf         PORTE,E_line        ; E line low
                bcf         PORTE,RS_line       ; RS line low
                bcf         PORTE,RW_line       ; Write mode
                call        delay_125           ; delay 125 microseconds
;**********************|
;      FUNCTION SET    |
;**********************|
                movlw       0x28        ; 0 0 1 0 1 0 0 0 (FUNCTION SET)
                                        ;         | | | |__ font select:
                                        ;         | | |     1 = 5x10 in 1/8 or 1/11
                                        ;         | | |     0 = 1/16 dc
                                        ;         | | |___ Duty cycle select
                                        ;         | |       0 = 1/8 or 1/11
                                        ;         | |       1 = 1/16)
                                        ;         | |___ Interface width
                                        ;         |       0 = 4 bits
                                        ;         |       1 = 8 bits
```

```
                           ;        |___ FUNCTION SET COMMAND
          call      send8  ; 4-bit send routine

; Set 4-bit mode command must be repeated
          movlw     0x28
          call      send8

;*********************|
; DISPLAY AND CURSOR ON |
;*********************|
          movlw     0x0e    ; 0 0 0 0 1 1 1 0 (DISPLAY ON/OFF)
                            ;         | | | |___ Blink character
                            ;         | | |      1 = on, 0 = off
                            ;         | | |___ Cursor on/off
                            ;         | |      1 = on, 0 = off
                            ;         | |____ Display on/off
                            ;         |      1 = on, 0 = off
                            ;         |____ COMMAND BIT
          call      send8
;*********************|
;   set entry mode    |
;*********************|
          movlw     0x06    ; 0 0 0 0 0 1 1 0 (ENTRY MODE SET)
                            ;           | | |___ display shift
                            ;           | |      1 = shift
                            ;           | |      0 = no shift
                            ;           | |____ increment mode
                            ;           |      1 = left-to-right
                            ;           |      0 = right-to-left
                            ;           |___ COMMAND BIT
          call      send8

;*********************|
; cursor/display shift |
;*********************|
          movlw     0x14    ; 0 0 0 1 0 1 0 0 (CURSOR/DISPLAY
SHIFT)
                            ;       | | | |_|___ don't care
                            ;       | |_|__ cursor/display shift
                            ;       |        00 = cursor shift left
                            ;       |        01 = cursor shift right
                            ;       |        10 = cursor and display
                            ;       |             shifted left
                            ;       |        11 = cursor and display
                            ;       |             shifted right
                            ;       |___ COMMAND BIT
          call      send8
;*********************|
```

```
;       clear display          |
;***********************|
        movlw       0x01        ; 0 0 0 0 0 0 0 1 (CLEAR DISPLAY)
                                ;                 |___ COMMAND BIT
        call        send8
; Per documentation
        call        delay_5     ; Test for busy
        return

;=======================
;   Procedure to delay
;     42 microseconds
;=======================
delay_125:
        movlw       D'42'       ; Repeat 42 machine cycles
        movwf       count1      ; Store value in counter
repeat:
        decfsz      count1,f    ; Decrement counter
        goto        repeat      ; Continue if not 0
        return                  ; End of delay

;=======================
;   Procedure to delay
;     5 milliseconds
;=======================
delay_5:
        movlw       D'42'       ; Counter = 41
        movwf       count2      ; Store in variable
delay:
        call        delay_125   ; Delay
        decfsz      count2,f    ; 40 times = 5 milliseconds
        goto        delay
        return                  ; End of delay
;=======================
;      pulse E line
;=======================
pulseE
        bsf         PORTE,E_line ; Pulse E line
        nop
        bcf         PORTE,E_line
        return

;============================
;   long delay sub-routine
;============================
long_delay
        movlw       D'200'      ; w delay count
        movwf       J           ; J = w
```

```
jloop:
        movwf   K                       ; K = w
kloop:
        decfsz  K,f                     ; K = K-1, skip next if zero
        goto    kloop
        decfsz  J,f                     ; J = J-1, skip next if zero
        goto    jloop
        return

;========================
;   send 2 nibbles in
;       4-bit mode
;========================
; Procedure to send two 4-bit values to Port-B lines
; 7, 6, 5, and 4. High-order nibble is sent first
; ON ENTRY:
;       w register holds 8-bit value to send
send8:
        movwf   store1                  ; Save original value
        call    merge4                  ; Merge with Port-B
; Now w has merged byte
        movwf   PORTD                   ; w to Port D
        call    pulseE                  ; Send data to LCD
; High nibble is sent
        movf    store1,w                ; Recover byte into w
        swapf   store1,w                ; Swap nibbles in w
        call    merge4
        movwf   PORTD
        call    pulseE                  ; Send data to LCD
        call    delay_125
        return
;==========================
;       merge bits
;==========================
; Routine to merge the 4 high-order bits of the
; value to send with the contents of Port-B
; so as to preserve the 4 low-bits in Port-B
; Logic:
;       AND value with 1111 0000 mask
;       AND Port-B with 0000 1111 mask
;       Now low nibble in value and high nibble in
;       Port-B are all 0 bits:
;           value = vvvv 0000
;           Port-B = 0000 bbbb
;       OR value and Port-B resulting in:
;               vvvv bbbb
; ON ENTRY:
;       w contains value bits
```

```
; ON EXIT:
;       w contains merged bits
merge4:
        andlw   b'11110000'         ; ANDing with 0 clears the
                                    ; bit. ANDing with 1 preserves
                                    ; the original value
        movwf   store2              ; Save result in variable
        movf    PORTD,w             ; Port-B to w register
        andlw   b'00001111'         ; Clear high nibble in Port-B
                                    ; and preserve low nibble
        iorwf   store2,w            ; OR two operands in w
        return

;==========================
;   Set address register
;       to LCD line 2
;==========================
; ON ENTRY:
;           Address of LCD line 2 in constant LCD_2
line2:
        bcf     PORTE,E_line        ; E line low
        bcf     PORTE,RS_line       ; RS line low, setup for
control
        call    delay_5             ; Busy?
; Set to second display line
        movlw   LCD_2               ; Address with high-bit set
        call    send8
; Set RS line for data
        bsf     PORTE,RS_line       ; RS = 1 for data
        call    delay_5             ; Busy?
        return

;==========================
;   Set address register
;       to LCD line 1
;==========================
; ON ENTRY:
;           Address of LCD line 1 in constant LCD_1
line1:
        bcf     PORTE,E_line        ; E line low
        bcf     PORTE,RS_line       ; RS line low, set up for
                                    ; control
        call    delay_5             ; busy?
; Set to second display line
        movlw   LCD_1               ; Address and command bit
        call    send8               ; 4-bit routine
; Set RS line for data
        bsf     PORTE,RS_line       ; Setup for data
```

```
                call    delay_5             ; Busy?
                return

;===========================
;   scroll to LCD line 2
;===========================
; Procedure to count the number of characters displayed on
; each LCD line. If the number reaches the value in the
; constant LCDlimit, then display is scrolled to the second
; LCD line. If at the end of the second line, then LCD is
; reset to the first line.
LCDscroll:
                incf    LCDcount,f          ; Bump counter
; Test for line limit
                movf    LCDcount,w
                sublw   LCDlimit            ; Count minus limit
                btfss   STATUS,Z            ; Is count - limit = 0
                goto    scrollExit          ; Go if not at end of line
; At this point the end of the LCD line was reached
; Test if this is also the end of the second line
                movf    LCDline,w
                sublw   0x01                ; Is it line 1?
                btfsc   STATUS,Z            ; Is LCDline minus 1 = 0?
                goto    line2End            ; Go if end of second line
; At this point it is the end of the top LCD line
                call    line2               ; Scroll to second line
                clrf    LCDcount ; Reset counter
                incf    LCDline,f           ; Bump line counter
                goto    scrollExit
; End of second LCD line
line2End:
                call    initLCD             ; Reset
                clrf    LCDcount ; Clear counters
                clrf    LCDline
                call    line1               ; Display to first line
scrollExit:
                return

;================================================================
;                   communications procedures
;================================================================
; Initialize serial port for 2400 baud, 8 bits, no parity,
; 1 stop
InitSerial:
                Bank1                       ; Macro to select bank1
; Bits 6 and 7 of Port C are multiplexed as TX/CK and RX/DT
; for USART operation. These bits must be set to input in the
; TRISC register
```

```
                movlw    b'11000000'          ; Bits for TX and RX
                iorwf    TRISC,f              ; OR into Trisc register
; The asynchronous baud rate is calculated as follows:
;                         Fosc
;               ABR = ---------
;                        S*(x+1)
; Where x is value in the SPBRG register and S is 64 if the high
; baud rate select bit (BRGH) in the TXSTA control register is
; clear, and 16 if the BRGH bit is set. For setting to 9600 baud
; using a 4Mhs oscillator at a high-speed baud rate the formula
; is:
;               4,000,000    4,000,000
;               ---------  = ---------  = 9,615 baud (0.16% error)
;               16*(25+1)       416
;
; At slow speed (BRGH = 0)
;               4,000,000    4,000,000
;               ---------  = ----------  = 2,403.85 (0.16% error)
;               64*(25+1)      1,664
;
                movlw    spbrgVal ; Value in spbrgVal = 25
                movwf    SPBRG                ; Place in baud rate generator
; TXSTA (Transmit Status and Control Register) bit map:
;    7  6  5  4  3  2  1  0   <== bits
;    |  |  |  |  |  |  |  |_____ TX9D 9nth data bit on
;    |  |  |  |  |  |  |            ? (used for parity)
;    |  |  |  |  |  |  |_____ TRMT Transmit Shift Register
;    |  |  |  |  |  |                 1 = TSR empty
;    |  |  |  |  |  |               * 0 = TSR full
;    |  |  |  |  |  |_____ BRGH High Speed Baud Rate
;    |  |  |  |  |                    (Asynchronous mode only)
;    |  |  |  |  |                    1 = high speed (* 4)
;    |  |  |  |  |                  * 0 = low speed
;    |  |  |  |  |_____ NOT USED
;    |  |  |  |_____ SYNC USART Mode Select
;    |  |  |                 1 = synchronous mode
;    |  |  |               * 0 = asynchronous mode
;    |  |  |_____ TXEN Transmit Enable
;    |  |                  * 1 = transmit enabled
;    |  |                    0 = transmit disabled
;    |  |_____ TX9 Enable 9-bit Transmit
;    |                       1 = 9-bit transmission mode
;    |                     * 0 = 8-bit mode
;    |_____ CSRC Clock Source Select
;                            Not used in asynchronous mode
;                            Synchronous mode:
;                               1 = Master Mode (internal clock)
;                             * 0 = Slave mode (external clock)
```

```
; Setup value: 0010 0000 = 0x20
        movlw   0x20        ; Enable transmission and low baud rate
        movwf   TXSTA
        Bank0                                   ; Bank 0
; RCSTA (Receive Status and Control Register) bit map:
;    7  6  5  4  3  2  1  0  <== bits
;    |  |  |  |  |  |  |  |_____ RX9D 9th data bit received
;    |  |  |  |  |  |  |         ? (can be parity bit)
;    |  |  |  |  |  |  |_____ OERR Overrun errror
;    |  |  |  |  |  |            ? 1 = error (cleared by software)
;    |  |  |  |  |  |_____ FERR Framing Error
;    |  |  |  |  |               ? 1 = error
;    |  |  |  |  |_____ NOT USED
;    |  |  |  |_____ CREN Continuous Receive Enable
;    |  |  |                      Asynchronous mode:
;    |  |  |                  *   1 = Enable continuous receive
;    |  |  |                      0 = Disables continuous receive
;    |  |  |                      Synchronous mode:
;    |  |  |                      1 = Enables until CREN cleared
;    |  |  |                      0 = Disables continuous receive
;    |  |  |_____ SREN Single Receive Enable
;    |  |                         ? Asynchronous mode =  don't care
;    |  |                         Synchronous master mode:
;    |  |                         1 = Enable single receive
;    |  |                         0 = Disable single receive
;    |  |_____ RX9 9th-bit Receive Enable
;    |                             1 = 9-bit reception
;    |                          *  0 = 8-bit reception
;    |_____ SPEN Serial Port Enable
;                                *  1 = RX/DT and TX/CK are serial pins
;
;                                   0 = Serial port disabled
; Setup value: 1001 0000 = 0x90
        movlw   0x90        ; Enable serial port and
                            ; continuous reception
        movwf   RCSTA
; Enable glocal and peripheral interrupts
;    7  6  5  4  3  2  1  0  <= INTCON bitmap
;    |  |  -  unrelated  --
;    |  |_____ Peripheral interrupts enable
;    |_____ Global interrupts enable
        movlw   b'11000000'
        movwf   INTCON
; Enable receive interrupt in PIE1 register
;    7  6  5  4  3  2  1  0  <= PIE1 bitmap
;          |_____ USART receive interrupt enable
        Bank1
        movlw   b'00100000'
```

```
                movwf       PIE1
; Clear error flags register
        Bank0
        clrf        errorFlags
        return

;==============================
;       transmit data
;==============================
; Test for Transmit Register Empty and transmit data in w
SerialSend:
        Bank0                   ; Select bank 0
        btfss       PIR1,TXIF   ; check if transmitter busy
        goto        $-1         ;wait until transmitter is not busy
        movwf       TXREG       ;and transmit the data
        return


;==============================================================
;==============================================================
;           interrupt handler for received characters
;==============================================================
;==============================================================
IntServ:
        movwf       tempW           ; Save W
        movf        STATUS,W ;  Store STATUS in W
        clrf        STATUS          ; Select bank0
        movwf       tempStatus      ; Save STATUS
        movf        PCLATH,W ;  Store PCLATH in W
        movwf       tempPclath      ; Save PCLATH
        clrf        PCLATH          ; Select program memory page 0
        movf        FSR,W           ; Store FSR in W
        movwf       tempFsr         ; Save FSR value

; Test for received data interrupt
        Bank0                               ; select bank0
;   7   6   5   4   3   2   1   0   <= PIR1
;               |_____ (RCIF) USART receive interrupt
;                                   flag
        btfsc       PIR1,RCIF       ; Test bit 5
        bsf                 STATUS,RP0      ; Bank 1 if RCIF set
;   7   6   5   4   3   2   1   0   <= PIE1
;               |_____ (RCIE) Receive interrupt enable
;                                   bit
        btfss       PIE1,RCIE       ; Test if interrupt is enabled
        goto        IntExit         ; Go if not enabled
;==============================
;       received data
;==============================
```

```
; Routine to handler received data. Overrun and framing
; errors are detected and remembered in the variable
; errorFlags, as follows:
;       7 6 5 4 3 2 1 0    <== errorFlags
;       — not used —  |  |___ overrun error
;                        |_____ framing error
        Bank0                               ; Select bank 0
; Test for overrun and framing errors.
; Bit 1 (OERR) of the RCSTA register detects overrun
; Bit 2 (FERR) of the RCSTA register detects framing error
        btfsc   RCSTA,OERR      ; Test for overrun error
        goto    OverErr         ; Error handler
        btfsc   RCSTA,FERR      ; Test for framing error
        goto    FrameErr ; Error handler
; At this point no error was detected
; Received data is in the USART RCREG register
        movf    RCREG,w         ; Received data into w
        call    send8           ; Display in LCD
        call    LCDscroll       ; Scroll at end of line
; Clear error flags
        clrf    errorFlags
        goto    IntExit
;===========================
;       error handlers
;===========================
; Errors are returned as bits in the errorFlags register
;       7 6 5 4 3 2 1 0   <= errorFlags
;       -- not used --  |  |____ overrun error
;                          |_____ framing error
; Error responses to be made by main code
OverErr:
        bsf     errorFlags,0    ; Bit 0 is overrun error
; Reset system
        bcf     RCSTA,CREN      ; Clear continuous receive bit
        bsf     RCSTA,CREN      ; Set to re-enable reception
        goto    IntExit
FrameErr:
        bsf     errorFlags,1; Bit 1 is framing error
        movf    RCREG,W         ; Read and throw away bad data
;=============================
;       interrupt handler exit
;=============================
IntExit:
        Bank0
        movf    tempFsr,w       ; Recover FSR value
        movwf   FSR             ; Restore in register
        movf    tempPclath,w    ; Recover PCLATH value
        movwf   PCLATH          ; Restore in register
```

```
        movf    tempStatus,W    ; Recover STATUS
        movwf   STATUS          ; Restore in register
        swapf   tempW,F         ; Swap file register in itself
        swapf   tempW,W         ; Restore in register
        retfie

        end
```

Chapter 15

Data EEPROM Programming

EEPROM stands for *Electrically-Erasable Programmable Read-Only Memory*. EEPROM is used in computers and digital devices as non-volatile storage. EEPROM is found in flash drives, BIOS chips, and in flash memory and EEPROM data storage memory in PICs and other microcontrollers.

EEPROM memory can be erased and programmed electrically without removing the chip. EPROM, the predecessor of EEPROM, required chip removal from the circuit and ultraviolet light exposure in order to erase the chip. In addition, EPROM requires higher-than-TTL voltages for reprogramming while EEPROM does not.

The PIC programmer regards EEPROM data memory as *onboard EEPROM memory* and EEPROM memory ICs as separate circuit components. In general, EEPROM elements are classified according to their electrical interfaces into serial and parallel. In this context we deal only with *serial EEPROMs*. The storage capacity of Serial EEPROMs ranges from a few bytes to 128 kilobytes. In PIC technology, the typical use of serial EEPROM onboard memory and EEPROM ICs is to store passwords, codes, configuration settings, and other information to be remembered after the system is turned off. For example, a PIC-based automated environment sensor can use EEPROM memory (onboard or independent) to store daily temperatures, humidity, air pressure, and other values. Later, this information could be downloaded to a PC and the EEPROM storage erased and reused for new data. In personal computers, EEPROM memory is used to store BIOS code and other system data.

Some early EEPROM could only be erased and rewritten about 100 times, while modern EEPROM tolerate thousands of erase-write cycles. EEPROM memory is different from *RAM (Random Access Memory)* in that RAM can be rewritten millions of times. Also, RAM is generally faster to write than EEPROM and considerably cheaper per unit of storage. On the other hand, RAM is volatile, so the contents are lost when power is removed.

PICs use EEPROM-type memory internally as flash program memory and as data memory. EEPROM data memory is covered in this chapter. Serial EEPROM memory is available as separate ICs that can be placed on the circuit board and accessed through PIC ports. For example, the Microchip 24LC04B EEPROM IC is a 4K electrically erasable PROM with a 2-wire serial interface that follows the I^{2}C convention. Programming serial EEPROM ICs is also covered in this chapter.

15.0 PIC Internal EEPROM Memory

Some PICs contain internal EEPROM data memory that is accessible to code. The amount of memory and the access mechanism varies from PIC to PIC. In fact, the mapping and access mechanisms varies even in devices belonging to the same family. In the sections that follow, we describe EEPROM data memory in the context of two different PICs of the mid-range family: the 16F84 and the 16F877.

15.0.1 EEPROM Programming on the 16F84

The 16F84 and 16F84A contain 64 bytes of EEPROM data memory. This memory is both readable and writable during normal operation. It is not mapped in the register file space, but is indirectly addressed through the *Special Function Registers EECON1, EECON2, EEDATA,* and *EEADR*. The address of EEPROM memory starts at location 0x00 and extends to the maximum contained in the PIC, in this case, 0x3f. The following registers relate to EEPROM operations:

1. EEDATA holds the data byte to be read or written.
2. EEADR contains the EEPROM address to be accessed by the read or write operation.
3. EECON1 contains the control bits for EEPROM operations.
4. EECON2 protects EEPROM memory from accidental access. This is not a physical register.

Figure 15-1 shows the bitmap of the EECON1 register in the 16F84.

The CPU continues to access EEPROM memory even if the device is code protected, but in this case the device programmer can not access EEPROM memory.

Reading EEPROM Data Memory on the 16F84

Reading an EEPROM data memory location in the 16F84 requires the following operations:

1. Bank 0 is selected and the address of the memory to be read is stored in the EEADR register.
2. Bank 1 is selected and the RD bit is set in the EECON1 register.
3. Bank 0 is selected and data is read from the EEDATA register.

The following procedure returns in the w register the data stored at the specified EEPROM memory address.

```
;==============================
;      read EEPROM 16F84
;==============================
; Procedure to read EEPROM memory. Address of memory
; location to read is stored in local register EEMemAdd
```

Data EEPROM Programming

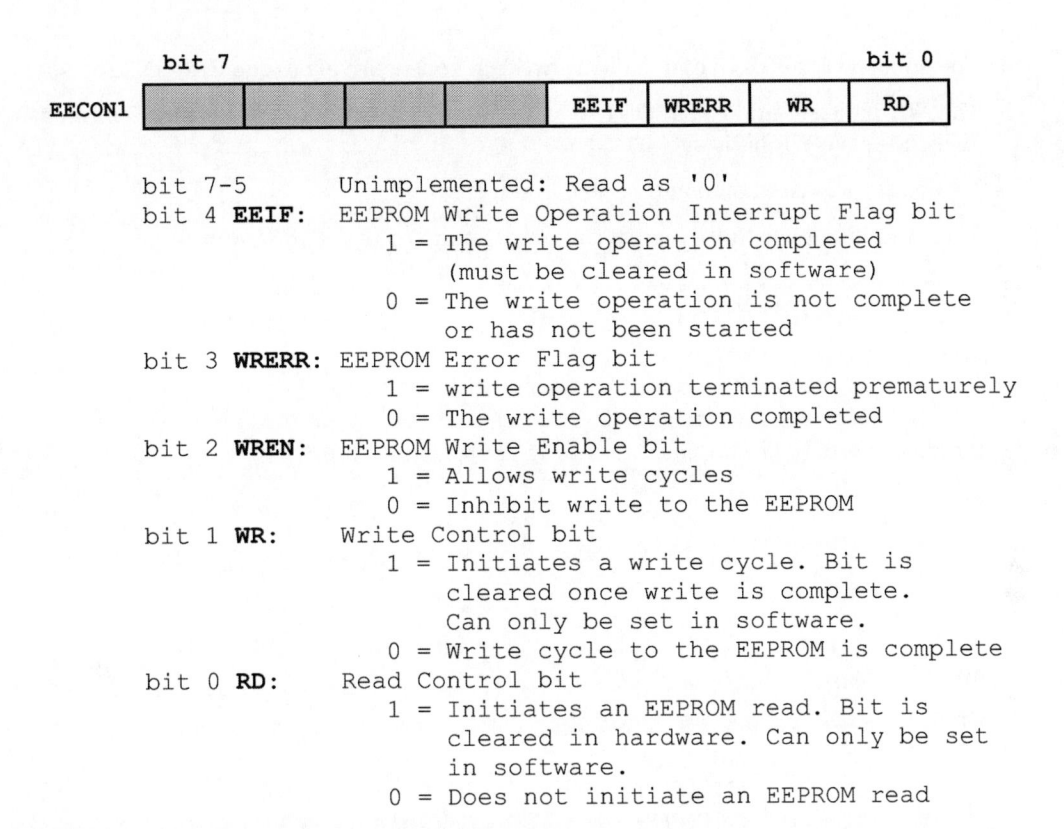

	bit 7							bit 0
EECON1					EEIF	WRERR	WR	RD

    bit 7-5   Unimplemented: Read as '0'
    bit 4 **EEIF**: EEPROM Write Operation Interrupt Flag bit
                        1 = The write operation completed
                            (must be cleared in software)
                        0 = The write operation is not complete
                            or has not been started
    bit 3 **WRERR**: EEPROM Error Flag bit
                        1 = write operation terminated prematurely
                        0 = The write operation completed
    bit 2 **WREN**: EEPROM Write Enable bit
                        1 = Allows write cycles
                        0 = Inhibit write to the EEPROM
    bit 1 **WR**:   Write Control bit
                        1 = Initiates a write cycle. Bit is
                            cleared once write is complete.
                            Can only be set in software.
                        0 = Write cycle to the EEPROM is complete
    bit 0 **RD**:   Read Control bit
                        1 = Initiates an EEPROM read. Bit is
                            cleared in hardware. Can only be set
                            in software.
                        0 = Does not initiate an EEPROM read

Figure 15-1 16F84 EECON1 Register Bit Map

```
; On exit: read data in w
EERead:
        bcf     STATUS,RP0      ; Bank 0
        movf    EEMemAdd,w      ; Address to w
        movwf   EEADR           ; w to address register
        bsf     STATUS,RP0      ; Bank 1
        bsf     EECON1,RD       ; EE Read
        bcf     STATUS,RP0      ; Bank 0
        movf    EEDATA,w        ; W = EEDATA
        return
```

16F84 EEPROM Data Memory Write

Writing to 16F84 EEPROM data memory consists of the following operations:

1. Bank 0 is selected and the address of the desired memory location is stored in the EEADR register.
2. The value to be written is stored in the EEDATA register.
3. Bank 1 is selected, interrupts are disabled, and the write enable bit (WREN) is set in the EECON1 register.

4. The special values 0x55 and 0xaa are written consecutively to the EECON2 register.
5. The WR bit is set in the EECON1 register. The EEPROM write takes place automatically after the WR bit is set.
6. Interrupts are re-enabled and bank 0 is selected.

The following procedure shows the processing for the EEPROM write.

```
;===============================
;          write EEPROM
;===============================
; Procedure to write ascl byte to EEPROM memory
; Address to write stored in local register EEMemAdd
; Data byte to write is in local register EEByte
EEWrite:
; Load byte to write into EE data register
        movf    EEByte,w     ; Data to w
        movwf   EEDATA       ; Write
; Set write address in EE address register
        movf    EEMemAdd,w   ; Address to w
        movwf   EEADR        ; w to address register
; Write data to EEPROM memory
        bsf     STATUS,RP0   ; Bank 1
        bcf     INTCON,GIE   ; Disable INTs.
        bsf     EECON1,WREN  ; Enable Write
        movlw   0x55         ; Code # 1
        movwf   EECON2       ; Write 0x55
        movlw   0xaa         ; Code # 2
        movwf   EECON2       ; Write 0xaa
        bsf     EECON1,WR    ; Set WR bit
; Write operation now takes place automatically
        bsf     INTCON,GIE   ; Re-enable interrupts
        bcf     STATUS,RP0   ; Bank 0
        return
```

Microchip documentation recommends that critical applications should verify the write operation by reading EEPROM memory after the write operation has taken place in order to make sure that the correct value was stored.

16F84 EEPROM Demonstration Program

The program EECounter, in the book's online software, is a demonstration of EEPROM memory access on the 16F84 PIC. The program keeps track of the number of times that the code has executed by storing each iteration in EEPROM data memory. The program uses the circuit shown in Figure 15-2 (see following page).

The EECounter program increments the value stored at EEPROM address 0x00 at every iteration and displays the result on the first LCD line. The following procedure is used to convert the binary value in EEPROM to 3 ASCII digits for display.

Data EEPROM Programming

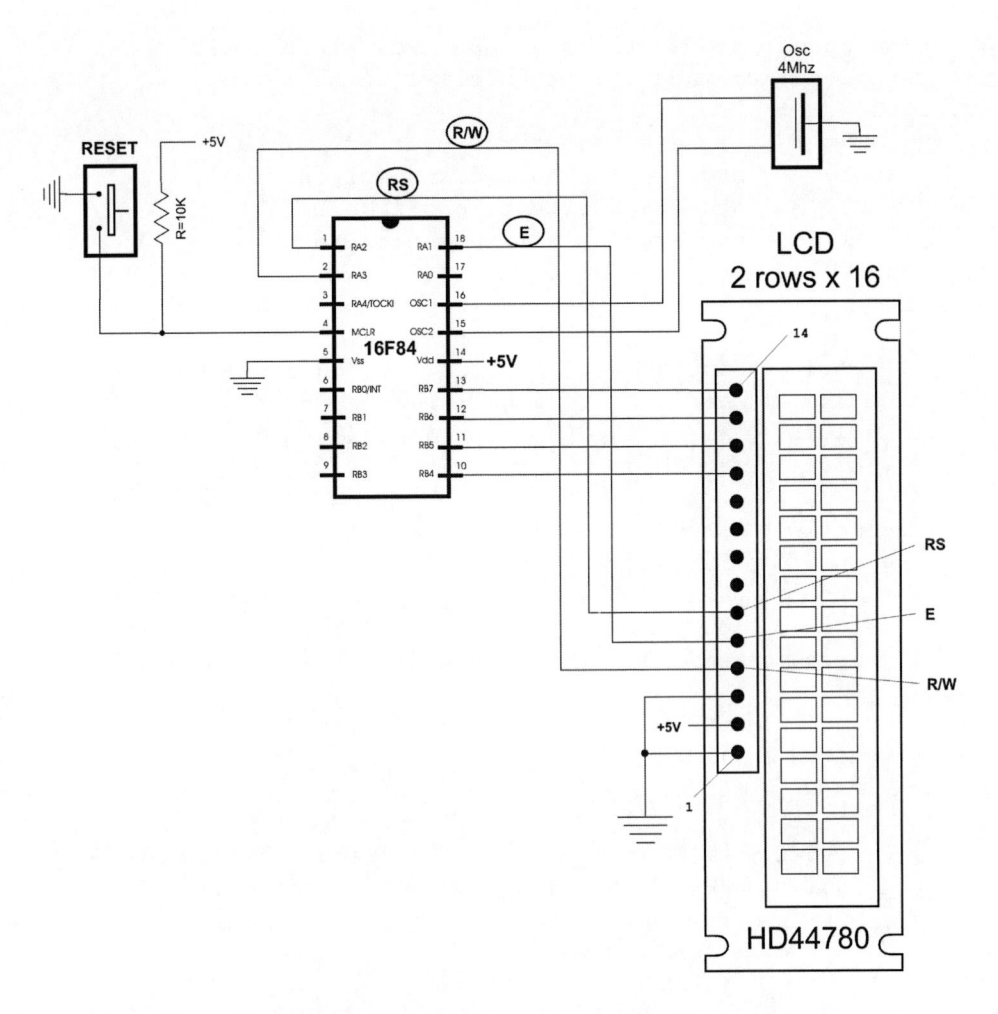

Figure 15-2 Circuit for 16F84 EEPROM Demonstration Program

```
;===============================
;   binary to ASCII decimal
;        conversion
;===============================
; ON ENTRY:
;        w register has binary value in range 0 to 255
; ON EXIT:
;        output variables asc100, asc10, and asc1 have
;        three ASCII decimal digits
; Routine logic:
;   The value 100 is subtracted from the source operand
;   until the remainder is < 0 (carry cleared). The number
;   of subtractions is the decimal hundreds result. 100 is
;   then added back to the subtrahend to compensate
;   for the last subtraction. Now 10 is subtracted in the
```

```
;       same manner to determine the decimal tenths result.
;       The final remainder is the decimal units result.
; Variables:
;       inNum       storage for source operand
;       asc100      storage for hundreds position result
;       asc10       storage for tenth position result
;       asc1        storage for unit position result
;       thisDig     Digit counter
bin2asc:
        movwf   inNum           ; Save copy of source value
        clrf    asc100          ; Clear hundreds storage
        clrf    asc10           ; Tens
        clrf    asc1            ; Units
        clrf    thisDig
sub100:
        movlw   .100
        subwf   inNum,f         ; Subtract 100
        btfsc   STATUS,C        ; Did subtract overflow?
        goto    bump100         ; No. Count subtraction
        goto    end100
bump100:
        incf    thisDig,f       ;increment digit counter
        goto    sub100
; Store 100th digit
end100:
        movf    thisDig,w       ; Adjusted digit counter
        addlw   0x30            ; Convert to ASCII
        movwf   asc100          ; Store it
; Calculate tenth position value
        clrf    thisDig
; Adjust minuend
        movlw   .100            ; Minuend
        addwf   inNum,f         ; Add value to minuend to
                                ; compensate for last
                                ; operation
sub10:
        movlw   .10
        subwf   inNum,f         ; Subtract 10
        btfsc   STATUS,C        ; Did subtract overflow?
        goto    bump10     ; No. Count subtraction
        goto    end10
bump10:
        incf    thisDig,f       ; Increment digit counter
        goto    sub10
; Store 10th digit
end10:
        movlw   .10
        addwf   inNum,f  ; Adjust for last subtraction
```

Data EEPROM Programming

```
        movf    thisDig,w       ; Digit counter contents
        addlw   0x30            ; Convert to ASCII
        movwf   asc10           ; Store it
; Calculate and store units digit
        movf    inNum,w         ; Store units value
        addlw   0x30            ; Convert to ASCII
        movwf   asc1            ; Store digit
        Return
```

15.0.2 EEPROM Programming on the 16F87x

The 16F87x PICs contain 128 or 256 bytes of EEPROM data memory. As in the 16F84, this memory is both readable and writable during normal operation. It is not mapped in the register file space but is indirectly addressed through Special Function Registers, as described later in this section.

In the 16F87x PICs both data EEPROM and flash Program Memory are readable and writable during normal operation. For data EEPROM memory, read and write operations take place one byte at a time. The write operation performs an *erase-then-write cycle*. No bulk erase function is available to user code.

The following Special Function Registers are used in 16F87x EEPROM data read and write operations:

1. EEDATA holds the data byte to be read or written.
2. EEADR contains the EEPROM address to be accessed by the read or write operation.
3. EECON1 contains the control bits for EEPROM operations.
4. EECON2 protects EEPROM memory from accidental access. This is not a physical register.

EEPROM data memory read and write operations do not interfere with normal PIC operations. The 16F873 and 16F874 PICs have 128 bytes of EEPROM data memory. These ICs require that the most-significant-bit of EEADR remains clear. The EEPROM data memory on these devices does not wrap around; that is, accessing EEPROM address 0x80 does not map to 0x00. The 16F876 and 16F877 devices have 256 bytes of EEPROM data memory. In these devices all 8-bits of the EEADR are used. Figure 15-3 (in the following page) is a bit map of the EECON1 register in the 16F87x.

The 16F87x EECON1 register contains one additional bit that is not present in the 16F84, named EEPGD. This bit determines if the EEPROM operation accesses program or data memory. When clear, any subsequent operations relate to EEPROM data. Otherwise, the operation accesses program memory. Read operations only require the RD bit, which initiates the read from the selected memory location. The RD bit is automatically cleared at the end of the read operation. The data in the selected memory location can be read in the EEDATA register as soon as the RD bit is set.

```
 bit 7                                              bit 0
┌───────┬───┬───┬───┬──────┬───────┬────┬────┐
│ EEPGD │   │   │   │ EEIF │ WRERR │ WR │ RD │
└───────┴───┴───┴───┴──────┴───────┴────┴────┘
```

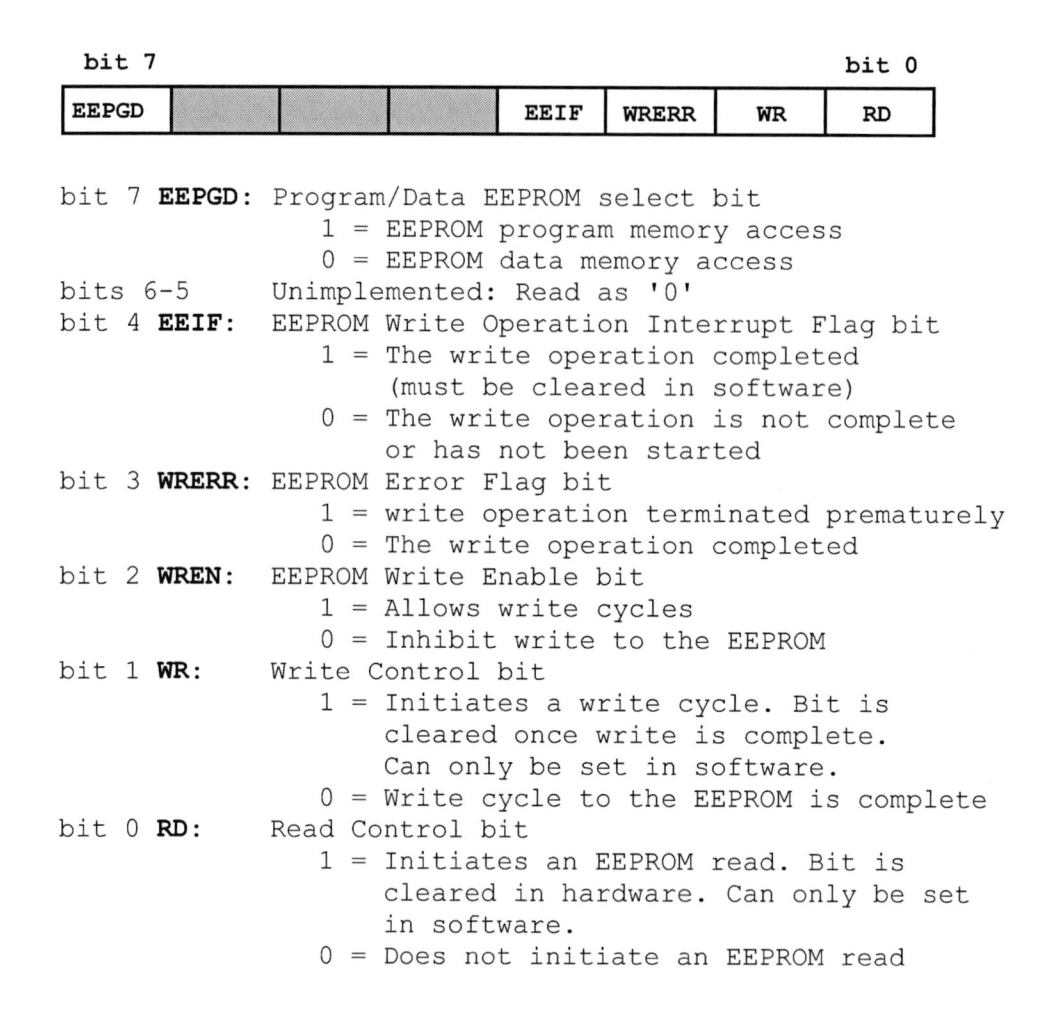

```
bit 7 EEPGD:  Program/Data EEPROM select bit
              1 = EEPROM program memory access
              0 = EEPROM data memory access
bits 6-5      Unimplemented: Read as '0'
bit 4 EEIF:   EEPROM Write Operation Interrupt Flag bit
              1 = The write operation completed
                  (must be cleared in software)
              0 = The write operation is not complete
                  or has not been started
bit 3 WRERR:  EEPROM Error Flag bit
              1 = write operation terminated prematurely
              0 = The write operation completed
bit 2 WREN:   EEPROM Write Enable bit
              1 = Allows write cycles
              0 = Inhibit write to the EEPROM
bit 1 WR:     Write Control bit
              1 = Initiates a write cycle. Bit is
                  cleared once write is complete.
                  Can only be set in software.
              0 = Write cycle to the EEPROM is complete
bit 0 RD:     Read Control bit
              1 = Initiates an EEPROM read. Bit is
                  cleared in hardware. Can only be set
                  in software.
              0 = Does not initiate an EEPROM read
```

Figure 15-3 16F87x EECON1 Register Bitmap

Write operations require two control bits, WR and WREN, and two status bits, WRERR and EEIF. The purpose of these bits is shown in Figure 15-3. Since the WREN bit enables or disables the write operation, it must be set before executing a write. The WR bit is used to initiate the write operation. This bit is automatically cleared at the end of the write. The interrupt flag bit EEIF can be use to detect when the *memory write* completes. The EEIF bit must be cleared by software before setting the WR bit. As soon as the WREN bit and the WR bit have been set, the desired memory address in EEADR is erased and the value in EEDATA written to the selected address.

The WRERR bit indicates when the 16F87x has been reset during a write operation. This bit should be cleared after power-on reset. The WRERR bit is set when a write operation is interrupted by a MCLR reset, or a Watchdog time-out.

Reading EEPROM Data Memory on the 16F87x

Reading an EEPROM data memory location in the 16F87x requires the following operations:

1. Write the address of the EEPROM location to be read to the EEDATA register. The address should be within the device's memory capacity.
2. The EEPGD bit in the EECON1 registered is cleared so as to access data memory.
3. The RD bit in the EECON1 register is set to start the read operation.
4. The data can now be read from the EEDATA register.

The following code fragment shows reading EEPROM data memory in the 16F877 PIC:

```
;===============================
;  16F877 read EEPROM data
;===============================
; Procedure to read EEPROM on-board memory
; ON ENTRY:
; Address of EEPROM memory location to read is stored in
; local register EEMemAdd
; ON EXIT:
; Read data in w
EERead:
        Bank2
        movf    EEMemAdd,W      ; EEPROM address
        movwf   EEADR           ; to read from
        Bank3
        bcf     EECON1,EEPGD    ; Point to Data memory
        bsf     EECON1,RD       ; Start read
        Bank2
        movf    EEDATA,W        ; Data to w register
        Bank0
        Return
```

Writing to EEPROM Data Memory in the 16F87x

Writing to 16F87x EEPROM data memory is more complex than on the 16F84 and much more complex than the read operation. The process consists of the following operations:

1. Make sure that a previous write operation is not in progress. This step is not necessary if write completion is checked at the end of the write routine.
2. The address to be accessed is stored in the EEADR register. Code should make certain that the address is within the device's range.
3. The data to be written is stored in the EEDATA register.
4. The EEPGD bit in the EECON1 register is cleared to select data memory access.
5. The WREN bit in the EECON1 register is set to enable the write function.

6. Interrupts are disabled to make sure the operation is not interrupted.
7. Three special operations are now executed:

    The value 0x55 is written to the EECON2 register.

    The value 0xaa is written to the EECON2 register.

    The WR bit in the EECON1 register is set.
8. Interrupts are enabled if the application uses interrupts.
9. The WREN bit is cleared to prevent accidental write operations.
10. The completion of the write operation can be ascertained either by checking that the WR bit is clear or that EEIF interrupt flag bit is set.

    The following code fragment is a procedure to write EEPROM data:

```
;===============================
;   16F87x write EEPROM data
;===============================
; Procedure to write data byte to EEPROM memory
; ON ENTRY:
; Address to write stored in local register EEMemAdd
; Data byte to write is in local register EEByte
EEWrite:
          Bank3
Wait2Start:
          btfsc     EECON1,WR         ; Wait for
          goto      Wait2Start        ; write to finish
          Bank2
          movf      EEMemAdd,w        ; Address to
          movwf     EEADR             ; SFR
          movf      EEByte,w          ; Data to
          movwf     EEDATA            ; SFR
          Bank3
          bcf       EECON1,EEPGD      ; Point to Data memory
          bsf       EECON1,WREN       ; and enable writes
; Disable interrupts. Can be done in any case
          bcf       INTCON,GIE
; Write special codes
          movlw     0x55              ; First code is 0x55
          movwf     EECON2
          movlw     0xaa              ; Second code is 0xaa
          movwf     EECON2
          bsf       EECON1,WR         ; Start write operation
          nop                         ; Time for write
          nop
; Test for end of write operation
wait2End:
          btfsc     EECON1,WR         ; Wait until WR clear
```

```
            goto      wait2End
; Re-enable interrupts if program uses interrupts
; If not, comment out next line
;           bsf       INTCON,GIE
            bcf       EECON1,WREN      ; Prevent accidental writes
            Bank0
            Return
```

GFR Access Issue in the 16F87x

Data memory space in the 16F87x is partitioned into four separate banks, labeled bank 0 to bank 3. The RP1 and RP0 bits in the Status register are used to select which one of the banks is currently accessible. In programming the *Special Functions Registers* it is necessary to find out on which bank a register is located to select it. The previous code fragment (the write EEPROM data procedure) requires three bank shifts since EEPROM special function registers are located in several banks.

In this context, we sometimes forget that in some PICs the *user registers*, also called the *General Purpose Registers*, can also be located in any one of the banks, although most applications locate the GPRs in bank 0. In the 16F87x there are 96 bytes of available space in bank 0 that can be used. So if the registers used by the application are allocated in this first bank (actually in the first 80 bytes of bank 0) then code must select bank 0 before accessing this data. Had this been the case, the preceding code fragment would have required four additional bank changes.

We were able to avoid this difficulty by placing the program GPRs in the bank 0 memory space from 0x70 to 0x7f. In the 16F87x, addresses 0x70 to 0x7f (15 bytes) are mirrored in the other three banks. In contrast, any GPR allocated below address 0x70 in bank 0 can be accessed only when bank 0 is selected. The 16x84 programmer may not be aware of this fact, since in the 16F84, the memory area available for GPRs, although physically located in bank 0, is mirrored in bank 1.

In the 16F87x, it is good programming practice to locate the most used GPRs in the 0x70 to 0x7f area so as to avoid unnecessary bank changes. Since the space is limited to 15 bytes, the programmer must exercise good judgment in deciding which registers to place in this area.

15.0.3 16F87x EEPROM Circuit and Program

The program Ser2EEP, in the book's online software, is a demonstration of EEPROM memory access on the 16F877 PIC. The program receives character data through the RS-232 line and stores them in EEPROM data memory. Received characters are echoed on the second LCD line. When the **<Enter>** key is detected, the text stored in EEPROM memory is displayed on the LCD. On startup, the top LCD line displays the prompt: "Receiving:". At that time, a message "Rdy-" is sent through the serial line so as to test the connection. Serial communications run at 2400 baud, no parity, 1 stop bit, and 8 character bits. Figure 15-4 (in the following page) shows the circuit used by the Ser2EEP program.

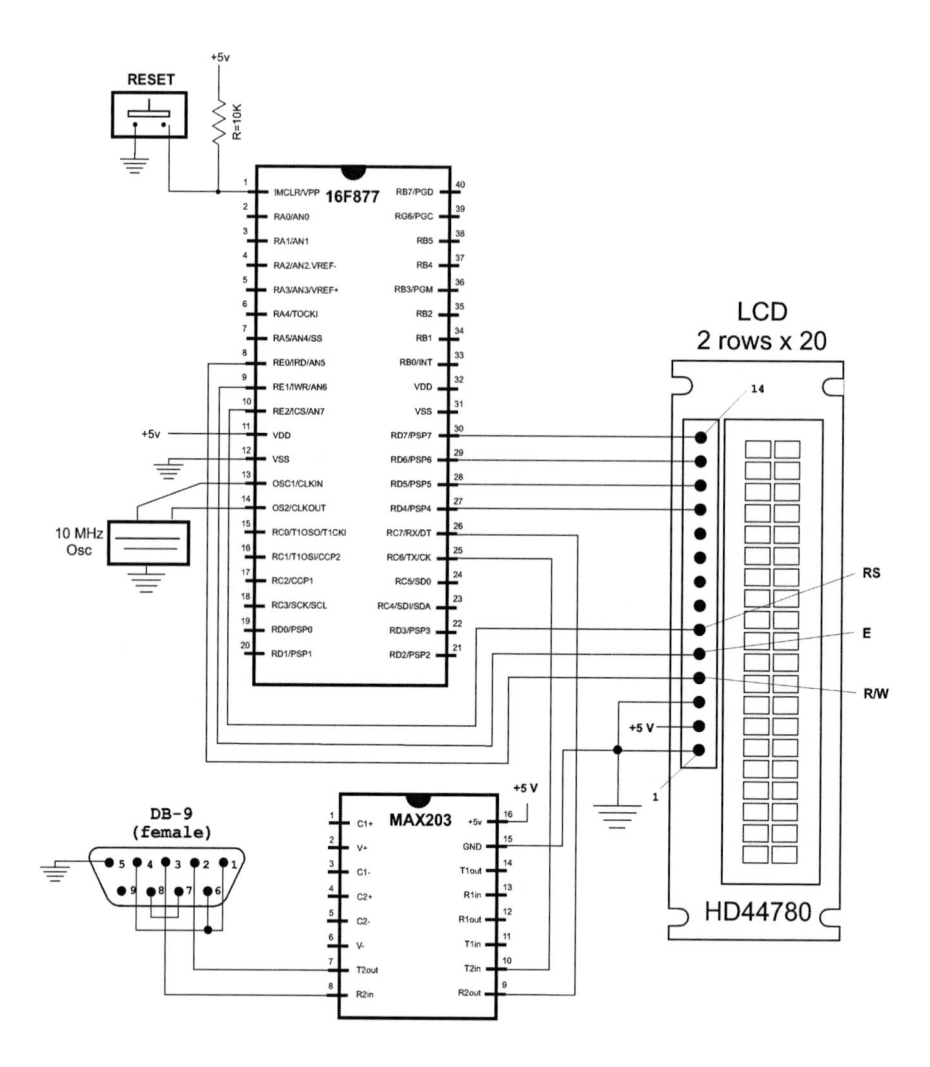

Figure 15-4 Circuit for the Ser2EEP Demonstration Program

The program's main driver routine, constants, and user registers are as follows:

```
;============================================================
;                       M A C R O S
;============================================================
; Macros to select the register banks
Bank0   MACRO                           ; Select RAM bank 0
        bcf     STATUS,RP0
        bcf     STATUS,RP1
        ENDM

Bank1   MACRO                           ; Select RAM bank 1
        bsf     STATUS,RP0
        bcf     STATUS,RP1
        ENDM
```

```
Bank2       MACRO                       ; Select RAM bank 2
            bcf         STATUS,RP0
            bsf         STATUS,RP1
            ENDM

Bank3       MACRO                       ; Select RAM bank 3
            bsf         STATUS,RP0
            bsf         STATUS,RP1
            ENDM
;=======================================================
;                   constant definitions
;   for PIC-to-LCD pin wiring and LCD line addresses
;=======================================================
#define E_line 1            ; |
#define RS_line 0           ; | - from wiring diagram
#define RW_line 2           ; |
; LCD line addresses (from LCD data sheet)
#define LCD_1 0x80          ; First LCD line constant
#define LCD_2 0xc0          ; Second LCD line constant
#define LCDlimit .20; Number of characters per line
#define spbrgVal .64; For 2400 baud on 10Mhz clock
; Note: The constants that define the LCD display
;       line addresses have the high-order bit set
;       so as to meet the requirements of controller
;       commands.
;
;=============================================================
;                   General Purpose Variables
;=============================================================
; Local variables
; Reserve 20 bytes for string buffer
            cblock      0x20
            strData
            endc
; Other data
            cblock      0x34        ; Start of block
            count1                  ; Counter # 1
            count2                  ; Counter # 2
            count3                  ; Counter # 3
            J                       ; counter J
            K                       ; counter K
            bufAdd
            index
            store1                  ; Local storage
            store2
            Endc
```

```
;===============================
;       Common RAM area
;===============================
; These GPRs can be accessed from any bank.
; 15 bytes are available, from 0x70 to 0x7f
        cblock  0x70
; For LCDscroll procedure
        LCDcount            ; Counter for characters per line
        LCDline             ; Current display line (0 or 1)
; Communications variables
        newData             ; not 0 if new data received
        ascVal
        errorFlags
; EEPROM-related variables
        EEMemAdd            ; EEPROM address to access
        EEByte              ; Data byte to write
        endc
;=============================================================
;                       P R O G R A M
;=============================================================
        org     0                   ; start at address
        goto    main
; Space for interrupt handlers
        org     0x08
main:
; Wiring:
;       LCD data to Port-D, lines 0 to 7
;       E line -> Port-E, 1
;       RW line -> Port-E, 2
;       RS line -> Port-E, 0
; Set PORTE D and E for output
; First, initialize Port-B by clearing latches
        clrf    STATUS
        clrf    PORTB
; Select bank 1 to TRIS Port-D for output
        Bank1
; TRIS Port-D for output. Port-D lines 4 to 7 are wired
; to LCD data lines. Port-D lines 0 to 4 are wired to LEDs.
        movlw   b'00000000'
        movwf   TRISD               ; and Port-D
; By default Port-A lines are analog. To configure them
; as digital code must set bits 1 and 2 of the ADCON1
; register (in bank 1)
        movlw   0x06                ; binary 0000 0110  is code to
                                    ; make all Port-A lines
digital
        movwf   ADCON1
; Port-B, lines are wired to keypad switches, as follows:
```

Data EEPROM Programming

```
;         7 6 5 4 3 2 1 0
;         | | | | |_|_|_|_____  switch rows (output)
;         |_|_|_|_____  switch columns (input)
; rows must be defined as output and columns as input
        movlw           b'11110000'
        movwf           TRISB
; TRIS Port-E for output
        movlw           b'00000000'
        movwf           TRISE                   ; TRIS Port-E
; Enable Port-B pullups for switches in OPTION register
        movlw           b'00001000'
        movwf           OPTION_REG
; Back to bank 0
        Bank0
; Initialize serial Port-for 2400 baud, 8 bits, no parity
; 1 stop
        call            InitSerial
; Test serial transmission by sending "RDY-"
        movlw           'R'
        call            SerialSend
        movlw           'D'
        call            SerialSend
        movlw           'Y'
        call            SerialSend
        movlw           '-'
        call            SerialSend
        movlw           0x20
        call            SerialSend
; Clear all output lines
        movlw           b'00000000'
        movwf           PORTD
        movwf           PORTE
; Wait and initialize HD44780
        call            delay_5         ; Allow LCD time to initialize
        call            initLCD         ; Then do forced
                                        ; initialization
        call            delay_5
; Clear character counter and line counter variables
        clrf            LCDcount
        clrf            LCDline
; Set display address to start of first LCD line
        call            line1
; Store address of display buffer
        movlw           0x20
        movwf           bufAdd
; Display "Receiving:" message prompt
        call            blank20  ; Clear buffer
        movlw           0x00                    ; Offset in buffer
```

```
                call    storeMS1 ; Store message at offset
                call    display20       ; Display message
; Start address of EEPROM
                clrf    EEMemAdd
; Setup for display in second line
                call    line2
                clrf    LCDline
                incf    LCDline,f; Set scroll control for line 2
;===============================================================
;           receive serial data, store, and display
;===============================================================
receive:
; Call serial receive procedure
                call    SerialRcv
; HOB of newData register is set if new data
; received
                btfss   newData,7
                goto    scanExit
; At this point new data was received.
                movwf   EEByte          ; Save received character
; Display character on LCD
                movf    EEByte,w ; Recover character
                call    send8           ; Display in LCD
                call    LCDscroll       ; Scroll at end of line
; Store character in EEPROM at location in EEMemAdd
                call    EEWrite ; Local procedure
                incf    EEMemAdd,f      ; Bump to next EEPROM
; Check for <Enter> key (0x0d) and execute display function
                movf    EEByte,w ; Recover last received
                sublw   0x0d
                btfsc   STATUS,Z ; Test if <Enter> key
                goto    isEnter ; Go if <Enter>
; Not <Enter> key, continue processing
scanExit:
                goto    receive         ; Continue
;============================
;    display EEPROM data
;============================
; This routine receives control when the <Enter> key is
; received.
; Action:
;       1. Clear LCD
;       2. Output is set to top LCD line
;       3. Characters stored in EEPROM are displayed
;          until 0x0d code is detected
isEnter:
                call    clearLCD
; Clear character counter and line counter variables
```

Data EEPROM Programming

```
            clrf        LCDcount
            clrf        LCDline
; Read data from EEPROM memory, starting at address 0
; and display on LCD until 0x0d terminator
            call        line1
            clrf        EEMemAdd    ; Start at EEPROM 0
readOne:
            call        EERead      ; Get character
; Store character
            movwf       EEByte      ; Save character
; Test for terminator
            sublw       0x0d
            btfsc       STATUS,Z    ; Test if 0x0d
            goto        atEnd       ; Go if 0x0d
; At this point character read is not 0x0d
; Display on LCD
            movf        EEByte,w    ; Recover character
; Display character on LCD
            call        send8       ; Display in LCD
            call        LCDscroll   ; Scroll at end of line
            incf        EEMemAdd,f  ; Next EEPROM byte
            goto        readOne
; End of execution
atEnd:
            goto        atEnd
```

The Ser2EEP program can be tested with any PC serial communications program set for the program's protocol parameters. We developed and tested the program using Windows Hyperterminal.

15.1 EEPROM Devices and Interfaces

In addition to onboard EEPROM memory that is available in many PICs, a circuit can contain EEPROM memory in separate integrated circuits. The reason for using separate EEPROM is the need for storing more data, since access to onboard EEPROM memory is usually faster, simpler, and requires less interface elements.

EEPROM devices are furnished in two different interface types: serial and parallel. Devices that use the parallel bus require an 8-bit data bus and an address bus wide enough to cover its entire memory space. Although parallel EEPROMS are faster than serial ones, in PIC and microcontroller technology, parallel devices are usually out of the question.

Serial EEPROMS also come in various flavors. In the PIC environment, the most used ones are *I2C (Inter-Integrated Circuit)*, *SPI (Serial Peripheral Interface)*, and *Microwire* which is a subset of SPI. Another interface called *1-Wire*, similar to I2C, finds some use in PIC systems.

Although of different design and having unique architectures, the several types of EEPROM devices share many features. For example, they all operate on a three-phase system that includes an *Opcode*, an *Address*, and a *Data* phase. Although each type of device has a unique instruction set, the basic operations perform similar functions: enable write, enable read, get read status, get write status, read data, and write data. For this reason we have selected a single one of these interfaces: I2C. This interface, probably because of its minimal use of communications lines, seems to be the most popular one in the PIC environment.

15.1.1 The I2C Serial Interface

I²C (or I2C) is a serial computer bus and interface developed by Philips Electronics for use in TV receivers. I2C has found considerable use in embedded systems and is supported by many types of devices; including EEPROMs, thermal sensors, real-time clocks, RF tuners, video decoders, etc. I2C devices are made by Philips, National Semiconductors, Microchip, and many others. The popularity of I2C is often attributed to its simplicity, low implementation cost, and minimal use of communications resources.

The typical use of I2C is for interfacing devices on a single board or in a closed system. The interface uses a two-wire bus and two signals: *SDA (serial data line)* and *SCL (serial clock line)*. These signals support serial transmission 8-bits at a time with 7-bit address-space devices. In the I2C protocol the device that initiates a transaction is called the master and the device being addressed is the slave. Normally the master controls the clock signal, but the slave can hold-off the master in the middle of a transaction by pulling the SCL line low. This is called ".". Not all I2C slave devices support this feature.

The presence of a clock signal makes I2C a synchronous protocol. Since the clock signal is part of the transmission, it can vary without disrupting data. For this reason, I2C is used in systems with imprecise clocks, such as the PIC RC oscillator.

Every I2C hardware slave device has a predefined device address, although some part of this address can be defined at the board level. At the start of every transaction, the master sends the device address of the slave it intends to access. The slave device monitors the bus and responds only to commands that include its own address. The number of available user-configurable address bits limits the number of identical devices that coexist on the same bus.

15.1.2 I2C Communications

SDA and SCL I2C signals are *open-drain*, that is, the master and slave devices can only drive the lines low, or leave them open (high). In operation, a *termination resistor* pulls the line up to Vcc if no I2C device is pulling it down. It is this mechanism that allows a slave device to suspend communications by holding down the SCL line. Furthermore, I2C lines can only be in one of two states, called *"float high"* and *"drive low."* Here again, it is the pull-up resistors that ensure that the line does not float in an unknown state.

Data EEPROM Programming

The pull-up resistors used in I2C hardware vary according to communication speed. Typically, lines at 100 kbps require 4.7K pull-up resistors and lines at 400 kbps 1K resistors.

15.1.3 EEPROM Communications Conditions

The descriptions and examples that follow are limited to EEPROM IC2 device acess. I2C is a bidirectional interface, implemented by an *Acknowledge* or *Ack* system. This system allows data to be sent in one direction to one device on the I2C bus. The device indicates that data has been received by issuing an Ack signal. This action from the receiver eliminates any doubts about whether the transmission was received or not.

Several so-called "conditions" serve to explain I2C communications. The conditions refer to the various bus states during transmission, such as start, stop, data, and acknowledge.

The *START* condition (represented by the letter S) indicates that an R2C device is ready to transfer data on the bus. The device initializes the transmission by pulling the SDA line low. Recall that both lines are high in the normal state. So the S condition is detected by a low SDA and a high SCL.

The *STOP* condition (represented by the letter P) indicates that a device has finished transferring and is releasing the bus. The P condition is detected when the SDA line is released while the SCL line remains high. Thus, by action of the pull-up resistor, the P conditions places both lines high.

The *RESTART* condition (represented by the letter R) indicates that a device is ready to transmit more data without releasing the line. This condition is called a *Repeated Start* (condition *Rs*) in the technical literature. The typical scenario for an R condition is when a START must be sent, but a STOP has not occurred. The RESTART condition issues a new START without releasing the line, as is the case when another data item must be sent. In the R condition, the SCL line is momentarily released while the SDA is held high.

The *DATA TRANSFER* condition (or just *DATA* condition) represents the transmission of 8 data bits by pulsing the SDA line while the SCL is high. The CLK signal and the SDA signal must be aligned so that the high and low bits on the SCL line coincide with the high state of the CLK line. The fact that the SCL line is meaningful only when the CLK signal is high allows the SCL line to change. A DATA byte can be a control code, an address, or an information element.

The *ACK* condition (represented by the letter A) is used to acknowledge a data reception. This condition is furnished by the device by bringing the SDS line low during the 9th clock pulse of a transmission sequence. The sequence starts with the S or R condition (one bit), followed by 8 data bits, and the 9th bit on the line requires that the SDA line be brought low, since the SDA line floats high.

The lack of the ACK signal is interpreted as *NACK* (represented by the letter *N*). NACK represents a negative acknowledge. The NACK signal is a passive response since the SDA line is normally held high. Both ACK (A) and NACK (N) refer to the previous byte of data.

15.1.4 EEPROM Write Operation

Figure 15-5 represents the I2C action sequence that takes place during a write to a small EEPROM, that is, one that requires a single address byte. Later we will see operations that access I2C EEPROMS with a 2-byte address space.

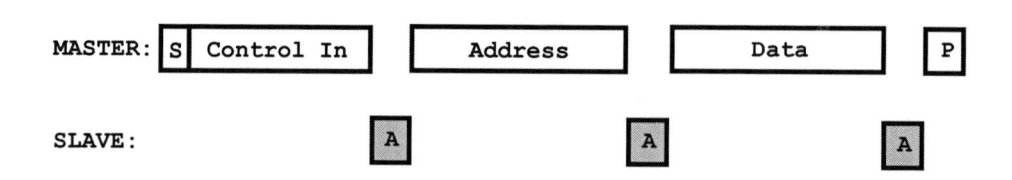

Figure 15-5 I2C Write Sequence to Small EEPROM

Note in Figure 15-5 that three bytes of information are required in the data transfer. The transmission starts with the S condition issued by the master, followed by a control byte. In the case of an EEPROM, the control byte indicates either a read or a write operation. In Figure 15-5, the control byte is labeled *Control In*, since it places the EEPROM device in input mode required for the write to take place. The EEPROM acknowledges the control byte by issuing the A condition. At this point, the master proceeds to transmit the address byte, which defines the EEPROM memory location at which the write operation is to take place. The slave (in this case the EEPROM) acknowledges reception of the address byte by issuing another A condition. Next, data is sent by the master and "ACKed" by the EEPROM. Finally, the master transmits the stop signal (condition P) which concludes the operation. The EEPROM does not proceed to write the data until the P signal is received.

15.1.5 EEPROM Read Operation

The read operation to a small EEPROM is similar to the write. In this case, 4 bytes of information must be exchanged: 3 from the master to the slave and one (the data item read) from the slave to the master. Figure 15-6 shows the sequence.

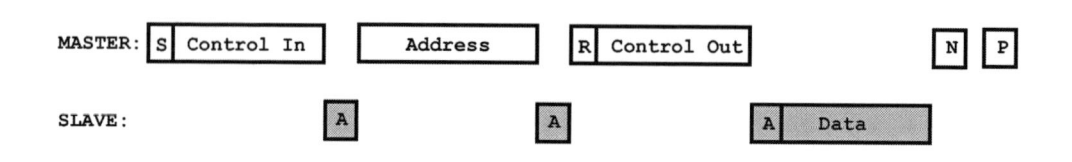

Figure 15-6 Read Sequence to a Small EEPROM

Note in Figure 15-6 that the first command from master to slave is a *Control In* byte to indicate that an address follows. Once the address is acknowledged, the master sends the restart command (R condition) and a *Control Out* command indicating that the master is requesting a read operation. The EEPROM then acknowledges and sends the data, to which the master responds with NACK to instruct the EEPROM that no more data is required.

In both read and write sequences, each byte transmitted requires a response from the other element. This response can be either an ACK (condition A) or a NACK (condition N). The RESTART bit that preceded the Control Out command is necessary since the P condition (STOP bit) has not been sent. The I2C protocol requires that the START condition be sent only on an idle bus, and never in the middle of a transmission.

Read and write operations to large EEPROMS, those with a two-byte address space, are identical to the ones described, except that there are two address bytes. The first one holds the high-order element and the second one the low-order. Here again, each byte transmitted must be acknowledged by the receiver.

15.1.6 I2C EEPROM Devices

EEPROM ICs that conform to the I2C specification are available to the PIC circuit designer. Microchip (the same company that manufactures the PIC microcontrollers) markets a series of I2C chips for this purpose. The line is designated as the 24XXX series of serial EEPROM devices. Table 15.1 lists several I2C EEPROM devices available from Microchip.

Table 15.1
I2C Compatible Serial EEPROM Devices from Microchip

DESIGNATION	MAX CLOCK	CAPACITY BITS	CAPACITY BYTES	PAGE SIZE (IN BYTES)	CASCADE
24xx00	400kHz	128	16	0	No
24xx01	400kHz	1K	128	8/16	No/8
24xx02	400kHz	2K	256	8/16/22	No/8
24xx04	400KHz	4K	512	16	No
24xx08	400Khz	8K	1K	16	No
24xx16	400Khz	16K	2K	16	No
24xx32	400Khz	32K	4K	32	8
24xx64	400Khz	64K	8K	8/32	8
24xx128	400kHz	128K	16K	64	8
24xx256	400kHz	256K	32K	64	8
24xx512	400kHz	512K	64K	64/128	No/4/8
24xx1025	1MHz	1024	128K	128	4

The memory capacity of the various EEPROM ICs ranges from 16 bytes to 128K. Since up to four 128K devices can be *cascaded*, the total accessible memory goes up to 512K. In selecting a particular IC one must take into account several parameters, since the page size, the maximum clock speed, and the number of similar devices that can be grouped changes within the same device type. In the example developed later in this chapter, we used the 24LC04B EEPROM with a total of 512 bytes located in two memory banks.

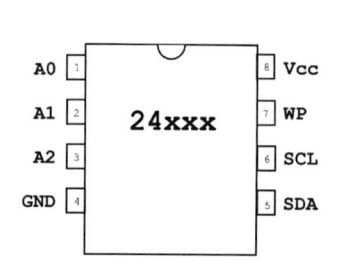

```
              24xxx EEPROM PINOUT

    A0-A2 - Chip address input
    GND   - Ground
    Vcc   - +1.8 to 5.5V power supply
    WP    - Write protect
    SCL   - Serial clock
    SDA   - Serial address/data I/O
```

Figure 15-7 Pin Out of the 24xxx EEPROM Line

The standard DIP package of the 24xxx EEPROMS consists of eight pins, as shown in Figure 15-7.

Lines A0 to A2 are used to encode the chip's address when supported by the device. The three lines allow up to eight possible combinations to identify up to eight similar cascaded devices, as shown in the corresponding column of Table 15.1. Also note in Table 15.1 that several EEPROMs do not support more than one device per address bus. In these cases, pins A0 to A2 are not meaningful. In devices that support this function, the pins must be hardwired to logic 0 or logic 1. If the pins are left floating the device could malfunction.

The SDA pin is bidirectional and is used to transfer addresses and data into and out of the device. Since it is an open drain it requires a pull-up resistor to Vcc. The resistor is typically 10 kΩ for 100 kHz, 2 kΩ for 400 kHz. During data transfer the SDA line is allowed to change only while SCL is low. SDA line changes while SCL is high are used for indicating the START and STOP conditions.

The SCL line is used to synchronize the data transfer to and from the device. The *WP (Write-Protect)* pin provides this function when tied to ground. For normal read/write operation the WP pin is tied to the Vcc line. Read operations are not affected by this pin. The write protect function allows using the EEPROM as a serial ROM.

15.1.7 PIC Master Synchronous Serial Port (MSSP)

Although I2C interfaces can and have been implemented in software, this emulation is not an attractive option now that more efficient and simpler hardware versions of I2C are available in many PICs. For this reason we do not discuss the software emulation of I2C in this book.

Some PIC microcontrollers come equipped with hardware modules to implement EEPROM serial protocols, including SPI and I2C. The module that provides these interfaces is named the *Master Synchronous Serial Port*, or *MSSP*. Although the MSSP module operates in slave or master mode, in the context of EEPROM programming the MSSP is set in master mode. The MSSP module can operate in a *free bus* mode,

Data EEPROM Programming

also called the multi-master function. In this section we discuss MSSP master mode operations.

I2C uses two communications lines, labeled the SDA or data line, and the SCL or clock line. The PICs that contain I2C hardware implementation multiplex two pins for these functions. In the case of the 16F877 (which we use in the forthcoming examples) the SCL line is attached to bit 3 in Port-C (16F877 pin number 18) and the SDA line to bit 4 in Port-C (16F877 pin number 23). When the PIC is used in MSSP mode, these two pins must be initialized for input by setting the corresponding TRIS register bits. The pull-up resistors for these lines must be provided externally. Figure 15-8 shows the minimal wiring diagram between a 16F877 PIC and a 24LC04B EEPROM IC.

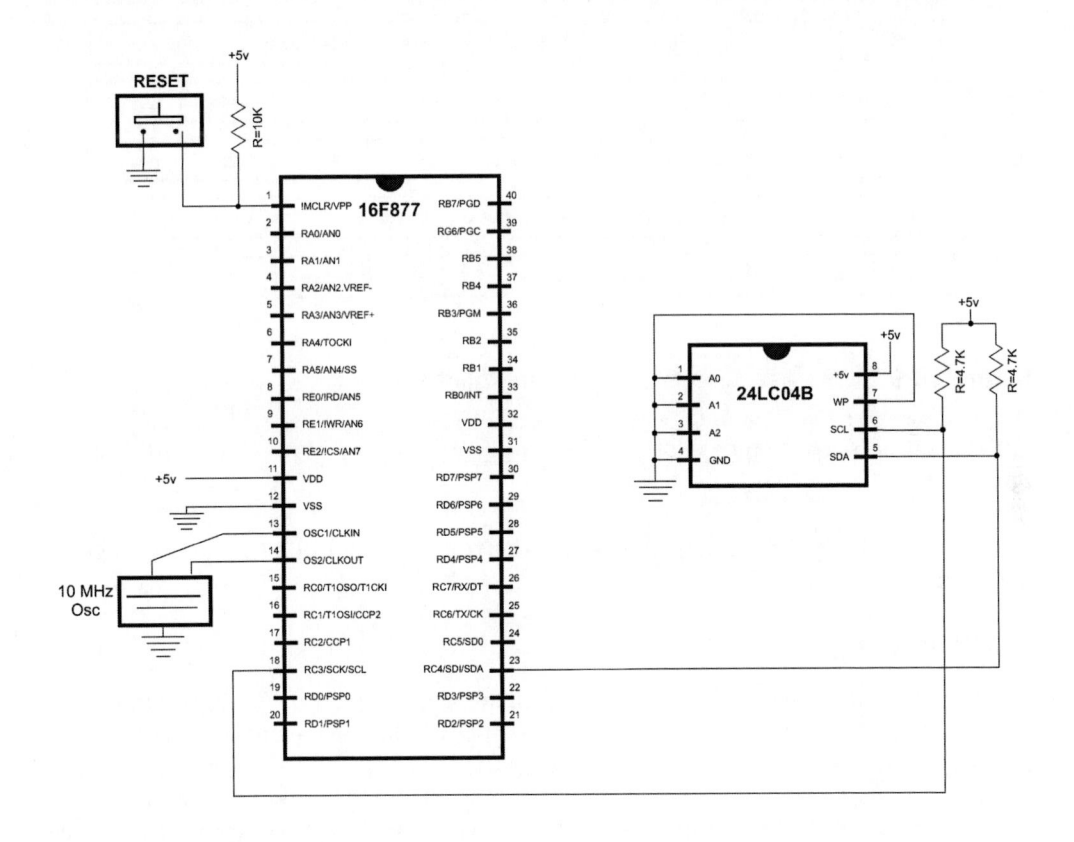

Figure 15-8 Wiring Diagram between a 16F877 PIC and 24LC04B EEPROM.

Note in Figure 15-8 that the address lines (A0 to A2) in the 24LC04B IC are wired to ground. The reason is that these lines are not used in this particular EEPROM (see Table 15.1). Also wired to ground is the *write protect* line. This allows read and write operations. Only two connections are required between the PIC and the EEPROM: the SLK and SDL lines. The 4.7K resistors are pull-ups to implement the open drain operation on these lines.

REGISTER NAME	7	6	5	4	3	2	1	0	bits
INTCON	GIE	PEIE							
PIR1					SSPIF				
PIE1					SSPIE				
PIR2					BCLIF				
PIE2					BCLIE				
SSPBUF	(Receive Buffer/Transmit Register)								
SSPCON	WCOL	SSPOV	SSPEN	CKP	SSPM3	SSPM2	SSPM1	SSPM0	
SSPCON2	GCEN	ACKSTAT	ACKDT	ACKEN	RCEN	PEN	RSEN	SEN	
SSPADD	(I2C Slave Address/Master Baud Rate Register)								
SSPSTAT	SMP	CKE	D/_A	P	S	R/_W	UA	BF	

Figure 15-9 SFRs Associated with I2C Operations

Several 16F87x registers relate to MSSP operation in I2C mode. Figure 15-9 shows these SFRs.

In the following subsection we discuss the registers and bits that apply to MSSP operation in Master Mode.

MSSP in Master Mode

In the context of accessing EEPROM circuits, the MSSP is operated in master mode. At this point we should consider that although the EEPROM device operates as a slave, it is a "smart" slave since it has a control engine capable of performing operations on its own, including reading and writing to its address space, recognizing commands, and issuing the corresponding responses. For example, in a data write operation the master sends the corresponding command code, followed by the address to which the data is to be written, followed by the data itself. The peripheral (in this case the EEPROM IC) receives and acknowledges the various bytes and executes the requested operations. In the case of a read command the EEPROM fetches and returns the data from the memory address requested in the command.

One of the special function registers most used in MSSP master mode operations is the SSPCON. Figure 15-10 is a bitmap of this register when operating in I2C master mode.

Data EEPROM Programming

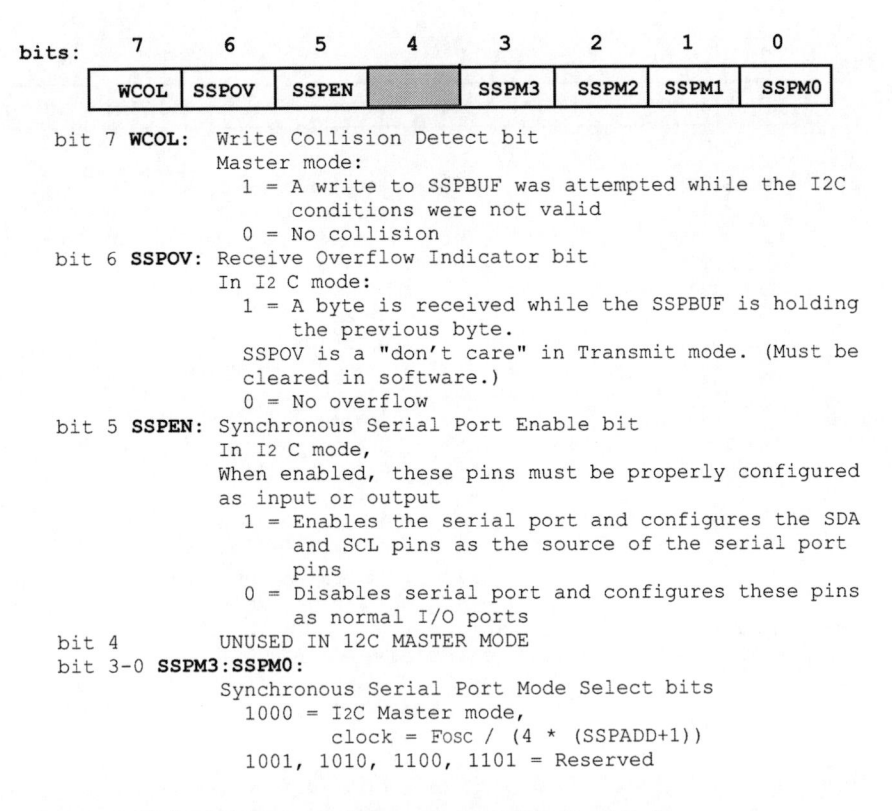

```
bits:    7      6      5      4      3      2      1      0
       ┌──────┬──────┬──────┬──────┬──────┬──────┬──────┬──────┐
       │ WCOL │ SSPOV│ SSPEN│▓▓▓▓▓▓│ SSPM3│ SSPM2│ SSPM1│ SSPM0│
       └──────┴──────┴──────┴──────┴──────┴──────┴──────┴──────┘

bit 7  WCOL:   Write Collision Detect bit
               Master mode:
                   1 = A write to SSPBUF was attempted while the I2C
                       conditions were not valid
                   0 = No collision
bit 6  SSPOV:  Receive Overflow Indicator bit
               In I2 C mode:
                   1 = A byte is received while the SSPBUF is holding
                       the previous byte.
                   SSPOV is a "don't care" in Transmit mode. (Must be
                   cleared in software.)
                   0 = No overflow
bit 5  SSPEN:  Synchronous Serial Port Enable bit
               In I2 C mode,
               When enabled, these pins must be properly configured
               as input or output
                   1 = Enables the serial port and configures the SDA
                       and SCL pins as the source of the serial port
                       pins
                   0 = Disables serial port and configures these pins
                       as normal I/O ports
bit 4          UNUSED IN I2C MASTER MODE
bit 3-0 SSPM3:SSPM0:
               Synchronous Serial Port Mode Select bits
                   1000 = I2C Master mode,
                       clock = Fosc / (4 * (SSPADD+1))
                   1001, 1010, 1100, 1101 = Reserved
```

Figure 15-10 SSPCON Register Bitmap in I2C Master Mode

The *WCOL* bit is an error flag that indicates that a *Write Collision* has occurred. Write collisions do not take place when programming an EEPROM device. This bit is useful in multi-master systems since it can detect when more than one master device is attempting to write to the bus.

The *SSPOV* bit (*Synchronous Serial Port Overflow*) is set by the microcontroller whenever there is an overflow error. An overflow occurs whenever an I2C transfer finishes but the previous data has not been read from SSPBUF. If SSPOV bit is set, it must be cleared by application code. Data in SSPBUF is not updated until the overflow condition is cleared.

The *SSPEN* bit (*Synchronous Serial Port Enable*) is set to turn on the SSP module, as is the case in I2C communications.

The bits SSPM0 through SSPM3 (*Synchronous Serial Port mode* bits) determine whether the MSSP module is configured for SPI or I2C and whether it is in slave or master mode. In the master mode, the MSSP module handles all details of I2C communications, such as generating the various conditions and sending and receiving data. The Master Mode is enabled by entering the binary value 1000 in this bit field.

Another frequently used register in I2C communications is SSPCON2. Figure 15-11 is a bitmap of this register in the I2C master mode.

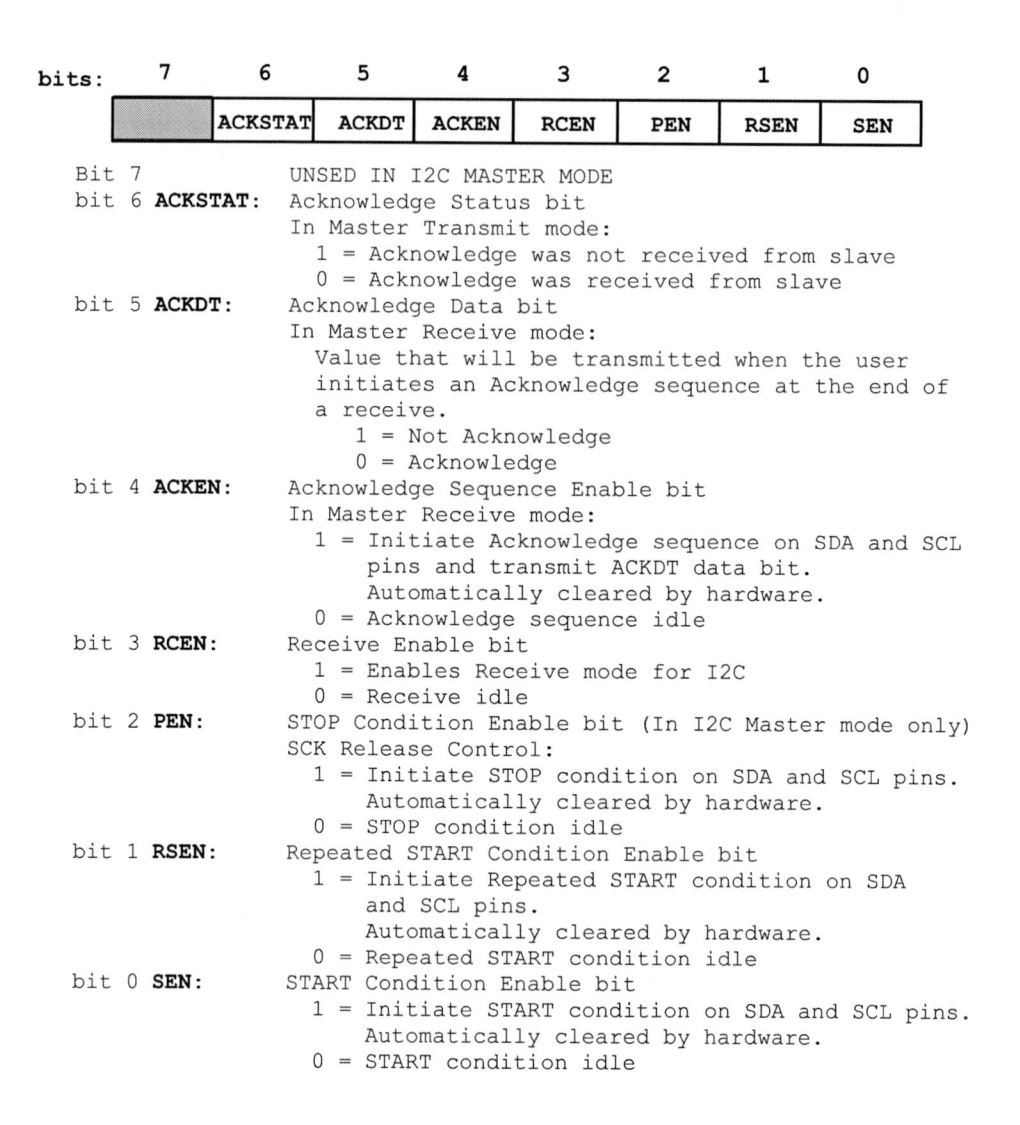

bits:	7	6	5	4	3	2	1	0
		ACKSTAT	ACKDT	ACKEN	RCEN	PEN	RSEN	SEN

Bit 7   UNSED IN I2C MASTER MODE
bit 6 **ACKSTAT**:   Acknowledge Status bit
                          In Master Transmit mode:
                              1 = Acknowledge was not received from slave
                              0 = Acknowledge was received from slave
bit 5 **ACKDT**:   Acknowledge Data bit
                          In Master Receive mode:
                              Value that will be transmitted when the user
                              initiates an Acknowledge sequence at the end of
                              a receive.
                                  1 = Not Acknowledge
                                  0 = Acknowledge
bit 4 **ACKEN**:   Acknowledge Sequence Enable bit
                          In Master Receive mode:
                              1 = Initiate Acknowledge sequence on SDA and SCL
                                  pins and transmit ACKDT data bit.
                                  Automatically cleared by hardware.
                              0 = Acknowledge sequence idle
bit 3 **RCEN**:   Receive Enable bit
                              1 = Enables Receive mode for I2C
                              0 = Receive idle
bit 2 **PEN**:   STOP Condition Enable bit (In I2C Master mode only)
                          SCK Release Control:
                              1 = Initiate STOP condition on SDA and SCL pins.
                                  Automatically cleared by hardware.
                              0 = STOP condition idle
bit 1 **RSEN**:   Repeated START Condition Enable bit
                              1 = Initiate Repeated START condition on SDA
                                  and SCL pins.
                                  Automatically cleared by hardware.
                              0 = Repeated START condition idle
bit 0 **SEN**:   START Condition Enable bit
                              1 = Initiate START condition on SDA and SCL pins.
                                  Automatically cleared by hardware.
                              0 = START condition idle

Figure 15-11 SSPON2 Register Bitmap in I2C Master Mode

The ACKSTAT bit is set when an ACK or NACK has been received. This bit can be tested by application code to determine if an ACK or NACK condition was received.

When the master reads data from a device, it must acknowledge the transfer by sending an ACK or NACK condition. The ACKDT bit determines the value of the condition to be sent: if it is clear an ACK is sent; otherwise a NACK is sent.

The ACKEN bit determines when the acknowledge condition is sent.

The RCEN bit places the MSSP module into I2C receive mode. When one byte of data is received, this bit automatically clears and the PIC returns to transmit mode. Code must ACK or NACK the data then reset this bit.

Data EEPROM Programming

Setting the PEN bit automatically sends a stop condition. This bit is automatically cleared at the end of the start condition.

The RSEN bit sends a restart condition. After the bit is set, application code must wait for the transfer to complete. This bit is reset automatically when the condition or data transfer finishes.

The SEN bit (for *Start condition Enable*) is equivalent to sending a start or restart condition. The SEN bit is reset after the start condition completes.

The *SSPSTAT* (*Synchronous Serial Port Status*) register contains three bits related to IC2 communications in master mode. The SMP bit controls the *slew rate*. The slew rate is a squelch filter for the I2C waveform that improves performance when transmission takes place at 400 kbps. This bit should be set at the 400 kbps transmission rate and reset at any slower rate. The CKE bit is used to allow the MSSP module to handle *SMBus peripherals*. Normally, this bit should be cleared. The *BF* bit (*buffer full*) indicates the SSPBUF contains unread data. In either the master or slave mode this data must be read before any other data is sent or received. The BF flag is set and cleared by the PIC. If SSPBUF is not read before another byte is received the buffer overflows and the SSPOV bit will be set.

Finally, the *SSPADD* (*Synchronous Serial Port Address*) register has a unique function in the I2C master mode: it controls the bus speed. The value entered into the SSPADD register determines the Baud Rate according to the following formula:

$$BaudRate = \frac{Fosc}{4 \bullet (SSPADD_{VAL} + 1)}$$

where *Fosc* is the oscillator speed in MHz. Solving this formula in terms of the value to be entered into *SSPADD*, we have:

$$SSPADD_{VAL} = \frac{Fosc}{4 \bullet Baud\ Rate} - 1$$

For a baud rate of 100 kbps (equal to 100,000Mhz) the formula is:

$$SSPADD_{VAL} = \frac{10,000,000}{4 \bullet 100,000} - 1 = \frac{100}{4} - 1 = 24$$

In this case, the value to be entered into the SSPADD register while using a communications speed of 100 kbps, in a PIC with a 10 KHz oscillator, is 24. The calculations can be checked by substituting into the original formula:

$$Baud\ Rate = \frac{Fosc}{4 \bullet (24+1)} = \frac{10,000,000}{100} = 100,000\ Mhz = 100\ kbps$$

15.1.8 I2C Serial EEPROM Programming on the 16F877

The 16F87x PIC family contains the Master Synchronous Serial Port module, which can be set in either Serial Peripheral Interface or Inter-Integrated Circuit mode. In the I2C mode the module performs either as a master, a multi-master, or a slave. In the context of driving an I2C EEPROM device, the MSSP module is initialized in the master mode. I2C firmware modes are provided for compatibility with other mid-range products.

The demonstration program named I2CEEP in the book's on line software receives character data from a PC through the RS-232 line and stores these characters in a 24LC04B EEPROM IC. The program uses the I2C serial interface facilities provided by the PIC's MSSP module. An on-board LCD echoes the received characters. When the PC user presses **<Enter>** text stored in the EEPROM IC is retrieved and displayed on the LCD.

On startup, the top LCD line displays the prompt: "Receiving:". At that time, a message "Rdy-" is sent through the serial line so as to test the connection. The program's serial communications run at 2400 baud, no parity, 1 stop bit, and 8 character bits. The 24LC04B SDA line is wired to PIC RC4 (MSSP SDA) and the SCL line is wired to PIC RC3 (MSSP SCL). In the 24LC04B the A0-A2 are not used. In the demonstration circuit, the WP lines are wired to ground. Program provides little error checking. The circuit in Figure 15-12 is used with the demonstration program.

The I2CEEP program includes three I2C-related functions:

1. **SetupI2C**. Initializes MSSP module for I2C mode in hardware master mode, configures the I2C lines, sets the slew rate for 100kbps, and sets the baud rate for 10Mhz
2. **WriteI2C**. Writes one byte to I2C EEPROM device. Data and address are stored in local variables.
3. **ReadI2C**. Reads one byte from I2C EEPROM device. Address is stored in a local variable and read data is returned in the w register.

As in previous 16F877 examples, we have placed the most used variables in the common RAM area, that is, in GPRs located from 0x70 to 0x7f. All three procedures use bank changing macros described and listed previously.

IC2 Initialization Procedure

The following procedure from the I2CEEP program initializes the MSSP module for operation in I2C mode with a 24LC04B EEPROM IC. The module is initialized for master mode operation on a PIC with a 10MhZ baud rate. For use with a faster or slower oscillator the value stored in the SSPADD register must be modified according to the formula.

```
;============================
;    I2C setup procedure
;============================
SetupI2C:
        Bank1
```

Data EEPROM Programming

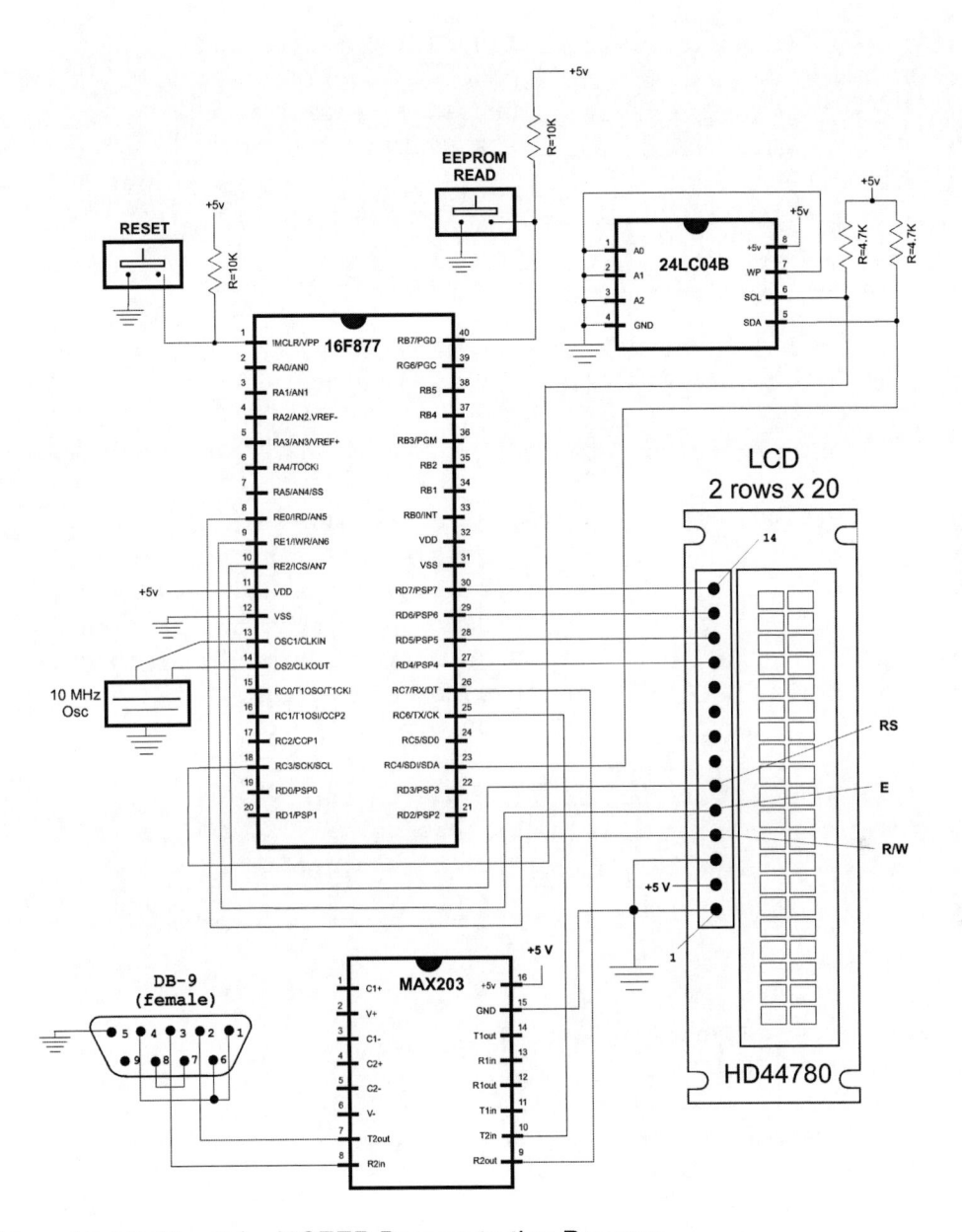

Figure 15-12 Circuit for I2CEEP Demonstration Program

```
        movlw    b'00011000'
        iorwf    TRISC,f           ; OR into TRISC
; Setup MSSP module for Master Mode operation
        Bank0
        movlw    B'00101000'; Enables MSSP and uses appropriate
;   0  0  1  0  1  0  0  0   Value to install
;   7  6  5  4  3  2  1  0   <== SSPCON bits in this operation
;   |  |  |  |  |__|__|__|___ Serial port select bits
```

```
;    |  |  |  |                  1000 = I2C master mode
;    |  |  |  |                  Clock = Fosc/(4*(SSPAD+1))
;    |  |  |  |_____  UNUSED IN MASTER MODE
;    |  |  |_____  SSP Enable
;    |  |                        1 = SDA and SCL pins as serial
;    |  |_____  Receive Overflow indicator
;    |                           0 = no overflow
;    |_____  Write collision detect
;                                0 = no collision detected
            movwf    SSPCON     ; Loaded into SSPCON
; Input levels and slew rate as standard I2C
            Bank1
            movlw    B'10000000'
;
; 1 0 0 0 0 0 0 0  Value to install
; 7 6 5 4 3 2 1 0  <== SSPSTAT bits in this operation
; | | | | | | | |___ Buffer full status bit READ ONLY
; | | | | | | |_____ UNUSED in present application
; | | | | | |_____ Read/write information READ ONLY
; | | | | |_____ UNUSED IN MASTER MODE
; | | | |_____ STOP bit READ ONLY
; | | |_____ Data address READ ONLY
; | |_____ SMP bus select
; |                  0 = use normal I2C specs
; |_____ Slew rate control
;                    0 = disabled
;
            movwf    SSPSTAT
; Setup Baud Rate
; Baud Rate = Fosc/(4*(SSPADD+1))
;    Fosc = 10Mhz
;    Baud Rate = 24 for 100 kbps
            movlw    .24        ; Value to use
            movwf    SSPADD     ; Store in SSPADD
            Bank0
            return
```

The procedures Send1I2c, WaitI2C, and the label FailI2C are listed in the subsection on the read procedure.

I2C Write Byte Procedure

The following procedure, from the I2CEEP program, writes one byte of data to an 24LC04B EEPROM IC, at the memory address stored in the variable EEMemAdd. The value to write is stored in the local variable EEByte.

Data EEPROM Programming

```
;==============================
;      I2C write procedure
;==============================
; Write one byte to I2C EEPROM 24LC04B
; Steps:
;                 1. Send START
;                 2. Send control. Wait for ACK
;                 3. Send address. Wait for ACK
;                 4. Send data. Wait for ACK
;                 5. Send STOP
; STEP 1:
WriteI2C:
        Bank1
        bsf     SSPCON2,SEN      ; Produce START Condition
        call    WaitI2C          ; Wait for I2C to complete
; STEP 2:
; Send control byte. Wait for ACK
        movlw   LC04READ         ; Control byte
        call    Send1I2C         ; Send Byte
        call    WaitI2C          ; Wait for I2C to complete
        btfsc   SSPCON2,ACKSTAT  ; Check ACK bit to see if
                                 ; I2C failed,
skip if not
        goto    FailI2C
; STEP 3:
; Send address. Wait for ACK
        Bank0
        movf    EEMemAdd,w  ; Load Address Byte
        call    Send1I2C         ; Send Byte
        call    WaitI2C  ; Wait for I2C operation to complete
        Bank1
        btfsc   SSPCON2,ACKSTAT ; Check ACK Status bit to see
                                 ; if I2C failed, skip if not
        goto    FailI2C
; STEP 4:
; Send data. Wait for ACK
        Bank0
        movf    EEByte,w         ; Load Data Byte
        call    Send1I2C    ; Send Byte
        call    WaitI2C     ; Wait for I2C operation to complete
        Bank1
        btfsc   SSPCON2,ACKSTAT ; Check ACK Status bit to see
                                 ; if I2C failed, skip if not
        goto    FailI2C
; STEP 5:
; Send STOP. Wait for ACK
        bsf     SSPCON2,PEN ; Send STOP condition
        call    WaitI2C  ; Wait for I2C operation to complete
```

; WRITE operation has completed successfully.
```
        Bank0
        return
```

The procedures Send1I2c, WaitI2C, and the label FailI2C are listed in the sub-section on the read procedure.

I2C Read Byte Procedure

The following procedure, from the I2CEEP program, reads a byte of data from the 24LC04B device. The address read is stored in the local variable EEMemAdd. The value read is returned in the w register. The listing also includes the support routines used by all three I2C procedures listed.

```
;==============================
;    I2C read procedure
;==============================
; Procedure to read one byte from 24LC04B EEPROM
; Steps:
;               1. Send START
;               2. Send control. Wait for ACK
;               3. Send address. Wait for ACK
;               4. Send RESTART + control. Wait for ACK
;               5. Switch to receive mode. Get data.
;               6. Send NACK
;               7. Send STOP
;               8. Retreive data into w register
; STEP 1:
ReadI2C
; Send RESTART. Wait for ACK
        Bank1
        bsf     SSPCON2,RSEN ; RESTART Condition
        call    WaitI2C ; Wait for I2C operation
; STEP 2:
; Send control byte. Wait for ACK
        movlw   LC04READ ; Control byte
        call    Send1I2C ; Send Byte
        call    WaitI2C ; Wait for I2C operation
; Now check to see if I2C EEPROM is ready
        Bank1
        btfsc   SSPCON2,ACKSTAT ; Check ACK Status bit
        goto    ReadI2C ; ACK Poll waiting for EEPROM
                                ; write to complete
; STEP 3:
; Send address. Wait for ACK
        Bank0
        movf    EEMemAdd,w      ; Load from address register
        call    Send1I2C        ; Send Byte
        call    WaitI2C         ; Wait for I2C operation
        Bank1
```

Data EEPROM Programming

```
                btfsc   SSPCON2,ACKSTAT ; Check ACK Status bit
                goto    FailI2C    ; failed, skipped if successful
; STEP 4:
; Send RESTART. Wait for ACK
                bsf     SSPCON2,RSEN ; Generate RESTART Condition
                call    WaitI2C ; Wait for I2C operation
; Send output control. Wait for ACK
                movlw   LC04WRITE ; Load CONTROL BYTE (output)
                call    Send1I2C          ; Send Byte
                call    WaitI2C ; Wait for I2C operation
                Bank1
                btfsc   SSPCON2,ACKSTAT ; Check ACK Status bit
                goto    FailI2C ; failed, skipped if successful
; STEP 5:
; Switch MSSP to I2C Receive mode
                bsf     SSPCON2,RCEN ; Enable Receive Mode (I2C)
; Get the data. Wait for ACK
                call    WaitI2C ; Wait for I2C operation
; STEP 6:
; Send NACK to acknowledge
                Bank1
                bsf     SSPCON2,ACKDT ; ACK DATA to send is 1 (NACK)
                bsf     SSPCON2,ACKEN ; Send ACK DATA now.
; Once ACK or NACK is sent, ACKEN is automatically cleared
; STEP 7:
; Send STOP. Wait for ACK
                bsf     SSPCON2,PEN ; Send STOP condition
                call    WaitI2C ; Wait for I2C operation
; STEP 8:
; Read operation has finished
                Bank0
                movf    SSPBUF,W ; Get data from SSPBUF into W
; Procedure has finished and completed successfully.
                return

;=============================
;    I2C support procedures
;=============================
; I2C Operation failed code sequence
; Procedure hangs up. User should provide error handling.
FailI2C
                Bank1
                bsf     SSPCON2,PEN ; Send STOP condition
                call    WaitI2C ; Wait for I2C operation
fail:
                goto    fail
; Procedure to transmit one byte
Send1I2C
```

```
                Bank0
                movwf   SSPBUF  ; Value to send to SSPBUF
                return
; Procedure to wait for the last I2C operation to complete.
; Code polls the SSPIF flag in PIR1.
WaitI2C
                Bank0
                btfss   PIR1,SSPIF  ; Check if I2C operation done
                goto    $-1 +               ; I2C module is not ready yet
                bcf     PIR1,SSPIF          ; I2C ready, clear flag
                return
```

15.2 Sample Programs

The following sections contain the code listing for the programs discussed in this chapter.

15.2.1 EECounter Program

```
; File name: EECounter.asm
; Last Update: May 22, 2006
; Author: Julio Sanchez
; Processor: 16F84A
;
; Description:
; Program to demonstrate on chip EEPROM data memory read
; and write operation. Program uses LCD display to output
; results.
; Operation:
; The program keeps track and displays the inNum of times
; the code has been started.
; For LCD display parameters see the LCDTest2 program.
; WARNING:
; Code assumes 4Mhz clock. Delay routines must be
; edited for faster clock
;
;===========================
;         switches
;===========================
; Switches used in __config directive:
;    _CP_ON         Code protection ON/OFF
; *  _CP_OFF
; *  _PWRTE_ON      Power-up timer ON/OFF
;    _PWRTE_OFF
;    _WDT_ON        Watchdog timer ON/OFF
; *  _WDT_OFF
;    _LP_OSC        Low power crystal oscillator
```

```
;  *  _XT_OSC         External parallel resonator/crystal oscillator
;
;     _HS_OSC         High speed crystal resonator (8 to 10 MHz)
;                     Resonator: Murate Erie CSA8.00MG = 8 MHz
;     _RC_OSC         Resistor/capacitor oscillator
;  |                  (simplest, 20% error)
;  |
;  |_____  * indicates setup values presently selected

;===========================
; setup and configuration
;===========================
          processor 16f84A
          include   <p16f84A.inc>
          __config  _XT_OSC & _WDT_OFF & _PWRTE_ON & _CP_OFF


;========================================================
;                  constant definitions
;   for PIC-to-LCD pin wiring and LCD line addresses
;========================================================
#define E_line 1            ; |
#define RS_line 2           ; | — from wiring diagram
#define RW_line 3           ; |
; LCD line addresses (from LCD data sheet)
#define LCD_1 0x80          ; First LCD line constant
#define LCD_2 0xc0          ; Second LCD line constant
; Note: The constants that define the LCD display line
;       addresses have the high-order bit set in
;       order to facilitate the controller command
;
;========================================================
;                  variables in PIC RAM
;========================================================
; Reserve 16 bytes for string buffer
          cblock    0x0c
          strData
          endc
; Reserve three bytes for ASCII digits
          cblock    0x1d
          asc100
          asc10
          asc1
          endc
; Continue with local variables
          cblock    0x20              ; Start of block
          count1                      ; Counter # 1
          count2                      ; Counter # 2
          count3                      ; Counter # 3
```

```
            pic_ad              ; Storage for start of text area
                                ; (labeled strData) in PIC RAM
            J                   ; counter J
            K                   ; counter K
            index               ; Index into text table (also used
                                ; for auxiliary storage)
            store1              ; Local temporary storage
            store2              ; Storage # 2
; EEPROM-related variables
            EEMemAdd ; EEPROM address to access
            EEByte              ; Data byte to write
; Storage for ASCII decimal conversion and digits
            inNum               ; Source operand
            thisDig             ; Digit counter
            endc


;============================================================
;                          program
;============================================================
            org     0           ; start at address
            goto    main
; Space for interrupt handlers
            org     0x08

main:
            movlw   b'00000000' ; All lines to output
            tris    PORTA       ; in Port-A
            tris    PORTB       ; and Port-B
            movlw   b'00000000' ; All outputs ports low
            movwf   PORTA
            movwf   PORTB
; Wait and initialize HD44780
            call    delay_5     ; Allow LCD time to initialize
                                ; itself
            call    delay_5
            call    initLCD     ; Then do forced
initialization
            call    delay_5     ; Wait again
; Store base address of text buffer in PIC RAM

            movlw   0x0c        ; Start address for buffer
            movwf   pic_ad      ; to local variable
; Initialize EEPROM data to 0x0
            clrf    EEMemAdd ; Set address to 0
;======================
;   first LCD line
;======================
; Store 16 blanks in PIC RAM, starting at address stored
```

Data EEPROM Programming

```
;   in variable pic_ad
            call        blank16
;   Call procedure to store ASCII characters for message
;   in text buffer
            movlw       d'0'                ; Offset into buffer
            call        storeMS1
;========================
;   Read EEPROM memory
;========================
;   EEPROM memory address to use is at 10 (0x0a). Variable
;   EEMemAdd is already initialized.
;   Fill data for EEPROM is 0xff. This value indicates
;   the first iteration
            call        EERead              ; Local procedure. Value in w
            movwf       EEByte              ; Save result
;   EEPROM data still in w
            incf        EEByte,f
            call        EEWrite
;   At this point iteration inNum is stored in EEByte
;   This value must be displayed on the LCD at offset 11
;   of the first line. This means it must be stored at offset
;   11 in the buffer. Since the buffer starts at 0x0c the
;   iteration digit must be stored at offset 0x0c+11=0x17
ShowEEData:
;   Binary data in EEByte
            movf        EEByte,w ; Value to w
            call        bin2asc             ; Conversion routine
;   At this point three ASCII digits are stored in local
;   variables. Move digits to display area
            movf        asc1,w              ; Unit digit
            movwf       0x18                ; Store in buffer
            movf        asc10,w             ; same with other digits
            movwf       0x17
            movf        asc100,w
            movwf       0x16
;   Display line
;   Set DDRAM address to start of first line
showLine:
            call        line1
;   Call procedure to display 16 characters in LCD
            call        display16
loopHere:
            goto        loopHere    ;done


;===============================================================
;                   initialize LCD for 4-bit mode
;===============================================================
initLCD:
```

```
; Initialization for Densitron LCD module as follows:
;    4-bit interface
;    2 display lines of 16 characters each
;    cursor on
;    left-to-right increment
;    cursor shift right
;    no display shift
;======================|
;   set command mode   |
;======================|
        bcf     PORTA,E_line     ; E line low
        bcf     PORTA,RS_line    ; RS line low
        bcf     PORTA,RW_line    ; Write mode
        call    delay_125        ; delay 125 microseconds
;*********************|
;     FUNCTION SET    |
;*********************|
        movlw   0x28     ; 0 0 1 0 1 0 0 0 (FUNCTION SET)
        call    send8    ; 4-bit send routine

; Set 4-bit mode command must be repeated
        movlw   0x28
        call    send8

;*********************|
; DISPLAY AND CURSOR ON |
;*********************|
        movlw   0x0e     ; 0 0 0 0 1 1 1 0 (DISPLAY ON/OFF)
        call    send8
;*********************|
;   set entry mode    |
;*********************|
        movlw   0x06     ; 0 0 0 0 0 1 1 0 (ENTRY MODE SET)
        call    send8

;*********************|
; cursor/display shift |
;*********************|
        movlw   0x14     ; 0 0 0 1 0 1 0 0 (CURSOR/DISPLAY
SHIFT)
        call    send8
;*********************|
;    clear display    |
;*********************|
        movlw   0x01     ; 0 0 0 0 0 0 0 1 (CLEAR DISPLAY)
        call    send8
; Per documentation
        call    delay_5  ; Test for busy
```

Data EEPROM Programming

```
        return

;=======================
;   Procedure to delay
;     42 microseconds
;=======================
delay_125
        movlw    D'42'              ; Repeat 42 machine cycles
        movwf    count1             ; Store value in counter
repeat
        decfsz   count1,f           ; Decrement counter
        goto     repeat             ; Continue if not 0
        return                      ; End of delay

;=======================
;   Procedure to delay
;     5 milliseconds
;=======================
delay_5
        movlw    D'41'              ; Counter = 41
        movwf    count2             ; Store in variable
delay
        call     delay_125          ; Delay
        decfsz   count2,f           ; 40 times = 5 milliseconds
        goto     delay
        return                      ; End of delay
;=======================
;     pulse E line
;=======================
pulseE
        bsf      PORTA,E_line       ; Pulse E line
        nop
        bcf      PORTA,E_line
        return

;=============================
;    long delay sub-routine
;       (for debugging)
;=============================
long_delay
        movlw    D'200'      ; w = 200 decimal
        movwf    J           ; J = w
jloop:
        movwf    K           ; K = w
kloop:
        decfsz   K,f         ; K = K-1, skip next if zero
        goto     kloop
        decfsz   J,f         ; J = J-1, skip next if zero
```

```
                goto    jloop
                return
;==============================
;    LCD display procedure
;==============================
; Sends 16 characters from PIC buffer with address stored
; in variable pic_ad to LCD line previously selected
display16
                call    delay_5                 ; Make sure not busy
; Set up for data
                bcf     PORTA,E_line            ; E line low
                bsf     PORTA,RS_line           ; RS line high for data
; Set up counter for 16 characters
                movlw   D'16'                   ; Counter = 16
                movwf   count3
; Get display address from local variable pic_ad
                movf    pic_ad,w ; First display RAM address to W
                movwf   FSR                     ; W to FSR
getchar
                movf    INDF,w   ; get character from display RAM
                                 ; location pointed to by file select
                                 ; register
                call    send8                   ; 4-bit interface routine
; Test for 16 characters displayed
                decfsz  count3,f                ; Decrement counter
                goto    nextchar ; Skipped if done
                return
nextchar:
                incf    FSR,f                   ; Bump pointer
                goto    getchar


;========================
;    send 2 nibbles in
;       4-bit mode
;========================
; Procedure to send two 4-bit values to Port-B lines
; 7, 6, 5, and 4. High-order nibble is sent first
; ON ENTRY:
;           w register holds 8-bit value to send
send8:
                movwf   store1                  ; Save original value
                call    merge4                  ; Merge with Port-B
; Now w has merged byte
                movwf   PORTB                   ; w to Port-B
                call    pulseE                  ; Send data to LCD
; High nibble is sent
                movf    store1,w ; Recover byte into w
                swapf   store1,w ; Swap nibbles in w
```

Data EEPROM Programming

```
            call    merge4
            movwf   PORTB
            call    pulseE           ; Send data to LCD
            call    delay_125
            return
;=================
;  merge bits
;=================
; Routine to merge the 4 high-order bits of the
; value to send with the contents of Port-B
; so as to preserve the 4 low-bits in Port-B
; Logic:
;       AND value with 1111 0000 mask
;       AND Port-B with 0000 1111 mask
;       Now low nibble in value and high nibble in
;       Port-B are all 0 bits:
;            value  = vvvv 0000
;            Port-B = 0000 bbbb
;       OR value and Port-B resulting in:
;                    vvvv bbbb
; ON ENTRY:
;       w contains value bits
; ON EXIT:
;       w contains merged bits
merge4:
            andlw   b'11110000'      ; ANDing with 0 clears the
                                     ; bit. ANDing with 1 preserves
                                     ; the original value
            movwf   store2           ; Save result in variable
            movf    PORTB,w          ; Port-B to w register
            andlw   b'00001111'      ; Clear high nibble in Port-b
                                     ; and preserve low nibble
            iorwf   store2,w         ; OR two operands in w
            return

;========================
;      blank buffer
;========================
; Procedure to store 16 blank characters in PIC RAM
; buffer starting at address stored in the variable
; pic_ad
blank16
            movlw   D'16'            ; Setup counter
            movwf   count1
            movf    pic_ad,w         ; First PIC RAM address
            movwf   FSR              ; Indexed addressing
            movlw   0x20             ; ASCII space character
storeit
```

```
                movwf     INDF        ; Store blank character in PIC RAM
                                      ; buffer using FSR register
                decfsz    count1,f    ; Done?
                goto      incfsr      ; no
                return                ; yes
incfsr:
                incf      FSR,f       ; Bump FSR to next buffer space
                goto      storeit

;========================
; Set address register
;     to LCD line 1
;========================
; ON ENTRY:
;          Address of LCD line 1 in constant LCD_1
line1:
                bcf       PORTA,E_line    ; E line low
                bcf       PORTA,RS_line   ; RS line low, set up for
control
                call      delay_5         ; busy?
; Set to second display line
                movlw     LCD_1           ; Address and command bit
                call      send8           ; 4-bit routine
; Set RS line for data
                bsf       PORTA,RS_line   ; Setup for data
                call      delay_5         ; Busy?
                return

;================================
;   first text string procedure
;================================
storeMS1:
; Procedure to store in PIC RAM buffer the message
; contained in the code area labeled msg1
; ON ENTRY:
;           variable pic_ad holds address of text buffer
;           in PIC RAM
;           w register hold offset into storage area
;           msg1 is routine that returns the string characters
;           and a zero terminator
;           index is local variable that hold offset into
;           text table. This variable is also used for
;           temporary storage of offset into buffer
; ON EXIT:
;           Text message stored in buffer
;
; Store offset into text buffer (passed in the w register)
; in temporary variable
```

Data EEPROM Programming

```
                movwf   index           ; Store w in index
; Store base address of text buffer in FSR
                movf    pic_ad,w ; first display RAM address to W
                addwf   index,w         ; Add offset to address
                movwf   FSR             ; W to FSR
; Initialize index for text string access
                movlw   0               ; Start at 0
                movwf   index           ; Store index in variable
; w still = 0
get_msg_char:
                call    msg1            ; Get character from table
; Test for zero terminator
                andlw   0x0ff
                btfsc   STATUS,Z ; Test zero flag
                goto    endstr1         ; End of string
; ASSERT: valid string character in w
;         store character in text buffer (by FSR)
                movwf   INDF            ; store in buffer by FSR
                incf    FSR,f           ; increment buffer pointer
; Restore table character counter from variable
                movf    index,w         ; Get value into w
                addlw   1               ; Bump to next character
                movwf   index   ; Store table index in variable
                goto    get_msg_char    ; Continue
endstr1:
                return

; Routine for returning message stored in program area
; Message has 10 characters
msg1:
                addwf   PCL,f           ; Access table
                retlw   'I'
                retlw   't'
                retlw   'e'
                retlw   'r'
                retlw   '.'
                retlw   0x20
                retlw   'N'
                retlw   'o'
                retlw   '.'
                retlw   0x20
                retlw   0


;===============================
;    binary to ASCII decimal
;           conversion
;===============================
; ON ENTRY:
```

```
;               w register has binary value in range 0 to 255
; ON EXIT:
;               output variables asc100, asc10, and asc1 have
;               three ASCII decimal digits
; Routine logic:
;   The value 100 is subtracted from the source operand
;   until the remainder is < 0 (carry cleared). The number
;   of subtractions is the decimal hundreds result. 100 is
;   then added back to the subtrahend to compensate
;   for the last subtraction. Now 10 is subracted in the
;   same manner to determine the decimal tenths result.
;   The final remainder is the decimal units result.
; Variables:
;      inNum      storage for source operand
;      asc100     storage for hundreds position result
;      asc10      storage for tenth position result
;      asc1       storage for unit position result
;      thisDig    Digit counter
bin2asc:
        movwf      inNum              ; Save copy of source value
        clrf       asc100     ; Clear hundreds storage
        clrf       asc10              ; Tens
        clrf       asc1               ; Units
        clrf       thisDig
sub100:
        movlw      .100
        subwf      inNum,f            ; Subtract 100
        btfsc      STATUS,C           ; Did subtract overflow?
        goto       bump100            ; No. Count subtraction
        goto       end100
bump100:
        incf       thisDig,f          ;increment digit counter
        goto       sub100
; Store 100th digit
end100:
        movf       thisDig,w          ; Adjusted digit counter
        addlw      0x30               ; Convert to ASCII
        movwf      asc100             ; Store it
; Calculate tenth position value
        clrf       thisDig
; Adjust minuend
        movlw      .100
        addwf      inNum,f            ; Minuend
                                      ; Add value to minuend to
                                                    ; compensate
for last operation
sub10:
        movlw      .10
        subwf      inNum,f            ; Subtract 10
```

Data EEPROM Programming

```
            btfsc    STATUS,C        ; Did subtract overflow?
            goto     bump10          ; No. Count subtraction
            goto     end10
bump10:
            incf     thisDig,f       ;increment digit counter
            goto     sub10
; Store 10th digit
end10:
            movlw    .10
            addwf    inNum,f         ; Adjust for last subtract
            movf     thisDig,w       ; get digit counter contents
            addlw    0x30            ; Conver to ASCII
            movwf    asc10           ; Store it
; Calculate and store units digit
            movf     inNum,w     ;   Store units value
            addlw    0x30            ; Convert to ASCII
            movwf    asc1            ; Store digit
            return


;=================================================================
;                       EEPROM procedures
;=================================================================
;==============================
;       read EEPROM
;==============================
; Procedure to read EEPROM memory. Address of memory
; location to read is stored in local register EEMemAdd
; On exit: read data in w
EERead:
            bcf      STATUS,RP0      ; Bank 0
            movf     EEMemAdd,w      ; Address to w
            movwf    EEADR           ; w to address register
            bsf      STATUS,RP0      ; Bank 1
            bsf      EECON1,RD       ; EE Read
            bcf      STATUS,RP0      ; Bank 0
            movf     EEDATA,w        ; W = EEDATA
            return

;==============================
;       write EEPROM
;==============================
; Procedure to write asc1 byte to EEPROM memory
; Address to write stored in local register EEMemAdd
; Data byte to write is in local register EEByte
EEWrite:
; Load byte to write into EE data register
            movf     EEByte,w ; Data to w
            movwf    EEDATA           ; Write
```

```
; Set write address in EE address register
        movf     EEMemAdd,w      ; Address to w
        movwf    EEADR           ; w to address register
; Write data to EEPROM memory
        bsf      STATUS,RP0      ; Bank 1
        bcf      INTCON,GIE      ; Disable INTs.
        bsf      EECON1,WREN     ; Enable Write
        movlw    0x55            ; Code # 1
        movwf    EECON2          ; Write 0x55
        movlw    0xaa            ; Code # 2
        movwf    EECON2          ; Write 0xaa
        bsf      EECON1,WR       ; Set WR bit
; Write operation now takes place automatically
        bsf      INTCON,GIE      ; Re-enable interrupts
        bcf              STATUS,RP0       ; Bank 0
        return

        End
```

15.2.2 Ser2EEP Program

```
; File name: Ser2EEP.asm
; Last revision: May 26, 2006
; Author: Julio Sanchez
; PIC: 16F877
;
; Description:
; Receive character data through RS-232 line and store in
; EEPROM data memory. Received characters are echoed on
; the second LCD line. When <Enter> key is detected (code
; 0x0d) the text stored in EEPROM memory is retrieved and
; displayed on the LCD. On startup the top LCD line displays
; the prompt: "Receiving:". At that time a message "Rdy- " is
; sent through the serial line so as to test the connection.
;
; Default serial line setting:
;               2400 baud
;               no parity
;               1 stop bit
;               8 character bits
;
; Program to use 4-bit PIC-to-LCD interface.
; Code assumes that LCD is driven by Hitachi HD44780
; controller and PIC 16F977. Display supports two lines
; each one with 20 characters. The length, wiring and base
; address of each display line is stored in #define
; statements. These statements can be edited to accommodate
; a different set-up.
```

Data EEPROM Programming

```
;
; WARNING:
; Code assumes 10 Mhz clock. Delay routines must be
; edited for a different clock. Clock speed also determines
; values for baud rate setting (see spbrgVal constant).
;
;===========================
;       16F877 switches
;===========================
; Switches used in __config directive:
;   _CP_ON           Code protection ON/OFF
; * _CP_OFF
; * _PWRTE_ON        Power-up timer ON/OFF
;   _PWRTE_OFF
;   _BODEN_ON        Brown-out reset enable ON/OFF
; * _BODEN_OFF
; * _PWRTE_ON        Power-up timer enable ON/OFF
;   _PWRTE_OFF
;   _WDT_ON          Watchdog timer ON/OFF
; * _WDT_OFF
;   _LPV_ON          Low voltage IC programming enable ON/OFF
; * _LPV_OFF
;   _CPD_ON          Data EE memory code protection ON/OFF
; * _CPD_OFF
; OSCILLATOR CONFIGURATIONS:
;   _LP_OSC          Low power crystal oscillator
;   _XT_OSC          External parallel resonator/crystal oscillator
;
; * _HS_OSC          High speed crystal resonator
;   _RC_OSC          Resistor/capacitor oscillator
; |                  (simplest, 20% error)
; |
; |_____ * indicates setup values presently selected

        processor        16f877           ; Define processor
        #include <p16f877.inc>
        __CONFIG _CP_OFF & _WDT_OFF & _BODEN_OFF & _PWRTE_ON &
_HS_OSC & _WDT_OFF & _LVP_OFF & _CPD_OFF

; __CONFIG directive is used to embed configuration data
; within the source file. The labels following the directive
; are located in the corresponding .inc file.
        errorlevel -302
; Suppress bank-related warning
;================================================================
;                       M A C R O S
;================================================================
; Macros to select the register banks
```

```
Bank0    MACRO              ; Select RAM bank 0
         bcf     STATUS,RP0
         bcf     STATUS,RP1
         ENDM

Bank1    MACRO              ; Select RAM bank 1
         bsf     STATUS,RP0
         bcf     STATUS,RP1
         ENDM

Bank2    MACRO              ; Select RAM bank 2
         bcf     STATUS,RP0
         bsf     STATUS,RP1
         ENDM

Bank3    MACRO              ; Select RAM bank 3
         bsf     STATUS,RP0
         bsf     STATUS,RP1
         ENDM
;========================================================
;                  constant definitions
;   for PIC-to-LCD pin wiring and LCD line addresses
;========================================================
#define E_line 1            ;|
#define RS_line 0           ;| - from wiring diagram
#define RW_line 2           ;|
; LCD line addresses (from LCD data sheet)
#define LCD_1 0x80          ; First LCD line constant
#define LCD_2 0xc0          ; Second LCD line constant
#define LCDlimit .20; Number of characters per line
#define spbrgVal .64; For 2400 baud on 10Mhz clock
; Note: The constants that define the LCD display
;       line addresses have the high-order bit set
;       so as to meet the requirements of controller
;       commands.
;
;========================================================
;                  General Purpose Variables
;========================================================
; Local variables
; Reserve 20 bytes for string buffer
         cblock   0x20
         strData
         endc
; Other data
         cblock   0x34              ; Start of block
         count1            ; Counter # 1
         count2            ; Counter # 2
```

Data EEPROM Programming

```
            count3              ; Counter # 3
            J                   ; counter J
            K                   ; counter K
            bufAdd
            index
            store1              ; Local storage
            store2
            endc
;==============================
;       Common RAM area
;==============================
; These GPRs can be accessed from any bank.
; 15 bytes are available, from 0x70 to 0x7f
            cblock   0x70
; For LCDscroll procedure
            LCDcount ; Counter for characters per line
            LCDline             ; Current display line (0 or 1)
; Communications variables
            newData             ; not 0 if new data received
            ascVal
            errorFlags
; EEPROM-related variables
            EEMemAdd ; EEPROM address to access
            EEByte              ; Data byte to write
            endc

;================================================================
;                       P R O G R A M
;================================================================
            org      0          ; start at address
            goto     main
; Space for interrupt handlers
            org                 0x08
main:
; Wiring:
;     LCD data to Port-D, lines 0 to 7
;     E line -> Port-E, 1
;     RW line -> Port-E, 2
;     RS line -> Port-E, 0
; Set PORTE D and E for output
; First, initialize Port-B by clearing latches
            clrf     STATUS
            clrf     PORTB
; Select bank 1 to TRIS Port-D for output
            Bank1
; TRIS Port-D for output. Port-D lines 4 to 7 are wired
; to LCD data lines. Port-D lines 0 to 4 are wired to LEDs.
            movlw    B'00000000'
```

```
                movwf   TRISD                   ; and Port-D
; By default Port-A lines are analog. To configure them
; as digital code must set bits 1 and 2 of the ADCON1
; register (in bank 1)
                movlw   0x06            ; binary 0000 0110  is code to
                                        ; make all Port-A lines digital
                movwf   ADCON1
; Port-B, lines are wired to keypad switches, as follows:
;       7 6 5 4 3 2 1 0
;       | | | | |_|_|_|_____ switch rows (output)
;       |_|_|_|_____ switch columns (input)
; rows must be defined as output and columns as input
                movlw   b'11110000'
                movwf   TRISB
; TRIS Port-E for output
                movlw   B'00000000'
                movwf   TRISE                   ; TRIS Port-E
; Enable Port-B pullups for switches in OPTION register
                movlw   b'00001000'
                movwf   OPTION_REG
; Back to bank 0
                Bank0
; Initialize serial Port-for 2400 baud, 8 bits, no parity
; 1 stop
                call    InitSerial
; Test serial transmission by sending "RDY-"
                movlw   'R'
                call    SerialSend
                movlw   'D'
                call    SerialSend
                movlw   'Y'
                call    SerialSend
                movlw   '-'
                call    SerialSend
                movlw   0x20
                call    SerialSend
; Clear all output lines
                movlw   b'00000000'
                movwf   PORTD
                movwf   PORTE
; Wait and initialize HD44780
                call    delay_5         ; Allow LCD time to initialize
itself
                call    initLCD         ; Then do forced
initialization
                call    delay_5         ; (Wait probably not
necessary)
; Clear character counter and line counter variables
```

Data EEPROM Programming

```
            clrf    LCDcount
            clrf    LCDline
; Set display address to start of first LCD line
            call    line1
; Store address of display buffer
            movlw   0x20
            movwf   bufAdd
; Display "Receiving:" message prompt
            call    blank20             ; Clear buffer
            movlw   0x00                ; Offset in buffer
            call    storeMS1            ; Store message at offset
            call    display20           ; Display message
; Start address of EEPROM
            clrf    EEMemAdd
; Setup for display in second line
            call    line2
            clrf    LCDline
            incf    LCDline,f           ; Set scroll control for
                                        ; line 2
;===============================================================
;           receive serial data, store, and display
;===============================================================
receive:
; Call serial receive procedure
            call    SerialRcv
; HOB of newData register is set if new data
; received
            btfss   newData,7
            goto    scanExit
; At this point new data was received.
            movwf   EEByte              ; Save received character
; Display character on LCD
            movf    EEByte,w ; Recover character
            call    send8               ; Display in LCD
            call    LCDscroll           ; Scroll at end of line
; Store character in EEPROM at location in EEMemAdd
            call    EEWrite             ; Local procedure
            incf    EEMemAdd,f          ; Bump to next EEPROM
; Check for <Enter> key (0x0d) and execute display function
            movf    EEByte,w ; Recover last received
            sublw   0x0d
            btfsc   STATUS,Z ; Test if <Enter> key
            goto    isEnter             ; Go if <Enter>
; Not <Enter> key, continue processing
scanExit:
            goto    receive             ; Continue
;=============================
;     display EEPROM data
```

```
;===============================
; This routine receives control when the <Enter> key is
; received.
; Action:
;         1. Clear LCD
;         2. Output is set to top LCD line
;         3. Characters stored in EEPROM are displayed
;            until 0x0d code is detected
isEnter:
        call    clearLCD
; Clear character counter and line counter variables
        clrf    LCDcount
        clrf    LCDline
; Read data from EEPROM memory, starting at address 0
; and display on LCD until 0x0d terminator
        call    line1
        clrf    EEMemAdd ; Start at EEPROM 0
readOne:
        call    EERead          ; Get character
; Store character
        movwf   EEByte          ; Save character
; Test for terminator
        sublw   0x0d
        btfsc   STATUS,Z ; Test if 0x0d
        goto    atEnd           ; Go if 0x0d
; At this point character read is not 0x0d
; Display on LCD
        movf    EEByte,w ; Recover character
; Display character on LCD
        call    send8           ; Display in LCD
        call    LCDscroll       ; Scroll at end of line
        incf    EEMemAdd,f      ; Next EEPROM byte
        goto    readOne

; End of execution
atEnd:
        goto    atEnd


;================================================================
;================================================================
;               L O C A L   P R O C E D U R E S
;================================================================
;================================================================
;==========================
; init LCD for 4-bit mode
;==========================
initLCD:
; Initialization for Densitron LCD module as follows:
```

```
;       4-bit interface
;       2 display lines of 16 characters each
;       cursor on
;       left-to-right increment
;       cursor shift right
;       no display shift
;=======================|
;   set command mode    |
;=======================|
        bcf     PORTE,E_line    ; E line low
        bcf     PORTE,RS_line   ; RS line low
        bcf     PORTE,RW_line   ; Write mode
        call    delay_125       ; delay 125 microseconds
        movlw   0x28     ; 0 0 1 0 1 0 0 0 (FUNCTION SET)
        call    send8    ; 4-bit send routine
; Set 4-bit mode command must be repeated
        movlw   0x28
        call    send8
        movlw   0x0e     ; 0 0 0 0 1 1 1 0 (DISPLAY ON/OFF)
        call    send8
        movlw   0x06     ; 0 0 0 0 0 1 1 0 (ENTRY MODE SET)
        call    send8
        movlw   0x14     ; 0 0 0 1 0 1 0 0 (CURSOR/DISPLAY
                         ;                       SHIFT)
        call    send8
        movlw   0x01     ; 0 0 0 0 0 0 0 1 (CLEAR DISPLAY)
                         ;               |___ COMMAND BIT
        call    send8
        call    delay_5  ; Test for busy
        return

.;============================
;   procedure to clear LCD
;============================
clearLCD:
        bcf     PORTE,E_line    ; E line low
        bcf     PORTE,RS_line   ; RS line low
        bcf     PORTE,RW_line   ; Write mode
        call    delay_125       ; delay 125 microseconds
        movlw   0x01     ; 0 0 0 0 0 0 0 1 (CLEAR DISPLAY)
                         ;               |___ COMMAND BIT
        call    send8
        call    delay_5  ; Test for busy
        return

;=======================
;   Procedure to delay
;     42 microseconds
```

```
;=======================
delay_125:
        movlw   .105            ; Repeat 105 machine cycles
        movwf   count1          ; Store value in counter
repeat
        decfsz  count1,f        ; Decrement counter
        goto    repeat          ; Continue if not 0
        return                  ; End of delay

;=======================
;   Procedure to delay
;     5 milliseconds
;=======================
delay_5:
        movlw   .105            ; Counter = 105 cycles
        movwf   count2          ; Store in variable
delay:
        call    delay_125       ; Delay
        decfsz  count2,f        ; 40 times = 5 milliseconds
        goto    delay
        return                  ; End of delay
;=======================
;       pulse E line
;=======================
pulseE
        bsf     PORTE,E_line    ; Pulse E line
        nop
        bcf     PORTE,E_line
        return

;============================
;   long delay sub-routine
;============================
long_delay:
        movlw   D'200'   ; w delay count
        movwf   J               ; J = w
jloop:
        movwf   K               ; K = w
kloop:
        decfsz  K,f             ; K = K-1, skip next if zero
        goto    kloop
        decfsz  J,f             ; J = J-1, skip next if zero
        goto    jloop
        return
;=======================
;   send 2 nibbles in
;     4-bit mode
;=======================
```

Data EEPROM Programming

```
;   Procedure to send two 4-bit values to Port-B lines
;   7, 6, 5, and 4. High-order nibble is sent first
;   ON ENTRY:
;           w register holds 8-bit value to send
send8:
          movwf     store1              ; Save original value
          call      merge4              ; Merge with Port-B
;   Now w has merged byte
          movwf     PORTD               ; w to Port-D
          call      pulseE              ; Send data to LCD
;   High nibble is sent
          movf      store1,w ; Recover byte into w
          swapf     store1,w ; Swap nibbles in w
          call      merge4
          movwf     PORTD
          call      pulseE              ; Send data to LCD
          call      delay_125
          return
;===========================
;         merge bits
;===========================
;   Routine to merge the 4 high-order bits of the
;   value to send with the contents of Port-B
;   so as to preserve the 4 low-bits in Port-B
;   Logic:
;       AND value with 1111 0000 mask
;       AND Port-B with 0000 1111 mask
;       Now low nibble in value and high nibble in
;       Port-B are all 0 bits:
;           value  = vvvv 0000
;           Port-B = 0000 bbbb
;       OR value and Port-B resulting in:
;                   vvvv bbbb
;   ON ENTRY:
;       w contain value bits
;   ON EXIT:
;       w contains merged bits
merge4:
          andlw     b'11110000'         ; ANDing with 0 clears the
                                        ; bit. ANDing with 1 preserves
                                        ; the original value
          movwf     store2              ; Save result in variable
          movf      PORTD,w             ; Port-B to w register
          andlw     b'00001111' ; Clear high nibble in Port-b
                                        ; and preserve low nibble
          iorwf     store2,w ; OR two operands in w
          return
;===========================
```

```
;     Set address register
;         to LCD line 2
;===========================
; ON ENTRY:
;          Address of LCD line 2 in constant LCD_2
line2:
        bcf     PORTE,E_line        ; E line low
        bcf     PORTE,RS_line       ; RS line low, setup for
                                    ;  control
        call    delay_5             ; Busy?
; Set to second display line
        movlw   LCD_2               ; Address with high-bit set
        call    send8
; Set RS line for data
        bsf     PORTE,RS_line       ; RS = 1 for data
        call    delay_5             ; Busy?
        return
;===========================
;     Set address register
;         to LCD line 1
;===========================
; ON ENTRY:
;          Address of LCD line 1 in constant LCD_1
line1:
        bcf     PORTE,E_line        ; E line low
        bcf     PORTE,RS_line       ; RS line low, set up for
                                    ;  control
        call    delay_5             ; busy?
; Set to second display line
        movlw   LCD_1               ; Address and command bit
        call    send8               ; 4-bit routine
; Set RS line for data
        bsf     PORTE,RS_line       ; Setup for data
        call    delay_5             ; Busy?
        return

;===========================
;    scroll to LCD line 2
;===========================
; Procedure to count the number of characters displayed on
; each LCD line. If the number reaches the value in the
; constant LCDlimit, then display is scrolled to the second
; LCD line. If at the end of the second line, then LCD is
; reset to the first line.
LCDscroll:
        incf    LCDcount,f          ; Bump counter
; Test for line limit
        movf    LCDcount,w
```

Data EEPROM Programming

```
            sublw    LCDlimit ; Count minus limit
            btfss    STATUS,Z ; Is count - limit = 0
            goto     scrollExit       ; Go if not at end of line
; At this point the end of the LCD line was reached
; Test if this is also the end of the second line
            movf     LCDline,w
            sublw    0x01              ; Is it line 1?
            btfsc    STATUS,Z ; Is LCDline minus 1 = 0?
            goto     line2End ; Go if end of second line
; At this point it is the end of the top LCD line
            call     line2             ; Scroll to second line
            clrf     LCDcount ; Reset counter
            incf     LCDline,f        ; Bump line counter
            goto     scrollExit
; End of second LCD line
line2End:
            call     initLCD          ; Reset
            clrf     LCDcount ; Clear counters
            clrf     LCDline
            call     line1             ; Display to first line
scrollExit:
            return


;==============================
;    LCD display procedure
;==============================
; Sends 20 characters from PIC buffer with address stored
; in variable bufAdd to LCD line previously selected
display20:
            call     delay_5          ; Make sure not busy
; Set up for data
            bcf      PORTA,E_line     ; E line low
            bsf      PORTA,RS_line    ; RS line high for data
; Set up counter for 20 characters
            movlw    D'20'
            movwf    count3
; Get display address from local variable bufAdd
            movf     bufAdd,w ; First display RAM address to W
            movwf    FSR               ; W to FSR
getchar:
            movf     INDF,w   ; get character from display RAM
                              ; location pointed to by file select
                              ; register
            call     send8    ; 4-bit interface routine
; Test for 20 characters displayed
            decfsz   count3,f         ; Decrement counter
            goto     nextchar ; Skipped if done
            return
```

```
nextchar:
        incf    FSR,f           ; Bump pointer
        goto    getchar

;================================
;   first text string procedure
;================================
storeMS1:
; Procedure to store in PIC RAM buffer the message
; contained in the code area labeled msg1
; ON ENTRY:
;         variable bufAdd holds address of text buffer
;         in PIC RAM
;         w register hold offset into storage area
;         msg1 is routine that returns the string characters
;         and a zero terminator
;         index is local variable that holds offset into
;         text table. This variable is also used for
;         temporary storage of offset into buffer
; ON EXIT:
;         Text message stored in buffer
;
; Store offset into text buffer (passed in the w register)
; in temporary variable
        movwf   index           ; Store w in index
; Store base address of text buffer in FSR
        movf    bufAdd,w ; first display RAM address to W
        addwf   index,w ; Add offset to address
        movwf   FSR     ; W to FSR
; Initialize index for text string access
        movlw   0               ; Start at 0
        movwf   index   ; Store index in variable
; w still = 0
get_msg_char:
        call    msg1    ; Get character from table
; Test for zero terminator
        andlw   0x0ff
        btfsc   STATUS,Z ; Test zero flag
        goto    endstr1  ; End of string
; ASSERT: valid string character in w
;         store character in text buffer (by FSR)
        movwf   INDF    ; store in buffer by FSR
        incf    FSR,f   ; increment buffer pointer
; Restore table character counter from variable
        movf    index,w ; Get value into w
        addlw   1       ; Bump to next character
        movwf   index   ; Store table index in variable
        goto    get_msg_char    ; Continue
```

Data EEPROM Programming

```
endstr1:
        return

; Routine for returning message stored in program area
; Message has 10 characters
msg1:
        addwf   PCL,f           ; Access table
        retlw   'R'
        retlw   'e'
        retlw   'C'
        retlw   'e'
        retlw   'i'
        retlw   'v'
        retlw   'i'
        retlw   'n'
        retlw   'g'
        retlw   ':'
        retlw   0


;========================
;       blank buffer
;========================
; Procedure to store 20 blank characters in PIC RAM
; buffer starting at address stored in the variable
; bufAdd
blank20:
        movlw   D'20'           ; Setup counter
        movwf   count1
        movf    bufAdd,w ; First PIC RAM address
        movwf   FSR             ; Indexed addressing
        movlw   0x20            ; ASCII space character
storeit:
        movwf   INDF    ; Store blank character in PIC RAM
                        ; buffer using FSR register
        decfsz  count1,f        ; Done?
        goto    incfsr  ; no
        return          ; yes
incfsr:
        incf    FSR,f   ; Bump FSR to next buffer space
        goto    storeit


;================================================================
;               communications procedures
;================================================================
; Initialize serial Port-for 2400 baud, 8 bits, no parity,
; 1 stop
InitSerial:
        Bank1                           ; Macro to select bank1
```

```
; Bits 6 and 7 of Port-C are multiplexed as TX/CK and RX/DT
; for USART operation. These bits must be set to input in the
; TRISC register
        movlw   b'11000000'     ; Bits for TX and RX
        iorwf   TRISC,f         ; OR into TRISc register
;
; The asynchronous baud rate is calculated as follows:
;                   Fosc
;          ABR = ---------
;                  S*(x+1)
;
; where x is value in the SPBRG register and S is 64 if the high
; baud rate select bit (BRGH) in the TXSTA control register is
; clear, and 16 if the BRGH bit is set. For setting to 2400 baud
; using a 10Mhs oscillator at a slow baud rate the formula
; is:
; At slow speed (BRGH = 0)
;           10,000,000    10,000,000
;          ----------- = ----------- = 2,403.84 (0.16% error)
;           64*(64+1)        4160
;
        movlw   spbrgVal ; Value in spbrgVal = 64
        movwf   SPBRG            ; Place in baud rate generator
; Setup value: 0010 0000 = 0x20
        movlw   0x20    ; Enable transmission and high baud
                        ; rate
        movwf   TXSTA
        Bank0                   ; Bank 0
; Setup value: 1001 0000 = 0x90
        movlw   0x90    ; Enable serial Port-and continuous
                        ; reception
        movwf   RCSTA
;
        clrf    errorFlags      ; Clear local error flags
                                ; register
        Return
;
;==============================
;       transmit data
;==============================
; Test for Transmit Register Empty and transmit data in w
SerialSend:
        Bank0                   ; Select bank 0
        btfss   PIR1,TXIF       ; check if transmitter busy
        goto    $-1             ; wait until transmitter is
not busy
        movwf   TXREG           ; and transmit the data
        return
```

Data EEPROM Programming

```asm
;===============================
;       receive data
;===============================
; Procedure to test line for data received and return value
; in w. Overrun and framing errors are detected and
; remembered in the variable errorFlags, as follows:
;        7 6 5 4 3 2 1 0    <== errorFlags
;        — not used —   |  |___ overrun error
;                          |_____ framing error
SerialRcv:
        clrf    newData         ; Clear new data received register
        Bank0                   ; Select bank 0
; Bit 5 (RCIF) of the PIR1 Register is clear if the USART
; receive buffer is empty. If so, no data has been received
        btfss   PIR1,RCIF       ; Check for received data
        return                  ; Exit if no data
; At this point data has been received. First eliminate
; possible errors: overrun and framing.
; Bit 1 (OERR) of the RCSTA register detects overrun
; Bit 2 (FERR( of the RCSTA register detects framing error
        btfsc   RCSTA,OERR      ; Test for overrun error
        goto    OverErr         ; Error handler
        btfsc   RCSTA,FERR      ; Test for framing error
        goto    FrameErr ; Error handler
; At this point no error was detected
; Received data is in the USART RCREG register
        movf    RCREG,w         ; get received data
        bsf     newData,7       ; Set bit 7 to indicate new
                                ; data
; Clear error flags
        clrf    errorFlags
        return
;==========================
;     error handlers
;==========================
OverErr:
        bsf     errorFlags,0    ; Bit 0 is overrun error
; Reset system
        bcf     RCSTA,CREN      ; Clear continuous receive bit
        bsf     RCSTA,CREN      ; Set to re-enable reception
        return
;error because FERR framing error bit is set
;can do special error handling here - this code simply clears
; and continues
FrameErr:
        bsf     errorFlags,1    ; Bit 1 is framing error
        movf    RCREG,W         ; Read and throw away bad data
```

```
            return
;================================================================
;                   local EEPROM data procedures
;================================================================
; GPRs used in EEPROM-related code are placed in the common
; RAM area (from 0x70 to 0x7f). This makes the registers
; accessible from any bank.
;==============================
;     read local EEPROM
;==============================
; Procedure to read EEPROM memory
; ON ENTRY:
; Address of EEPROM memory location to read is stored in
; local register EEMemAdd
; ON EXIT:
; Read data in w
EERead:
        Bank2
        movf       EEMemAdd,W      ; EEPROM address
        movwf      EEADR           ; to read from
        Bank3
        bcf        EECON1,EEPGD    ; Point to Data memory
        bsf        EECON1,RD       ; Start read
        Bank2
        movf       EEDATA,W        ; Data to w register
        Bank0
        return

;==============================
;     write local EEPROM
;==============================
; Procedure to write data byte to EEPROM memory
; ON ENTRY:
; Address to write stored in local register EEMemAdd
; Data byte to write is in local register EEByte
EEWrite:
        Bank3
Wait2Start:
        btfsc      EECON1,WR       ; Wait for
        GOTO       Wait2Start      ; write to finish
        Bank2
        movf       EEMemAdd,w      ; Address to
        movwf      EEADR           ; SFR
        movf       EEByte,w        ; Data to
        movwf      EEDATA          ; SFR
        Bank3
        bcf        EECON1,EEPGD    ; Point to Data memory
        bsf        EECON1,WREN     ; and enable writes
```

Data EEPROM Programming

```
; Disable interrupts. Can be done in any case
        bcf     INTCON,GIE
; Write special codes
        movlw   0x55             ; First code is 0x55
        movwf   EECON2
        movlw   0xaa             ; Second code is 0xaa
        movwf   EECON2
        bsf     EECON1,WR        ; Start write operation
        nop                      ; Time for write
        nop
; Test for end of write operation
wait2End:
        btfsc   EECON1,WR        ; Wait until WR clear
        goto    wait2End
;
; Re-enable interrupts if program uses interrupts
; If not, comment out next line
;       bsf     INTCON,GIE
;
        bcf     EECON1,WREN      ; Prevent accidental writes
        Bank0
        return


;===========================================================
        end              ; END OF PROGRAM
;===========================================================
```

15.2.3 I2CEEP Program

```
; File name: I2CEEP.asm
; Last revision: May 28, 2006
; Author: Julio Sanchez
; Processor: 16F877
;
; Description:
; Receive character data through RS-232 line and store in
; 24LC04B EEPROM IC, using the I2C serial protocol in the
; PIC's MSSP module. Received characters are echoed on
; the second LCD line. When <Enter> key is detected (code
; 0x0d) the text stored in EEPROM memory is retrieved and
; displayed on the LCD. On startup the top LCD line displays
; the prompt: "Receiving:". At that time a message "Rdy- " is
; sent through the serial line so as to test the connection.
;
; Default serial line setting:
;               2400 baud
;               no parity
;               1 stop bit
```

```
;                    8 character bits
;
; Wiring:
; 24LC04B SDA line is wired to PIC RC4 (MSSP SDA)
; 24LC04B SCL line is wired to PIC RC3 (MSSP SCL)
; 24LC04B A0-A2 and WP lines are not used (GND)
;
; Program to use 4-bit PIC-to-LCD interface.
; Code assumes that LCD is driven by Hitachi HD44780
; controller and PIC 16F977. Display supports two lines
; each one with 20 characters. The length, wiring and base
; address of each display line is stored in #define
; statements. These statements can be edited to accommodate
; a different set-up.
;
; WARNING:
; Code assumes 10 Mhz clock. Delay routines must be
; edited for a different clock. Clock speed also determines
; values for baud rate setting (see spbrgVal constant).
;
;===========================
;       16F877 switches
;===========================
; Switches used in __config directive:
;   _CP_ON           Code protection ON/OFF
; * _CP_OFF
; * _PWRTE_ON        Power-up timer ON/OFF
;   _PWRTE_OFF
;   _BODEN_ON        Brown-out reset enable ON/OFF
; * _BODEN_OFF
; * _PWRTE_ON        Power-up timer enable ON/OFF
;   _PWRTE_OFF
;   _WDT_ON          Watchdog timer ON/OFF
; * _WDT_OFF
;   _LPV_ON          Low voltage IC programming enable ON/OFF
; * _LPV_OFF
;   _CPD_ON          Data EE memory code protection ON/OFF
; * _CPD_OFF
; OSCILLATOR CONFIGURATIONS:
;   _LP_OSC          Low power crystal oscillator
;   _XT_OSC          External parallel resonator/crystal oscillator
;
; * _HS_OSC          High speed crystal resonator
;   _RC_OSC          Resistor/capacitor oscillator
;   |                (simplest, 20% error)
;   |
;   |_____ * indicates setup values presently selected
```

Data EEPROM Programming

```
        processor      16f877        ; Define processor
        #include <p16f877.inc>
            __CONFIG _CP_OFF & _WDT_OFF & _BODEN_OFF & _PWRTE_ON &
_HS_OSC & _WDT_OFF & _LVP_OFF & _CPD_OFF

; __CONFIG directive is used to embed configuration data
; within the source file. The labels following the directive
; are located in the corresponding .inc file.
        errorlevel -302
; Suppress bank-related warning
;================================================================
;                         M A C R O S
;================================================================
; Macros to select the register banks
Bank0   MACRO                ; Select RAM bank 0
        bcf       STATUS,RP0
        bcf       STATUS,RP1
        ENDM

Bank1   MACRO                ; Select RAM bank 1
        bsf       STATUS,RP0
        bcf       STATUS,RP1
        ENDM

Bank2   MACRO                ; Select RAM bank 2
        bcf       STATUS,RP0
        bsf       STATUS,RP1
        ENDM

Bank3   MACRO                ; Select RAM bank 3
        bsf       STATUS,RP0
        bsf       STATUS,RP1
        ENDM
;========================================================
;              constant definitions
;   for PIC-to-LCD pin wiring and LCD line addresses
;========================================================
#define E_line 1            ; |
#define RS_line 0           ; | - from wiring diagram
#define RW_line 2           ; |
; LCD line addresses (from LCD data sheet)
#define LCD_1 0x80          ; First LCD line constant
#define LCD_2 0xc0          ; Second LCD line constant
#define LCDlimit .20; Number of characters per line
#define spbrgVal .64; For 2400 baud on 10Mhz clock
; Note: The constants that define the LCD display
;       line addresses have the high-order bit set
;       so as to meet the requirements of controller
```

```
;               commands.
;================================================================
;            constants for I2C initialization
;================================================================
; I2C connected to 24LC04B EEPROM.
; The MSSP module is in I2C MASTER mode.
#define LC04READ  0xa0      ; I2C value for read control byte
#define LC04WRITE 0xa1      ; I2C value for write control byte


;================================================================
;                   General Purpose Variables
;================================================================
; Local variables
; Reserve 20 bytes for string buffer
        cblock   0x20
        strData
        endc
; Other data
        cblock   0x34        ; Start of block
        count1               ; Counter # 1
        count2               ; Counter # 2
        count3               ; Counter # 3
        J                    ; counter J
        K                    ; counter K
        bufAdd
        index
        store1               ; Local storage
        store2
; For LCDscroll procedure
        LCDcount ; Counter for characters per line
        LCDline              ; Current display line (0 or 1)
        Endc
;
;==============================
;     Common RAM area
;==============================
; These GPRs can be accessed from any bank.
; 15 bytes are available, from 0x70 to 0x7f
        cblock   0x70
; Communications variables
        newData              ; not 0 if new data received
        ascVal
        errorFlags
; EEPROM-related variables
        EEMemAdd ; EEPROM address to access
        EEByte               ; Data byte to write
        endc
```

Data EEPROM Programming

```
;=================================================================
;                        P R O G R A M
;=================================================================
         org         0               ; start at address
         goto        main
; Space for interrupt handlers
         org         0x08
main:
; Wiring:
;       LCD data to Port-D, lines 0 to 7
;       E line -> Port-E, 1
;       RW line -> Port-E, 2
;       RS line -> Port-E, 0
; Set PORTE D and E for output
; First, initialize Port-B by clearing latches
         clrf        STATUS
         clrf        PORTB
; Select bank 1 to TRIS Port-D for output
         Bank1
; TRIS Port-D for output. Port-D lines 4 to 7 are wired
; to LCD data lines. Port-D lines 0 to 4 are wired to LEDs.
         movlw       B'00000000'
         movwf       TRISD           ; and Port-D
; By default Port-A lines are analog. To configure them
; as digital code must set bits 1 and 2 of the ADCON1
; register (in bank 1)
         movlw       0x06            ; binary 0000 0110  is code to
                                     ; make all Port-A lines digital
         movwf       ADCON1
; Port-B, lines are wired to keypad switches, as follows:
;    7 6 5 4 3 2 1 0
;    | | | | |_|_|_|_____ switch rows (output)
;    |_|_|_|_____ switch columns (input)
; rows must be defined as output and columns as input
         movlw       b'11110000'
         movwf       TRISB
; TRIS Port-E for output
         movlw       B'00000000'
         movwf       TRISE           ; TRIS Port-E
; Enable Port-B pullups for switches in OPTION register
         movlw       b'00001000'
         movwf       OPTION_REG
; Back to bank 0
         Bank0
; Initialize serial port for 2400 baud, 8 bits, no parity
; 1 stop
         call        InitSerial
; Test serial transmission by sending "RDY-"
```

```
                movlw       'R'
                call        SerialSend
                movlw       'D'
                call        SerialSend
                movlw       'Y'
                call        SerialSend
                movlw       '-'
                call        SerialSend
                movlw       0x20
                call        SerialSend
; Clear all output lines
                movlw       b'00000000'
                movwf       PORTD
                movwf       PORTE
; Wait and initialize HD44780
                call        delay_5     ; Allow LCD time to initialize itself
                call        initLCD     ; Then do forced initialization
                call        delay_5
; Clear character counter and line counter variables
                clrf        LCDcount
                clrf        LCDline
; Set display address to start of first LCD line
                call        line1
; Store address of display buffer
                movlw       0x20
                movwf       bufAdd
; Display "Receiving:" message prompt
                call        blank20     ; Clear buffer
                movlw       0x00        ; Offset in buffer
                call        storeMS1    ; Store message at offset
                call        display20   ; Display message
; Start address of EEPROM
                clrf        EEMemAdd
; Setup for display in second line
                call        line2
                clrf        LCDline
                incf        LCDline,f   ; Set scroll control for line 2
; Initialize I2C EEPROM operation
                call        SetupI2C    ; Local procedure
;================================================================
;           receive serial data, store, and display
;================================================================
receive:
; Call serial receive procedure
                call        SerialRcv
; HOB of newData register is set if new data
; received
                btfss       newData,7
```

Data EEPROM Programming

```
                goto    scanExit
; At this point new data was received.
        movwf   EEByte          ; Save received character
; Display character on LCD
        movf    EEByte,w ; Recover character
        call    send8           ; Display in LCD
        call    LCDscroll       ; Scroll at end of line
; Store character in EEPROM at location in EEMemAdd
        call    WriteI2C ; Local procedure
        incf    EEMemAdd,f      ; Bump to next EEPROM
; Check for <Enter> key (0x0d) and execute display function
        movf    EEByte,w ; Recover last received
        sublw   0x0d
        btfsc   STATUS,Z ; Test if <Enter> key
        goto    isEnter         ; Go if <Enter>
; Not <Enter> key, continue processing
scanExit:
        goto    receive         ; Continue
;=============================
;    display EEPROM data
;=============================
; This routine receives control when the <Enter> key is
; received.
; Action:
;       1. Clear LCD
;       2. Output is set to top LCD line
;       3. Characters stored in EEPROM are displayed
;          until 0x0d code is detected
isEnter:
        call    clearLCD
; Clear character counter and line counter variables
        clrf    LCDcount
        clrf    LCDline
; Read data from EEPROM memory, starting at address 0
; and display on LCD until 0x0d terminator
        call    line1
        clrf    EEMemAdd ; Start at EEPROM 0
readOne:
        call    ReadI2C         ; Get character
; Store character
        movwf   EEByte          ; Save character
; Test for terminator
        sublw   0x0d
        btfsc   STATUS,Z ; Test if 0x0d
        goto    atEnd           ; Go if 0x0d
; At this point character read is not 0x0d
; Display on LCD
        movf    EEByte,w ; Recover character
```

```
; Display character on LCD
        call    send8           ; Display in LCD
        call    LCDscroll       ; Scroll at end of line
        incf    EEMemAdd,f      ; Next EEPROM byte
        goto    readOne
; End of execution
atEnd:
        goto    atEnd


;================================================================
;================================================================
;              L O C A L    P R O C E D U R E S
;================================================================
;================================================================
;==========================
; init LCD for 4-bit mode
;==========================
initLCD:
; Initialization for Densitron LCD module as follows:
;    4-bit interface
;    2 display lines of 16 characters each
;    cursor on
;    left-to-right increment
;    cursor shift right
;    no display shift
;=====================|
;   set command mode  |
;=====================|
        bcf     PORTE,E_line    ; E line low
        bcf     PORTE,RS_line   ; RS line low
        bcf     PORTE,RW_line   ; Write mode
        call    delay_125       ; delay 125 microseconds
        movlw   0x28            ; 0 0 1 0 1 0 0 0 (FUNCTION SET)
        call    send8           ; 4-bit send routine
; Set 4-bit mode command must be repeated
        movlw   0x28
        call    send8
        movlw   0x0e    ; 0 0 0 0 1 1 1 0 (DISPLAY ON/OFF)
        call    send8
        movlw   0x06    ; 0 0 0 0 0 1 1 0 (ENTRY MODE SET)
        call    send8
        movlw   0x14    ; 0 0 0 1 0 1 0 0 (CURSOR/DISPLAY SHIFT)
        call    send8
        movlw   0x01    ; 0 0 0 0 0 0 0 1 (CLEAR DISPLAY)
        call    send8
        call    delay_5 ; Test for busy
```

Data EEPROM Programming

```
            return

;;==========================
;   procedure to clear LCD
;==========================
clearLCD:
        bcf     PORTE,E_line        ; E line low
        bcf     PORTE,RS_line       ; RS line low
        bcf     PORTE,RW_line       ; Write mode
        call    delay_125           ; delay 125 microseconds
        movlw   0x01                ; 0 0 0 0 0 0 0 1
        call    send8
        call    delay_5  ; Test for busy
        return

;======================
;  Procedure to delay
;    42 microseconds
;======================
delay_125:
        movlw   .105                ; Repeat 105 machine cycles
        movwf   count1              ; Store value in counter
repeat:
        decfsz  count1,f            ; Decrement counter
        goto    repeat              ; Continue if not 0
        return                      ; End of delay

;======================
;  Procedure to delay
;    5 milliseconds
;======================
delay_5:
        movlw   .105                ; Counter = 105 cycles
        movwf   count2              ; Store in variable
delay:
        call    delay_125           ; Delay
        decfsz  count2,f            ; 40 times = 5 milliseconds
        goto    delay
        return                      ; End of delay
;======================
;     pulse E line
;======================
pulseE
        bsf     PORTE,E_line        ; Pulse E line
        nop
        bcf     PORTE,E_line
        return
```

```
;==============================
;   long delay sub-routine
;==============================
long_delay:
        movlw   D'200'          ; w delay count
        movwf   J               ; J = w
jloop:
        movwf   K               ; K = w
kloop:
        decfsz  K,f             ; K = K-1, skip next if zero
        goto    kloop
        decfsz  J,f             ; J = J-1, skip next if zero
        goto    jloop
        return
;========================
;   send 2 nibbles in
;       4-bit mode
;========================
; Procedure to send two 4-bit values to Port-B lines
; 7, 6, 5, and 4. High-order nibble is sent first
; ON ENTRY:
;           w register holds 8-bit value to send
send8:
        movwf   store1          ; Save original value
        call    merge4          ; Merge with Port-B
; Now w has merged byte
        movwf   PORTD           ; w to Port-D
        call    pulseE          ; Send data to LCD
; High nibble is sent
        movf    store1,w ; Recover byte into w
        swapf   store1,w ; Swap nibbles in w
        call    merge4
        movwf   PORTD
        call    pulseE          ; Send data to LCD
        call    delay_125
        return
;==========================
;       merge bits
;==========================
; Routine to merge the 4 high-order bits of the
; value to send with the contents of Port-B
; so as to preserve the 4 low-bits in Port-B
; Logic:
;       AND value with 1111 0000 mask
;       AND Port-B with 0000 1111 mask
;       Now low nibble in value and high nibble in
;       Port-B are all 0 bits:
;           value = vvvv 0000
```

```
;                 Port-B = 0000 bbbb
;       OR value and Port-B resulting in:
;                        vvvv bbbb
; ON ENTRY:
;       w contain value bits
; ON EXIT:
;       w contains merged bits
merge4:
          andlw     b'11110000'        ; ANDing with 0 clears the
                                       ; bit. ANDing with 1 preserves
                                       ; the original value
          movwf     store2             ; Save result in variable
          movf      PORTD,w            ; Port-B to w register
          andlw     b'00001111'        ; Clear high nibble in Port-b
                                       ; and preserve low nibble
          iorwf     store2,w ; OR two operands in w
          return
;==========================
;    Set address register
;       to LCD line 2
;==========================
; ON ENTRY:
;          Address of LCD line 2 in constant LCD_2
line2:
          bcf       PORTE,E_line       ; E line low
          bcf       PORTE,RS_line      ; RS line low, setup for
control
          call      delay_5            ; Busy?
; Set to second display line
          movlw     LCD_2
          call      send8              ; Address with high-bit set
; Set RS line for data
          bsf       PORTE,RS_line      ; RS = 1 for data
          call      delay_5            ; Busy?
          return
;==========================
;    Set address register
;       to LCD line 1
;==========================
; ON ENTRY:
;          Address of LCD line 1 in constant LCD_1
line1:
          bcf       PORTE,E_line       ; E line low
          bcf       PORTE,RS_line      ; RS line low, set up for
control
          call      delay_5            ; busy?
; Set to second display line
          movlw     LCD_1              ; Address and command bit
```

```
                call    send8           ; 4-bit routine
; Set RS line for data
                bsf     PORTE,RS_line   ; Setup for data
                call    delay_5         ; Busy?
                return

;===========================
;   scroll to LCD line 2
;===========================
; Procedure to count the number of characters displayed on
; each LCD line. If the number reaches the value in the
; constant LCDlimit, then display is scrolled to the second
; LCD line. If at the end of the second line, then LCD is
; reset to the first line.
LCDscroll:
                incf    LCDcount,f      ; Bump counter
; Test for line limit
                movf    LCDcount,w
                sublw   LCDlimit        ; Count minus limit
                btfss   STATUS,Z        ; Is count - limit = 0
                goto    scrollExit      ; Go if not at end of line
; At this point the end of the LCD line was reached
; Test if this is also the end of the second line
                movf    LCDline,w
                sublw   0x01            ; Is it line 1?
                btfsc   STATUS,Z        ; Is LCDline minus 1 = 0?
                goto    line2End        ; Go if end of second line
; At this point it is the end of the top LCD line
                call    line2           ; Scroll to second line
                clrf    LCDcount        ; Reset counter
                incf    LCDline,f       ; Bump line counter
                goto    scrollExit
; End of second LCD line
line2End:
                call    initLCD         ; Reset
                clrf    LCDcount        ; Clear counters
                clrf    LCDline
                call    line1           ; Display to first line
scrollExit:
                return

;==============================
;   LCD display procedure
;==============================
; Sends 20 characters from PIC buffer with address stored
; in variable bufAdd to LCD line previously selected
display20:
                call    delay_5         ; Make sure not busy
```

Data EEPROM Programming

```
; Set up for data
        bcf     PORTA,E_line    ; E line low
        bsf     PORTA,RS_line   ; RS line high for data
; Set up counter for 20 characters
        movlw   D'20'
        movwf   count3
; Get display address from local variable bufAdd
        movf    bufAdd,w ; First display RAM address to W
        movwf   FSR             ; W to FSR
getchar
        movf    INDF,w  ; get character from display RAM
                        ; location pointed to by file select
                        ; register
        call    send8   ; 4-bit interface routine
; Test for 20 characters displayed
        decfsz  count3,f        ; Decrement counter
        goto    nextchar ; Skipped if done
        return
nextchar:
        incf    FSR,f   ; Bump pointer
        goto    getchar


;================================
;   first text string procedure
;================================
storeMS1:
; Procedure to store in PIC RAM buffer the message
; contained in the code area labeled msg1
; ON ENTRY:
;       variable bufAdd holds address of text buffer
;       in PIC RAM
;       w register hold offset into storage area
;       msg1 is routine that returns the string characters
;       and a zero terminator
;       index is local variable that hold offset into
;       text table. This variable is also used for
;       temporary storage of offset into buffer
; ON EXIT:
;       Text message stored in buffer
;
; Store offset into text buffer (passed in the w register)
; in temporary variable
        movwf   index           ; Store w in index
; Store base address of text buffer in FSR
        movf    bufAdd,w ; first display RAM address to W
        addwf   index,w         ; Add offset to address
        movwf   FSR             ; W to FSR
; Initialize index for text string access
```

```
                movlw    0              ; Start at 0
                movwf    index          ; Store index in variable
; w still = 0
get_msg_char:
                call     msg1           ; Get character from table
; Test for zero terminator
                andlw    0x0ff
                btfsc    STATUS,Z       ; Test zero flag
                goto     endstr1        ; End of string
; ASSERT: valid string character in w
;         store character in text buffer (by FSR)
                movwf    INDF           ; store in buffer by FSR
                incf     FSR,f          ; increment buffer pointer
; Restore table character counter from variable
                movf     index,w        ; Get value into w
                addlw    1              ; Bump to next character
                movwf    index          ; Store table index in variable
                goto     get_msg_char   ; Continue
endstr1:
                return

; Routine for returning message stored in program area
; Message has 10 characters
msg1:
                addwf    PCL,f          ; Access table
                retlw    'R'
                retlw    'e'
                retlw    'c'
                retlw    'e'
                retlw    'i'
                retlw    'v'
                retlw    'i'
                retlw    'n'
                retlw    'g'
                retlw    ':'
                retlw    0


;========================
;    blank buffer
;========================
; Procedure to store 20 blank characters in PIC RAM
; buffer starting at address stored in the variable
; bufAdd
blank20:
                movlw    D'20'          ; Setup counter
                movwf    count1
```

Data EEPROM Programming

```
            movf    bufAdd,w  ; First PIC RAM address
            movwf   FSR       ; Indexed addressing
            movlw   0x20      ; ASCII space character
storeit
            movwf   INDF      ; Store blank character in PIC RAM
                              ; buffer using FSR register
            decfsz  count1,f  ; Done?
            goto    incfsr    ; no
            return            ; yes
incfsr:
            incf    FSR,f     ; Bump FSR to next buffer space
            goto    storeit


;=================================================================
;                   communications procedures
;=================================================================
; Initialize serial port for 2400 baud, 8 bits, no parity,
; 1 stop
InitSerial:
            Bank1             ; Macro to select bank1
; Bits 6 and 7 of Port-C are multiplexed as TX/CK and RX/DT
; for USART operation. These bits must be set to input in the
; TRISC register
            movlw   b'11000000'     ; Bits for TX and RX
            iorwf   TRISC,f         ; OR into TRISc register
; The asynchronous baud rate is calculated as follows:
;                       Fosc
;           ABR =      ------
;                      S*(x+1)
; where x is value in the SPBRG register and S is 64 if the high
; baud rate select bit (BRGH) in the TXSTA control register is
; clear, and 16 if the BRGH bit is set. For setting to 2400 baud
; using a 10Mhs oscillator at a slow baud rate the formula
; is:
; At slow speed (BRGH = 0)
;           10,000,000     10,000,000
;           ----------  =  ----------  = 2,403.84 (0.16% error)
;           64*(64+1)         4160
;
            movlw   spbrgVal  ; Value in spbrgVal = 64
            movwf   SPBRG     ; Place in baud rate generator
; Setup value: 0010 0000 = 0x20
            movlw   0x20      ; Enable transmission and high
                              ; baud rate
            movwf   TXSTA
            Bank0             ; Bank 0
; Setup value: 1001 0000 = 0x90
            movlw   0x90      ; Enable serial port and
```

```
                              ; continuous reception
        movwf    RCSTA
;
        clrf     errorFlags ; Clear local error flags register
        return
;===============================
;      transmit data
;===============================
; Test for Transmit Register Empty and transmit data in w
SerialSend:
        Bank0                ; Select bank 0
        btfss    PIR1,TXIF   ; check if transmitter busy
        goto     $-1         ; wait until transmitter is not busy
        movwf    TXREG       ; and transmit the data
        return


;===============================
;      receive data
;===============================
; Procedure to test line for data received and return value
; in w. Overrun and framing errors are detected and
; remembered in the variable errorFlags, as follows:
;       7 6 5 4 3 2 1 0   <== errorFlags
;       — not used —   |  |___ overrun error
;                         |_____ framing error
SerialRcv:
        clrf     newData     ; Clear new data received register
        Bank0                ; Select bank 0
; Bit 5 (RCIF) of the PIR1 Register is clear if the USART
; receive buffer is empty. If so, no data has been received
        btfss    PIR1,RCIF   ; Check for received data
        return               ; Exit if no data
; At this point data has been received. First eliminate
; possible errors: overrun and framing.
; Bit 1 (OERR) of the RCSTA register detects overrun
; Bit 2 (FERR( of the RCSTA register detects framing error
        btfsc    RCSTA,OERR  ; Test for overrun error
        goto     OverErr     ; Error handler
        btfsc    RCSTA,FERR  ; Test for framing error
        goto     FrameErr ; Error handler
; At this point no error was detected
; Received data is in the USART RCREG register
        movf     RCREG,w     ; get received data
        bsf      newData,7   ; Set bit 7 to indicate new
data
; Clear error flags
        clrf     errorFlags
        return
```

Data EEPROM Programming

```
;===========================
;      error handlers
;===========================
OverErr:
        bsf     errorFlags,0      ; Bit 0 is overrun error
; Reset system
        bcf     RCSTA,CREN        ; Clear continuous receive bit
        bsf     RCSTA,CREN        ; Set to re-enable reception
        return
;error because FERR framing error bit is set
;can do special error handling here - this code simply clears
; and continues
FrameErr:
        bsf     errorFlags,1      ; Bit 1 is framing error
        movf    RCREG,W           ; Read and throw away bad data
        return
;================================================================
;              I2C EEPROM data procedures
;================================================================
; GPRs used in EEPROM-related code are placed in the common
; RAM area (from 0x70 to 0x7f). This makes the registers
; accessible from any bank.
;=============================
;       LIST OF PROCEDURES
;=============================
; SetupI2C   --  Initialize MSSP module for I2C mode
;                in hardware master mode
;                Configure I2C lines
;                Set slew rate for 100kbps
;                Set baud rate for 10Mhz
; WriteI2C   --  Write byte to I2C EEPROM device
;                Data is stored in EEByte variable
;                Address is stored in EEMemAdd
; ReadI2C    --  Read byte from I2C EEPROM device
;                Address stored in EEMemAdd
;                Read data returned in w register
;=============================
;      I2C setup procedure
;=============================
SetupI2C:
        Bank1
        movlw   b'00011000'
        iorwf   TRISC,f           ; OR into TRISC
; Setup MSSP module for Master Mode operation
        Bank0
        movlw   B'00101000'; Enables MSSP and uses appropriate
;  0  0  1  0  1  0  0  0   Value to install
;  7  6  5  4  3  2  1  0   <== SSPCON bits in this operation
```

```
;   |   |   |   |   |__|__|__|___  Serial port select bits
;   |   |   |   |                  1000 = I2C master mode
;   |   |   |   |                  Clock = Fosc/(4*(SSPAD+1))
;   |   |   |   |_____  UNUSED IN MASTER MODE
;   |   |   |_____  SSP Enable
;   |   |                          1 = SDA and SCL pins as serial
;   |   |_____  Receive Overflow indicator
;   |                              0 = no overflow
;   |_____  Write collision detect
;                                  0 = no collision detected
            movwf   SSPCON  ; This is loaded into SSPCON
; Input levels and slew rate as standard I2C
            Bank1
            movlw   B'10000000'
;   1   0   0   0   0   0   0   0  Value to install
;   7   6   5   4   3   2   1   0  <== SSPSTAT bits in this operation
;   |   |   |   |   |   |   |   |___ Buffer full status bit READ ONLY
;   |   |   |   |   |   |   |_____  UNUSED in present application
;   |   |   |   |   |   |_____  Read/write information READ ONLY
;   |   |   |   |   |_____  UNUSED IN MASTER MODE
;   |   |   |   |_____  STOP bit READ ONLY
;   |   |   |_____  Data address READ ONLY
;   |   |_____  SMP bus select
;   |                                0 = use normal I2C specs
;   |_____  Slew rate control
;                                    0 = disabled
            movwf   SSPSTAT
; Setup Baud Rate
; Baud Rate = Fosc/(4*(SSPADD+1))
;     Fosc = 10Mhz
;     Baud Rate = 24 for 100 kbps
            movlw   .24             ; Value to use
            movwf   SSPADD  ; Store in SSPADD
            Bank0
            return


;==============================
;     I2C write procedure
;==============================
; Write one byte to I2C EEPROM 24LC04B
; Steps:
;               1. Send START
;               2. Send control. Wait for ACK
;               3. Send address. Wait for ACK
;               4. Send data. Wait for ACK
;               5. Send STOP
; STEP 1:
WriteI2C:
```

Data EEPROM Programming

```
                Bank1
                bsf     SSPCON2,SEN         ; Produce START Condition
                call    WaitI2C ; Wait for I2C to complete
; STEP 2:
; Send control byte. Wait for ACK
                movlw   LC04READ            ; Control byte
                call    Send1I2C            ; Send Byte
                call    WaitI2C ; Wait for I2C to complete
                btfsc   SSPCON2,ACKSTAT ; Check ACK bit to see if
                                            ; I2C failed, skip if not
                goto    FailI2C
; STEP 3:
; Send address. Wait for ACK
                Bank0
                movf    EEMemAdd,w          ; Load Address Byte
                call    Send1I2C            ; Send Byte
                call    WaitI2C ; Wait for I2C operation to complete
                Bank1
                btfsc   SSPCON2,ACKSTAT ; Check ACK Status bit to see
                                            ; If I2C failed, skip if not
                goto    FailI2C
; STEP 4:
; Send data. Wait for ACK
                Bank0
                movf    EEByte,w            ; Load Data Byte
                call    Send1I2C            ; Send Byte
                call    WaitI2C ; Wait for I2C operation to complete
                Bank1
                btfsc   SSPCON2,ACKSTAT ; Check ACK Status bit to see
                                            ; if I2C failed, skip if not
                goto    FailI2C
; STEP 5:
; Send STOP. Wait for ACK
                bsf     SSPCON2,PEN         ; Send STOP condition
                call    WaitI2C ; Wait for I2C operation to complete
; WRITE operation has completed successfully.
                Bank0
                return

;==============================
;    I2C read procedure
;==============================
; Procedure to read one byte from 24LC04B EEPROM
; Steps:
;       1. Send START
;       2. Send control. Wait for ACK
;       3. Send address. Wait for ACK
;       4. Send RESTART + control. Wait for ACK
```

```
;           5. Switch to receive mode. Get data.
;           6. Send NACK
;           7. Send STOP
;           8. Retrieve data into w register
; STEP 1:
ReadI2C
; Send RESTART. Wait for ACK
          Bank1
          bsf       SSPCON2,RSEN  ; RESTART Condition
          call      WaitI2C ; Wait for I2C operation
; STEP 2:
; Send control byte. Wait for ACK
          movlw     LC04READ           ; Control byte
          call      Send1I2C           ; Send Byte
          call      WaitI2C ; Wait for I2C operation
; Now check to see if I2C EEPROM is ready
          Bank1
          btfsc     SSPCON2,ACKSTAT ; Check ACK Status bit
          goto      ReadI2C ; ACK Poll waiting for EEPROM
                            ; write to complete
; STEP 3:
; Send address. Wait for ACK
          Bank0
          movf      EEMemAdd,w         ; Load from address register
          call      Send1I2C           ; Send Byte
          call      WaitI2C            ; Wait for I2C operation
          Bank1
          btfsc     SSPCON2,ACKSTAT   ; Check ACK Status bit
          goto      FailI2C ; failed, skipped if successful
; STEP 4:
; Send RESTART. Wait for ACK
          bsf       SSPCON2,RSEN       ; Generate RESTART Condition
          call      WaitI2C            ; Wait for I2C operation
; Send output control. Wait for ACK
          movlw     LC04WRITE          ; Load CONTROL BYTE (output)
          call      Send1I2C           ; Send Byte
          call      WaitI2C            ; Wait for I2C operation
          Bank1
          btfsc     SSPCON2,ACKSTAT ; Check ACK Status bit
          goto      FailI2C ; failed, skipped if successful
; STEP 5:
; Switch MSSP to I2C Receive mode
          bsf       SSPCON2,RCEN       ; Enable Receive Mode (I2C)
; Get the data. Wait for ACK
          call      WaitI2C            ; Wait for I2C operation
; STEP 6:
; Send NACK to acknowledge
          Bank1
```

Data EEPROM Programming

```
            bsf     SSPCON2,ACKDT   ; ACK DATA to send is 1 (NACK)
            bsf     SSPCON2,ACKEN   ; Send ACK DATA now.
; Once ACK or NACK is sent, ACKEN is automatically cleared
; STEP 7:
; Send STOP. Wait for ACK
            bsf     SSPCON2,PEN     ; Send STOP condition
            call    WaitI2C         ; Wait for I2C operation
; STEP 8:
; Read operation has finished
            Bank0
            movf    SSPBUF,W        ; Get data from SSPBUF into W
; Procedure has finished and completed successfully.
            return


;=============================
;   I2C support procedures
;=============================
; I2C Operation failed code sequence
; Procedure hangs up. User should provide error handling.
FailI2C
            Bank1
            bsf     SSPCON2,PEN     ; Send STOP condition
            call    WaitI2C         ; Wait for I2C operation
fail:
            goto    fail

; Procedure to transmit one byte
Send1I2C
            Bank0
            movwf   SSPBUF          ; Value to send to SSPBUF
            return

; Procedure to wait for the last I2C operation to complete.
; Code polls the SSPIF flag in PIR1.
WaitI2C
            Bank0
            btfss   PIR1,SSPIF      ; Check if I2C operation done
            goto    $-1             ; I2C module is not ready yet
            bcf     PIR1,SSPIF      ; I2C ready, clear flag
            return


;=============================================================
            end                     ; END OF PROGRAM
;=============================================================
```

Chapter 16

Analog to Digital and Realtime Clocks

Digits are a human invention; nature does not count or measure using numbers. We measure natural forces and phenomena using digital representations, but the forces and phenomena themselves are continuous. Time, pressure, voltage, current, temperature, humidity, gravitational attraction, all exist as continuous entities which we measure in volts, pounds, hours, amperes, or degrees, so as to better understand them and to be able to perform numerical calculations.

In this sense, natural phenomena occur in analog quantities. Sometimes they are digitized so as to facilitate measurements and manipulations. For example, a potentiometer in an electrical circuit allows reducing the voltage level from the circuit maximum to ground, or zero level. In order to measure and control the action of the potentiometer, we need to quantify its action by producing a digital value within the physical range of the circuit; that is, we need to convert an *analog* quantity that varies continuously between 0 and 5 volts, to a discrete *digital* value range. If, in this case, the voltage range of the potentiometer is from 5 to 0 volts, we can digitize its action into a numeric range of 0 to 500 units, or measure the angle or rotation of the potentiometer disk in degrees from 0 to 180. The device that performs either conversion is called an A/D or *analog-to-digital converter*. The reverse process, digital-to-analog, is also necessary, although not as often as A/D. In this chapter we explore A/D conversions in PIC software and hardware.

The second topic of this chapter is the measurement of time in discrete (albeit, digital) units. In this context we speak of "realtime" as years, days, hours, minutes, and so on. So a realtime clock measures time in hours, minutes, and seconds, and a realtime calendar measures it in years, months, weeks, and days. Since time is a continuum that escapes our comprehension, we must divide it into measurable chunks that can be manipulated and calculated. However, not all time units are in proportional relation with one another. There are 60 seconds in a minute and 60 minutes in an hour, but 24 hours in a day and 28, 29, 30, or 31 days in a month. Furthermore, the months and the days of the week have traditional names. Finally, the *Gregorian calendar* requires adding a 29th day to February on any year that is evenly divisible by 4. The device or software to perform all of these time calculations is referred to as a realtime clock. In this chapter we discuss the use of realtime clocks in PIC circuits.

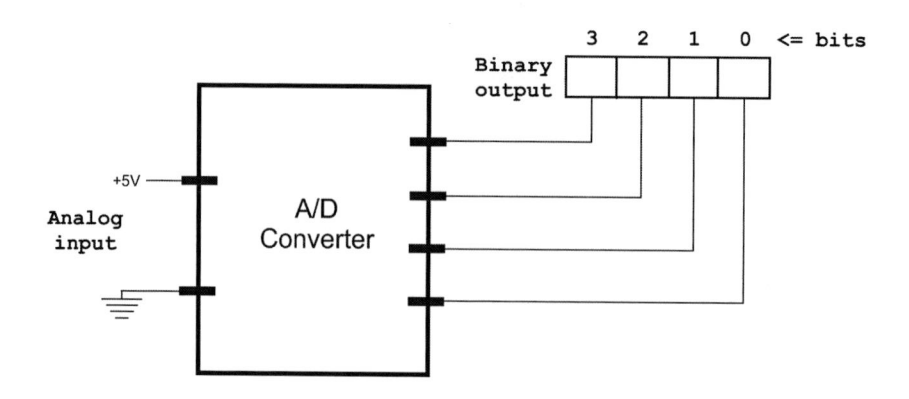

```
                                    3  2  1  0  <= bits
                          Binary  ┌──┬──┬──┬──┐
                          output  └──┴──┴──┴──┘
         +5V
Analog          A/D
input         Converter
```

Figure 16-1 A/D Converter Block Diagram

16.0 A/D Converters

In electronics, the typical *A/D* or *ADC converter* is a device that takes a voltage input and returns a binary digital number. Figure 16-1 is a block diagram of an A/D converter.

The electronic A/C converter requires an input in the form of an electrical voltage. Non-electric quantities must be changed into a voltage level before the conversion can be performed. The device that performs this conversion is called a *transducer*. For example, a digital barometer must be equipped with a transducer that converts the measurement into voltage levels. The voltage levels can then be fed into an A/D converter and the result output in digital form.

16.0.1 Converter Resolution

An ideal A/D converter outputs into an infinite number of discrete steps that exactly represent the analog quantity. Needless to say, such a device cannot exist, and a real A/D converter must be limited to a numeric range. For example, the device in Figure 16-1 outputs a voltage range of 0 to +5 volts in four binary digits that represent values between 0 and 15. Another A/D converter may produce output in eight binary digits, and another in sixteen binary digits. The number of discrete values in the conversion is called the *resolution*. The converter's resolution is usually expressed in bits. Figure 16-2 represents an A/C converter with a voltage range of 0 to +5 volts and a resolution of three bits.

Suppose that a value of 2.5 volts were input into the A/D converter in Figure 16-2. Since the output has a resolution in the range 0 to 7, the converter's output would be either 4 or 5. The non-linear characteristic of the output determines a *quantization error* that increases as the converter resolution decreases. Converters used in PIC circuits have a resolution of either 8, 10, or 12 bits. In each case the *output range*, or *quantization level*, is 0 to 255, 0 to 1023, or 0 to 4095. The voltage resolution of the converter is its maximum voltage range divided by the number of quantization levels. A device with a voltage range of 5 volts and a range of 255 levels has a voltage resolution of:

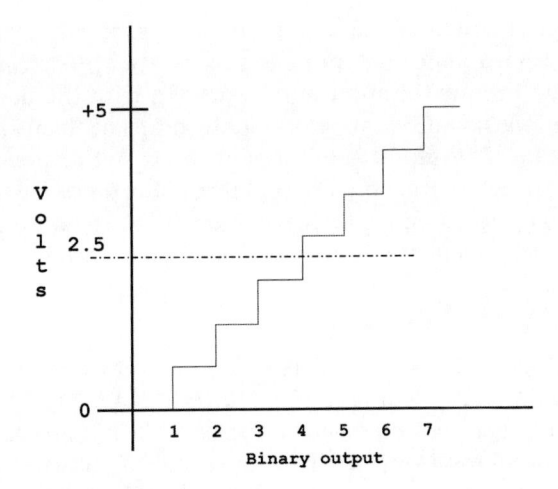

Figure 16-2 Converter Quantization Error

$$voltage\ resolution = \frac{5}{255} = 0.01960\ volts = 19.60 mV$$

16.0.2 ADC Implementation

The analog-to-digital converter performs accurately only if the input voltage is within the converter's valid range. This range is usually selected by setting high and low voltage references on converter pins. For example, if +4 volts is input into the converter's *positive reference pin* and +2 volts into the *negative reference pin*, then the converter's voltage range lies between these values. In many PIC applications the converter range is selected as the system's supply voltage and ground, that is, +5 and 0 volts. When a different range is externally referenced, there is a general restriction that the range cannot exceed the system's positive and negative limits (Vdd and Vss). Also, a minimum difference is required between the high and low voltage references.

The output of the ADC is a digital representation of the original analog signal. In this context, the term *quantization* refers to subdividing a range into small but measurable increments. The quantization process can introduce a quantization error, which is similar to a rounding error.

The time required for the holding capacitor on the ADC to charge is called the *acquisition time*. The holding capacitor on the ADC must be given sufficient time to settle to the analog input voltage level before the actual conversion is initiated. Otherwise, the conversion is not accurate. The acquisition time is determined by the impedance of the internal multiplexer and that of the analog source. The exact acquisition time can be determined from the device's data sheet, although 10K ohms is the maximum recommended source impedance for 8- and 10-bit converters and 2.5K ohms for 12-bit converters.

Most analog-to-digital converters in PIC applications, either internal or external, are of the successive approximation type. The *successive approximation algorithm* performs a conversion on one bit at a time, beginning with the most significant bit and ending with the least significant bit. To determine each bit in the range, the value of the input signal is tested to see if it is in the upper or lower portion of this range. If in the upper portion, the conversion bit is a 1, otherwise it is a 0. The next most significant bit is then tested in the lower half of the remaining range. The process is continued until the least-significant bit has been determined.

16.1 A/D Integrated Circuits

Several popular integrated circuits are used to perform as A/D converters, among them the ADC0831, the LTC1298, and the MAX 190 and MAX 191. The variations consist in the resolution and interfacing of the different ICs. Of these, the ADC0831, from National Semiconductor, is an 8-bit resolution, serial interface A/D quite suited to applications for small, mid-range PICs such as the 16F84. The input range of the 0831 is 0 to 5 volts, which matches the TTL voltage levels used in PIC circuits. The 0831 pin diagram is shown in Figure 16-3.

```
              ADC0831 PINOUT

_CS  [1]      8] Vcc      _CS   - Chip Select (active low)
Vin+ [2]      7] CLK      Vin+  - Analog voltage input +
     ADC0831              Vin-  - Analog voltage input -
Vin- [3]      6] DO       GND   - Ground
GND  [4]      5] Vref     Vref  - Voltage reference
                          DO    - Data out
                          CLK   - Clock signal
                          Vcc   - +5V power
```

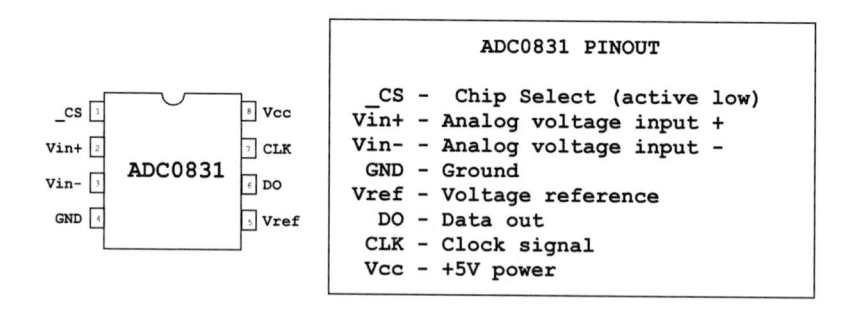

Figure 16-3 ADC0831 Pin Diagram

The ADC0831 uses three control lines, labeled *DO* (*data out*), *CLK* (*clock*), and *_CS* (*chip select*) in Figure 16-3. Interfacing the ADC0831 requires three I/O lines. Of these, two can be multiplexed with other functions or with other ADC0831. Actually, only the chip-select (CS) pin requires a dedicated line. This allows for several ADCs to be multiplexed on the CLK and DO lines as long as each one has its own CS connection to the microcontroller. In this case, the controller determines which device is being read by the port to which CS line is connected.

The input voltage range of the ADC0831 is determined by the *Vref* (*positive voltage reference* line) and *Vin-* (*negative voltage reference line*) pins. Vref is used to set the maximum level and Vin- the minimum. Since the ADC0831 has an 8-bit range, the voltage reading that matches the Vref value is read as 255 and the one that matches the Vin- value is read as 0. The minimum difference between the voltage limits is of 1 volt.

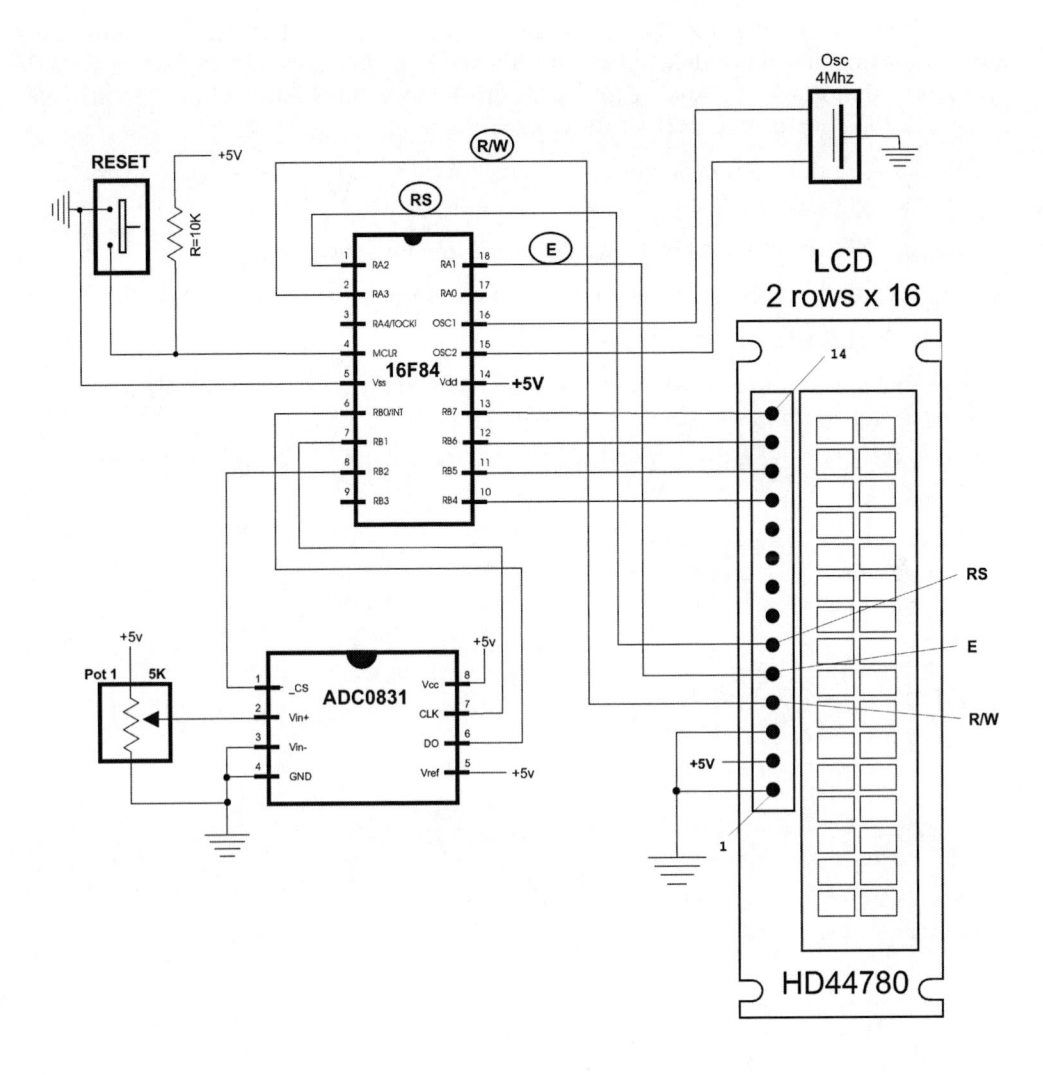

Figure 16-4 ADC0831 Demonstration Circuit

16.1.1 ADC0331 Sample Circuit and Program

A simple circuit to illustrate the action of an analog-to-digital converter consists of connecting a potentiometer with the positive voltage reference line, as shown in Figure 16-4. In the circuit the potentiometer was selected so as to produce a voltage range between 0 and +5 volts. Vref was wired to the circuit's +5 V source and Vin- was wired to ground. The potentiometer variable line was connected to the ADC0831 Vin+ line and the other ADC lines to the corresponding 16F84 Port-B pins.

The sample program is named ADF84, and can be found in the book's online software. The ADF84 program uses the ADC0831 to convert the analog voltage from the potentiometer, in the range +5 to 0 volts, into a digital value in the range 0 to 255. The value read is then displayed on the LCD. The initialization routine defines

Port-B, line 0 as input since this is the one connected to the DO line. The remaining lines in ports A and B are defined as output. ADC0831 processing consists of a single procedure that reads the analog line and returns an 8-bit digital value. The processing required is performed in the following steps:

1. The data return register (named rcvdata) is cleared and the bit counter register is initialized to count 8 bits.
2. The ADC0831 is prepared by bringing the CS line low and pulsing the CLK line.
3. The CLK line is pulsed and one bit is read from the low-order bit (DO line) of Port-B.
4. The bit is shifted into the data return register and the bit counter is decremented.
5. If the bit counter is exhausted, execution ends and the ADC is turned off. Otherwise processing continues at step 3.

The following procedure, from the ADF84 program, reads digital data from the ADC0831:

```
;=============================
;    procedure to read and
;      convert analog line
;=============================
; ON ENTRY:
; Code assumes that the ADC0831 DO line is initialized for
; input, while CLK and CS lines are output
; From ADC0831 wiring diagram. All lines in Port-B
;         DO          =       RB0     ==> INPUT
;         CLK         =       RB1     <== OUTPUT
;         CS          =       RB2     <== OUTPUT
; ON EXIT:
; Returns 8-bit digital value in the register rcvdata
;
ana2dig:
; Clear data register and init counter for 8 bits
         clrf      rcvdata   ; Clear register
         movlw     0x08                ; Initialize counter
         movwf     bitCount
; Prepare to read analog line
         bcf       PORTB,CS  ; CS pin low to enable ADC
         nop                 ; Delay for 4MHz clock
         bsf       PORTB,CLK ; Set CLK high
         Nop
         bcf       PORTB,CLK ; Reset CLK to start conversion
         nop
nextB:
; Pulse CLK line to read bit from ADC
         bsf       PORTB,CLK           ; CLK high
         nop       bcf       PORTB,CLK           ; CLK low
         Nop
; Read analog line and store data, bit by bit
```

```
            movf     PORTB,w    ; Read all Port-B bits
            movwf    store1     ; Store value for later
            rrf      store1,f   ; Rotate bit into carry flag
            rlf      rcvdata,f  ; Rotate carry flag into result
                                ; register
            decfsz   bitCount,f ; Bump counter, skip next
                                ; if counter zero
            goto     nextB
; Value read is stored in rcvdata register
            bsf      PORTB,CLK  ; Final clock pulse
            Nop
            bcf      PORTB,CLK
            nop
            bsf      PORTB,CS   ; Turn off ADC
            call     long_delay ; Time to settle
            Return
```

16.2 PIC On-Board A/D Hardware

A few years ago, A/D conversions always required the use of devices such as the ones described in the previous sections. Nowadays, many PIC microcontrollers come with onboard A/D hardware. One of the advantages of using onboard A/D converters is saving interface lines. The circuit shown in Figure 16-4 requires devoting three lines to the interface between the ADC0831 and the PIC 16F84. On the other hand, a similar circuit can be implemented in a PIC with internal A/C conversion by simply connecting the analog device to the corresponding PIC port. In the PIC world, where I/O lines are often in short supply, this advantage is not insignificant.

At the time we are writing, PICs equipped with A/D converters have either 8- or 10-bit resolution and can receive analog input in 2 to 16 different channels. The 16F877 with eight analog input channels at a 10-bit resolution is discussed. Nowadays, these PICs are easy to obtain. On the other hand, if the resolution required exceeds 10-bits then the designer has to resort to an independent A/D IC, such as the LTC1298, which has a 12-bit resolution, or to others with even higher numbers of output bits.

16.2.1 A/D Module on the 16F87x

The PICs of the 16F87x family are equipped with an analog-to-digital converter module. The number of lines depends on the specific version of the device: 28-pin devices have five A/D lines and all others have eight lines. The converter *uses a sample and hold* capacitor to store the analog charge and performs a successive approximation algorithm to produce the digital result. The converter resolution is 10 bits, which are stored in two 8-bit registers. One of the registers has only four significant bits.

The A/D module has high- and low-voltage reference inputs that are selected by software. The module can operate while the processor is in SLEEP mode, but only if the A/D clock pulse is derived from its internal RC oscillator. The module contains four registers accessible to the application:

- ADRESH - Result High Register
- ADRESL - Result Low Register
- ADCON0 - Control Register 0
- ADCON1 - Control Register 1

Of these, it is the ADCON0 register that controls most of the operations of the A/C module. Port-A pins RA0 to RA5 and Port-E pins RE0 to RE2 are multiplexed as analog input pins into the A/C module. In the 28-pin versions of the 16F87x, port pins RA0 to RA5 provide the five input channels. In all other implementations of the 16F87X, Port-E pins RE0 to RE2 provide the three additional channels.

Figure 16-5 shows the registers associated with A/D module operations.

REGISTER NAME	7	6	5	4	3	2	1	0 bits
INTCON	GIE	PEIE						
PIR1		ADIF						
PIE1		ADIE						
ADRESH	A/D Result Register High Byte							
ADRESL	A/D Result Register Low Byte							
ADCON0	ADSC1	ADSC0	CHS2	CHS1	CHS0	GO/DONE		ADON
ADCON1	ADFM				PCFG3	PCFG2	PCFG1	PCFG0

Figure 16-5 Registers Related to A/C Module Operations

The ADCON0 Register

The ADCON0 register is located in bank 0, at address 0x1f. Seven of the eight bits are meaningful in A/D control and status operations. Figure 16-6 is a bitmap of the ADCON0 register.

In Figure 16-6, bits 7 and 6, labeled ADSC1 and ADSC0, are the selection bits for the *A/D conversion clock*. The conversion time per bit is defined as *TAD* in PIC documentation. A/D conversion requires a minimum of 12 TAD in a 10-bit ADC. The source of the A/D conversion clock is software selected. The four possible options for TAD are:

1. Fosc/2
2. Fosc/8
3. Fosc/32
4. Internal A/D module RC oscillator (varies between 2 and 6 µs)

Analog to Digital and Realtime Clocks

bits:	7	6	5	4	3	2	1	0
	ADSC1	ADSC0	CHS2	CHS1	CHS0	GO/DONE		ADON

```
bit 7-6 ADCS1:ADCS0: A/D Conversion Clock Select bits
        00 = FOSC/2
        01 = FOSC/8
        10 = FOSC/32
        11 = FRC (internal A/D module RC oscillator)
bit 5-3 CHS2:CHS0: Analog Channel Select bits
        000 = channel 0, (RA0=AN0)
        001 = channel 1, (RA1=AN1)
        010 = channel 2, (RA2=AN2)
        011 = channel 3, (RA3=AN3)
        100 = channel 4, (RA5=AN4)
        101 = channel 5, (RE0=AN5) | not active
        110 = channel 6, (RE1=AN6) | in 28-pin
        111 = channel 7, (RE2=AN7) | 16F87x PICS
bit 2 GO/DONE: A/D Conversion Status bit
      If ADON = 1:
      1 = A/D conversion in progress (setting this
          bit starts the A/D conversion)
      0 = A/D conversion not in progress (this bit
          is automatically cleared by hardware when
          the A/D conversion is complete)
bit 1 Unimplemented: Read as '0'
bit 0 ADON: A/D On bit
      1 = A/D converter module is operating
      0 = A/D converter module is shut-off and
          consumes no power
```

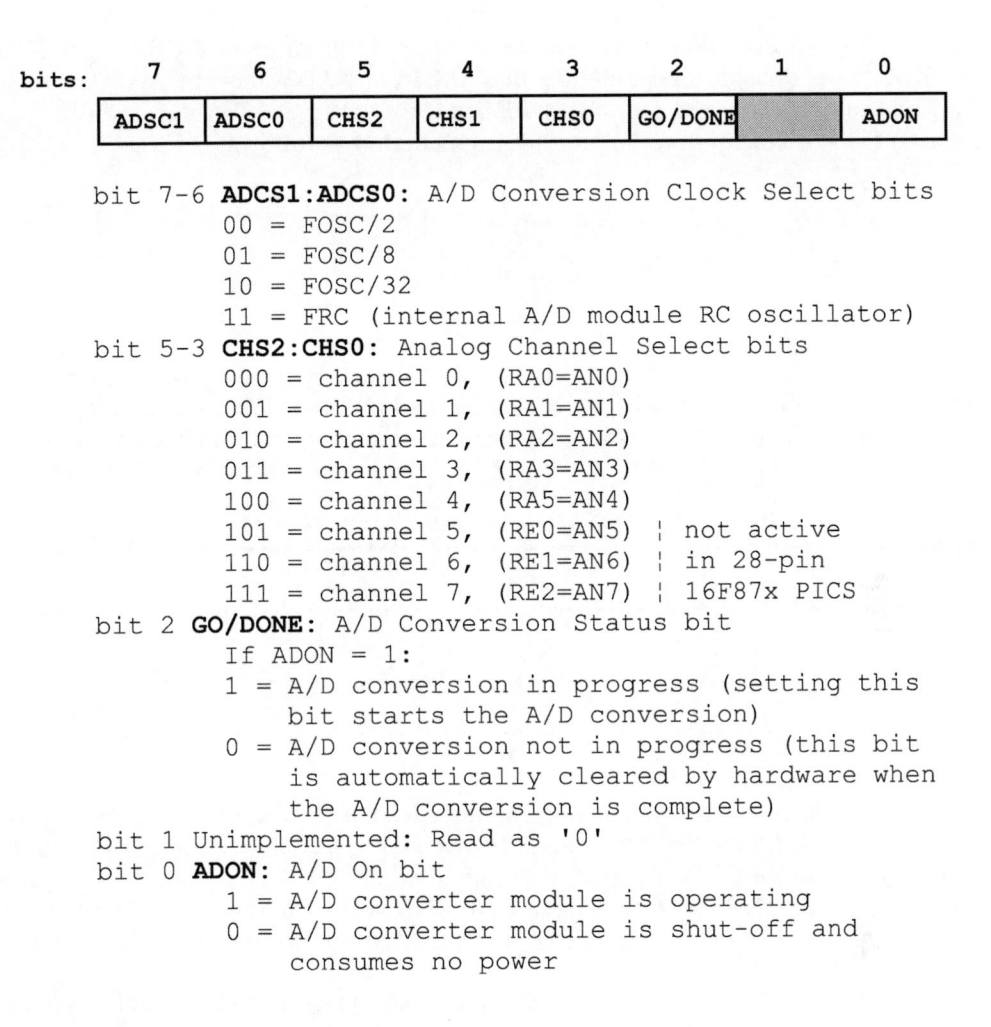

Figure 16-6 ADCON0 Register Bitmap

The conversion time is the analog-to-digital clock period multiplied by the number of bits of resolution in the converter, plus the two to three additional clock periods for settling time, as specified in the data sheet of the specific device. The various sources for the analog-to-digital converter clock represent the main oscillator frequency divided by 2, 8, or 32. The third choice is the use of a dedicated internal RC clock that has a typical period of 2 to 6 µs. Since the conversion time is determined by the system clock, a faster clock results in a faster conversion time.

The A/D conversion clock must be selected to ensure a minimum Tad time of 1.6 µs. The formula for converting processor speed (in MHz) into Tad microseconds is as follows:

$$Tad = \frac{1}{\frac{Tosc}{Tdiv}}$$

Where *Tad* is A/D conversion time, *Tosc* is the oscillator clock frequency in MHz, and *Tdiv* is the divisor determined by bits ADSC1 and ADSC0 of the ADCON0 register. For example, in a PIC running at 10MHz if we select the Tosc/8 option (divisor equal 8) the A/D conversion time per bit is calculated as follows:

$$Tad = \frac{1}{\frac{5Mhz}{8}} = 1.6$$

In this case, the minimum recommended conversion speed of 1.6 μs is achieved. However, in a PIC with an oscillator speed of 10MHz, this option produces a conversion speed of 0.8 μs, less than the recommended minimum. In this case we would have to select the divisor 32 option, giving a conversion speed of 3.2 μs.

Table 16.1
A/C Converter Tad at Various Oscillator Speeds

OPERATION	ADCS1:ADCS0	TAD IN MICROSECONDS			
		20MHZ	10MHZ	5MHZ	1.25MHZ
Fosc/2	00	0.1	0.2	0.4	**1.6**
Fosc/8	01	0.4	0.8	**1.6**	**6.4**
Fosc/32	10	**1.6**	**3.2**	**6.4**	25.6
RC	11	**2-6**	**2-6**	**2-6**	**2-6**

Note: values in bold are within the recommended limits

In Table 16.1, converter speeds of less than 1.6 μs or higher than 10 μs are not recommended. Recall that the Tad speed of the converter is calculated per bit, so the total conversion time in a 10-bit device (such as the 16F87x) is approximately the Tad speed multiplied by 10 bits, plus 3 additional cycles. Therefore, a device operating at a Tad speed of 1.6 μs requires 1.6 μs * 13, or 20.8 μs for the entire conversion.

Bits CHS2 to CHS0 in the ADCON0 register (see Figure 16-6) determine which of the analog channels is selected. This is required, since there are several channels for analog input but only one A/2 converter circuitry. So the setting of this bit field determines which of six or eight possible channels is currently read by the A/C converter. An application can change the setting of these bits in order to read several analog inputs in succession.

Bit 2 of the ADCON0 register, labeled GO/DONE, is both a control and a status bit. Setting the GO/DONE bit starts A/D conversion. Once conversion has started, the bit indicates if it is still in progress. Code can test the status of the GO/DONE bit in order to determine if conversion has concluded.

Bit 0 of the ADCON0 register turns the A/D module on and off. The initialization routine of an A/D-enabled application turns on this bit. Programs that do not use the A/D conversion module leave the bit off to conserve power.

The ADCON1 Register

The ADCON1 register also plays an important role in programming the A/D module. Bit 7 of the ADCON1 register is used to determine the *bit justification* of the digital result. This is possible because the 10-bit result is returned in two 8-bit registers; therefore, the six unused bits can be placed either on the left- or the right-hand side of the 16-bit result. If ADCON1 bit 7 is set then the result is right-justified; otherwise it is left-justified. Figure 16-7 shows the location of the significant bits.

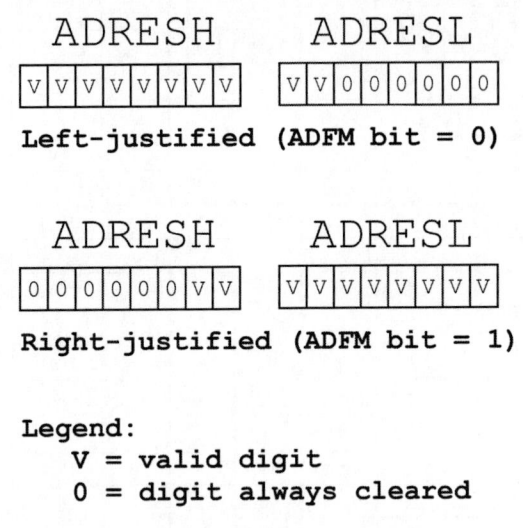

```
         ADRESH              ADRESL
       |V|V|V|V|V|V|V|V|   |V|V|0|0|0|0|0|0|
       Left-justified (ADFM bit = 0)


         ADRESH              ADRESL
       |0|0|0|0|0|0|V|V|   |V|V|V|V|V|V|V|V|
       Right-justified (ADFM bit = 1)


       Legend:
           V = valid digit
           0 = digit always cleared
```

Figure 16-7 Left- and Right-justification of A/D Result

One common use of right justification is to reduce the number of significant bits in the conversion result. For example, an application on the 16F877 that uses the A/D conversion module requires only 8-bit accuracy in the result. In this case, code can left-justify the conversion result, read the ADRESH register, and ignore the low-order bits in the ADRESL register. By ignoring the two low-order bits, the 10-bit accuracy of the A/D hardware is reduced to eight bits and the converter performs as an 8-bit accuracy unit.

The bit field labeled PCFG3 to PCFG0 in the ADCON1 register determines port configuration as analog or digital and the mapping of the positive and negative voltage reference pins. The number of possible combinations is limited by the four bits allocated to this field, so the programmer and circuit designer must select the option that is most suited to the application when the ideal one is not available. Table 16.2 (in the following page) shows the port configuration options.

For example, there is a circuit that calls for two analog inputs, wired to ports RA0 and RA1, with no reference voltages. In Table 16.2 we can find two options that select ports RA0 and RA1 and are analog inputs: these are the ones selected with PCFG bits 0100 and 0101. The first option also selects port RA3 as analog input, even though not required in this case. The second one also selects port RA3 as a positive voltage reference, also not required.

Either option works in this case; however, any pin configured for analog input produces incorrect results if used as a digital source. Therefore, a channel configured for analog input cannot be used for non-analog purposes. On the other hand, a

Table 16.2
A/D Converter Port Configuration Options

PCFG3:PCFG0	An7 Re2	An6 Re1	An5 Re0	An4 Ra5	An3 Ra3	An2 Ra2	An1 Ra1	An0 Ra0	Vref+	Vref-	CHAN/Refs
0000	A	A	A	A	A	A	A	A	VDD	VSS	8/0
0001	A	A	A	A	Vre+	A	A	A	RA3	VSS	7/1
0010	D	D	D	A	A	A	A	A	VDD	VSS	5/0
0011	D	D	D	A	Vre+	A	A	A	RA3	VSS	4/1
0100	D	D	D	D	A	D	A	A	VDD	VSS	3/0
0101	D	D	D	D	Vre+	D	A	A	RA3	VSS	2/1
011x	D	D	D	D	D	D	D	D	VDD	VSS	0/0
1000	A	A	A	A	Vre+	Vre-	A	A	RA3	RA2	6/2
1001	D	D	A	A	A	A	A	A	VDD	VSS	6/0
1010	D	D	A	A	Vre+	A	A	A	RA3	VSS	5/1
1011	D	D	A	A	Vre+	Vre-	A	A	RA3	RA2	4/2
1100	D	D	D	A	Vre+	Vre-	A	A	RA3	RA2	3/2
1101	D	D	D	D	Vre+	Vre-	A	A	RA3	RA2	2/2
1110	D	D	D	D	D	D	D	A	VDD	VSS	1/0
1111	D	D	D	D	Vre+	Vre-	D	A	RA3	RA2	1/2

```
Legend:
    D = digital input
    A = analog input
    CHAN/Refs = analog channels/voltage reference inputs
```

channel configured for digital input should not be used for analog data since extra current is consumed by the hardware. Finally, channels to be used for analog-to-digital conversion must be configured for input in the corresponding TRIS register.

SLEEP Mode Operation

The A/D module can be made to operate in SLEEP mode. As mentioned previously, SLEEP mode operation requires that the A/D clock source be set to RC by setting both ADCS bits in the ADCON0 register. When the RC clock source is selected, the A/D module waits one instruction cycle before starting the conversion. During this period, the SLEEP instruction is executed, thus eliminating all *digital switching noise* from the conversion. The completion of the conversion is detected by testing the GO/DONE bit. If a different clock source is selected, then a SLEEP instruction causes the conversion-in-progress to be aborted and the A/D module to be turned off.

16.2.2 A/D Module Sample Circuit and Program

The circuit in Figure 16-8 is designed to demonstrate the use of the A/D converter module in PICs of the 16F87x family.

Analog to Digital and Realtime Clocks

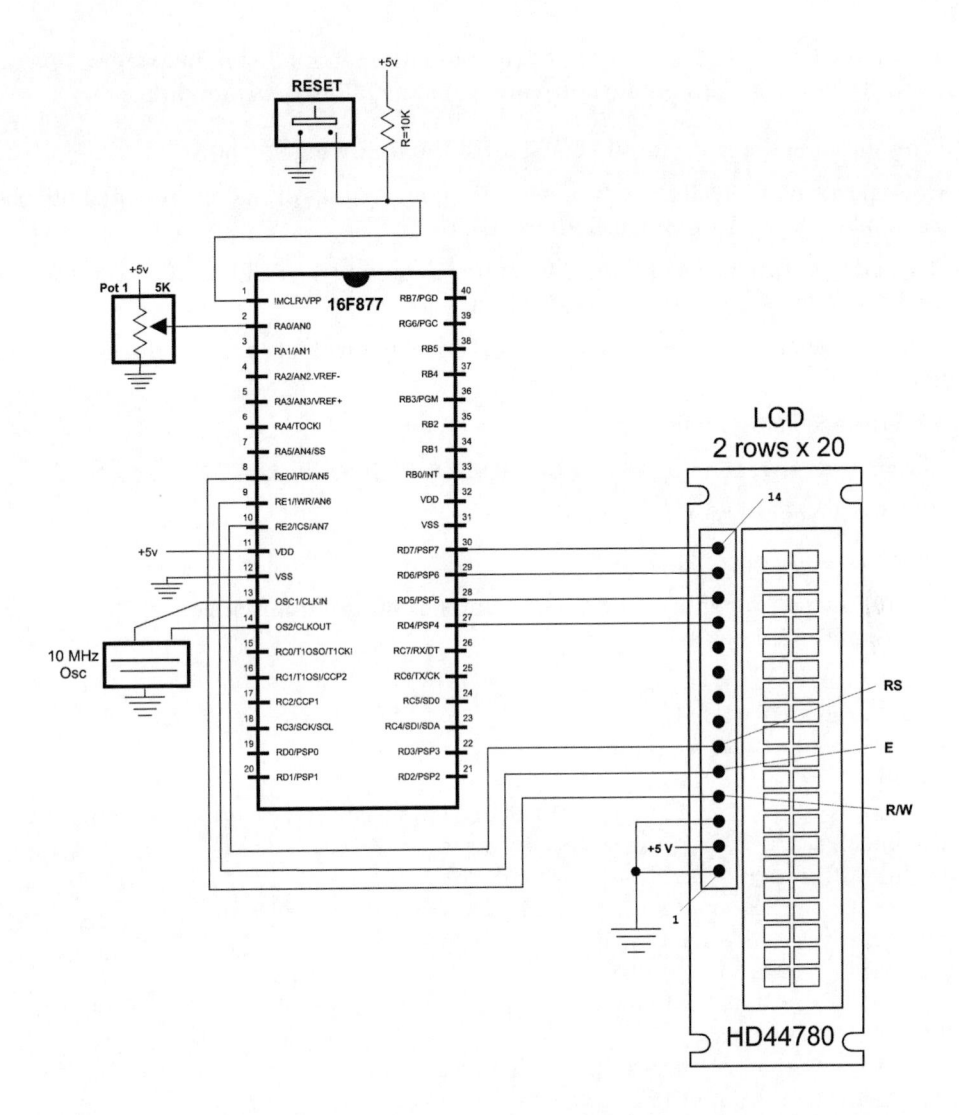

Figure 16-8 Demonstration Circuit for A/D Conversion Module

Comparing Figure 16-8 with Figure 16-4, which uses the ADC0831 IC, we notice the economy of resources that results from selecting a PIC with an onboard A/D module. In the circuit of Figure 16-4 three microcontroller I/O ports must be used to connect the converter IC to the PIC. In the circuit of Figure 16-8, the potentiometer is connected directly to a single PIC port, saving two I/O lines. Considering the number of different PIC architectures that are equipped with onboard A/D converters, the circuit designer should explore this possibility before deciding on using a separate converter IC. At the same time, recall that two of the three input lines used by converter ICs can be shared. In a design with more than one converter IC the use of input lines is not a 3 to 1 ratio.

The circuit in Figure 16-8 consists of a 5K potentiometer wired to analog port RA0 of a 16F877 PIC. The LCD display is used to show three digits, in the range 0 to 255, that represent the relative position of the potentiometer's disk. The program named A2DinLCD, in the book's online software, uses the built-in A/D module.

Programming the A/D module consists of the following steps:

1. Configure the PIC I/O lines to be used in the conversion. All analog lines are initialized as input in the corresponding TRIS registers.
2. Select the ports to be used in the conversion by setting the PCFGx bits in the ADCON1 register. Selects right- or left-justification.
3. Select the analog channels, select the A/D conversion clock, and enable the A/D module.
4. Wait the acquisition time.
5. Initiate the conversion by setting the GO/DONE bit in the ADCON0 register.
6. Wait for the conversion to complete.
7. Read and store the digital result.

The following procedure from the A2DinLCD program initialized the A/D module for the required processing:

```
;=============================
;     init A/D module
;=============================
; 1. Procedure to initialize the A/D module, as follows:
;    Configure the PIC I/O lines. Init analog lines as input
; 2. Select ports to be used by setting the PCFGx bits in the
;    ADCON1 register. Selects right- or left-justification.
; 3. Select the analog channels, select the A/D conversion
;    clock, and enable the A/D module.
; 4. Wait the acquisition time.
; 5. Initiate the conversion by setting the GO/DONE bit in the
;    ADCON0 register.
; 6. Wait for the conversion to complete.
; 7. Read and store the digital result.
InitA2D:
        Bank1                ; Select bank for TRISA register
        movlw    b'00000001'
        movwf    TRISA       ; Set Port-A, line 0, as input
; Select the format and A/D port configuration bits in
; the ADCON1 register
; Format is left-justified so that ADRESH bits are the
; most significant
;   0  x  x  x  1  1  0   <== value installed in ADCON1
;   7  6  5  4  3  2  1  0   <== ADCON1 bits
;   |           |__|__|__|____ RA0 is analog.
;   |                          Vref+ = Vdd
;   |                          Vref- = Vss
;   |_____  0 = left-justified
; ADCON1 is in bank 1
```

Analog to Digital and Realtime Clocks

```
            movlw     b'00001110'
            movwf     ADCON1    ; RA0 is analog. All others digital
                                ; Vref+ = Vdd
; Select D/A options in ADCON0 register
; For a 10Mhz clock the Fosc32 option produces a conversion
; speed of 1/(10/32) = 3.2 microseconds, which is within the
; recommended range of 1.6 to 10 microseconds.
;   1  0  0  0  0  0  0  1  <== value installed in ADCON0
;   7  6  5  4  3  2  1  0  <== ADCON0 bits
;   |  |  |  |  |  |  |  |____ A/D function select
;   |  |  |  |  |  |             1 = A/D ON
;   |  |  |  |  |  |_____ A/D status bit
;   |  |  |__|__|_____ Analog Channel Select
;   |  |                         000 = Chanel 0 (RA0)
;   |__|_____ A/D Clock Select
;                                10 = Fosc/32
; ADCON0 is in bank 0
            Bank0
            movlw     b'10000001'
            movwf     ADCON0    ; Channel 0, Fosc/32, A/D enabled
; Delay for selection to complete
            call      delayAD   ; Local procedure
            return
```

Once the module is initialized, the analog line is read by the following procedure:

```
;=============================
;        read A/D line
;=============================
; Procedure to read the value in the A/D line and convert
; to digital
ReadA2D:
; Initiate conversion
            Bank0                ; Bank for ADCON0 register
            bsf       ADCON0,GO  ; Set the GO/DONE bit
; GO/DONE bit is cleared automatically when conversion ends
convWait:
            btfsc     ADCON0,GO  ; Test bit
            goto      convWait   ; Wait if not clear
; At this point conversion has concluded
; ADRESH register (bank 0) holds 8 MSBs of result
; ADRESL register (bank 1) holds 4 LSBs.
; In this application value is left-justified. Only the
; MSBs are read
            movf      ADRESH,W ; Digital value to w register
            return
```

The delay routine required in this case is coded as follows:

```
;=======================
;     delay procedure
;=======================
; For a 10Mhz clock the Fosc32 option produces a conversion
; speed of 1/(10/32) = 3.2 microseconds. At 3.2 ms per bit
; 13 bits require approximately 41 ms. The instruction time
; at 10Mhz is 10 ms. 4/10 = 0.4 ms per instruction. To delay
; 41 ms a 10Mhz PIC must execute 11 instructions. Add one
; more for safety.
delayAD:
        movlw   .12                ; Repeat 12 machine cycles
        movwf   count1   ; Store value in counter
repeat11:
        decfsz  count1,f ; Decrement counter
        goto    repeat11 ; Continue if not 0
        return
```

16.3 Realtime Clocks

In the context of microcontrollers and embedded systems, *realtime clocks* (also called *RTC*s) are integrated circuits designed to keep track of time in conventional hours, that is, in years, days, hours, minutes, and seconds. Many realtime clock ICs are available with various characteristics, data formats, modes of operation, and interfaces. Most of the ones used in PIC circuits have a serial interface in order to save access ports. Most RTC chips provide a battery connection so that time can be kept when the system is turned off.

In the sections that follow, we discuss one popular RTC chip: the NJU6355, but this is by no means the only option for embedded systems.

16.3.1 The NJU6355 Realtime Clock

The NJU6355 series is a serial I/O realtime clock used in microcontroller-based embedded systems. The IC includes its own quartz crystal oscillator, counter, shift register, voltage regulator, and interface controller. The PIC interface requires four lines. Operating voltage is TTL level so it can be wired directly on the typical PIC circuit. The output data includes year, month, day-of-week, hour, minutes, and seconds. Figure 16-9 is the pin diagram for the NJU6355.

NJU6355 output is in packed BCD format, that is, each decimal digit is represented by a 4-bit binary number. The chip's logic correctly calculates the number of days in each month as well as the leap years. All unused bits are reported as binary 0. Figure 16-10 is a bitmap of the formatted timer data.

Analog to Digital and Realtime Clocks 559

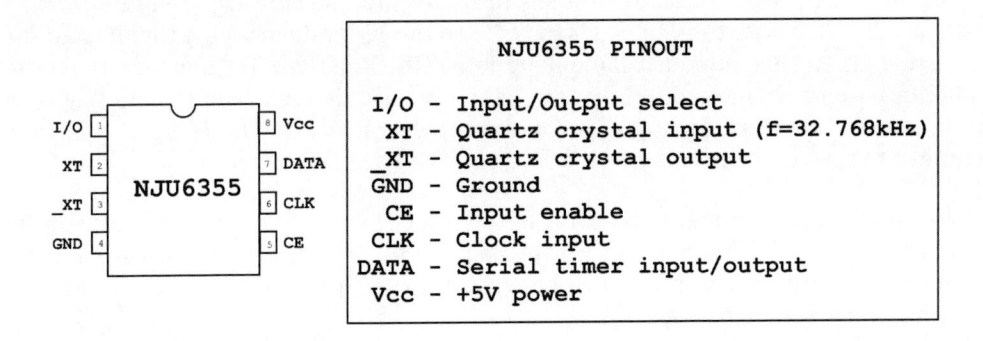

```
          NJU6355 PINOUT

  I/O  - Input/Output select
   XT  - Quartz crystal input (f=32.768kHz)
  _XT  - Quartz crystal output
  GND  - Ground
   CE  - Input enable
  CLK  - Clock input
  DATA - Serial timer input/output
  Vcc  - +5V power
```

Pin diagram: I/O (1), XT (2), _XT (3), GND (4), CE (5), CLK (6), DATA (7), Vcc (8)

Figure 16-9 NJU6355 Pin Diagram

Timer data is read when the I/O line is low and the CE line is high. Output from the 6355 is LSB first. A total of 52 significant bits are read in bottom-up order for data as shown in Figure 16-10. That is, the first bit received is the least-significant bit of the year, then the month, then the day, and so forth. All date items are eight bits, except the day of week which is four bits. Non-significant bits in each field are reported as zero; this means that the value for the 10th month (October) is encoded as binary digits 00001010. Reporting unused digits as zero simplifies the conversion into BCD and ASCII.

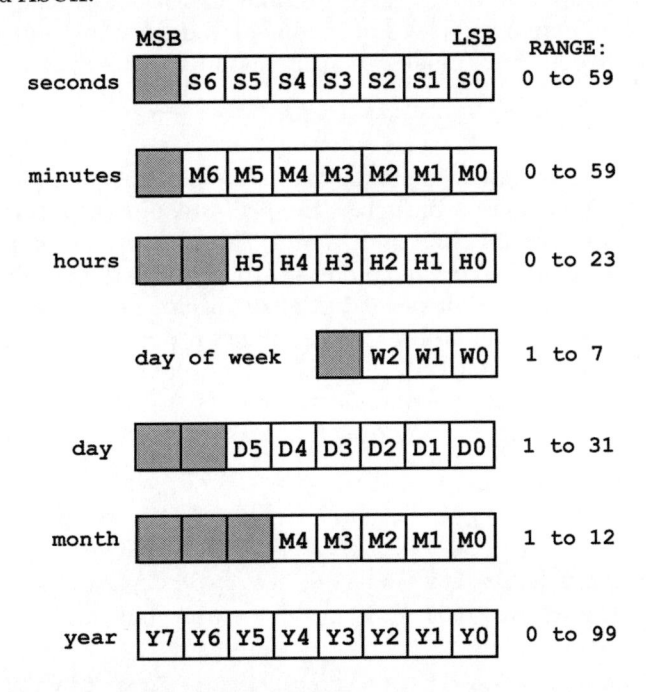

Field	Bits (MSB → LSB)	RANGE
seconds	S6 S5 S4 S3 S2 S1 S0	0 to 59
minutes	M6 M5 M4 M3 M2 M1 M0	0 to 59
hours	H5 H4 H3 H2 H1 H0	0 to 23
day of week	W2 W1 W0	1 to 7
day	D5 D4 D3 D2 D1 D0	1 to 31
month	M4 M3 M2 M1 M0	1 to 12
year	Y7 Y6 Y5 Y4 Y3 Y2 Y1 Y0	0 to 99

Figure 16-10 NJU6355 Timer Data Format

The NJU6355 does not report valid time data until after it has been initialized, even if there are power and clock signals into the chip. Initialization requires writing data into the 6355 registers. In order to write to the IC, code must set the I/O and the CE lines high. At this moment all clock updates stop and the RTC goes into the write mode. Input data is latched in LSB first, starting with the year and concluding with the minutes. There is no provision for writing seconds into the RTC, so the total number of bits written is 44.

The 6355 contains a mechanism for detecting conditions that could compromise the clock's operation, such as low power. In this case, the special value 0xee is written into each digit of the internal registers to inform processing routines that the timer has been compromised.

The NJU6355 requires the installation of an external crystal oscillator. The crystal must have a frequency of 32.768 kHz. The time-keeping accuracy of the RTC is determined by the quartz oscillator. The capacity of the oscillator must match that of the RTC and of the circuit. A standard crystal with a capacitance of 12.5pF works well for applications that do not demand high clock accuracy. For more exacting applications the 6355 can be programmed to check the clock frequency and determine its error. The chip's frequency-checking mode is described in an NJU6355 Application Note available from New Japan Radio Co., Ltd.

16.3.2 RTC Demonstration Circuit and Program

The circuit shown in Figure 16-11 is a simple application of the 6355 RTC. The circuit uses a NJU6355 in conjunction with a 16F86 PIC and an LCD. The demonstration program, named RTC2LCD, sets up RTC and reads clock data in an endless loop. The hours, minutes, and seconds are displayed at the top line of the LCD as follows:

```
H:xx  M:xx  S:xx
```

where xx represents the two BCD digits read from the clock and converted to ASCII decimal for display. The program initializes the 6355 to some arbitrary values contained in the corresponding **#define** statements. These values are copied into program variables by a local procedure and then used to initialize the RTC registers. Two procedures relate to RTC operation: one to initialize the clock hardware and the other one to read the current time. In addition, two auxiliary procedures are implemented: one to read clock data and one to write clock data. Since clock data can be in 8- or 4-bit formats each procedure contains a separate entry point to handle the 4-bit option. The procedure to initialize and the one to write clock data are coded as follows:

```
;=============================
;           init RTC
;=============================
; Procedure to initialize the real time clock chip. If chip
; is not initialized it will not operate and the values
; read will be invalid.
; Since the 6355 operates in BCD format the stored values must
; be converted to packed BCD.
; According to wiring diagram
```

Analog to Digital and Realtime Clocks

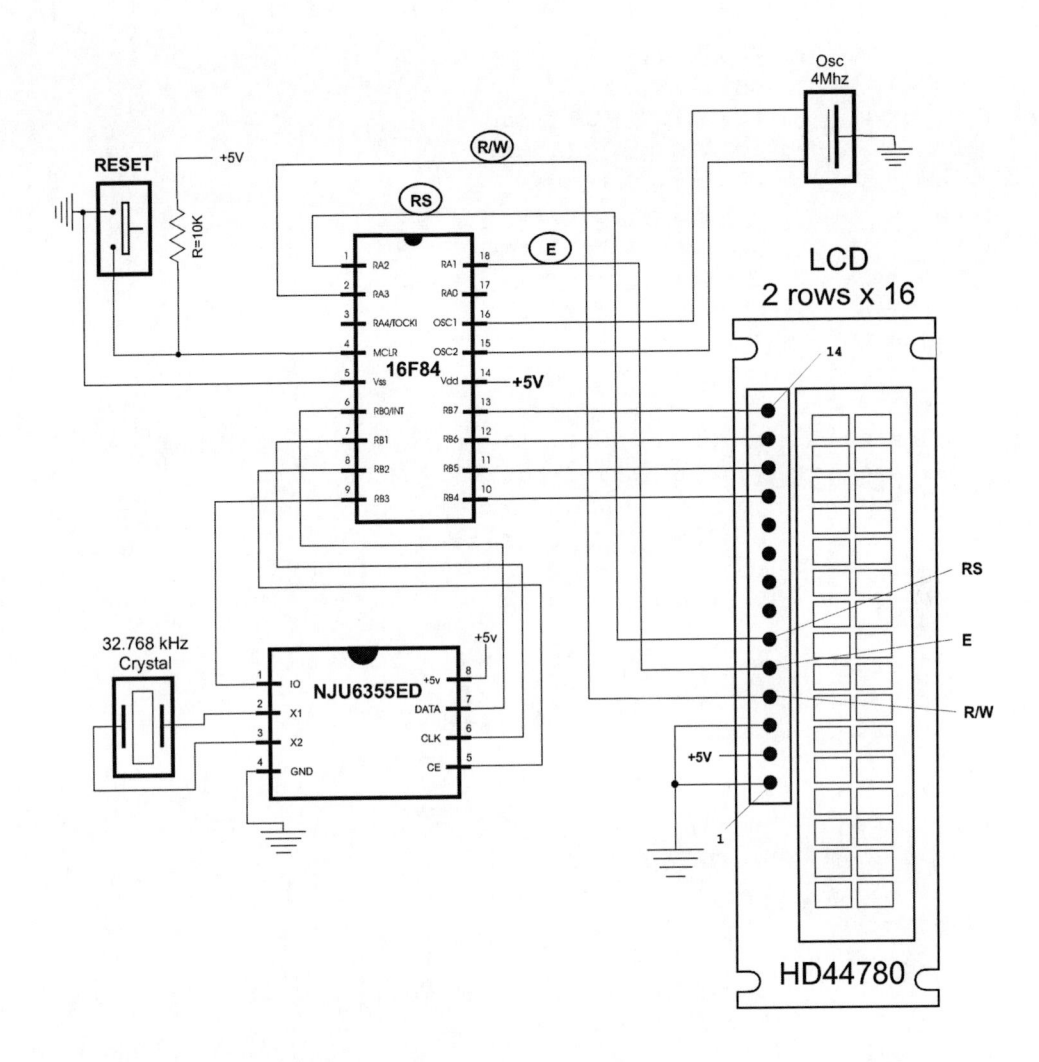

Figure 16-11 Real-time Clock Demonstration Circuit

```
; NJU6355 Interface for setting time:
; DAT           PORTB,0         Output
; CLK           PORTB,1         Output
; CE            PORTB,2         Output
; IO            PORTB,3         Output
setRTC:
        Bank1
        movlw   b'00000000'     ; All lines are output
        movlw   TRISB
        Bank0
; Writing to the 6355 requires that the CLK bit be held
; low while the IO and CE lines are high
        bcf     PORTB,CLK       ; CLK low
        call    delay_5
```

```
              bsf       PORTB,IO ; IO high
              call      delay_5
              bsf       PORTB,CE ; CE high
; Data is stored in RTC as follows:
;   year              8 bits (0 to 99)
;   month             8 bits (1 to 12)
;   day               8 bits (1 to 31)
;   dayOfWeek         4 bits (1 to 7)
;   hour              8 bits (0 to 23)
;   minutes           8 bits (0 to 59)
;                            ======
;   Total                    44 bits
; Seconds cannot be written to RTC. RTC seconds register
; is automatically initialized to zero
              movf      year,w    ; Get item from storage
              call      bin2bcd   ; Convert to BCD
              movwf     temp1
              call      writeRTC

              movf      month,w
              call      bin2bcd
              movwf     temp1
              call      writeRTC

              movf      day,w
              call      bin2bcd
              movwf     temp1
              call      writeRTC

              movf      dayOfWeek,w        ; dayOfWeek of week is 4-bits
              call      bin2bcd
              movwf     temp1
              call      write4RTC

              movf      hour,w
              call      bin2bcd
              movwf     temp1
              call      writeRTC

              movf      minutes,w
              call      bin2bcd
              movwf     temp1
              call      writeRTC
; Done
              bcf       PORTB,CLK          ; Hold CLK line low
              call      delay_5
              bcf       PORTB,CE ; and the CE line
                                           ; to the RTC
```

Analog to Digital and Realtime Clocks

```
            call      delay_5
            bcf       PORTB,IO        ; RTC in output mode
            return
;=============================
;   write 4/8 bits to RTC
;=============================
; Procedure to write 4 or 8 bits to the RTC registers
; ON ENTRY:
;     temp1 register holds value to be written
; ON EXIT:
;     nothing
write4RTC
            movlw     .4              ; Init for 4 bits
            goto      allBits
writeRTC:
            movlw     .8              ; Init for 8 bits
allBits:
            movwf     counter         ; Store in bit counter
writeBits:
            bcf       PORTB,CLK       ; Clear the CLK line
            call      delay_5         ; Wait
            bsf       PORTB,DAT       ; Set the data line to RTC
            btfss     temp1,0         ; Send LSB
            bcf       PORTB,DAT       ; Clear data line
            call      delay_5         ; Wait for operation to
complete
            bsf       PORTB,CLK       ; Bring CLK line high to
validate
            rrf       temp1,f         ; Rotate bits in storage
            decfsz    counter,1       ; Decrement bit counter
            goto      writeBits       ; Continue if not last bit
            return
```

The following procedures are used by the RTC2LCD program to read the data in the RTC registers:

```
;=============================
;       read RTC data
;=============================
; Procedure to read the current time from the RTC and store
; data (in packed BCD format) in local time registers.
; According to wiring diagram
; NJU6355 Interface for read operations:
; DAT           PORTB,0         Input
; CLK           PORTB,1         Output
; CE            PORTB,2         Output
; IO            PORTB,3         Output
Get_Time
; Clear Port-B
```

```
        movlw    b'00000000'
        movwf    PORTB
; Make data line input
        Bank1
        movlw    b'00000001'
        movwf    TRISB
        Bank0
; Reading RTC data requires that the IO line be low and the
; CE line be high. CLK line is held low
        bcf      PORTB,CLK      ; CLK low
        call     delay_125
        bcf      PORTB,IO ; IO line low
        call     delay_125
        bsf      PORTB,CE ; and CE line high
; Data is read from RTC as follows:
;  year           8 bits (0 to 99)
;  month          8 bits (1 to 12)
;  day            8 bits (1 to 31)
;  dayOfWeek      4 bits (1 to 7)
;  hour           8 bits (0 to 23)
;  minutes        8 bits (0 to 59)
;  seconds        8 bits (0 to 59)
;                        ======
;   Total                52 bits
;
        call     readRTC
        movwf    year
        call     delay_125

        call     readRTC
        movwf    month
        call     delay_125

        call     readRTC
        movwf    day
        call     delay_125

; dayOfWeek of week is a 4-bit value
        call     read4RTC
        movwf    dayOfWeek
        call     delay_125

        call     readRTC
        movwf    hour
        call     delay_125

        call     readRTC
        movwf    minutes
```

Analog to Digital and Realtime Clocks

```
            call     delay_125

            call     readRTC
            movwf    seconds
            bcf      PORTB,CE ; CE line low to end output
            return

;==============================
;  read 4/8 bits from RTC
;==============================
; Procedure to read 4/8 bits stored in 6355 registers
; Value returned in w register
read4RTC
            movlw    .4                ; 4 bit read
            goto     anyBits
readRTC:
            movlw    .8                ; 8 bits read
anyBits:
            movwf    counter
; Read 6355 read operation requires the IO line be set low
; and the CE line high. Data is read in the following order:
; year, month, day, day-of-week, hour, minutes, seconds
readBits:
            bsf      PORTB,CLK; Set CLK high to validate data
            bsf      STATUS,C ; Set the carry flag (bit = 1)
; Operation:
;   If data line is high, then bit read is a 1-bit
;   otherwise bit read is a 0-bit
            btfss    PORTB,DAT         ; Is data line high?
                                       ; Leave carry set (1 bit) if high
            bcf      STATUS,C ; Clear the carry bit (make bit 0)
; At this point the carry bit matches the data line
            bcf      PORTB,CLK         ; Set CLK low to end read
; The carry bit is now rotated into the temp1 register
            rrf      temp1,1
            decfsz   counter,1         ; Decrement the bit counter
            goto     readBits ; Continue if not last bit
; At this point all bits have been read (8 or 4)
            movf     temp1,0           ; Result to w
            return
```

BCD Conversion Procedures

In addition to the RTC procedures to initialize the clock registers and to read clock data, the application requires auxiliary procedures to manipulate and display data in BCD format. BCD encodings, covered in Section 3.4, are a way of representing decimal digits in binary form. Two common BCD formats are used: packed and unpacked. In the unpacked format each byte encodes a single BCD value. In packed form two BCD digits are encoded per byte. The 6355 uses the packed BCD format.

Since program data is usually in binary form, it is useful to have a routine to convert binary data into BCD form. A simple algorithm for converting binary to BCD is as follows:

1. The value 10 is subtracted from the source operand until the remainder is less than 0 (carry cleared). The number of subtractions is the high-order BCD digit.
2. The value 10 is then added back to the subtrahend to compensate for the last subtraction.
3. The final remainder is the low-order BCD digit.

The binary to BCD conversion procedure is coded as follows:

```
;==============================
;   binary to BCD conversion
;==============================
; Convert a binary number into two packed BCD digits
; ON ENTRY:
;         w register has binary value in range 0 to 99
; ON EXIT:
;         output variables bcdLow and bcdHigh contain two
;         unpacked BCD digits
;         w contains two packed BCD digits
; Routine logic:
;   The value 10 is subtracted from the source operand
;   until the remainder is < 0 (carry cleared). The number
;   of subtractions is the high-order BCD digit. 10 is
;   then added back to the subtrahend to compensate
;   for the last subtraction. The final remainder is the
;   low-order BCD digit
; Variables:
;     inNum      storage for source operand
;     bcdHigh    storage for high-order nibble
;     bcdLow     storage for low-order nibble
;     thisDig    Digit counter
bin2bcd:
        movwf    inNum              ; Save copy of source value
        clrf     bcdHigh  ; Clear storage
        clrf     bcdLow
        clrf     thisDig
min10:
        movlw    .10
        subwf    inNum,f            ; Subtract 10
        btfsc    STATUS,C           ; Did subtract overflow?
        goto     sum10              ; No. Count subtraction
        goto     fin10
sum10:
```

Analog to Digital and Realtime Clocks

```
                incf    thisDig,f       ; Increment digit counter
                goto    min10
; Store 10th digit
fin10:
                movlw   .10
                addwf   inNum,f         ; Adjust
                movf    thisDig,w       ; Get digit counter contents
                movwf   bcdHigh         ; Store it
; Calculate and store low-order BCD digit
                movf    inNum,w         ; Store units value
                movwf   bcdLow          ; Store digit
; Combine both digits
                swapf   bcdHigh,w       ; High nibble to HOBs
                iorwf   bcdLow,w ; ORin low nibble
                return
```

Since the program requires displaying values encoded in BCD format, a routine is necessary to convert two packed BCD digits into two ASCII decimal digits. The conversion logic is quite simple since the BCD digit is converted to ASCII by adding 0x30 to its value. All that is necessary is to shift bits in the packed BCD operand so as to isolate each digit and then add 0x30 to each one. The routine's code is as follows:

```
;===============================
;    BCD to ASCII decimal
;          conversion
;===============================
; ON ENTRY:
;         w register has two packed BCD digits
; ON EXIT:
;         output variables asc10, and asc1 have
;         two ASCII decimal digits
; Routine logic:
;   The low order nibble is isolated and the value 30H
;   added to convert to ASCII. The result is stored in
;   the variable asc1. Then the same is done to the
;   high-order nibble and the result is stored in the
;   variable asc10

Bcd2asc:
                movwf   store1          ; Save input
                andlw   b'00001111'     ; Clear high nibble
                addlw   0x30            ; Convert to ASCII
                movwf   asc1            ; Store result
                swapf   store1,w ; Recover input and swap digits
                andlw   b'00001111' ; Clear high nibble
                addlw   0x30            ; Convert to ASCII
                movwf   asc10           ; Store result
                return
```

16.4 Sample Programs

The following sections contain the sample programs discussed in this chapter.

16.4.1 ADF84 Program

```
; File name: ADCF84.asm
; Last Update: June 8, 2006
; Author: Julio Sanchez
; Processor: 16F84A
;
; Description:
; Program to demonstrate use of the ADC0831 Analog to
; Digital converter with the 16F84 PIC. Program reads the
; value of a potentionmeter connected to Port-A, line 0
; and displays resistance in the range 0 to 255 on the
; attached LCD.
; Circuit:
;     ADC0831             16F84              CIRCUIT
; PIN   LINE
;  6     DO  ------------- RB0
;  7     CLK ------------- RB1
;  1     CS  ------------- RB2
;  2     Vin+ ------------------------------- POT2
;  3     Vin- ------------------------------- GND
;  5     Vref ------------------------------- +5v
;  8     Vcc  ------------------------------- +5v
;
; For LCD display parameters see the LCDTest2 program.
; WARNING:
; Code assumes 4Mhz clock. Delay routines must be
; edited for faster clock
;
;==========================
;          switches
;==========================
; Switches used in __config directive:
;   _CP_ON          Code protection ON/OFF
; * _CP_OFF
; * _PWRTE_ON       Power-up timer ON/OFF
;   _PWRTE_OFF
;   _WDT_ON         Watchdog timer ON/OFF
; * _WDT_OFF
;   _LP_OSC         Low power crystal oscillator
; * _XT_OSC         External parallel resonator/crystal oscillator

;   _HS_OSC         High speed crystal resonator (8 to 10 MHz)
;                   Resonator: Murate Erie CSA8.00MG = 8 MHz
;   _RC_OSC         Resistor/capacitor oscillator
```

```
;   |                 (simplest, 20% error)
;   |
;   |_____  * indicates setup values presently selected


;=========================
; setup and configuration
;=========================
        processor 16f84A
        include   <p16f84A.inc>
        __config  _XT_OSC & _WDT_OFF & _PWRTE_ON & _CP_OFF

        errorlevel -302
; Suppress bank-related warning
;===============================================================
;                        M A C R O S
;===============================================================
; Macros to select the register banks in 16F84
Bank0   MACRO                   ; Select RAM bank 0
        bcf     STATUS,RP0
        ENDM

Bank1   MACRO                           ; Select RAM bank 1
        bsf     STATUS,RP0
        ENDM
;=========================================================
;               constant definitions
;   for PIC-to-LCD pin wiring and LCD line addresses
;=========================================================
#define E_line 1                ; |
#define RS_line 2               ; | - from circuit wiring diagram
#define RW_line 3               ; |
; LCD line addresses (from LCD data sheet)
#define LCD_1 0x80              ; First LCD line constant
#define LCD_2 0xc0              ; Second LCD line constant
; Note: The constants that define the LCD display line
;       addresses have the high-order bit set in
;       order to facilitate the controller command

; Defines from ADC0831 wiring diagram
; all lines in Port-A
#define  DO              0    ; |
#define  CLK             1    ; | - from circuit wiring diagram
#define  CS              2    ; |
;
;=========================================================
;               variables in PIC RAM
;=========================================================
; Reserve 16 bytes for string buffer
```

```
            cblock  0x0c
            strData
            endc
; Reserve three bytes for ASCII digits
            cblock  0x1d
            asc100
            asc10
            asc1
            endc
; Continue with local variables
            cblock  0x20            ; Start of block
            count1                  ; Counter # 1
            count2                  ; Counter # 2
            count3                  ; Counter # 3
            pic_ad                  ; Storage for start of text area
            J                       ; counter J
            K                       ; counter K
            index                   ; Index into text table (also used
                                    ; for auxiliary storage)
            store1                  ; Local temporary storage
            store2                  ; Storage # 2
            rcvdata                 ; Received data
            bitCount
; Storage for ASCII decimal conversion and digits
            inNum                   ; Source operand
            thisDig                 ; Digit counter
            endc


;============================================================
;                        program
;============================================================
            org     0               ; start at address
            goto    main
; Space for interrupt handlers
            org     0x08
main:
            Bank1
            movlw   b'00000000'     ; All lines to output
            movwf   TRISA           ; in Port-A
            movlw   b'00000001'     ; B line 0 to input
            movwf   TRISB
            Bank0
            movlw   b'00000000'     ; All outputs ports low
            movwf   PORTA
            movwf   PORTB
; Wait and initialize HD44780
            call    delay_5         ; Allow LCD time to initialize
                                    ; itself
```

Analog to Digital and Realtime Clocks

```
            call       delay_5
            call       initLCD          ; Then do forced
initialization
            call       delay_5          ; Wait again
; Store base address of text buffer in PIC RAM
            movlw      0x0c             ; Start address for buffer
            movwf      pic_ad           ; to local variable
;======================
;   first LCD line
;======================
; Store 16 blanks in PIC RAM, starting at address stored
; in variable pic_ad
            call       blank16
; Call procedure to store ASCII characters for message
; in text buffer
            movlw      d'0'             ; Offset into buffer
            call       storeMS1 ; Store message text in buffer
; Initialize ADC0831
nextAna:
            call       ana2dig          ; Read analog line
            call       delay_125
; Display result
            movf       rcvdata,w
            call       bin2asc          ; Conversion routine
; At this point three ASCII digits are stored in local
; variables. Move digits to display area
            movf       asc1,w           ; Unit digit
            movwf      .26              ; Store in buffer
            movf       asc10,w          ; same with other digits
            movwf      .25
            movf       asc100,w
            movwf      .24
; Display line
; Set DDRAM address to start of first line
            call       line1
; Call procedure to display 16 characters in LCD
            call       display16
            call       long_delay
            goto       nextAna


;=================================================================
;                initialize LCD for 4-bit mode
;=================================================================
initLCD:
; Initialization for Densitron LCD module as follows:
;         4-bit interface
;     2 display lines of 16 characters each
;     cursor on
```

```
;     left-to-right increment
;     cursor shift right
;     no display shift
;======================|
;   set command mode   |
;======================|
         bcf      PORTA,E_line      ; E line low
         bcf      PORTA,RS_line     ; RS line low
         bcf      PORTA,RW_line     ; Write mode
         call     delay_125         ; delay 125 microseconds
;*********************|
;      FUNCTION SET    |
;*********************|
         movlw    0x28     ; 0 0 1 0 1 0 0 0 (FUNCTION SET)
         call     send8    ; 4-bit send routine

; Set 4-bit mode command must be repeated
         movlw    0x28
         call     send8


;*********************|
; DISPLAY AND CURSOR ON |
;*********************|
         movlw    0x0e     ; 0 0 0 0 1 1 1 0 (DISPLAY ON/OFF)
         call     send8
;*********************|
;    set entry mode    |
;*********************|
         movlw    0x06     ; 0 0 0 0 0 1 1 0 (ENTRY MODE SET)
         call     send8


;*********************|
; cursor/display shift |
;*********************|
         movlw    0x14     ; 0 0 0 1 0 1 0 0 (CURSOR/DISPLAY
SHIFT)
         call     send8
;*********************|
;    clear display     |
;*********************|
         movlw    0x01     ; 0 0 0 0 0 0 0 1 (CLEAR DISPLAY)
         call     send8
; Per documentation
         call     delay_5  ; Test for busy
         return


;======================
;   Procedure to delay
```

```
;       42 microseconds
;========================
delay_125
        movlw   D'42'               ; Repeat 42 machine cycles
        movwf   count1              ; Store value in counter
repeat
        decfsz  count1,f            ; Decrement counter
        goto    repeat              ; Continue if not 0
        return                      ; End of delay

;========================
;   Procedure to delay
;    5 milliseconds
;========================
delay_5:
        movlw   D'41'               ; Counter = 41
        movwf   count2              ; Store in variable
delay:
        call    delay_125           ; Delay
        decfsz  count2,f            ; 40 times = 5 milliseconds
        goto    delay
        return                      ; End of delay
;========================
;      pulse E line
;========================
pulseE
        bsf     PORTA,E_line        ; Pulse E line
        nop
        bcf     PORTA,E_line
        return

;=============================
;    long delay sub-routine
;        (for debugging)
;=============================
long_delay
        movlw   D'200'              ; w = 200 decimal
        movwf   J                   ; J = w
jloop:
        movwf   K                   ; K = w
kloop:
        decfsz  K,f                 ; K = K-1, skip next if zero
        goto    kloop
        decfsz  J,f                 ; J = J-1, skip next if zero
        goto    jloop
        return
;=============================
;    LCD display procedure
```

```
;===============================
; Sends 16 characters from PIC buffer with address stored
; in variable pic_ad to LCD line previously selected
display16
        call    delay_5             ; Make sure not busy
; Set up for data
        bcf     PORTA,E_line        ; E line low
        bsf     PORTA,RS_line       ; RS line high for data
; Set up counter for 16 characters
        movlw   D'16'               ; Counter = 16
        movwf   count3
; Get display address from local variable pic_ad
        movf    pic_ad,w            ; First display RAM address to W
        movwf   FSR                 ; W to FSR
getchar:
        movf    INDF,w              ; get character from display RAM
                                    ; location pointed to by file select
                                    ; register
        call    send8               ; 4-bit interface routine
; Test for 16 characters displayed
        decfsz  count3,f            ; Decrement counter
        goto    nextchar            ; Skipped if done
        return
nextchar:
        incf    FSR,f               ; Bump pointer
        goto    getchar

;=======================
;   send 2 nibbles in
;       4-bit mode
;=======================
; Procedure to send two 4-bit values to Port-B lines
; 7, 6, 5, and 4. High-order nibble is sent first
; ON ENTRY:
;       w register holds 8-bit value to send
send8:
        movwf   store1              ; Save original value
        call    merge4              ; Merge with Port-B
; Now w has merged byte
        movwf   PORTB               ; w to Port-B
        call    pulseE              ; Send data to LCD
; High nibble is sent
        movf    store1,w            ; Recover byte into w
        swapf   store1,w            ; Swap nibbles in w
        call    merge4
        movwf   PORTB
        call    pulseE              ; Send data to LCD
        call    delay_125
```

Analog to Digital and Realtime Clocks

```
            return
;==================
;    merge bits
;==================
; Routine to merge the 4 high-order bits of the
; value to send with the contents of Port-B
; so as to preserve the 4 low-bits in Port-B
; Logic:
;       AND value with 1111 0000 mask
;       AND Port-B with 0000 1111 mask
;       Now low nibble in value and high nibble in
;       Port-B are all 0 bits:
;           value  = vvvv 0000
;           Port-B = 0000 bbbb
;       OR value and Port-B resulting in:
;                  vvvv bbbb
; ON ENTRY:
;       w contain value bits
; ON EXIT:
;       w contains merged bits
merge4:
            andlw    b'11110000'      ; ANDing with 0 clears the
                                      ; bit. ANDing with 1 preserves
                                      ; the original value
            movwf    store2           ; Save result in variable
            movf     PORTB,w          ; Port-B to w register
            andlw    b'00001111'      ; Clear high nibble in Port-B
                                      ; and preserve low nibble
            iorwf    store2,w         ; OR two operands in w
            return

;=========================
;    blank buffer
;=========================
; Procedure to store 16 blank characters in PIC RAM
; buffer starting at address stored in the variable
; pic_ad
blank16:
            movlw    D'16'            ; Setup counter
            movwf    count1
            movf     pic_ad,w ; First PIC RAM address
            movwf    FSR              ; Indexed addressing
            movlw    0x20             ; ASCII space character
storeit:
            movwf    INDF             ; Store blank character in PIC
RAM
                                      ; buffer using FSR register
            decfsz   count1,f         ; Done?
```

```
                goto     incfsr                  ; no
                return                           ; yes
incfsr:
                incf     FSR,f                   ; Bump FSR to next buffer
space
                goto     storeit

;========================
; Set address register
;     to LCD line 1
;========================
; ON ENTRY:
;           Address of LCD line 1 in constant LCD_1
line1:
                bcf      PORTA,E_line            ; E line low
                bcf      PORTA,RS_line           ; RS line low, set up for
                                                 ; control
                call     delay_5                 ; busy?
; Set to second display line
                movlw    LCD_1                   ; Address and command bit
                call     send8                   ; 4-bit routine
; Set RS line for data
                bsf      PORTA,RS_line           ; Setup for data
                call     delay_5                 ; Busy?
                return

;===============================
;   first text string procedure
;===============================
storeMS1:
; Procedure to store in PIC RAM buffer the message
; contained in the code area labeled msg1
; ON ENTRY:
;           variable pic_ad holds address of text buffer
;           in PIC RAM
;           w register hold offset into storage area
;           msg1 is routine that returns the string characters
;           and a zero terminator
;           index is local variable that hold offset into
;           text table. This variable is also used for
;           temporary storage of offset into buffer
; ON EXIT:
;           Text message stored in buffer
;
; Store offset into text buffer (passed in the w register)
; in temporary variable
                movwf    index                   ; Store w in index
; Store base address of text buffer in FSR
```

```
                movf    pic_ad,w ; first display RAM address to W
                addwf   index,w         ; Add offset to address
                movwf   FSR             ; W to FSR
; Initialize index for text string access
                movlw   0               ; Start at 0
                movwf   index           ; Store index in variable
; w still = 0
get_msg_char:
                call    msg1            ; Get character from table
; Test for zero terminator
                andlw   0x0ff
                btfsc   STATUS,Z ; Test zero flag
                goto    endstr1         ; End of string
; ASSERT: valid string character in w
;       store character in text buffer (by FSR)
                movwf   INDF            ; store in buffer by FSR
                incf    FSR,f           ; increment buffer pointer
; Restore table character counter from variable
                movf    index,w         ; Get value into w
                addlw   1               ; Bump to next character
                movwf   index           ; Store table index in
variable
                goto    get_msg_char    ; Continue
endstr1:
                return

; Routine for returning message stored in program area
; Message has 10 characters
msg1:
                addwf   PCL,f           ; Access table
                retlw   'P'
                retlw   'o'
                retlw   't'
                retlw   ' '
                retlw   'R'
                retlw   'e'
                retlw   's'
                retlw   'i'
                retlw   's'
                retlw   't'
                retlw   ':'
                retlw   0
;==============================
;    binary to ASCII decimal
;           conversion
;==============================
; ON ENTRY:
;           w register has binary value in range 0 to 255
```

```
;  ON EXIT:
;           output variables asc100, asc10, and asc1 have
;           three ASCII decimal digits
;  Routine logic:
;    The value 100 is subtracted from the source operand
;    until the remainder is < 0 (carry cleared). The number
;    of subtractions is the decimal hundreds result. 100 is
;    then added back to the subtrahend to compensate
;    for the last subtraction. Now 10 is subtracted in the
;    same manner to determine the decimal tenths result.
;    The final remainder is the decimal units result.
;  Variables:
;      inNum     storage for source operand
;      asc100    storage for hundreds position result
;      asc10     storage for tenth position result
;      asc1      storage for unit position result
;      thisDig   Digit counter
bin2asc:
          movwf     inNum     ; Save copy of source value
          clrf      asc100    ; Clear hundreds storage
          clrf      asc10     ; Tens
          clrf      asc1      ; Units
          clrf      thisDig
sub100:
          movlw     .100
          subwf     inNum,f          ; Subtract 100
          btfsc     STATUS,C         ; Did subtract overflow?
          goto      bump100          ; No. Count subtraction
          goto      end100
bump100:
          incf      thisDig,f        ;increment digit counter
          goto      sub100
;  Store 100th digit
end100:
          movf      thisDig,w        ; Adjusted digit counter
          addlw     0x30             ; Convert to ASCII
          movwf     asc100           ; Store it
;  Calculate tenth position value
          clrf      thisDig
;  Adjust minuend
          movlw     .100             ; Minuend
          addwf     inNum,f          ; Add value to minuend to
                                     ; compensate for last
operation
sub10:
          movlw     .10
          subwf     inNum,f          ; Subtract 10
          btfsc     STATUS,C         ; Did subtract overflow?
```

Analog to Digital and Realtime Clocks

```
            goto     bump10              ; No. Count subtraction
            goto     end10
bump10:
            incf     thisDig,f           ;increment digit counter
            goto     sub10
; Store 10th digit
end10:
            movlw    .10
            addwf    inNum,f             ; Adjust for last subtract
            movf     thisDig,w           ; get digit counter contents
            addlw    0x30                ; Convert to ASCII
            movwf    asc10               ; Store it
; Calculate and store units digit
            movf     inNum,w             ; Store units value
            addlw    0x30                ; Convert to ASCII
            movwf    asc1                ; Store digit
            return


;================================================================
;                        ADC0831 procedures
;================================================================
;============================
;    procedure to read and
;      convert analog line
;============================
; ON ENTRY:
; Code assumes that the ADC0831 DO line is initialized for
; input, while CLK and CS lines are output
; From ADC0831 wiring diagram. All lines in Port-B
;         DO     =     RB0    ==> INPUT
;         CLK    =     RB1    <== OUTPUT
;         CS     =     RB2    <== OUTPUT
; ON EXIT:
; Returns 8-bit digital value in the register rcvdata
;
ana2dig:
; Clear data register and init counter for 8 bits
            clrf     rcvdata             ; Clear register
            movlw    0x08                ; Initialize counter
            movwf    bitCount
; Prepare to read analog line
            bcf      PORTB,CS ; CS pin low to enable ADC
            nop                          ; Delay for 4Mhz clock
            bsf      PORTB,CLK           ; Set CLK high
            nop
            bcf      PORTB,CLK           ; Reset CLK to start
conversion
            nop
```

nextB:
; Pulse CLK line to read bit from ADC
        bsf   PORTB,CLK   ; CLK high
        nop
        bcf   PORTB,CLK   ; CLK low
        nop
; Read analog line and store data, bit by bit
        movf   PORTB,w ; Read all Port-B bits
        movwf   store1   ; Store value for later
        rrf   store1,f ; Rotate bit into carry flag
        rlf   rcvdata,f   ; Rotate carry flag into result
                                ; register
        decfsz bitCount,f   ; Bump counter, skip next
                                ; if counter zero
        goto   nextB
; Value read is stored in rcvdata register
        bsf   PORTB,CLK   ; Final clock pulse
        nop
        bcf   PORTB,CLK
        nop
        bsf   PORTB,CS ; Turn off ADC
        call   long_delay   ; Time to settle
        return

        end
```

## 16.4.2 A2DinLCD Program

```
; File name: A2DinLCD.asm
; Last revision: June 2, 2006
; Author: Julio Sanchez
; Processor: 16F877
;
; Description:
; Program to demonstrate use of the Analog to Digital
; Converter (A/D) module on the 16F877. Program reads the
; value of a potentionmeter connected to Port-A, line 0
; and displays resistance in the range 0 to 255 on the
; attached LCD.
;
; WARNING:
; Code assumes 10Mhz clock. Delay routines must be
; edited for faster clock. Clock speed is also used to
; set up the A/D converter clock.
;
;===========================
;   16F877 switches
```
```

```
;============================
; Switches used in __config directive:
;   _CP_ON          Code protection ON/OFF
; * _CP_OFF
; * _PWRTE_ON       Power-up timer ON/OFF
;   _PWRTE_OFF
;   _BODEN_ON       Brown-out reset enable ON/OFF
; * _BODEN_OFF
; * _PWRTE_ON       Power-up timer enable ON/OFF
;   _PWRTE_OFF
;   _WDT_ON         Watchdog timer ON/OFF
; * _WDT_OFF
;   _LPV_ON         Low voltage IC programming enable ON/OFF
; * _LPV_OFF
;   _CPD_ON         Data EE memory code protection ON/OFF
; * _CPD_OFF
; OSCILLATOR CONFIGURATIONS:
;   _LP_OSC         Low power crystal oscillator
;   _XT_OSC         External parallel resonator/crystal oscillator
;
; * _HS_OSC         High speed crystal resonator
;   _RC_OSC         Resistor/capacitor oscillator
;   |               (simplest, 20% error)
;   |
;   |_____ * indicates setup values presently selected

        processor       16f877           ; Define processor
        #include <p16f877.inc>
        __CONFIG _CP_OFF & _WDT_OFF & _BODEN_OFF & _PWRTE_ON &
_HS_OSC & _WDT_OFF & _LVP_OFF & _CPD_OFF
; __CONFIG directive is used to embed configuration data
; within the source file. The labels following the directive
; are located in the corresponding .inc file.

        errorlevel -302
; Suppress bank-related warning
;================================================================
;                       M A C R O S
;================================================================
; Macros to select the register banks
Bank0   MACRO                   ; Select RAM bank 0
        bcf     STATUS,RP0
        bcf     STATUS,RP1
        ENDM

Bank1   MACRO                   ; Select RAM bank 1
        bsf     STATUS,RP0
        bcf     STATUS,RP1
```

```
                ENDM

Bank2   MACRO                   ; Select RAM bank 2
        bcf         STATUS,RP0
        bsf         STATUS,RP1
        ENDM

Bank3   MACRO                   ; Select RAM bank 3
        bsf         STATUS,RP0
        bsf         STATUS,RP1
        ENDM
;=========================================================
;                  constant definitions
;   for PIC-to-LCD pin wiring and LCD line addresses
;=========================================================
#define E_line 1            ;|
#define RS_line 0           ;| — from wiring diagram
#define RW_line 2           ;|
; LCD line addresses (from LCD data sheet)
#define LCD_1 0x80          ; First LCD line constant
#define LCD_2 0xc0          ; Second LCD line constant
#define LCDlimit .20; Number of characters per line
#define  spbrgVal .64; For 2400 baud on 10Mhz clock
; Note: The constants that define the LCD display
;       line addresses have the high-order bit set
;       so as to meet the requirements of controller
;       commands.
;=========================================================
;              variables in PIC RAM
;=========================================================
; Reserve 20 bytes for string buffer
        cblock      0x20
        strData
        endc

; Reserve three bytes for ASCII digits
        cblock      0x34
        asc100
        asc10
        asc1
        endc

; Data
        cblock      0x37                ; Start of block
        count1                  ; Counter # 1
        count2                  ; Counter # 2
        count3                  ; Counter # 3
        pic_ad
```

```
            J                   ; counter J
            K                   ; counter K
            index
            store1              ; Local storage
            store2
; For LCDscroll procedure
            LCDcount ; Counter for characters per line
            LCDline             ; Current display line (0 or 1)
            endc

; Common RAM area for most critical variables
            cblock  0x70
; Storage for ASCII decimal conversion and digits
            inNum               ; Source operand
            thisDig             ; Digit counter
            endc


;================================================================
;                       P R O G R A M
;================================================================
            org     0           ; start at address
            goto    main
; Space for interrupt handlers
            org     0x08
main:
; Wiring:
;       LCD data to Port-D, lines 0 to 7
;       E line -> Port-E, 1
;       RW line -> Port-E, 2
;       RS line -> Port-E, 0
; Set PORTE D and E for output
; First, initialize Port-B by clearing latches
            clrf    STATUS
            clrf    PORTB
; Select bank 1 to tris Port-D for output
            Bank1
; Tris Port-D for output. Port-D lines 4 to 7 are wired
; to LCD data lines. Port-D lines 0 to 4 are wired to LEDs.
            movlw   B'00000000'
            movwf   TRISD               ; and Port-D
; By default Port-A lines are analog. To configure them
; as digital code must set bits 1 and 2 of the ADCON1
; register (in bank 1)
            movlw   0x06                ; binary 0000 0110  is code to
                                        ; make all Port-A lines
digital
            movwf   ADCON1
; Port-B, lines are not used by this application. Init
```

```
; to output
        movlw   b'00000000'
        movwf   TRISB
; Tris Port-E for output. LCD lines are in Port-E
        movwf   TRISE           ; Tris Port-E
; Enable Port-B pullups for switches in OPTION register
;     7  6  5  4  3  2  1  0 <= OPTION bits
;     |  |  |  |  |  |__|__|_____ PS2-PS0 (prescaler bits)
;     |  |  |  |  |               Values for Timer0
;     |  |  |  |  |               000 = 1:2    001 = 1:4
;     |  |  |  |  |               010 = 1:8    011 = 1:16
;     |  |  |  |  |               100 = 1:32   101 = 1:64
;     |  |  |  |  |               110 = 1:128 *111 = 1:256
;     |  |  |  |  |_____ PSA (prescaler assign)
;     |  |  |  |                 *1 = to WDT
;     |  |  |  |                  0 = to Timer0
;     |  |  |  |_____ TOSE (Timer0 edge select)
;     |  |  |                    *0 = increment on low-to-high
;     |  |  |                     1 = increment in high-to-low
;     |  |  |_____ TOCS (TMR0 clock source)
;     |  |                       *0 = internal clock
;     |  |                        1 = RA4/TOCKI bit source
;     |  |_____ INTEDG (Edge select)
;     |                          *0 = falling edge
;     |_____ RBPU (Pullup enable)
;                                *0 = enabled
;                                 1 = disabled
        movlw   b'00001000'
        movwf   OPTION_REG
; Back to bank 0
        Bank0
; Clear all output lines
        movlw   b'00000000'
        movwf   PORTD
        movwf   PORTE
; Wait and initialize HD44780
        call    delay_5         ; Allow LCD time to initialize itself
        call    initLCD         ; Then do forced initialization
        call    delay_5         ; (Wait probably not necessary)
; Clear character counter and line counter variables
        clrf    LCDcount
        clrf    LCDline
; Initialize A/D conversion lines
        call    InitA2D         ; Local procedure
; Store base address of text buffer in PIC RAM
```

Analog to Digital and Realtime Clocks

```
            movlw   0x20              ; Start address for buffer
            movwf   pic_ad            ; to local variable
; Store 20 blanks in PIC RAM, starting at address stored
; in variable pic_ad
            call    blank20
; Call procedure to store ASCII characters for message
; in text buffer
            movlw   d'0'              ; Offset into buffer
            call    storeMS1
;=============================
;       read POT digital value
;=============================
readPOT:
            call    ReadA2D           ; Local procedure
; w has digital value read from analog line RA0
; Display result
            call    bin2asc           ; Conversion routine
; At this point three ASCII digits are stored in local
; variables. Move digits to display area
            movf    asc1,w            ; Unit digit
            movwf   0x2e              ; Store in buffer
            movf    asc10,w           ; same with other digits
            movwf   0x2d
            movf    asc100,w
            movwf   0x2c
; Display line
; Set DDRAM address to start of first line
showLine:
            call    line1
; Call procedure to display 16 characters in LCD
            call    display20
            goto    readPOT


;================================================================
;================================================================
;              L O C A L    P R O C E D U R E S
;================================================================
;================================================================
;=========================
; init LCD for 4-bit mode
;=========================
initLCD:
; Initialization for Densitron LCD module as follows:
;    4-bit interface
;    2 display lines of 20 characters each
;    cursor on
;    left-to-right increment
;    cursor shift right
```

```
;    no display shift
;======================|
;   set command mode   |
;======================|
        bcf     PORTE,E_line        ; E line low
        bcf     PORTE,RS_line       ; RS line low
        bcf     PORTE,RW_line       ; Write mode
        call    delay_125           ; delay 125 microseconds
;**********************|
;     FUNCTION SET     |
;**********************|
        movlw   0x28        ; 0 0 1 0 1 0 0 0 (FUNCTION SET)
        call    send8       ; 4-bit send routine

; Set 4-bit mode command must be repeated
        movlw   0x28
        call    send8
;**********************|
; DISPLAY AND CURSOR ON |
;**********************|
        movlw   0x0e        ; 0 0 0 0 1 1 1 0 (DISPLAY ON/OFF)
        call    send8
;**********************|
;   set entry mode     |
;**********************|
        movlw   0x06        ; 0 0 0 0 0 1 1 0 (ENTRY MODE SET)
        call    send8
;**********************|
; cursor/display shift |
;**********************|
        movlw   0x14        ; 0 0 0 1 0 1 0 0 (CURSOR/DISPLAY
SHIFT)
        call    send8
;**********************|
;    clear display     |
;**********************|
        movlw   0x01        ; 0 0 0 0 0 0 0 1 (CLEAR DISPLAY)
        call    send8
; Per documentation
        call    delay_5 ; Test for busy
        return

;======================
;   Procedure to delay
;   125ms. at 10Mhz
;======================
delay_125:
        movlw   .110                ; Repeat 110 machine cycles
```

```
                movwf      count1            ; Store value in counter
repeat:
                decfsz     count1,f          ; Decrement counter
                goto       repeat            ; Continue if not 0
                return                       ; End of delay

;=======================
;   Procedure to delay
;   5 milliseconds
;=======================
delay_5:
                movlw      .110              ; Counter = 110
                movwf      count2            ; Store in variable
delay:
                call       delay_125         ; Delay
                decfsz     count2,f          ; 40 times = 5 milliseconds
                goto       delay
                return                       ; End of delay
;=======================
;     pulse E line
;=======================
pulseE:
                bsf        PORTE,E_line      ; Pulse E line
                nop
                bcf        PORTE,E_line
                return

;=============================
;   long delay sub-routine
;=============================
long_delay:
                movlw      .200              ; w delay count
                movwf      J                 ; J = w
jloop:
                movwf      K                 ; K = w
kloop:
                decfsz     K,f               ; K = K-1, skip next if zero
                goto       kloop
                decfsz     J,f               ; J = J-1, skip next if zero
                goto       jloop
                return


;=============================
;    display buffer on LCD
;=============================
; Sends 20 characters from PIC buffer with address stored
; in variable pic_ad to LCD line previously selected
display20:
```

```
            call      delay_5               ; Make sure not busy
; Set up for data
            bcf       PORTA,E_line          ; E line low
            bsf       PORTA,RS_line         ; RS line high for data
; Set up counter for 20 characters
            movlw     D'20'
            movwf     count3
; Get display address from local variable pic_ad
            movf      pic_ad,w ; First display RAM address to W
            movwf     FSR                   ; W to FSR
getchar
            movf      INDF,w   ; get character from display RAM
                                ; location pointed to by file select
                                ; register
            call      send8                 ; 4-bit interface routine
; Test for 16 characters displayed
            decfsz    count3,f              ; Decrement counter
            goto      nextchar              ; Skipped if done
            return
nextchar:
            incf      FSR,f                 ; Bump pointer
            goto      getchar

;=========================
;    send 2 nibbles in
;       4-bit mode
;=========================
; Procedure to send two 4-bit values to Port-B lines
; 7, 6, 5, and 4. High-order nibble is sent first
; ON ENTRY:
;            w register holds 8-bit value to send
send8:
            movwf     store1                ; Save original value
            call      merge4                ; Merge with Port-B
; Now w has merged byte
            movwf     PORTD                 ; w to Port-D
            call      pulseE                ; Send data to LCD
; High nibble is sent
            movf      store1,w ; Recover byte into w
            swapf     store1,w ; Swap nibbles in w
            call      merge4
            movwf     PORTD
            call      pulseE                ; Send data to LCD
            call      delay_125
            return
;==========================
;      merge bits
;==========================
```

Analog to Digital and Realtime Clocks

```
; Routine to merge the 4 high-order bits of the
; value to send with the contents of Port-B
; so as to preserve the 4 low-bits in Port-B
; Logic:
;       AND value with 1111 0000 mask
;       AND Port-B with 0000 1111 mask
;       Now low nibble in value and high nibble in
;       Port-B are all 0 bits:
;            value  = vvvv 0000
;            Port-B = 0000 bbbb
;       OR value and Port-B resulting in:
;                   vvvv bbbb
; ON ENTRY:
;       w contain value bits
; ON EXIT:
;       w contains merged bits
merge4:
        andlw     b'11110000'       ; ANDing with 0 clears the
                                    ; bit. ANDing with 1 preserves
                                    ; the original value
        movwf     store2            ; Save result in variable
        movf      PORTD,w           ; Port-B to w register
        andlw     b'00001111'       ; Clear high nibble in Port-B
                                    ; and preserve low nibble
        iorwf     store2,w ; OR two operands in w
        return
;===========================
;   Set address register
;        to LCD line 1
;===========================
; ON ENTRY:
;       Address of LCD line 1 in constant LCD_1
line1:
        bcf       PORTE,E_line      ; E line low
        bcf       PORTE,RS_line     ; RS line low, set up for
                                    ; control
        call      delay_5           ; busy?
; Set to second display line
        movlw     LCD_1             ; Address and command bit
        call      send8             ; 4-bit routine
; Set RS line for data
        bsf       PORTE,RS_line     ; Setup for data
        call      delay_5           ; Busy?
        return
;=================================
;   first text string procedure
;=================================
storeMS1:
```

```
; Procedure to store in PIC RAM buffer the message
; contained in the code area labeled msg1
; ON ENTRY:
;              variable pic_ad holds address of text buffer
;              in PIC RAM
;              w register hold offset into storage area
;              msg1 is routine that returns the string characters
;              and a zero terminator
;              index is local variable that hold offset into
;              text table. This variable is also used for
;              temporary storage of offset into buffer
; ON EXIT:
;              Text message stored in buffer
;
; Store offset into text buffer (passed in the w register)
; in temporary variable
         movwf   index              ; Store w in index
; Store base address of text buffer in FSR
         movf    pic_ad,w ; first display RAM address to W
         addwf   index,w            ; Add offset to address
         movwf   FSR                ; W to FSR
; Initialize index for text string access
         movlw   0                  ; Start at 0
         movwf   index              ; Store index in variable
; w still = 0
get_msg_char:
         call    msg1               ; Get character from table
; Test for zero terminator
         andlw   0x0ff
         btfsc   STATUS,Z ; Test zero flag
         goto    endstr1            ; End of string
; ASSERT: valid string character in w
;         store character in text buffer (by FSR)
         movwf   INDF               ; store in buffer by FSR
         incf    FSR,f              ; increment buffer pointer
; Restore table character counter from variable
         movf    index,w            ; Get value into w
         addlw   1                  ; Bump to next character
         movwf   index    ; Store table index in variable
         goto    get_msg_char       ; Continue
endstr1:
         return
; Routine for returning message stored in program area
; Message has 10 characters
msg1:
         addwf   PCL,f              ; Access table
         retlw   'P'
         retlw   'o'
```

Analog to Digital and Realtime Clocks

```
            retlw    't'
            retlw    ' '
            retlw    'R'
            retlw    'e'
            retlw    's'
            retlw    'i'
            retlw    's'
            retlw    't'
            retlw    ':'
            retlw    0
;========================
;     blank buffer
;========================
; Procedure to store 20 blank characters in PIC RAM
; buffer starting at address stored in the variable
; pic_ad
blank20:
            movlw    D'20'              ; Setup counter
            movwf    count1
            movf     pic_ad,w ; First PIC RAM address
            movwf    FSR                ; Indexed addressing
            movlw    0x20               ; ASCII space character
storeit:
            movwf    INDF     ; Store blank character in PIC RAM
                              ; buffer using FSR register
            decfsz   count1,f           ; Done?
            goto     incfsr   ; no
            return            ; yes
incfsr:
            incf     FSR,f    ; Bump FSR to next buffer space
            goto     storeit

;===============================
;    binary to ASCII decimal
;           conversion
;===============================
; ON ENTRY:
;         w register has binary value in range 0 to 255
; ON EXIT:
;         output variables asc100, asc10, and asc1 have
;         three ASCII decimal digits
; Routine logic:
;   The value 100 is subtracted from the source operand
;   until the remainder is < 0 (carry cleared). The number
;   of subtractions is the decimal hundreds result. 100 is
;   then added back to the subtrahend to compensate
;   for the last subtraction. Now 10 is subtracted in the
;   same manner to determine the decimal tenths result.
```

```
;    The final remainder is the decimal units result.
; Variables:
;      inNum       storage for source operand
;      asc100      storage for hundreds position result
;      asc10       storage for tenth position result
;      asc1        storage for unit position result
;      thisDig     Digit counter
bin2asc:
        movwf   inNum           ; Save copy of source value
        clrf    asc100          ; Clear hundreds storage
        clrf    asc10           ; Tens
        clrf    asc1            ; Units
        clrf    thisDig
sub100:
        movlw   .100
        subwf   inNum,f         ; Subtract 100
        btfsc   STATUS,C        ; Did subtract overflow?
        goto    bump100         ; No. Count subtraction
        goto    end100
bump100:
        incf    thisDig,f       ; Increment digit counter
        goto    sub100
; Store 100th digit
end100:
        movf    thisDig,w       ; Adjusted digit counter
        addlw   0x30            ; Convert to ASCII
        movwf   asc100          ; Store it
; Calculate tenth position value
        clrf    thisDig
; Adjust minuend
        movlw   .100    ; Minuend
        addwf   inNum,f ; Add value to minuend to
                        ; Compensate for last operation
sub10:
        movlw   .10
        subwf   inNum,f         ; Subtract 10
        btfsc   STATUS,C        ; Did subtract overflow?
        goto    bump10          ; No. Count subtraction
        goto    end10
bump10:
        incf    thisDig,f       ;increment digit counter
        goto    sub10
; Store 10th digit
end10:
        movlw   .10
        addwf   inNum,f         ; Adjust for last subtract
        movf    thisDig,w       ; get digit counter contents
        addlw   0x30            ; Convert to ASCII
```

Analog to Digital and Realtime Clocks

```
                movwf     asc10               ; Store it
; Calculate and store units digit
                movf      inNum,w             ; Store units value
                addlw     0x30                ; Convert to ASCII
                movwf     asc1                ; Store digit
                return

;================================================================
;                 Analog to Digital Procedures
;================================================================
;============================
;       init A/D module
;============================
; 1. Procedure to initialize the A/D module, as follows:
;    Configure the PIC I/O lines. Init analog lines as input
; 2. Select ports to be used by setting the PCFGx bits in the
;    ADCON1 register. Selects right- or left-justification.
; 3. Select the analog channels, select the A/D conversion
;    clock, and enable the A/D module.
; 4. Wait the acquisition time.
; 5. Initiate the conversion by setting the GO/DONE bit in the
;    ADCON0 register.
; 6. Wait for the conversion to complete.
; 7. Read and store the digital result.
InitA2D:
                Bank1                         ; Select bank for TRISA register
                movlw     b'00000001'
                movwf     TRISA               ; Set Port-A, line 0, as input
; Select the format and A/D port configuration bits in
; the ADCON1 register
; Format is left-justified so that ADRESH bits are the
; most significant
;     0  x  x  x  1  1  1  0   <== value installed in ADCON1
;     7  6  5  4  3  2  1  0   <== ADCON1 bits
;     |           |__|__|__|____ RA0 is analog.
;     |                          Vref+ = Vdd
;     |                          Vref- = Vss
;     |_____ 0 = left-justified
; ADCON1 is in bank 1
                movlw     b'00001110'
                movwf     ADCON1              ; RA0 is analog. All others ;
                                              ; digital
                                              ; Vref+ = Vdd
; Select D/A options in ADCON0 register
; For a 10Mhz clock the Fosc32 option produces a conversion
; speed of 1/(10/32) = 3.2 microseconds, which is within the
; recommended range of 1.6 to 10 microseconds.
;     1  0  0  0  0  0  0  1   <== value installed in ADCON0
```

```
;       7  6  5  4  3  2  1  0   <== ADCON0 bits
;       |  |  |  |  |  |  |  |____ A/D function select
;       |  |  |  |  |  |  |       1 = A/D ON
;       |  |  |  |  |  |  |_____ A/D status bit
;       |  |  |__|__|_____ Analog Channel Select
;       |  |                        000 = Chanel 0 (RA0)
;       |__|_____ A/D Clock Select
;                                   10 = Fosc/32
; ADCON0 is in bank 0
        Bank0
        movlw   b'10000001'
        movwf   ADCON0              ; Channel 0, Fosc/32, A/D
                                    ; enabled
; Delay for selection to complete. (Existing routine provides
; more than 20 microseconds required)
        call    delayAD             ; Local procedure
        return
;============================
;       read A/D line
;============================
; Procedure to read the value in the A/D line and convert
; to digital
ReadA2D:
; Initiate conversion
        Bank0                       ; Bank for ADCON0 register
        bsf     ADCON0,GO           ; Set the GO/DONE bit
; GO/DONE bit is cleared automatically when conversion ends
convWait:
        btfsc   ADCON0,GO           ; Test bit
        goto    convWait ; Wait if not clear
; At this point conversion has concluded
; ADRESH register (bank 0) holds 8 MSBs of result
; ADRESL register (bank 1) holds 4 LSBs.
; In this application value is left-justified. Only the
; MSBs are read
        movf    ADRESH,W ; Digital value to w register
        return
;======================
;   delay procedure
;======================
; For a 10Mhz clock the Fosc32 option produces a conversion
; speed of 1/(10/32) = 3.2 microseconds. At 3.2 ms per bit
; 13 bits require approximately 41 ms. The instruction time
; at 10Mhz is 10 ms. 4/10 = 0.4 ms per instruction. To delay
; 41 ms a 10Mhz PIC must execute 11 instructions. Add one
; more for safety.
delayAD:
        movlw   .12                 ; Repeat 12 machine cycles
```

```
            movwf    count1            ; Store value in counter
repeat11:
            decfsz   count1,f          ; Decrement counter
            goto     repeat11 ; Continue if not 0
            return

;================================================================
            end             ; END OF PROGRAM
;================================================================
```

16.4.3 RTC2LCD Program

```
; File name: RTC2LCD.asm
; Last Update: June 6, 2006
; Author: Julio Sanchez
; Processor: 16F84A
;
; Description:
; Program to demonstrate use of the NJU6355 Real Time Clock
; IC. Program uses LCD to display results of hours, minutes,
; and seconds, as follows:
;
; Top LCD line:    H:xx M:yy S:zz
;
; Initialization values are in #define statements that start
; with i, such as iYear, iMonth, etc.
;
; For LCD display parameters see the LCDTest2 program.
; WARNING:
; Code assumes 4Mhz clock. Delay routines must be
; edited for faster clock
;
;===========================
;          switches
;===========================
; Switches used in __config directive:
;    _CP_ON        Code protection ON/OFF
; *  _CP_OFF
; *  _PWRTE_ON     Power-up timer ON/OFF
;    _PWRTE_OFF
;    _WDT_ON       Watchdog timer ON/OFF
; *  _WDT_OFF
;    _LP_OSC       Low power crystal oscillator
; *  _XT_OSC       External parallel resonator/crystal oscillator
;
;    _HS_OSC       High speed crystal resonator (8 to 10 MHz)
;                  Resonator: Murate Erie CSA8.00MG = 8 MHz
;    _RC_OSC       Resistor/capacitor oscillator
```

```
;   |                 (simplest, 20% error)
;   |
;   |_____ * indicates setup values presently selected

;==========================
; setup and configuration
;==========================
        processor 16f84A
        include   <p16f84A.inc>
        __config  _XT_OSC & _WDT_OFF & _PWRTE_ON & _CP_OFF

        errorlevel -302
; Suppress bank-related warning
;===============================================================
;                       M A C R O S
;===============================================================
; Macros to select the register banks in 16F84
Bank0   MACRO                   ; Select RAM bank 0
        bcf       STATUS,RP0
        ENDM

Bank1   MACRO                   ; Select RAM bank 1
        bsf       STATUS,RP0
        ENDM
;======================================================
;               constant definitions
;   for PIC-to-LCD pin wiring and LCD line addresses
;======================================================
#define E_line 1                ; |
#define RS_line 2               ; | - from circuit wiring diagram
#define RW_line 3               ; |
; LCD line addresses (from LCD data sheet)
#define LCD_1 0x80              ; First LCD line constant
#define LCD_2 0xc0              ; Second LCD line constant
; Note: The constants that define the LCD display line
;       addresses have the high-order bit set in
;       order to facilitate the controller command

; Defines from realtime clock wiring diagram
; all lines in Port-B
#define DAT             0       ; |
#define CLK             1       ; | - from circuit wiring diagram
#define CE              2       ; |
#define IO              3       ; |
;
; Defines for RTC initialization (values are arbitrary)
#define iYear     .7
#define iMonth    .6
```

```
        #define iDay    .5
        #define iDoW    .4
        #define iHour   .3
        #define iMin    .2
        #define iSec    .1
;========================================================
;                PIC register equates
;========================================================
;========================================================
;              variables in PIC RAM
;========================================================
; Reserve 16 bytes for string buffer
        cblock  0x0c
        strData
        endc
; Reserve three bytes for ASCII digits
        cblock  0x1d
        asc100
        asc10
        asc1
        endc
; Continue with local variables
        cblock  0x20            ; Start of block
        count1                  ; Counter # 1
        count2                  ; Counter # 2
        count3                  ; Counter # 3
        pic_ad                  ; Storage for start of text area
        J                       ; counter J
        K                       ; counter K
        index                   ; Index into text table (also used
                                ;       for auxiliary storage)
        store1                  ; Local temporary storage
        store2                  ; Storage # 2
; Storage for BCD digits
        bcdLow                  ; low-order nibble of packed BCD
        bcdHigh                 ; High-order nibble
; Variables for Real-Time Clock
        year
        month
        day
        dayOfWeek               ; Sunday to Saturday (1 to 7)
        hour
        minutes
        seconds
        temp1
        counter
; Storage for BCD conversion routine
        inNum                   ; Source operand
```

```
            thisDig              ; Digit counter
            endc

;============================================================
;                        program
;============================================================
            org         0                ; start at address
            goto        main
; Space for interrupt handlers
            org         0x08

main:
            movlw       b'00000000'      ; All lines to output
            Bank1
            movwf       TRISA            ; in Port-A
            movwf       TRISB            ; and Port-B
            Bank0
            movlw       b'00000000'      ; All outputs ports low
            movwf       PORTA
            movwf       PORTB
; Wait and initialize HD44780
            call        delay_5          ; Allow LCD time to initialize
                                         ; itself
            call        delay_5
            call        initLCD          ; Then do forced
initialization
            call        delay_5          ; Wait again
; Store base address of text buffer in PIC RAM
            movlw       0x0c             ; Start address for buffer
            movwf       pic_ad           ; to local variable
;=====================
;    first LCD line
;=====================
; Store 16 blanks in PIC RAM, starting at address stored
; in variable pic_ad
            call        blank16
; Call procedure to store ASCII characters for message
; in text buffer
            movlw       d'0'             ; Offset into buffer
            call        storeMS1 ; Store message text in buffer
; Initialize real time clock
            call        initRTC          ; Initialize variables
            call        setRTC           ; Start clock
            call        delay_5          ; Wait for operation to
                                         ; conclude
newTime:
; Get variables from RTC
            call        Get_Time
```

Analog to Digital and Realtime Clocks

```
            call    delay_5             ; Wait
            movf    hour,w              ; Get hours
            call    Bcd2asc             ; Conversion routine
; At this point three ASCII digits are stored in local
; variables. Move digits to display area
            movf    asc1,w              ; Unit digit
            movwf   .15                 ; Store in buffer
            movf    asc10,w             ; Same with other digit
            movwf   .14
            call    delay_5
            movf    minutes,w
            call    Bcd2asc             ; Conversion routine
; At this point three ASCII digits are stored in local
; variables. Move two digits to display area
            movf    asc1,w              ; Unit digit
            movwf   .20                 ; Store in buffer
            movf    asc10,w             ; same with other digit
            movwf   .19
            call    delay_5

            movf    seconds,w
            call    Bcd2asc             ; Conversion routine
; Move digits to display area
            movf    asc1,w              ; Unit digit
            movwf   .25                 ; Store in buffer
            movf    asc10,w             ; same with other digit
            movwf   .24
            call    delay_5
; Set DDRAM address to start of first line
            call    line1
; Call procedure to display 16 characters in LCD
            call    display16
            goto    newTime


;===============================================================
;                  initialize LCD for 4-bit mode
;===============================================================
initLCD:
; Initialization for Densitron LCD module as follows:
;        4-bit interface
;    2 display lines of 16 characters each
;    cursor on
;    left-to-right increment
;    cursor shift right
;    no display shift
;======================|
;   set command mode   |
;======================|
```

```
        bcf             PORTA,E_line        ; E line low
        bcf             PORTA,RS_line       ; RS line low
        bcf             PORTA,RW_line       ; Write mode
        call            delay_125           ; delay 125 microseconds
;*********************|
;      FUNCTION SET   |
;*********************|
        movlw   0x28    ; 0 0 1 0 1 0 0 0 (FUNCTION SET)
        call    send8   ; 4-bit send routine

; Set 4-bit mode command must be repeated
        movlw   0x28
        call    send8

;*********************|
; DISPLAY AND CURSOR ON |
;*********************|
        movlw   0x0e    ; 0 0 0 0 1 1 1 0 (DISPLAY ON/OFF)
        call    send8
;*********************|
;    set entry mode   |
;*********************|
        movlw   0x06    ; 0 0 0 0 0 1 1 0 (ENTRY MODE SET)
        call    send8

;*********************|
; cursor/display shift |
;*********************|
        movlw   0x14    ; 0 0 0 1 0 1 0 0 (CURSOR/DISPLAY SHIFT)
        call    send8
;*********************|
;    clear display    |
;*********************|
        movlw   0x01    ; 0 0 0 0 0 0 0 1 (CLEAR DISPLAY)
        call    send8
; Per documentation
        call    delay_5 ; Test for busy
        return

;======================
;  Procedure to delay
;    42 microseconds
;======================
delay_125
        movlw   D'42'           ; Repeat 42 machine cycles
        movwf   count1          ; Store value in counter
```

Analog to Digital and Realtime Clocks

```
repeat:
        decfsz  count1,f        ; Decrement counter
        goto    repeat          ; Continue if not 0
        return                  ; End of delay

;=======================
;   Procedure to delay
;      5 milliseconds
;=======================
delay_5:
        movlw   D'41'           ; Counter = 41
        movwf   count2          ; Store in variable
delay:
        call    delay_125       ; Delay
        decfsz  count2,f        ; 40 times = 5 milliseconds
        goto    delay
        return                  ; End of delay
;=======================
;      pulse E line
;=======================
pulseE:
        bsf     PORTA,E_line    ; Pulse E line
        nop
        bcf     PORTA,E_line
        return

;=============================
;   long delay sub-routine
;        (for debugging)
;=============================
long_delay:
                movlw   D'200'  ; w = 200 decimal
                movwf   J       ; J = w
jloop:
        movwf   K               ; K = w
kloop:
        decfsz  K,f             ; K = K-1, skip next if zero
        goto    kloop
        decfsz  J,f             ; J = J-1, skip next if zero
        goto    jloop
        return
;=============================
;    LCD display procedure
;=============================
; Sends 16 characters from PIC buffer with address stored
; in variable pic_ad to LCD line previously selected
display16
        call    delay_5         ; Make sure not busy
```

```
; Set up for data
        bcf     PORTA,E_line    ; E line low
        bsf     PORTA,RS_line   ; RS line high for data
; Set up counter for 16 characters
        movlw   D'16'           ; Counter = 16
        movwf   count3
; Get display address from local variable pic_ad
        movf    pic_ad,w        ; First display RAM address to W
        movwf   FSR             ; W to FSR
getchar:
        movf    INDF,w          ; get character from display RAM
                                ; location pointed to by file select
                                ; register
        call    send8           ; 4-bit interface routine
; Test for 16 characters displayed
        decfsz  count3,f        ; Decrement counter
        goto    nextchar        ; Skipped if done
        return
nextchar:
        incf    FSR,f           ; Bump pointer
        goto    getchar

;========================
;   send 2 nibbles in
;      4-bit mode
;========================
; Procedure to send two 4-bit values to Port-B lines
; 7, 6, 5, and 4. High-order nibble is sent first
; ON ENTRY:
;       w register holds 8-bit value to send
send8:
        movwf   store1          ; Save original value
        call    merge4          ; Merge with Port-B
; Now w has merged byte
        movwf   PORTB           ; w to Port-B
        call    pulseE          ; Send data to LCD
; High nibble is sent
        movf    store1,w        ; Recover byte into w
        swapf   store1,w        ; Swap nibbles in w
        call    merge4
        movwf   PORTB
        call    pulseE          ; Send data to LCD
        call    delay_125
        return
;==================
;   merge bits
;==================
; Routine to merge the 4 high-order bits of the
```

```
; value to send with the contents of Port-B
; so as to preserve the 4 low-bits in Port-B
; Logic:
;       AND value with 1111 0000 mask
;       AND Port-B with 0000 1111 mask
;       Now low nibble in value and high nibble in
;       Port-B are all 0 bits:
;           value = vvvv 0000
;           Port-B = 0000 bbbb
;       OR value and Port-B resulting in:
;                    vvvv bbbb
; ON ENTRY:
;       w contain value bits
; ON EXIT:
;       w contains merged bits
merge4:
        andlw   b'11110000'      ; ANDing with 0 clears the
                                 ; bit. ANDing with 1 preserves
                                 ; the original value
        movwf   store2           ; Save result in variable
        movf    PORTB,w          ; Port-B to w register
        andlw   b'00001111' ; Clear high nibble in Port-b
                                 ; and preserve low nibble
        iorwf   store2,w ; OR two operands in w
        return

;=========================
;       blank buffer
;=========================
; Procedure to store 16 blank characters in PIC RAM
; buffer starting at address stored in the variable
; pic_ad
blank16
        movlw   D'16'            ; Setup counter
        movwf   count1
        movf    pic_ad,w ; First PIC RAM address
        movwf   FSR              ; Indexed addressing
        movlw   0x20             ; ASCII space character
storeit
        movwf   INDF     ; Store blank character in PIC RAM
                         ; buffer using FSR register
        decfsz  count1,f         ; Done?
        goto    incfsr           ; no
        return                   ; yes
incfsr
        incf    FSR,f    ; Bump FSR to next buffer space
        goto    storeit
```

```
;========================
; Set address register
;    to LCD line 1
;========================
; ON ENTRY:
;         Address of LCD line 1 in constant LCD_1
line1:
        bcf     PORTA,E_line        ; E line low
        bcf     PORTA,RS_line       ; RS line low, set up for
                                    ; control
        call    delay_5             ; busy?
; Set to second display line
        movlw   LCD_1               ; Address and command bit
        call    send8               ; 4-bit routine
; Set RS line for data
        bsf     PORTA,RS_line       ; Setup for data
        call    delay_5             ; Busy?
        return

;===============================
;   first text string procedure
;===============================
storeMS1:
; Procedure to store in PIC RAM buffer the message
; contained in the code area labeled msg1
; ON ENTRY:
;         variable pic_ad holds address of text buffer
;         in PIC RAM
;         w register hold offset into storage area
;         msg1 is routine that returns the string characters
;         and a zero terminator
;         index is local variable that hold offset into
;         text table. This variable is also used for
;         temporary storage of offset into buffer
; ON EXIT:
;         Text message stored in buffer
;
; Store offset into text buffer (passed in the w register)
; in temporary variable
        movwf   index               ; Store w in index
; Store base address of text buffer in FSR
        movf    pic_ad,w ; first display RAM address to W
        addwf   index,w             ; Add offset to address
        movwf   FSR                 ; W to FSR
; Initialize index for text string access
        movlw   0                   ; Start at 0
        movwf   index               ; Store index in variable
; w still = 0
```

```
get_msg_char:
        call    msg1            ; Get character from table
; Test for zero terminator
        andlw   0x0ff
        btfsc   STATUS,Z ; Test zero flag
        goto    endstr1         ; End of string
; ASSERT: valid string character in w
;        store character in text buffer (by FSR)
        movwf   INDF            ; store in buffer by FSR
        incf    FSR,f           ; increment buffer pointer
; Restore table character counter from variable
        movf    index,w         ; Get value into w
        addlw   1               ; Bump to next character
        movwf   index           ; Store table index in
variable
        goto    get_msg_char    ; Continue
endstr1:
        return

; Routine for returning message stored in program area
; Message has 10 characters
msg1:
        addwf   PCL,f           ; Access table
        retlw   'H'
        retlw   ':'
        retlw   ' '
        retlw   ' '
        retlw   ' '
        retlw   'M'
        retlw   ':'
        retlw   ' '
        retlw   ' '
        retlw   ' '
        retlw   'S'
        retlw   ':'
        retlw   0

;==============================
;    BCD to ASCII decimal
;          conversion
;==============================
; ON ENTRY:
;        w register has two packed BCD digits
; ON EXIT:
;        output variables asc10, and asc1 have
;        two ASCII decimal digits
; Routine logic:
;    The low order nibble is isolated and the value 30H
```

```
;          added to convert to ASCII. The result is stored in
;          the variable asc1. Then the same is done to the
;          high-order nibble and the result is stored in the
;          variable asc10

Bcd2asc:
           movwf    store1            ; Save input
           andlw    b'00001111'  ; Clear high nibble
           addlw    0x30              ; Convert to ASCII
           movwf    asc1              ; Store result
           swapf    store1,w ; Recover input and swap digits
           andlw    b'00001111'  ; Clear high nibble
           addlw    0x30              ; Convert to ASCII
           movwf    asc10             ; Store result
           return


;=============================================================
;                       6355 RTC procedures
;=============================================================
;==============================
;          init RTC
;==============================
; Procedure to initialize the real time clock chip. If chip
; is not initialized it will not operate and the values
; read will be invalid.
; Since the 6355 operates in BCD format the stored values must
; be converted to packed BCD.
; According to wiring diagram
; NJU6355 Interface for setting time:
; DAT             PORTB,0            Output
; CLK             PORTB,1            Output
; CE              PORTB,2            Output
; IO              PORTB,3            Output
setRTC:
           Bank1
           movlw    b'00000000'       ; All output bits
           movlw    TRISB
           Bank0
; Writing to the 6355 requires that the CLK bit be held
; low while the IO and CE lines are high
           bcf      PORTB,CLK         ; CLK low
           call     delay_5
           bsf      PORTB,IO ; IO high
           call     delay_5
           bsf      PORTB,CE ; CE high
; Data is stored in RTC as follows:
;    year                   8 bits (0 to 99)
;    month                  8 bits (1 to 12)
```

```
;       day                     8 bits (1 to 31)
;       dayOfWeek               4 bits (1 to 7)
;       hour                    8 bits (0 to 23)
;       minutes                 8 bits (0 to 59)
;                               ======
;       Total                   44 bits
; Seconds cannot be written to RTC. RTC seconds register
; is automatically initialized to zero
        movf    year,w          ; Get item from storage
        call    bin2bcd         ; Convert to BCD
        movwf   temp1
        call    writeRTC

        movf    month,w
        call    bin2bcd
        movwf   temp1
        call    writeRTC

        movf    day,w
        call    bin2bcd
        movwf   temp1
        call    writeRTC

        movf    dayOfWeek,w     ; dayOfWeek of week is 4-bits
        call    bin2bcd
        movwf   temp1
        call    write4RTC

        movf    hour,w
        call    bin2bcd
        movwf   temp1
        call    writeRTC

        movf    minutes,w
        call    bin2bcd
        movwf   temp1
        call    writeRTC
; Done
        bcf     PORTB,CLK       ; Hold CLK line low
        call    delay_5
        bcf     PORTB,CE        ; and the CE line
; to the RTC
        call    delay_5
        bcf     PORTB,IO        ; RTC in output mode
        return
;============================
;       read RTC data
;============================
```

```
; Procedure to read the current time from the RTC and store
; data (in packed BCD format) in local time registers.
; According to wiring diagram
; NJU6355 Interface for read operations:
; DAT               PORTB,0         Input
; CLK               PORTB,1         Output
; CE                PORTB,2         Output
; IO                PORTB,3         Output
Get_Time
; Clear Port-B
          movlw     b'00000000'
          movwf     PORTB
; Make data line input
          Bank1
          movlw     b'00000001'
          movwf     TRISB
          Bank0
; Reading RTC data requires that the IO line be low and the
; CE line be high. CLK line is held low
          bcf       PORTB,CLK       ; CLK low
          call      delay_125
          bcf       PORTB,IO ; IO line low
          call      delay_125
          bsf       PORTB,CE ; and CE line high
; Data is read from RTC as follows:
;   year                  8 bits (0 to 99)
;   month                 8 bits (1 to 12)
;   day                   8 bits (1 to 31)
;   dayOfWeek             4 bits (1 to 7)
;   hour                  8 bits (0 to 23)
;   minutes               8 bits (0 to 59)
;   seconds               8 bits (0 to 59)
;                         ======
;   Total                 52 bits
          call      readRTC
          movwf     year
          call      delay_125

          call      readRTC
          movwf     month
          call      delay_125

          call      readRTC
          movwf     day
          call      delay_125

; dayOfWeek of week is a 4-bit value
          call      read4RTC
```

```
            movwf     dayOfWeek
            call      delay_125

            call      readRTC
            movwf     hour
            call      delay_125

            call      readRTC
            movwf     minutes
            call      delay_125

            call      readRTC
            movwf     seconds

            bcf       PORTB,CE ; CE line low to end output
            return

;=============================
;  read 4/8 bits from RTC
;=============================
; Procedure to read 4/8 bits stored in 6355 registers
; Value returned in w register
read4RTC
            movlw     .4               ; 4 bit read
            goto      anyBits
readRTC
            movlw     .8               ; 8 bits read
anyBits:
            movwf     counter
; Read 6355 read operation requires the IO line be set low
; and the CE line high. Data is read in the following order:
; year, month, day, day-of-week, hour, minutes, seconds
readBits:
            bsf       PORTB,CLK; Set CLK high to validate data
            bsf       STATUS,C         ; Set the carry flag (bit = 1)
; Operation:
;   If data line is high, then bit read is a 1-bit
;   otherwise bit read is a 0-bit
            btfss     PORTB,DAT        ; Is data line high?
                                       ; Leave carry set (1 bit) if high
            bcf       STATUS,C ; Clear the carry bit (make bit 0)
; At this point the carry bit matches the data line
            bcf       PORTB,CLK        ; Set CLK low to end read
; The carry bit is now rotated into the temp1 register
            rrf       temp1,1
            decfsz    counter,1        ; Decrement the bit counter
            goto      readBits ; Continue if not last bit
; At this point all bits have been read (8 or 4)
```

```
                movf        temp1,0             ; Result to w
                return

;==============================
;    write 4/8 bits to RTC
;==============================
; Procedure to write 4 or 8 bits to the RTC registers
; ON ENTRY:
;     temp1 register holds value to be written
; ON EXIT:
;     nothing
write4RTC
                movlw       .4                  ; Init for 4 bits
                goto        allBits
writeRTC
                movlw       .8                  ; Init for 8 bits
allBits:
                movwf       counter             ; Store in bit counter
writeBits:
                bcf         PORTB,CLK           ; Clear the CLK line
                call        delay_5             ; Wait
                bsf         PORTB,DAT           ; Set the data line to RTC
                btfss       temp1,0             ; Send LSB
                bcf         PORTB,DAT           ; Clear data line
                call        delay_5             ; Wait for operation to
                                                ; complete
                bsf         PORTB,CLK           ; Bring CLK line high to
                                                ; validate
                rrf         temp1,f             ; Rotate bits in storage
                decfsz      counter,1           ; Decrement bit counter
                goto        writeBits           ; Continue if not last bit
                return

;==============================
;    init time variables
;==============================
; Procedure to initialize time variables for testing
; Constants used in initialization are located in
; #define statements.
initRTC:
                movlw       iYear
                movwf       year
                movlw       iMonth
                movwf       month
                movlw       iDay
                movwf       day
                movlw       iDoW
                movwf       dayOfWeek
```

```
            movlw       iHour
            movwf       hour
            movlw       iMin
            movwf       minutes
            movlw       iSec
            movwf       seconds
            return
;=============================
;  binary to BCD conversion
;=============================
; Convert a binary number into two packed BCD digits
; ON ENTRY:
;       w register has binary value in range 0 to 99
; ON EXIT:
;       output variables bcdLow and bcdHigh contain two
;       packed unpacked BCD digits
;       w contains two packed BCD digits
; Routine logic:
;   The value 10 is subtracted from the source operand
;   until the reminder is < 0 (carry cleared). The number
;   of subtractions is the high-order BCD digit. 10 is
;   then added back to the subtrahend to compensate
;   for the last subtraction. The final reminder is the
;   low-order BCD digit
; Variables:
;       inNum       storage for source operand
;       bcdHigh     storage for high-order nibble
;       bcdLow      storage for low-order nibble
;       thisDig     Digit counter
bin2bcd:
            movwf       inNum               ; Save copy of source value
            clrf        bcdHigh    ; Clear storage
            clrf        bcdLow
            clrf        thisDig
min10:
            movlw       .10
            subwf       inNum,f             ; Subtract 10
            btfsc       STATUS,C            ; Did subtract overflow?
            goto        sum10               ; No. Count subtraction
            goto        fin10
sum10:
            incf        thisDig,f           ;increment digit counter
            goto        min10
; Store 10th digit
fin10:
            movlw       .10
            addwf       inNum,f             ; Adjust for last subtract
            movf        thisDig,w           ; get digit counter contents
```

```
        movwf   bcdHigh             ; Store it
; Calculate and store low-order BCD digit
        movf    inNum,w             ; Store units value
        movwf   bcdLow              ; Store digit
; Combine both digits
        swapf   bcdHigh,w           ; High nibble to HOBs
        iorwf   bcdLow,w ; ORin low nibble
        return
;=============================================================
        end     ; END OF PROGRAM
;=============================================================
```

Appendix A

Resistor Color Codes

The resistor color coding system applies to carbon film, metal oxide film, fusible, precision metal film, and wirewound resistors of the axial lead type. This system is employed when the surface area is not sufficient to print the actual resistance value. Several color codes are used, the most common ones are the 4-band and 5-band codes. In the 4-band code the first two bands represent the magnitude of the resistance, the third band is a multiplier for this value, and the fourth one encodes the error tolerance. In the 5-band code the first three bands represent the magnitude, the fourth one serves as a multiplier, and the fifth one is the error tolerance.

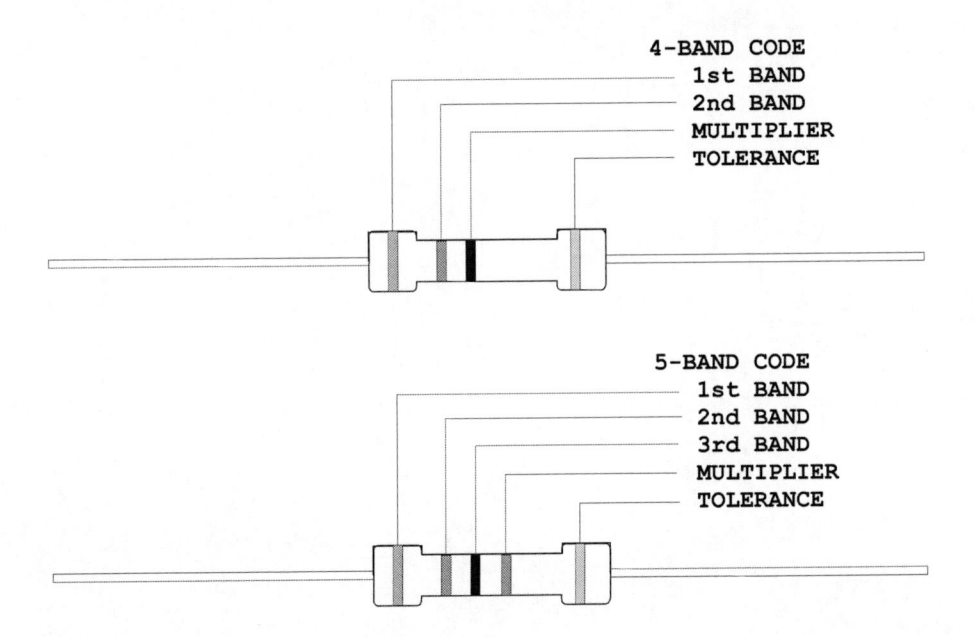

```
4-BAND CODE
        1st BAND
        2nd BAND
        MULTIPLIER
        TOLERANCE
```

```
5-BAND CODE
        1st BAND
        2nd BAND
        3rd BAND
        MULTIPLIER
        TOLERANCE
```

The color codes for the various bands are as follows:

COLOR	MAGNITUDE	MULTIPLIER	TOLERANCE
Black	0	1	
Brown	1	10	1%
Red	2	100	2%
Orange	3	1K	
Yellow	4	10K	
Green	5	100K	0.5%
Blue	6	1M	0.25%
Violet	7	10M	0.10%
Grey	8		0.05%
White	9		
Gold		0.1	5%
Silver		0.01	10%

To read the resistance value first determine if it is a 4-band or a 5-band encoding. Then proceed to identify the tolerance band, which is usually either gold or silver. Starting at the opposite end, read the two or three magnitude bands and multiply this value by the multiplier band. For example, a resistor with four color bands: red, orange, brown, and gold is a 230 Ohm resistor with a 5% error tolerance.

There are several calculators on line that allow you to easily find the resistance value. You can locate these calculators by searching for the keywords: resistor color codes.

Appendix B

Building Your Own Circuit Boards

Several methods have been developed for making printed circuit boards on a small scale, as would be convenient for the experimenter and prototype developer. If you look through the pages of any electronics supply catalog you will find kits and components based on different technologies of various levels of complexity. The method we describe in this appendix is perhaps the simplest one since it does not require a photographic process.

The process consists of the following steps:

1. The circuit diagram is drawn on the PC using a general-purpose or a specialized drawing program.
2. A printout is made of the circuit drawing on photographic paper.
3. The printout is transferred to a copper-clad circuit board blank by ironing over the backside with a household clothes iron.
4. The resulting board is placed in an etching bath that eats away all the copper, except the circuit image ironed onto the board surface.
5. The board is washed of etchant, cleaned, drilled, and the components soldered to it in the conventional manner.
6. Optionally another image can be ironed onto the backside of the board to provide component identification, logos, etc.

The following URL contains detailed information on making your own PCBs :

```
Http://www.fullnet.com/u/tomg/gooteedr.htm
```

Drawing the Circuit Diagram

Any computer drawing program serves this purpose. We have used CorelDraw but there are several specialized PCB drawing programs available on the Internet. The following is a circuit board drawing used by us for a PIC flasher circuit described in the text:

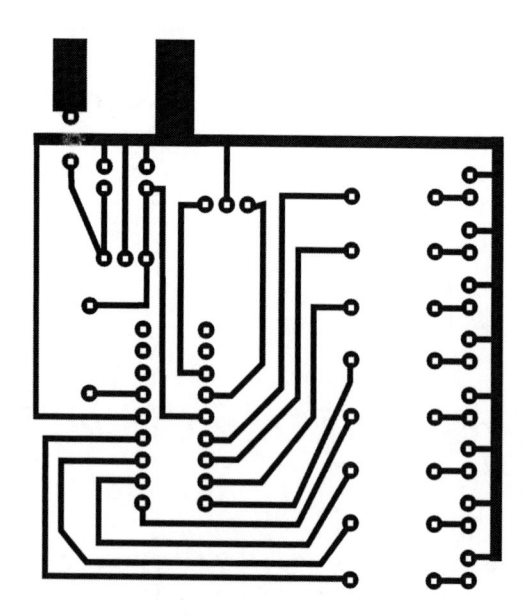

Figure B-1 PIC Flasher Circuit Board Drawing

Note in the drawing that the circuit locations where the components are to be soldered consist of small circular pads, usually called *solder pads*. The following illustration zooms into the lower corner of the drawing to show the details of the solder pads.

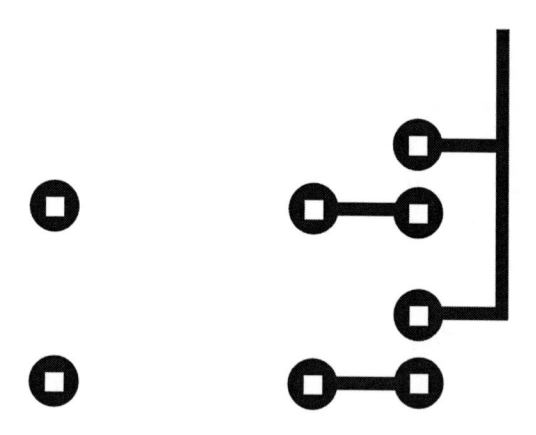

Figure B-2 Detail of Circuit Board Pads

Quite often it is necessary for a circuit line to cross between two standard pads. In this case the pads can be modified so as to allow it. The modified pads are shown in Figure B-3.

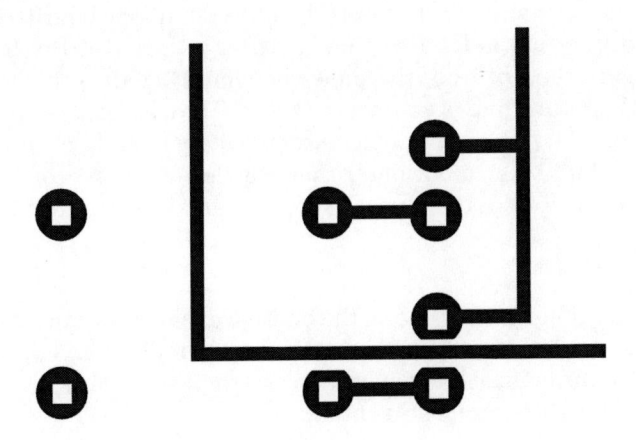

Figure B-3 Modified Circuit Boards Pads

Printing the PCB Diagram

The circuit diagram must be printed using a laser printer. Inkjet toners do not produce an image that resist the action of the etchant. Although in our experiments we used LaserJet printers it is well documented that virtually any laser printer will work. Laser copiers have also been used successfully for creating the PCB circuit image.

With the method we are describing, the width of the traces can become an issue. The traces in the PCB image of Figure B-1 are 2 points, which is 0.027". Traces half that width and less have been used successfully with this method but as the traces become thinner the entire process becomes more critical. For most simple circuits 0.020" traces should be a useful limit. Also be careful not to touch the glossy side of the paper or the printed image with fingers.

Note that the pattern is drawn as if you were looking from the component side of the board.

Transferring the PCB Image

Users of this method state that one of the most critical elements is the paper used in printing the circuit. Pinholes in some papers can degrade the image to the point that the circuit lines (especially if they are very thin) do not etch correctly. Another problem relates to removing the ironed-on paper from the board without damaging the board surface.

Glossy, coated inkjet-printer paper works well. Even better results can be obtained with glossy photo paper. We use a common high-gloss photographic paper available from Staples and sold under the name of "picture paper". The 30 sheets, 8-by-10 size, have the Staples number B031420197 1713. The UPC barcode is: 7 18103 02238 5.

Transferring the image onto the board blank is done by applying heat from a common clothes iron, set on the hottest setting, onto the paper/board sandwich. In most irons the hottest setting is labeled "linen." After going over the back of the paper several times with the hot iron, the paper becomes fused to the copper side of the blank board. The board/paper sandwich is then allowed to soak in water for about 10 minutes, after which the paper can be removed by peeling or light scrubbing with a toothbrush. It has been mentioned that Hewlett-Packard toner cartridges with microfine particles work better than the store-brand toner cartridges.

Etching the Board

Once the paper has been removed and the board washed it is time to prepare the board for etching. The preliminary operations consist of rubbing the copper surface of the board with Scotchbrite plastic abrasive pad and then scrubbing the surface with a paper towel soaked with Acetone solvent.

When the board is rubbed and clean, it is time to etch the circuit. The etching solution contains Ferric Chloride and is available from Radio Shack as a solution and from Jameco Electronics as a powder to be mixed by the user. PCB Ferric Chloride etchant should be handled with rubber gloves and rubber apron since it stains the skin and utensils. Also, concentrated acid fumes from Ferric Chloride solution are toxic and can cause severe burns. These chemicals should be handled according to cautions and warnings posted in the containers.

The Ferric Chloride solution should be stored and used in a plastic or glass container, never metal. Faster etching is accomplished if the etching solution is first warmed by placing the bottle in a tub of hot water. Once the board is in the solution, face up, the container is rocked back and forth. It is also possible to aid in the copper removal by rubbing the surface with a rubber-gloved finger.

Finishing the Board

The etched board should be washed well, first in water and then in Lacquer Thinner or Acetone; either solvent works. It is better to just rub the board surface with a paper towel soaked in the solvent. Keep in mind that most solvents are flammable and explosive, and also toxic.

After the board is clean the mounting holes can be drilled using the solder pads as a guide. A small electric drill at high revolutions, such as a Dremmel tool, works well for this operation. The standard drill size for the mounting holes is 0.035". A #60 drill (0.040") also works well. Once all the holes are drilled, the components can be mounted from the backside and soldered at the pads.

The Backside Image

The component side (backside) of the PCB can be printed with an image of the components to be mounted or with logos or other text. A single-sided blank board has no copper coating on the backside so the image is just ironed on without etching. Probably the best time to print the backside image is after the board has been etched and drilled but before mounting the components.

Building Your Own Circuit Boards

Since the image is to be transferred directly to the board, it must be a mirror image of the desired graphics and text. Most drawing programs contain a mirroring transformation so the backside image can be drawn using the component side as a guide, and then mirrored horizontally before ironing it on the backside of the board. Figure B-4 shows the backside image of the sample circuit board, before and after mirroring.

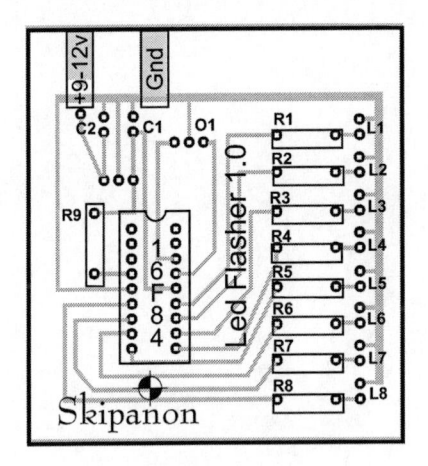

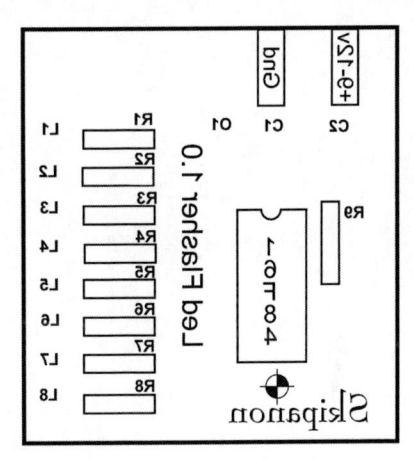

Figure B-4 Graphics and Text for Board Backside Image

Note on the left-side image in Figure B-4 that a lighter copy of the circuit diagram was used to lay out the image of the backside. Once drawn, the backside drawing was mirrored horizontally, as shown in the right-side image.

Appendix C

Mid-range Instruction Set

This Appendix describes the instructions in the PIC mid-range family. Not all instructions are implemented in all devices but all of them work in the specific PICs discussed in the text, that is, the 16F84A and the 16F877.

Table C.1

Mid-range PIC Instruction Set

MNEMONIC	OPERAND	DESCRIPTION	CYCLES	BITS AFFECTED
		BYTE-ORIENTED OPERATIONS:		
ADDWF	f,d	Add w and f	1	C,DC,Z
ANDWF	f,d	AND w with f	1	Z
CLRF	f	Clear f	1	Z
CLRW	-	Clear w	1	Z
COMF	f,d	Complement f	1	Z
DECF	f,d	Decrement f	1	Z
DECFSZ	f,d	Decrement, skip if 0	1(2)	-
INCF	f,d	Increment f	1	Z
INCFSZ	f,d	Increment, skip if 0	1(2)	-
IORWF	f,d	Inclusive OR w and f	1	Z
MOVF	f,d	Move f	1	Z
MOVWF	f	Move w to f	1	-
NOP	-	No operation	1	-
RLF	f,d	Rotate left through carry	1	C
RRF	f,d	Rotate right through carry	1	C
SUBWF	f,d	Subtract w from f	1	C,DC,Z
SWAPF	f,d	Swap nibbles in f	1	-
XORWF				
		BIT-ORIENTED OPERATIONS		
BCF	f,b	Bit clear in f	1	-
BSF	f,b	Bit set in f	1	-
BTFSC	f,b	Bit test, skip if clear	1	-
BTFSS	f,b	Bit test, skip if set	1	-
		LITERAL AND CONTROL OPERATIONS		
ADDLW	k	Add literal and w	1	C,DC,Z

(continues)

Table C.1

Mid-range PIC Instruction Set (continued)

MNEMONIC	OPERAND	DESCRIPTION	CYCLES	BITS AFFECTED
		LITERAL AND CONTROL OPERATIONS		
ANDLW	k	AND literal and w	1	Z
CALL	k	Call procedure	2	-
CLRWDT	-	Clear watchdog timer	1	TO,PD
GOTO	k	Go to address	2	-
IORLW	k	Inclusive OR literal with w	1	Z
MOVLW	k	Move literal to w	1	-
RETFIE	-	Return from interrupt	2	-
RETLWk	-	Return literal in w	2	-
RETURN	-	Return from procedure	2	-
SLEEP	-	Go into SLEEP mode	1	TO,PD
SUBLW	k	Subtract literal and w	1	C,DC,Z
XORLW	k	Exclusive OR literal with w	1	Z

Legend:
  f = file register
  d = destination: 0 = w register
                             1 = file register
  b = bit position
  k = 8-bit constant

Table C.2

Conventions used in Instruction Descriptions

FIELD	DESCRIPTION
f	Register file address (0x00 to 0x7F)
w	Working register (accumulator)
b	Bit address within an 8-bit file register (0 to 7)
k	Literal field, constant data or label (may be either an 8-bit or an 11-bit value)
x	Don't care (0 or 1)

d	Destination select;
	    d = 0: store result in W,
	    d = 1: store result in file register f.

dest	Destination either the W register or the specified register file location
label	Label name
TOS	Top of Stack
PC	Program Counter
PCLATH	Program Counter High Latch
GIE	Global Interrupt Enable bit
WDT	Watchdog Timer
!TO	Time-out bit
!PD	Power-down bit
[ ]	Optional element
[XXX]	Contents of memory location pointed at by XXX register
( )	Contents
->	Assigned to
< >	Register bit field
italics	User defined term

ADDLW Add Literal and w

Syntax:	[label] ADDLW k
Operands:	k in range 0 to 255
Operation:	(w) + k -> w
Status Affected:	C, DC, Z
Description:	The contents of the w register are added to the eight bit literal 'k' and the result is placed in the w register
Words:	1
Cycles:	1

Example1:

```
ADDLW    0x15
Before Instruction:      w = 0x10
After Instruction:       w = 0x25
```

Example 2:

```
ADDLW    var1
Before Instruction:      w = 0x10
var1 is data memory variable
var1 = 0x37
After Instruction        w = 0x47
```

ADDWF Add w and f

Syntax: [label] ADDWF   f,d
Operands: f in range 0 to 127
d = 0 / 1
Operation: (W) + (f) -> destination
Status Affected: C, DC, Z
Description: Add the contents of the W register with register 'f'. If 'd' is 0 the result is stored in the w register. If 'd' is 1 the result is stored back in register 'f'.
Words: 1
Cycles: 1

Example 1:

```
ADDWF    FSR,0
Before Instruction:
    w = 0x17
    FSR = 0xc2
After Instruction
    W = 0xd9
    FSR = 0xc2
```

Example 2:

```
ADDWF    INDF, 1
before Instruction:
    W = 0x17
    FSR = 0xC2
    Contents of Address (FSR) = 0x20
After Instruction
    W = 0x17
    FSR = 0xC2
    Contents of Address (FSR) = 0x37
```

BCF   Bit Clear f

Syntax: [label] BCF f,b
Operands: f in range 0 to 127
b in range 0 to 7
Operation: 0 ->f<b>
Status Affected: None
Description: Bit 'b' in register 'f' is cleared.
Words: 1
Cycles: 1

Example 1:
```
BCF     reg1,7
Before Instruction: reg1 = 0xc7 (1100 0111)
After Instruction: reg1 = 0x47 (0100 0111)
```

Example 2:
```
BCF     INDF,3
Before Instruction:  w   = 0x17
                     FSR = 0xc2
                     [FSR]= 0x2f
After Instruction
                     w   = 0x17
                     FSR = 0xc2
                     [FSR] = 0x27
```

BSF Bit Set f

Syntax: [label] BSF f,b
Operands: f in range 0 to 127
b in range 0 to 7
Operation: 1-> f<b>
Status Affected: None
Description: Bit 'b' in register 'f' is set.
Words: 1
Cycles: 1
Example 1:

```
BSF    reg1,6
Before Instruction :    reg1 = 0011 1010
After Instruction:      reg1 = 0111 1010
```

Example 2:

```
BSF    INDF,3
Before Instruction:     w = 0x17
                        FSR = 0xc2
                        [FSR] = 0x20

After Instruction
                        w = 0x17
                        FSR = 0xc2
                        [FSR] = 0x28
```

BTFSC   Bit Test f, Skip if Clear

Syntax:	[label] BTFSC f,b
Operands:	f in range 0 to 127
	b in range 0 to 7
Operation:	skip next instruction if (f<b>) = 0
Status Affected:	None
Description:	If bit 'b' in register 'f' is '0' then the next instruction is skipped. If bit 'b' is '0' then the next instruction (fetched during the current instruction execution) is discarded, and a NOP is executed instead, making this a 2 cycle instruction.
Words:	1

Example 1:

```
repeat:
        btfsc   reg1,4
        goto    repeat
Case 1: Before Instruction
            PC = $
            reg1 = xxx0 xxxx
        After Instruction
            Since reg1<4>= 0,
            PC = $ + 2 (goto skipped)
Case 2: Before Instruction
            PC = $
            reg1= xxx1 xxxx
        After Instruction
            Since FLAG<4>=1,
            PC = $ + 1 (goto executed)
```

BTFSS   Bit Test f, Skip if Set

Syntax:	[label] BTFSC f,b
Operands:	f in range 0 to 127
	b in range 0 to 7
Operation:	skip next instruction if (f<b>) = 1
Status Affected:	None
Description:	If bit 'b' in register 'f' is '1' then the next instruction is skipped. If bit 'b' is '0' then the next instruction (fetched during the current instruction execution) is discarded, and a NOP is executed instead, making this a 2 cycle instruction.
Words:	1
Cycles:	1(2)

```
repeat:
        btfss   reg1,4
        goto    repeat
Case 1: Before Instruction
            PC = $
            Reg1 = xxx1 xxxx
        After Instruction
            Since Reg1<4>= 1,
            PC = $ + 2 (goto skipped)
Case 2: Before Instruction
            PC = $
            Reg1 = xxx0 xxxx
        After Instruction
            Since Reg1<4>=0,
            PC = $ + 1 (goto executed)
```

CALL Call Subroutine

Syntax:	[label] CALL k
Operands:	k in range 0 to 2047
Operation:	(PC) + -> TOS,
	k-> PC<10:0>,
	(PCLATH<4:3>)-> PC<12:11>
Status Affected:	None
Description:	Call Subroutine. First, the 13-bit return address (PC+1) is pushed onto the stack. The eleven bit immediate address is loaded into PC bits <10:0>. The upper bits of the PC are loaded from PCLATH<4:3>. CALL is a two cycle instruction.
Words:	1
Cycles:	2
Example 1:	

```
Here:
        call    There
Before Instruction
        PC = AddressHere
        After Instruction
        TOS = Address Here + 1
        PC = Address There
```

CLRF — Clear f

Syntax:	[label] CLRF f
Operands:	f in range 0 to 127
Operation:	00h ->f 1-> Z
Status Affected:	Z
Description:	The contents of register 'f' are cleared and the Z bit is set.
Words:	1
Cycles:	1

Example 1:

```
clrf    reg1
Before Instruction:     reg1 = 0x5a
After Instruction:      reg1 = 0x00
                        Z = 1
```

Example 2:

```
Clrf    INDF
Before Instruction:     FSR = 0xc2
                        [FSR]= 0xAA
After Instruction:      FSR = 0xc2
                        [FSR] = 0x00
                        Z = 1
```

CLRW   Clear w

Syntax:	[label] CLRW
Operands:	None
Operation:	00h -> w
	1-> Z
Status Affected:	Z
Description:	w register is cleared. Zero bit (Z) is set.
Words:	1
Cycles:	1
Example 1:	

```
CLRW
Before Instruction:    w = 0x5A
After Instruction:     w = 0x00
                       Z = 1
```

CLRWDT   Clear Watchdog Timer

Syntax:	[label] CLRWDT
Operands:	None
Operation:	00h -> WDT
	0 -> WDT prescaler count,
	1 -> TO
	1 -> PD
Status Affected:	TO, PD
Description:	CLRWDT instruction clears the Watchdog Timer. It also clears the prescaler count of the WDT. Status bits TO and PD are set. The instruction does not change the assignment of the WDT prescaler.
Words:	1
Cycles:	1
Example 1:	

```
          CLRWDT
          Before Instruction:     WDT counter= x
                                  WDT prescaler = 1:128
          After Instruction:      WDT counter=0x00
                                  WDT prescaler count=0
                                  TO = 1
                                  PD = 1
                                  WDT prescaler = 1:128
```

COMF   Complement f

Syntax:	[label] COMF f,d
Operands:	f in range 0 to 127
	d is 0 or 1
Operation:	(f) -> destination
Status Affected:	Z
Description:	The contents of register 'f' are 1's complemented. If 'd' is 0 the result is stored in w. If 'd' is 1 the result is stored back in register 'f'.
Words:	1
Cycles:	1

Example 1:
```
        comf    reg1,0
    Before Instruction:     reg1 = 0x13
    After Instruction:      reg1 = 0x13
                            w    = 0xEC
```

Example 2:
```
        comf    INDF,1
    Before Instruction:     FSR   = 0xc2
                            [FSR] = 0xAA
    After Instruction:      FSR   = 0xc2
                            [FSR] = 0x55
```

Example 3:
```
        comf    reg1,1
    Before Instruction:     reg1 = 0xff
    After Instruction:      reg1 = 0x00
```

DECF — Decrement f

Syntax:	[label] DECF f,d
Operands:	f in range 0 to 127 d is either 0 or 1
Operation:	(f) - 1 -> destination
Status Affected:	Z
Description:	Decrement register 'f'. If 'd' is 0 the result is stored in the w register. If 'd' is 1 the result is stored back in register 'f'.
Words:	1
Cycles:	1

Example 1:
```
decf    count,1
Before Instruction:        count = 0x01
                           Z = 0
After Instruction:         count = 0x00
                           Z = 1
```

Example 2:
```
decf    INDF,1
Before Instruction:        FSR = 0xc2
                           [FSR] = 0x01
                           Z = 0
After Instruction:         FSR = 0xc2
                           [FSR] = 0x00
                           Z = 1
```

Example 3:
```
decf    count,0
Before Instruction:        count = 0x10
                           w = x
                           Z = 0
After Instruction:         count = 0x10
                           w = 0x0f
```

DECFSZ   Decrement f, Skip if 0

Syntax:	[label] DECFSZ f,d
Operands:	f in the range 0 to 127 d is either 0 or 1
Operation:	(f) - 1 -> destination; skip if result = 0
Status Affected:	None
Description:	The contents of register 'f' are decremented. If 'd' is 0 the result is placed in the w register. If 'd' is 1 the result is placed back in register 'f'. If the result is 0, then the next instruction (fetched during the current instruction execution) is discarded and a NOP is executed instead, making this a 2 cycle instruction.
Words:	1
Cycles:	1(2)
Example	

```
          here:
                  decfsz  count,1
                  goto    here
          Case 1:
          Before Instruction:      PC = $
                                   count = 0x01
                  After Instruction:       count = 0x00
                                   PC = $ + 2 (goto skipped)
          Case 2:
          Before Instruction:      PC = $
                                   count = 0x04
                  After Instruction:       count = 0x03
                                   PC = $ + 1 (goto executed)
```

GOTO — Unconditional Branch

Syntax:	[label] GOTO k
Operands:	$0 \le k \le 2047$
Operation:	k -> PC<10:0>
	PCLATH<4:3> ->PC<12:11>
Status Affected:	None
Description:	GOTO is an unconditional branch. The eleven bit immediate value is loaded into PC bits <10:0>. The upper bits of PC are loaded from PCLATH<4:3>.
	GOTO is a two cycle instruction.
Words:	1
Cycles:	2
Example	

```
goto     There
After Instruction:      PC = address of There
```

INCF Increment f

Syntax: [label] INCF f,d
Operands: f in range 0 to 127
d is either 0 or 1
Operation: (f) + 1 -> destination
Status Affected: Z
Description: The contents of register 'f' are incremented. If 'd' is 0 the result is placed in the w register. If 'd' is 1 the result is placed back in register 'f'.
Words: 1
Cycles: 1

Example 1

```
incf    count,1
Before Instruction:     count = 0xff
                        Z = 0
After Instruction:      count = 0x00
                        Z = 1
```

Example 2

```
incf    INDF,1
Before Instruction:     FSR = 0xC2
                        [FSR] = 0xff
                        Z = 0
After Instruction:      FSR = 0xc2
                        [FSR] = 0x00
                        Z = 1
```

Example 3

```
incf    count,0
Before Instruction:     count = 0x10
                        w = x
                        Z = 0
After Instruction:      count = 0x10
                        w = 0x11
                        Z = 0
```

INCFSZ   Increment f, Skip if 0

Syntax:	[label] INCFSZ f,d
Operands:	f in range 0 to 127 d is either 0 or 1
Operation:	(f) + 1 -> destination, skip if result = 0
Status Affected:	None
Description:	The contents of register 'f' are incremented. If 'd' is 0 the result is placed in the w register. If 'd' is 1 the result is placed back in register 'f'. If the result is 0, then the next instruction (fetched during the current instruction execution) is discarded and a NOP is executed instead, making this a 2 cycle instruction.
Words:	1
Cycles:	1(2)

Example

```
        Here:
                incfsz  count,1
                goto    Here
Case 1:
        Before Instruction:     PC = $
                                count = 0x10
        After Instruction:      count = 0x11
                                PC = $ + 1 (goto executed)
Case 2:
        Before Instruction:     PC = $
                                count = 0x00
        After Instruction:      count = 0x01
                                PC = $ + 2 (goto skipped)
```

IORLW — Inclusive OR Literal with w

Syntax: [label] IORLW k
Operands: k is in range 0 to 255
Operation: (w).OR. k -> w
Status Affected: Z
Description: The contents of the w register is OR'ed with the eight bit literal 'k'. The result is placed in the w register.
Words: 1
Cycles: 1

Example 1

```
iorlw   0x35
Before Instruction:     w = 0x9a
After Instruction:      w = 0xbfF
                        Z = 0
```

Example 2

```
iorlw   myreg
Before Instruction:     w = 0x9a
Myreg is a variable representing a location
in PIC RAM.             [Myreg] = 0x37
After Instruction:      w = 0x9F
                        Z = 0
```

Example 3

```
iorlw   0x00
Before Instruction:     w = 0x00
After Instruction:      w = 0x00
```

IORWF Inclusive OR w with f

Syntax: [label] IORWF f,d
Operands: f is in range 0 to 127
d is either 0 or 1
Operation: (W).OR. (f) -> destination
Status Affected: Z
Description: Inclusive OR the w register with register 'f'. If 'd' is 0 the result is placed in the w register. If 'd' is 1 the result is placed back in register 'f'.
Words: 1
Cycles: 1

Example 1

```
        iorwf   result,0
        Before Instruction:     result = 0x13
                                w = 0x91

        After Instruction:      result = 0x13
                                w = 0x93
                                Z = 0
```

Example 2

```
        iorwf   INDF,1
        Before Instruction:     w = 0x17
                                FSR = 0xc2
                                [FSR] = 0x30

        After Instruction:      w = 0x17
                                FSR = 0xc2
                                [FSR] = 0x37
                                Z = 0
```

Example 3

```
                iorwf   result,1
Case 1: Before Instruction:     result = 0x13
                                w = 0x91

        After Instruction:      result = 0x93
                                w = 0x91
                                Z = 0

Case 2: Before Instruction:     result = 0x00
                                w = 0x00

        After Instruction:      result = 0x00
                                w = 0x00
                                Z = 1
```

MOVF   Move f

Syntax:	[label] MOVF f,d
Operands:	f is in range 0 to 127
	d is either 0 or 1
Operation:	(f) -> destination
Status Affected:	Z
Description:	The contents of register 'f' is moved to a destination dependent upon the status of 'd'. If 'd' = 0, destination is W register. If 'd' = 1, the destination is file register 'f' itself. 'd' = 1 is useful to test a file register since status flag Z is affected.
Words:	1
Cycles:	1

Example 1

```
            movf    FSR,0
            Before Instruction:         w   = 0x00
                                        FSR = 0xc2
            After Instruction:          w   = 0xc2
                                        Z   = 0
```

Example 2

```
            movf    INDF,0
            Before Instruction:         w     = 0x17
                                        FSR   = 0xc2
                                        [FSR] = 0x00
            After Instruction:          w     = 0x17
                                        FSR   = 0xc2
                                        [FSR] = 0x00
                                        Z     = 1
```

Example 3

```
            movf    FSR,1
    Case 1: Before Instruction:         FSR = 0x43
            After Instruction:          FSR = 0x43
                                        Z   = 0
    Case 2: Before Instruction:         FSR = 0x00
            After Instruction:          FSR = 0x00
                                        Z   = 1
```

MOVLW   Move Literal to w

Syntax:	[label] MOVLW k
Operands:	k in range 0 to 255
Operation:	k- > w
Status Affected:	None
Description:	The eight bit literal 'k' is loaded into W register. The don't cares will assemble as 0's.
Words:	1
Cycles:	1

Example 1

```
movlw   0x5a
After Instruction:        w = 0x5A
```

Example 2

```
movlw   myreg
Before Instruction:       w = 0x10
                          [myreg] = 0x37
After Instruction:        w = 0x37
```

MOVWF   Move w to f

Syntax: [label] MOVWF f
Operands: f in range 0 to 127
Operation: (w) -> f
Status Affected: None
Description: Move data from W register to register 'f'.
Words: 1
Cycles: 1

Example 1

```
        movwf    OPTION_REG
        Before Instruction:     OPTION_REG = 0xff
                                w = 0x4f
        After Instruction:      OPTION_REG = 0x4f
                                w = 0x4f
```

Example 2

```
        movwf    INDF
        Before Instruction:     w = 0x17
                                FSR = 0xC2
                                [FSR] = 0x00
        After Instruction:      w = 0x17
                                FSR = 0xC2
                                [FSR] = 0x17
```

NOP   No Operation

Syntax:	[label] NOP
Operands:	None
Operation:	No operation
Status Affected:	None
Description:	No operation.
Words:	1
Cycles:	1
Example	

```
nop
Before Instruction:     PC = $
fter Instruction:       PC = $ + 1
```

OPTION Load Option Register

Syntax:	[label] OPTION
Operands:	None
Operation:	(w) -> OPTION_REG
Status Affected:	None
Description:	The contents of the w register are loaded in the OPTION_REG register. This instruction is supported for code compatibility with PIC16C5X products. Since OPTION_REG is a Readable/writable register, code can directly address it without using this instruction.

Words: 1
Cycles: 1
Example:

```
        movlw   b'01011100'
        option
```

RETFIE — Return from Interrupt

Syntax:	[label] RETFIE
Operands:	None
Operation:	TOS -> PC, 1 -> GIE
Status Affected:	None
Description:	Return from Interrupt. The 13-bit address at the Top of Stack (TOS) is loaded in the PC. The Global Interrupt Enable bit, GIE (INTCON<7>), Is automatically set, enabling Interrupts. This is a two cycle instruction.
Words:	1
Cycles:	2
Example	

```
retfie
After Instruction:      PC  = TOS
                        GIE = 1
```

RETLW   Return with Literal in W

Syntax:	[label] RETLW k
Operands:	k in range 0 to 255
Operation:	k -> w; TOS -> PC
Status Affected:	None
Description:	The w register is loaded with the eight bit literal 'k'. The program counter is loaded 13-bit address at the Top of Stack (the return address). This is a two cycle instruction.
Words:	1
Cycles:	2

Example

```
        movlw    2          ; Load w with desired
                            ; Table offset
        call     table      ; When call returns w
                            ; contains value stored
                            ; in table
Table:
        addwf    pc         ; w = offset
        retlw    .22        ; First table entry
        retlw    .23        ; Second table entry
        retlw    .24
         .
         .
         .
        retlw    .29        ; Last table entry
        Before Instruction:    w = 0x02
        After Instruction:     w = .24
```

RETURN Return from Subroutine

Syntax: [label] RETURN
Operands: None
Operation: TOS -> PC
Status Affected: None
Description: Return from subroutine. The stack is POPed and the top of the stack (TOS) is loaded into the program counter. This is a two cycle instruction.
Words: 1
Cycles: 2
Example

```
return
After Instruction:      PC = TOS
```

RLF   Rotate Left f through Carry

Syntax: [label] RLF f,d
Operands: f in range 0 to 127
                  d is either 0 or 1
Operation: See description below
Status Affected: C
Description: The contents of register 'f' are rotated one bit to the left through the Carry Flag. If 'd' is 0 the result is placed in the W register. If 'd' is 1 the result is stored back in register 'f'.
Words: 1
Cycles: 1

Example 1

```
          rlf      reg1,0
          Before Instruction:        reg1 = 1110 0110
                                     C = 0

          After Instruction:         reg1 = 1110 0110
                                     w = 1100 1100
                                     C = 1
```

Example 2

```
          rlf      INDF,1
Case 1:   Before Instruction:        w = xxxx xxxx
                                     FSR = 0xc2
                                     [FSR] = 0011 1010
                                     C = 1

          After Instruction:         w = 0x17
                                     FSR = 0xc2
                                     [FSR] = 0111 0101
                                     C = 0

Case 2:   Before Instruction:        w = xxxx xxxx
                                     FSR = 0xC2
                                     [FSR] = 1011 1001
                                     C = 0

          After Instruction:         w = 0x17
                                     FSR = 0xC2
                                     [FSR] = 0111 0010
                                     C = 1
```

RRF Rotate Right f through Carry

Syntax: [label] RRF f,d
Operands: f in range 0 to 127
d is either 0 or 1
Operation: See description below
Status Affected: C
Description: The contents of register 'f' are rotated one bit to the right through the Carry Flag. If 'd' is 0 the result is placed in the w register. If 'd' is 1 the result is placed back in register 'f'.
Words: 1
Cycles: 1

Example 1

```
          rrf     reg1,0
          Before Instruction:     reg1= 1110 0110
                                  w = xxxx xxxx
                                  C = 0

          After Instruction:      reg1= 1110 0110
                                  w = 0111 0011
                                  C = 0
```

Example 2

```
          rrf     INDF,1
Case 1:   Before Instruction:     w = xxxx xxxx
                                  FSR = 0xc2
                                  [FSR] = 0011 1010
                                  C = 1

          After Instruction:      w = 0x17
                                  FSR = 0xC2
                                  [FSR] = 1001 1101
                                  C = 0

Case 2:   Before Instruction:     w = xxxx xxxx
                                  FSR = 0xC2
                                  [FSR] = 0011 1001
                                  C = 0

          After Instruction:      w = 0x17
                                  FSR = 0xc2
                                  [FSR] = 0001 1100
                                  C = 1
```

SLEEP

Syntax:	[label] SLEEP
Operands:	None
Operation:	00h -> WDT,
	0 -> WDT prescaler count,
	1 -> TO,
	0 -> PD
Status Affected:	TO, PD
Description:	The power-down status bit, PD is cleared. Time-out status bit, TO is set. Watchdog Timer and its prescaler count are cleared. The processor is put into SLEEP mode with the oscillator stopped. The SLEEP instruction does not affect the assignment of the WDT prescaler.
Words:	1
Cycles:	1
Example:	

```
SLEEP
```

SUBLW — Subtract w from Literal

Syntax:	[label] SUBLW k
Operands:	k in range 0 to 255
Operation:	k - (W) -> W
Status Affected:	C, DC, Z
Description:	The w register is subtracted (2's complement method) from the eight bit literal 'k'. The result is placed in the w register.
Words:	1
Cycles:	1

Example 1

```
        sublw   0x02
Case 1: Before Instruction:     w = 0x01
                                C = x
                                Z = x

        After Instruction:      w = 0x01
                                C = 1 if result +
                                Z = 0

Case 2: Before Instruction:     w = 0x02
                                C = x
                                Z = x

        After Instruction:      w = 0x00
                                C = 1 ; result = 0
                                Z = 1

Case 3: Before Instruction:     w = 0x03
                                C = x
                                Z = x

        After Instruction:      w = 0xff
                                C = 0 ; result -
                                Z = 0
```

Example 2

```
        sublw   myreg
        Before Instruction:     w = 0x10
                                [myreg] = 0x37
        After Instruction       w = 0x27
                                C = 1 ; result +
```

SUBWF   Subtract w from f

Syntax: [label] SUBWF   f,d
Operands: f in range 0 to 127
               d is either 0 or 1
Operation: (f) - (W) -> destination
Status Affected: C, DC, Z
Description: Subtract (2's complement method) w register from register 'f'. If 'd' is 0 the result is stored in the w register. If 'd' is 1 the result is stored back in register 'f'.
Words: 1
Cycles: 1
Example 1

```
            subwf   reg1,1
    Case 1: Before Instruction:     reg1 = 3
                                    w = 2
                                    C = x
                                    Z = x
            After Instruction:      reg1 = 1
                                    w = 2
                                    C = 1 ; result +
                                    Z = 0
    Case 2: Before Instruction:     reg1 = 2
                                    w = 2
                                    C = x
                                    Z = x
            After Instruction:      reg1 = 0
                                    w = 2
                                    C = 1 ; result = 0
                                    Z = 1
    Case 3: Before Instruction:     reg1 = 1
                                    w = 2
                                    C = x
                                    Z = x
            After Instruction:      reg1 = 0xff
                                    w = 2
                                    C = 0 ; result is -
                                    Z = 0
```

SWAPF Swap Nibbles in f

Syntax:	[label] SWAPF f,d
Operands:	f in range 0 to 127 d is either 0 or 1
Operation:	(f<3:0>) -> destination<7:4>, (f<7:4>) -> destination<3:0>
Status Affected:	None
Description:	The upper and lower nibbles of register 'f' are exchanged. If 'd' is 0 the result is placed in w register. If 'd' is 1 the result is placed in register 'f'.
Words:	1
Cycles:	1

Example 1
```
          swapf    reg,0
          Before Instruction:          reg1 = 0xa5
          After Instruction:           reg1 = 0xa5
                                       W = 0x5a
```

Example 2
```
          Swapf    INDF,1
          Before Instruction:          w = 0x17
                                       FSR = 0xc2
                                       [FSR] = 0x20
          After Instruction:           w = 0x17
                                       FSR = 0xC2
                                       [FSR] = 0x02
```

Example 3
```
          swapf    reg,1
          Before Instruction:          reg1 = 0xa5
          After Instruction:           reg1 = 0x5a
```

TRIS   Load TRIS Register

Syntax:	[label] TRIS f
Operands:	f in range 5 to 7
Operation:	(W) -> TRIS register f;
Status Affected:	None
Description:	The instruction is supported for code compatibility with the PIC16C5X products. Since TRIS registers are readable and writable, code can address these registers directly
Words:	1
Cycles:	1

Example

```
movlw   B'00000000'
tris    PORTB
```

XORLW Exclusive OR Literal with W

Syntax:	[label] XORLW k
Operands:	k in range 0 to 255
Operation:	(w).XOR. k -> W
Status Affected:	Z
Description:	The contents of the w register are XOR'ed with the eight bit literal 'k'. The result is placed in the w register.
Words:	1
Cycles:	1

Example 1

```
          xorlw    b'10101111'
          Before Instruction:      w = 1011 0101
          After Instruction :      w = 0001 1010
                                   Z = 0
```

Example 2

```
          xorlw    myreg
          Before Instruction:      w = 0xaf
                                   [Myreg] = 0x37
          After Instruction:       w = 0x18
                                   Z = 0
```

XORWF Exclusive OR w with f

Syntax: [label] XORWF f,d
Operands: f in range 0 to 127
d is either 0 or 1
Operation: (W).XOR. (f) -> destination
Status Affected: Z
Description: Exclusive OR the contents of the w register with register 'f'. If 'd' is 0 the result is stored in the w register. If 'd' is 1 the result is stored back in register 'f'.
Words: 1
Cycles: 1

Example 1

```
        xorwf   reg,1
        Before Instruction:         w   = 1011 0101
                                    reg = 1010 1111
        After Instruction:          reg = 0001 1010
                                    w   = 1011 0101
```

Example 2

```
        xorwf   reg,0
        Before Instruction          w   = 1011 0101
                                    reg = 1010 1111
        After Instruction:          reg = 1010 1111
                                    w   = 0001 1010
```

Example 3

```
        xorwf   INDF,1
        Before Instruction:         w     = 1011 0101
                                    FSR   = 0xc2
                                    [FSR] = 1010 1111
        After Instruction:          w     = 1011 0101
                                    FSR   = 0xc2
                                    [FSR] = 0001 1010
```

Appendix D

Supplementary Programs

In this Appendix we have listed several programs that were developed while writing this book and for some reason were not used in the text. They are provided to the reader as a code grab-bag in the hope some may find a useful fragment or routine among those listed. Each program contains a description of its purpose and functionality. The code for the supplementary programs is available in the book's on-line software package.

```
; File: SecondCnt.ASM
; Date: April 29, 2006
; Author: Julio Sanchez
;
; Description:
; Using timer0 to delay one second at a signal
; rate of 1,000,000 beats per second
;===========================
;        switches
;===========================
; Switches used in __config directive:
;   _CP_ON          Code protection ON/OFF
; * _CP_OFF
; * _PWRTE_ON       Power-up timer ON/OFF
;   _PWRTE_OFF
;   _WDT_ON         Watchdog timer ON/OFF
; * _WDT_OFF
;   _LP_OSC         Low power crystal occilator
; * _XT_OSC         External parallel resonator/crystal oscillator
;
;   _HS_OSC         High speed crystal resonator (8 to 10 MHz)
;                   Resonator: Murate Erie CSA8.00MG = 8 MHz
;   _RC_OSC         Resistor/capacitor oscillator (simplest, 20%
; error)
; |
; |_____ * indicates setup values
```

```
            processor 16f84A
            include     <p16f84A.inc>
            __config    _XT_OSC & _WDT_OFF & _PWRTE_ON & _CP_OFF

;=======================================================
;                   PIC register equates
;=======================================================
porta       equ                 0x05
portb       equ         0x06
status      equ                 0x03
z                   equ                 0x02
c                   equ                 0x00
tmr0        equ                 0x01
; countL    equ                 0x01        ; Alias for tmr0
;
;=======================================================
;               variables in PIC RAM
;=======================================================
; Local variables
            cblock  0x0d                    ; Start of block
            J                               ; counter J
            K                               ; counter K
; 3-byte auxiliary counter. Low order byte is kept
; in the timer0 register
            countM      ; Medium byte
            countH      ; High byte

            endc
;=======================================================
;               m a i n    p r o g r a m
;=======================================================
            org     0           ; start at address 0
            goto    main
;
;==============================
;       interrupt handler
;==============================
            org             0x04
;           goto    IntServ
;==============================
;       main program
;==============================
main:
; Clear the watchdog timer and reset prescaler
            clrf    tmr0
            clrwdt
; Set up the OPTION register bit map
```

Supplementary Programs

```
                movlw    b'11011000'
;       7   6   5   4   3   2   1   0   <= OPTION bits
;       |   |   |   |   |   |__|__|_____ PS2-PS0 (prescaler bits)
;       |   |   |   |   |                Values for Timer0
;       |   |   |   |   |               *000 = 1:2    001 = 1:4
;       |   |   |   |   |                010 = 1:8    011 = 1:16
;       |   |   |   |   |                100 = 1:32   101 = 1:64
;       |   |   |   |   |                110 = 1:128 *111 = 1:256
;       |   |   |   |   |_____  PSA (prescaler assign)
;       |   |   |   |                    1 = to WDT
;       |   |   |   |                   *0 = to Timer0
;       |   |   |   |_____  TOSE (Timer0 edge select)
;       |   |   |                        0 = increment on low-to-high
;       |   |   |                       *1 = increment in high-to-low
;       |   |   |_____  TOCS (TMR0 clock source)
;       |   |                           *0 = internal clock
;       |   |                            1 = RA4/TOCKI bit source
;       |   |_____  INTEDG (Edge select)
;       |                               *0 = falling edge
;       |_____  RBPU (Pullup enable)
;                                        0 = enabled
;                                       *1 = disabled
                option
; Setup ports
                movlw    0x00            ; Set port B to output
                tris     portb
                clrf     portb           ; All port B to 0
; Port A is not used in this program
mloop:
                bsf      portb,0
                call     TM0delay
                bcf      portb,0
                call     TM0delay
                goto     mloop
;********************************
;   one second delay sub-routine
;           using Timer0
;********************************
; Routine logic:
; The prescaler is assigned to timer0 and setup so
; that the timer runs at 1:2 rate. This means that
; every time the counter reaches 128 (0x80) a total
; of 256 machine cycles have elapsed. The value 0x80
; is detected by testing bit 7 of the counter
; register. This method gives the routine a total of
; 128 machine cycles before the next counter beat must
; be acknowledged.
TM0delay:
```

```
; Timer is designed to count from 0 to 1,000,000
; 1,000,000 = 0x0f 0x42 0x40
;              ---- ---- ----
;               |    |    |___ (see note)
;               |    |_____ countM
;               |_____ countH
; Note:
;    The initial count of 0x40 (64 decimal) is ensured
; by initializing the tmr0 register to count 32 timer
; beats at the 1:2 prescaler rate. 128 - 32 = 96 = 0x60
; Initialize the counters:
        movlw   0x0f
        movwf   countH
        movlw   0x42
        movwf   countM
        movlw   0x60
        movwf   tmr0
; Routine tests timer overflow by testing bit 7 of
; the tmr0 register.
cycle:
        movlw   3
        subwf   tmr0,w
        btfsc   status,c
        goto    cycle
; Subtract 256 from beat counter by decrementing the
; mid-order byte
        decfsz  countM,f
        goto    cycle               ; Continue if mid-byte not zero
; At this point the mid-order byte has overflowed.
; High-order byte must be decremented.
        decfsz  countH,f
        goto    cycle
; At this point one second has elapsed
        return
        end
```

```
; File name: SevenSeg.asm
; Date: April 19, 2006
; Author: Julio Sanchez
;
; Reference: SevenSeg Circuit and Board
;
; Description:
; Test program for reading four toggle switches and
; displaying the represented hex number on seven-segment
; LED. Also contains a pushbutton switch to activate a
; piezo buzzer. Switches are wired active low.
;
; Switches used in __config directive:
;    _CP_ON         Code protection ON/OFF
; *  _CP_OFF
; *  _PWRTE_ON      Power-up timer ON/OFF
;    _PWRTE_OFF
;    _WDT_ON        Watchdog timer ON/OFF
; *  _WDT_OFF
;    _LP_OSC        Low power crystal occilator
; *  _XT_OSC        External parallel resonator/crystal oscillator

;    _HS_OSC        High speed crystal resonator (8 to 10 MHz)
;                   Resonator: Murate Erie CSA8.00MG = 8 MHz
;    _RC_OSC        Resistor/capacitor oscillator (simplest, 20%
error)
;  |
;  |_____ * indicates setup values
;
;=========================
; setup and configuration
;=========================
        processor 16f84A
        include   <p16f84A.inc>
        __config  _XT_OSC & _WDT_OFF & _PWRTE_ON & _CP_OFF
;
;========================================================
;                  constant definitions
;              (per circuit wiring diagram)
;========================================================
#define Pb_sw 4 ; Port A line 4 to push button switch
;
;========================================================
;                  PIC register equates
;========================================================
Porta    equ     0x05
Portb    equ     0x06
;=============================
```

```
;       local variables
;==============================
        cblock  0x0c            ; Start of block
        J                       ; counter J
        K                       ; counter K
        endc
;===============================================================
;                       program
;===============================================================
        org     0               ; start at address 0
        goto    main
;
; Space for interrupt handlers
        org     0x08

main:
; Port A. Five low-order lines set for input
        movlw   B'00011111'     ; w = 00011111 binary
        tris    porta           ; port A (lines 0 to 4) to
input
; Port B. All eight lines for output
        movlw   B'00000000'     ; w := 00000000 binary
        tris    portb           ; port B to output
;================================
; Pushbutton switch processing
;================================
pbutton:
; Push button switch on demo board is wired to port A bit 4
; Switch logic is active low
        btfss   porta,Pb_sw     ; Test and skip if switch bit
                                ; set
        goto    buzzit          ; Buzz if switch ON,
; At this point port A bit 4 is set (switch is off)
        call    buzoff          ; Buzzer off
        goto    readdip         ; Read DIP switches
buzzit:
        call    buzon           ; Turn on buzzer
        goto    pbutton
;============================
;   dip switch monitoring
;============================
readdip:
; Read port A switches
        movf    porta,w         ; Port A bits to w
; Since board is wired active low then all switch bits
; must be negated. This is done by XORing with 1-bits
        xorlw   b'11111111'     ; Invert all bits in w
; Mask off 4 high-order bits
```

Supplementary Programs

```
        andlw   b'00001111'             ; And with mask
; At this point the w register contains a 4-bit value
; in the range 0 to 0xf. Use this value (in w) to
; obtain seven-segment display code
        call    segment
        movwf   portb                   ; Display switch bits
        goto    pbutton
;=================================
;   routine to returns 7-segment
;           codes
;=================================
segment:
        addwf   PCL,f           ; PCL is program counter latch
        retlw   0x3f            ; 0 code
        retlw   0x06            ; 1
        retlw   0x5b            ; 2
        retlw   0x4f            ; 3
        retlw   0x66            ; 4
        retlw   0x6d            ; 5
        retlw   0x7d            ; 6
        retlw   0x07            ; 7
        retlw   0x7f            ; 8
        retlw   0x6f            ; 9
        retlw   0x77            ; A
        retlw   0x7c            ; B
        retlw   0x39            ; C
        retlw   0x5b            ; D
        retlw   0x79            ; E
        retlw   0x71            ; F
        retlw   0x7f            ; Just in case all on

;=============================
;   piezo buzzer ON
;=============================
; Routine to turn on piezo buzzer on port B bit 7
buzon:
        bsf     portb,7         ; Tune on bit 7, port B
        return
;
;=============================
;   piezo buzzer OFF
;=============================
; Routine to turn off piezo buzzer on port B bit 7
buzoff:
        bcf     portb,7         ; Bit 7 port b clear
        return
;=============================
;   long delay sub-routine
```

```
;       (for code testing)
;==============================
long_delay
                movlw   D'200'  ; w = 200 decimal
                movwf   J               ; J = w
jloop:  movwf   K               ; K = w
kloop:  decfsz  K,f             ; K = K-1, skip next if zero
                goto    kloop
                decfsz  J,f                     ; J = J-1, skip next if zero
                goto    jloop
                return
                end
```

```
;   File name: TestStr.asm
;   Date: April 19, 2006
;   Author: Julio Sanchez
;
;   Description:
;   Program to test sending strings to LCD memory directly
;   Program uses delay loops for interface timing.
;   WARNING:
;   Code assumes 4Mhz clock. Delay routines must be
;   edited for faster clock

;   Displays: Minnesota State, Mankato
;
;===========================
;         switches
;===========================
; Switches used in __config directive:
;   _CP_ON         Code protection ON/OFF
; * _CP_OFF
; * _PWRTE_ON      Power-up timer ON/OFF
;   _PWRTE_OFF
;   _WDT_ON        Watchdog timer ON/OFF
; * _WDT_OFF
;   _LP_OSC        Low power crystal occilator
; * _XT_OSC        External parallel resonator/crystal oscillator

;   _HS_OSC        High speed crystal resonator (8 to 10 MHz)
;                  Resonator: Murate Erie CSA8.00MG = 8 MHz
;   _RC_OSC        Resistor/capacitor oscillator (simplest, 20%
error)
;  |
;  |_____ * indicates setup values

;=========================
; setup and configuration
;=========================
        processor 16f84A
        include    <p16f84A.inc>
        __config   _XT_OSC & _WDT_OFF & _PWRTE_ON & _CP_OFF

;========================================================
;                constant definitions
;   for PIC-to-LCD pin wiring and LCD line addresses
;========================================================
#define E_line  1              ; |
#define RS_line 2              ; | -- from wiring diagram
#define RW_line 3              ; |
; LCD line addresses (from LCD data sheet)
```

```
#define LCD_1 0x80         ; First LCD line constant
#define LCD_2 0xc0         ; Second LCD line constant
; Note: The constants that define the LCD display line
;       addresses have the high-order bit set in
;       order to facilitate the controller command
;
;========================================================
;                   PIC register equates
;========================================================
        porta    equ     0x05
        Portb    equ     0x06
        fsr      equ     0x04
        Status   equ     0x03
        indf     equ     0x00
        z        equ     2
;========================================================
;                 variables in PIC RAM
;========================================================
; Reserve 16 bytes for string buffer
        cblock   0x0c
        strData
        endc
; Leave 16 bytes and Continue with local variables
        cblock   0x1d              ; Start of block
        count1   ; Counter # 1
        count2   ; Counter # 2
        count3   ; Counter # 3
        pic_ad   ; Storage for start of text area
                 ; (labeled strData) in PIC RAM
        J        ; counter J
        K        ; counter K
        index    ; Index into text table (also used
                 ; for auxiliary storage)
        endc
;================================================================
;                         program
;================================================================
                org       0           ; start at address
                goto      main
; Space for interrupt handlers
                org       0x08

main:
        movlw    b'00000000' ; All lines to output
        tris     porta                ; in port A
        tris     portb                ; and port B
        movlw    b'00000000' ; All outputs ports low
        movwf    porta
```

Supplementary Programs

```
                movwf   portb
; Wait and initialize HD44780
                call    delay_5             ; Allow LCD time to initialize itself
                call    initLCD             ; Then do forced initialization
                call    delay_5             ; (Wait probably not necessary)
; Store base address of text buffer in PIC RAM
                movlw   0x0c                ; Start address of text buffer
                movwf   pic_ad              ; to local variable
;=========================
;       test routine
;=========================
; Set DDRAM address to start of first line
                call    line1
; Store characters and send directly
                movlw   'H'
                movwf   portb
                call    pulseE
                movlw   'e'
                movwf   portb
                call    pulseE
                movlw   'l'
                movwf   portb
                call    pulseE
                movlw   'l'
                movwf   portb
                call    pulseE
                movlw   'o'
                movwf   portb
                call    pulseE
                call    delay_5
;======================
;       done!
;======================
loopHere:
                goto    loopHere   ;done

;***************************************************************
;                    INITIALIZE LCD PROCEDURE
;***************************************************************
initLCD
; Initialization for Densitron LCD module as follows:
;       8-bit interface
;   2 display lines of 16 characters each
;   cursor on
;   left-to-right increment
```

```
;       cursor shift right
;       no display shift
;*********************|
;       COMMAND MODE      |
;*********************|
        bcf         porta,E_line     ; E line low
        bcf         porta,RS_line    ; RS line low for command
        bcf         porta,RW_line    ; Write mode
        call        delay_125                     ;delay 125 microseconds
;*********************|
;       FUNCTION SET     |
;*********************|
        movlw       0x38      ; 0 0 1 1 1 0 0 0 (FUNCTION SET)
                              ;         | | | |__ font select:
                              ;         | | |    1 = 5x10 in 1/8 or 1/11
                              ;         | | |    0 = 1/16 dc
                              ;         | | |___ Duty cycle select
                              ;         | |     0 = 1/8 or 1/11
                              ;         | |     1 = 1/16
                              ;         | |___ Interface width
                              ;         |     0 = 4 bits
                              ;         |     1 = 8 bits
                              ;         |___ FUNCTION SET COMMAND
        movwf       portb     ;0011 1000
        call        pulseE    ;pulseE and delay

;*********************|
;       DISPLAY OFF       |
;*********************|
        movlw       0x08      ; 0 0 0 0 1 0 0 0 (DISPLAY ON/OFF)
                              ;         | | | |___ Blink character
                              ;         | | |     1 = on, 0 = off
                              ;         | | |___ Cursor on/off
                              ;         | |     1 = on, 0 = off
                              ;         | |___ Display on/off
                              ;         |     1 = on, 0 = off
                              ;         |___ COMMAND BIT

        movwf       portb
        call        pulseE    ;pulseE and delay

;*********************|
; DISPLAY AND CURSOR ON |
;*********************|
        movlw       0x0e      ; 0 0 0 0 1 1 1 0 (DISPLAY ON/OFF)
                              ;         | | | |___ Blink character
                              ;         | | |     1 = on, 0 = off
```

Supplementary Programs

```
                            ;            | | |___ Cursor on/off
                            ;            | |      1 = on, 0 = off
                            ;            | |___ Display on/off
                            ;            |      1 = on, 0 = off
                            ;            |___ COMMAND BIT
        movwf   portb
        call    pulseE      ;pulseE and delay

;*********************|
;    ENTRY MODE SET   |
;*********************|
        movlw   0x06        ; 0 0 0 0 0 1 1 0 (ENTRY MODE SET)
                            ;           | | |___ display shift
                            ;           | |      1 = shift
                            ;           | |      0 = no shift
                            ;           | |___ cursor increment
                            ;           |      1 = left-to-right
                            ;           |      0 = right-to-left
                            ;           |___ COMMAND BIT
        movwf   portb       ;00000110
        call    pulseE

;*********************|
; CURSOR/DISPLAY SHIFT |
;*********************|
        movlw   0x14        ; 0 0 0 1 0 1 0 0 (CURSOR/DISPLAY
                            ;                   SHIFT)
                            ;         | | | |_|___ don't care
                            ;         | |_|__ cursor/display shift
                            ;         |         00 = cursor shift left
                            ;         |         01 = cursor shift right
                            ;         |         10 = cursor and display
                            ;         |              shifted left
                            ;         |         11 = cursor and display
                            ;         |              shifted right
                            ;         |___ COMMAND BIT
        movwf   portb       ;0001 1111
        call    pulseE

;*********************|
;    CLEAR DISPLAY    |
;*********************|
        movlw   0x01        ; 0 0 0 0 0 0 0 1 (CLEAR DISPLAY)
                            ;                         |___
COMMAND BIT
        movwf   portb       ;0000 0001
;
        call    pulseE
```

```
           call     delay_5   ;delay 5 milliseconds after init
           return
;***************************************************************
;                  DELAY AND PULSE PROCEDURES
;***************************************************************
;=======================
;   Procedure to delay
;      42 microseconds
;=======================
delay_125
           movlw    D'42'     ; Repeat 42 machine cycles
           movwf    count1    ; Store value in counter
repeat
           decfsz   count1,f  ; Decrement counter
           goto     repeat    ; Continue if not 0
           return             ; End of delay
;---------------------------------------------------------------
;=======================
;   Procedure to delay
;      5 milliseconds
;=======================
delay_5
           movlw    D'41'     ; Counter = 41
           movwf    count2    ; Store in variable
delay
           call     delay_125          ; Delay
           decfsz   count2,f           ; 40 times = 5 milliseconds
           goto     delay
           return                      ; End of delay
;=======================
;       pulseE E line
;=======================
pulseE
           bsf      porta,E_line       ;pulse E line
           bcf      porta,E_line
           call     delay_125          ;delay 125 microseconds
           return
;=============================
;    long delay sub-routine
;       (for debugging)
;=============================
long_delay
           movlw    D'200'    ; w = 200 decimal
           movwf    J         ; J = w
jloop:     movwf    K         ; K = w
kloop:     decfsz   K,f       ; K = K-1, skip next if zero
                    goto      kloop
           decfsz   J,f       ; J = J-1, skip next if zero
```

Supplementary Programs

```
                goto    jloop
                return
;==============================
;   LCD display procedure
;==============================
; Sends 16 characters from PIC buffer with address stored
; in variable pic_ad to LCD line previously selected
display16:
; Set up for data
                bcf     porta,E_line        ; E line low
                bsf     porta,RS_line       ; RS line low for control
                call    delay_125           ; Delay
; Set up counter for 16 characters
                movlw   D'16'               ; Counter = 16
                movwf   count3
; Get display address from local variable pic_ad
                movf    pic_ad,w ; First display RAM address to W
                movwf   fsr      ; W to FSR
getchar:
                movf    indf,w   ; get character from display RAM
                                 ; location pointed to by file select
                                 ; register
                movwf   portb
                call    pulseE   ;send data to display
; Test for 16 characters displayed
                decfsz  count3,f            ; Decrement counter
                goto    nextchar            ; Skipped if done
                return
nextchar:
                incf    fsr,f               ; Bump pointer
                goto    getchar
;========================
;    blank buffer
;========================
; Procedure to store 16 blank characters in PIC RAM
; buffer starting at address stored in the variable
; pic_ad
blank16:
                movlw   D'16'    ; Setup counter
                movwf   count1
                movf    pic_ad,w ; First PIC RAM address
                movwf   fsr      ; Indexed addressing
                movlw   0x20     ; ASCII space character
storeit:
                movwf   indf     ; Store blank character in PIC RAM
                                 ; buffer using fsr register
                decfsz  count1,f ; Done?
                goto    incfsr              ; no
```

```
            return                  ; yes
incfsr:
        incf    fsr,f               ; Bump fsr to next buffer
space
        goto    storeit
;=========================
; Set address register
;    to LCD line 1
;=========================
; ON ENTRY:
;         Address of LCD line 1 in constant LCD_1
line1:
        bcf     porta,E_line        ; E line low
        bcf     porta,RS_line       ; RS line low, set up for
                                    ; control
        call    delay_125           ; delay 125 microseconds
; Set to second display line
        movlw   LCD_1               ; Address and command bit
        movwf   portb
        call    pulseE              ; Pulse and delay
; Set RS line for data
        bsf     porta,RS_line       ; Setup for data
        call    delay_125           ; Delay
        return
;=========================
; Set address register
;    to LCD line 2
;=========================
; ON ENTRY:
;         Address of LCD line 2 in constant LCD_2
line2:
        bcf     porta,E_line        ; E line low
        bcf     porta,RS_line       ; RS line low, setup for
                                    ; control
        call    delay_125           ; delay
; Set to second display line
        movlw   LCD_2               ; Address with high-bit set
        movwf   portb
        call    pulseE              ; Pulse and delay
; Set RS line for data
        bsf     porta,RS_line       ; RS = 1 for data
        call    delay_125           ; delay
        return


;================================
;   first text string procedure
;================================
storeMSU:
```

```
; Procedure to store in PIC RAM buffer the message
; contained in the code area labeled msg1
; ON ENTRY:
;          variable pic_ad holds address of text buffer
;          in PIC RAM
;          w register hold offset into storage area
;          msg1 is routine that returns the string characters
;          and a zero terminator
;          index is local variable that hold offset into
;          text table. This variable is also used for
;          temporary storage of offset into buffer
; ON EXIT:
;          Text message stored in buffer
;
; Store offset into text buffer (passed in the w register)
; in temporary variable
         movwf   index            ; Store w in index
; Store base address of text buffer in fsr
         movf    pic_ad,w ; first display RAM address to W
         addwf   index,w          ; Add offset to address
         movwf   fsr              ; W to FSR
; Initialize index for text string access
         movlw   0        ; Start at 0
         movwf   index    ; Store index in variable
; w still = 0
get_msg_char:
         call    msg1     ; Get character from table
; Test for zero terminator
         andlw   0x0ff
         btfsc   status,z ; Test zero flag
         goto    endstr1  ; End of string
; ASSERT: valid string character in w
;         store character in text buffer (by fsr)
         movwf   indf     ; store in buffer by fsr
         incf    fsr,f    ; increment buffer pointer
; Restore table character counter from variable
         movf    index,w  ; Get value into w
         addlw   1        ; Bump to next character
         movwf   index    ; Store table index in variable
         goto    get_msg_char    ; Continue
endstr1:
         return

; Routine for returning message stored in program area
msg1:
         addwf   PCL,f            ; Access table
         retlw   'M'
         retlw   'i'
```

```
        retlw    'n'
        retlw    'n'
        retlw    'e'
        retlw    's'
        retlw    'o'
        retlw    't'
        retlw    'a'
        retlw    0

;=================================
;   second text string procedure
;=================================
storeUniv:
; Processing identical to procedure StoreMSU
        movwf    index    ; Store w in index
; Store base address of text buffer in fsr
        movf     pic_ad,0 ; first display RAM address to W
        addwf    index,0  ; Add offset to address
        movwf    fsr      ; W to FSR
; Initialize index for text string access
        movlw    0        ; Start at 0
        movwf    index    ; Store index in variable
; w still = 0
get_msg_char2:
        call     msg2     ; Get character from table
; Test for zero terminator
        andlw    0x0ff
        btfsc    status,z ; Test zero flag
        goto     endstr2  ; End of string
; ASSERT: valid string character in w
;         store character in text buffer (by fsr)
        movwf    indf     ; Store in buffer by fsr
        incf     fsr,f    ; Increment buffer pointer
; Restore table character counter from variable
        movf     index,w  ; Get value into w
        addlw    1        ; Bump to next character
        movwf    index    ; Store table index in variable
        goto     get_msg_char2   ; Continue
endstr2:
        return

; Routine for returning message stored in program area
msg2:
        addwf    PCL,f            ; Access table
        retlw    'S'
        retlw    't'
        retlw    'a'
        retlw    't'
```

```
        retlw    'e'
        retlw    ','
        retlw    0x20
        retlw    'M'
        retlw    'a'
        retlw    'n'
        retlw    'k'
        retlw    'a'
        retlw    't'
        retlw    'o'
        retlw    0

        end
```

```
; File: TestDemo1.asm
; Date: June 2, 2006
; Author: Julio Sanchez
; Processor: 16F84A
;
; Description:
; Program to exercise the demonstration circuit and board
; number 1
;============================
;         switches
;============================
; Switches used in __config directive:
;    _CP_ON        Code protection ON/OFF
; *  _CP_OFF
; *  _PWRTE_ON     Power-up timer ON/OFF
;    _PWRTE_OFF
;    _WDT_ON       Watchdog timer ON/OFF
; *  _WDT_OFF
;    _LP_OSC       Low power crystal occilator
; *  _XT_OSC       External parallel resonator/crystal
;                  oscillator
;    _HS_OSC       High speed crystal resonator (8 to 10 MHz)
;                  Resonator: Murate Erie CSA8.00MG = 8 MHz
;    _RC_OSC       Resistor/capacitor oscillator
; |
; |_____ * indicates setup values

        processor 16f84A
        include   <p16f84A.inc>
        __config  _XT_OSC & _WDT_OFF & _PWRTE_ON & _CP_OFF
;=======================================================
;                      constants
;=======================================================
;#define dummy    100
;=======================================================
;                 variables in PIC RAM
;=======================================================

; with local variables
        cblock  0x0c        ; Start of block
;       hexDig              ; Hex digit counter
        count1              ; Counter # 1
        j                   ; counter J
        k                   ; counter K
        endc


;=======================================================
;                      P R O G R A M
```

```
;================================================
          org     0              ; start at address 0
          goto    main
;
; Space for interrupt handlers
          org     0x08
main:
; Port A (5 lob) for input
          movlw   B'00011111'    ; w := 00001111 binary
          tris    PORTA          ; port A (lines 0 to 4) to
input
; Port bit (8 lines) for output
          movlw   B'00000000'    ; w := 00000000 binary
          tris    PORTB          ; port B to output
;=============================
; Pushbutton switch processing
;=============================
pbutton:
; Push button switch on demo board is wired to RA4
; Switch logic is active low
          btfss   PORTA,4        ; Test and skip if bit is set
          goto    buzzit         ; Buzz if switch ON
; At this point port A bit 4 is set (switch is off)
          call    buzoff         ; Buzzer off
          goto    readdip        ; Read DIP switches
buzzit:
          call    buzon          ; Turn on buzzer
          goto    pbutton
;=============================
;     DIP switch processing
;=============================
; Read all bits of port A
readdip:
          movf    PORTA,w        ; Port A bits to w
; If board uses active low then all switch bits must be negated
; This is done by XORing with 1-bits
          xorlw   b'11111111'    ; Invert all bits in w
; Eliminate all 4 high order bits (just in case)
          andlw   b'00001111'              ; And with mask
; Get digit into w
          call    segment        ; get digit code
          movwf   PORTB          ; Display digit
          call    delay          ; Give time
; Update digit and loop counter
          goto    pbutton

;******************************
;   7-segment table of hex codes
```

```
;*****************************
segment:
        addwf   PCL,f       ; PCL is program counter latch
        retlw   0x3f        ; 0 code
        retlw   0x06        ; 1
        retlw   0x5b        ; 2
        retlw   0x4f        ; 3
        retlw   0x66        ; 4
        retlw   0x6d        ; 5
        retlw   0x7d        ; 6
        retlw   0x07        ; 7
        retlw   0x7f        ; 8
        retlw   0x6f        ; 9
        retlw   0x77        ; A
        retlw   0x7c        ; B
        retlw   0x39        ; C
        retlw   0x5b        ; D
        retlw   0x79        ; E
        retlw   0x71        ; F
        retlw   0x7f        ; Just in case all on

;***************************
;   piezo buzzer ON
;***************************
; Routine to turn on piezo buzzer on port B bit 7
buzon:
        bsf     PORTB,7     ; Tune on bit 7, port B
        return

;***************************
;   piezo buzzer OFF
;***************************
; Routine to turn off piezo buzzer on port B bit 7
buzoff:
        bcf     PORTB,7     ; Bit 7 port b clear
        return
;================================
;       delay sub-routine
;================================
delay:
        movlw   .200        ; w = 200 decimal
        movwf   j           ; j = w
jloop:
        movwf   k           ; k = w
kloop:
        decfsz  k,f         ; k = k-1, skip next if zero
        goto    kloop
        decfsz  j,f         ; j = j-1, skip next if zero
```

```
        goto    jloop
        return

        end
```

```
; File: Timer0.ASM
; Date: April 27, 2006
; Author: Julio Sanchez
; Processor: a6F84A
;
; Description:
; Program to demonstrate programming of the 16F84A
; TIMER0 module. Program flashes eight LEDs in sequence
; counting from 0 to 0xff. Timer0 is used to delay
; the count.
;============================
;          switches
;============================
; Switches used in __config directive:
;    _CP_ON          Code protection ON/OFF
; *  _CP_OFF
; *  _PWRTE_ON       Power-up timer ON/OFF
;    _PWRTE_OFF
;    _WDT_ON         Watchdog timer ON/OFF
; *  _WDT_OFF
;    _LP_OSC         Low power crystal occilator
; *  _XT_OSC         External parallel resonator/crystal oscillator

;    _HS_OSC         High speed crystal resonator (8 to 10 MHz)
;                    Resonator: Murate Erie CSA8.00MG = 8 MHz
;    _RC_OSC         Resistor/capacitor oscillator (simplest, 20%
error)
; |
; |____ * indicates setup values

        processor 16f84A
        include   <p16f84A.inc>
        __config  _XT_OSC & _WDT_OFF & _PWRTE_ON & _CP_OFF
;=======================================================
;              variables in PIC RAM
;=======================================================
; None in this application
;
;=======================================================
;              m a i n   p r o g r a m
;=======================================================
        org     0           ; start at address 0
        goto    main
;
;============================
;      interrupt handler
;============================
        org             0x08
```

```
;===============================
;       main program
;===============================
main:
; Clear the watchdog timer and reset prescaler
          clrwdt
; Set up the OPTION register bit map
          movlw     b'11010111'
;    7   6   5   4   3   2   1   0  <= OPTION bits
;    |   |   |   |   |   |___|___|_____ PS2-PS0 (prescaler bits)
;    |   |   |   |   |                  Values for Timer0
;    |   |   |   |   |                  000 = 1:2     001 = 1:4
;    |   |   |   |   |                  010 = 1:8     011 = 1:16
;    |   |   |   |   |                  100 = 1:32    101 = 1:64
;    |   |   |   |   |                  110 = 1:128  *111 = 1:256
;    |   |   |   |   |_____ PSA (prescaler assign)
;    |   |   |   |                      1 = to WDT
;    |   |   |   |                     *0 = to Timer0
;    |   |   |   |_____ TOSE (Timer0 edge select)
;    |   |   |                          0 = increment on low-to-high
;    |   |   |                         *1 = increment in high-to-low
;    |   |   |_____ TOCS (TMR0 clock source)
;    |   |                             *0 = internal clock
;    |   |                              1 = RA4/TOCKI bit source
;    |   |_____ INTEDG (Edge select)
;    |                                 *0 = falling edge
;    |_____ RBPU (Pullup enable)
;                                      0 = enabled
;                                     *1 = disabled
          option
; Setup ports
          movlw     0x00            ; Set port B to output
          tris      PORTB
          clrf      PORTB           ; All port B to 0
; Port A is not used in this program
mloop:
          incf      PORTB,f         ; Add 1 to register value
          call      TM0delay
          goto      mloop
;*****************************
;     delay sub-routine
;         uses Timer0
;*****************************
TM0delay:
; Initialize the timer register
          clrf      TMR0            ; Clear SFR for Timer0
; Routine tests the value in the TMR0 register by
; subtracting 0xff from the value in TMR0. The zero flag
```

```
; is set if TMR0 = 0xff
cycle:
        movf    TMR0,w              ; Timer to w
; w has TMR0 register value
        sublw   0xff                ; Subtract max value
; Zero flag is set if value in TMR0 = 0xff
        btfss   STATUS,Z ; Test for zero
        goto    cycle               ; Repeat
        return

                End
```

```
; File: TimerTst.ASM
; Date: April 27, 2006
; Author: Julio Sanchez
;
; Description:
; Using the timer to generate a signal at 1 Mhz
;============================
;          switches
;============================
; Switches used in __config directive:
;    _CP_ON         Code protection ON/OFF
; *  _CP_OFF
; *  _PWRTE_ON      Power-up timer ON/OFF
;    _PWRTE_OFF
;    _WDT_ON        Watchdog timer ON/OFF
; *  _WDT_OFF
;    _LP_OSC        Low power crystal occilator
; *  _XT_OSC        External parallel resonator/crystal oscillator
;
;    _HS_OSC        High speed crystal resonator (8 to 10 MHz)
;                   Resonator: Murate Erie CSA8.00MG = 8 MHz
;    _RC_OSC        Resistor/capacitor oscillator (simplest, 20%
;                   error)
; |
; |_____ * indicates setup values

        processor 16f84A
        include   <p16f84A.inc>
        __config  _XT_OSC & _WDT_OFF & _PWRTE_ON & _CP_OFF

;========================================================
;                  PIC register equates
;========================================================
        porta    equ     0x05
        Portb    equ     0x06
        Status   equ     0x03
        z        equ     0x02
        tmr0     equ     0x01
;
;========================================================
;                  variables in PIC RAM
;========================================================
; Local variables
        cblock   0x0d        ; Start of block
        J                    ; counter J
        K                    ; counter K
        countL               ; Auxiliary counter
        countH               ; ISR counter
```

```
                endc
;===========================================================
;                 m a i n   p r o g r a m
;===========================================================
          org       0               ; start at address 0
          goto      main
;
;============================
;     interrupt handler
;============================
          org                 0x04
;         goto      IntServ
;============================
;       main program
;============================
main:
; Clear the watchdog timer and reset prescaler
          clrwdt
; Set up the OPTION register bit map
          movlw     b'11010011'
;     7   6   5   4   3   2   1   0 <= OPTION bits
;     |   |   |   |   |   |__|__|_____ PS2-PS0 (prescaler bits)
;     |   |   |   |   |                 Values for Timer0
;     |   |   |   |   |                 000 = 1:2    001 = 1:4
;     |   |   |   |   |                 010 = 1:8    011 = 1:16
;     |   |   |   |   |                 100 = 1:32   101 = 1:64
;     |   |   |   |   |                 110 = 1:128 *111 = 1:256
;     |   |   |   |   |_____ PSA (prescaler assign)
;     |   |   |   |                     1 = to WDT
;     |   |   |   |                    *0 = to Timer0
;     |   |   |   |_____ TOSE (Timer0 edge select)
;     |   |   |                         0 = increment on low-to-high
;     |   |   |                        *1 = increment in high-to-low
;     |   |   |_____ TOCS (TMR0 clock source)
;     |   |                            *0 = internal clock
;     |   |                             1 = RA4/TOCKI bit source
;     |   |_____ INTEDG (Edge select)
;     |                                *0 = falling edge
;     |_____ RBPU (Pullup enable)
;                                       0 = enabled
;                                      *1 = disabled
          option
; Setup ports
          movlw     0x00            ; Set port B to output
          tris      portb
          clrf      portb           ; All port B to 0
; Port A is not used in this program
mloop:
```

```
            bsf             portb,0
            call    TM0delay
            bcf             portb,0
            call    TM0delay
            goto    mloop
;*****************************
;     delay sub-routine
;         uses Timer0
;*****************************
TM0delay:
; Initialize the timer register
            clrf    tmr0    ; Clear SFR for Timer0
; Routine tests the value in the tmr0 register by
; xoring with a mask of all ones. The operation sets
; the zero flag if tmr0 is zero.
cycle:
            movf    tmr0,w          ; Timer to w
; w has tmr0 register value
            sublw   0xff            ; Subtract max value
; Zero flag is set if value in tmr0 = 0xff
            btfss   status,z ; Test for zero
            goto    cycle           ; Repeat
            return

            end
```

```
; File name: TTYUsart.asm
; Last update: May, 2006
; Author: Julio Sanchez
; Processor: 16F84A
;
; Description:
; Program to emulate USART operation in PIC code. Uses
; PIC-to-LCD interface. Display has 2 lines, each with
; 16 characters.
; Program operation:
; Characters received from the RS232 line are displayed on
; the LCD. LCD lines scroll automatically. A pushbutton
; activates the send operation by transmitting the text
; string: Ready- which is also displayed on the LCD.
;
; Program communications and LCD parameters are stored in
; #define statements. These statements can be edited to
; accommodate a different set-up. Program uses delay loops
; for interface timing.
;
; WARNING:
; Code assumes 4Mhz clock. Delay routines must be
; edited for faster clock
;
; BAUD RATE CALCULATIONS:
; A 4Mhz clock oscillator has a clock frequency of 1 Mhz:
; Since the baud rate is the number of clock cycles per
; second, for a 4Mhz clock it is:
;                   1
; bit time = ------ sec. = 208.33 microseconds
;                 4,800
; Calculating one half the baud rate allows resetting the
; clock from the edge to the center of a time pulse:
;
;            |<======== falling edge of start bit
;            |         |<======== center of bit time
;          >|          |< one-half baud rate
;            |          |
;_____.          |         ._____.
;            |_____|         |_____
;                208/2 = 104
; The PIC clock counts up from 0 to 255. So to implement
; a 104 microsecond delay we must start counting at
; clock beat:
;                   255 - 104 = 151
; plus one microsecond for movlw instruction used to
; initialize the clock:
;                   151 + 1 = 152
```

```
; For one full baud rate delay:
;                 255 - 208 = 47 + 1 = 48
; The following two constants are stored in #define
; statements:
;                 halfBaud = 152
;                 fullBaud = 48
; Setting the prescaler to TMR0 reduces the baud rate
; to one-half. Other prescaler values will reduce the
; baud rate accordingly.
;
; Wiring diagram:
;    RB4-RB7 ===> LCD data lines 4 to 7 (output)
;    RB0 =======> MAX202 T2in line (output)
;    RA0 =======> MAX202 R2out line (input)
;    RA1 =======> LCD E line (output)
;    RA2 =======> LCD RS line (output)
;    RA3 =======> LCD R/W line (output - not used)
;    RA4 =======> Pushbutton switch 1
;                 (input - active low)
;
;===========================
;        switches
;===========================
; Switches used in __config directive:
;   _CP_ON         Code protection ON/OFF
; * _CP_OFF
; * _PWRTE_ON     Power-up timer ON/OFF
;   _PWRTE_OFF
;   _WDT_ON        Watchdog timer ON/OFF
; * _WDT_OFF
;   _LP_OSC        Low power crystal occilator
; * _XT_OSC        External parallel resonator/crystal oscillator

;   _HS_OSC        High speed crystal resonator (8 to 10 MHz)
;                  Resonator: Murate Erie CSA8.00MG = 8 MHz
;   _RC_OSC        Resistor/capacitor oscillator
;   |              (simplest, 20% error)
;   |
;   |_____ * indicates setup values presently selected

;=========================
; setup and configuration
;=========================
        processor 16f84A
        include   <p16f84A.inc>
        __config  _XT_OSC & _WDT_OFF & _PWRTE_ON & _CP_OFF


;================================================================
```

```
;                       M A C R O S
;===============================================================
; Macros to select the register banks
Bank0    MACRO                           ; Select RAM bank 0
         bcf       STATUS,RP0
         ENDM

Bank1    MACRO                           ; Select RAM bank 1
         bsf       STATUS,RP0
         ENDM
;===============================================================
;                   constant definitions
;       for PIC-to-LCD pin wiring and LCD line addresses
;===============================================================
#define E_line 1            ; |
#define RS_line 2           ; | -- from wiring diagram
#define RW_line 3           ; |
; LCD line addresses (from LCD data sheet)
#define LCD_1 0x80          ; First LCD line constant
#define LCD_2 0xc0          ; Second LCD line constant
#define LCDlimit .16; Number of characters per line
; 4800 baud clock countdown values
; Code reduces rate to 2400 baud by entering a minimal
; prescaler to TRM0
#define halfBaud .152  ; For one-half bit time
#define fullBaud .48   ; For one full bit time
;
; Note: The constants that define the LCD display line
;       addresses have the high-order bit set in
;       order to facilitate the controller command
;
;========================================================
;            PIC register and flag equates
;========================================================
z        equ      2         ; Zero flag
c        equ      0         ; Carry flag
;========================================================
;          buffer and variables in PIC RAM
;========================================================
; Create a 16-byte storage area
         cblock   0x0c      ; Start of first data block
         lineBuf                     ; buffer for text storage
         endc
; Leave 16 bytes and Continue with local variables
         cblock   0x1c      ; Second data block
         count1             ; Counter # 1
         count2             ; Counter # 2
         J                           ; counter J
```

Supplementary Programs

```
            K                ; counter K
            store1           ; Local temporary storage
            store2           ; Storage # 2
; For LCDscroll procedure
            LCDcount ; Counter for characters per line
            LCDline          ; Current display line (0 or 1)
            bufPtr           ; Buffer pointer
; Variables for serial communications
            tempData ; Temporary storage for bit manipulations
            rcvData          ; Final storage for received character
            bitCount ; Bit counter
            sendData ; Character to send
            endc


;===========================================================
;                m a i n    p r o g r a m
;===========================================================
            org              0              ; start at address
            goto    main
; Space for interrupt handlers
        org     0x08

main:
        Bank1
        movlw   b'00010001' ; Port A lines I/O setup
                            ; RA0 = RS232 input (R2out)
                            ; RA4 = Pushbutton SW # 1
        movwf   TRISA
        movlw   b'00000000' ; Port B lines as follow:
;   RB4-RB7 ===> LCD data lines 4 to 7 (output)
;   RB0 ========> MAX202 T2in line (output)
; RB0 =
        movwf   TRISB
        Bank0
; Clear bits in port A output lines
        bcf     PORTA,1
        bcf     PORTA,2
        bcf     PORTA,3
        movlw   b'00000000' ; All outputs ports low
        movwf   PORTB
; Wait and initialize HD44780
        call    delay_5 ; Allow LCD time to initialize
                        ; itself
        call    delay_5
        call    initLCD ; Then do forced initialization
        call    delay_5 ; Wait again
; Set port B, line 0 high so start bit is detected
        bsf              PORTB,0
```

```
;============================
;   wait for start command
;============================
; Program waits until pushbutton number 1 is pressed
; to continue execution. Pushbutton 1 is active low
; and wired to RA4
pb1Wait:
        btfsc   PORTA,4         ; Test port A, line 4
        goto    pb1Wait         ; Loop if not clear
;============================
;   display and send "Ready-"
;============================
; Set LCD base address
        call    line1
; Initialize system for UART emulation at 2400 baud
        call    initTTY
; Display on LCD and test serial transmission by sending
; the string "Ready-"
        movlw   'R'
        movwf   sendData ; Store in send register
        call    send8    ; Local LCD display procedure
        call    sendTTY  ; Local send procedure
        movlw   'e'
        movwf   sendData ; Store in send register
        call    send8    ; Local LCD display procedure
        call    sendTTY  ; Local send procedure
        movlw   'a'
        movwf   sendData ; Store in send register
        call    send8    ; Local LCD display procedure
        call    sendTTY  ; Local send procedure
        movlw   'd'
        movwf   sendData ; Store in send register
        call    send8    ; Local LCD display procedure
        call    sendTTY  ; Local send procedure
        movlw   'y'
        movwf   sendData ; Store in send register
        call    send8    ; Local LCD display procedure
        call    sendTTY  ; Local send procedure
        movlw   '-'
        movwf   sendData ; Store in send register
        call    send8    ; Local LCD display procedure
        call    sendTTY  ; Local send procedure
; Init character counter and line counter variables for
; LCD line scroll procedure
        movlw   0x06             ; 6 characters already
displayed
        movwf   LCDcount
        clrf    LCDline          ; LCD line counter
```

```
;============================
;     monitor RS232 line
;============================
nextChar:
        call    rcvTTY                  ; Receive character
; Store character in local line buffer using indirect
; addressing
; 16-byte buffer named lineBuf starts at address 0x0c
; Register variable bufPtr holds offset into buffer
        movlw   0x0c            ; Buffer base address
        addwf   bufPtr,w        ; Add pointer in w
        movwf   FSR             ; Value to index register
        movf    rcvData,        ; Character into w
        movwf   INDF            ; Store w in [FSR]
        incf    bufPtr,f        ; Bump pointer
; Send character (still in w)
        call    send8           ; Display it
        call    LCDscroll       ; Scroll display lines
        goto    nextChar        ; Continue


;================================================================
;               initialize LCD for 4-bit mode
;================================================================
initLCD:
; Initialization for Densitron LCD module as follows:
;       4-bit interface
;   2 display lines of 16 characters each
;   cursor on
;   left-to-right increment
;   cursor shift right
;   no display shift
;======================|
;   set command mode   |
;======================|
        bcf     PORTA,E_line    ; E line low
        bcf     PORTA,RS_line   ; RS line low
        bcf     PORTA,RW_line   ; Write mode
        call    delay_125               ; delay 125 microseconds
;**********************|
;       FUNCTION SET   |
;**********************|
        movlw   0x28    ; 0 0 1 0 1 0 0 0 (FUNCTION SET)
                        ;         | | | |__ font select:
                        ;         | | |   1 = 5x10 in 1/8 or 1/11
                        ;         | | |   0 = 1/16 dc
                        ;         | | |___ Duty cycle select
```

```
                            ;        | |       0 = 1/8 or 1/11
                            ;        | |       1 = 1/16
                            ;        | |___ Interface width
                            ;        |         0 = 4 bits
                            ;        |         1 = 8 bits
                            ;        |___ FUNCTION SET COMMAND
        call     send8      ; 4-bit send routine

; Set 4-bit mode command must be repeated
        movlw    0x28
        call     send8

;*********************|
; DISPLAY AND CURSOR ON |
;*********************|
        movlw    0x0e      ; 0 0 0 0 1 1 1 0 (DISPLAY ON/OFF)
                           ;         | | | |___ Blink character
                           ;         | | |        1 = on, 0 = off
                           ;         | | |___ Cursor on/off
                           ;         | |        1 = on, 0 = off
                           ;         | |____ Display on/off
                           ;         |          1 = on, 0 = off
                           ;         |____ COMMAND BIT
        call     send8
;*********************|
;   set entry mode    |
;*********************|
        movlw    0x06      ; 0 0 0 0 0 1 1 0 (ENTRY MODE SET)
                           ;           | | |___ display shift
                           ;           | |       1 = shift
                           ;           | |       0 = no shift
                           ;           | |____ increment mode
                           ;           |        1 = left-to-right
                           ;           |        0 = right-to-left
                           ;           |___ COMMAND BIT
        call     send8

;*********************|
; cursor/display shift |
;*********************|
        movlw    0x14      ; 0 0 0 1 0 1 0 0 (CURSOR/DISPLAY
                           ;                    SHIFT)
                           ;      | | | |_|___ don't care
                           ;      | |_|__ cursor/display shift
                           ;      |         00 = cursor shift left
                           ;      |         01 = cursor shift right
                           ;      |         10 = cursor and display
                           ;      |                shifted left
```

```
                              ;       |         11 = cursor and display
                              ;       |              shifted right
                              ;       |___ COMMAND BIT
         call      send8
;*********************|
;    clear display         |
;*********************|
         movlw     0x01       ; 0 0 0 0 0 0 0 1 (CLEAR DISPLAY)
                              ;                 |___ COMMAND BIT
         call      send8
; Per documentation
         call      delay_5    ; Test for busy
         return


;=====================
;  Procedure to delay
;   42 microseconds
;=====================
delay_125
         movlw     D'42'      ; Repeat 42 machine cycles
         movwf     count1     ; Store value in counter
repeat
         decfsz    count1,f   ; Decrement counter
         goto      repeat     ; Continue if not 0
         return               ; End of delay


;=====================
;  Procedure to delay
;   5 milliseconds
;=====================
delay_5
         movlw     D'41'      ; Counter = 41
         movwf     count2     ; Store in variable
delay
         call      delay_125         ; Delay
         decfsz    count2,f   ; 40 times = 5 milliseconds
         goto      delay
         return               ; End of delay
;=====================
;    pulse E line
;=====================
pulseE
         bsf       PORTA,E_line       ; Pulse E line
         nop
         bcf       PORTA,E_line
         return
;============================
;   long delay sub-routine
```

```
;       (for debugging)
;==============================
long_delay
        movlw   D'200'      ; w = 200 decimal
        movwf   J           ; J = w
jloop:  movwf   K           ; K = w
kloop:  decfsz  K,f         ; K = K-1, skip next if zero
        goto    kloop
        decfsz  J,f         ; J = J-1, skip next if zero
        goto    jloop
        return
;========================
;   send 2 nibbles in
;      4-bit mode
;========================
; Procedure to send two 4-bit values to port B lines
; 7, 6, 5, and 4. High-order nibble is sent first
; ON ENTRY:
;         w register holds 8-bit value to send
send8:
        movwf   store1      ; Save original value
        call    merge4      ; Merge with port B
; Now w has merged byte
        movwf   PORTB       ; w to port B
        call    pulseE      ; Send data to LCD
; High nibble is sent
        movf    store1,w    ; Recover byte into w
        swapf   store1,w    ; Swap nibbles in w
        call    merge4
        movwf   PORTB
        call    pulseE      ; Send data to LCD
        call    delay_125
        return
;=================
;   merge bits
;=================
; Routine to merge the 4 high-order bits of the
; value to send with the contents of port B
; so as to preserve the 4 low-bits in port B
; Logic:
;       AND value with 1111 0000 mask
;       AND port B with 0000 1111 mask
;       Now low nibble in value and high nibble in
;       port B are all 0 bits:
;           value  = vvvv 0000
;           port B = 0000 bbbb
;       OR value and port B resulting in:
;                   vvvv bbbb
```

```
; ON ENTRY:
;       w contain value bits
; ON EXIT:
;       w contains merged bits
merge4:
        andlw   b'11110000'     ; ANDing with 0 clears the
                                ; bit. ANDing with 1 preserves
                                ; the original value
        movwf   store2          ; Save result in variable
        movf    PORTB,w         ; port B to w register
        andlw   b'00001111'     ; Clear high nibble in port b
                                ; and preserve low nibble
        iorwf   store2,w        ; OR two operands in w
        return

;========================
; Set address register
;    to LCD line 1
;========================
; ON ENTRY:
;       Address of LCD line 1 in constant LCD_1
line1:
        bcf     PORTA,E_line    ; E line low
        bcf     PORTA,RS_line   ; RS line low, set up for
                                ; control
        call    delay_5         ; busy?
; Set to second display line
        movlw   LCD_1           ; Address and command bit
        call    send8           ; 4-bit routine
; Set RS line for data
        bsf     PORTA,RS_line   ; Setup for data
        call    delay_5         ; Busy?
; Clear buffer and pointer
        call    blankBuf
        clrf    bufPtr          ; Clear
        return
;========================
; Set address register
;    to LCD line 2
;========================
; ON ENTRY:
;       Address of LCD line 2 in constant LCD_2
line2:
        bcf     PORTA,E_line    ; E line low
        bcf     PORTA,RS_line   ; RS line low, setup for
control
        call    delay_5         ; Busy?
; Set to second display line
```

```
            movlw    LCD_2               ; Address with high-bit set
            call     send8
; Set RS line for data
            bsf      PORTA,RS_line       ; RS = 1 for data
            call     delay_5             ; Busy?
; Clear buffer and pointer
            call     blankBuf
            clrf     bufPtr
            return

;===========================
;    scroll LCD line 2
;===========================
; Procedure to count the number of characters displayed on
; each LCD line. If the number reaches the value in the
; constant LCDlimit, then display is scrolled to the second
; LCD line. If at the end of the second line, then the
; second line is scrolled to the first line and display
; continues at the start of the second line
; reset to the first line.
LCDscroll:
            incf     LCDcount,f          ; Bump counter
; Test for line limit
            movf     LCDcount,w
            sublw    LCDlimit            ; Count minus limit
            btfss    STATUS,z            ; Is count - limit = 0
            goto     scrollExit          ; Go if not at end of line
; At this point the end of the LCD line was reached
; Test if this is also the end of the second line
            movf     LCDline,w
            sublw    0x01                ; Is it line 1?
            btfsc    STATUS,z            ; Is LCDline minus 1 = 0?
            goto     line2End            ; Go if end of second line
; At this point it is the end of the top LCD line
            call     line2               ; Scroll to second line
            clrf     LCDcount            ; Reset counter
            incf     LCDline,f           ; Bump line counter
            goto     scrollExit
; End of second LCD line
line2End:
; Scroll second line to first line. Characters to be
; scrolled are stored in buffer starting at address 0x0c.
; 16 characters are to be moved
; First clear LCD
            call     initLCD
            call     delay_5             ; Make sure not busy
; Set up for data
            bcf      PORTA,E_line        ; E line low
```

```
                bsf     PORTA,RS_line   ; RS line high for data
; Set up counter for 16 characters
                movlw   D'16'           ; Counter = 16
                movwf   count2
; Get address of storage buffer
                movlw   0x0c
                movwf   FSR             ; W to FSR
getchar:
                movf    INDF,w          ; get character from display RAM
                                        ; location pointed to by file select
                                        ; register
                call    send8           ; 4-bit interface routine
; Test for 16 characters displayed
                decfsz  count2,f        ; Decrement counter
                goto    nextchar        ; Skipped if done
; At this point scroll operation has concluded
                clrf    LCDcount        ; Clear counters
; Stay at line 2
                clrf    LCDline
                incf    LCDline,f
                call    line2           ; Set for second line
scrollExit:
                return
nextchar:
                incf    FSR,f           ; Bump pointer
                goto    getchar


;============================
;   clear line buffer
;============================
; Use indirect addressing to store 16 blanks in the
; buffer located at 0x0c
blankBuf:
                Bank0
                movlw   0x0c            ; Pointer to RAM
                movwf   FSR             ; To index register
blank16:
                clrf    INDF            ; Clear memory pointed at by FSR
                incf    FSR,f           ; Bump pointer
                btfss   FSR,4           ; 000x0000 when bit 4 is set
                                        ; count reached 16
                goto    blank16
                return


;================================================================
;                       initialize for TTY
;================================================================
; Procedure to initialize RS232 reception
```

```
; Assumes:
;                           2400 baud
;                           8 data bits
;                           no parity
;                           one stop bit
initTTY:
; First initialize receiver to RS-232 line parameters
; Disable global and peripheral interrupts
;   7  6  5  4  3  2  1  0   <= INTCON bitmap
;   |  ?  |  ?  ?  ?  ?  ?   (? = unrelated bits)
;   |     |_____ Timer0 interrupt on overflow
;   |_____ Global interrupts
            bcf     INTCON,5        ; Disable TMR0 interrupts
            bcf     INTCON,7        ; Disable global interrupts
            clrf    TMR0            ; Reset timer
            clrwdt                  ; Clear WDT for prescaler
                                    ; assign
            Bank1
; Set up the OPTION register bit map
;   7  6  5  4  3  2  1  0   <= OPTION bits
;   1  1  0  1  1  0  0  0   <= setup
;   |  |  |  |  |  |__|__|_____ PS2-PS0 (prescaler bits)
;   |  |  |  |  |              Values for Timer0
;   |  |  |  |  |              *000 = 1:2    001 = 1:4
;   |  |  |  |  |               010 = 1:8    011 = 1:16
;   |  |  |  |  |               100 = 1:32   101 = 1:64
;   |  |  |  |  |               110 = 1:128  111 = 1:256
;   |  |  |  |  |_____ PSA (prescaler assign)
;   |  |  |  |                  1 = to WDT
;   |  |  |  |                 *0 = to Timer0
;   |  |  |  |_____ TOSE (Timer0 edge select)
;   |  |  |                     0 = increment on low-to-high
;   |  |  |                    *1 = increment in high-to-low
;   |  |  |_____ TOCS (TMR0 clock source)
;   |  |                       *0 = internal clock
;   |  |                        1 = RA4/TOCKI bit source
;   |  |_____ INTEDG (Edge select)
;   |                           0 = falling edge
;   |                          *1 = raising edge
;   |_____ RBPU (Pullup enable)
;                               0 = enabled
;                              *1 = disabled
            movlw   b'11010000'     ; set up timer/counter
            movwf   OPTION_REG
            Bank0
            return
;===============================================================
;                    receive character
```

Supplementary Programs

```
;================================================================
; Receive a single character through the serial port.
; Assumes: 4800 baud, 8 data bits, no parity, 1 stop bit.
; Receiving line is Port A, line 0
rcvTTY:
        movlw   0x08            ; Counter for 8 bits
        movwf   bitCount
; The start of character transmission is signaled by
; the sender by setting the line low
startBit:
        btfsc   PORTA,0         ; Test for low on line
        goto    startBit ; Go if not low
;==========================
;  offset to data bit
;==========================
; At this point the receiver has found the falling
; edge of the start bit. It must now wait one and
; one-half the baud rate to synchronize in the center
; of the sender's first data bit
;, as follows:
;       |<========= falling edge of START bit
;       |       |<========== center of start bit
;       |       |           |<====== center of data bit
;       |-----------|-----|
;_____                _____              _____
;       |            |                 |            |      <== SIGNAL
;       -----------               ----------
;       |<-- 208 -->|<104>| <====== ms. for 4800 baud
;
; Clock start count for one-half bit = 255 - 104 = 151
; Clock start count for one full bit =  255 - 208 = 47
; One clock cycle is added for the movwf intruction:
;    clkHalf = 152 (for one-half bit countdown)
;    clkFull = 48 (for one full bit countdown)
        movlw   halfBaud ; Skip one-half bit
        movwf   TMR0     ; Initialize tmr0 and start count
        bcf     INTCON,2        ; Clear overflow flag
;============================
;           start bit
;============================
wait1:
        btfss   INTCON,2        ; Timer count overflow?
        goto    wait1           ; No, keep waiting
; At this point we are at the center of the start bit
        btfsc   PORTA,0         ; Check to see it is still low
        goto    startBit ; No, it is high. False start
; At this point the clock is at the center of the start
; bit. The first data bit must be read one full baud
```

```
; period later
        movlw   fullBaud            ; One full bit delay
        movwf   TMR0                ; Start timer
        bcf     INTCON,2            ; clear tmr0 overflow flag
wait2:
        btfss   INTCON,2            ; End of one full baud period?
        goto    wait2               ; Wait if not end of period
; Timer is now at the center of the first/next data bit
; Timer must be reset immediately so that code will not
; lose synchronization with sender
        movlw   fullBaud ; Skip to next data bit
        movwf   TMR0                ; Restart timer
        bcf     INTCON,2            ; Reset overflow flag
; Now the data bit can be read and stored
        movf    PORTA,w             ; Read port B
        movwf   tempData ; Store in temporary variable
        rrf     tempData,f          ; Rotate bit 0 into carry flag
        rrf     rcvData,f   ; Rotate carry flag into storage
                                    ; register high-order bit
        decfsz  bitCount,f          ; End of data?
        goto    wait2    ; Continue until 8 bits received
;=============================
;           stop bit
;=============================
stopWait:
        btfss   INTCON,2            ; Test time
        goto    stopWait ; Wait
        return                      ; Exit


;===============================================================
;                        send character
;===============================================================
; Procedure to send one character through the RS232 line.
; Assumes: 2400 baud, 8 data bits, no parity, one stop bit
; Sending line is Port B, line 0
; ON ENTRY:
;         variable sendData holds character to send
sendTTY:
        movlw   0x08                ; Init bit counter
        movwf   bitCount
        bcf     PORTB,0             ; Low for start bit
        movlw   fullBaud            ; For one baud space
        movwf   TMR0                ; Start timer
        bcf     INTCON,2            ; Clear timer flag
start2snd:
        btfss   INTCON,2            ; Full baud done?
        goto    start2snd           ; No
        movlw   fullBaud            ; Reset for one full bit
```

```
                              ; period
        movwf   TMR0          ; Start timer
        bcf     INTCON,2      ; Clear flag
; At this point the start bit has been sent
; Data follows
sendOut:
        rrf     sendData,f  ; Rotate bit into carry
        bcf     PORTB,0  ; Assume data bit is 0
        btfsc   STATUS,c ; Test if carry set
        bsf     PORTB,0  ; Change bit to 1 if clear
; Hold bit for 1 baud period
timeBit:
        btfss   INTCON,2      ; Wait for baud period to end
        goto    timeBit       ; Loop if not yet
        movlw   fullBaud      ; Reset timer
        movwf   TMR0          ; Start timer
        bcf     INTCON,2      ; Clear flag
; Test for last bit
        decfsz  bitCount,f    ; Count this bit
        goto    sendOut       ; Continue if not last bit
; Done. Send stop bit
        bsf     PORTB,0       ; High for stop bit
stopBit:
        btfss   INTCON,2      ; Timer done?
        goto    stopBit       ; No
; Set port B line 0 high back again
        bsf     PORTB,0
        call    delay_5       ; And hold
        return

        End
```

```
; File: Watchdog.asm
; Date: May 2, 2006
; Author: Julio Sanchez
;
; Description:
; Program to demonstrate the use of the watchdog timer
; in breaking out of an endless loop.
; A LED on port B, line 1, flashes on and off at 1/2
; second intervals for 20 iterations. At that time the
; program enters an endless loop. The watchdog timer
; times-out and restarts the program
;===========================
;           switches
;===========================
; Switches used in __config directive:
;   _CP_ON          Code protection ON/OFF
; * _CP_OFF
; * _PWRTE_ON       Power-up timer ON/OFF
;   _PWRTE_OFF
; * _WDT_ON         Watchdog timer ON/OFF
;   _WDT_OFF
;   _LP_OSC         Low power crystal occilator
; * _XT_OSC         External parallel resonator/crystal oscillator
;
;   _HS_OSC         High speed crystal resonator (8 to 10 MHz)
;                   Resonator: Murate Erie CSA8.00MG = 8 MHz
;   _RC_OSC         Resistor/capacitor oscillator
; |
; |_____ * indicates setup values

;=========================
; setup and configuration
;=========================
        processor 16f84A
        include    <p16f84A.inc>
        __config   _XT_OSC & _WDT_ON & _PWRTE_ON & _CP_OFF

;====================================================
;                  PIC register equates
;====================================================
porta   equ     0x05
Portb   equ     0x06
status  equ     0x03
z       equ     0x02
tmr0    equ     0x01
;
;====================================================
;               variables in PIC RAM
```

```
;=========================================================
; Local variables
        cblock   0x0d      ; Start of block
        J                  ; counter J
        K                  ; counter K
        count1             ; Auxiliary counter
        count2             ; Second auxiliary counter
        old_w              ; Context saving
        old_status  ; Idem
        endc

;=========================================================
;                m a i n   p r o g r a m
;=========================================================
        org      0         ; start at address 0
        goto     main
;
;=============================
;      interrupt handler
;=============================
        org      0x04
;=============================
;      main program
;=============================
main:
; Setting the prescaler to the watchdog timer following
; the sequence recommended by Microchip
        movlw    b'10010101'       ; Clock source and some
prescaler
        option
        clrf     tmr0              ; Clear timer and prescaler
        movlw    b'10111101' ; WDT, do not change prescale
        option                              ; again
; Reset watchdog timer
        clrwdt
; Final setting of OPTION register
        movlw    b'10111000'
; Set up the OPTION register bit map
;   7  6  5  4  3  2  1  0 <= OPTION bits
;   |  |  |  |  |  |__|__|_____ PS2-PS0 (prescaler bits)
;   |  |  |  |  |              Values for WDT
;   |  |  |  |  |              000 = 1:1    001 = 1:2
;   |  |  |  |  |              010 = 1:4    011 = 1:8
;   |  |  |  |  |              100 = 1:16   101 = 1:32
;   |  |  |  |  |              110 = 1:64 *111 = 1:128
;   |  |  |  |  |_____ PSA (prescaler assign)
;   |  |  |  |                 *1 = to WDT
;   |  |  |  |                  0 = to Timer0
```

```
;       |   |   |   |_____ TOSE (Timer0 edge select)
;       |   |   |                       0 = increment on low-to-high
;       |   |   |                      *1 = increment in high-to-low
;       |   |   |_____ TOCS (TMR0 clock source)
;       |   |                           0 = internal clock
;       |   |                          *1 = RA4/TOCKI bit source
;       |   |_____ INTEDG (Edge select)
;       |                              *0 = falling edge
;       |_____ RBPU (Pullup enable)
;                                       0 = enabled
;                                      *1 = disabled
        option
        movlw   b'00000000'     ; Port B is output
        tris    portb           ; all others are output
        clrf    portb           ; All port B to 0
; Port A is not used by this program
;============================
;     flash LED 20 times
;============================
; Program flashes LED wired to port B, line 2
; 5 times before entering the endless loop
        movlw   D'5'            ; Number of iterations
        movwf   count2          ; To counter
lights:
        movlw   b'00000010'     ; Mask with bit 1 set
        xorwf   portb,f         ; Complement bit 1
        call    long_delay
        call    long_delay
        call    long_delay
        decfsz  count2,f        ; Decrement counter
        goto    lights
        clrwdt                  ; Clear watchdog
;============================
;         endless loop
;============================
endless:
        goto    endless


;============================
;     delay sub-routine
;============================
long_delay
                movlw   D'200'  ; w = 20 decimal
                movwf   J               ; J = w
jloop:  movwf   K               ; K = w
kloop:  decfsz  K,f             ; K = K-1, skip next if zero
                clrwdt
                goto    kloop
```

```
                decfsz  J,f             ; J = J-1, skip next if zero
                clrwdt
                goto    jloop
                return

                end
```

```
; File name: I2CEEP.asm
; Last revision: May 28, 2006
; Author: Julio Sanchez
; Processor: 16F877
;
; Description:
; Receive character data through RS-232 line and store in
; 24LC04B EEPROM IC, using the I2C serial protocol in the
; PIC's MSSP module. Received characters are echoed on
; the second LCD line. When <Enter> key is detected (code
; 0x0d) the text stored in EEPROM memory is retrieved and
; displayed on the LCD. On startup the top LCD line displays
; the prompt: "Receiving:". At that time a message "Rdy- " is
; sent through the serial line so as to test the connection.
;
; Default serial line setting:
;              2400 baud
;              no parity
;              1 stop bit
;              8 character bits
;
; Wiring:
; 24LC04B SDA line is wired to PIC RC4 (MSSP SDA)
; 24LC04B SCL line is wired to PIC RC3 (MSSP SCL)
; 24LC04B A0-A2 and WP lines are not used (GND)
;
; Program to use 4-bit PIC-to-LCD interface.
; Code assumes that LCD is driven by Hitachi HD44780
; controller and PIC 16F977. Display supports two lines
; each one with 20 characters. The length, wiring and base
; address of each display line is stored in #define
; statements. These statements can be edited to accommodate
; a different set-up.
;
; WARNING:
; Code assumes 10 Mhz clock. Delay routines must be
; edited for a different clock. Clock speed also determines
; values for baud rate setting (see spbrgVal constant).
;
;===========================
;       16F877 switches
;===========================
; Switches used in __config directive:
;   _CP_ON           Code protection ON/OFF
; * _CP_OFF
; * _PWRTE_ON        Power-up timer ON/OFF
;   _PWRTE_OFF
;   _BODEN_ON        Brown-out reset enable ON/OFF
```

```
;  *  _BODEN_OFF
;  *  _PWRTE_ON      Power-up timer enable ON/OFF
;     _PWRTE_OFF
;     _WDT_ON        Watchdog timer ON/OFF
;  *  _WDT_OFF
;     _LPV_ON        Low voltage IC programming enable ON/OFF
;  *  _LPV_OFF
;     _CPD_ON        Data EE memory code protection ON/OFF
;  *  _CPD_OFF
; OSCILLATOR CONFIGURATIONS:
;     _LP_OSC        Low power crystal oscillator
;     _XT_OSC        External parallel resonator/crystal oscillator
;
;  *  _HS_OSC        High speed crystal resonator
;     _RC_OSC        Resistor/capacitor oscillator
;  |                 (simplest, 20% error)
;  |
;  |_____ * indicates setup values presently selected

        processor       16f877          ; Define processor
        #include <p16f877.inc>
        __CONFIG _CP_OFF & _WDT_OFF & _BODEN_OFF & _PWRTE_ON &
_HS_OSC & _WDT_OFF & _LVP_OFF & _CPD_OFF

; __CONFIG directive is used to embed configuration data
; within the source file. The labels following the directive
; are located in the corresponding .inc file.
        errorlevel -302
; Suppress bank-related warning
;=============================================================
;                         M A C R O S
;=============================================================
; Macros to select the register banks
Bank0   MACRO                           ; Select RAM bank 0
        bcf     STATUS,RP0
        bcf     STATUS,RP1
        ENDM

Bank1   MACRO                           ; Select RAM bank 1
        bsf     STATUS,RP0
        bcf     STATUS,RP1
        ENDM

Bank2   MACRO                           ; Select RAM bank 2
        bcf     STATUS,RP0
        bsf     STATUS,RP1
        ENDM
```

```
Bank3    MACRO                        ; Select RAM bank 3
         bsf      STATUS,RP0
         bsf      STATUS,RP1
         ENDM
;========================================================
;              constant definitions
;   for PIC-to-LCD pin wiring and LCD line addresses
;========================================================
#define E_line 1             ; |
#define RS_line 0            ; | -- from wiring diagram
#define RW_line 2            ; |
; LCD line addresses (from LCD data sheet)
#define LCD_1 0x80           ; First LCD line constant
#define LCD_2 0xc0           ; Second LCD line constant
#define LCDlimit .20; Number of characters per line
#define spbrgVal .64; For 2400 baud on 10Mhz clock
; Note: The constants that define the LCD display
;       line addresses have the high-order bit set
;       so as to meet the requirements of controller
;       commands.
;========================================================
;        constants for I2C initialization
;========================================================
; I2C connected to 24LC04B EEPROM.
; The MSSP module is in I2C MASTER mode.
#define LC04READ 0xa0        ; I2C value for read control byte
#define LC04WRITE 0xa1       ; I2C value for write control byte


;========================================================
;                General Purpose Variables
;========================================================
; Local variables
; Reserve 20 bytes for string buffer
         cblock   0x20
         strData
         endc
; Other data
         cblock   0x34                ; Start of block
         count1                       ; Counter # 1
         count2                       ; Counter # 2
         count3                       ; Counter # 3
         J                            ; counter J
         K                            ; counter K
         bufAdd
         index
         store1                       ; Local storage
         store2
; For LCDscroll procedure
```

```
        LCDcount ; Counter for characters per line
        LCDline          ; Current display line (0 or 1)
        endc
;===============================
;       Common RAM area
;===============================
; These GPRs can be accessed from any bank.
; 15 bytes are available, from 0x70 to 0x7f
        cblock   0x70
; Communications variables
        newData          ; not 0 if new data received
        ascVal
        errorFlags
; EEPROM-related variables
        EEMemAdd ; EEPROM address to access
        EEByte           ; Data byte to write
        endc


;================================================================
;                       P R O G R A M
;================================================================
        org      0       ; start at address
        goto     main
; Space for interrupt handlers
        org      0x08
main:
; Wiring:
;     LCD data to port D, lines 0 to 7
;     E line -> port E, 1
;     RW line -> port E, 2
;     RS line -> port E, 0
; Set PORTE D and E for output
; First, initialize port B by clearing latches
        clrf     STATUS
        clrf     PORTB
; Select bank 1 to tris port D for output
        Bank1
; Tris port D for output. Port D lines 4 to 7 are wired
; to LCD data lines. Port D lines 0 to 4 are wired to LEDs.
        movlw    B'00000000'
        movwf    TRISD             ; and port D
; By default port A lines are analog. To configure them
; as digital code must set bits 1 and 2 of the ADCON1
; register (in bank 1)
        movlw    0x06              ; binary 0000 0110 is code to
                                   ; make all
port A lines digital
        movwf    ADCON1
```

```
; Port B, lines are wired to keypad switches, as follows:
;    7 6 5 4 3 2 1 0
;    | | | | |_|_|_|_____ switch rows (output)
;    |_|_|_|_____ switch columns (input)
; rows must be defined as output and columns as input
         movlw    b'11110000'
         movwf    TRISB
; Tris port E for output
         movlw    B'00000000'
         movwf    TRISE             ; Tris port E
; Enable port B pullups for switches in OPTION register
         movlw    b'00001000'
         movwf    OPTION_REG
; Back to bank 0
         Bank0
; Initialize serial port for 2400 baud, 8 bits, no parity
; 1 stop
         call     InitSerial
; Test serial transmission by sending "RDY-"
         movlw    'R'
         call     SerialSend
         movlw    'D'
         call     SerialSend
         movlw    'Y'
         call     SerialSend
         movlw    '-'
         call     SerialSend
         movlw    0x20
         call     SerialSend
; Clear all output lines
         movlw    b'00000000'
         movwf    PORTD
         movwf    PORTE
; Wait and initialize HD44780
         call     delay_5   ; Allow LCD time to initialize itself
         call     initLCD   ; Then do forced initialization
         call     delay_5   ; (Wait probably not necessary)
; Clear character counter and line counter variables
         clrf     LCDcount
         clrf     LCDline
; Set display address to start of first LCD line
         call     line1
; Store address of display buffer
         movlw    0x20
         movwf    bufAdd
; Display "Receiving:" message prompt
         call     blank20          ; Clear buffer
         movlw    0x00             ; Offset in buffer
```

```
        call    storeMS1 ; Store message at offset
        call    display20       ; Display message
; Start address of EEPROM
        clrf    EEMemAdd
; Setup for display in second line
        call    line2
        clrf    LCDline
        incf    LCDline,f ; Set scroll control for line 2
; Initialize I2C EEPROM operation
        call    SetupI2C        ; Local procedure
;===============================================================
;          receive serial data, store, and display
;===============================================================
receive:
; Call serial receive procedure
        call    SerialRcv
; HOB of newData register is set if new data
; received
        btfss   newData,7
        goto    scanExit
; At this point new data was received.
        movwf   EEByte          ; Save received character
; Display character on LCD
        movf    EEByte,w ; Recover character
        call    send8           ; Display in LCD
        call    LCDscroll       ; Scroll at end of line
; Store character in EEPROM at location in EEMemAdd
        call    WriteI2C        ; Local procedure
        incf    EEMemAdd,f      ; Bump to next EEPROM
; Check for <Enter> key (0x0d) and execute display function
        movf    EEByte,w ; Recover last received
        sublw   0x0d
        btfsc   STATUS,Z ; Test if <Enter> key
        goto    isEnter         ; Go if <Enter>
; Not <Enter> key, continue processing
scanExit:
        goto    receive         ; Continue
;=============================
;    display EEPROM data
;=============================
; This routine receives control when the <Enter> key is
; received.
; Action:
;       1. Clear LCD
;       2. Output is set to top LCD line
;       3. Characters stored in EEPROM are displayed
;          until 0x0d code is detected
isEnter:
```

```
        call    clearLCD
; Clear character counter and line counter variables
        clrf    LCDcount
        clrf    LCDline
; Read data from EEPROM memory, starting at address 0
; and display on LCD until 0x0d terminator
        call    line1
        clrf    EEMemAdd ; Start at EEPROM 0
readOne:
        call    ReadI2C         ; Get character
; Store character
        movwf   EEByte          ; Save character
; Test for terminator
        sublw   0x0d
        btfsc   STATUS,Z ; Test if 0x0d
        goto    atEnd           ; Go if 0x0d
; At this point character read is not 0x0d
; Display on LCD
        movf    EEByte,w ; Recover character
; Display character on LCD
        call    send8           ; Display in LCD
        call    LCDscroll       ; Scroll at end of line
        incf    EEMemAdd,f      ; Next EEPROM byte
        goto    readOne
; End of execution
atEnd:
        goto    atEnd


;=============================================================
;=============================================================
;                 L O C A L    P R O C E D U R E S
;=============================================================
;=============================================================
;=========================
; init LCD for 4-bit mode
;=========================
initLCD:
; Initialization for Densitron LCD module as follows:
;       4-bit interface
;    2 display lines of 16 characters each
;    cursor on
;    left-to-right increment
;    cursor shift right
;    no display shift
;=====================|
;    set command mode  |
;=====================|
        bcf             PORTE,E_line    ; E line low
```

```
            bcf       PORTE,RS_line    ; RS line low
            bcf       PORTE,RW_line    ; Write mode
            call      delay_125                  ; delay 125
microseconds
            movlw     0x28     ; 0 0 1 0 1 0 0 0 (FUNCTION SET)
            call      send8    ; 4-bit send routine
; Set 4-bit mode command must be repeated
            movlw     0x28
            call      send8
            movlw     0x0e     ; 0 0 0 0 1 1 1 0 (DISPLAY ON/OFF)
            call      send8
            movlw     0x06     ; 0 0 0 0 0 1 1 0 (ENTRY MODE SET)
            call      send8
            movlw     0x14     ; 0 0 0 1 0 1 0 0 (CURSOR/DISPLAY
                               ;                       SHIFT)
            call      send8
            movlw     0x01     ; 0 0 0 0 0 0 0 1 (CLEAR DISPLAY)
                               ;               |___ COMMAND BIT
            call      send8
            call      delay_5  ; Test for busy
            return
.;===========================
;   procedure to clear LCD
;===========================
clearLCD:
            bcf               PORTE,E_line     ; E line low
            bcf               PORTE,RS_line    ; RS line low
            bcf               PORTE,RW_line    ; Write mode
            call      delay_125                ; delay 125
microseconds
            movlw     0x01     ; 0 0 0 0 0 0 0 1 (CLEAR DISPLAY)
                               ;               |___ COMMAND BIT
            call      send8
            call      delay_5  ; Test for busy
            return

;=======================
;   Procedure to delay
;     42 microseconds
;=======================
delay_125:
            movlw     .105             ; Repeat 105 machine cycles
            movwf     count1           ; Store value in counter
repeat
            decfsz    count1,f         ; Decrement counter
            goto      repeat           ; Continue if not 0
            return                     ; End of delay
```

```
;=======================
;   Procedure to delay
;   5 milliseconds
;=======================
delay_5:
        movlw    .105              ; Counter = 105 cycles
        movwf    count2            ; Store in variable
delay
        call     delay_125         ; Delay
        decfsz   count2,f          ; 40 times = 5 milliseconds
        goto     delay
        return                     ; End of delay
;=======================
;     pulse E line
;=======================
pulseE
        bsf      PORTE,E_line      ; Pulse E line
        nop
        bcf      PORTE,E_line
        return

;=============================
;   long delay sub-routine
;=============================
long_delay
                 movlw    D'200'   ; w delay count
                 movwf    J        ; J = w
jloop:  movwf    K                 ; K = w
kloop:  decfsz   K,f               ; K = K-1, skip next if zero
                 goto     kloop
                 decfsz   J,f      ; J = J-1, skip next if zero
                 goto     jloop
                 return
;=======================
;   send 2 nibbles in
;      4-bit mode
;=======================
; Procedure to send two 4-bit values to port B lines
; 7, 6, 5, and 4. High-order nibble is sent first
; ON ENTRY:
;         w register holds 8-bit value to send
send8:
        movwf    store1            ; Save original value
        call     merge4            ; Merge with port B
; Now w has merged byte
        movwf    PORTD             ; w to port D
        call     pulseE            ; Send data to LCD
; High nibble is sent
```

Supplementary Programs

```
            movf    store1,w        ; Recover byte into w
            swapf   store1,w        ; Swap nibbles in w
            call    merge4
            movwf   PORTD
            call    pulseE          ; Send data to LCD
            call    delay_125
            return
;===========================
;       merge bits
;===========================
; Routine to merge the 4 high-order bits of the
; value to send with the contents of port B
; so as to preserve the 4 low-bits in port B
; Logic:
;       AND value with 1111 0000 mask
;       AND port B with 0000 1111 mask
;       Now low nibble in value and high nibble in
;       port B are all 0 bits:
;           value  = vvvv 0000
;           port B = 0000 bbbb
;       OR value and port B resulting in:
;                   vvvv bbbb
; ON ENTRY:
;       w contain value bits
; ON EXIT:
;       w contains merged bits
merge4:
            andlw   b'11110000'     ; ANDing with 0 clears the
                                    ; bit. ANDing with 1 preserves
                                    ; the original value
            movwf   store2          ; Save result in variable
            movf    PORTD,w         ; port B to w register
            andlw   b'00001111' ; Clear high nibble in port b
                                    ; and preserve low nibble
            iorwf   store2,w ; OR two operands in w
            return
;===========================
;   Set address register
;       to LCD line 2
;===========================
; ON ENTRY:
;       Address of LCD line 2 in constant LCD_2
line2:
            bcf     PORTE,E_line    ; E line low
            bcf     PORTE,RS_line   ; RS line low, setup for
                                    ; control
            call    delay_5         ; Busy?
; Set to second display line
```

```
            movlw   LCD_2                   ; Address with high-bit set
            call    send8
; Set RS line for data
            bsf     PORTE,RS_line           ; RS = 1 for data
            call    delay_5                 ; Busy?
            return
;===========================
;    Set address register
;        to LCD line 1
;===========================
; ON ENTRY:
;           Address of LCD line 1 in constant LCD_1
line1:
            bcf     PORTE,E_line            ; E line low
            bcf     PORTE,RS_line           ; RS line low, set up for
                                            ; control
            call    delay_5                 ; busy?
; Set to second display line
            movlw   LCD_1                   ; Address and command bit
            call    send8                   ; 4-bit routine
; Set RS line for data
            bsf     PORTE,RS_line           ; Setup for data
            call    delay_5                 ; Busy?
            return

;===========================
;    scroll to LCD line 2
;===========================
; Procedure to count the number of characters displayed on
; each LCD line. If the number reaches the value in the
; constant LCDlimit, then display is scrolled to the second
; LCD line. If at the end of the second line, then LCD is
; reset to the first line.
LCDscroll:
            incf    LCDcount,f              ; Bump counter
; Test for line limit
            movf    LCDcount,w
            sublw   LCDlimit                ; Count minus limit
            btfss   STATUS,Z                ; Is count - limit = 0
            goto    scrollExit              ; Go if not at end of line
; At this point the end of the LCD line was reached
; Test if this is also the end of the second line
            movf    LCDline,w
            sublw   0x01                    ; Is it line 1?
            btfsc   STATUS,Z                ; Is LCDline minus 1 = 0?
            goto    line2End                ; Go if end of second line
; At this point it is the end of the top LCD line
            call    line2                   ; Scroll to second line
```

Supplementary Programs

```
                clrf    LCDcount        ; Reset counter
                incf    LCDline,f       ; Bump line counter
                goto    scrollExit
; End of second LCD line
line2End:
                call    initLCD         ; Reset
                clrf    LCDcount        ; Clear counters
                clrf    LCDline
                call    line1           ; Display to first line
scrollExit:
                return

;==============================
;   LCD display procedure
;==============================
; Sends 20 characters from PIC buffer with address stored
; in variable bufAdd to LCD line previously selected
display20:
                call    delay_5         ; Make sure not busy
; Set up for data
                bcf     PORTA,E_line    ; E line low
                bsf     PORTA,RS_line   ; RS line high for data
; Set up counter for 20 characters
                movlw   D'20'
                movwf   count3
; Get display address from local variable bufAdd
                movf    bufAdd,w        ; First display RAM address to W
                movwf   FSR             ; W to FSR
getchar
                movf    INDF,w          ; get character from display RAM
                                        ; location pointed to by file select
                                        ; register
                call    send8           ; 4-bit interface routine
; Test for 20 characters displayed
                decfsz  count3,f        ; Decrement counter
                goto    nextchar        ; Skipped if done
                return
nextchar:
                incf    FSR,f           ; Bump pointer
                goto    getchar

;================================
;   first text string procedure
;================================
storeMS1:
; Procedure to store in PIC RAM buffer the message
; contained in the code area labeled msg1
; ON ENTRY:
```

```
;               variable bufAdd holds address of text buffer
;               in PIC RAM
;               w register hold offset into storage area
;               msg1 is routine that returns the string characters
;               and a zero terminator
;               index is local variable that hold offset into
;               text table. This variable is also used for
;               temporary storage of offset into buffer
; ON EXIT:
;               Text message stored in buffer
;
; Store offset into text buffer (passed in the w register)
; in temporary variable
        movwf   index                   ; Store w in index
; Store base address of text buffer in FSR
        movf    bufAdd,w ; first display RAM address to W
        addwf   index,w                 ; Add offset to address
        movwf   FSR                     ; W to FSR
; Initialize index for text string access
        movlw   0                       ; Start at 0
        movwf   index   ; Store index in variable
; w still = 0
get_msg_char:
        call    msg1                    ; Get character from table
; Test for zero terminator
        andlw   0x0ff
        btfsc   STATUS,Z ; Test zero flag
        goto    endstr1                 ; End of string
; ASSERT: valid string character in w
;         store character in text buffer (by FSR)
        movwf   INDF    ; store in buffer by FSR
        incf    FSR,f   ; increment buffer pointer
; Restore table character counter from variable
        movf    index,w ; Get value into w
        addlw   1       ; Bump to next character
        movwf   index   ; Store table index in variable
        goto    get_msg_char    ; Continue
endstr1:
        return

; Routine for returning message stored in program area
; Message has 10 characters
msg1:
        addwf   PCL,f           ; Access table
        retlw   'R'
        retlw   'e'
        retlw   'c'
        retlw   'e'
```

```
        retlw   'i'
        retlw   'v'
        retlw   'i'
        retlw   'n'
        retlw   'g'
        retlw   ':'
        retlw   0
```

```
;========================
;       blank buffer
;========================
; Procedure to store 20 blank characters in PIC RAM
; buffer starting at address stored in the variable
; bufAdd
blank20:
        movlw   D'20'       ; Setup counter
        movwf   count1
        movf    bufAdd,w    ; First PIC RAM address
        movwf   FSR         ; Indexed addressing
        movlw   0x20        ; ASCII space character
storeit
        movwf   INDF        ; Store blank character in PIC RAM
                            ; buffer using FSR register
        decfsz  count1,f    ; Done?
        goto    incfsr      ; no
        return              ; yes
incfsr
        incf    FSR,f       ; Bump FSR to next buffer space
        goto    storeit
```

```
;================================================================
;               communications procedures
;================================================================
; Initialize serial port for 2400 baud, 8 bits, no parity,
; 1 stop
InitSerial:
        Bank1               ; Macro to select bank1
; Bits 6 and 7 of Port C are multiplexed as TX/CK and RX/DT
; for USART operation. These bits must be set to input in the
; TRISC register
        movlw   b'11000000' ; Bits for TX and RX
        iorwf   TRISC,f     ; OR into Trisc register
; The asynchronous baud rate is calculated as follows:
;                   Fosc
;           ABR = -----------
;                   S*(x+1)
; where x is value in the SPBRG register and S is 64 if the high
; baud rate select bit (BRGH) in the TXSTA control register is
```

```
; clear, and 16 if the BRGH bit is set. For setting to 2400 baud
; using a 10Mhs oscillator at a slow baud rate the formula
; is:
; At slow speed (BRGH = 0)
;           10,000,000    10,000,000
;           ----------  = ----------  = 2,403.84 (0.16% error)
;           64*(64+1)       4160
;
        movlw    spbrgVal ; Value in spbrgVal = 64
        movwf    SPBRG    ; Place in baud rate generator
; Setup value: 0010 0000 = 0x20
        movlw    0x20     ; Enable transmission and high baud
                          ; rate
        movwf    TXSTA
        Bank0             ; Bank 0
; Setup value: 1001 0000 = 0x90
        movlw    0x90     ; Enable serial port and continuous
                          ; reception
        movwf    RCSTA
;
        clrf     errorFlags; Clear local error flags register
        return
;===============================
;       transmit data
;===============================
; Test for Transmit Register Empty and transmit data in w
SerialSend:
        Bank0                     ; Select bank 0
        btfss    PIR1,TXIF        ; check if transmitter busy
        goto     $-1     ; wait until transmitter is not busy
        movwf    TXREG   ; and transmit the data
        return

;===============================
;       receive data
;===============================
; Procedure to test line for data received and return value
; in w. Overrun and framing errors are detected and
; remembered in the variable errorFlags, as follows:
;       7 6 5 4 3 2 1 0    <== errorFlags
;       -- not used ----  |  |___ overrun error
;                         |_____ framing error
SerialRcv:
        clrf    newData ; Clear new data received register
        Bank0           ; Select bank 0
; Bit 5 (RCIF) of the PIR1 Register is clear if the USART
; receive buffer is empty. If so, no data has been received
        btfss    PIR1,RCIF        ; Check for received data
```

```
                return                          ; Exit if no data
; At this point data has been received. First eliminate
; possible errors: overrun and framing.
; Bit 1 (OERR) of the RCSTA register detects overrun
; Bit 2 (FERR) of the RCSTA register detects framing error
        btfsc       RCSTA,OERR          ; Test for overrun error
        goto        OverErr             ; Error handler
        btfsc       RCSTA,FERR          ; Test for framing error
        goto        FrameErr ; Error handler
; At this point no error was detected
; Received data is in the USART RCREG register
        movf        RCREG,w             ; get received data
        bsf         newData,7           ; Set bit 7 to indicate new data
; Clear error flags
        clrf        errorFlags
        return
;===========================
;     error handlers
;===========================
OverErr:
        bsf         errorFlags,0        ; Bit 0 is overrun error
; Reset system
        bcf         RCSTA,CREN          ; Clear continuous receive bit
        bsf         RCSTA,CREN          ; Set to re-enable reception
        return
; error because FERR framing error bit is set
; can do special error handling here - this code simply clears
; and continues
FrameErr:
        bsf                 errorFlags,1; Bit 1 is framing error
        movf        RCREG,W             ; Read and throw away bad data
        return
;================================================================
;                I2C EEPROM data procedures
;================================================================
; GPRs used in EEPROM-related code are placed in the common
; RAM area (from 0x70 to 0x7f). This makes the registers
; accessible from any bank.
;============================
;     LIST OF PROCEDURES
;============================
; SetupI2C    ---    Initialize MSSP module for I2C mode
;                    in hardware master mode
;                    Configure I2C lines
;                    Set slew rate for 100kbps
;                    Set baud rate for 10Mhz
; WriteI2C    ---    Write byte to I2C EEPROM device
;                    Data is stored in EEByte variable
```

```
;                      Address is stored in EEMemAdd
; ReadI2C    ---   Read byte from I2C EEPROM device
;                      Address stored in EEMemAdd
;                      Read data returned in w register
;=============================
;     I2C setup procedure
;=============================
SetupI2C:
          Bank1
          movlw     b'00011000'
          iorwf     TRISC,f             ; OR into TRISC
; Setup MSSP module for Master Mode operation
          Bank0
          movlw     B'00101000'; Enables MSSP and uses appropriate
;  0  0  1  0  1  0  0  0  Value to install
;  7  6  5  4  3  2  1  0  <== SSPCON bits in this operation
;  |  |  |  |  |__|__|__|___ Serial port select bits
;  |  |  |  |              1000 = I2C master mode
;  |  |  |  |              Clock = Fosc/(4*(SSPAD+1))
;  |  |  |  |_____ UNUSED IN MASTER MODE
;  |  |  |_____ SSP Enable
;  |  |                    1 = SDA and SCL pins as serial
;  |  |_____ Receive 0verflow indicator
;  |                       0 = no overflow
;  |_____ Write collision detect
;                          0 = no collision detected
          movwf     SSPCON    ; This is loaded into SSPCON
; Input levels and slew rate as standard I2C
          Bank1
          movlw     B'10000000'
;  1  0  0  0  0  0  0  0  Value to install
;  7  6  5  4  3  2  1  0  <== SSPSTAT bits in this operation
;  |  |  |  |  |  |  |  |___ Buffer full status bit READ ONLY
;  |  |  |  |  |  |  |_____ UNUSED in present application
;  |  |  |  |  |  |_____ Read/write information READ ONLY
;  |  |  |  |  |_____ UNUSED IN MASTER MODE
;  |  |  |  |_____ STOP bit READ ONLY
;  |  |  |_____ Data address READ ONLY
;  |  |_____ SMP bus select
;  |                       0 = use normal I2C specs
;  |_____ Slew rate control
;                          0 = disabled
          movwf     SSPSTAT
; Setup Baud Rate
; Baud Rate = Fosc/(4*(SSPADD+1))
;    Fosc = 10Mhz
;    Baud Rate = 24 for 100 kbps
          movlw     .24                 ; Value to use
```

Supplementary Programs

```
                movwf   SSPADD  ; Store in SSPADD
        Bank0
        return

;=============================
;       I2C write procedure
;=============================
; Write one byte to I2C EEPROM 24LC04B
; Steps:
;               1. Send START
;               2. Send control. Wait for ACK
;               3. Send address. Wait for ACK
;               4. Send data. Wait for ACK
;               5. Send STOP
; STEP 1:
WriteI2C:
        Bank1
        bsf     SSPCON2,SEN ; Produce START Condition
        call    WaitI2C ; Wait for I2C to complete
; STEP 2:
; Send control byte. Wait for ACK
        movlw   LC04READ        ; Control byte
        call    Send1I2C        ; Send Byte
        call    WaitI2C ; Wait for I2C to complete
        btfsc   SSPCON2,ACKSTAT ; Check ACK bit to see if
                                ; I2C failed, skip if not
        goto    FailI2C
; STEP 3:
; Send address. Wait for ACK
        Bank0
        movf    EEMemAdd,w      ; Load Address Byte
        call    Send1I2C        ; Send Byte
        call    WaitI2C ; Wait for I2C operation to complete
        Bank1
        btfsc   SSPCON2,ACKSTAT ; Check ACK Status bit to see
                                ; if I2C failed, skip if not
        goto    FailI2C
; STEP 4:
; Send data. Wait for ACK
        Bank0
        movf    EEByte,w        ; Load Data Byte
        call    Send1I2C ; Send Byte
        call    WaitI2C ; Wait for I2C operation to complete
        Bank1
        btfsc   SSPCON2,ACKSTAT ; Check ACK Status bit to see
                                ; if I2C failed, skip if not
        goto            FailI2C
; STEP 5:
```

```
; Send STOP. Wait for ACK
        bsf     SSPCON2,PEN ; Send STOP condition
        call    WaitI2C ; Wait for I2C operation to complete
; WRITE operation has completed successfully.
        Bank0
        return


;============================
;    I2C read procedure
;============================
; Procedure to read one byte from 24LC04B EEPROM
; Steps:
;                 1. Send START
;                 2. Send control. Wait for ACK
;                 3. Send address. Wait for ACK
;                 4. Send RESTART + control. Wait for ACK
;                 5. Switch to receive mode. Get data.
;                 6. Send NACK
;                 7. Send STOP
;                 8. Retreive data into w register
; STEP 1:
ReadI2C
; Send RESTART. Wait for ACK
        Bank1
        bsf     SSPCON2,RSEN ; RESTART Condition
        call    WaitI2C ; Wait for I2C operation
; STEP 2:
; Send control byte. Wait for ACK
        movlw   LC04READ ; Control byte
        call    Send1I2C ; Send Byte
        call    WaitI2C ; Wait for I2C operation
; Now check to see if I2C EEPROM is ready
        Bank1
        btfsc   SSPCON2,ACKSTAT ; Check ACK Status bit
        goto    ReadI2C ; ACK Poll waiting for EEPROM
                        ; write to complete
; STEP 3:
; Send address. Wait for ACK
        Bank0
        movf    EEMemAdd,w      ; Load from address register
        call    Send1I2C        ; Send Byte
        call    WaitI2C ; Wait for I2C operation
        Bank1
        btfsc   SSPCON2,ACKSTAT ; Check ACK Status bit
        goto    FailI2C ; failed, skipped if successful
; STEP 4:
; Send RESTART. Wait for ACK
        bsf     SSPCON2,RSEN ; Generate RESTART Condition
```

```
                call    WaitI2C ; Wait for I2C operation
; Send output control. Wait for ACK
        movlw   LC04WRITE ; Load CONTROL BYTE (output)
        call    Send1I2C        ; Send Byte
        call    WaitI2C ; Wait for I2C operation
        Bank1
        btfsc   SSPCON2,ACKSTAT ; Check ACK Status bit
        goto    FailI2C ; failed, skipped if successful
; STEP 5:
; Switch MSSP to I2C Receive mode
        bsf     SSPCON2,RCEN ; Enable Receive Mode (I2C)
; Get the data. Wait for ACK
        call    WaitI2C ; Wait for I2C operation
; STEP 6:
; Send NACK to acknowledge
        Bank1
        bsf     SSPCON2,ACKDT ; ACK DATA to send is 1 (NACK)
        bsf     SSPCON2,ACKEN ; Send ACK DATA now.
; Once ACK or NACK is sent, ACKEN is automatically cleared
; STEP 7:
; Send STOP. Wait for ACK
        bsf     SSPCON2,PEN ; Send STOP condition
        call    WaitI2C ; Wait for I2C operation
; STEP 8:
; Read operation has finished
        Bank0
        movf    SSPBUF,W ; Get data from SSPBUF into W
; Procedure has finished and completed successfully.
        return


;=============================
;   I2C support procedures
;=============================
; I2C Operation failed code sequence
; Procedure hangs up. User should provide error handling.
FailI2C
        Bank1
        bsf     SSPCON2,PEN ; Send STOP condition
        call    WaitI2C ; Wait for I2C operation
fail:
        goto    fail

; Procedure to transmit one byte
Send1I2C
        Bank0
        movwf   SSPBUF ; Value to send to SSPBUF
        return
```

```
; Procedure to wait for the last I2C operation to complete.
; Code polls the SSPIF flag in PIR1.
WaitI2C
        Bank0
        btfss   PIR1,SSPIF ; Check if I2C operation done
        goto    $-1        ; I2C module is not ready yet
        bcf     PIR1,SSPIF ; I2C ready, clear flag
        return

;=============================================================
        end             ; END OF PROGRAM
;=============================================================
```

```
; File name: Key2LCD.asm
; Date: May 11, 2006
; Author: Julio Sanchez
;
; Description:
; Decode 4 x 4 keypad and display scan code in LCD.
; Program to use 4-bit PIC-to-LCD interface.
; Code assumes that LCD is driven by Hitachi HD44780
; controller and PIC 16F977. Display supports two lines
; each one with 20 characters. The wiring and base
; address of each display line is stored in #define
; statements. These statements can be edited to
; accommodate a different set-up.
; Keypad switch wiring (values are scan codes):
;        --- KEYPAD --
;         0   1   2   3    <= port B0 |
;         4   5   6   7    <= port B1 |--- ROWS = OUTPUTS
;         8   9   A   B    <= port B2 |
;         C   D   E   F    <= port B3 |
;         |   |   |   |
;         |   |   |   |_____ port B4 |
;         |   |   |_____ port B5 |--- COLUMNS = INPUTS
;         |   |_____ port B6 |
;         |_____ port B7 |
;
; Program operations:
; 1. Key press action generates a scan code in the range
;    0x0 to 0xf.
; 2. Program converts scan code to ASCII digit and displays
;    the digit on the LCD.
; 3. When the end of the first LCD line is reached, display
;    continues in the second line. When the end of the
;    second line is reached, LCD is reset to the first line
;
; Program uses delay loops for interface timing.
; WARNING:
; Code assumes 4Mhz clock. Delay routines must be
; edited for faster clock
;
;===========================
;     16F877 switches
;===========================
; Switches used in __config directive:
;    _CP_ON           Code protection ON/OFF
; *  _CP_OFF
; *  _PWRTE_ON        Power-up timer ON/OFF
;    _PWRTE_OFF
;    _BODEN_ON        Brown-out reset enable ON/OFF
```

```
;   *   _BODEN_OFF
;   *   _PWRTE_ON        Power-up timer enable ON/OFF
;       _PWRTE_OFF
;       _WDT_ON          Watchdog timer ON/OFF
;   *   _WDT_OFF
;       _LPV_ON          Low voltage IC programming enable ON/OFF
;   *   _LPV_OFF
;       _CPD_ON          Data EE memory code protection ON/OFF
;   *   _CPD_OFF
; OSCILLATOR CONFIGURATIONS:
;       _LP_OSC          Low power crystal occilator
;       _XT_OSC          External parallel resonator/crystal oscillator
;
;   *   _HS_OSC          High speed crystal resonator
;       _RC_OSC          Resistor/capacitor oscillator
;   |                    (simplest, 20% error)
;   |
;   |_____ * indicates setup values presently selected

        processor       16f877          ; Define processor
        #include <p16f877.inc>
        __CONFIG _CP_OFF & _WDT_OFF & _BODEN_OFF & _PWRTE_ON &
_HS_OSC & _WDT_OFF & _LVP_OFF & _CPD_OFF

; __CONFIG directive is used to embed configuration data
;  within the source file. The labels following the directive
;  are located in the corresponding .inc file.

;================================================================
;                    constant definitions
;       for PIC-to-LCD pin wiring and LCD line addresses
;================================================================
#define E_line 1             ; |
#define RS_line 0            ; | -- from wiring diagram
#define RW_line 2            ; |
; LCD line addresses (from LCD data sheet)
#define LCD_1 0x80           ; First LCD line constant
#define LCD_2 0xc0           ; Second LCD line constant
#define LCDlimit .20; Number of characters per line
; Note: The constants that define the LCD display line
;       addresses have the high-order bit set in
;       order to facilitate the controller command
;
;================================================================
;                    PIC register equates
;================================================================
portd   equ             0x08
porte   equ     0x09
```

```
fsr             equ                     0x04
status    equ               0x03
indf      equ     0x00
z               equ                     2
c               equ                     0
;================================================================
;                       variables in PIC RAM
;================================================================
; Reserve 20 bytes for string buffer
        cblock  0x20
        strData
        endc
; Leave 16 bytes and Continue with local variables
        cblock  0x34    ; Start of block
        count1          ; Counter # 1
        count2          ; Counter # 2
        count3          ; Counter # 3
        pic_ad          ; Storage for start of text area
                        ; (labeled strData) in PIC RAM
        J               ; counter J
        K               ; counter K
        index           ; Index into text table (also used
                        ; for auxiliary storage)
        store1          ; Local storage
        store2
; For LCDscroll procedure
        LCDcount ; Counter for characters per line
        LCDline         ; Current display line (0 or 1)
; Keypad processing variables
        keyMask         ; For keypad processing
        rowMask         ; For masking-off key rows
        rowCode         ; Row addend for calculating scan code
        rowCount ; Counter for key rows (0 to 3)
        scanCode ; Final key code
        newScan         ; 0 if no new scan code detected
        endc
;================================================================
;                       M A C R O S
;================================================================
; Macros to select the register banks
; Data memory bank selection bits:
; RP1:RP0         Bank
;   0:0             0    Ports A,B,C,D, and E
;   0:1             1    Tris A,B,C,D, and E
;   1:0             2
;   1:1             3
Bank0   MACRO                   ; Select RAM bank 0
        bcf     STATUS,RP0
```

```
                    bcf     STATUS,RP1
                    ENDM

Bank1   MACRO                       ; Select RAM bank 1
                    bsf     STATUS,RP0
                    bcf     STATUS,RP1
                    ENDM

Bank2   MACRO                       ; Select RAM bank 2
                    bcf     STATUS,RP0
                    bsf     STATUS,RP1
                    ENDM

Bank3   MACRO                       ; Select RAM bank 3
                    bsf     STATUS,RP0
                    bsf     STATUS,RP1
                    ENDM
;=============================================================
;                   M A I N    P R O G R A M
;=============================================================
                    org     0               ; start at address
                    goto    main
; Space for interrupt handlers
        org     0x08
main:
; Wiring:
;       LCD data to port D, lines 0 to 7
;       E line -> port E, 1
;       RW line -> port E, 2
;       RS line -> port E, 0
; Set porte D and E for output
; First, initialize port B by clearing latches
        clrf    STATUS
        clrf    PORTB
; Select bank 1 to tris port D for output
        Bank1
; Tris port D for output. Port D lines 4 to 7 are wired
; to LCD data lines. Port D lines 0 to 4 are wired to LEDs.
        movlw   B'00000000'
        movwf   TRISD           ; and port D
; By default port A lines are analog. To configure them
; as digital code must set bits 1 and 2 of the ADCON1
; register (in bank 1)
        movlw   0x06            ; binary 0000 0110  is code to
                                ; make all
port A lines digital
        movwf   ADCON1
; Port B, lines are wired to keypad switches, as follows:
```

```
;       7 6 5 4 3 2 1 0
;       | | | | |_|_|_|_____ switch rows (output)
;       |_|_|_|_____ switch columns (input)
; rows must be defined as output and columns as input
        movlw   b'11110000'
        movwf   TRISB
; Tris port E for output
        movlw   B'00000000'
        movwf   TRISE               ; Tris port E
; Enable port B pullups for switches in OPTION register
;       7 6 5 4 3 2 1 0 <= OPTION bits
;       | | | | | |__|__|_____ PS2-PS0 (prescaler bits)
;       | | | | |             Values for Timer0
;       | | | | |             000 = 1:2    001 = 1:4
;       | | | | |             010 = 1:8    011 = 1:16
;       | | | | |             100 = 1:32   101 = 1:64
;       | | | | |             110 = 1:128 *111 = 1:256
;       | | | | |_____ PSA (prescaler assign)
;       | | | |              *1 = to WDT
;       | | | |               0 = to Timer0
;       | | | |_____ TOSE (Timer0 edge select)
;       | | |                *0 = increment on low-to-high
;       | | |                 1 = increment in high-to-low
;       | | |_____ TOCS (TMR0 clock source)
;       | |                  *0 = internal clock
;       | |                   1 = RA4/TOCKI bit source
;       | |_____ INTEDG (Edge select)
;       |                    *0 = falling edge
;       |_____ RBPU (Pullup enable)
;                            *0 = enabled
;                             1 = disabled
        movlw   b'00001000'
        movwf   OPTION_REG
; Back to bank 0
        Bank0
; Clear all output lines
        movlw   b'00000000'
        movwf   portd
        movwf   porte
; Wait and initialize HD44780
        call    delay_5   ; Allow LCD time to initialize itself
        call    initLCD   ; Then do forced initialization
        call    delay_5   ; (Wait probably not necessary)
; Set display address to start of second LCD line
        call    line1
; Clear character counter and line counter variables
        clrf    LCDcount
        clrf    LCDline
```

```
;========================
;      scan keypad
;========================
; Keypad switch wiring:
;          x    x    x    x    <= port B0 |
;          x    x    x    x    <= port B1 |--- ROWS = OUTPUTS
;          x    x    x    x    <= port B2 |
;          x    x    x    x    <= port B3 |
;          |    |    |    |
;          |    |    |    |_____ port B4 |
;          |    |    |_____ port B5 |--- COLUMNS = INPUTS
;          |    |_____ port B6 |
;          |_____ port B7 |
; Switches are connected to port B lines
; Clear scan code register
         clrf     scanCode
;============================
;  scan keypad and display
;============================
keyScan:
; Port B, lines are wired to pushbutton switches, as follows:
;   7 6 5 4 3 2 1 0
;   | | | | |_|_|_|_____ switch rows (output)
;   |_|_|_|_____ switch columns (input)
; Keypad processing:
; switch rows are successively grounded (row = 0)
; Then column values are tested. If a column returns 0
; in a 0 row, that switch is down.
; Initialize row code addend
         clrf     rowCode              ; First row is code 0
         clrf     newScan              ; No new scan code detected
; Initialize row count
         movlw    D'4'                 ; Four rows
         movwf    rowCount             ; Register variable
         movlw    b'11111110'          ; All set but LOB
         movwf    rowMask
keyLoop:
; Initialize row eliminator mask:
; The row mask is ANDed with the key mask to successively
; mask-off each row, for example:
;
;                        |------- row 3
;                        ||------ row 2
;                        |||----- row 1
;                        ||||---- row 0
;                   0000 1111 <= key mask
;           AND     1111 1101 <= mask for row 1
;                   ---------
```

```
;                 0000 1101 <= row 1 is masked off
;
; The row mask, which is initially 1111 1110, is rotated left
; through the carry in order to mask off the next row
        movlw   b'00001111'       ; Mask off all lines
        movwf   keyMask           ; To local register
; Set row mask for current row
        movf    rowMask,w         ; Mask to w
        andwf   keyMask,f         ; Update key mask
        movf    keyMask,w         ; Key mask to w
        movwf   PORTB             ; Mask-off port B lines
; Read port B lines 4 to 7 (columns are input)
        btfss   PORTB,4
        call    col0              ; Key column procedures
        btfss   PORTB,5
        call    col1
        btfss   PORTB,6
        call    col2
        btfss   PORTB,7
        call    col3
; Index to next row by adding 4 to row code
        movf    rowCode,w         ; Code to w
        addlw   D'4'
        movwf   rowCode
;=========================
;     shift row mask
;=========================
; Set the carry flag
        bsf     STATUS,c
        rlf     rowMask,f         ; Rotate mask bits in storage
;=========================
;     end of keypad?
;=========================
; Test for last key row (maximum count is 4)
        decfsz  rowCount,f        ; Decrement counter
        goto    keyLoop
;===============================================================
;===============================================================
;                 display scan code
;===============================================================
;===============================================================
; At this point all keys have been tested.
; Variable newScan = 0 if no new scan code detected, else
; variable scanCode holds scan code
        movf    newScan,f         ; Copy onto intsef (sets z flag)
        btfsc   STATUS,z          ; Is it zero
        goto    scanExit
; At this point a new scan code is detected
```

```
                movf    scanCode,w          ; To w
; If scan code is in the range 0 to 9, that is, a decimal
; digit, then ASCII conversion consists of adding 0x30.
; If the scan code represents one of the hex letters
; (0xa to 0xf) then ASCII conversion requires adding
; 0x37
                sublw   0x09                ; 9 - w
; if w from 0 to 9 then 9 - w = positive (c flag = 1)
; if w = 0xa then 9 - 10 = -1 (c flag = 0)
; if w = 0xc then 9 - 12 = -2 (c flag = 0)
                btfss   STATUS,c ; Test carry flag
                goto    hexLetter ; Carry clear, must be a letter
; At this point scan code is a decimal digit in the
; range 0 to 9. Conver to ASCII by adding 0x30
                movf    scanCode,w          ; Recover scan code
                addlw   0x30                ; Convert to ASCII
                goto    displayDig
hexLetter:
                movf    scanCode,w          ; Recover scan code
                addlw   0x37                ; Convert to ASCII
displayDig:
                call    send8       ; Display routine
                call    LCDscroll ; Auto line scrolling procedure
scanExit:
                call    long_delay          ; Debounce
                goto    keyScan             ; Continue

;===========================
;   calculate scan code
;===========================
; The column position is added to the row code (stored
; in rowCode register). Sum is the scan code
col0:
                movf    rowCode,w ; Row code to w
                addlw   0x00      ; Add 0 (clearly not necessary)
                movwf   scanCode ; Final value
                incf    newScan,f           ; New scan code
                return

col1:
                movf    rowCode,w           ; Row code to w
                addlw   0x01                ; Add 1
                movwf   scanCode
                incf    newScan,f
                return

col2:
                movf    rowCode,w           ; Row code to w
```

Supplementary Programs

```
            addlw    0x02              ; Add 2
            movwf    scanCode
            incf     newScan,f
            return

col3:
            movf     rowCode,w         ; Row code to w
            addlw    0x03              ; Add 3
            movwf    scanCode
            incf     newScan,f
            return

;================================================================
;                initialize LCD for 4-bit mode
;================================================================
initLCD:
; Initialization for Densitron LCD module as follows:
;       4-bit interface
;    2 display lines of 16 characters each
;    cursor on
;    left-to-right increment
;    cursor shift right
;    no display shift
;======================|
;    set command mode  |
;======================|
            bcf      porte,E_line      ; E line low
            bcf      porte,RS_line     ; RS line low
            bcf      porte,RW_line     ; Write mode
            call     delay_125         ; delay 125 microseconds
;**********************|
;      FUNCTION SET    |
;**********************|
            movlw    0x28     ; 0 0 1 0 1 0 0 0 (FUNCTION SET)
                              ;         | | | |__ font select:
                              ;         | | |     1 = 5x10 in 1/8 or 1/11
                              ;         | | |     0 = 1/16 dc
                              ;         | | |___ Duty cycle select
                              ;         | |      0 = 1/8 or 1/11
                              ;         | |      1 = 1/16
                              ;         | |___ Interface width
                              ;         |      0 = 4 bits
                              ;         |      1 = 8 bits
                              ;         |___ FUNCTION SET COMMAND
            call     send8    ; 4-bit send routine

; Set 4-bit mode command must be repeated
            movlw    0x28
```

```
        call    send8

;**********************|
; DISPLAY AND CURSOR ON |
;**********************|
        movlw   0x0e    ; 0 0 0 0 1 1 1 0 (DISPLAY ON/OFF)
                        ;         | | | |___ Blink character
                        ;         | | |       1 = on, 0 = off
                        ;         | | |___ Cursor on/off
                        ;         | |         1 = on, 0 = off
                        ;         | |____ Display on/off
                        ;         |           1 = on, 0 = off
                        ;         |____ COMMAND BIT
        call    send8
;**********************|
;   set entry mode     |
;**********************|
        movlw   0x06    ; 0 0 0 0 0 1 1 0 (ENTRY MODE SET)
                        ;         | | |___ display shift
                        ;         | |       1 = shift
                        ;         | |       0 = no shift
                        ;         | |____ cursor increment
                        ;         |         1 = left-to-right
                        ;         |         0 = right-to-left
                        ;         |____ COMMAND BIT
        call    send8

;**********************|
; cursor/display shift |
;**********************|
        movlw   0x14    ; 0 0 0 1 0 1 0 0 (CURSOR/DISPLAY
                        ;                   SHIFT)
                        ;     | | | |_|___ don't care
                        ;     | |_|__ cursor/display shift
                        ;     |         00 = cursor shift left
                        ;     |         01 = cursor shift right
                        ;     |         10 = cursor and display
                        ;     |              shifted left
                        ;     |         11 = cursor and display
                        ;     |              shifted right
                        ;     |___ COMMAND BIT
        call    send8
;**********************|
;   clear display      |
;**********************|
        movlw   0x01    ; 0 0 0 0 0 0 0 1 (CLEAR DISPLAY)
                        ;                 |___ COMMAND BIT
```

```
                call    send8
; Per documentation
                call    delay_5     ; Test for busy
                return


;======================
;   Procedure to delay
;      42 microseconds
;======================
delay_125:
                movlw   D'42'       ; Repeat 42 machine cycles
                movwf   count1      ; Store value in counter
repeat
                decfsz  count1,f    ; Decrement counter
                goto    repeat      ; Continue if not 0
                return              ; End of delay


;======================
;   Procedure to delay
;      5 milliseconds
;======================
delay_5:
                movlw   D'42'               ; Counter = 41
                movwf   count2              ; Store in variable
delay
                call    delay_125           ; Delay
                decfsz  count2,f            ; 40 times = 5 milliseconds
                goto    delay
                return                      ; End of delay
;======================
;      pulse E line
;======================
pulseE
                bsf     porte,E_line        ; Pulse E line
                Nop
                bcf     porte,E_line
                return


;============================
;    long delay sub-routine
;        (for debugging)
;============================
long_delay
                movlw   D'200'      ; w delay count
                movwf   J           ; J = w
jloop:          movwf   K           ; K = w
kloop:          decfsz  K,f         ; K = K-1, skip next if zero
                goto    kloop
```

```
        decfsz  J,f             ; J = J-1, skip next if zero
        goto    jloop
        return

;========================
;   send 2 nibbles in
;     4-bit mode
;========================
; Procedure to send two 4-bit values to port B lines
; 7, 6, 5, and 4. High-order nibble is sent first
; ON ENTRY:
;       w register holds 8-bit value to send
send8:
        movwf   store1          ; Save original value
        call    merge4          ; Merge with port B
; Now w has merged byte
        movwf   portd           ; w to port D
        call    pulseE          ; Send data to LCD
; High nibble is sent
        movf    store1,w ; Recover byte into w
        swapf   store1,w ; Swap nibbles in w
        call    merge4
        movwf   portd
        call    pulseE          ; Send data to LCD
        call    delay_125
        return
;=================
;   merge bits
;=================
; Routine to merge the 4 high-order bits of the
; value to send with the contents of port B
; so as to preserve the 4 low-bits in port B
; Logic:
;       AND value with 1111 0000 mask
;       AND port B with 0000 1111 mask
;       Now low nibble in value and high nibble in
;       port B are all 0 bits:
;           value = vvvv 0000
;           port B = 0000 bbbb
;       OR value and port B resulting in:
;                  vvvv bbbb
; ON ENTRY:
;       w contain value bits
; ON EXIT:
;       w contains merged bits
merge4:
        andlw   b'11110000'     ; ANDing with 0 clears the
                                ; bit. ANDing with 1 preserves
```

Supplementary Programs

```
                                    ; the original value
        movwf   store2              ; Save result in variable
        movf    portd,w             ; port B to w register
        andlw   b'00001111'         ; Clear high nibble in port b
                                    ; and preserve low nibble
        iorwf   store2,w ; OR two operands in w
        return

;========================
; Set address register
;    to LCD line 1
;========================
; ON ENTRY:
;       Address of LCD line 2 in constant LCD_2
line1:
        bcf     porte,E_line        ; E line low
        bcf     porte,RS_line       ; RS line low, setup for
                                    ; control
        call    delay_5             ; Busy?
; Set to second display line
        movlw   LCD_1               ; Address with
high-bit set
        call    send8
; Set RS line for data
        bsf     porte,RS_line       ; RS = 1 for data
        call    delay_5             ; Busy?
        return

;========================
; Set address register
;    to LCD line 2
;========================
; ON ENTRY:
;       Address of LCD line 2 in constant LCD_2
line2:
        bcf     porte,E_line        ; E line low
        bcf     porte,RS_line       ; RS line low, setup for
                                    ; control
        call    delay_5             ; Busy?
; Set to second display line
        movlw   LCD_2               ; Address with high-bit set
        call    send8
; Set RS line for data
        bsf     porte,RS_line       ; RS = 1 for data
        call    delay_5             ; Busy?
        return

;===========================
```

```
;   scroll to LCD line 2
;===========================
; Procedure to count the number of characters displayed on
; each LCD line. If the number reaches the value in the
; constant LCDlimit, then display is scrolled to the second
; LCD line. If at the end of the second line, then LCD is
; reset to the first line.
LCDscroll:
        incf    LCDcount,f              ; Bump counter
; Test for line limit
        movf    LCDcount,w
        sublw   LCDlimit                ; Count minus limit
        btfss   STATUS,z                ; Is count - limit = 0
        goto    scrollExit              ; Go if not at end of line
; At this point the end of the LCD line was reached
; Test if this is also the end of the second line
        movf    LCDline,w
        sublw   0x01                    ; Is it line 1?
        btfsc   STATUS,z                ; Is LCDline minus 1 = 0?
        goto    line2End                ; Go if end of second line
; At this point it is the end of the top LCD line
        call    line2                   ; Scroll to second line
        clrf    LCDcount                ; Reset counter
        incf    LCDline,f               ; Bump line counter
        goto    scrollExit
; End of second LCD line
line2End:
        call    initLCD                 ; Reset
        clrf    LCDcount                ; Clear counters
        clrf    LCDline
        call    line1                   ; Display to first line
scrollExit:
        return
        end
```

```
;   File name: KeyPad.asm
;   Last revision: May 12, 2006
;   Author: Julio Sanchez
;
;   Description:
;   Program to scan a 4 x 4 keypad
;   Keypad switch wiring (values are scan codes):
;          --- KEYPAD --
;          0   1   2   3    <= port B0 |
;          4   5   6   7    <= port B1 |--- ROWS = OUTPUTS
;          8   9   A   B    <= port B2 |
;          C   D   E   F    <= port B3 |
;          |   |   |   |
;          |   |   |   |_____ port B4 |
;          |   |   |_____ port B5 |--- COLUMNS = INPUTS
;          |   |_____ port B6 |
;          |_____ port B7 |
;
;   Key press action generates a scan code in the range
;   0x0 to 0xf, starting at the top-left pushbutton and
;   increasing left-to-right and from the top down.
;
;   Scan code is displayed in LEDs wired to port D,
;   lines 0, 1, 2, and 3
;
;   WARNING:
;   Code assumes 4Mhz clock. Delay routines must be
;   edited for faster clock
;
;===========================
;       16F877 switches
;===========================
;   Switches used in __config directive:
;     _CP_ON              Code protection ON/OFF
; *   _CP_OFF
; *   _PWRTE_ON           Power-up timer ON/OFF
;     _PWRTE_OFF
;     _BODEN_ON           Brown-out reset enable ON/OFF
; *   _BODEN_OFF
; *   _PWRTE_ON           Power-up timer enable ON/OFF
;     _PWRTE_OFF
;     _WDT_ON             Watchdog timer ON/OFF
; *   _WDT_OFF
;     _LPV_ON             Low voltage IC programming enable ON/OFF
; *   _LPV_OFF
;     _CPD_ON             Data EE memory code protection ON/OFF
; *   _CPD_OFF
;   OSCILLATOR CONFIGURATIONS:
```

```
;       _LP_OSC          Low power crystal occilator
;       _XT_OSC          External parallel resonator/crystal oscillator
;
; *     _HS_OSC          High speed crystal resonator
;       _RC_OSC          Resistor/capacitor oscillator
; |                      (simplest, 20% error)
; |
; |_____ * indicates setup values presently selected

        processor       16f877          ; Define processor
        #include <p16f877.inc>
        __CONFIG _CP_OFF & _WDT_OFF & _BODEN_OFF & _PWRTE_ON &
_HS_OSC & _WDT_OFF & _LVP_OFF & _CPD_OFF

;========================================================
;                   PIC register equates
;========================================================
Status  equ     0x03
c       equ     0
;========================================================
;                   variables in PIC RAM
;========================================================
        cblock  0x20
        J                       ; Counters
        K
        keyMask                 ; For keypad processing
        rowMask                 ; For masking-off key rows
        rowCode                 ; Row addend for calculating scan code
        RowCount                ; Counter for key rows (0 to 3)
        ScanCode                ; Final key code
        endc

;========================================================
;                           program
;========================================================
        org             0       ; start at address
        goto    main
; Space for interrupt handlers
        org             0x08
main:
; First, initialize port B by clearing latches
        clrf    STATUS
        clrf    PORTB
; Select bank 1 to tris port D for output
        bcf     STATUS,RP1      ; Clear banks 2/3 selector
        bsf     STATUS,RP0      ; Select bank 1 for tris
registers
; Tris port D for output. Port D is wired to LEDs.
```

```
                movlw       B'00000000'
                movwf       TRISD               ; and port D
; Port B, lines are wired to pushbutton switches, as follows:
;   7 6 5 4 3 2 1 0
;   | | | | |_|_|_|_____ switch rows (output)
;   |_|_|_|_____ switch columns (input)
; rows must be defined as output and columns as input
                movlw       b'11110000'
                movwf       TRISB
; Enable port B pullups for switches in OPTION register
                movlw       b'00001000'
                movwf       OPTION_REG
; Back to bank 0
                bcf                     STATUS,RP0
;========================
; Monitor switches and
;     toggle LEDS
;========================
; Keypad switch wiring:
;           x   x   x   x   <= port B0 |
;           x   x   x   x   <= port B1 |--- ROWS = OUTPUTS
;           x   x   x   x   <= port B2 |
;           x   x   x   x   <= port B3 |
;           |   |   |   |
;           |   |   |   |_____ port B4 |
;           |   |   |_____ port B5 |--- COLUMNS = INPUTS
;           |   |_____ port B6 |
;           |_____ port B7 |
; LEDS are wired to port D, lines 0, 1, 2, and 3
; Test switches
; First, all LEDs off
                movlw       b'00000000'
                movwf       PORTD               ; Place in port
; And clear scan code register
                clrf        scanCode
;=============================
;  scan keypad and display
;=============================
keyScan:
; Port B, lines are wired to pushbutton switches, as follows:
;   7 6 5 4 3 2 1 0
;   | | | | |_|_|_|_____ switch rows (output)
;   |_|_|_|_____ switch columns (input)
; Keypad processing:
; switch rows are successively grounded (row = 0)
; Then column values are tested. If a column returns 0
; in a 0 row, that switch is down.
; Initialize row code addend
```

```
                clrf    rowCode         ; First row is code 0
; Initialize row count
                movlw   D'4'            ; Four rows
                movwf   rowCount        ; Register variable
                movlw   b'11111110'     ; All set but LOB
                movwf   rowMask
keyLoop:
; Initialize row eliminator mask:
; The row mask is ANDed with the key mask to successively
; mask-off each row, for example:
;
;                       |------- row 3
;                       ||------ row 2
;                       |||----- row 1
;                       ||||---- row 0
;               0000 1111 <= key mask
;       AND     1111 1101 <= mask for row 1
;               ---------
;               0000 1101 <= row 1 is masked off
;
; The row mask, which is initially 1111 1110, is rotated left
; through the carry in order to mask off the next row
                movlw   b'00001111'     ; Mask off all lines
                movwf   keyMask         ; To local register
; Set row mask for current row
                movf    rowMask,w       ; Mask to w
                andwf   keyMask,f       ; Update key mask
                movf    keyMask,w       ; Key mask to w
                movwf   PORTB           ; Mask-off port B lines
; Read port B lines 4 to 7 (columns are input)
                btfss   PORTB,4
                call    col0            ; Key column procedures
                btfss   PORTB,5
                call    col1
                btfss   PORTB,6
                call    col2
                btfss   PORTB,7
                call    col3
; Index to next row by adding 4 to row code
                movf    rowCode,w       ; Code to w
                addlw   D'4'
                movwf   rowCode
;========================
;     shift row mask
;========================
; Set the carry flag
                bsf             STATUS,c
                rlf     rowMask,f       ; Rotate mask bits in storage
```

```
;=========================
;      end of keypad?
;=========================
; Test for last key row (maximum count is 4)
        decfsz   rowCount,f   ; Decrement counter
        goto     keyLoop
;=========================
;    display scan code
;=========================
; At this point all keys have been tested
; variable scanCode holds scan code
        movf     scanCode,w
        movwf    PORTD
        call     long_delay      ; Debounce
        goto     keyScan         ; Continue

;===========================
;    calculate scan code
;===========================
; The column position is added to the row code (stored
; in rowCode register). Sum is the scan code
col0:
        movf     rowCode,w       ; Row code to w
        addlw    0x00    ; Add 0 (clearly not necessary)
        movwf    scanCode ; Final value
        return

col1:m
        movf     rowCode,w       ; Row code to w
        addlw    0x01            ; Add 1
        movwf    scanCode
        return

col2:
        movf     rowCode,w       ; Row code to w
        addlw    0x02            ; Add 2
        movwf    scanCode
        return

col3:
        movf     rowCode,w       ; Row code to w
        addlw    0x03            ; Add 3
        movwf    scanCode
        return

;=============================
;    long delay sub-routine
;=============================
```

```
long_delay
                movlw   D'200'      ; w delay count
                movwf   J                       ; J = w
jloop:  movwf   K                   ; K = w
kloop:  decfsz  K,f                 ; K = K-1, skip next if zero
                goto    kloop
                decfsz  J,f         ; J = J-1, skip next if zero
                goto    jloop
                return
        End
```

```
; File name: RamDemo.asm
; Last revision: May 27, 2006
; Author: Julio Sanchez
; PIC: 16F877
;
; Description:
; Program to demonstrate access to General Purpose Register
; located in different banks. 16F877 GPR space is as
; follows:
; BANK 0          BANK 1          BANK 2          BANK 3
;  0x20            0xa0            0x110           0x190  |
;   --              --              --              --    | bank
;   --              --              --              --    | related
;   --             0xef            0x16f           0x1ef  |
;   --             0xf0            0x170           0x1f0  |
;  0x70          <==    access registers at 0x70-0x7f
;  0x7f                 these GPRs are common to all banks
;
; Program is designed to be used with a debugger so that
; access to different registers can be tested
; GPR registers are named according to the following format:
;
;               REGx_yyy
;                 |  |||
;                 |  |||_____ hex address (up to 3 digits)
;                 |_____ bank number (0 to 3)
;
; CONCLUSIONS:
; GPRs located between 0x70 and 0x7f can be accessed from
; any bank. GPRs at other addresses require previous bank
; selection
;
;===========================
;       16F877 switches
;===========================
; Switches used in __config directive:
;    _CP_ON           Code protection ON/OFF
; *  _CP_OFF
; *  _PWRTE_ON        Power-up timer ON/OFF
;    _PWRTE_OFF
;    _BODEN_ON        Brown-out reset enable ON/OFF
; *  _BODEN_OFF
; *  _PWRTE_ON        Power-up timer enable ON/OFF
;    _PWRTE_OFF
;    _WDT_ON          Watchdog timer ON/OFF
; *  _WDT_OFF
;    _LPV_ON          Low voltage IC programming enable ON/OFF
; *  _LPV_OFF
```

```
;    _CPD_ON         Data EE memory code protection ON/OFF
; *  _CPD_OFF
; OSCILLATOR CONFIGURATIONS:
;    _LP_OSC         Low power crystal occilator
;    _XT_OSC         External parallel resonator/crystal oscillator
; *  _HS_OSC         High speed crystal resonator
;    _RC_OSC         Resistor/capacitor oscillator
;    |               (simplest, 20% error)
;    |
;    |_____ * indicates setup values presently selected

        processor       16f877          ; Define processor
        #include <p16f877.inc>
        __CONFIG _CP_OFF & _WDT_OFF & _BODEN_OFF & _PWRTE_ON &
_HS_OSC & _WDT_OFF & _LVP_OFF & _CPD_OFF
; __CONFIG directive is used to embed configuration data
; within the source file. The labels following the directive
; are located in the corresponding .inc file.
        errorlevel -302
; Supress bank-related warning
;================================================================
;                       M A C R O S
;================================================================
; Macros to select the register banks
Bank0   MACRO                           ; Select RAM bank 0
        bcf     STATUS,RP0
        bcf     STATUS,RP1
        ENDM

Bank1   MACRO                           ; Select RAM bank 1
        bsf     STATUS,RP0
        bcf     STATUS,RP1
        ENDM

Bank2   MACRO                           ; Select RAM bank 2
        bcf     STATUS,RP0
        bsf     STATUS,RP1
        ENDM

Bank3   MACRO                           ; Select RAM bank 3
        bsf     STATUS,RP0
        bsf     STATUS,RP1
        ENDM
;========================================================
;                   PIC register equates
;========================================================
reg0_20 equ             0x20
```

Supplementary Programs

```
reg0_50   equ                 0x50
reg0_7e   equ                 0x7e
reg1_a0   equ                 0xa0
;==============================================================
;                      P R O G R A M
;==============================================================
          org             0           ; start at address
          goto    main
; Space for interrupt handlers
          org             0x08
main:
          Bank0
          nop
          movlw   0xee                ; Test value
          movwf   reg0_20
          movwf   reg0_7e
;
          Bank1
; Register at address 0x20 in bank 0 cannot be accessed
          nop
          movf    reg0_20,w
          nop
; However, the register at 0x7e CAN be accessed
          movf    reg0_7e,w
          nop
          Bank0
          movlw   0x0
          movf    reg0_7e,w
          nop
; How about from bank 3
          Bank3
          movlw   0x0
          movf    reg0_7e,w
          nop

loopHere:
          goto    loopHere
;==============================================================
          end               ; END OF PROGRAM
;==============================================================
```

```
; File name: RTC6355.asm
; Last revision: June 4, 2006
; Author: Julio Sanchez
; PIC: 16F877
;
; Description:
; Program to demonstrate programming of the NJU6355ED
; realtime clock IC.
;
; Operation:
;
; WARNING:
; Code assumes 10Mhz clock. Delay routines must be
; edited for faster clock. Clock speed also determines
; values for baud rate setting (see spbrgVal constant).
;
;===========================
;       16F877 switches
;===========================
; Switches used in __config directive:
;    _CP_ON          Code protection ON/OFF
; *  _CP_OFF
; *  _PWRTE_ON       Power-up timer ON/OFF
;    _PWRTE_OFF
;    _BODEN_ON       Brown-out reset enable ON/OFF
; *  _BODEN_OFF
; *  _PWRTE_ON       Power-up timer enable ON/OFF
;    _PWRTE_OFF
;    _WDT_ON         Watchdog timer ON/OFF
; *  _WDT_OFF
;    _LPV_ON         Low voltage IC programming enable ON/OFF
; *  _LPV_OFF
;    _CPD_ON         Data EE memory code protection ON/OFF
; *  _CPD_OFF
; OSCILLATOR CONFIGURATIONS:
;    _LP_OSC         Low power crystal occilator
;    _XT_OSC         External parallel resonator/crystal oscillator

; *  _HS_OSC         High speed crystal resonator
;    _RC_OSC         Resistor/capacitor oscillator
;    |               (simplest, 20% error)
;    |
;    |_____ * indicates setup values presently selected

        processor       16f877          ; Define processor
        #include <p16f877.inc>
        __CONFIG _CP_OFF & _WDT_OFF & _BODEN_OFF & _PWRTE_ON &
_HS_OSC & _WDT_OFF & _LVP_OFF & _CPD_OFF
```

Supplementary Programs

```
; __CONFIG directive is used to embed configuration data
; within the source file. The labels following the directive
; are located in the corresponding .inc file.

        errorlevel -302
; Supress bank-related warning
;===============================================================
;                       M A C R O S
;===============================================================
; Macros to select the register banks
Bank0   MACRO                           ; Select RAM bank 0
        bcf     STATUS,RP0
        bcf     STATUS,RP1
        ENDM

Bank1   MACRO                           ; Select RAM bank 1
        bsf     STATUS,RP0
        bcf     STATUS,RP1
        ENDM

Bank2   MACRO                           ; Select RAM bank 2
        bcf     STATUS,RP0
        bsf     STATUS,RP1
        ENDM

Bank3   MACRO                           ; Select RAM bank 3
        bsf     STATUS,RP0
        bsf     STATUS,RP1
        ENDM
;========================================================
;               constant definitions
;  for PIC-to-LCD pin wiring and LCD line addresses
;========================================================
#define E_line  1               ;|
#define RS_line 0               ;| -- from wiring diagram
#define RW_line 2               ;|
; LCD line addresses (from LCD data sheet)
#define LCD_1 0x80              ; First LCD line constant
#define LCD_2 0xc0              ; Second LCD line constant
#define LCDlimit .20; Number of characters per line
#define spbrgVal .64; For 2400 baud on 10Mhz clock

; Defines from real-time clock wiring diagram
#define CLK     PORTC,1
#define DAT     PORTC,3
#define IO      PORTC,5         ; input/output select
#define CE      PORTA,2         ; chip enable bit
```

```
;=========================================================
;               variables in PIC RAM
;=========================================================
; Local variables
; Reserve 20 bytes for string buffer
        cblock  0x20
        strData
        endc
; Reserve three bytes for ASCII digits
        cblock  0x34
        asc100
        asc10
        asc1
        endc
; Data
        cblock  0x37       ; Start of block
        count1             ; Counter # 1
        count2             ; Counter # 2
        count3             ; Counter # 3
        pic_ad
        J                  ; counter J
        K                  ; counter K
        index
        store1             ; Local storage
        store2
; For LCDscroll procedure
        LCDcount ; Counter for characters per line
        LCDline            ; Current display line (0 or 1)
; Communications variables
        newData            ; not 0 if new data received
        ascVal
        errorFlags
; Variables for Real-Time Clock
        year      ; Year (00h-99h)
        month     ; Month (01h-12h)
        day       ; Day of Month (01h-31h)
        dow       ; Day of Week (01h-07h)
        hour      ; Hour (00h-23h)
        min       ; Minute (00h-59h)
        sec       ; Second (00h-59h)
        T0        ; Temporary storage
        T1        ; Temporary storage
        endc

; EEPROM-related variables are placed in common area so
; they may be accessed from any bank
        cblock  0x70
        EEMemAdd ; EEPROM address to access
```

```
            EEByte              ; Data byte to write
; Storage for ASCII decimal conversion and digits
            inNum               ; Source operand
            thisDig             ; Digit counter
            endc

;================================================================
;                       P R O G R A M
;================================================================
                org             0           ; start at address
                goto    main
; Space for interrupt handlers
            org                 0x08
main:
; Wiring:
;       LCD data to port D, lines 0 to 7
;       E line -> port E, 1
;       RW line -> port E, 2
;       RS line -> port E, 0
; Set PORTE D and E for output
; Data memory bank selection bits:
; RP1:RP0           Bank
;   0:0              0      Ports A,B,C,D, and E
;   0:1              1      Tris A,B,C,D, and E
;   1:0              2
;   1:1              3
; First, initialize port B by clearing latches
            clrf    STATUS
            clrf    PORTB
; Select bank 1 to tris port D for output
            Bank1
; Tris port D for output. Port D lines 4 to 7 are wired
; to LCD data lines. Port D lines 0 to 4 are wired to LEDs.
            movlw   B'00000000'
            movwf   TRISD               ; and port D
; By default port A lines are analog. To configure them
; as digital code must set bits 1 and 2 of the ADCON1
; register (in bank 1)
            movlw   0x06        ; binary 0000 0110 is code to
                                ; make all port A lines digital
            movwf   ADCON1
; Port B, lines are wired to keypad switches, as follows:
;    7 6 5 4 3 2 1 0
;    | | | | |_|_|_|_____ switch rows (output)
;    |_|_|_|_____ switch columns (input)
; rows must be defined as output and columns as input
            movlw   b'11110000'
            movwf   TRISB
```

```
; Tris port E for output
        movlw   B'00000000'
        movwf   TRISE           ; Tris port E
; NJU6355 Interface:
; CLK           PORTC,1         Output
; DAT           PORTC,3         Output
; IO            PORTC,5         Output
; CE            PORTA,2         Output
        movlw   b'00000000'
        movwf   TRISC
        movlw   TRISA
; Enable port B pullups for switches in OPTION register
;       7  6  5  4  3  2  1  0 <= OPTION bits
;       |  |  |  |  |  |__|__|_____ PS2-PS0 (prescaler bits)
;       |  |  |  |  |               Values for Timer0
;       |  |  |  |  |               000 = 1:2    001 = 1:4
;       |  |  |  |  |               010 = 1:8    011 = 1:16
;       |  |  |  |  |               100 = 1:32   101 = 1:64
;       |  |  |  |  |               110 = 1:128 *111 = 1:256
;       |  |  |  |  |_____ PSA (prescaler assign)
;       |  |  |  |                 *1 = to WDT
;       |  |  |  |                  0 = to Timer0
;       |  |  |  |_____ TOSE (Timer0 edge select)
;       |  |  |                    *0 = increment on low-to-high
;       |  |  |                     1 = increment in high-to-low
;       |  |  |_____ TOCS (TMR0 clock source)
;       |  |                       *0 = internal clock
;       |  |                        1 = RA4/TOCKI bit source
;       |  |_____ INTEDG (Edge select)
;       |                          *0 = falling edge
;       |_____ RBPU (Pullup enable)
;                                  *0 = enabled
;                                   1 = disabled
        movlw   b'00001000'
        movwf   OPTION_REG
; Clear the write error flag (WRERR) in EECON1
        Bank3
        bcf     EECON1,WRERR
; Back to bank 0
        Bank0
; Clear all output lines
        movlw   b'00000000'
        movwf   PORTD
        movwf   PORTE
        movwf   PORTA
        movwf   PORTC
; Wait and initialize HD44780
        call    delay_5 ; Allow LCD time to initialize itself
```

Supplementary Programs

```
        call    initLCD   ; Then do forced initialization
        call    delay_5   ; (Wait probably not necessary)
; Clear character counter and line counter variables
        clrf    LCDcount
        clrf    LCDline
; Store base address of text buffer in PIC RAM
        movlw   0x20      ; Start address for buffer
        movwf   pic_ad    ; to local variable
; Initialize EEPROM address and data
        clrf    EEMemAdd  ; Set address to 0
        clrf    EEByte
;======================
;   first LCD line
;======================
; Store 20 blanks in PIC RAM, starting at address stored
; in variable pic_ad
        call    blank20
; Call procedure to store ASCII characters for message
; in text buffer
        movlw   d'0'      ; Offset into buffer
        call    storeMS1
;======================
;   Read EEPROM memory
;======================
; EEPROM memory address to use is at 10 (0x0a). Variable
; EEMemAdd is already initialized.
; Fill data for EEPROM is 0xff. This value indicates
; the first iteration
        call    EERead    ; Local procedure. Value in w
        movwf   EEByte    ; Store value
; At this point w must be 0
; EEPROM data still in w
        clrf    EEMemAdd  ; Address 0
        incf    EEByte,f
        call    EEWrite
; At this point iteration number is stored in EEByte
; This value must be displayed on the LCD at offset 11
; of the first line. This means it must be stored at offset
; 11 in the buffer. Since the buffer starts at 0x20 the
; iteration digit must be stored at offset 0x20+11=0x2b
ShowEEData:
; Binary data in EEByte
        movf    EEByte,w  ; Value to w
        call    bin2asc            ; Conversion routine
; At this point three ASCII digits are stored in local
; variables. Move digits to display area
        movf    asc1,w             ; Unit digit
        movwf   0x2b               ; Store in buffer
```

```
            movf    asc10,w         ; same with other digits
            movwf   0x2a
            movf    asc100,w
            movwf   0x29
; Display line
; Set DDRAM address to start of first line

showLine:

; Testing real time clock
;           call    initRTC         ; Initialize variables
;           call    Set_Time

; Wait
            movlw   .50
            movwf   count3
longWait:
            call    long_delay
            decfsz  count3,f
            goto    longWait

; Get variables from RTC
;           call    Get_Time
over1:
; Store data in EEPROM
            clrf    EEMemAdd        ; Address 0
            incf    EEMemAdd,f      ; Address 1
;           movf    Hours,w
            movlw   0xee
            movwf   EEByte
            call    EEWrite

            incf    EEMemAdd,f      ; Address 2
            movf    min,w
            movwf   EEByte
            call    EEWrite

            incf    EEMemAdd,f      ; Address 3
            movf    sec,w
            movwf   EEByte
            call    EEWrite

            call    line1
; Call procedure to display 16 characters in LCD
            call    display20
loopHere:
            goto    loopHere        ;done
```

```
;================================================================
;================================================================
;                  L O C A L    P R O C E D U R E S
;================================================================
;================================================================
;==========================
; init LCD for 4-bit mode
;==========================
initLCD:
; Initialization for Densitron LCD module as follows:
;       4-bit interface
;   2 display lines of 16 characters each
;   cursor on
;   left-to-right increment
;   cursor shift right
;   no display shift
;=======================|
;   set command mode    |
;=======================|
        bcf     PORTE,E_line        ; E line low
        bcf     PORTE,RS_line       ; RS line low
        bcf     PORTE,RW_line       ; Write mode
        call    delay_125           ; delay 125 microseconds
;***********************|
;     FUNCTION SET      |
;***********************|
        movlw   0x28       ; 0 0 1 0 1 0 0 0 (FUNCTION SET)
                           ;         | | | |__ font select:
                           ;         | | |     1 = 5x10 in 1/8 or 1/11
                           ;         | | |     0 = 1/16 dc
                           ;         | | |___ Duty cycle select
                           ;         | |      0 = 1/8 or 1/11
                           ;         | |      1 = 1/16
                           ;         | |___ Interface width
                           ;         |      0 = 4 bits
                           ;         |      1 = 8 bits
                           ;         |___ FUNCTION SET COMMAND
        call    send8      ; 4-bit send routine

; Set 4-bit mode command must be repeated
        movlw   0x28
        call    send8

;***********************|
; DISPLAY AND CURSOR ON |
;***********************|
        movlw   0x0e       ; 0 0 0 0 1 1 1 0 (DISPLAY ON/OFF)
                           ;         | | | |___ Blink character
```

```
                              ;        | | |      1 = on, 0 = off
                              ;        | | |___ Cursor on/off
                              ;        | |      1 = on, 0 = off
                              ;        | |____ Display on/off
                              ;        |        1 = on, 0 = off
                              ;        |____ COMMAND BIT
        call    send8
;*********************|
;   set entry mode    |
;*********************|
        movlw   0x06  ; 0 0 0 0 0 1 1 0 (ENTRY MODE SET)
                      ;           | | |___ display shift
                      ;           | |      1 = shift
                      ;           | |      0 = no shift
                      ;           | |____ increment mode
                      ;           |        1 = left-to-right
                      ;           |        0 = right-to-left
                      ;           |____ COMMAND BIT
        call    send8

;*********************|
; cursor/display shift|
;*********************|
        movlw   0x14  ; 0 0 0 1 0 1 0 0 (CURSOR/DISPLAY
                      ;         | | | | |   SHIFT)
                      ;         | | | |_|___ don't care
                      ;         | |_|__ cursor/display shift
                      ;         |        00 = cursor shift left
                      ;         |        01 = cursor shift right
                      ;         |        10 = cursor and display
                      ;         |             shifted left
                      ;         |        11 = cursor and display
                      ;         |             shifted right
                      ;         |____ COMMAND BIT
        call    send8
;*********************|
;    clear display    |
;*********************|
        movlw   0x01  ; 0 0 0 0 0 0 0 1 (CLEAR DISPLAY)
                      ;                 |___ COMMAND BIT
        call    send8
; Per documentation
        call    delay_5 ; Test for busy
        return

;=====================
;  Procedure to delay
;    42 microseconds
```

Supplementary Programs

```
;=======================
delay_125:
        movlw   D'42'           ; Repeat 42 machine cycles
        movwf   count1          ; Store value in counter
repeat
        decfsz  count1,f        ; Decrement counter
        goto    repeat          ; Continue if not 0
        return                  ; End of delay

;=======================
;   Procedure to delay
;   5 milliseconds
;=======================
delay_5:
        movlw   D'42'           ; Counter = 41
        movwf   count2          ; Store in variable
delay
        call    delay_125       ; Delay
        decfsz  count2,f        ; 40 times = 5 milliseconds
        goto    delay
        return                  ; End of delay
;=======================
;       pulse E line
;=======================
pulseE
        bsf     PORTE,E_line    ; Pulse E line
        nop
        bcf     PORTE,E_line
        return

;==============================
;   long delay sub-routine
;==============================
long_delay
                movlw   D'200'          ; w delay count
                movwf   J               ; J = w
jloop:  movwf   K                       ; K = w
kloop:  decfsz  K,f                     ; K = K-1, skip next if zero
                goto    kloop
                decfsz  J,f             ; J = J-1, skip next
if zero
                goto    jloop
                return


;==============================
;   LCD display procedure
;==============================
; Sends 20 characters from PIC buffer with address stored
```

```
; in variable pic_ad to LCD line previously selected
display20:
        call    delay_5                 ; Make sure not busy
; Set up for data
        bcf     PORTA,E_line    ; E line low
        bsf     PORTA,RS_line   ; RS line high for data
; Set up counter for 20 characters
        movlw   D'20'
        movwf   count3
; Get display address from local variable pic_ad
        movf    pic_ad,w ; First display RAM address to W
        movwf   FSR             ; W to FSR
getchar
        movf    INDF,w  ; get character from display RAM
                        ; location pointed to by file select
                        ; register
        call    send8           ; 4-bit interface routine
; Test for 16 characters displayed
        decfsz  count3,f        ; Decrement counter
        goto    nextchar        ; Skipped if done
        return
nextchar:
        incf    FSR,f           ; Bump pointer
        goto    getchar


;========================
;    send 2 nibbles in
;       4-bit mode
;========================
; Procedure to send two 4-bit values to port B lines
; 7, 6, 5, and 4. High-order nibble is sent first
; ON ENTRY:
;       w register holds 8-bit value to send
send8:
        movwf   store1          ; Save original value
        call    merge4          ; Merge with port B
; Now w has merged byte
        movwf   PORTD
        call    pulseE          ; Send data to LCD
; High nibble is sent
        movf    store1,w ; Recover byte into w
        swapf   store1,w ; Swap nibbles in w
        call    merge4
        movwf   PORTD
        call    pulseE          ; Send data to LCD
        call    delay_125
        return
```

```
;===========================
;       merge bits
;===========================
; Routine to merge the 4 high-order bits of the
; value to send with the contents of port B
; so as to preserve the 4 low-bits in port B
; Logic:
;       AND value with 1111 0000 mask
;       AND port B with 0000 1111 mask
;       Now low nibble in value and high nibble in
;       port B are all 0 bits:
;            value = vvvv 0000
;           port B = 0000 bbbb
;       OR value and port B resulting in:
;                    vvvv bbbb
; ON ENTRY:
;       w contains value bits
; ON EXIT:
;       w contains merged bits
merge4:
        andlw   b'11110000'     ; ANDing with 0 clears the
                                ; bit. ANDing with 1 preserves
                                ; the original value
        movwf   store2          ; Save result in variable
        movf    PORTD,w         ; port B to w register
        andlw   b'00001111'     ; Clear high nibble in port b
                                ; and preserve low nibble
        iorwf   store2,w ;      OR two operands in w
        return

;===========================
;   Set address register
;       to LCD line 2
;===========================
; ON ENTRY:
;       Address of LCD line 2 in constant LCD_2
line2:
        bcf     PORTE,E_line    ; E line low
        bcf     PORTE,RS_line   ; RS line low, setup for
                                ; control
        call    delay_5         ; Busy?
; Set to second display line
        movlw   LCD_2           ; Address with high-bit set
        call    send8
; Set RS line for data
        bsf     PORTE,RS_line   ; RS = 1 for data
        call    delay_5         ; Busy?
        return
```

```
;===========================
;    Set address register
;       to LCD line 1
;===========================
; ON ENTRY:
;          Address of LCD line 1 in constant LCD_1
line1:
        bcf     PORTE,E_line     ; E line low
        bcf     PORTE,RS_line    ; RS line low, set up for
                                 ; control
        call    delay_5          ; busy?
; Set to second display line
        movlw   LCD_1            ; Address and command bit
        call    send8            ; 4-bit routine
; Set RS line for data
        bsf     PORTE,RS_line    ; Setup for data
        call    delay_5          ; Busy?
        return

;===========================
;    scroll to LCD line 2
;===========================
; Procedure to count the number of characters displayed on
; each LCD line. If the number reaches the value in the
; constant LCDlimit, then display is scrolled to the second
; LCD line. If at the end of the second line, then LCD is
; reset to the first line.
LCDscroll:
        incf    LCDcount,f       ; Bump counter
; Test for line limit
        movf    LCDcount,w
        sublw   LCDlimit         ; Count minus limit
        btfss   STATUS,Z         ; Is count - limit = 0
        goto    scrollExit       ; Go if not at end of line
; At this point the end of the LCD line was reached
; Test if this is also the end of the second line
        movf    LCDline,w
        sublw   0x01             ; Is it line 1?
        btfsc   STATUS,Z         ; Is LCDline minus 1 = 0?
        goto    line2End         ; Go if end of second line
; At this point it is the end of the top LCD line
        call    line2            ; Scroll to second line
        clrf    LCDcount         ; Reset counter
        incf    LCDline,f        ; Bump line counter
        goto    scrollExit
; End of second LCD line
line2End:
```

```
            call    initLCD             ; Reset
            clrf    LCDcount            ; Clear counters
            clrf    LCDline
            call    line1               ; Display to first line
scrollExit:
            return

;================================
;   first text string procedure
;================================
storeMS1:
; Procedure to store in PIC RAM buffer the message
; contained in the code area labeled msg1
; ON ENTRY:
;           variable pic_ad holds address of text buffer
;           in PIC RAM
;           w register hold offset into storage area
;           msg1 is routine that returns the string characters
;           and a zero terminator
;           index is local variable that hold offset into
;           text table. This variable is also used for
;           temporary storage of offset into buffer
; ON EXIT:
;           Text message stored in buffer
;
; Store offset into text buffer (passed in the w register)
; in temporary variable
            movwf   index               ; Store w in index
; Store base address of text buffer in FSR
            movf    pic_ad,w ; first display RAM address to W
            addwf   index,w             ; Add offset to address
            movwf   FSR                 ; W to FSR
; Initialize index for text string access
            movlw   0                   ; Start at 0
            movwf   index               ; Store index in variable
; w still = 0
get_msg_char:
            call    msg1                ; Get character from table
; Test for zero terminator
            andlw   0x0ff
            btfsc   STATUS,Z            ; Test zero flag
            goto    endstr1             ; End of string
; ASSERT: valid string character in w
;         store character in text buffer (by FSR)
            movwf   INDF                ; store in buffer by FSR
            incf    FSR,f               ; increment buffer pointer
; Restore table character counter from variable
            movf    index,w             ; Get value into w
```

```
                addlw    1              ; Bump to next character
                movwf    index          ; Store table index in
variable
                goto     get_msg_char   ; Continue
endstr1:
                return
; Routine for returning message stored in program area
; Message has 10 characters
msg1:
                addwf    PCL,f          ; Access table
                retlw    'I'
                retlw    't'
                retlw    'e'
                retlw    'r'
                retlw    '.'
                retlw    0x20
                retlw    'N'
                retlw    'o'
                retlw    '.'
                retlw    0x20
                retlw    0
;========================
;     blank buffer
;========================
; Procedure to store 20 blank characters in PIC RAM
; buffer starting at address stored in the variable
; pic_ad
blank20:
                movlw    D'20'          ; Setup counter
                movwf    count1
                movf     pic_ad,w ; First PIC RAM address
                movwf    FSR            ; Indexed addressing
                movlw    0x20           ; ASCII space character
storeit
                movwf    INDF     ; Store blank character in PIC RAM
                                  ; buffer using FSR register
                decfsz   count1,f       ; Done?
                goto     incfsr         ; no
                return                  ; yes
incfsr
                incf     FSR,f    ; Bump FSR to next buffer space
                goto     storeit


;===============================
;   binary to ASCII decimal
;           conversion
;===============================
; ON ENTRY:
```

Supplementary Programs

```
;              w register has binary value in range 0 to 255
; ON EXIT:
;              output variables asc100, asc10, and asc1 have
;              three ASCII decimal digits
; Routine logic:
;   The value 100 is subtracted from the source operand
;   until the remainder is < 0 (carry cleared). The number
;   of subtractions is the decimal hundreds result. 100 is
;   then added back to the subtrahend to compensate
;   for the last subtraction. Now 10 is subtracted in the
;   same manner to determine the decimal tenths result.
;   The final remainder is the decimal units result.
; Variables:
;       inNum       storage for source operand
;       asc100      storage for hundreds position result
;       asc10       storage for tenth position result
;       asc1        storage for unit position reslt
;       thisDig     Digit counter
bin2asc:
        movwf   inNum       ; Save copy of source value
        clrf    asc100      ; Clear hundreds storage
        clrf    asc10       ; Tens
        clrf    asc1        ; Units
        clrf    thisDig
sub100:
        movlw   .100
        subwf   inNum,f     ; Subtract 100
        btfsc   STATUS,C    ; Did subtract overflow?
        goto    bump100     ; No. Count subtraction
        goto    end100
bump100:
        incf    thisDig,f   ;increment digit counter
        goto    sub100
; Store 100th digit
end100:
        movf    thisDig,w   ; Adjusted digit counter
        addlw   0x30        ; Convert to ASCII
        movwf   asc100      ; Store it
; Calculate tenth position value
        clrf    thisDig
; Adjust minuend
        movlw   .100        ; Minuend
        addwf   inNum,f     ; Add value to minuend to
                            ; compensate for last
                            ; operation
sub10:
        movlw   .10
        subwf   inNum,f     ; Subtract 10
```

```
            btfsc   STATUS,C          ; Did subtract overflow?
            goto    bump10            ; No. Count subtraction
            goto    end10
bump10:
            incf    thisDig,f         ;increment digit counter
            goto    sub10
; Store 10th digit
end10:
            movlw   .10
            addwf   inNum,f           ; Adjust for last subtract
            movf    thisDig,w         ; get digit counter contents
            addlw   0x30              ; Convert to ASCII
            movwf   asc10             ; Store it
; Calculate and store units digit
            movf    inNum,w           ; Store units value
            addlw   0x30              ; Convert to ASCII
            movwf   asc1              ; Store digit
            return
;================================================================
;                     local EEPROM data procedures
;================================================================
; GPRs used in EEPROM-related code are placed in the common
; RAM area (from 0x70 to 0x7f). This makes the registers
; accessible from any bank.
;==============================
;      read local EEPROM
;==============================
; Procedure to read EEPROM memory
; ON ENTRY:
; Address of EEPROM memory location to read is stored in
; local register EEMemAdd
; ON EXIT:
; Read data in w
EERead:
            Bank2
            movf    EEMemAdd,W        ; EEPROM address
            movwf   EEADR             ; to read from
            Bank3
            bcf     EECON1,EEPGD      ; Point to Data memory
            bsf     EECON1,RD         ; Start read
            Bank2
            movf    EEDATA,W          ; Data to w register
            Bank0
            return

;==============================
;      write local EEPROM
;==============================
```

```
; Procedure to write data byte to EEPROM memory
; ON ENTRY:
; Address to write stored in local register EEMemAdd
; Data byte to write is in local register EEByte
EEWrite:
        Bank3
Wait2Start:
        btfsc   EECON1,WR       ; Wait for
        GOTO    Wait2Start      ; write to finish
        Bank2
        movf    EEMemAdd,w      ; Address to
        movwf   EEADR           ; SFR
        movf    EEByte,w        ; Data to
        movwf   EEDATA          ; SFR
        Bank3
        bcf     EECON1,EEPGD    ; Point to Data memory
        bsf     EECON1,WREN     ; and enable writes
; Disable interrupts. Can be done in any case
        bcf     INTCON,GIE
; Write special codes
        movlw   0x55            ; First code is 0x55
        movwf   EECON2
        movlw   0xaa            ; Second code is 0xaa
        movwf   EECON2
        bsf     EECON1,WR       ; Start write operation
        nop                     ; Time for write
        nop
; Test for end of write operation
wait2End:
        btfsc   EECON1,WR       ; Wait until WR clear
        goto    wait2End
;
; Reenable interrupts if program uses interrupts
; If not, comment out next line
;       bsf     INTCON,GIE
;
        bcf     EECON1,WREN     ; Prevent accidental writes
        Bank0
        return
;================================================================
;                      6355 RTC procedures
;================================================================
; Write time/date to real-time clock
;

setclk
        movlw   08h             ; CE=low, IO=high
        movwf   PORTA
```

```
                movlw       10h                 ; CE,IO,CLK,DATA=out
                tris        PORTA
                movlw       year                ; INDF=year
                movwf       FSR
                bsf         IO          ; Enable clock I/O
                call        CS3                 ; Set year, month, and day
                movf        INDF,W              ; Set day of week
                call        CSNI
                call        CS2                 ; Set hr, min (zero second)
                bcf         IO          ; Disable clock I/O
                return
;
CS3             call        CSB                 ; Set next byte and advance
CS2             call        CSB                 ; Set next byte and advance
;
CSB             movf        INDF,W              ; Shift out LSN
                call        CSN
                swapf       INDF,W              ; Shift out MSN
CSNI            incf        FSR,f               ; Bump storage pointer
;
CSN             movwf       T1                  ; T1=data
                movlw       4                   ; T0=bit count
                movwf       T0
next
                movb        PORTA.0,T1.0        ; Output next bit
                rrf         T1,f                ; Ready next bit
                bsf         CLK                 ; Clock bit out of clock
                bcf         CLK
                decfsz      T0,f                ; Continue until byte done
                goto        next
                return                          ; Done
;
;=====================================
; Read time/date from real-time clock
;=====================================
getclk
                movlw       00h                 ; CE,IO=low
                movwf       PORTA
                movlw       11h                 ; CE,IO,CLK=out, DATA=in
                tris        PORTA
                movlw       year                ; INDF=year
                movwf       FSR
                bsf         IO                  ; Enable clock I/O
                call        CG3                 ; Get year, month, and day
                call        CGNI                ; Get day of week
                swapf       dow
                call        CG3                 ; Get hr, min, and sec
                bcf         IO          ; Disable clock I/O
```

Supplementary Programs

```
                return              ; Done
;
CG3     call    CGB         ; Get next byte and advance
        call    CGB         ; Get next byte and advance
;
CGB     call    CGN         ; Shift LSN in
CGNI    call    CGN         ; Shift MSN in
        call    CGMask      ; Mask unused bits
        andwf   INDF
        incf    FSR         ; Bump storage pointer
        return              ; Done
;
CGN     movlw   4           ; T0=bit count
        movwf   T0
next1   rrf     INDF        ; Shift accumulator
        movb    INDF.7,PORTA.0   ; Store bit to acc.
        bsf     CLK         ; Clock next bit
        bcf     CLK
        decfsz  T0          ; Continue until done
        goto    next1
        return              ; Done
;
CGMask  movf    FSR,W       ; W=RTC reg (0=year, ...)
        addlw   -year
        addwf   PCL         ; W=bit mask for register
        retlw   0FFh        ; Year (8 bits)
        retlw   01Fh        ; Month (5 bits)
        retlw   03Fh        ; Day of month (6 bits)
        retlw   070h        ; Day of week (3 bits)
        retlw   03Fh        ; Hour (6 bits)
        retlw   07Fh        ; Minute (7 bits)
        retlw   07Fh        ; Second (7 bits)

;===============================================================
        end                 ; END OF PROGRAM
;===============================================================
```

```
; File name: SerEEI2C.asm
; Last revision: May 28, 2006
; Author: Julio Sanchez
; PIC: 16F877
;
; Description:
; Receive character data through RS-232 line and store in
; 24LC04B EEPROM IC, using the I2C serial protocol in the
; PIC's MSSP module. Received characters are echoed on
; the second LCD line. When <Enter> key is detected (code
; 0x0d) the text stored in EEPROM memory is retrieved and
; displayed on the LCD. On startup the top LCD line displays
; the prompt: "Receiving:". At that time a message "Rdy- " is
; sent through the serial line so as to test the connection.
;
; Default serial line setting:
;               2400 baud
;               no parity
;               1 stop bit
;               8 character bits
;
; Wiring:
; 24LC04B SDA line is wired to PIC RC4 (MSSP SDA)
; 24LC04B SCL line is wired to PIC RC3 (MSSP SCL)
; 24LC04B A0-A2 and WP lines are not used (GND)
;
; Program to use 4-bit PIC-to-LCD interface.
; Code assumes that LCD is driven by Hitachi HD44780
; controller and PIC 16F977. Display supports two lines
; each one with 20 characters. The length, wiring and base
; address of each display line is stored in #define
; statements. These statements can be edited to accommodate
; a different set-up.
;
; WARNING:
; Code assumes 10 Mhz clock. Delay routines must be
; edited for a different clock. Clock speed also determines
; values for baud rate setting (see spbrgVal constant).
;
;===========================
;      16F877 switches
;===========================
; Switches used in __config directive:
;   _CP_ON          Code protection ON/OFF
; * _CP_OFF
; * _PWRTE_ON       Power-up timer ON/OFF
;   _PWRTE_OFF
;   _BODEN_ON       Brown-out reset enable ON/OFF
```

```
;  *  _BODEN_OFF
;  *  _PWRTE_ON      Power-up timer enable ON/OFF
;     _PWRTE_OFF
;     _WDT_ON        Watchdog timer ON/OFF
;  *  _WDT_OFF
;     _LPV_ON        Low voltage IC programming enable ON/OFF
;  *  _LPV_OFF
;     _CPD_ON        Data EE memory code protection ON/OFF
;  *  _CPD_OFF
; OSCILLATOR CONFIGURATIONS:
;     _LP_OSC        Low power crystal oscillator
;     _XT_OSC        External parallel resonator/crystal oscillator
;
;  *  _HS_OSC        High speed crystal resonator
;     _RC_OSC        Resistor/capacitor oscillator
;     |              (simplest, 20% error)
;     |
;     |_____ * indicates setup values presently selected

        processor       16f877          ; Define processor
        #include <p16f877.inc>
        __CONFIG _CP_OFF & _WDT_OFF & _BODEN_OFF & _PWRTE_ON & _HS_OSC & _WDT_OFF & _LVP_OFF & _CPD_OFF

; __CONFIG directive is used to embed configuration data
; within the source file. The labels following the directive
; are located in the corresponding .inc file.
        errorlevel -302
; Suppress bank-related warning
;===============================================================
;                        M A C R O S
;===============================================================
; Macros to select the register banks
Bank0   MACRO                           ; Select RAM bank 0
        bcf     STATUS,RP0
        bcf     STATUS,RP1
        ENDM

Bank1   MACRO                           ; Select RAM bank 1
        bsf     STATUS,RP0
        bcf     STATUS,RP1
        ENDM

Bank2   MACRO                           ; Select RAM bank 2
        bcf     STATUS,RP0
        bsf     STATUS,RP1
        ENDM
```

```
Bank3     MACRO                          ; Select RAM bank 3
          bsf       STATUS,RP0
          bsf       STATUS,RP1
          ENDM
;==========================================================
;                 constant definitions
;   for PIC-to-LCD pin wiring and LCD line addresses
;==========================================================
#define E_line 1           ; |
#define RS_line 0          ; | -- from wiring diagram
#define RW_line 2          ; |
; LCD line addresses (from LCD data sheet)
#define LCD_1 0x80         ; First LCD line constant
#define LCD_2 0xc0         ; Second LCD line constant
#define LCDlimit .20; Number of characters per line
#define spbrgVal .64; For 2400 baud on 10Mhz clock
; Note: The constants that define the LCD display
;         line addresses have the high-order bit set
;         so as to meet the requirements of controller
;         commands.


;==========================================================
;                    local equates
;==========================================================
WRITE_ADDR    equ    b'10100000' ; Control byte for write
READ_ADDR     equ    b'10100001' ; Control byte for read
;==========================================================
;                 General Purpose Variables
;==========================================================
; Local variables
; Reserve 20 bytes for string buffer
          cblock    0x20
          strData
          endc
; Other data
          cblock    0x34      ; Start of block
          count1              ; Counter # 1
          count2              ; Counter # 2
          count3              ; Counter # 3
          J                   ; counter J
          K                   ; counter K
          bufAdd
          index
          store1              ; Local storage
          store2
; For LCDscroll procedure
          LCDcount ; Counter for characters per line
          LCDline             ; Current display line (0 or 1)
```

Supplementary Programs

```
                endc
;===============================
;       Common RAM area
;===============================
; These GPRs can be accessed from any bank.
; 15 bytes are available, from 0x70 to 0x7f
        cblock  0x70
; Communications variables
        newData             ; not 0 if new data received
        ascVal
        errorFlags
; EEPROM-related variables
        datai               ; Data input byte buffer
        datao               ; Data output byte buffer
        bytecount           ; Counter for byte loops
        pollcnt             ; Counter for polling loops
        loops               ; Delay loop counter
        loops2              ; Delay loop counter
        EEMemAdd            ; EEPROM address to access
        EEByte              ; Data byte to write
        endc


;============================================================
;                       P R O G R A M
;============================================================
        org     0           ; start at address
        goto    main
; Space for interrupt handlers
        org             0x08
main:
; Wiring:
;       LCD data to port D, lines 0 to 7
;       E line -> port E, 1
;       RW line -> port E, 2
;       RS line -> port E, 0
; Set PORTE D and E for output
; First, initialize port B by clearing latches
        clrf    STATUS
        clrf    PORTB
; Select bank 1 to tris port D for output
        Bank1
; Tris port D for output. Port D lines 4 to 7 are wired
; to LCD data lines. Port D lines 0 to 4 are wired to LEDs.
        movlw   B'00000000'
        movwf   TRISD              ; and port D
; By default port A lines are analog. To configure them
; as digital code must set bits 1 and 2 of the ADCON1
; register (in bank 1)
```

```
        movlw   0x06                ; binary 0000 0110  is code to
                                    ; make all port A lines digital
        movwf   ADCON1
; Port B, lines are wired to keypad switches, as follows:
;   7 6 5 4 3 2 1 0
;   | | | | |_|_|_|_____ switch rows (output)
;   |_|_|_|_____ switch columns (input)
; rows must be defined as output and columns as input
        movlw   b'11110000'
        movwf   TRISB
; Tris port E for output
        movlw   B'00000000'
        movwf   TRISE               ; Tris port E
; Enable port B pullups for switches in OPTION register
        movlw   b'00001000'
        movwf   OPTION_REG
; Back to bank 0
        Bank0
; Initialize serial port for 2400 baud, 8 bits, no parity
; 1 stop
        call    InitSerial
; Test serial transmission by sending "RDY-"
        movlw   'R'
        call    SerialSend
        movlw   'D'
        call    SerialSend
        movlw   'Y'
        call    SerialSend
        movlw   '-'
        call    SerialSend
        movlw   0x20
        call    SerialSend
; Clear all output lines
        movlw   b'00000000'
        movwf   PORTD
        movwf   PORTE
; Wait and initialize HD44780
        call    delay_5  ; Allow LCD time to initialize itself
        call    initLCD  ; Then do forced initialization
        call    delay_5  ; (Wait probably not necessary)
; Clear character counter and line counter variables
        clrf    LCDcount
        clrf    LCDline
; Set display address to start of first LCD line
        call    line1
; Store address of display buffer
        movlw   0x20
        movwf   bufAdd
```

```
; Display "Receiving:" message prompt
        call    blank20         ; Clear buffer
        movlw   0x00            ; Offset in buffer
        call    storeMS1 ; Store message at offset
        call    display20       ; Display message
; Start address of EEPROM
        clrf    EEMemAdd
; Setup for display in second line
        call    line2
        clrf    LCDline
        incf    LCDline,f ; Set scroll control for line 2
;================================================================
;          receive serial data, store, and display
;================================================================
receive:
; Call serial receive procedure
        call    SerialRcv
; HOB of newData register is set if new data
; received
        btfss   newData,7
        goto    scanExit
; At this point new data was received.
        movwf   EEByte          ; Save received character
; Display character on LCD
        movf    EEByte,w ; Recover character
        call    send8           ; Display in LCD
        call    LCDscroll       ; Scroll at end of line
; Store character in EEPROM at location in EEMemAdd
        call    ic2Write        ; Local procedure
        incf    EEMemAdd,f      ; Bump to next EEPROM
; Check for <Enter> key (0x0d) and execute display function
        movf    EEByte,w ; Recover last received
        sublw   0x0d
        btfsc   STATUS,Z ; Test if <Enter> key
        goto    isEnter         ; Go if <Enter>
; Not <Enter> key, continue processing
scanExit:
        goto    receive         ; Continue
;=============================
;    display EEPROM data
;=============================
; This routine receives control when the <Enter> key is
; received.
; Action:
;          1. Clear LCD
;          2. Output is set to top LCD line
;          3. Characters stored in EEPROM are displayed
;             until 0x0d code is detected
```

```
isEnter:
        call    clearLCD
; Clear character counter and line counter variables
        clrf    LCDcount
        clrf    LCDline
; Read data from EEPROM memory, starting at address 0
; and display on LCD until 0x0d terminator
        call    line1
        clrf    EEMemAdd ; Start at EEPROM 0
readOne:
        call    ic2Read         ; Get character
; Store character
        movwf   EEByte          ; Save character
; Test for terminator
        sublw   0x0d
        btfsc   STATUS,Z ; Test if 0x0d
        goto    atEnd           ; Go if 0x0d
; At this point character read is not 0x0d
; Display on LCD
        movf    EEByte,w ; Recover character
; Display character on LCD
        call    send8           ; Display in LCD
        call    LCDscroll       ; Scroll at end of line
        incf    EEMemAdd,f      ; Next EEPROM byte
        goto    readOne

; End of execution
atEnd:
        goto    atEnd


;================================================================
;================================================================
;                 L O C A L    P R O C E D U R E S
;================================================================
;================================================================
;=========================
; init LCD for 4-bit mode
;=========================
initLCD:
; Initialization for Densitron LCD module as follows:
;       4-bit interface
;    2 display lines of 16 characters each
;    cursor on
;    left-to-right increment
;    cursor shift right
;    no display shift
;======================|
;    set command mode  |
```

```
;=======================|
        bcf     PORTE,E_line        ; E line low
        bcf     PORTE,RS_line       ; RS line low
        bcf     PORTE,RW_line       ; Write mode
        call    delay_125                   ; delay 125 microseconds
        movlw   0x28        ; 0 0 1 0 1 0 0 0 (FUNCTION SET)
        call    send8       ; 4-bit send routine
; Set 4-bit mode command must be repeated
        movlw   0x28
        call    send8
        movlw   0x0e        ; 0 0 0 0 1 1 1 0 (DISPLAY ON/OFF)
        call    send8
        movlw   0x06        ; 0 0 0 0 0 1 1 0 (ENTRY MODE SET)
        call    send8
        movlw   0x14        ; 0 0 0 1 0 1 0 0 (CURSOR/DISPLAY
                            ;                           SHIFT)
        call    send8
        movlw   0x01        ; 0 0 0 0 0 0 0 1 (CLEAR DISPLAY)
                            ;                 |___ COMMAND BIT
        call    send8
        call    delay_5  ; Test for busy
        return

.;===========================
;    procedure to clear LCD
;===========================
clearLCD:
        bcf             PORTE,E_line        ; E line low
        bcf             PORTE,RS_line       ; RS line low
        bcf             PORTE,RW_line       ; Write mode
        call    delay_125                   ; delay 125 microseconds
        movlw   0x01        ; 0 0 0 0 0 0 0 1 (CLEAR DISPLAY)
                            ;                 |___ COMMAND BIT
        call    send8
        call    delay_5  ; Test for busy
        return

;======================
;   Procedure to delay
;    42 microseconds
;======================
delay_125:
        movlw   .105                        ; Repeat 105 machine cycles
        movwf   count1                      ; Store value in counter
repeat
```

```
            decfsz    count1,f          ; Decrement counter
            goto      repeat            ; Continue if not 0
            return                      ; End of delay

;=======================
;    Procedure to delay
;    5 milliseconds
;=======================
delay_5:
            movlw     .105              ; Counter = 105 cycles
            movwf     count2            ; Store in variable
delay
            call      delay_125         ; Delay
            decfsz    count2,f          ; 40 times = 5 milliseconds
            goto      delay
            return                      ; End of delay
;========================
;      pulse E line
;========================
pulseE
            bsf       PORTE,E_line      ; Pulse E line
            nop
            bcf       PORTE,E_line
            return

;=============================
;    long delay sub-routine
;=============================
long_delay
            movlw     D'200'     ; w delay count
            movwf     J                 ; J = w
jloop:      movwf     K                 ; K = w
kloop:      decfsz    K,f               ; K = K-1, skip next if zero
            goto      kloop
            decfsz    J,f               ; J = J-1, skip next if zero
            goto      jloop
            return
;=======================
;    send 2 nibbles in
;      4-bit mode
;=======================
; Procedure to send two 4-bit values to port B lines
; 7, 6, 5, and 4. High-order nibble is sent first
; ON ENTRY:
;        w register holds 8-bit value to send
send8:
            movwf     store1            ; Save original value
            call      merge4            ; Merge with port B
```

```
; Now w has merged byte
        movwf   PORTD           ; w to port D
        call    pulseE          ; Send data to LCD
; High nibble is sent
        movf    store1,w ; Recover byte into w
        swapf   store1,w ; Swap nibbles in w
        call    merge4
        movwf   PORTD
        call    pulseE          ; Send data to LCD
        call    delay_125
        return
;===========================
;       merge bits
;===========================
; Routine to merge the 4 high-order bits of the
; value to send with the contents of port B
; so as to preserve the 4 low-bits in port B
; Logic:
;       AND value with 1111 0000 mask
;       AND port B with 0000 1111 mask
;       Now low nibble in value and high nibble in
;       port B are all 0 bits:
;           value  = vvvv 0000
;           port B = 0000 bbbb
;       OR value and port B resulting in:
;                  vvvv bbbb
; ON ENTRY:
;       w contain value bits
; ON EXIT:
;       w contains merged bits
merge4:
        andlw   b'11110000'     ; ANDing with 0 clears the
                                ; bit. ANDing with 1 preserves
                                ; the original value
        movwf   store2          ; Save result in variable
        movf    PORTD,w         ; port B to w register
        andlw   b'00001111' ; Clear high nibble in port b
                                ; and preserve low nibble
        iorwf   store2,w ; OR two operands in w
        return
;===========================
;   Set address register
;       to LCD line 2
;===========================
; ON ENTRY:
;       Address of LCD line 2 in constant LCD_2
line2:
        bcf     PORTE,E_line    ; E line low
```

```
            bcf     PORTE,RS_line     ; RS line low, setup for
                                      ; control
            call    delay_5           ; Busy?
; Set to second display line
            movlw   LCD_2             ; Address with high-bit set
            call    send8
; Set RS line for data
            bsf     PORTE,RS_line     ; RS = 1 for data
            call    delay_5           ; Busy?
            return
;===========================
;   Set address register
;       to LCD line 1
;===========================
; ON ENTRY:
;       Address of LCD line 1 in constant LCD_1
line1:
            bcf     PORTE,E_line      ; E line low
            bcf     PORTE,RS_line     ; RS line low, set up for
                                      ; control
            call    delay_5           ; busy?
; Set to second display line
            movlw   LCD_1             ; Address and command bit
            call    send8             ; 4-bit routine
; Set RS line for data
            bsf     PORTE,RS_line     ; Setup for data
            call    delay_5           ; Busy?
            return

;===========================
;   scroll to LCD line 2
;===========================
; Procedure to count the number of characters displayed on
; each LCD line. If the number reaches the value in the
; constant LCDlimit, then display is scrolled to the second
; LCD line. If at the end of the second line, then LCD is
; reset to the first line.
LCDscroll:
            incf    LCDcount,f        ; Bump counter
; Test for line limit
            movf    LCDcount,w
            sublw   LCDlimit          ; Count minus limit
            btfss   STATUS,Z          ; Is count - limit = 0
            goto    scrollExit        ; Go if not at end of line
; At this point the end of the LCD line was reached
; Test if this is also the end of the second line
            movf    LCDline,w
            sublw   0x01              ; Is it line 1?
```

```
            btfsc    STATUS,Z          ; Is LCDline minus 1 = 0?
            goto     line2End          ; Go if end of second line
; At this point it is the end of the top LCD line
            call     line2             ; Scroll to second line
            clrf     LCDcount          ; Reset counter
            incf     LCDline,f         ; Bump line counter
            goto     scrollExit
; End of second LCD line
line2End:
            call     initLCD           ; Reset
            clrf     LCDcount          ; Clear counters
            clrf     LCDline
            call     line1             ; Display to first line
scrollExit:
            return

;==============================
;    LCD display procedure
;==============================
; Sends 20 characters from PIC buffer with address stored
; in variable bufAdd to LCD line previously selected
display20:
            call     delay_5           ; Make sure not busy
; Set up for data
            bcf      PORTA,E_line      ; E line low
            bsf      PORTA,RS_line     ; RS line high for data
; Set up counter for 20 characters
            movlw    D'20'
            movwf    count3
; Get display address from local variable bufAdd
            movf     bufAdd,w          ; First display RAM address to W
            movwf    FSR               ; W to FSR
getchar
            movf     INDF,w            ; get character from display RAM
                                       ; location pointed to by file select
                                       ; register
            call     send8             ; 4-bit interface routine
; Test for 20 characters displayed
            decfsz   count3,f          ; Decrement counter
            goto     nextchar          ; Skipped if done
            return
nextchar:
            incf     FSR,f             ; Bump pointer
            goto     getchar

;==============================
;   first text string procedure
;==============================
```

```
storeMS1:
; Procedure to store in PIC RAM buffer the message
; contained in the code area labeled msg1
; ON ENTRY:
;           variable bufAdd holds address of text buffer
;           in PIC RAM
;           w register hold offset into storage area
;           msg1 is routine that returns the string characters
;           and a zero terminator
;           index is local variable that hold offset into
;           text table. This variable is also used for
;           temporary storage of offset into buffer
; ON EXIT:
;           Text message stored in buffer
;
; Store offset into text buffer (passed in the w register)
; in temporary variable
         movwf    index              ; Store w in index
; Store base address of text buffer in FSR
         movf     bufAdd,w ; first display RAM address to W
         addwf    index,w             ; Add offset to address
         movwf    FSR                 ; W to FSR
; Initialize index for text string access
         movlw    0                   ; Start at 0
         movwf    index               ; Store index in variable
; w still = 0
get_msg_char:
         call     msg1                ; Get character from table
; Test for zero terminator
         andlw    0x0ff
         btfsc    STATUS,Z ; Test zero flag
         goto     endstr1             ; End of string
; ASSERT: valid string character in w
;         store character in text buffer (by FSR)
         movwf    INDF                ; store in buffer by FSR
         incf     FSR,f               ; increment buffer pointer
; Restore table character counter from variable
         movf     index,w             ; Get value into w
         addlw    1                   ; Bump to next character
         movwf    index               ; Store table index in
                                      ; variable
         goto     get_msg_char        ; Continue
endstr1:
         return

; Routine for returning message stored in program area
; Message has 10 characters
msg1:
```

```
        addwf   PCL,f           ; Access table
        retlw   'R'
        retlw   'e'
        retlw   'C'
        retlw   'e'
        retlw   'i'
        retlw   'v'
        retlw   'i'
        retlw   'n'
        retlw   'g'
        retlw   ':'
        retlw   0

;========================
;     blank buffer
;========================
; Procedure to store 20 blank characters in PIC RAM
; buffer starting at address stored in the variable
; bufAdd
blank20:
        movlw   D'20'           ; Setup counter
        movwf   count1
        movf    bufAdd,w        ; First PIC RAM address
        movwf   FSR             ; Indexed addressing
        movlw   0x20            ; ASCII space character
storeit
        movwf   INDF    ; Store blank character in PIC RAM
                        ; buffer using FSR register
        decfsz  count1,f        ; Done?
        goto    incfsr          ; no
        return                  ; yes
incfsr
        incf    FSR,f           ; Bump FSR to next buffer space
        goto    storeit


;================================================================
;              communications procedures
;================================================================
; Initialize serial port for 2400 baud, 8 bits, no parity,
; 1 stop
InitSerial:
        Bank1                                   ; Macro to select bank1
; Bits 6 and 7 of Port C are multiplexed as TX/CK and RX/DT
; for USART operation. These bits must be set to input in the
; TRISC register
        movlw   b'11000000'     ; Bits for TX and RX
```

```
              iorwf    TRISC,f            ; OR into Trisc register
; The asynchronous baud rate is calculated as follows:
;                     Fosc
;            ABR = -----------
;                    S*(x+1)
; where x is value in the SPBRG register and S is 64 if the high
; baud rate select bit (BRGH) in the TXSTA control register is
; clear, and 16 if the BRGH bit is set. For setting to 2400 baud
; using a 10Mhs oscillator at a slow baud rate the formula
; is:
; At slow speed (BRGH = 0)
;            10,000,000     10,000,000
;           ----------  =  ----------  = 2,403.84 (0.16% error)
;            64*(64+1)        4160
;
          movlw    spbrgVal         ; Value in spbrgVal = 64
          movwf    SPBRG            ; Place in baud rate generator
; Setup value: 0010 0000 = 0x20
          movlw    0x20             ; Enable transmission and high
                                    ; baud rate
          movwf    TXSTA
          Bank0                     ; Bank 0
; Setup value: 1001 0000 = 0x90
          movlw    0x90             ; Enable serial port and
                                    ; continuous reception
          movwf    RCSTA
;
          clrf     errorFlags       ; Clear local error flags
                                    ; register
          return
;==============================
;       transmit data
;==============================
; Test for Transmit Register Empty and transmit data in w
SerialSend:
          Bank0                     ; Select bank 0
          btfss    PIR1,TXIF        ; check if transmitter busy
          goto     $-1              ; wait until transmitter is
                                    ; not busy
          movwf    TXREG            ; and transmit the data
          return

;==============================
;       receive data
;==============================
; Procedure to test line for data received and return value
; in w. Overrun and framing errors are detected and
; remembered in the variable errorFlags, as follows:
```

```
;              7  6  5  4  3  2  1  0    <== errorFlags
;              -- not used ----  |  |___ overrun error
;                                |_____ framing error
SerialRcv:
        clrf    newData                 ; Clear new data received register
        Bank0                           ; Select bank 0
; Bit 5 (RCIF) of the PIR1 Register is clear if the USART
; receive buffer is empty. If so, no data has been received
        btfss   PIR1,RCIF               ; Check for received data
        return                          ; Exit if no data
; At this point data has been received. First eliminate
; possible errors: overrun and framing.
; Bit 1 (OERR) of the RCSTA register detects overrun
; Bit 2 (FERR( of the RCSTA register detects framing error
        btfsc   RCSTA,OERR              ; Test for overrun error
        goto    OverErr                 ; Error handler
        btfsc   RCSTA,FERR              ; Test for framing error
        goto    FrameErr ; Error handler
; At this point no error was detected
; Received data is in the USART RCREG register
        movf    RCREG,w                 ; get received data
        bsf     newData,7               ; Set bit 7 to indicate new data
; Clear error flags
        clrf    errorFlags
        return
;=========================
;     error handlers
;=========================
OverErr:
        bsf     errorFlags,0    ; Bit 0 is overrun error
; Reset system
        bcf     RCSTA,CREN      ; Clear continuous receive bit
        bsf     RCSTA,CREN      ; Set to re-enable reception
        return
;error because FERR framing error bit is set
;can do special error handling here - this code simply clears
; and continues
FrameErr:
        bsf     errorFlags,1;   Bit 1 is framing error
        movf    RCREG,W         ; Read and throw away bad data
        return
;================================================================
;                    I2C EEPROM data procedures
;================================================================
; GPRs used in EEPROM-related code are placed in the common
; RAM area (from 0x70 to 0x7f). This makes the registers
; accessible from any bank.
```

```
;******************Byte read test subroutine ******************
;           This routine tests the byte read feature
;           of the serial EEPROM device.  It will read
;           1 byte of data at address 0x5AA5 from the device.
;**************************************************************
ic2Read
    call    BSTART              ; Generate Start condition
                                ; Send control byte
    bcf     STATUS,RP0          ; Select Bank 00
    movlw   WRITE_ADDR          ; Load control byte for write
    movwf   datao               ; Copy to datao for output
    call    TX                  ; Send control byte to device
                                ; Send word address high byte
    bcf     STATUS,RP0          ; Select Bank 00
    movlw   0x5A                ; Load 0x5A for word address
    movwf   datao               ; Copy to datao for output
    call    TX                  ; Send high byte to device
                                ; Send word address low byte
    bcf     STATUS,RP0          ; Select Bank 00
    movlw   0xA5                ; Load 0xA5 for word address
    movwf   datao               ; Copy to datao for output
    call    TX                  ; Send word address to device
    call    BRESTART            ; Generate Restart condition
                                ; Send control byte
    bcf     STATUS,RP0          ; Select Bank 00
    movlw   READ_ADDR           ; Load control byte for read
    movwf   datao               ; Copy to datao for output
    call    TX                  ; Send control byte to device
                                ; Read data byte
    bsf     STATUS,RP0          ; Select Bank 01
    bsf     SSPCON2,ACKDT       ; Select to send NO ACK bit
    call    RX                  ; Read data byte from device
    call    BSTOP               ; Generate Stop condition
    retlw   0

;******************Byte write test subroutine*****************
;           This routine tests the byte write feature
;           of the serial EEPROM device.  It will write
;           1 byte of data to the device at address 0x5AA5.
;**************************************************************
ic2Write
    call    BSTART              ; Generate Start condition
                                ; Send control byte
    bcf     STATUS,RP0          ; Select Bank 00
    movlw   WRITE_ADDR          ; Load control byte for write
    movwf   datao               ; Copy to datao for output
    call    TX                  ; Send control byte to device
                                ; Send word address high byte
```

Supplementary Programs

```
        bcf     STATUS,RP0          ; Select Bank 00
        movlw   0x5A                ; Load 0x5A for word address
        movwf   datao               ; Copy to datao for output
        call    TX
                                    ; Send word address low byte
        bcf     STATUS,RP0          ; Select Bank 00
        movlw   0xA5                ; Load 0xA5 for word address
        movwf   datao               ; Copy to datao for output
        call    TX                  ; Send word address to device
                                    ; Send data byte
        bcf     STATUS,RP0          ; Select Bank 00
        movlw   0xAA                ; Load 0xAA for data byte
        movwf   datao               ; Copy to datao for output
        call    TX                  ; Send data byte to device
        call    BSTOP               ; Generate Stop condition
        call    Poll                ; Poll for write completion
        retlw   0

;*****************Acknowledge Polling subroutine**************
;           This subroutine polls the EEPROM device
;           for an ACK bit, which indicates that the
;           internal write cycle has completed. Code
;           is in place for a timeout routine, just
;           uncomment the 'goto TimedOut' line, and
;           provide a 'TimedOut' label.
;*************************************************************
Poll
        bcf     STATUS,RP0          ; Select Bank 00
        movlw   .40
        movwf   pollcnt             ; Set max polling times to 40
polling
        call    BRESTART            ; Generate start bit
        bcf     STATUS,RP0          ; Select Bank 00
        movlw   WRITE_ADDR          ; Now send the control byte
        movwf   datao               ; Copy control byte to buffer
        call    TX                  ; Output control byte to device
        bsf     STATUS,RP0          ; Select Bank 01
        btfss   SSPCON2,ACKSTAT     ; Was the ACK bit low?
        goto    exitpoll            ; If yes, stop polling
                                    ; If no, check if polled 40 times
        bcf     STATUS,RP0          ; Select Bank 00
        decfsz  pollcnt,F           ; Is poll counter down to zero?
        goto    polling             ; If no, poll again
;       goto    TimedOut            ; If yes, part didn't respond
                                    ; in time, so take action
exitpoll
        call    BSTOP               ; Generate stop bit
```

```
        movlw   .40
        movwf   pollcnt             ; Set max polling times to 40
polling
        call    BRESTART            ; Generate start bit
        bcf     STATUS,RP0          ; Select Bank 00
        movlw   WRITE_ADDR          ; Now send the control byte
        movwf   datao               ; Copy control byte to buffer
        call    TX                  ; Output control byte to device
        bsf     STATUS,RP0          ; Select Bank 01
        btfss   SSPCON2,ACKSTAT     ; Was the ACK bit low?
        goto    exitpoll            ; If yes, stop polling
                                    ; If no, check if polled 40
times
        bcf     STATUS,RP0          ; Select Bank 00
        decfsz  pollcnt,F           ; Is poll counter down to zero?
        goto    polling             ; If no, poll again
;       goto    TimedOut            ; If yes, part didn't respond
                                    ; in time, so take action
exitpoll
        call    BSTOP               ; Generate stop bit
        retlw   0
;*****************Initialization subroutine******************
;           This routine initializes the MSSP module
;           for I2C Master mode, with a 100 kHz clock.
;************************************************************
Init
        bcf     STATUS,RP1          ; Select Bank 01
        bsf     STATUS,RP0
        movlw   b'11111111'
        movwf   TRISC               ; Set PORTC to all inputs
        clrf    SSPSTAT             ; Disable SMBus inputs
        bsf     SSPSTAT,SMP         ; Disable slew rate control
        movlw   0x18                ; Load 0x18 into WREG
        movwf   SSPADD              ; Setup 100 kHz I2C clock
        clrf    SSPCON2             ; Clear control bits
        bcf     STATUS,RP0          ; Select Bank 00
        movlw   b'00101000'
        movwf   SSPCON              ; Enable SSP, select I2C Master
mode
        bcf     PIR1,SSPIF          ; Clear SSP interrupt flag
        bcf     PIR2,BCLIF          ; Clear Bit Collision flag
        retlw   0

;*****************Start bit subroutine**********************
;           This routine generates a Start condition
;           (high-to-low transition of SDA while SCL
;           is still high).
;************************************************************
```

```
BSTART
    bcf     STATUS,RP1
    bcf     STATUS,RP0          ; Select Bank 00
    bcf     PIR1,SSPIF          ; Clear SSP interrupt flag
    bsf     STATUS,RP0          ; Select Bank 01
    bsf     SSPCON2,SEN         ; Generate Start condition
    bcf     STATUS,RP0          ; Select Bank 00
bstart_wait
    btfss   PIR1,SSPIF          ; Check if operation completed
    goto    bstart_wait         ; If not, keep checking
    retlw   0

;*****************Restart bit subroutine**********************
;           This routine generates a Repeated Start
;           condition (high-to-low transition of SDA
;           while SCL is still high).
;*************************************************************
BRESTART
    bcf     STATUS,RP1
    bcf     STATUS,RP0          ; Select Bank 00
    bcf     PIR1,SSPIF          ; Clear SSP interrupt flag
    bsf     STATUS,RP0          ; Select Bank 01
    bsf     SSPCON2,RSEN        ; Generate Restart condition
    bcf     STATUS,RP0          ; Select Bank 00
brestart_wait
    btfss   PIR1,SSPIF          ; Check if operation completed
    goto    brestart_wait       ; If not, keep checking
    retlw   0

;*****************Stop bit subroutine***********************
;           This routine generates a Stop condition
;           (low-to-high transition of SDA while SCL
;           is still high).
;*************************************************************
BSTOP
    bcf     STATUS,RP1
    bcf     STATUS,RP0          ; Select Bank 00
    bcf     PIR1,SSPIF          ; Clear SSP interrupt flag
    bsf     STATUS,RP0          ; Select Bank 01
    bsf     SSPCON2,PEN         ; Generate Stop condition
    bcf     STATUS,RP0          ; Select Bank 00
bstop_wait
    btfss   PIR1,SSPIF          ; Check if operation completed
    goto    bstop_wait          ; If not, keep checking
    retlw   0

;*****************Data transmit subroutine********************
```

```
;               This routine transmits the byte of data
;               stored in 'datao' to the serial EEPROM
;               device. Instructions are also in place
;               to check for an ACK bit, if desired.
;               Just replace the 'goto' instruction,
;               or create an 'ackfailed' label, to provide
;               the functionality.
;****************************************************************
TX
    bcf     STATUS,RP1
    bcf     STATUS,RP0              ; Select Bank 00
    bcf     PIR1,SSPIF              ; Clear SSP interrupt flag
    movf    datao,W                 ; Copy datao to WREG
    movwf   SSPBUF                  ; Write byte out to device
tx_wait
    btfss   PIR1,SSPIF              ; Check if operation completed
    goto    tx_wait                 ; If not, keep checking
;   bsf     STATUS,RP0              ; Select Bank 01
;   btfsc   SSPCON2,ACKSTAT         ; Check if ACK bit was received
;   goto    ackfailed               ; This executes if no ACK
received
    retlw   0

;******************Data receive subroutine*********************
;               This routine reads in one byte of data from
;               the serial EEPROM device, and stores it in
;               'datai'. It then responds with either an
;               ACK or a NO ACK bit, depending on the value
;               of 'ACKDT' in 'SSPCON2'.
;****************************************************************
RX
    bcf     STATUS,RP1
    bcf     STATUS,RP0              ; Select Bank 00
    bcf     PIR1,SSPIF              ; Clear SSP interrupt flag
    bsf     STATUS,RP0              ; Select Bank 01
    bsf     SSPCON2,RCEN            ; Initiate reception of byte
    bcf     STATUS,RP0              ; Select Bank 00
rx_wait
    btfss   PIR1,SSPIF              ; Check if operation completed
    goto    rx_wait                 ; If not, keep checking
    movf    SSPBUF,W                ; Copy byte to WREG
    movwf   datai                   ; Copy WREG to datai
    bcf     PIR1,SSPIF              ; Clear SSP interrupt flag
    bsf     STATUS,RP0              ; Select Bank 01
    bsf     SSPCON2,ACKEN           ; Generate ACK/NO ACK bit
    bcf     STATUS,RP0              ; Select Bank 00
rx_wait2
    btfss   PIR1,SSPIF              ; Check if operation completed
```

```
        goto    rx_wait2                ; If not, keep checking
        retlw   0

;================================================================
        end                     ; END OF PROGRAM
;================================================================
```

Index

!MCLR pin 147,223
#define directive 288
__config 82-183,189,195,198,201,203,208,
   233,236,260,263,265,269,306,317,328,
   390,395,400,406,421,439,492,505,522,
   568,581,595,659,663,667,678,682,685,
   689,704,708,729,743,749,752,772
16F84A UART Emulation 369
16F877 PIC Initialization Code ,381
16F87x USART
   Asynchronous Receiver 380
   Asynchronous Transmitter ,379
   Module 376
24LC04B EEPROM 460,479,481,486,
   488,490,521,524,539,708,710,726,772
4050 hex buffer IC 121
4-bit data transfer mode 302,304
555 timer 106
7400 NAND gate 87
7402 NOR quad 2-input NOR gate 88
7404 hex inverter 86,105
7408 AND gate 87
74138 118
7414 hex Schmitt trigger 93
74165 IC 114
7432 quad 2-input OR gate 88
7486 quad 2-input XOR gate 90
7805 voltage regulator 96,192

A
A/D
   channels 141
   conversion clock 550-551
   conversions 161,543,549
   converters 161,549
   module 161,549-550,552,554-556,593
abacists 21
   and algorists 21
ACK condition 477,484

acquisition time 545,556,593
active-high 199
active-low 147,199,215,389,395
ADC0831 546-549,555,568-569,571,579
ADCON0 550-552,554,556-557,593-594
ADCON1 180-181,550,552-553,556,593
address counter 276,278,283-285
ADRESH 550,553,556-557,593-594
ADRESL 550,553,557,594
al-Khowarizmi 21
ALU 59,149,151,243
American Standard Code for Information
   Interchange 34
Ampere, Andre 5
analog-to-digital converters 95,129
AND gate 84,87-90,97-98,101,108,110,
   114,116
ANSI/IEEE Standard 754
arithmetic
   instructions 53
   -logic unit 149
asynchronous
   inputs 101
   logic 98
   serial communications 341,350
   signal 211
   transmission 379
atomic number 2
auxiliary operations 55

B
Babbage, Charles 84
bank selection macros 181
banking techniques 145
banksel directive 180-181
base-emitter current 82
battery types 191
battery-operated devices 222
Baudot, Jean Maurice Emile 340

baud period 340,344,418,420,702-703
Benson, David 293
BCD
    digits to ASCII decimal 72
    to ASCII decimal 567,605
Bell Telephone System 343
biased switch 120
BiCMOS 94
big-endian format 40
binary
    arithmetic 37,42,44,51,59,62
    to ASCII decimal conversion 29
    to BCD conversion 566,611
binary-coded decimals 22,33
bipolar transistor 81-82,84-85,93-94
bistable device 96
bit stream 339-341
bit-banging 340
Black, Roman 245
Black-Ammerman method 245-246,253
Bohr, Niels 1
Boole, George 37
bootloaders 131
breadboard 132,164,194,206,278
breakover point 79
breakpoint 165,170-172
    modes 172
Bothe, Walther 84
buffer pointer variable 372-373
build process 170
busy flag 276,280,283,285,287,290,293,
    295-296,317,321-322
buzzers 77,95,122
byte ordering 40

C
capacitor circuit 121
carrier detect 347
carry/overflow flag 56
cathode 78,80-81,123
cblock directive 179,289,372
Centronics interface 350
ceramic resonator 146
character
    bits 341-342,344,420,438,469,
        486,504,522,708,772
    representations 34
circuit breakers 6
circuit tester 349
CISC 143,154,193

clear display 332,411,431,449,496,572,
    586,600,695,738,760
        to send 347
clearing the display 275
CLK 363,365,400,403-404,477,546,548,
    561-565,568,579-580,606-610,770-771
clock rate 147,242,244,246,253,265,269,
    343,351,354,370
clocks 195,102,105,107,122-123,174,280,
    341,344,476,543,558
CMOS 93-94,121,135,138,199,368
    4HCT 94
    logic gates 94
    transistor 94
code protec-
    tion 135,182-183,421,439,505,522,581,
    709, 730,743,750,752,773
collector-emitter current 82
comment symbol 176
commented bitmaps 178
common
    ground 6,343
    -cathode seven-segment LED 123
compare operator 193
comparison operations 63
conductors 4-5,9,13,15-16,77,367,476
configuration bits 145,182-183,556,593
contrast
    control 281
    control line 281
converter resolution 161,544,549
Coulomb, Charles Agustin 4
counter mode 246,248
counting 19,21,23,109-111,113,160,241,
    243-245,252,255,259,263,293,370,405,
    682,688
CP1600 processor 130
crystal
    displays 77,95,123,159,275
    oscillator 195,198,201,203,208,229,233,
        236,260,263,266,269,306,317,328,
        390,395,405,406,421-422,439,492-
        493,505,522,558,560,568,581,595,
        659,663,667,682,685,689,704,709,
        730,744,750,752,773
    resonator 146,183,190,195,198,201,
        203,208,229,233,236,260,263,266,
        269,306,317,328,390,395,401,406,
        422,439,493,505,522,568,581,595,
        659,663,667,678,682,685,689,704,
        709,730,744,750,752,773
current-limiting resistor 281

Index

cur-
   sor/dislay shift 292,310,321,331,411,430-431,448,496,572,586,600,671,694,738,760

D
D flip-flop 99-101
Dallas Semiconductors 367
DCE 344-347
data
   registers 154
   set ready 347
   stream 342,344
   terminal ready 347
DATA TRANSFER condition 477
data-logging 340
DB-9 connector 344,369
DC power supply 86,95
DDRAM 276-277,281-288,292,298-299,301-302,308,319,330,495,571,585,599,669,758
   address mapping 285
debouncing
   routine 222
   the switch 222
decimal-to-binary conversion 30
decoding gates 110
default radix 169
De Forest, Lee 84
delay loop 196-197,243-244,247,250,260,287,293-295,306,317,327,405,667,688,729
demo board 147, 202,206,208-209,664,679
demultiplexers 115
denormal numbers 49
denormals 48-49
detecting overflow 70
development boards 131
differential signaling 349
digit carry 61
digital
   circuits 85,95-96,101,115
   switching noise 554
   technology 103
diode 15,17-18,77-82,84,93-95
direct addressing 154-156,158,372-374,416,699
display
   data RAM 276,281,285
   shift 178,283-286,292,309-310,320-321,330-331,410-411,429-431,444-448,496,511,528,572,586,599-
display shift *(continued)*
   600,670-671,693-694,714,737-738,759-760,778
   mode 285
divide-by-two circuit 108-109
dopants 15
double-counter loop 197
data register 276,362,365,403,548,579
dual inline package 86

E
EBCDIC 36
EEADR 153,185,460-462,465-468,503-504,520,768-769
EECON1 153,185,217,460-462,465-469,503-504,520-521,756,768-769
   register 217,460-462,465,467-468
EECON2 153,460,462,465,468,504,521,769
EEDATA 153,185,460-462,465-468,503,520,768-769
EEPROM
   data memory 138,153,158,161,217,459-462,465,467,469,492,504
   programming 153,343,366,376,480
EIA/TIA-561 standard 346
EIA232E 340
EIA-485 339,349-350
   in PIC-based Systems xii,350
   standard xii,349
electrical
   charge 2-5,12,15-16,38
   circuits 6,8
   current 3,17,78,120,124
Electrically-Erasable Programmable Read-Only Memory (see EEPROM)
embedded system XV-XVI,164,476,558
entry mode set 285-286
EPROM 130,135-136,138,145,149,153,158,161,172,174,182,213,217-218,223,343,366,376,459-463,465,467-469,471-483,485-495,497,499,501,503-505,507,509-511,513,515,517,519-521,523-529,531,533,535,537-541,708,710-711,713-714,723-726,754,757-758,768-769,772,775,777-778,787-789,792
error-recovery mechanism 260
ESD 367
Ethernet 339,346
   cables 346

excess-n 48
external interrupt 139,148,214,220,224, 226,230,234,237,256,271
    flag 220,226,230,234,237,256,271
    source 214,224
external oscillator 135

F

Fairchild Semiconductor 16
farad 13
Faraday, Michael 13
flip-flops 77,85,95,98-101,103-104,109-111,113-115,121,360
float switch 120
floating-point
    BCD 52
    numbers 33,47-48,50
four-line decoder 118
Franklin, Benjamin 78
free running timer 243,245
FSR register 155,158,313,324,335,373, 387,500,517,535,575,591,603,721,766, 785
full-duplex 344,349,376,386
full-handshake 348
function set 285

G

general purpose registers 154-155,157, 289
Global Interrupt Enable bit 211,242, 623,647
global interrupts 220,226,230,234,237, 257,271,386,417,700
GPR 148,154-155,157,172,179-180,184, 197,469,472,486,507,520,524,537,711, 723,749,768,775,787
GPSIM 130
GPUTILS 130

H

half-duplex 161,344,349,376
handshake 347-349,360
hardware debuggers 170,174
Harvard architecture 33,142-143
HD44780 275-281,283,287,295,306,308, 316,318,321,327,329,408,420,425,438, 443,473,494,504,508,522,526,570,584,
HD44780 *(continued)*
598,669,691,708,712,729,733,756,772, 776
    controller: 276,283
    instruction set 283
heavy water 2
helium atom 2
high-end PICs 141
Hindu notation 23
Hindu-Arabic numerals 20-21
Hoerni, Jean 16
hysteresis 92-93,121

I

I/O ports 141,158-159,280,555
I/V curve 79
I2C
    communications 477,483
    EEPROM devices 479
    master mode 482-483,485,488,538,724
    serial interface 486
ICD 2 130,170,174
ideal waveform 104
IEEE 38,47-51
    754 Single Format 48
in-circuit debugger 130,172,174
include file 168,184,186,191,195
INDF register 158,372-373
indirect addressing 155-156,158,372-374,416,699
inductors 8,14,77
input devices 118,122,159
Institute of Electrical and Electronics Engineers 38
instruction pipeline 144
insulators 4
INTCON register 149,155,211,214,242, 255-256,259,356,358,386,393,398
INTEDG bit 214
integrated
    circuits 16,38,85-86,89,94,161,339, 360,376,475,546,558
    development environment 163,175
Inter-Integrated Circuit (see I2C)
internal
    clock 129,178,241,246,249-251, 253,256,259,261,264,267,271,343, 355,370,383,392,397,417,425,436, 443,453,584,661,683,686,700,706, 733,756

Index

registers 130,276,560
International Standards Organization 34
interrupt
   handlers 190,209,218,261,308,
      318,329,391,396,402,408,423,441,
      472,494,507,525,570,583,598,664,
      668,679,691,711,732,744,751,755,
      775
   mechanism 148,211,214,216,218
   request line 103
      sources 148,211,215,217,224
         -driven counter 247,255
         -driven timer 247,255
interrupts on the 16F84 211
inverted borrow 194
IRQ 103
isotopes 2

J
Japanese Kana characters 275,277

K
keypads 121-122,159

L
LAB-X1 development board 131
large EEPROMS 479
LCD
   initialization 287,290,303
   programming 287,293
LED 77,79-81,95,103,122-124,126,159,
   189,194-200,202-206,219-223,225-226,
   228-234,236-239,248-250,257-258,260,
   262-263,267,273,279,281,342-343,349,
   351-352,356,358,360,364,389,391-392,
   395-398,400,424,442,472,507,525,583,
   663,682,704,706,711,732,743-745,755,
   775
LED/pushbutton circuit 199
left-justification. 556,593
Leibniz, Gottfried 22,37
liquid crystal
   display driver circuit 276,278
   displays 77,95,123,275
little-endian format 40
logic gates 77,84-86,89,94-95,97,113,119
logical

AND 57,90
logical *(continued)*
   instructions 56
   NOT 58
   OR 57,90
low voltage programming 131
low-power Schottky 93
LTC1298 546,549

M
make-before-break action 120
marking state 340-341
Master Asynchronous Serial Port (see
   MSSP)
Master Synchronous Serial Port 161,
   376,480,486
Mauchly, John 37
MAX 190 546
MAX202 367-369,406,408,689,691
MAX232 367-368
megacandela 80
memory
   addressing 27
   banks 154,180,479
   storage 27,51,342
mercury switch 120
metal oxide semiconductor transistor 83
Microchip Technology 129,184
microcontroller clocks 107
microcontrollers 34,39,55-56,60-65,70,77,
   103,107,118,122,129-130,134-135,141-
   142,153,163,182,191,211,289,304,459,
   479-480,549,558
MicroPro 131,175
Microwire 475
mid-range
   instruction set 181
   PIC family 138
mismatch period 215
models of the atom 1
MOS transistor 83-84,94
motors 6,77,95,349
MPLAB 130,163-167,170-173,184,186,
   288-289
   assembler 165,184
   debuggers 165,288
   documentation 170
   editor 165
   IDE 164-166,171
   in-circuit emulators 165
   linker 165

SIM 170-173
MPU 276
MSSP 158,161,343,366,376,480-487,491,
    521-522,524,537,540,708,710,723-724,
    727,772,790
multi-byte counters 245
multiplexers 115,118
multipoint connection 349
multi-throw switch 120

N
NACK signal 478
NAND gate 84,87-90,97-98,114
n-channel MOS 83
negative-to-positive charges 5
nematic crystals 124
nested interrupts 216
NJU6355 177,558-561,563,595,606,608,
    752,756
NMOS 83-84
non-maskable interrupts 148
NOR gate 88-89,91,110,116
normalized form 49-50
Noyce, Robers 16
NPN transistor 81,83
n-type 78,81-83,93
    silicon 78,82-83,93
null modem 347-349,369
    cable 347-348,369
number systems 19,22

O
Ohm's Law v,5-6,9-10
Ohm, Georg Simon 5
on-board A/D hardware 161
OPTION register 152,211-212,214,220,
    226,230,234,241-243,246-247,250-251,
    253,256,259,261,264,266,270,355,391,
    397,417,424,442,473,508,525,584,660,
    683,686,700,705,712,733,745,756,776
OR gate 84,88-91,97,101,110,116
orbital model 1-2,16
oscillating signal 95
oscillator 105-107,135-136,144-147,161,
    183,190-192,195-196,198,201,203,208,
    222,229,233,236,242-243,260,263,266,
    269-270,294,306-307,317,328,354,370,
    377-378,383,390,395,401,405-406,421-
    422,435,439,453,476,485-486,492-493,
    505,518,522,535,549-552,558,560,568,
oscillator (continued)
    581,595,652,659,663,667,678,682,685,
    688-689,704,709,722,730,744,750,752,
    773,786
    type 195
output devices 122,159,189
overflow 42-44,56,59,61,64,67,69-71,160,
    213-214,218,241-242,245-247,255-258,
    268,271-272,355-360,387,391-394,396-
    399,417-419,464,483,485,488,502-503,
    538,566,578,592,611,662,700-702,724,
    767-768

P
packed BCD format 51,60-61,558,563,
    566,608
parallel
    communications 305-306,339-340,
        350-351
    data transmission 286,351
    port 174
    slave port 158,351
parity bit 341,344-345,370-371,384,436,
    454
passive matrix display 126
PCB 133-134,206-207,615,617-618
PCON register 148
PIC
    architecture 141,144,149,158,179,555
    clocking system 144
    programmers 175
    serial communications 352,366,368
PIC/LCD
    circuits 296
    port access 296
PIC10 devices 135
PIC1650 129
PIC-driven LCDs 275
PICMicro viii,129
picofarads 13
Pingala 37
pixilated output device 124
PMOS 83-84
p-n junction 78,80,82
PNP transistor 83
polled routines 255,386
positional
    system 21,24,27,66
    weights 22,29,45
positive and negative logic 89-90

Index

potentiometer 6,9,543,547,555-556
power supplies 81,95
power-down state 223,233
printed circuit board 86,133,206,615
programmable prescaler 160,214,241
programmers 131,174-176,206
protocol-based programming 366
prototyping 86,133-134
p-type 78,81-83,93
pulsing the E line 298
pushbutton circuit 199
pushbutton switch 121,164,191,199-201, 206,215,217,219,221,223,225,229,231, 233,248-249,260,262,352,355-356,358, 389-390,395,400,426,444,663,734,745
push-to-break switch 120
push-to-make switch 120
Pythagoras 24

Q

quantization level 544
quartz oscillator 560

R

R/W line 280,287,290,296,302,322,371-372,406,689
RA4/TOCKI pin 212,241,246-247
radix complement 41-43,47-49
  representation 42,48-49
RB0 interrupt 212,219-221,223-224,226, 230-231,234,237,257,271
RC
  network 106-107
  oscillator 107,146,161,183,294,476, 549-550
RCSTA register 376,380-381,385,387-388,437,456,519,536,723,787
read data 342,358,398,461,476,485-486, 503
reading the busy flag 280,287
receive data 346
receiver circuit 352
re-enabling interrupts. 211
reflective LCDs 125-126
register variables 218,257,268-269,271
request to send 347
reset switch 147
resistor color codes 614
resistor/capacitor 100,107,146

combination 100
resistor/capacitor *(continued)*
  circuit 121
resistors 8-14,77,85,94,106,121,212, 476-477,481,613
  in parallel 11
RESTART condition 477
right justification 553
rheostat (see variable resistor)
ring indicator 346
ripple counter 108-110,112-113
RISC architecture 135
ROM-based 135
rotary switch 97
rotate operation 62-63,113
round-off errors 244
RS flip-flop 98,102,121
RS-232 131,161,174-175,339-340,343-347, 350,352,354,366-371,375-376,379, 381,416,469,486,504,521,700,708, 772
RS-232-C ,339-340,343-347,350,352,354, 366-371,375-376,379,381,416
  communications on the 16F84 366
RS-422 340
RS-423 340
RX/TX pin 380

S

sample and hold capacitor 549
Schmitt trigger inverter 93,121
Schottky diodes 93
semiconducting material 79-80
semiconductor device 8,15-16,79,81
semiconductors 4,15,77
sender circuit 352,360
serial bit stream 339-340
serial
  communications 114-115,215,241, 306,339-341,343-344,347,350,352, 366,368,375-376,381,386,389,395, 400,408,475,486,691
  data transmission 342
  EEPROMS 161,376
  LCDs 275
Serial Peripheral Interface (see SPI)
series-parallel circuit 7
set
  CGRAM address 285-286
  entry mode 331,411,430,448,496,572,

586,600,694,738,760
seven-segment
  displays 77,95,122-123,159
  LED 122-123,126,189,204-206,248-250, 262,351
SFR 148-149,151,153-155,184,252,265, 468,482,520,683,687,769
  registers 148,155
Shannon, Claude E. 37,84
shift registers 85,113-115,161,360,362, 400
sign extension 71
signed numbers 24,40-41,48,59,67
sign-magnitude representation 40-41,67
simulators and debuggers 171
single-step mode 172
single-word instructions 142-143
sink the current 199
SLEEP mode 135-136,147,150,161,219, 549,554,622,652
  operation 554
small EEPROM 478
software timers, 293
source current 199-200
spacing state 340-341,344
special function registers 154-155,172, 469,482
SPI 161,475,480,483,492,541,728,790-792
square wave generator 106
SSP Interrupt 149
SSPADD register 485-486
start bit 341-342,344,354,356,358-359, 370-371,379,393,399,405,408,418-419, 688,691,701-703,789-791
START condition 477,479
static
  electrical charge 3
  electricity 94
STATUS register 145,149,151-152,154- 155,180,217-218,222-223,228,232,239, 258-259,273
Stibitz, George 37
stop bits 341,356,358,370,393,398
STOP condition 477,480,489,491,539, 541,726-727
stripboards 133
strobing 295
successive approximation algorithm 161,546,549
switch debouncing 121-122
synchronous
  communications 342-343,354
  idle characters 342

logic 98
synchronous *(continued)*
  serial transmission 343
system clock 223,244,376,551

T
tally system 19-20
template circuit 191
TFT displays 127
three-input AND gates 108
time delay routine 280,293
timer
  beats per second 243-244
  interrupt flag 258-259,272
  malfunction 260
  register 241-243,245-247,252,265,683, 687
timer0 module 214
timing
  generation circuit 276-277
  techniques 243
TMR0 overflow interrupt 213
transceiver IC 367-368
transducer 345,544
transformers 77
transistor 16,77,81-85,93-94,127
transistor-transistor logic 84
transmissive LCD 125
transmit data 384,437,455,518,536,722, 786
TRIS register 159,481,554,556,656
truth table 87,90-91,97,108,115
TTL 84-86,92-94,105-106,112,115,350, 367-369,459,546,558
  clock 105
  -compatible clock 105
  board 368
twisted-pair cables 346
two-bit ripple counter 108
TXSTA register 376-377,381
types of numbers 23

U
UART 115,340,342-343,349,366-369,381, 409,692
  module 367,381
unpacked BCD format 51
unsigned
  division 67

Index

iinteger 33,37,56,60
unsigned *(continued)*
   multiplication 64
up- and down-counter 112
USART
   138,148,158,161,340,342-343,349,366-368,376,379-386,388,405,435,437,452-456,518-519,535-536,688,721-723,785,
     module 161,343,366,368,376
     receive interrupt 388,454-455
USB 130-131,138,153,174-175,339,343
UTF-16 37
UTF-32 37
UTF-8 37

V
variable
   resistors. 9
   time-lapse routine 252
variable-lapse delay 253,265,267,269
voltage regulator 95-96,192,558
von Neumann bottleneck 142
   Burks, and Goldstine 22
von Neumann, John 37

W
watchdog circuits 129
watch-dog timer 150,152,622,660,683,686,704-705
Watt, James 5
Weiner, Norbert 37
whole numbers 24-25
wire wrap 133
word size 39,44,56,59,64
WRERR bit 466
write data 280,468,476,520,769

X
XNOR 84
XOR gate 84,90-91

Z
zero flag 56,63-64,193,218,245,252,265,300,314-315,325-326,336-337,501,516,534,577,590,605,675-676,683,687,720,765,784
Zuse, Konrad 22